chemistry

———— FIFTH EDITION

bju press®
Greenville, South Carolina

NOTE: The fact that materials produced by other publishers may be referred to in this volume does not constitute an endorsement of the content or theological position of materials produced by such publishers. Any references and ancillary materials are listed as an aid to the student or the teacher and in an attempt to maintain the accepted academic standards of the publishing industry.

CHEMISTRY Student Edition
Fifth Edition

Coordinating Writer
David M. Quigley, MEd

Writer
Christopher D. Coyle

Consultants
Kelly Driskell
Ben Zuehlke

Biblical Worldview
Renton Rathbun, ThM
Bryan Smith, PhD

Academic Oversight
Jeff Heath, EdD
Rachel Santopietro, MEd
Michael Winningham, MA

Project Editor
Rick Vasso, MDiv

Cover, Design, and Interior Concept Design
Sarah Lompe

Page Layout
Carrie Walker

Illustration
Leib Chigrin
David Lompe
Sarah Lompe
Frank Ramspott
Ina Stanimirova
   c/o Lemonadeillustration agency

Permissions
Katie Jackson
Lily Kielmeyer
Rita Mitchell
Elizabeth Walker

Project Coordinator
Chris Daniels

Photo credits appear on pages 616–21.

The text for this book is set in Adobe Gil Sans, Adobe Minion Pro, Adobe Minion Variable Concept, Adobe Myriad Pro, Helvetica, IM FELL Great Primer by Igino Marini, Minion® Math by typoma® GmbH, Playfair Display by Claus Eggers Sørensen, Raleway, STIX, and Symbol.

All trademarks are the registered and unregistered marks of their respective owners. BJU Press is in no way affiliated with these companies. No rights are granted by BJU Press to use such marks, whether by implication, estoppel, or otherwise.

The cover photo shows a close-up of a cluster of bubbles on a black background.

© 2021 BJU Press
Greenville, South Carolina 29609
Fourth Edition © 2015 BJU Press
First Edition © 1984 BJU Press

Printed in the United States of America
All rights reserved

ISBN 978-1-62856-686-4

15 14 13 12 11 10 9 8 7 6 5 4 3

Contents

Chemistry and Beyond   *vii*

01 FOUNDATIONS OF CHEMISTRY   *XII*

- **1.1 Chemistry: Modeling Matter**   1
  - *Mini Lab*: Seeing Is Believing?   5
- **1.2 Chemistry Helps People**   6
  - *Ethics*: Christian Ethics and Chemistry   8
  - *Case Study*: The Cost of Vaccine Research   10
- **1.3 Doing Chemistry**   10
  - *Case Study*: Experiment in Orbit   13
  - *Case Study*: The Search for a Malaria Vaccine   15
  - *Case Study*: Nanoscience   15
  - *Case Study*: Quinine in Time   16
  - *Ethics*: Pesticides   19

02 MATTER   *20*

- **2.1 The Classification of Matter**   21
  - *Mini Lab*: Paper Chromatography   26
  - *Case Study*: Understanding Trihydrogen   29
- **2.2 Energy and Matter**   31
  - *Worldview Investigation*: The Big Bang   32
- **2.3 The States of Matter**   40
  - *Serving as a Materials Scientist*: Informative Fibers   43

03 MEASUREMENTS IN CHEMISTRY   *48*

- **3.1 Measurement Systems**   49
  - *How It Works*: Speedometer   54
  - *Case Study*: Redefining the Kilogram   56
- **3.2 Measurements**   58
  - *Mini Lab*: Accuracy and Precision   62
- **3.3 Problem Solving in Chemistry**   65
  - *Serving as a Climatologist*: Weather Watchers   70
  - *Ethics*: Ethical Medical Testing   74

04 ATOMIC STRUCTURE   *76*

- **4.1 Early Thoughts about Matter**   77
- **4.2 The Development of Atomic Models**   80
  - *Mini Lab*: Indirect Observation   84
- **4.3 Useful Notations**   85
  - *Serving as a Science Teacher*: Encouraging Inquiring Minds   87
  - *Ethics*: Radium Girls   93

05 ELECTRON ARRANGEMENT   *94*

- **5.1 Bohr Model**   95
  - *Worldview Investigation*: Exoplanets   97
  - *How It Works*: Spectroscopy: Fingerprinting Atoms   98
  - *Mini Lab*: Lights, Spectroscope, Action!   99
- **5.2 Quantum-Mechanical Model**   100
- **5.3 Electron Configurations**   106

06 PERIODIC TABLE & ELEMENTS   *116*

- **6.1 The Periodic Table**   117
  - *Worldview Investigation*: Element Origins   126
- **6.2 Periodic Trends**   127
  - *Case Study*: What's in a Name?   130
- **6.3 Elements by Their Groups**   133
  - *Mini Lab*: Dense, Denser, Densest?   133
  - *How It Works*: Hydrogen Fuel Cell   135
  - *Serving as a Chemical Engineer*: Chemistry Everywhere   147
  - *Ethics*: Rare-Earth Elements and Risks   151

07 CHEMICAL BONDS   *152*

- **7.1 Bonding Basics**   153
  - *How It Works*: Recycled Plastic   157
- **7.2 Types of Bonds**   158
- **7.3 Properties of Compounds**   167
  - *Worldview Investigation*: Biodegradable Plastic   169
  - *Mini Lab*: Pie Pan Predictions   170
  - *Ethics*: Plastic—Wonder Product or Destroyer of Worlds?   173

08 BOND THEORIES & MOLECULAR GEOMETRY   *174*

- **8.1 Bond Theories**   175
  - *Serving as a Patent Attorney*: Protecting the Giving of Proper Credit   179
- **8.2 Molecular Geometry**   182
  - *Worldview Investigation*: Refreshing Water   186
  - *How It Works*: Water Striders   188
  - *Mini Lab*: A Pile of Water   190
  - *Ethics*: The Law of the River   195

Contents

09 CHEMICAL COMPOUNDS — 196

- **9.1 Ionic Compounds** — 197
  - *Worldview Investigation*: IUPAC — 206
  - *Serving as an Anesthesiologist*: The Comfort of Not Feeling a Thing — 208
- **9.2 Covalent Compounds** — 210
  - *Mini Lab*: Same Stuff, Different Name? — 212
- **9.3 Acids** — 213
  - *Ethics*: Drug Testing — 217

10 CHEMICAL REACTIONS & EQUATIONS — 218

- **10.1 Chemical Equations** — 219
  - *Mini Lab*: Conserving Atoms — 220
  - *Case Study*: Waste Not, Want Not — 222
- **10.2 Types of Reactions** — 228
  - *How It Works*: Dynamite! — 230
  - *Serving as an Explosive Ordnance Disposal (EOD) Technician*: The Bomb Squad — 233
  - *Ethics*: Explosives Development — 239

11 CHEMICAL CALCULATIONS — 240

- **11.1 The Mole** — 241
  - *How It Works*: Carbon Monoxide Detector — 242
- **11.2 Stoichiometry** — 252
  - *Serving as a Chemical Abatement Specialist*: Not Your Average Cleanup Crew — 255
  - *Mini Lab*: Blowup — 256
- **11.3 Real-World Stoichiometry** — 259
  - *Case Study*: Sulfuric Acid — 262
  - *Ethics*: Mandatory Detectors — 267

12 GASES — 268

- **12.1 Properties of Gases** — 269
- **12.2 Gas Laws** — 274
  - *Mini Lab*: Changing Volume — 276
  - *Case Study*: When Oxygen Is Bad — 280
- **12.3 Gas Stoichiometry** — 283
  - *How It Works*: Airbags — 290
  - *Worldview Investigation*: Greenhouse Gases — 292
  - *Ethics*: Deadly Safety Device? — 297

13 SOLIDS & LIQUIDS — 298

- **13.1 Intermolecular Forces** — 299
- **13.2 Solids** — 304
  - *How It Works*: Cryogenics — 307
- **13.3 Liquids** — 312
  - *Mini Lab*: Through the Void — 313
  - *Ethics*: Cryonics — 325

14 SOLUTIONS — 326

- **14.1 The Dissolving Process** — 327
  - *Case Study*: Pharmaceutical Pollution — 329
  - *Mini Lab*: Off to the Races — 331
- **14.2 Measures of Concentration** — 338
- **14.3 Colligative Properties** — 343
  - *How It Works*: Reverse Osmosis — 348
- **14.4 Suspensions and Colloids** — 349
  - *Serving as an Environmental Scientist*: Water Watcher — 351
  - *Ethics*: Wastewater Management — 355

15 THERMOCHEMISTRY — 356

- **15.1 Thermodynamics and Phase Changes** — 357
  - *Mini Lab*: Comparing Thermal Energy Transfer — 364
- **15.2 Thermodynamics and Chemical Changes** — 365
- **15.3 Reaction Tendency** — 371
  - *Case Study*: Entropy and Life — 374
  - *Worldview Investigation*: Heat Death — 380

16 CHEMICAL KINETICS — 386

- **16.1 Reaction Rates** — 387
  - *Mini Lab*: Changing Reaction Rates — 389
  - *Case Study*: Spontaneous Combustion — 395
- **16.2 Reaction Mechanisms** — 396
  - *Serving as a Pharmacologist*: Lifesaving Chemistry — 401
  - *How It Works*: Sustained-Release Medication — 403
  - *Ethics*: Medical Marijuana — 407

Contents

17
CHEMICAL EQUILIBRIUM — 408

- **17.1 Equilibrium** — 409
  - *Mini Lab*: Mix, Change, Repeat — 415
- **17.2 Le Châtelier's Principle** — 416
  - *How It Works*: The Haber Process — 421
  - *Worldview Investigation*: Ethanol — 422
- **17.3 Solution Equilibrium** — 424

18
ACIDS, BASES, AND SALTS — 434

- **18.1 Defining Acids and Bases** — 435
  - *Case Study*: Royal Acid to the Rescue — 439
- **18.2 Acid-Base Equilibria** — 440
  - *How It Works*: Breathalyzer® — 448
  - *Mini Lab*: Acid or Base? — 452
- **18.3 Neutralization** — 453
  - *Worldview Investigation*: Influencing Others — 455

19
REDOX REACTIONS — 462

- **19.1 Redox Reactions** — 463
  - *Serving as an Electrochemist*: Power the Future — 465
- **19.2 Electrochemical Reactions** — 471
  - *Worldview Investigation*: Battery Recycling — 474
  - *Mini Lab*: Observing a Voltaic Cell — 479
  - *Case Study*: Smaller, Safer Batteries — 480
  - *Ethics*: Electric Cars — 483

20
ORGANIC CHEMISTRY — 484

- **20.1 Organic Compounds** — 485
  - *Serving as an Odor Tester*: Improving Life One Smell at at Time — 487
- **20.2 Hydrocarbons** — 488
- **20.3 Substituted Hydrocarbons** — 497
  - *Mini Lab*: Isomerism in Substituted Hydrocarbons — 498
  - *Case Study*: Bugs, Be Gone! — 505
- **20.4 Organic Reactions** — 506
  - *Worldview Investigation*: Aromatherapy — 508

21
BIOCHEMISTRY — 514

- **21.1 Chemistry of Life** — 515
- **21.2 Carbohydrates** — 517
  - *How It Works*: New Bone from Sugar? — 520
  - *Mini Lab*: *Simple* Sugars? — 521
- **21.3 Lipids** — 522
  - *Serving as a Biomedical Researcher*: Making a Miracle Molecule — 524
- **21.4 Proteins** — 525
- **21.5 Nucleic Acids** — 528
  - *Worldview Investigation*: Abiogenesis — 532
  - *Ethics*: Paleo Diets—Ancient Key to Modern Health? — 535

22
NUCLEAR CHEMISTRY — 536

- **22.1 Inside the Nucleus** — 537
  - *Serving as a Nuclear Engineer*: Harnessing the Power of the Atom — 541
- **22.2 Nuclear Decay** — 545
  - *Worldview Investigation*: Radiometric Dating — 555
- **22.3 Using Nuclear Chemistry** — 556
  - *Mini Lab*: Inquiring into Chain Reactions — 561
  - *Ethics*: Nuclear Power — 569

CASE STUDIES

- The Cost of Vaccine Research — 10
- Experiment in Orbit — 13
- The Search for a Malaria Vaccine — 15
- Nanoscience — 15
- Quinine in Time — 16
- Understanding Trihydrogen — 29
- Redefining the Kilogram — 56
- What's in a Name? — 130
- Waste Not, Want Not — 222
- Sulfuric Acid — 262
- When Oxygen Is Bad — 280
- Pharmaceutical Pollution — 329
- Entropy and Life — 374
- Spontaneous Combustion — 395
- Royal Acid to the Rescue — 439
- Smaller, Safer Batteries — 480
- Bugs, Be Gone! — 505

ETHICS

- Christian Ethics and Chemistry — 8
- Pesticides — 19
- Ethical Medical Testing — 74
- Radium Girls — 93
- Rare-Earth Elements and Risks — 151

Contents

Plastic—Wonder Product or Destroyer of Worlds?	*173*
The Law of the River	195
Drug Testing	217
Explosives Development	239
Mandatory Detectors	267
Deadly Safety Device?	297
Cryonics	325
Wastewater Management	355
Medical Marijuana	407
Electric Cars	483
Paleo Diets—Ancient Key to Modern Health?	535
Nuclear Power	569

HOW IT WORKS

Speedometer	54
Spectroscopy: Fingerprinting Atoms	98
Hydrogen Fuel Cell	135
Recycled Plastic	157
Water Striders	188
Dynamite!	230
Carbon Monoxide Detector	242
Airbags	290
Cryogenics	307
Reverse Osmosis	348
Sustained-Release Medication	403
The Haber Process	421
Breathalyzer	448
New Bone from Sugar?	520

MINI LABS

Seeing Is Believing?	5
Paper Chromatography	26
Accuracy and Precision	62
Indirect Observation	84
Lights, Spectroscope, Action!	99
Dense, Denser, Densest?	133
Pie Pan Predictions	170
A Pile of Water	190
Same Stuff, Different Name?	212
Conserving Atoms	220
Blowup	256
Changing Volume	276
Through the Void	313
Off to the Races	331
Comparing Thermal Energy Transfer	364
Changing Reaction Rates	389
Mix, Change, Repeat	415
Acid or Base?	452
Observing a Voltaic Cell	479
Isomerism in Substituted Hydrocarbons	498
Simple Sugars?	521
Inquiring into Chain Reactions	561

SERVING AS A(N) . . .

Materials Scientist: Informative Fibers	43
Climatologist: Weather Watchers	70
Science Teacher: Encouraging Inquiring Minds	87
Chemical Engineer: Chemistry Everywhere	147
Patent Attorney: Protecting the Giving of Proper Credit	179
Anesthesiologist: The Comfort of Not Feeling a Thing	208
Explosive Ordnance Disposal (EOD) Technician: The Bomb Squad	233
Chemical Abatement Specialist: Not Your Average Cleanup Crew	255
Environmental Scientist: Water Watcher	351
Pharmacologist: Lifesaving Chemistry	401
Electrochemist: Power the Future	465
Odor Tester: Improving Life One Smell at a Time	487
Biomedical Researcher: Making a Miracle Molecule	524
Nuclear Engineer: Harnessing the Power of the Atom	541

WORLDVIEW INVESTIGATIONS

The Big Bang	32
Exoplanets	97
Element Origins	126
Biodegradable Plastic	169
Refreshing Water	186
IUPAC	206
Greenhouse Gases	292
Heat Death	380
Ethanol	422
Influencing Others	455
Battery Recycling	474
Aromatherapy	508
Abiogenesis	532
Radiometric Dating	555

APPENDIXES

A	Understanding Scientific Terms	570
B	Math Helps	572
C	Fundamental and Derived Units of the SI	578
D	Metric Prefixes	579
E	Physical Constants	580
F	Common Abbreviations and Symbols	580
G	Common Ions	583
H	Standard Thermodynamic Property Values	584
I	K_{sp} for Minimally Soluble Substances	586
J	Families of Organic Compounds	587

GLOSSARY — 588

INDEX — 606

PERIODIC TABLE OF THE ELEMENTS — 626

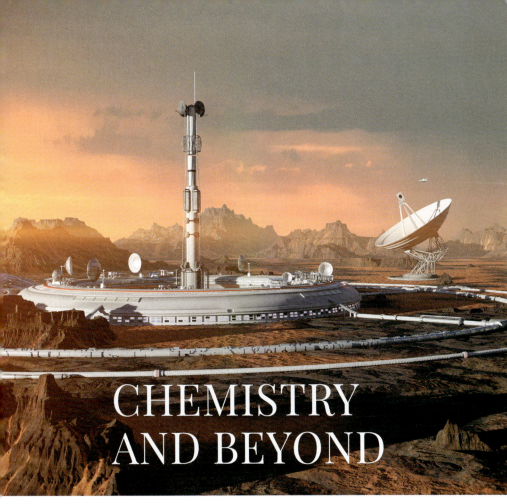

Left: Mockup of a futuristic base on Mars; *Top*: zinnia plant grown on the International Space Station (ISS) floating in microgravity; *Bottom*: astronauts outside the ISS.

CHEMISTRY AND BEYOND

People have been studying chemistry since Creation. There are numerous chemicals mentioned throughout Scripture. While we do know a lot about chemistry, we certainly don't know it all. As NASA looks to visit Mars and beyond, we need to expand our thinking about chemistry. We have a good grasp on much of chemistry here on Earth, but does that change in space?

A number of scientists are working on questions related to how various science fields differ in space. How do candles burn in reduced gravity? Do chemical reactions behave differently away from Earth's gravity? Could the science of photosynthesis help us in space exploration? Are biological processes different in space?

While there are numerous ways to investigate microgravity here on Earth, there is nothing better than experimenting about the effects of space than to actually be there. The International Space Station (ISS) has proven itself to be a vital research facility for all the nations involved in the program.

Investigating combustion in microgravity was one of the first experiments done on the ISS. NASA is currently conducting a series of combustion experiments on the ISS. They hope to learn more about how the process occurs so that they can find ways to burn fuels more efficiently and with less pollution. Combustion is one of the many chemistry concepts that scientists investigate on the ISS.

But chemistry isn't just for exploring space. Chemistry affects you all day every day—the clothes you wear, the food you eat, the car you drive, the medicine you take. You'll see in this textbook that chemistry is a powerful tool for us to use in God's world.

Chemistry could someday lead to cures for some of the diseases that plague us now, such as cancer, HIV, malaria, and others. It could revolutionize the cars we drive, the technology we use, and the way we generate electricity. It might even take us to Mars.

But *should* we try to do all these things? Should we try to establish a space station on Mars, isolating people from the earth that God created as a haven for life? Is chemistry the ultimate solution for all the problems we have now?

Chemistry alone can't answer these questions. But God's Word gives us the perspective we need to use chemistry in ways that glorify God and are truly helpful for His image-bearers. Let's explore how we can wisely use chemistry to take us above the clouds.

Features of This Textbook

This textbook is for you. We've designed it to help you learn. In the back of the textbook you will see other features, including an appendix section, glossary, index, and periodic table. We've designed this textbook with you in mind. We hope it will help you appreciate the wonders of God's creation even more. Flip through the following pages to see the features that we've included to lead you to success in chemistry.

❶ **Chapter Opener**
a short article that highlights issues and developments in chemistry that demonstrate how science intersects with your life

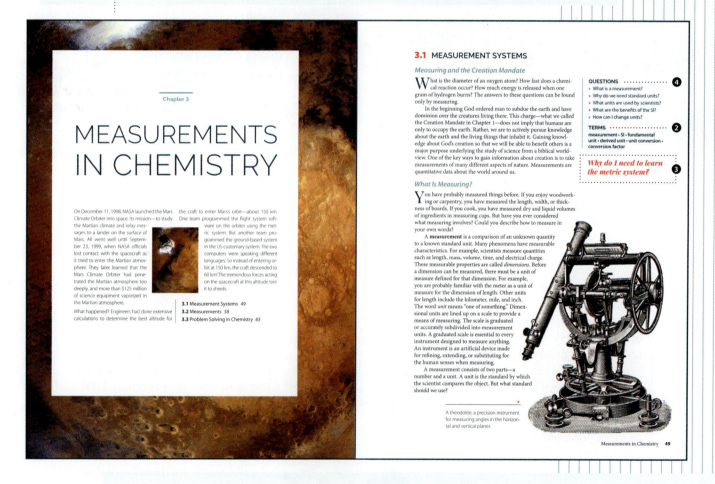

❷ **Vocabulary Terms**
the key terms that will be introduced in a section

❸ **Essential Question**
the big question that you will learn about in a section

❹ **Key Questions**
the smaller questions that you can ask along the way through a section to help you answer the essential question

viii Chemistry and Beyond

Element names come from many different sources. Latin names serve as the basis for many of the symbols for some elements that were known in ancient times. For example, *cuprum*, the Latin name for copper (Cu), means "from the island of Cyprus" and is the source for its symbol. Elements are named after many things.

color: iridium (Ir) from the Latin for "rainbow"

people: curium (Cm) for Pierre and Marie Curie, early researchers of radioactivity

places: californium (Cf) for California

heavenly bodies: helium (He) from *helios*, the Greek word for "sun"

miscellaneous words: bromine (Br) from the Greek word for "stench"

Today it is generally accepted that whoever discovers a new element has the honor of naming it, subject to the approval of the International Union of Pure and Applied Chemistry (IUPAC), an international organization responsible for standardization in chemistry. The periodic table of the elements in the back of this textbook displays the symbols for all the known elements.

Compounds

A pure substance that consists of two or more elements chemically combined in a fixed ratio is called a **compound**. Some compounds form distinct particles—molecules—while other compounds form vast crystalline arrays with repeating arrangements of ions. As you can imagine, there are many more compounds than elements. Just as chemical symbols are used to represent elements, chemists use **chemical formulas** to represent the millions of kinds of compounds.

The chemical formulas are made up of the chemical symbols for each element present. Numbers written at the lower right of a chemical symbol are called *chemical subscripts*. They indicate the number of atoms or groups of atoms in a chemical formula. A molecule of water contains two hydrogen atoms and a single oxygen atom and so is written H_2O (the 1 for oxygen is assumed). When a subscript follows a group of symbols that are surrounded by parentheses, it refers to the entire group. The formula $Ca(HCO_3)_2$ consists of a calcium atom and two HCO_3 groups, for a total of one calcium atom, two hydrogen atoms, two carbon atoms, and six oxygen atoms. Notice that the subscripts act very much like exponents in mathematical expressions. The six oxygen atoms comes from the three oxygen atoms multiplied by two.

CASE STUDY

UNDERSTANDING TRIHYDROGEN

Trihydrogen is an interesting molecular ion consisting of three hydrogen atoms held together by a single chemical bond. Initially discovered over one hundred years ago in 1911, trihydrogen has recently become of particular interest to astrochemists. In the early 1960s it was hypothesized that trihydrogen could exist in the interstellar medium, the material located between stars. During the 1990s its existence in the interstellar medium was confirmed, and by the early 2000s it was noted that it existed in large quantities. Today naturalistic scientists believe that trihydrogen was a key material in the transition from the origin of matter to the matter we see today.

Questions to Consider

1. Of what is trihydrogen made?
2. What is an ion?
3. What does the existence of trihydrogen in the interstellar medium imply about the origin of matter?

---

❺ **Case Studies**

opportunities to investigate specific areas in chemistry to apply what you have learned in a chapter

❻ **Bold-Faced Terms**

vocabulary terms that you need to know

❼ **Italicized Terms**

terms that will be defined later in the textbook or that are important terms in other scientific fields

---

10 CHAPTER REVIEW

Chapter Summary

TERMS
precipitate
chemical equation
reactant
product

10.1 CHEMICAL EQUATIONS

- Chemical reactions are a rearranging of the atoms in chemical substances to form new substances. Four signs that could indicate that a chemical reaction has occurred are a change in energy, the production of a gas, the formation of a precipitate, and a change of color.
- A balanced chemical equation shows the formulas and symbols for all substances involved in a chemical reaction and accounts for all atoms in both the reactants and the products to comply with the law of conservation of matter.
- Coefficients, subscripts, and special symbols provide more details on the quantities and conditions of a reaction.
- As a model of chemical reactions, chemical equations provide a lot of critical information but do not tell us everything about a reaction.

10.2 TYPES OF REACTIONS

- Five major categories of chemical reactions are synthesis, decomposition, combustion, single-replacement, and double-replacement.
- We can predict whether a single-replacement reaction will occur on the basis of the reactivity of the substances in the reaction. Double-replacement reactions will occur if at least one of the products is a precipitate, gas, or water.
- We can more accurately model what is going on in a double-replacement reaction with complete ionic and net ionic equations.

TERMS
synthesis reaction
decomposition reaction
combustion reaction
single-replacement reaction
double-replacement reaction
complete ionic equation
spectator ion
net ionic equation

Chapter Review Questions

RECALLING FACTS

1. What happens to chemical bonds in chemical reactions?
2. What is the only definitive way to know whether a chemical reaction has occurred?
3. Why is there typically a change in energy during a chemical reaction?
4. What is a precipitate?
5. Explain the steps to writing a balanced chemical equation.
6. Consider the reaction below.
   $HCl(aq) + H_2O(l) \rightleftharpoons Cl^-(aq) + H_3O^+(aq)$
   a. Why are two opposite arrows shown?
   b. Why is (*aq*) written after several of the substances?

---

❽ **Chapter Summary**

handy statements of the big ideas of the chapter, including vocabulary lists

❾ **Review Questions**

questions at the end of each section and chapter that will have you recall facts, demonstrate your understanding of concepts, and cause you to use critical thinking

HOW IT WORKS

New Bone from Sugar?

If you've ever taken a wilderness first aid course, you may have learned how to make a splint for a broken bone by using a tree branch. But medical researchers are taking this idea one step further—they are helping bones heal with help from a surprising source. Researchers in Canada have recently had much success in quickening new bone growth by injecting test subjects with an aerogel derived from plant cellulose. The nanocrystals of the cellulose form a scaffolded structure in which osteoblasts can deposit new bone. And unlike commonly used ceramic implant materials, which are hard and brittle, the new aerogel is soft enough to flex and completely fill all the gaps in a host bone. Those gaps hinder the growth of new bone when using ceramic implants; the new aerogel has demonstrated 50% more bone growth after twelve weeks compared with ceramic. Our God-given ability to think of creative ways to use the resources around us is part of exercising godly dominion over His world.

Polysaccharides

Polysaccharides, literally "many sugars," are molecules that contain many sugar units. This is evident from their large molecular masses (up to several million atomic mass units). Multitudes of sugar units are bonded into long chains. Polysaccharides may be built from several different monosaccharides, but we will concentrate on the polymers of glucose.

Plants such as rice, potatoes, wheat, and oats store food in polysaccharide deposits called starch. *Starch* exists as one of two glucose polymers. Most of the mass of starch found in plants exists as the branched isomer called *amylopectin*. But most of the actual starch molecules are the linear form known as *amylose*. Because the amylose molecules are very much smaller than amylopectin molecules, the more numerous amylose molecules make up only about a quarter of the mass of starch. Together, these two polysaccharides supply nearly three-fourths of the world's food energy.

Humans and animals also store glucose as a polysaccharide known as *glycogen*. Like amylopectin, glycogen is branched, but to an even greater degree. In humans, most glycogen is stored in the liver and skeletal muscles. It serves as a readily accessible energy reserve since it can be quickly broken down and converted into the molecules needed for cellular respiration.

Humans can't digest much of the plant matter that herbivorous animals can because plants contain lots of *cellulose*, a polysaccharide that is similar to starch and glycogen. In the polysaccharides that humans can digest, such as starch, each glucose unit is oriented in the same direction. But the glucose units in cellulose alternate directions. Thus, the difference between the starch in a tasty potato and the cellulose in grass, cotton, or a splintery piece of wood is the manner in which the glucose units are arranged. This may seem like a minor difference at first, but it isn't. Humans can digest the uniformly oriented starch, but they lack the enzyme needed to digest cellulose. Cows and other herbivores can't actually digest cellulose either—but symbiotic bacteria that live in their guts can. This mutualistic relationship allows both bacteria and herbivores to thrive.

MINI LAB

SIMPLE SUGARS?

You've seen that the ring forms of simple sugars have many isomers, but what about the open-chain forms? Can they also exist as many isomers?

Are simple sugars, you know, like, really simple?

Procedure

A Use your molecular modeling set to build a model of D-glucose as shown at right. When finished, you should be able to view your model from above in a manner that places the hydroxyl (–OH) groups on either the left or right side of the carbon backbone. Keep the double-bonded oxygen of the carbonyl group to the right to help you maintain your orientation when viewing.

B Swap the positions of the hydroxyl group and hydrogen atom on any internal carbon. When viewed from above, that hydroxyl group should now appear to be on the opposite side of the carbon backbone from its original position. Make sure that all the other substituents on the model remain in their original locations.

1. Can the affected carbon atom be rotated so that the hydroxyl group and hydrogen atom are returned to their original positions?

C The affected carbon atom in Step B is an example of a *stereocenter*. Isomers created around a stereocenter are called *stereoisomers*. The three remaining internal carbons are also stereocenters. Repeat Step B for each of those carbons. The resulting model is L-glucose.

D Try to make as many different aldohexoses as you can by continuing to swap substituents on the internal carbons, either individually or in various combinations.

2. Write the structural formula for each aldohexose you come up with for Step D.

3. How many different aldohexoses are possible in addition to D-glucose?

Conclusion

4. So, are open-chain simple sugars simple? Defend your answer.

Going Further

5. Search the internet to find the names of the stereoisomers that you created for Question 2. Label each of your structural formulas with the appropriate name.

EQUIPMENT
- molecular modeling set

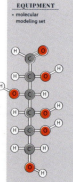

D-glucose

21.2 SECTION REVIEW

1. What are carbohydrates?
2. How does the structure of one kind of monosaccharide differ from another?
3. What can we know about a compound when told that it is an aldohexose?
4. What products result from a condensation reaction between glucose and fructose?
5. How are monosaccharides, disaccharides, and polysaccharides related?
6. Describe at least two roles of carbohydrates in living things.

CAREERS

SERVING AS A CHEMICAL ABATEMENT SPECIALIST: *NOT YOUR AVERAGE CLEANUP CREW*

Asbestos, lead, mold, PCBs—what do these have in common? Each is a type of contaminant that might exist in a home, business, factory, or job site. Chemical agents like these can pose serious health risks to people, even if the contamination occurred many years in the past. The threat is often not readily apparent, either, since the problems usually exist in out-of-the-way places such as inside walls and ceilings, beneath floors, and even inside lamps, thermostats, and switches. Because of this, inspections take place whenever buildings are scheduled for demolition or renovation.

Chemical abatement specialists are specially trained technicians who identify chemical hazards and safely neutralize them. They use their knowledge of hazardous chemicals and the technologies needed to deal with them to help make safer working and living environments. As a chemical abatement specialist, you could demonstrate God's care for others by improving their lives through the wise remediation of chemical health risks.

EXAMPLE 11-12: MASS-TO-MOLE CONVERSIONS

How many moles of phosphoric acid can be formed from 3.550×10^3 g of diphosphorus pentoxide?

$$P_2O_5(s) + 3H_2O(l) \longrightarrow 2H_3PO_4(aq)$$

Write what you know and what you want to know.

3.550×10^3 g P_2O_5
? mol H_3PO_4

Note that the given value is in grams and the desired value is in moles. Since the conversion factors from the balanced equation are based on a mole ratio, the mass of diphosphorus pentoxide must first be converted to moles. Then the moles of diphosphorus pentoxide can be converted to moles of phosphoric acid according to the mole ratio from the equation above. These two steps should be done together as follows.

Convert g P_2O_5 to mol H_3PO_4.

$$3.550 \times 10^3 \text{ g } P_2O_5 \left(\frac{1 \text{ mol } P_2O_5}{141.94 \text{ g } P_2O_5}\right)\left(\frac{2 \text{ mol } H_3PO_4}{1 \text{ mol } P_2O_5}\right) = 50.02 \text{ mol } H_3PO_4$$

---

⑩ How It Works Boxes
descriptions of how an everyday technology works and its relationship to the content of the chapter

⑪ Mini Labs
short hands-on activities to get you thinking and working like a scientist

⑫ Career Boxes
information about careers in chemistry (that could be yours!) that can be followed to wisely use God's world and help people

⓭ Worldview Investigation Boxes
inquiry-based investigations that help you think through controversial areas of chemistry through the lens of Scripture

⓮ Ethics Boxes
opportunities to apply a biblical worldview to ethical issues in chemistry

Chemistry and Beyond xi

Chapter 1

FOUNDATIONS OF CHEMISTRY

It all begins with a seemingly harmless mosquito bite. But if the mosquito is a host for microorganisms in the genus *Plasmodium*, which causes malaria, the person bitten could end up in a coma. A few days later, he might even be dead. Malaria is a scourge of the world's tropical and subtropical regions, threatening half the world's population. In 2017, about 219 million people were infected, leading to an estimated 435,000 deaths, mostly among children. Sub-Saharan Africa suffers the most due to weather conditions and bad sanitation practices that favor the most aggressive forms of malaria. But scientists, millionaires, politicians, and physicians from around the world have united in the fight against malaria. Their arsenal for vanquishing malaria is expanding with the help of chemistry. In 2015, regulators approved a vaccine for malaria—the first vaccine ever licensed for use against a human parasitic disease. Though the vaccine's success rate is low, it is still considered a major milestone toward eliminating the threat of malaria.

1.1 Chemistry: Modeling Matter *1*
1.2 Chemistry Helps People *6*
1.3 Doing Chemistry *10*

1.1 CHEMISTRY: MODELING MATTER

Chemistry and Worldview

Our world is full of significant problems that need answers—things like malaria, contaminated drinking water, pollution, and the need to increase crop production. Other problems are more a matter of convenience, like developing alternative fuels for our cars or making smaller and more lightweight computers. Chemistry can enable us to solve such real-world problems. The use of chemistry in this manner is known as *applied science*. But some chemists are engaged in *pure science*, which is an effort to understand how and why things work the way they do. Of course, the two kinds of science are not wholly separate from one another. Scientists must first discover the *how* and *why* of something before they can apply the knowledge gained to problem solving.

But what exactly is chemistry? **Chemistry** is the study of matter and the changes that it undergoes. **Matter**, as you recall from physical science, is anything that takes up space and has mass. You are sitting on, using, wearing, and eating chemicals all the time! This just shows you the scope and impact that chemistry can have on you for life.

Christians can get excited about applying chemistry to solve problems like malaria. We can also find it very rewarding just to satisfy our curiosity about God's marvelous creation and how it works. The Bible gives us a distinct **worldview**, a perspective from which to see and interpret all of life. Your worldview is made up of what you believe about the most important things in life. Like a pair of glasses, the Bible brings every part of the world into focus, including chemistry. In this textbook, you'll be presented with opportunities to examine how chemistry and a biblical worldview fit together.

Chemistry as Modeling

Chemistry is all about *modeling*. Like all other areas of science, it relies on our observations of the world. But we are limited in our abilities, especially when it comes to relating observations to explanations. To understand complicated things about the world, we have to simplify a problem by leaving some information out. This means that sometimes explanations may not be fully exhaustive—they provide an overall picture of a phenomenon but often omit some finer details.

Scientists have created models for every area of science. A **model** is a workable explanation, description, or representation of a phenomenon. A model of a quinine molecule (right), for instance, is a simplified representation of an actual molecule. People make models all the time to predict the path that a hurricane will take, the reaction when an acid and base are mixed, and changes in climate. Scientific knowledge can be really useful in the framework of an explanatory model.

QUESTIONS
» What do chemists study?
» How does worldview affect a chemist's work?
» What is modeling?
» Do scientific models reveal what is true about the world?
» How can Christians tackle ethical issues?

TERMS
chemistry • matter • worldview • model

What is chemistry?

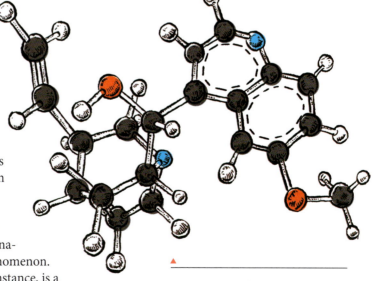

Quinine is one of the oldest medicines used to treat malaria. Each ball in this model represents a different element. This model relies on another model: the atomic model.

Foundations of Chemistry 1

Chemistry
IN HISTORY

Physicist and Nobel Prize recipient Richard Feynman said, "What I cannot create, I do not understand." He also said, "Scientific knowledge is a body of statements of varying degrees of certainty—some most unsure, some nearly sure, but none absolutely certain. Now, we scientists are used to this, and we take it for granted that it is perfectly consistent to be unsure—that it is possible to live and *not know* [italics added]." Feynman summed up an important aspect of science: Science is not about establishing what is true. It is about producing workable models. Only God's Word can tell us what is true.

So how are scientific models useful? Models are essential to solving problems in the world for two reasons. First, they are workable—they help us make sense of data. Second, they have predictive power. They can help us to see what could happen in the future. Models like the atomic model are the foundation of modern chemistry. Feynman used the atomic model when he participated in the Manhattan Project, the secret program that developed the atomic bomb during World War II.

Chemistry developed over time, often for purely practical reasons. People needed clothing, food, tools, and medicines to make life easier in a world full of discomfort and disease. But people have not always viewed and studied chemistry the way we do now. They had very different ideas about the nature of matter and the changes that it undergoes.

Ancient Chemistry

The Bible mentions that even as early as five generations after Cain men began to work with metals. Egypt during Joseph's time had a culture that understood and used the medicines of early pharmacists called *apothecaries*. The "science" of chemistry in ancient times involved a lot of trial and error!

The Philosophers

The ancient Greeks went a step beyond just learning how to use matter—they also applied reasoning to think about the nature of matter. Eventually, the idea that matter was made of atoms rose from this Greek culture. But the Greeks never made it to the laboratory. Their experiments were only in their minds.

The Alchemists

Chemistry as practiced by ancient civilizations, though robustly scientific in many ways, included a strong element of superstition. Eventually, a new emphasis on experimentation led to a surge of scientific discoveries. The odd blend of astrology and mysticism with observation and experimentation was called *alchemy*. Alchemists searched for immortality through medicines, tried to change common metals to gold, and experimented to discover the nature of the elements. They even developed secret languages to encode their recipes for different concoctions. Alchemists like Tycho Brahe, Isaac Newton, Robert Boyle, and Francis Bacon revolutionized science.

Classical Chemistry

By the mid-1600s, widespread acceptance of a new method of scientific inquiry led to rapid developments in chemistry. Elements were redefined and chemists began isolating them in the laboratory. By 1800 chemistry had become an established academic discipline. In America, several colleges made chemistry a part of their curriculum. Benjamin Rush (above) of the College of Philadelphia was the first professor of chemistry in the United States. He provided the medications that Lewis and Clark took on their expedition of discovery in the American West. Branches of chemistry began to develop around the world as chemists used atomic models to interpret the results of their experiments.

Modern Chemistry

The study of chemistry thrives today. People still need clothing, food, tools, and medicines to make life easier in a world full of discomfort and diseases like malaria. But though we do chemistry for the same reasons, we use different models to interpret what we observe about the nature of matter and the changes that it undergoes.

Models and Worldview

Scientists use the models they create within their worldviews, and worldview affects how they interpret evidence and make sense of things. Many scientific models fit within a biblical worldview, but some do not. Some scientists create and promote models that describe the world in ways that do not agree with the Bible. They may even try to use their models to show that the Bible is not true. For example, many scientists believe that life evolved from nonliving compounds. People who believe the Bible cannot accept this because the Bible traces all living things to God's special Creation by His Word. Scientists with a biblical worldview try to create models that accurately model the world and are also faithful to the Bible. An example of this is the catastrophic plate tectonics model, which models how the earth's surface changed very quickly after the Flood.

Chemists throughout history have developed and refined the models that we use now to understand matter and the changes that it undergoes. Without useful models, scientists are powerless to understand the world and solve problems that really matter, like the scourge of malaria.

1.1 SECTION REVIEW

1. State a definition of chemistry.

2. What reasons can you give for people to pursue discoveries in chemistry?

3. How does a scientist's worldview affect the way that he does science?

4. Evaluate the statement, "Scientific models reveal what is true about the world."

5. How has human history affected the development of chemistry? Think about the places where chemistry has advanced over time.

6. Do an online search for "history of chemistry." Use this information to create a timeline of key dates in chemistry's history. For historical context, include a few other significant dates in your timeline such as the birth of Christ, fall of Rome, Protestant Reformation, American Revolution, Civil War, end of World Wars I and II, or fall of the Soviet Union.

7. Research the following fields in chemistry and fit them into the Venn diagram below: inorganic chemistry, organic chemistry, biochemistry, nuclear chemistry, physical chemistry, and analytical chemistry. Biophysics has been done as an example for you.

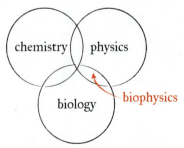

8. Many scientists have a worldview that rejects any supernatural explanation for either the original creation or the development of life on Earth. Scientists who hold to this *naturalistic worldview* usually believe that the universe is the result of the big bang and that all life on Earth has evolved from primitive to more complex forms. Can such a worldview be used as the basis for justifying the expense of developing vaccines? Explain.

MINI LAB

SEEING IS BELIEVING?

You may have heard that the process of science involves making observations. One type of observation is simply seeing something with our own eyes. But can we always rely on what our eyes seem to be telling us about something?

Can we rely on our senses to discover truth?

Procedure

A Allow your eyes to scan back and forth over the image at right.

1. What color(s) do the dots appear to be while you are scanning?

2. As you are scanning, do the dots appear to remain the same color?

B Now focus on a single dot.

3. What color is the dot that you focused on?

C Now focus on another dot.

4. What color is the second dot?

5. What color are *all* of the dots in the image?

Conclusion

6. How did your initial observation of dot color compare with a more careful examination of each dot?

Going Further

7. In what way(s) do you think this activity is similar to the modeling nature of science?

Optical illusions work through a variety of methods, but each produces a perception that is different than reality. Natural phenomena can fool our sense of perception as well, which is why good science relies on multiple strands of evidence. Mathematical tools such as statistical analysis also help us find order in situations that our natural senses can't perceive.

EQUIPMENT
- none

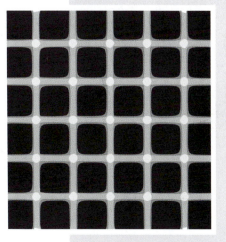

Foundations of Chemistry   5

QUESTIONS
» Why do naturalists and Christians see things differently?
» Do chemists have an obligation to serve others?
» Does worldview affect the way scientists do chemistry?

TERMS
presupposition • naturalism • Creation Mandate

Why is chemistry important?

1.2 CHEMISTRY HELPS PEOPLE

A Biblical Worldview of Chemistry

When we think about a biblical worldview of anything, we begin with the story in which God has placed us. The story is composed of God's good creation, man's fall into sin, and Christ's work in redeeming man. When we apply this story to a worldview, we mean: Creation (how things were meant to be), Fall (how things went wrong), and Redemption (how Christians are to respond to a fallen world). Chemistry has a place in our worldview as well.

CREATION

Science is not a list of absolute truths. Renowned scientist Karl Popper stated, "The mistake usually made in this field can be explained historically: science was considered to be a system of knowledge…as certain as it could be made….Later it became clear that absolutely certain truth was not attainable." To understand science we need a reliable view of the world. We need something certain. Because reality is revealed to us by our God, we interpret chemistry as something created by God. It is sustained every moment by His wisdom and power (Col. 1:16–17; Heb. 1:3). We begin our work in chemistry already expecting to see similarity and relationship between the physical world and the Creator.

FALL

Sadly man, in his sin, has come to believe that the world is the way it is, ultimately, because of the chemical properties of matter and energy. For those holding this view, human beings are basically chemical beings, not image-bearers of God. These assumptions color every point of their inquiry and interpretation of their observations. This line of thinking has infected some Christian thinkers as well. In having accepted the idea that science is the ultimate truth-teller, they hold that it is the work of science to "confirm" the reliability of Scripture.

REDEMPTION

But Christians should understand that all our thoughts and actions are to be governed by God's Word. Therefore, chemistry is to be limited to an explanation of our observations, not the ultimate grounding of this world. Christians should defend the claim that it is God who reveals His interpretation of the world to us through His Word. Though Scripture is not a chemistry textbook, only Scripture can account for the existence, practice, and perspective of every subject in creation, including chemistry. Part of what makes chemistry so important, then, is that we have the privilege of studying God's atomic handiwork as He holds the universe together by the word of His power.

Modeling and a Biblical Worldview

One of the challenges in chemistry and in science in general is that scientists don't always share the same view of science. All scientists approach their work with certain **presuppositions**, or assumptions, about the world, which are the basis of their worldview. Presuppositions don't come from proof; they are based on belief. We are all pressed to have faith in *something*.

Naturalism is a worldview that is based on a belief that matter is all that exists and human reasoning informed by science is the only reliable path to truth. Astronomer Carl Sagan expressed the essence of this worldview when he said, "The Cosmos [the material universe] is all that is or ever was or ever will be." Naturalists have faith in themselves.

Naturalists have developed the theory of evolution to explain the chance existence of elements, plants, animals, people, and ultimately the universe. Since they believe that there is no Creator, humans are not accountable to God. There is no absolute code of morality. People are tiny specks in a huge, impersonal world with little to elevate them above the rest of an evolved universe. A person is nothing more than a bag of chemicals. Love, happiness, justice, peace, and life itself are nothing more than chemical reactions.

Why do people with a naturalistic worldview study chemistry? To them, chemistry is a method of survival, a way to preserve or improve the current human way of life. Helping people is just preserving one's kind as produced by evolution. But naturalism cannot defend this way of life as good. Goodness implies a moral standard independent of science. The naturalistic worldview is thus unable to defend the work of chemistry as a morally good thing.

But Christians have faith in God and His Word. God's Word helps us to see not only the fallen nature of the world but also the fallen nature of our minds. And those who believe the Creator's Word are in the best position to understand His world. When science contradicts the Bible, it is not nature contradicting the Bible. Instead, it is a scientist's model affected by his worldview that contradicts the Bible. The Bible and nature both have the same divine Source, so they cannot contradict each other.

When we study chemistry from a biblical worldview, we have the perspective to expose error in science. We can fight against a dominant secular worldview in chemistry and rebuild it for the glory of God.

We have seen that one reason chemistry is important is that we are able to study God's handiwork. Another reason chemistry is so important is that we are able to use it to obey God's command to love our neighbor. Throughout this textbook we will look at examples of how the study of chemistry intersects with human activity.

Chemistry is a tool, and like any tool it can be used for either good or evil. The application of moral principles to how we live our lives, including the use of chemistry, is known as ethics. Christian ethics is unique because we base our ethical decisions on God's Word, the Bible.

For each issue we look at, we will consider three ways, sometimes referred to as an ethics triad (biblical principles, biblical outcomes, and biblical motivations), in which the Bible guides ethical decision-making.

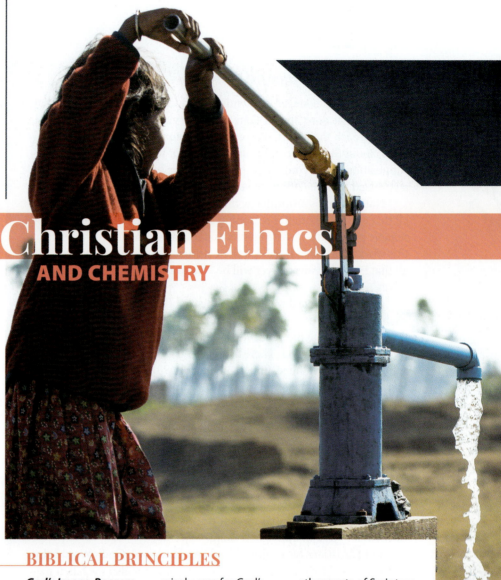

Christian Ethics
AND CHEMISTRY

BIBLICAL PRINCIPLES

God's Image-Bearers. Foundational to our ethical decision-making is the understanding that we all bear God's image. Therefore, we must make decisions out of respect for all people and for their protection (Gen. 1:26–27).

Creation Mandate. God's first commandment to us is to have dominion over the world that He created. Therefore, we must wisely care for God's creation according to this **Creation Mandate**. We must balance the appropriate use of the world's resources with the needs of people around the world. Nothing belongs to us; we are stewards of God's world. (Gen. 1:28).

God's Whole Truth. God's image in people and the Creation Mandate touch on many ethical issues. But other parts of Scripture also give us helpful insights into what God wants us to do. Part of making ethical decisions requires that we understand what His Word teaches. We cannot live any part of our lives separated from God and His Word (2 Tim. 3:15–16).

What does God's Word say?

BIBLICAL OUTCOMES

Human Flourishing. As soon as God created mankind, He blessed him (Gen. 1:28). Throughout Scripture (Ps. 1; Matt. 5), we see that God's desire is for all people to be blessed and to prosper. Jesus came to give us life and to have us live that life abundantly (John 10:10). Our ethical decisions must align with God's will to maximize human flourishing.

A Thriving Creation. Part of our obligation to the Creation Mandate is to ensure that creation thrives (Gen. 1:28; 2:5, 15). God has assigned us an important role as stewards of the world that He created. We must wisely use and develop the earth's resources to ensure that it flourishes.

Glorifying God. Just as Jesus came to glorify God, everything we do should glorify God (Matt. 5:16; 1 Cor. 6:20; 10:31). Our decisions should show God that we love and honor Him. This obligation includes every aspect of our lives: school, work, and play. So it is not enough that our decisions help others or that creation thrives. Our decisions must always give God the honor that He is due.

What results are right?

BIBLICAL MOTIVATIONS

Faith in God. The Bible discusses works versus faith (Jam. 2:14–26). The passage concludes that we are to live out our faith in God through our works. We are motivated to act because of our faith in God. Good works can stem only from our faith in God (Rom. 14:23; Heb. 11:6).

Hope in God's Promises. In the Bible, hope is not something that we wish for; it is something that God has promised. Biblical hope is confident expectation. The promises of God allow us to take action without fear (2 Tim. 1:7–9). Scripture teaches us that God can never lie, so we can act with the assurance that God will follow through on His promises.

Love for God and Others. As stated in 1 Corinthians 13, our greatest motivation for doing right is love. We have the love of God in us, and we do right when we are motivated by our love for God and our love for others. John 13:34–35 teaches that love is the outward sign of a transformed life.

How can I grow through this decision?

Foundations of Chemistry 9

1.2 SECTION REVIEW

1. Summarize a biblical worldview of chemistry in one sentence.
2. Are Christian chemists obligated to help protect the environment? Explain.
3. Give one presupposition of a biblical worldview and one of a naturalistic worldview.
4. How would a naturalist view each of the subjects below? How would a Christian view each one?
    a. God
    b. fighting malaria
    c. the origin of man
    d. the reliability of man's reasoning
    e. the reliability of the Bible
    f. conserving the environment
5. What should a Christian do when science seems to contradict what the Bible says?
6. Suggest at least one way that a Christian can use chemistry to return glory to God. Give an example not mentioned in the textbook.

CASE STUDY

THE COST OF VACCINE RESEARCH

Developing a new vaccine is not cheap, especially in the case of malaria, in which the targeted pathogen is not the usual bacteria or virus. Cost estimates for developing any vaccine range from $500–$800 million. This represents a significant investment for companies doing the research. And even though a vaccine has been approved for use against malaria, it's not yet known just how effective it will be if it is ever put into widespread use.

Use the case study above to answer Questions 7–8.

7. What three questions, each a part of the ethics triad, should Christians consider when thinking about applying biblical ethics?
8. Use the three parts of the ethics triad to justify the expense of developing a malaria vaccine.

QUESTIONS
» What is the difference between a theory and a law?
» How is scientific inquiry done?
» What should a scientific question look like?
» What are some examples of scientific inquiry?

TERMS
quantitative data • qualitative data • hypothesis • scientific inquiry • theory • law

1.3 DOING CHEMISTRY

Observation and Reasoning

Malaria has deeply shaped human culture. It has plagued mankind since ancient times. Chinese emperors, Egyptian pharaohs, citizens of the Roman Empire, and even George Washington all had malaria. It may have determined the outcome of wars and the decline of empires. But what causes malaria? It wasn't until the 1800s that scientists began to make discoveries about how people are infected with malaria. Answering a *scientific question* like this begins with observation.

How do chemists solve problems?

We use our senses to collect data or information about the world. But sometimes our senses need a little help! Scientific tools like telescopes, microscopes, or mini video cameras such as those used in arthroscopy extend our ability to collect data about the world. This data can be in the form of numbers determined through measuring, called **quantitative data**, but it can also be in the form of words used to describe something—**qualitative data**.

After obtaining data, scientists use models and reasoning to make connections and explain the data. *Deductive reasoning* proceeds from general statements called *premises* to a specific conclusion. *Inductive reasoning* proceeds from known data to an unknown general conclusion. It uses facts or data to draw general conclusions, but it cannot prove these conclusions to be completely certain. Instead, it shows only the probability that a conclusion is true on the basis of the evidence under consideration. Most scientific conclusions are based on data, so they use inductive reasoning, but deduction also plays a part. Let's look at some examples and see whether you can figure out which type of reasoning is being used.

EXAMPLE 1-1: IDENTIFYING TYPES OF REASONING

Reasoning

Determine whether the following arguments use deductive or inductive reasoning.

a. When potassium chloride, sodium chloride, magnesium chloride, or calcium chloride is mixed in water, it dissolves readily. Because barium chloride is a similar salt, it also should dissolve in water.

b. Any substance that floats on top of water has a lower density than water. Vegetable oil floats on top of water. Therefore, vegetable oil is less dense than water.

Solution

a. This argument attempts to show that the conclusion is probable or most likely based on similar situations, but it cannot prove that this salt or any other salt will dissolve in the same manner. Therefore, the argument uses inductive reasoning. As an example, silver chloride is an insoluble salt, showing that inductive reasoning is not completely conclusive.

b. This argument compares premises and arrives at a conclusion that must be true if the premises are true. Therefore, the argument uses deductive reasoning.

Sir Ronald Ross, a British doctor, used reasoning to learn about how people get malaria. He grew up in India, an area of the world where malaria is widespread. His grandfather contracted the disease when Ross was a child, and Ross purposed to fight malaria when he grew older. He studied it in the late 1800s, connecting malaria to mosquitoes after deliberately exposing uninfected mosquitoes to a malaria patient and then finding the malaria parasite in the mosquitoes' intestines. This led to a new approach to fighting malaria—controlling mosquito populations. We use this approach in the fight to eradicate malaria today.

Scientific Inquiry

When you think of scientists, you probably picture people in a laboratory performing experiments. Scientific inquiry actually starts with what many consider to be in the middle of the process—observation. After a scientist makes an observation, he may recognize that there is a scientific problem or question that needs answering. He may then do some initial research to determine what is already known about the problem and what investigations others have done on it. On the basis of his research, he may then develop a suggested explanation for the scientific question. This suggested explanation is called a **hypothesis**, and it should be as simple as possible. It should be reasonably based on current scientific models related to what is being studied. Most importantly, it must be testable.

Testing hypotheses about currently observable phenomena is referred to as *operational science*. Confirming the effectiveness of a new vaccine is an example of operational science. Hypotheses regarding phenomena that occurred in the unobservable past are part of *historical science*. A good example of historical science is a hypothesis regarding the origins of matter. Both creationists and naturalists have ideas and beliefs about how matter came into being, but the actual origin of matter is an event that happened in the past and cannot be repeated. It is therefore outside the realm of operational science.

After defining a hypothesis, scientists then collect data to test the hypothesis. Some people think that this is done only in a laboratory, but in reality scientists gather data in many different ways and in a variety of locations, often outdoors (called *field research*, or *fieldwork*). Scientists next use reasoning to analyze data that they have collected. The analysis is used to evaluate the hypothesis and form a conclusion as to whether the data supports or refutes the hypothesis. In the event that the data refutes a hypothesis, all is not lost—the scientist may simply need to modify the hypothesis. Sometimes, though, the data will suggest that a hypothesis needs to be discarded and replaced by a better one.

This process of identifying a problem or question, forming a hypothesis, testing it, and evaluating the results is collectively referred to as **scientific inquiry**. This may be more familiar to you as the *scientific method*. The scientific method is usually taught in a way that leaves students thinking of a series of steps to rigidly follow. In reality, the scientific inquiry process is much less rigid and linear. It is cyclical, with numerous instances of returning to prior tasks to modify, update, or refine them, or perhaps even start over.

EXPERIMENTATION

Experiments are a major part of scientific inquiry. They rely on a simple, reasonable, testable hypothesis. An *experiment* is a way to observe a natural process, sometimes under controlled conditions, so that a scientist can know whether her hypothesis is right or wrong. Experiments are an important part of the model-making process. In a controlled experiment, only one condition changes at a time so that a scientist can isolate its effect on the experiment's outcome. The different factors that change are called *variables*. The scientist changes one of these variables, called the *independent variable*. This can generate changes in other factors that she did not directly cause. These are called *dependent variables*. The group of samples exposed to the independent variable is the *experimental group*. The group of samples not exposed to the variable is called the *control group*. What a scientist observes or determines using data gathered from an experiment is said to be *empirical*.

A key element of scientific inquiry is repetition. To arrive at the best possible conclusions with the least uncertainty, scientists repeat experiments using multiple samples called *trials*. Each trial provides more data. A single trial that yields data that supports a hypothesis is not as conclusive as hundreds or perhaps even thousands of trials that produce the same results.

Confused by all these terms? Don't worry! Read the example below to help you sort out how scientists use them.

EXAMPLE 1-2: A CONTROLLED EXPERIMENT

Case Study: Experiment in Orbit
One of the questions that NASA has explored aboard the International Space Station is how plants grow in the absence of gravity. In a microgravity environment like space, how do plants know which way is up or down? In one particular experiment, scientists used magnets on flax seeds in zero gravity to see whether magnetism affected the direction of plant growth. They hypothesized that if certain slightly magnetic parts of a plant called starch grains were exposed to a highly magnetic environment, then the starch grains would move along the magnetic field lines. This movement could control the direction in which the roots grow.

Here's an example of how they might have set up their experiment. A group of fifty seeds could be grown in the dark while in orbit around the earth. They could be subjected to strong magnetic fields, but because they would be in orbit, they would not experience gravity. Another group of fifty seeds could be grown under these same conditions, but without the magnetic fields. Now suppose that after the seeds sprout, root growth in forty-two plants from the first group seems to follow magnetic lines. The rest of them either have no recognizable pattern in their root growth or do not germinate. The scientists could conclude that the starch grains do help plants know which way is down in this case, though there might be other factors at work.

Using the information from the scenario just described, identify the independent variable, dependent variable, experimental group, and control group. Then check your answers below.

Solution
The independent variable is the magnetic environment. The dependent variable is the direction of root growth. The experimental group is the fifty seeds subjected to the strong magnetic fields. And the control group is the fifty seeds grown without the magnetic fields.

Hypotheses are sometimes tested using an experiment in which scientists cannot control the conditions. These are called *natural experiments*. In natural experiments, scientists usually try to determine what variables are interacting, even though they cannot control them. Research in ecology, meteorology, and astronomy often involves natural experiments.

SURVEYS

Though experiments are really important to scientific inquiry, there is much more to the scientific process. Sometimes the only way to test a hypothesis is to conduct a survey. A *scientific survey* is a process that involves randomly selecting representative samples from a larger population. Surveys rely on inductive reasoning.

Surveys and experiments can work together. A scientist might take a sample of a population and experiment on the sample to find out characteristics of the population. For example, in 1993 the Mississippi and Missouri River basins flooded, resulting in $15 billion of damage and fifty deaths in the upper Midwestern states. These floodwaters appeared to have contaminated private water supplies such as wells, springs, and cisterns. In nine states in that region, about 18% of the water supply came from these sources. A survey of water quality was conducted in these nine states by collecting samples from some of the water supplies and analyzing them. The survey found that many of the water supplies were contaminated with bacteria and fertilizers. The results of this survey had the potential to protect thousands of people from illness.

USING THE RESULTS

After a scientist has experimented or conducted a survey, he must formulate his results and have his work examined by other experts in a *peer review* process. If his work is accepted, he can communicate his findings to the world so that they can be useful. He might do this through publishing papers in scientific journals, which other scientists read to stay up-to-date on current research. Scientific work helps us refine our models and advance our understanding of God's creation.

At times scientists test an idea that consistently *explains* a certain phenomenon. The scientific community calls such an idea a **theory**. Since theories offer explanations of what we observe, we can think of them as scientific models. They prove their reliability if they are able to make accurate predictions to explain several different bodies of data. Sometimes they can attempt to connect several different phenomena. The relationship between space and time as described by the theory of relativity is an example of this.

A scientific **law** is a statement that *describes* a recognizable, repeating pattern in nature. Laws often take the form of a mathematical equation. Scientific laws are based on observations made under many different conditions. Generally, scientists consider laws to be very reliable statements of science. Examples include Newton's laws of motion, Kepler's laws of planetary motion, and the laws of thermodynamics. You may hear someone say that scientific laws have no known exceptions, but this is not always the case. Newton's law of universal gravitation, for example, has been superseded by Einstein's theory of general relativity. But we still use Newton's law because it remains workable and useful for many situations, even if it lacks the explanatory power of Einstein's later theory.

Thinking like a Scientist

Right now you are somewhat of an inexperienced chemistry student. But you don't have to stay that way! If you keep an open mind, you can be transformed into a student chemist. The key to becoming one is to learn to think and work like a scientist. This textbook is your guide to help you along the way. To get you started, take a look at the case studies below of real scientists who made significant breakthroughs. You will see how scientific research leads to more scientific research and that ideas must be tested and modified to help us better understand our universe.

You never know—maybe God will call you to glorify Him through the work of a scientist!

Case Study 1:
THE SEARCH FOR A MALARIA VACCINE

In 2011–12, the Bill and Melinda Gates Foundation financed the development of a new vaccine for malaria called RTS,S or Mosquirix®. This vaccine includes a chemical found on the surface of malaria parasites that triggers a human immune response to the parasite. Researchers first conducted trials of the vaccine on adults in Gambia, then on children in Mozambique, then on younger African children and infants. Early results of this testing were mixed, but it was a start. The vaccine didn't prevent malaria entirely, but it did reduce severe malaria in children. Although many others still got malaria after the vaccine, those that did had less severe cases. This was likely due to the vaccination.

Though Mosquirix is now approved for use, researchers are still trying to improve its effectiveness. For now, Mosquirix is not the cure-all researchers were hoping for, but it is possible that further research will improve the vaccine's effectiveness.

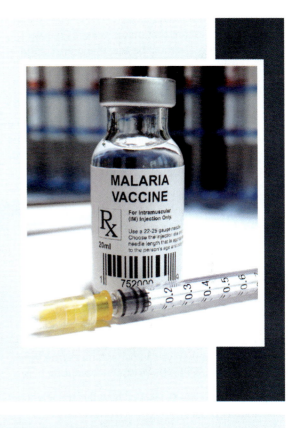

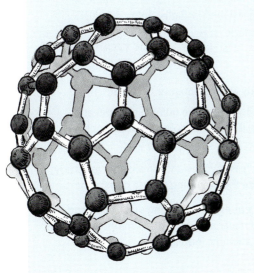

Case Study 2: NANOSCIENCE

One of the pioneering fields of science is known as nanoscience—science on the scale of one billionth of a meter. Carbon atoms can be linked to form many different substances, like graphite, coal, diamond, and even a soccer-ball-shaped molecule called a *buckminsterfullerene*. In 1991, Sumio Iijima discovered nanotubes, which form by linking carbon atoms or other atoms to make tubes only a few nanometers in diameter, or one fifty-thousandth that of a human hair. These tubes behave very differently according to their size and shape. Early models predicted how water would behave inside a nanotube, but in a paper published in July 2004, Russian author Alexander Kolesnikov and his research group found additional evidence that water in a nanotube will not freeze, even at temperatures of 8 K (–445 °F)! The team has committed to refining the model of nanotube water behavior and empirically determining its properties.

1.3 SECTION REVIEW

1. Your friend says that answering a scientific question usually involves collecting qualitative data or quantitative data but not both. Do you agree or disagree? Explain.

2. Determine whether the following statements use inductive or deductive reasoning.

   a. My grandpa's doctor has performed 100 cardiac bypass surgeries, and all were successful. When he performs this surgery on my grandpa, it will likely be successful.

   b. Everything that the Bible teaches is true. The Bible teaches that man is sinful by nature. Therefore, it must be true that man is sinful by nature.

   c. If humanism is correct, then war, treachery, and death are just the normal rhythms of an evolving world. But we all feel deep down that these things are unnatural. Thus, humanism is not a correct explanation of where our world has come from and where it is going.

3. Give an example of a scientific question not mentioned in the text that is best answered with an experiment. Describe an experiment that a scientist could do to answer that question.

4. Design a survey to test whether there are any gold deposits in the Grand Canyon.

5. Many Christians try to discount evolution by saying, "It's just a theory." And many naturalists use the word *theory* to mean that evolution is something that we can depend on as true. Why do you think that there is some confusion here?

6. Did either Case Study 1 or 2 involve the scientific inquiry process? If so, which one(s)? Are either of the examples not truly scientific because they did not strictly adhere to the process?

7. In Case Study 1, what was the independent variable in the malaria vaccine study? What was the dependent variable? What was the experimental group? Can you infer what the control group was like?

8. Name a few of the qualities that the case studies have in common. What can you deduce about the scientific inquiry process from these qualities?

9. Although it is not mentioned in Case Study 2, why do you think that the use of models in nanotechnology is essential?

CASE STUDY

QUININE IN TIME

For someone with a severe case of malaria, getting quinine in time could be a matter of life or death. But it has not always been easy to get it to the people who need it most. The Quechua people of Peru were the first to use quinine, derived from the bark of local cinchona trees. They ground the bark, mixed it with water, and drank it as a tonic to reduce chills from fevers. Quinine crossed the Atlantic with Jesuit missionaries, who brought this tonic water back to a malaria-ridden Europe. Quinine became known as Jesuit's bark, and it was one of the most precious cargos that traveled between Peru and Europe.

Demand grew for cheaper and more accessible quinine. Other forms of quinine with fewer bad side effects were developed. Bayer laboratories developed chloroquine in 1934. Cut off from supplies of the drug by World War II, American chemists took on the task of synthesizing quinine and succeeded in 1944.

In 1970, researchers found an ancient Chinese medical manual from AD 340. It shed light on a new weapon in the war against malaria, an herb called *Artemisia annua*. This herb's antimalarial chemical, called *artesunate*, is becoming one of the most promising medications to treat malaria. It can be injected, has fewer negative side effects, and could be more effective than any other medication in the war on malaria.

Use the case study above to answer Questions 10–12.

10. Were the efforts to make quinine cheaper and more accessible an example of pure science or applied science? Explain.

11. Describe at least two instances from the case study that illustrate the exercising of wise dominion.

12. Formulate a question regarding artesunate and its use against malaria that scientists might investigate using scientific inquiry.

01 CHAPTER REVIEW
Chapter Summary

TERMS
- chemistry
- matter
- worldview
- model

1.1 CHEMISTRY: MODELING MATTER

- Chemistry is the study of matter (anything that takes up space and has mass) and the changes that it undergoes.

- The Bible gives us a distinct worldview, or perspective, to see and interpret all of life, including chemistry.

- A model is a physical, conceptual, or mathematical representation of some aspect of the world.

- Chemistry relies on models that explain, describe, or represent what we observe in the universe, though they cannot give us the whole picture.

- People have been using chemistry for practical purposes since the beginning, though they have not always viewed and studied it the way that we do now.

1.2 CHEMISTRY HELPS PEOPLE

- A biblical worldview can be summed up by the Bible's account of the Creation, Fall, and Redemption.

- The Creation Mandate is God's command to the human race to exercise good and wise dominion over the works of God's hands.

- Christians can use chemistry to glorify God, to love their neighbors, and to expose error introduced by a naturalistic worldview.

- A naturalistic worldview relies on man's reasoning, while a biblical worldview relies on God's Word.

TERMS
- presupposition
- naturalism
- Creation Mandate

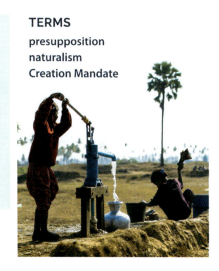

TERMS
- quantitative data
- qualitative data
- hypothesis
- scientific inquiry
- theory
- law

1.3 DOING CHEMISTRY

- Scientists answer scientific questions by observing and collecting both qualitative and quantitative data, then using reasoning to explain or describe the data.

- Scientific inquiry involves asking questions, making hypotheses, collecting data, evaluating hypotheses, drawing conclusions, and having the process reviewed by peers.

- The scientific process is not limited to controlled experiments but could also include natural experiments, surveys, and even accidental discoveries.

- Hypotheses, theories, and laws are all scientific models. Hypotheses and theories explain, while laws describe.

- The scientific process does not help us discover absolute truth but instead helps us develop and refine our models of the natural world.

- Applied science looks for specific applications of science, while pure science explores the natural world to learn new things about the universe.

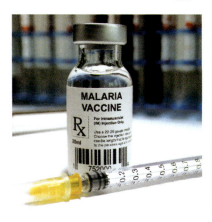

Foundations of Chemistry

01 CHAPTER REVIEW
Chapter Review Questions

RECALLING FACTS

1. What are the foundations of scientific models?
2. Is it possible to establish truth using science? Explain.
3. By what criteria did the Greeks judge the validity of their scientific ideas? What criteria do modern scientists use?
4. Though alchemists were often superstitious, they left a rich legacy for modern chemists. What was their main contribution?
5. Summarize a biblical worldview in one sentence.
6. What are the three parts of the ethics triad to be considered when thinking about ethical questions from a biblical perspective?
7. Is it possible to do science without direct observation? Explain.
8. Which type of reasoning do scientists use to develop theories and models?
9. Which type of reasoning is used to make predictions from theories and models?
10. How is a scientific theory different from either a hypothesis or scientific law?

UNDERSTANDING CONCEPTS

11. Classify each of the following scientific activities as pure chemistry or applied chemistry.
    a. measuring the energy released when iron forms rust
    b. exploring how the size of silver grains on photographic film affects the resolution of the finished print
    c. isolating a chemical compound from the leaves of a newly discovered tropical plant
    d. developing a method to purify polluted well water in a small town
    e. identifying elements in a distant star
12. How is the modern practice of chemistry both similar to and different from that of ancient times?
13. Chemists have developed chemical fertilizers to increase food yields for a growing world population. How is this an example of fulfilling the Creation Mandate?
14. Describe two ways that a Christian can serve God as a chemist.
15. Give examples of quantitative data and qualitative data.
16. Contrast inductive and deductive reasoning.
17. Give two examples of scientific questions.
18. A dermatologist (a scientist who studies skin) is doing a scientific survey of a population of people. Her goal is to see whether there is a connection between the rate of tanning bed usage and the incidence of skin cancer. Which is the best hypothesis that she could propose before making observations? Why?
    a. If tanning bed usage is the largest factor in the recent increase in skin cancer, then I expect to see the most recent cases of skin cancer in younger people who spend significant amounts of time in tanning beds. The rate of skin cancer in active elderly people, construction workers, park rangers, groundskeepers, and other people whose occupations require them to spend a great deal of time outside should be relatively stable.
    b. If tanning bed usage is the largest factor in the recent increase in skin cancer, then I expect to see an increasing rate of elderly people with skin cancer.
    c. If tanning bed usage is the largest factor in the incidence of skin cancer, then I expect to see a direct correlation between hours spent tanning and the occurrence of skin cancer.
19. Look back at Example 1-2 on page 13. Identify steps of scientific inquiry used in the proposed experiment with plants on the International Space Station.
20. If scientific laws are generally reliable statements of science, can they be proved wrong? If so, how?

CRITICAL THINKING

21. Do you agree or disagree with this statement: "The pursuit of pure science is less important than doing applied science"? Explain.
22. How does chemistry affect your life? Describe at least three examples.
23. Are chemists obligated to serve others and the environment through their work? Explain.
24. Why do you think that the Lord Jesus closely associated the command to love others with the command to love God (Mark 12:30–31)?
25. Identify the core problem in the following statement:

    People's ideas of right and wrong are completely the result of the development of our culture and the result of chemical reactions in the brain. When a person does something he views as right, his body produces melatonin, which gives him a good feeling. The increased levels of melatonin can be measured in the blood. The good feeling reinforces his behavior. Morality is just chemistry, and therefore there is no compelling reason to associate ideas of right and wrong with any notion of God. It is all the result of chemical processes and their effect on the body.

26. Design a controlled experiment to test the effect of acid rain on plants.

27. Review the example of the scientific survey discussed on page 14 to monitor the pollutants of private wells in the Midwest. Why are the results of this survey somewhat questionable? What would make them more useful?

28. Can Christians ever agree with naturalists regarding the process of science?

29. Choose a career involving chemistry that interests you. Look online for information to write a one-paragraph summary about that career.

Use the ethics triad on pages 8–9 and the information below to answer Questions 30–34.

30. Does God's Word provide any insight about using pesticides? Consider the three biblical principles.

31. What are the acceptable and unacceptable outcomes of using or not using pesticides?

32. How may the biblical motivations of faith in God, hope in God's promises, and love for God and others impact your decision to use or not use pesticides?

33. From the perspective of a biblical worldview, what is the most important thing to consider when determining whether to use pesticides?

34. What is your opinion about using pesticides? Describe how you would respond to someone who has an opposing opinion.

ethics

PESTICIDES

Part of exercising good and wise dominion is weighing the consequences of an action. Good and wise dominion keeps in mind that people are valuable because they are made in God's image. For example, during World War II, the chemical DDT was used with great success as an insecticide to control diseases like malaria and typhus. As its use became widespread for agriculture and for disease prevention in the United States, malaria cases plummeted. In some instances it was applied in amounts much greater than what was needed. But the results were good—malaria virtually disappeared in the United States.

But in 1962, American marine biologist Rachel Carson published a book called *Silent Spring*, which claimed that animals were being driven to the brink of extinction by DDT. It generated a public outcry against the insecticide. DDT was soon banned, and malaria cases worldwide skyrocketed. Using DDT improperly certainly caused problems, but not using it at all caused even greater problems. Animals are a resource that God has entrusted to us, but God values people's lives more than those of animals. A wise approach to dominion is to seek a balance: saving people's lives while caring for the rest of creation.

Aerial spraying of DDT

Now scientists are bringing back DDT in limited ways, such as by treating mosquito nets with DDT and spraying indoors. They are also exploring other insecticides to control malaria. But the damage has already been done. Good and wise dominion involves planning ahead and examining all the possibilities to save and improve people's lives, often using science.

Chapter 2

MATTER

On a clear night in 1572, Danish astronomer Tycho Brahe had the opportunity of a lifetime: he noticed a "new star" in the sky. Over the following two weeks, the star grew brighter. Surprisingly, within two years the star had faded from view. We now know that Tycho Brahe had observed a supernova, the explosive event that marks the end of the luminous phase of a star's existence.

While a supernova is of immense interest to astronomers and astrophysicists, some chemists are also interested in these events. Many chemists believe that all of the elements in nature have come from the nuclear reactions in stars. Scientists understand that a star's energy is produced by nuclear fusion. Initially, hydrogen nuclei fuse to form helium. As a star ages, the nuclear fuel and products change until it produces iron. Iron has the most stable nucleus of any element and can't undergo nuclear fusion. At this point, a supernova occurs and propels both energy (up to billions of times the energy of the sun) and matter (the first twenty-six elements) into space.

Along with these elements, the supernova ejects billions of neutrons. Iron can capture these neutrons and form higher mass elements. Secular scientists now believe that all of the other naturally occurring elements have formed through a series of neutron capture and nuclear decay events. The elements formed from those supernova events in the past supposedly then went on to become planets, chemical compounds, and even living organisms. So are we just stardust?

2.1 The Classification of Matter *21*
2.2 Energy and Matter *31*
2.3 The States of Matter *40*

2.1 THE CLASSIFICATION OF MATTER

In the Chapter opener you read a current hypothesis in which secular scientists attempt a naturalistic explanation for the origin of matter. While their model explains most observations about matter, they have no observable evidence for its origin, but they accept their model on faith. According to the Bible, everything was created out of nothing by God. A creation model for the origin of matter also explains almost all our observations about matter. We accept the biblical account on faith, and the Genesis account is the foundational truth of a biblical understanding of chemistry.

Organizing Our Study of Matter

The world that God spoke into existence contains an amazing variety of material. Just look around you to see the different types of materials and the things made from them: paper, wood, metal, plastic, computers, and cellphones. And these are just a few of the many forms that matter takes. Scientists have identified millions of different materials, both natural and manmade. Some materials may be in the same form as they were when created. But nearly all accessible natural minerals have been partially or even significantly modified by the Genesis Flood. Humans also rework natural materials into other useful materials.

Recall from Chapter 1 that matter is defined as anything that has mass and takes up space. With so many different types of matter, it would be good to organize our study of matter with a good classification system. Although there are different ways that we could classify matter, this textbook divides it into two major categories: pure substances and mixtures. We can make distinctions between these two categories on the basis of the physical and chemical properties of matter.

QUESTIONS
» Where did matter come from?
» What are some properties of matter?
» What are the different types of matter?
» Where did the names and symbols for elements come from?
» What do chemical formulas tell us?

TERMS
physical property • physical change • chemical property • chemical change • pure substance • mixture • heterogeneous mixture • homogeneous mixture • element • atom • chemical symbol • molecule • compound • chemical formula

Isn't all matter the same?

Properties are the distinguishing characteristics of matter. Scientists divide these characteristics into two classes: physical and chemical.

PHYSICAL PROPERTIES AND CHANGES

A **physical property** is any property of matter that can be observed or measured without altering a substance's chemical composition. Color, shape, texture, state of matter, odor, and taste are examples of properties that we can observe without changing the composition of the material. Other common physical properties of matter follow.

» *What does it look like?*
» *How much is there?*
» *How closely is it packed together?*

Density

The amount of matter packed into a given volume is its density. Dense objects have a lot of matter packed into the space they occupy. Less dense objects have less matter in the same space. Density will be explored further in Chapter 3.

Usually rocks sink because they are denser than water, but pumice—a volcanic rock—floats due to air trapped within its pores.

Malleability

Materials that can be hammered easily into shapes or thin sheets are malleable. This property is possible because the connections between the particles that make up these materials are strong but allow some movement. Most metals are malleable. Thirty grams of gold can be hammered out to cover thirty square meters—more than the surface area of the family car!

Conductivity

The ability of a material to transfer heat or electricity between its particles represents its conductivity. For example, the handle of a metal spoon left in a hot pot on a stove becomes hot, but the handle of a wooden spoon does not. The metal conducts heat easily; the wood does not. Silver is the most electrically conductive metal, but because of silver's high cost, we usually use copper for electrical wiring.

Ductility

Materials that can be drawn into long, thin wires are ductile. Ductility is another property of most metals. The same properties that make gold the most malleable metal also make it the most ductile. A single ounce of gold can be drawn into a fine wire that is eighty kilometers long. Copper and aluminum are also extremely ductile.

The properties listed here are all properties that do not change no matter how much or little of the material is present. For example, a material is ductile regardless of how much material there is. Properties such as these that do not depend on the amount of material present are called *intensive properties*. Other properties, like mass and volume, do change depending on how much of the material is present. These properties are called *extensive properties*.

Physical changes are any changes in the appearance, shape, or state of matter in a material that do not cause a change in its chemical composition. Think about a ductile material like copper. As we draw copper into a long wire, it changes shape, but it is still copper. Boiling is a physical change in which a material changes from its liquid state to vapor, but its identity is not altered.

CHEMICAL PROPERTIES AND CHANGES

Chemical properties are those properties that describe how matter acts in the presence of other materials or how it changes composition under certain conditions. Some examples of chemical properties are reactivity, flammability, and toxicity.

Reactivity

Substances can range from being completely stable to being highly reactive depending on the presence of other substances or particular conditions. Sodium is stable when stored in oil, but it will react vigorously in the presence of water. Water itself is an example of a material that behaves differently under different conditions. Water, typically very stable, will decompose in the presence of electrical energy.

Flammability

While reactivity relates to how something responds to different substances or conditions, flammability is all about whether a substance will burn or not. We can burn hydrogen, demonstrating that it is flammable. When tested for flammability, hydrogen "changes" into water vapor through a chemical change. Also, flammability can depend on various conditions. For instance, metal zinc is not flammable unless it is in powdered form.

Toxicity

Some substances have an adverse effect on living organisms. Toxicity indicates how much damage a substance will do to a particular organism. Toxicity is often reported as its LD_{50} rating. LD stands for the lethal dose, and the LD_{50} of a particular substance is the amount that would be expected to kill half an exposed population. Arsenic is a metalloid that is toxic to humans in very low doses.

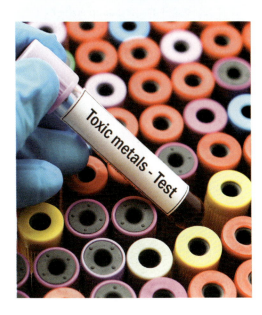

When a substance undergoes a change in which its chemical identity changes, we say that it has experienced a **chemical change**—also called a chemical reaction. When the composition of the substance changes, so do its chemical and physical properties. When iron rusts, it undergoes a chemical change. The iron particles combine with oxygen particles to form iron(III) oxide—rust. Not only is rust a new substance; it also has different properties from oxygen or iron.

Matter

Pure Substances and Mixtures

Scientists use chemical and physical properties to organize matter into pure substances and mixtures.

Pure Substances

A **pure substance** consists of only one type of matter. Scientists further categorize pure substances into *elements* and *compounds*. While elements cannot be broken down by chemical means, compounds can be broken down into elements through chemical changes. You will learn about elements and compounds later in this section.

Mixtures

A **mixture** is two or more substances that are physically combined in a variable ratio. There are two types of mixtures—heterogeneous and homogeneous.

Dissolved CO_2 gives soda its fizz.

A **heterogeneous mixture** consists of two or more materials in distinct regions, called *phases*. Consider oil and water: if left undisturbed, these materials will separate into distinct regions, the oil bubbles and water (see image at left). Many salad dressings contain a mixture of vinegar and oil. No matter how vigorously you shake the dressing, a microscopic examination will always show two separate phases—oil and vinegar. Because of the different phases, heterogeneous mixtures typically have a nonuniform appearance.

Scientists further categorize heterogeneous mixtures as colloids and suspensions depending on the size of the dispersed particles. We will discuss colloids and suspensions in Chapter 14.

A **homogeneous mixture** has only a single phase, which gives it a uniform appearance throughout. Homogeneous mixtures are also called *solutions*.

If you dissolve sugar in water and stir it until it completely dissolves, it appears as a single liquid phase—sugar water. Likewise, when molten gold and silver are thoroughly mixed, a single phase of white gold results. Air is an example of a gaseous solution. Notice the soda, which is a solution, in the bottle at left. While it is a mixture, there is only one liquid phase visible in the bottle.

Separating mixtures

Since mixtures are combined physically, they can be separated by physical means.

The specific method used for separating a mixture will depend on the physical properties of the substances that make up the mixture. Some common methods of separating mixtures include filtration, decanting, and distillation.

Filtration

When a mixture consists of a fluid and another substance with significantly larger particle size, we can separate the substances by filtration. As the small fluid particles move through the filter, the filter blocks the larger particles. Many of us are thankful for the common coffee filter that allows the liquid coffee to pass through while trapping the larger coffee grounds.

Distillation

If the substances in a homogeneous mixture have sufficiently different boiling points, we can separate them by distillation. In this process, the mixture is heated until the lower boiling point is reached. As the mixture boils, it is collected in a separate container as a pure substance. The mixture is further heated until all the individual substances are separated. In the setup below, the mixture in the flask on the left will be heated to the boiling point of one component. That component is condensed in the central part of the apparatus and collected in the flask on the right side. Distillation is commonly used to separate hydrocarbons out of crude oil or to produce alcohols.

Decanting

A heterogeneous mixture of two liquids or a liquid and a solid can be separated by decanting. We decant a mixture by pouring the less dense material off the top of the denser material. Traditionally, decanters were vessels from which wine was decanted, leaving solid sediments behind.

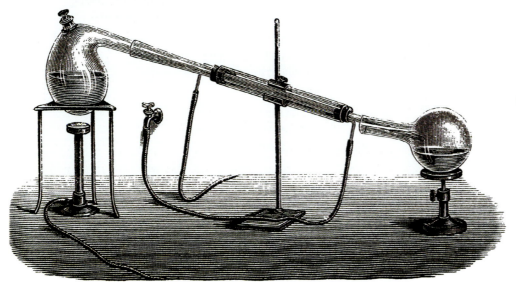

We regularly deal with mixtures in our daily lives, but most of our study of chemistry will involve pure substances.

MINI LAB

PAPER CHROMATOGRAPHY

Chromatography is another method for separating a mixture. In paper chromatography, a sample is placed on chromatography paper (the stationary phase). The paper is then dipped into a solvent (the mobile phase). The solvent dissolves the sample. If the sample is a mixture, the solvent will separate the sample. The components of the mixture, as well as the degree of separation, enable a scientist to identify the mixture.

Is the ink in a marker a mixture or a pure substance?

Procedure

A Straighten the paper clip so that it can span the beaker.

B Cut three 1.5 cm wide strips of chromatography paper. The paper should be slightly longer than the beaker height.

C Pour approximately 30 mL of water into the beaker.

D Fold the top of each piece of chromatography paper so that the bottom of the paper will be just hanging in the water. Don't allow the paper to get wet.

E With a pencil, draw a line 1.5 cm above the bottom edge of the paper.

F On the lines you just drew, put a dot with each of the different markers that you have.

G Hang the papers on the clip so that the paper is in the water but the marker dot is not under the water. Cover the beaker with the watch glass.

1. Why do you think the paper is the stationary phase and the solvent is the mobile phase?

H Remove the papers when the solvent stops moving upward or before it reaches the top of the paper. Mark the highest point of solvent travel.

Conclusion

2. Which of the marker colors was made up of a mixture of colors?

3. Did all the colors travel the same distance up the paper?

Going Further

4. Why do you think the different colors traveled different distances?

Scientists use a measure called the *Rf value* to compare the degree of separation of the different components. They calculate the Rf value for a component by dividing the distance the component traveled by the distance the solvent traveled.

5. Rf values provide quantitative data for the analysis of chromatography. How is this beneficial to scientists?

EQUIPMENT

- beaker, 300 mL
- watch glass
- paper clip
- chromatography paper
- water-based markers (3)

Elements

In the description of the Garden of Eden in Genesis 2, gold is mentioned. Throughout the Bible, there are numerous references to many materials (such as silver, sulfur, tin, and iron) that today we know as elements. Scientists discovered many other elements shortly after the rise of modern science in the 1800s. And other elements have recently been synthesized in the laboratories of nuclear physicists. An *element* is a pure substance that cannot be broken down into a simpler substance by ordinary chemical means. Each element has unique chemical and nuclear properties. Today there are 118 known elements that are represented on the periodic table (see table below).

We also define an **element** as a pure substance that consists of only one kind of atom. For example, a piece of pure copper metal is made up of only copper atoms. An **atom** is the smallest particle that makes up an element and is capable of chemical interactions. An atom consists of a nucleus containing positively charged *protons* and, in most cases, uncharged *neutrons*. A cloud of negatively charged *electrons* surrounds the nucleus. When the number of protons and electrons in an atom are equal, the atom is electrically neutral. However, when they are out of balance, the atom has a net electrical charge and is called an *ion*.

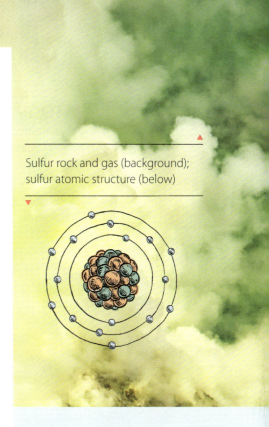

Sulfur rock and gas (background); sulfur atomic structure (below)

Each element has a special **chemical symbol** that represents its name. The first letter of the name of the element often serves as its symbol (e.g., H for hydrogen, N for nitrogen, and O for oxygen).

Frequently the names of more than one element have the same first letter. To avoid confusion in these cases, we use a second, lowercase letter. The second letter is usually related to the sound of the element's name.

Since the single letter C stands for carbon, we use Ca for calcium and Cd for cadmium. We always uppercase the first letter of an element's symbol, and the second letter if present is *always* lowercase. Careless writing of symbols can result in serious errors.

Consider the symbol for the element cobalt: Co. If written carelessly as CO, the referenced element could be confused for the compound carbon monoxide, a poisonous gas found in automobile exhaust.

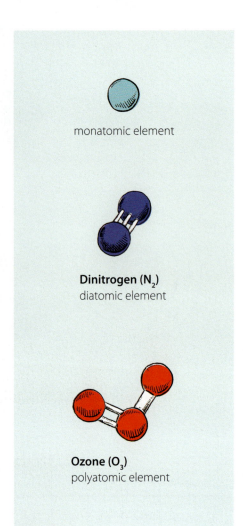

monatomic element

Dinitrogen (N$_2$)
diatomic element

Ozone (O$_3$)
polyatomic element

The particles of elements in their natural state may consist of individual atoms or groups of atoms. If an element occurs naturally as individual atoms, the element is called a *monatomic element*. Monatomic elements are rare because most atoms tend to bond with one or more other atoms. Examples of monatomic elements are neon, argon, and xenon. Elements whose atoms naturally bond into two-atom units are called *diatomic elements*. The seven diatomic elements are hydrogen, nitrogen, oxygen, fluorine, chlorine, bromine, and iodine. Elements whose particles are normally composed of groups of more than two identical atoms are called *polyatomic elements*. For instance, sulfur often occurs in the form of eight atoms bonded into a single unit. Oxygen can also exist as ozone high in the atmosphere.

The subscript number in each chemical symbol shown at left indicates the number of atoms of the element that are present. If there is no subscript, we understand that the element exists in the form of uncombined atoms. Distinct groups of atoms bonded together are called **molecules**. Molecules may consist of atoms from one element or from different elements.

Molecule
diamond molecular structure made up of carbon atoms

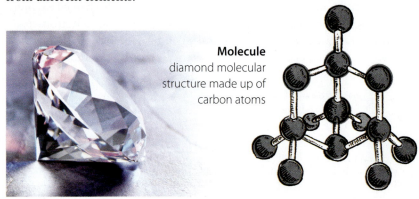

You can easily find the seven diatomic elements on a periodic table if you remember the Hydrogen Seven. Remember hydrogen, then go to element number 7 (nitrogen) and make a large number 7 on the periodic table (oxygen, fluorine, chlorine, bromine, and iodine).

Element names come from many different sources. Latin names serve as the basis for many of the symbols for some elements that were known in ancient times. For example, *cuprum*, the Latin name for copper (Cu), means "from the island of Cyprus" and is the source for its symbol. Elements are named after many things.

color: iridium (Ir) from the Latin for "rainbow"

people: curium (Cm) for Pierre and Marie Curie, early researchers of radioactivity

places: californium (Cf) for California

heavenly bodies: helium (He) from *helios*, the Greek word for "sun"

miscellaneous words: bromine (Br) from the Greek word for "stench"

Today it is generally accepted that whoever discovers a new element has the honor of naming it, subject to the approval of the International Union of Pure and Applied Chemistry (IUPAC), an international organization responsible for standardization in chemistry. The periodic table of the elements in the back of this textbook displays the symbols for all the known elements.

Compounds

A pure substance that consists of two or more elements chemically combined in a fixed ratio is called a **compound**. Some compounds form distinct particles—molecules—while other compounds form vast crystalline arrays with repeating arrangements of ions. As you can imagine, there are many more compounds than elements. Just as chemical symbols are used to represent elements, chemists use **chemical formulas** to represent the millions of kinds of compounds.

The chemical formulas are made up of the chemical symbols for each element present. Numbers written at the lower right of a chemical symbol are called *chemical subscripts*. They indicate the number of atoms or groups of atoms in a chemical formula. A molecule of water contains two hydrogen atoms and a single oxygen atom and so is written H_2O (the 1 for oxygen is assumed). When a subscript follows a group of symbols that are surrounded by parentheses, it refers to the entire group. The formula $Ca(HCO_3)_2$ consists of a calcium atom and two HCO_3 groups, for a total of one calcium atom, two hydrogen atoms, two carbon atoms, and six oxygen atoms. Notice that the subscripts act very much like exponents in mathematical expressions. The six oxygen atoms comes from the three oxygen atoms multiplied by two.

CASE STUDY

UNDERSTANDING TRIHYDROGEN

Trihydrogen is an interesting molecular ion consisting of three hydrogen atoms held together by a single chemical bond. Initially discovered over one hundred years ago in 1911, trihydrogen has recently become of particular interest to astrochemists. In the early 1960s it was hypothesized that trihydrogen could exist in the interstellar medium, the material located between stars. During the 1990s its existence in the interstellar medium was confirmed, and by the early 2000s it was noted that it existed in large quantities. Today naturalistic scientists believe that trihydrogen was a key material in the transition from the origin of matter to the matter we see today.

Questions to Consider

1. Of what is trihydrogen made?
2. What is an ion?
3. What does the existence of trihydrogen in the interstellar medium imply about the origin of matter?

TABLE 2-1
Some Compounds and Their Formulas

Compound	Formula	Atoms
ammonia	NH_3	one nitrogen, three hydrogens
rust	Fe_2O_3	two irons, three oxygens
table salt	$NaCl$	one sodium, one chlorine
cleansing lime	$Ca(HCO_3)_2$	one calcium, two hydrogens, two carbons, six oxygens
sucrose (table sugar)	$C_{12}H_{22}O_{11}$	twelve carbons, twenty-two hydrogens, eleven oxygens
water	H_2O	two hydrogens, one oxygen

In compounds that form molecules, each molecule has a definite number of atoms. A molecule is the smallest distinct particle of a compound. Therefore, molecular chemical formulas indicate the type and number of atoms of each element present in each molecule. To illustrate this, let's consider two compounds of oxygen and nitrogen. The formula NO represents nitrogen monoxide. This formula tells chemists that each molecule of this compound contains one nitrogen atom chemically combined with one oxygen atom. Another compound of nitrogen and oxygen is dinitrogen trioxide (N_2O_3). This compound's molecule consists of two nitrogen atoms combined with three oxygen atoms.

In compounds that form crystalline arrays, there are no distinct particles that make up the compound. Since there are no molecules, the chemical formula represents the smallest whole number ratio of the elements that are in the compound and is called a *formula unit*. Common table salt is a compound of sodium and chlorine ions in a crystalline structure. They combine in a one-to-one ratio, so the chemical formula for table salt is NaCl, indicating that for every sodium ion there is one chlorine ion. Some common compounds, their formulas, and the atoms that they contain are listed in Table 2-1.

EXAMPLE 2-1: DECIPHERING COMPOUND FORMULAS

How many atoms of each element are present in each of the following formulas?

a. $Na_2S_2O_3$ b. $Mg(NO_3)_2$ c. N_3O_5

Solution

a. The subscripts show that two sodium atoms, two sulfur atoms, and three oxygen atoms are present.

b. No subscript after the magnesium indicates that only one atom is present. The subscript 2 after the nitrate (NO_3) group means that two nitrate groups are present, for a total of two nitrogen atoms and six oxygen atoms.

c. The subscripts show that three nitrogen atoms and five oxygen atoms are present.

2.1 SECTION REVIEW

1. What is a physical change? Give three examples of physical changes.

2. Differentiate between physical and chemical properties of matter. Give two examples of each property.

3. What are the two main classes of matter, and what characteristics differentiate them?

4. Explain the difference between heterogeneous and homogeneous mixtures.

5. (True or False) Mixtures can be separated by either physical or chemical processes.

6. Create a hierarchy chart that relates the terms *compound, element, heterogeneous mixture, homogeneous mixture, mixture,* and *pure substance*.

7. Why do you think the chemical symbol for potassium is K?

8. Give the number of each kind of atom in the following compounds.

   a. Li_2O
   b. $Ca(OH)_2$
   c. $HC_2H_3O_2$
   d. $(NH_4)_2CO_3$

2.2 ENERGY AND MATTER

Work and Energy

We do all sorts of work in our daily lives, like when we rake leaves in the yard, move furniture, or ride our bikes. *Work* is done anytime an object moves while a force is applied along its direction of motion. Whenever we do work, we get hungry because we have to use energy to do work, and eating replenishes that energy. Scientists describe **energy** as the ability to do work. According to this description, energy is something that matter has—an ability—just like having a wallet full of cash gives you the ability to buy things. The value of what you purchase with cash should be equivalent to the amount of money you spend. Similarly, there is an equivalence between energy and work done. In fact, scientists use the same dimensional unit to measure them—the *joule* (J).

Since energy is the ability to do work, matter moves whenever work is done, and work makes changes to the energy in matter. We can see how closely connected energy and matter are. Albert Einstein made this connection even stronger with his famous equation, $E = mc^2$. This equation demonstrates the equivalence of matter and energy—they are two forms of the same thing. In the beginning, God spoke everything—matter and energy—into existence. Today we work to increase our understanding of these and other concepts to fulfill the Creation Mandate.

QUESTIONS
» What forms of energy are there?
» What is thermodynamics?
» Where did energy come from?
» Are heat and temperature the same?
» Why are there different temperature scales?

TERMS
energy • **kinetic energy** • **potential energy** • **thermal energy** • **thermodynamics** • **law of conservation of mass-energy** • **system** • **heat** • **temperature** • **exothermic process** • **endothermic process** • **Celsius scale** • **absolute zero** • **Kelvin scale** • **joule**

How can energy be lost?

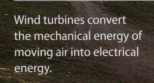

Wind turbines convert the mechanical energy of moving air into electrical energy.

WORLDVIEW INVESTIGATION

THE BIG BANG

You can't read much about origins in science publications without seeing a reference to the Big Bang theory, the currently accepted naturalistic theory regarding the origin of the universe. But what exactly does this theory say? There is much misinformation about what exactly the theory states and implies.

Task

Your Christian school and a homeschool cooperative have been working together to produce a monthly publication. You have been asked to write an article about the big bang, including a perspective from a biblical worldview.

Procedure

1. Research the big bang by doing an internet search using the keywords "big bang" and "big bang cosmology."
2. To get a biblical perspective, go to the Answers in Genesis website and search for "big bang" and "Big Bang theory."
3. Write your article, citing your sources properly. Have a classmate review your article and give you feedback.
4. Complete your article and submit it by the deadline.

Conclusion

We hear information from many sources, some reliable and others not as much. We can all fall into the trap of accepting what we have heard without doing the necessary research to verify the information. But we do a disservice to ourselves and others when we repeat incorrect information. We must be willing to seek out reliable resources and verify what we hear before we repeat information.

Forms of Energy

Energy is related to the force that matter generates and the resulting actions of that force. One of the most obvious forms of energy is *mechanical energy*. This kind of energy is commonly possessed by objects that are moving or have the potential to move. Mechanical energy is the sum of an object's kinetic energy and all forms of its potential energy. **Kinetic energy** (*KE*) is the energy due to an object's motion, while **potential energy** (*PE*) is the energy of an object due to its position or condition.

The sum of the kinetic energy of the particles in an object is called **thermal energy**. We perceive the transfer of thermal energy, known as *heat*, as feeling hot or cold. Sound moves in waves and possesses *acoustic (sound) energy*. Some sounds are loud enough (or we could say "highly energetic enough") that our skin and other organs can sense the energy (e.g., explosions).

The forces that move electrically charged particles involve *electrical energy*. The movement of charged particles creates magnetic fields, which in turn affect other charged particles or magnetic fields. Because electricity and magnetism are so closely related, the energy associated with these two kinds of forces is called *electromagnetic energy*. Visible light is one form of electromagnetic energy.

Energy is stored in compounds as *chemical energy*—a very important concept in chemistry. As chemical reactions occur, the bonds in compounds are broken, the atoms are rearranged, and new bonds form. These changes use and release chemical energy. And finally, when the nucleus of an atom breaks apart or when particles are added to or removed from a nucleus, *nuclear energy* is involved.

As you can see, every form of energy can change the motion of matter to do work. When energy is absorbed by matter, it has done some kind of work on that matter. More importantly, any one kind of energy can be converted into other forms.

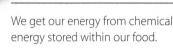

We get our energy from chemical energy stored within our food.

Conservation of Mass-Energy

Steam engines, like the one at right, were designed to convert energy from one form to another. In the early nineteenth century, scientists developed the science of **thermodynamics** to study the movement and conversion of energy, especially thermal energy. This new field of study began with the invention of the steam engine during the Industrial Revolution. Scientists were interested in being able to make this new invention more efficient. Scientists quickly recognized that the laws of thermodynamics seem to govern all areas of science. No exception to these laws has ever been observed in natural processes—they apply to every field, at all times, in all instances.

In the course of their studies, scientists discovered that in all energy conversions, energy is always conserved. In other words, in any energy conversion, the sum of all the kinds of energy produced always equals the total energy consumed. Many careful measurements have shown that energy cannot be created or destroyed, but can change from one form to another. This principle is known as the *law of energy conservation*. We often hear references to lost energy. What is being described is the loss of *useful* energy. The energy in gasoline is used to move a car, but most of the energy is "lost," not used to move the car, but is converted to thermal energy that leaves the car "system" with the exhaust.

Since Einstein's equation reveals the equivalence of matter and energy, the law of energy conservation is sometimes called the **law of conservation of mass-energy**. Matter and energy cannot be created or destroyed but can be transformed or transferred between objects. This principle is also known as the *first law of thermodynamics*. Apart from instances where God created matter or energy, the total amount of mass and energy has remained constant. All mass-energy changes that have occurred since Creation, by natural processes or by man, have merely changed one form of energy or matter into another form of energy or matter.

The first law of thermodynamics is problematic for the naturalist regarding the origins of the universe. It cannot explain the source of matter and energy from a purely naturalistic perspective. A creationist has no difficulty with this idea. Because all creation and scientific laws have their origin in God, they do not contradict one another.

A steam fire engine from the 1800s is a fantastic example of converting between different forms of energy.

The Law of Entropy

We know that water flows downhill, that warm coffee cools, and that wind blows from high pressure to low pressure. All these experiences teach us that things naturally flow from high energy states to lower energy states. Think about placing a drop of food coloring into a beaker of water. Even if you leave it undisturbed, eventually the color evenly disperses. The particles of color spread naturally from high concentration to lower concentration through a process called *diffusion*. You would not expect the color particles to return to form a drop-sized volume somewhere in the container. The probability of such an event is extremely remote—so remote that you would probably say that it is impossible. The measure of the dispersal of energy is called *entropy*. The entropy of the universe is continuously increasing as time marches on.

diffusion

After observing countless instances of entropy, scientists formed a conclusion. According to what is known as the *second law of thermodynamics*, the entropy of an isolated system can never decrease. All natural processes tend to progress toward increased entropy, and toward decreased usable energy. A significant consequence of the second law of thermodynamics is that during any energy conversion, some of the energy changes into an unusable form; that is, we often say that the energy is lost. We can see the second law of thermodynamics in action in the three examples stated above. As water flows downhill, moving from high to low potential energy, some of the energy is transformed to thermal energy as the water does work on the ground over which it moves. That cup of coffee transfers thermal energy to the environment. The energy spreads throughout the room and is less usable. The pressure differential in the atmosphere produces wind that does work throughout the atmosphere, and energy is spread out.

The universe is too big to study, so in order to simplify the analysis of a phenomenon, scientists will limit the scope of their investigation. They define the boundaries within which the analysis is limited. Everything inside the boundaries is called the system of interest, or just the **system**. A system could be a beaker in a laboratory, a rocket engine, or the entire rocket. A system can be a distinct part of an object, such as the burner of a stove. It can also be a small portion identified by imaginary boundaries within a larger volume of a substance, such as a volume of air within the atmosphere. Throughout this discussion of energy, the system is assumed to be something with distinct boundaries, an object, or a sample of a substance.

Every system is subject to the second law. While work can be done on a system to increase the energy and decrease the entropy within the system, the total entropy of the system and its surroundings will still increase. Any decrease in entropy in one part of the universe comes at the expense of increased entropy elsewhere.

Thermal Energy, Temperature, and Heat

As scientists worked to understand the new field of thermodynamics, they were interested in the concept of how thermal energy moved. The term we use today for the movement, or flow, of thermal energy, is **heat**, which always flows from warm to cold. The early theory of heat, that thermal energy moved as an invisible, self-repelling form of matter, was called the *caloric theory*. The caloric theory was workable at the time since it explained many observations about matter and thermal energy. Eventually there were some new observations that the caloric theory could not explain, and therefore, scientists needed to modify or replace their theory.

According to the *particle model of matter*, all matter is constructed from particles, whether atoms, molecules, or combinations of both. This model has been proposed in various forms since at least the fifth century BC. With the advances of quantitative chemistry in the early 1800s, scientists accepted the particle model of matter on experimental rather than philosophical grounds.

There is strong evidence for the continuous movement of particles in matter. For example, the odor of perfume diffuses from an open bottle, and water droplets evaporate. Under a high-powered light microscope, we can see minute bits of matter moving in random ways as water particles jostle them. English botanist Robert Brown observed this in 1827. The random jiggling of microscopic matter is now known as *Brownian motion*.

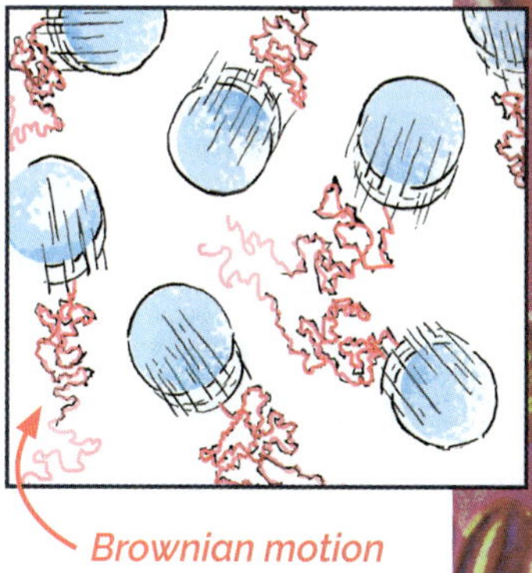

Brownian motion

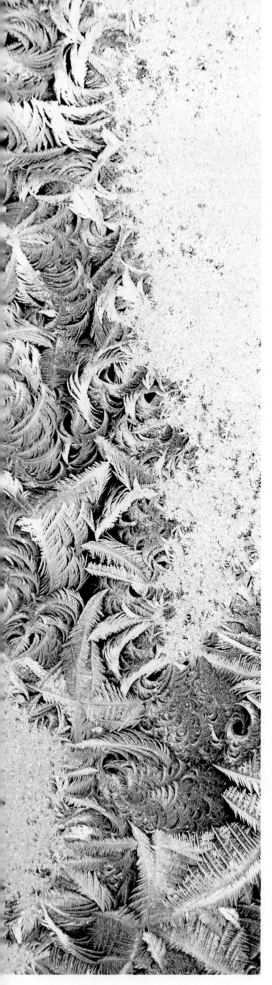

In the nineteenth century, thermodynamics proposed a new model for the movement of heat. Scientists agreed that particles of matter are in constant motion. At the particle level, this motion may occur in straight lines (translational motion); as small, random vibrations (vibrational motion); or as a spin (rotational motion). The sum of all these motions determines the total kinetic energy of the particle.

As stated previously, kinetic energy is the energy due to the motion of an object. The kinetic energy of a system is proportional to the mass of the system and the square of its velocity. The formula for kinetic energy is

$$KE = \frac{1}{2}mv^2,$$

where KE is the kinetic energy in joules, m is the mass of the object in kilograms, and v is the object's velocity in meters per second. You can see that for a given particle, kinetic energy varies with velocity. Because the velocity is squared, doubling a particle's speed quadruples its kinetic energy.

It is very difficult to measure the energy of each particle in a system, but we can study the energy of the entire object. As mentioned above, the thermal energy of a system is the sum of the kinetic energies of all particles in the system. While we can measure changes to thermal energy, it is easier to measure temperature. We tend to think of temperature as a measure of how hot or cold an object is, but scientists define **temperature** as the *average* kinetic energy of the system's particles.

To understand the relationship between temperature, thermal energy, and heat, consider a large pot full of hot soup. If you scoop out some of the soup to fill a small bowl, the soup in the pot and the bowl will have the same temperature because the particles have the same average kinetic energy. But the soup in the pot will have more thermal energy. This is due to the greater number of particles in the pot than in the bowl—more particles produce a greater sum of kinetic energies. As the soup sits in the bowl, it cools. The soup is transferring thermal energy to the air around it. Recall that this flow of thermal energy is called heat. And heat always moves thermal energy from a higher temperature object to an object with a lower temperature.

The word "heat" is most correctly used as a noun but can also be used as a verb to describe a process, such as "heating a beaker." In this sense, it means to add thermal energy to something and is the opposite of cooling, which is the removal of thermal energy.

Most chemical and physical processes either release or absorb energy. These processes can be described as *exergonic* (releasing energy) or *endergonic* (absorbing energy). In thermodynamics, we are typically interested in the movement of thermal energy. Processes that release thermal energy, such as burning natural gas, the explosion of a firecracker, or the dissolving of calcium chloride, are called **exothermic processes**. Changes that absorb thermal energy are called **endothermic processes**. Chemical cold packs used to treat sprains feel cold because the chemical change absorbs thermal energy from its surroundings. We will discuss exothermic and endothermic reactions in Chapter 15.

Measuring Temperature and Thermal Energy

Scientists have developed instruments to measure temperature by using *thermometric properties*—properties of matter that vary directly with changes in temperature. One such property is volume. In Chapter 13 you will study how the volumes of liquids and solids generally increase proportionally with temperature. If a liquid is enclosed within a thin rod of glass, a small temperature change can produce a measurable change in the distance that the liquid extends into the rod. In this instrument, called a *thermometer*, the change can be compared with a calibrated scale to determine the temperature.

The two temperature scales used most often in chemistry are the Celsius and Kelvin scales. The Celsius scale was named in honor of Swedish astronomer Anders Celsius, who first considered making a centigrade scale, one with one hundred degrees between its definition points. The Celsius scale measures temperature using the degree Celsius (°C) as its unit. Originally, the scale was calibrated to the freezing and boiling points of water. Today the **Celsius scale** uses absolute zero and the triple point of water as its two reference points. These two reference points, also called *fiduciary points*, provide verifiable points for instrument standardization. **Absolute zero** is the temperature at which particles have minimum motion. Absolute zero equates to −273.15 °C. The *third law of thermodynamics* states that we can never decrease a system's entropy to an absolute minimum. In other words, it is impossible to get any system to absolute zero. The *triple point* of a substance is the temperature and pressure at which the substance exists as a solid, liquid, and vapor in equilibrium.

The Celsius temperature scale contains both positive and negative temperatures. While this is not a problem for observing temperatures, negative values create problems when doing calculations with the gas laws (Chapter 12). For this reason, another temperature scale was proposed in the 1800s that consisted of only positive values, called the *absolute thermodynamic temperature*, or **Kelvin scale**. The lowest possible value on the Kelvin scale is absolute zero, the point of minimum particle motion. This scale was named after the prominent Scottish physicist William Thomson, more commonly known as Lord Kelvin. A temperature change of one kelvin is equal to one degree Celsius. The Kelvin scale uses the same two reference points as the Celsius scale and uses the kelvin (K) as its unit. In 2019, both the degree Celsius and the kelvin were redefined on the basis of the Boltzmann constant—a physical constant.

TABLE 2-2 *Select Temperatures on Temperature Scales*

Scale	Absolute Zero	Triple Point	Freezing Point of Water	Boiling Point of Water
Celsius (°C)	-273.15	0.01	0	100
Kelvin (K)	0	273.16	273.15	373.15
Fahrenheit (°F)	-459.67	32.02	32	212

As you can see from the relationships between the Kelvin and Celsius scales in Table 2-2, the numerical value of a Kelvin temperature is always 273.15 degrees more than the equivalent Celsius temperature. Kelvin temperatures are never negative, and therefore their numerical values are *always* greater than the corresponding Celsius temperature. This observation should help you when working on conversion problems. When required, you can make temperature measurements in degrees Celsius and then convert them to kelvins. The equations for conversion between Celsius and Kelvin temperatures are shown below.

$$T_K = T_C + 273.15 \text{ and } T_C = T_K - 273.15$$

T_K is the Kelvin temperature and T_C is the Celsius temperature. Note that we do not use the degree symbol when referring to Kelvin temperatures. So we say that the normal freezing point of water is "273.15 kelvins," not "273.15 degrees Kelvin."

Although the Fahrenheit scale is not usually used for scientific purposes, it is the scale that most Americans know best. Use the following formula to convert from Fahrenheit to Celsius readings.

$$T_C = \frac{5}{9}(T_F - 32)$$

The use of a little algebra will enable you to rearrange the formula to convert Celsius temperatures into Fahrenheit temperatures.

EXAMPLE 2-2: CONVERTING FROM KELVIN TO CELSIUS AND FAHRENHEIT

The temperature of seawater jetting from a black smoker on the bottom of the ocean at the Mid-Atlantic Ridge is 671.50 K. What is this temperature in degrees Celsius and Fahrenheit?

Solution
Write what you know.
$T_K = 671.50$ °C
$T_C = ?$
$T_F = ?$

Write the formula and solve for the unknown.
$T_C = T_K - 273.15$

Plug in known values and evaluate.
$T_C = 671.50 - 273.15$
$= 398.35$

Write the formula and solve for the unknown.

$T_C = \frac{5}{9}(T_F - 32)$
$\frac{9}{5}T_C = \frac{5}{9}(T_F - 32)\frac{9}{5}$
$\frac{9}{5}T_C + 32 = T_F - 32 + 32$
$T_F = \frac{9}{5}T_C + 32$

Plug in known values and evaluate.

$T_F = \frac{9}{5}(398.35) + 32$
$= 717.03 + 32$
$= 749.03$

The temperature of the seawater from the black smoker is 398.35 °C or 749.03 °F

EXAMPLE 2-3: CONVERT FROM CELSIUS TO KELVIN

While performing an experiment, you collected a sample of gas in a bottle immersed in 32.50 °C water. What is this temperature in kelvins?

Write what you know.

$T_C = 32.50\ °C$
$T_K = ?$

Write the formula and solve for the unknown.

$T_K = T_C + 273.15$

Plug in known values and evaluate.

$T_K = 32.50 + 273.15$
$ = 305.65\ K$

The water temperature is 305.65 K

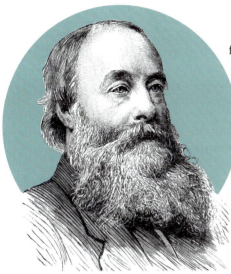

In most science applications, the standard unit of measure for work, energy, or heat in a system is the **joule** (J). The joule is named for the English physicist James Prescott Joule, who distinguished himself by his scientific integrity and his precise measurements. Another energy unit that is used less today than in the past is the *calorie* (cal)—the amount of heat transfer required to change the temperature of one gram of water one degree Celsius. Calories and joules relate to each other according to this definition: 1 cal = 4.184 J (exactly). Larger units are called kilocalories (kcal), which are a thousand times as large. The kilocalorie is equivalent to the Calorie (note the uppercase C), which we still use in reference to the energy content of foods.

James Prescott Joule

2.2 SECTION REVIEW

1. Is energy something a system does or something it has? Explain.

2. Name the principal type of energy involved in each of the following.
   a. vibrations from a loudspeaker
   b. a pot of near-boiling water
   c. a camera flash
   d. an explosion of dynamite
   e. changes occurring at the center of a star
   f. a rolling ball

3. State the first law of thermodynamics. What does this law imply for the origins of matter and energy?

4. What is entropy?

5. Compare the entropy of a firecracker before it is lighted with its entropy after it explodes. Explain the difference.

6. State three pieces of evidence for the idea that the particles of substances are in continuous motion.

7. Since heat transfers take place only between systems with different temperatures, explain why our current understanding of matter supports the third law of thermodynamics.

8. Convert the following temperatures as indicated.
   a. 22 °C (room temperature) to kelvins
   b. −218.30 °C (melting point of oxygen) to kelvins
   c. 496 K (temperature of rock at a depth of 5.9 km in the earth's crust) to degrees Celsius
   d. 2.726 K (average temperature of deep space) to degrees Celsius

QUESTIONS

» Do the particles in a solid move?
» How do states of matter differ?
» What causes solids to melt?
» Are the melting and freezing points always the same?

TERMS

kinetic-molecular theory • solid • liquid • gas • phase change • condensation • vaporization • freezing • melting • sublimation • deposition

2.3 THE STATES OF MATTER

According to the particle model of matter, all matter is composed of tiny particles (atoms, molecules, or ions) that are in constant motion. If we add energy to these particles, their motion increases. The greater the amount of energy added, the faster the resulting motions will be. If particle motions were not in some way limited, matter would fly apart. Instead, forces between these particles limit the movement of the particles toward and away from each other. Scientists call the model described by these ideas the **kinetic-molecular theory** since it describes the motion of particles in matter.

Why do ice cubes in the freezer disappear?

Common States of Matter

One of the physical properties mentioned in Section 2.1 was the states that matter can exist in. There are many different states of matter, but we will address the three most common—solid, liquid, and gas.

Many students believe that the state of matter is determined by the temperature of the material. While that is partially true, it is more accurate to say that state of matter is determined by the kinetic energy of the particles and the attractive forces between the particles as explained by the kinetic-molecular theory.

Solids

We may think of metals, plastic, or wood when we consider solids. Solids are commonly used to build things because of their properties. The particles in a **solid** have relatively little kinetic energy compared with the attractive forces that are present between particles. The attractive forces significantly limit the movement of the particles, which vibrate in relatively fixed positions. This produces several distinct properties.

For example, solids tend to have a fixed volume and a rigid, fixed shape. They are also relatively incompressible because the structure of atoms makes them resist being squeezed more compactly. We will discuss the structure of atoms in Chapters 4 and 5.

Solids have a definite shape and volume, are difficult to compress, are packed closely together, and vibrate in rigidly held positions.

40 Chapter 2

Liquids

Water is a common substance that people think of when they think about liquids. One of the most recognizable properties of liquids is that they can flow—they are *fluid*. Liquids can flow because their particles have a degree of freedom and can move about. The particles in a **liquid** have enough kinetic energy to partially overcome the attractive forces between them and are not held rigidly in fixed positions. While they can move about somewhat, the attractive forces have not been completely overcome, so the particles remain close to each other. Because they are still close to each other, the volume of a liquid is relatively constant—it is nearly incompressible. The shape of a liquid, however, changes and is determined by the container that holds it.

A liquid has a definite volume and assumes the shape of its container. Liquids are difficult to compress and are packed closely together, but particles are able to slide past each other.

Gases

If a substance has enough kinetic energy for its particles to completely overcome all the attractive forces between them, the material is a **gas**, or *vapor*. The particles are widely spaced and interact only occasionally. They move rapidly in random directions across great distances. At 25 °C, the average velocity of an oxygen molecule is 444 m/s. Because of the wide spacing of their particles, gases are highly compressible. They have no definite shape and expand to fill their container. Since gas particles are highly mobile, they are also fluids.

Other states of matter exist under extreme conditions. One of these, *plasma*, is actually the most abundant state of matter in the universe. Stars are made of plasma. The high temperatures and pressures in stars remove electrons from gases, leaving a collection of positive ions and free electrons.

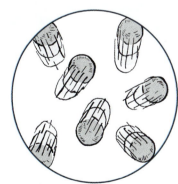

A gas has a variable shape and volume that is limited only by its container. Gases are easy to compress and have highly mobile particles with high speeds.

chlorine gas

condensation

Changes of State

A substance can experience a change of state as thermal energy is added to or removed from it. As thermal energy is added to a material, the particles either gain kinetic energy, resulting in an increased temperature or a change of state. The material will progress from a solid to a liquid, then to a gas. Each of these changes in state provides the particles with a greater degree of freedom of movement, resulting in different properties for the matter.

Conversely, as energy is removed from a material, it will progress from a gas, then to a liquid, and finally to a solid. Removing thermal energy results in either a change of state or a lowering of the material's temperature. Any change of state is a physical change because the substance changes form but is still chemically the same substance.

As a gas cools by transferring thermal energy or by converting it into other forms of energy, its particles slow down. Lacking the necessary energy to resist the attractive forces between them, the particles begin clumping and then finally draw together into drops recognizable as the liquid state of the substance. The change of state, or **phase change**, from gas to liquid is called **condensation**. This process occurs, for example, when fog droplets or dew form, which is why the temperature at which this occurs is called the *dew point*. The opposite process to condensation is **vaporization**, in which a liquid changes into a vapor, or gas. Boiling and evaporation are two different mechanisms. You will learn about these in Chapter 13.

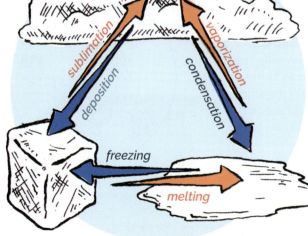

When particles in a liquid lose thermal energy, their motion slows to the point that they become locked in position by the attractive forces between them. The liquid transitions to the solid state in a process called **freezing**, or *solidification*. The opposite process, a solid **melting** to become a liquid, occurs as thermal energy is transferred into the material. Melting and freezing normally take place at the same temperature, called the *melting point* (if thermal energy is being added) or the *freezing point* (if thermal energy is being removed).

Under certain conditions, a solid may change directly into a gas, or a gas may change directly into a solid, without passing through the liquid phase. The change from a solid to a gas is called **sublimation**. The change from a gas directly to the solid state is called **deposition**. Sublimation occurs even at temperatures well below the melting point of most solids. Recall that the temperature of a system is related to the *average* kinetic energy of its particles. Some particles can individually gain kinetic energy well above the average. They then break away from the attractive forces at the surface of the solid, entering the gaseous state. This process removes material from the solid. Some common examples of substances that sublime directly from a solid to a gas include mothballs, solid air fresheners, and ice cubes in a freezer. These solids gradually change to a gas and seem to "disappear." The deposition of gaseous water vapor to solid ice is the principal process in the formation of snowflakes and frost.

CAREERS

SERVING AS A MATERIALS SCIENTIST: *INFORMATIVE FIBERS*

If you look up information about the field of materials science, you'll read about some amazing things that scientists are working on—microscopic sensors, room temperature superconductors, quantum dots, metal foams, amorphous metals, programmable materials, cloaking technology, and even flexible electronics that you could wear. Much of it sounds like science fiction, but materials scientists bring the completely fantastic closer to reality every day.

Today many people wear a wristband that monitors pulse, blood pressure, biochemical levels, calories burned, exercise levels, and sleep times. But materials scientists think that these devices could do so much more. They are currently working on devices that could monitor athletic performance, chronic diseases, and blood alcohol content. Some applications could also transmit this information to medical professionals.

One group of researchers in Japan is working on flexible ferroelectric polymers. These could potentially allow manufacturers to incorporate biometric-collection technology into the clothing we wear. The goal would be to collect needed data for sharing with doctors but to do it in a way with the least negative impact on a patient's daily activities. For some patients, that could be the difference between life or death. For other people, it could just be a matter of living a fuller life with fewer restrictions.

sublimation

Final Thoughts on Origins

Throughout this chapter, you have learned about matter. Most of the things that we know about matter are straightforward with little need for interpretation. Understand that all scientists look at the same evidence. We all study the same properties of matter, the same behavior of matter, and the same changes in matter. However, our worldview affects how we interpret that data. As evidenced in the Chapter opener, the origin of matter is subject to interpretation. Our worldview has a direct impact on our understanding of where matter came from.

Scientists with a naturalistic worldview claim that all matter and energy originated with the big bang. They believe that hydrogen formed early in the universe. Over time, under the influence of various forces, this hydrogen was changed into all of the now-known elements. Those elements formed compounds and with sufficient time and fortuitous circumstances formed complex organic compounds. These compounds then transformed into living organisms. Through mutations and natural selection, the organisms evolved into all the different forms of life that have existed, both now and in the past.

The law of cause and effect creates difficulty for scientists with a naturalistic worldview. They accept, on faith, the certainty of the Big Bang model, even though that model doesn't include any cause for what started it all. But scientists who view the world through the lens of the Bible believe that God created everything in the Creation as described in the Bible. All matter and energy were formed through the power of God. The forces and processes that we see today are effects of the properties that God created in matter and energy. Over time those forces and processes have changed some aspects of the world. We too accept the origin of matter on faith, but our faith is in a loving God and the written record of the origin of matter and energy.

2.3 SECTION REVIEW

1. List the common states of matter for a substance in order from the most to least energetic.

2. Use a graphic organizer to summarize the characteristics of the three common states of matter described in the chapter.

3. Why is temperature the most important factor in determining the state of a substance?

4. Name the change of state that is described by each of the following processes.

    a. Pond water solidifies on a cold winter day.

    b. The fragrance of an open perfume bottle is detected from across the room.

    c. Steam leaves a teakettle that is sitting on a hot stove.

    d. Frost patterns appear on a cold windowpane in winter.

    e. Water mist collects on the bathroom mirror after a shower.

    f. You smell the odor of mothballs when you open a clothes box.

    g. As a candle burns, wax drips down the sides of the candle.

02 CHAPTER REVIEW
Chapter Summary

TERMS
physical property
physical change
chemical property
chemical change
pure substance
mixture
heterogeneous mixture
homogeneous mixture
element
atom
chemical symbol
molecule
compound
chemical formula

2.1 THE CLASSIFICATION OF MATTER

- Scientists classify matter on the basis of its physical and chemical properties.

- Physical properties are any property of matter that can be observed without altering the material's chemical composition. Physical changes change a material's appearance or state without causing a change in its composition.

- Chemical properties describe how matter acts in the presence of other materials or how it changes composition under certain conditions. Chemical changes alter the composition of matter by rearranging its particles.

- Matter can be classified as pure substances (elements or compounds) or mixtures (heterogeneous or homogeneous).

- Atoms are the fundamental particles of elements. Molecules are two or more atoms chemically bonded.

- Chemical symbols are used to represent the names of elements in chemical expressions. Chemical formulas represent the composition of compounds.

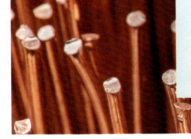

2.2 ENERGY AND MATTER

- Matter and energy are two fundamental quantities in the universe. Energy is the ability to do work. Work moves matter and alters the energy within matter.

- There are two major categories of energy: kinetic and potential.

- Thermodynamics is the study of the movement and transformation of energy. The laws of thermodynamics establish some basic principles regarding the origin and destiny of energy and matter.

- Thermal energy is the sum of the kinetic energy of the particles within a substance. Thermal energy naturally flows between two points at different temperatures. The flow of thermal energy is called heat and always moves from hotter to colder temperatures.

- Temperature is a measure of the average kinetic energy of the particles within a substance.

- An exothermic process results in a net transfer of thermal energy *out* of a system. An endothermic process results in a net transfer of thermal energy *into* a system.

- The Celsius and Kelvin (absolute) temperature scales are commonly used in chemistry.

TERMS
energy
kinetic energy
potential energy
thermal energy
thermodynamics
law of conservation of mass-energy
system
heat
temperature
exothermic process
endothermic process
Celsius scale
absolute zero
Kelvin scale
joule

Matter 45

02 CHAPTER REVIEW

Chapter Summary

2.3 THE STATES OF MATTER

- According to the kinetic-molecular theory, matter consists of tiny particles in constant motion.
- Three states of matter occur commonly on earth: solid, liquid, and gas.
- Changes of state occur as energy is added to or removed from matter.
- As energy is added to a substance, either the temperature of the substance will increase or the state of matter will change from a solid to a liquid and then to a gas.
- As energy is removed from a substance, either the temperature of the substance will decrease or the state of matter will change from a gas, then to a liquid, and finally to a solid.

TERMS
kinetic-molecular theory
solid
liquid
gas
phase change
condensation
vaporization
freezing
melting
sublimation
deposition

Chapter Review Questions

RECALLING FACTS

1. Define *matter*.
2. List three physical properties of matter, including at least one that is not mentioned in Chapter 2.
3. What is a heterogeneous mixture?
4. Describe the two kinds of pure substances. How do these differ from mixtures?
5. What do chemical formulas tell us?
6. What specifically does thermodynamics investigate? What invention spurred its development?
7. State the second law of thermodynamics. What does this law imply for the origins of matter and energy?
8. What are the two fiduciary points for the Celsius and Kelvin scales?
9. Summarize the kinetic-molecular theory of matter.
10. What are the two processes of vaporization?
11. How does the melting point of a substance compare with its freezing point?
12. Which meteorological conditions are caused by sublimation?
13. Discuss the two beliefs about the origin of matter outlined in Chapter 2.

UNDERSTANDING CONCEPTS

14. In the Chapter opener image, the matter and energy from the supernova is shown moving outward at high speed. What would have to happen for that matter to form a planet?
15. Indicate whether each of the following properties is a physical or chemical property.
    - **a.** color
    - **b.** density
    - **c.** magnetism
    - **d.** corrosiveness
    - **e.** insulating ability
    - **f.** flammability
16. Indicate whether each of the following processes involves a physical change, a chemical change, or both.
    - **a.** a wafer of silicon being scored and cut into chips to be used in computer microprocessors
    - **b.** a glacier melting
    - **c.** small water droplets forming on the outside surface of a cold can of soft drink
    - **d.** dynamite exploding and the resulting collapse of an old building
    - **e.** a burning candlewick melting wax
    - **f.** the rusting of a bicycle
    - **g.** the growth of a child
    - **h.** the erosion of a hillside

17. Which of the separation techniques mentioned in Chapter 2 would work well for separating a solution of alcohol and water? Explain.

18. How are the chemical symbols representing the names of elements written? Where do the names come from?

19. Classify each of the following as an element or a compound. Then identify each as a monatomic, diatomic, or polyatomic element or as a molecule.

    a. oxygen (O_2)   d. helium (He)

    b. methane (CH_4)   e. carbon monoxide (CO)

    c. ozone (O_3)   f. hydrogen peroxide (H_2O_2)

20. Compare elements and compounds.

21. Compare compounds and mixtures.

22. Use the classification scheme of matter presented in this chapter to classify the following substances as elements, compounds, homogeneous mixtures, or heterogeneous mixtures. Assume that the substances are uncontaminated.

    a. vegetable soup   d. crude oil

    b. air   e. sulfuric acid (H_2SO_4)

    c. oxygen (O_2)   f. granite

23. Identify the elements and the number of atoms of each element for each of the following.

    a. O_3

    b. $CaCl_2$

    c. $(NH_4)_3PO_4$

    d. $CH_3(CH_2)_3CH_2OH$

    e. $Al(C_2H_3O_2)_3$

    f. NH_4HCO_3

24. How are work, energy, and matter related?

25. How do the laws of thermodynamics relate to the theories about the origin of matter and energy?

26. Compare thermal energy, temperature, and heat.

27. What do you mean when you say, "This cup of cocoa that I am holding is hot."

28. Describe exothermic and endothermic processes and give an example of each.

29. Why does a chemical ice pack feel cold?

30. Why is the Kelvin scale called the absolute thermodynamic temperature scale?

31. Convert each of the following temperatures to the scale indicated.

    a. 313.0 K to Celsius

    b. –118.53 °C to Kelvin

    c. 87.2 °F to Kelvin

32. Explain what is happening to the particles in a solid as thermal energy is added. Assume that the material changes to a vapor. Include in your explanation the concepts of melting, melting point, vaporization, and boiling point.

33. Are the melting point and the freezing point always the same for a particular pure substance? Explain.

CRITICAL THINKING

34. If a new element were being named kelvinium to honor Lord Kelvin, what would be a good chemical symbol? Explain.

35. People's ideas are not always in harmony with the way that the universe operates. Identify the law of thermodynamics that indicates that each of the following is impossible.

    a. A perpetual motion machine, once started with an initial input of energy, will continue running indefinitely.

    b. A machine produces more energy than it consumes.

    c. The universe spontaneously came into existence.

36. Derive a formula to convert Kelvin temperatures to Fahrenheit temperatures.

37. A classmate tells you that since a solid has a fixed shape and volume, its particles must not be moving. Is he correct? Explain.

38. A classmate tells you that the origin of matter and energy in the universe by naturalistic processes is a fact beyond scientific debate. Do you agree or disagree? Why?

Chapter 3

MEASUREMENTS IN CHEMISTRY

On December 11, 1998, NASA launched the Mars Climate Orbiter into space. Its mission—to study the Martian climate and relay messages to a lander on the surface of Mars. All went well until September 23, 1999, when NASA officials lost contact with the spacecraft as it tried to enter the Martian atmosphere. They later learned that the Mars Climate Orbiter had penetrated the Martian atmosphere too deeply, and more than $125 million of science equipment vaporized in the Martian atmosphere.

What happened? Engineers had done extensive calculations to determine the best altitude for the craft to enter Mars's orbit—about 150 km. One team programmed the flight system software on the orbiter using the metric system. But another team programmed the ground-based system in the US customary system. The two computers were speaking different languages. So instead of entering orbit at 150 km, the craft descended to 60 km! The tremendous forces acting on the spacecraft at this altitude tore it to shreds.

3.1 Measurement Systems *49*
3.2 Measurements *58*
3.3 Problem Solving in Chemistry *65*

3.1 MEASUREMENT SYSTEMS

Measuring and the Creation Mandate

What is the diameter of an oxygen atom? How fast does a chemical reaction occur? How much energy is released when one gram of hydrogen burns? The answers to these questions can be found only by measuring.

In the beginning God ordered man to subdue the earth and have dominion over the creatures living there. This charge—what we called the Creation Mandate in Chapter 1—does not imply that humans are only to occupy the earth. Rather, we are to actively pursue knowledge about the earth and the living things that inhabit it. Gaining knowledge about God's creation so that we will be able to benefit others is a major purpose underlying the study of science from a biblical worldview. One of the key ways to gain information about creation is to take measurements of many different aspects of nature. Measurements are quantitative data about the world around us.

What Is Measuring?

You have probably measured things before. If you enjoy woodworking or carpentry, you have measured the length, width, or thickness of boards. If you cook, you have measured dry and liquid volumes of ingredients in measuring cups. But have you ever considered what measuring involves? Could you describe how to measure in your own words?

A **measurement** is a comparison of an unknown quantity to a known standard unit. Many phenomena have measurable characteristics. For example, scientists measure quantities such as length, mass, volume, time, and electrical charge. These measurable properties are called *dimensions*. Before a dimension can be measured, there must be a unit of measure defined for that dimension. For example, you are probably familiar with the meter as a unit of measure for the dimension of length. Other units for length include the kilometer, mile, and inch. The word *unit* means "one of something." Dimensional units are lined up on a scale to provide a means of measuring. The scale is graduated or accurately subdivided into measurement units. A graduated scale is essential to every instrument designed to measure anything. An instrument is an artificial device made for refining, extending, or substituting for the human senses when measuring.

A measurement consists of two parts—a number and a unit. A unit is the standard by which the scientist compares the object. But what standard should we use?

QUESTIONS
» What is a measurement?
» Why do we need standard units?
» What units are used by scientists?
» What are the benefits of the SI?
» How can I change units?

TERMS
measurement • SI • fundamental unit • derived unit • unit conversion • conversion factor

Why do I need to learn the metric system?

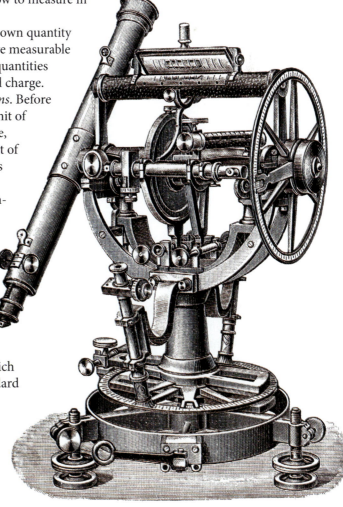

▶ A theodolite, a precision instrument for measuring angles in the horizontal and vertical planes

Metric System of Measurement

With the increased emphasis on experimental science during the seventeenth and eighteenth centuries came a desire to share research across Europe. Scientists quickly recognized a problem in that scientists were not using the same measurement units. They needed standardized units. In 1670, French astronomer Gabriel Mouton proposed a system that would become the metric system. The system was based on a decimal scale, that is, a scale that consisted of base units multiplied by powers of ten.

In 1799 France became the first European nation to adopt the metric measurement system, which consisted of the kilogram and the meter. These standard units were defined by the mass of a cylinder and the length of a bar, respectively, each made of durable platinum metal.

In 1960 many nations adopted the modern metric system, called the Système International d'Unités, which is French for the International System of Units. We commonly refer to it as the **SI**. Worldwide, scientists use the SI as the standard measuring system for their research. It is also the system used throughout most of the world for everyday measurements. The United States is one of only three countries that have not adopted the SI.

There are three reasons for the widespread usage of the SI. First, it has only one unit for each measurable dimension. For instance, the meter is the unit used for the dimension of length. The units in the SI are also based on natural phenomena, and most are based on physical constants, meaning that the units can be verified easily. Finally, the SI is a decimal system, which means that units can be scaled by factors of ten to measure larger and smaller objects. The decimal nature of the SI makes converting between units easier.

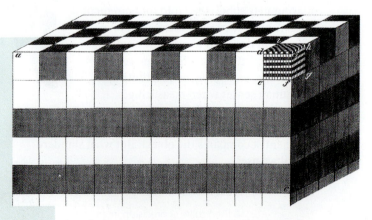

In a decimal system, dimensions can be divided into smaller and smaller increments, which form the basis of the prefixes in the SI.

THE SEVEN SI FUNDAMENTAL UNITS

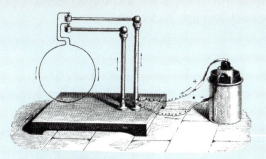

Unit: ampere (A)
Dimension: electrical current (I)
Basis: elementary charge, vibrational frequency of cesium-133

Unit: candela (cd)
Dimension: light intensity (I_v)
Basis: luminous efficacy of 540 Thz light, Planck constant, vibrational frequency of cesium-133

Unit: kelvin (K)
Dimension: temperature (T)
Basis: Boltzmann constant Planck constant, vibrational frequency of cesium-133

Unit: kilogram (kg)
Dimension: mass (m)
Basis: Planck constant, speed of light, vibrational frequency of cesium-133

Unit: meter (m)
Dimension: length (ℓ)
Basis: speed of light, vibrational frequency of cesium-133

Unit: mole (mol)
Dimension: number of particles (n)
Basis: Avagadro's number

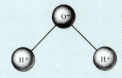

Unit: second (s)
Dimension: time (t)
Basis: vibrational frequency of cesium-133

SI Units

There are seven *fundamental units* in the SI, each of which measures a basic dimension. Scientists originally based these units on physical objects that they maintained in France. Over time we have redefined each of these seven units so that we now define them on the basis of physical constants, which is why they are the fundamental units. You may hear these referred to as base units, but keep in mind that the term *base unit* also refers to the part of the metric system that is multiplied by a power of ten to form a prefixed unit. A summary of the fundamental units along with their symbols is shown at left.

This spark plug gap gauge is a set of precisely measured metal plates that allow mechanics to properly adjust the spacing between a spark plug's central and lateral electrodes.

Measurements in Chemistry   51

Some SI Derived Units

Unit: square meter (m²)
Dimension: area (A)

Unit: cubic meter (m³)
Dimension: volume (V)

Unit: kilogram per cubic meter (kg/m³)
Dimension: density (ρ)

Unit: hertz (Hz)
Dimension: frequency (f)

Unit: newton (N)
Dimension: force (F)
Expressed as Fundamental Units: kg·m/s²

Unit: pascal (Pa)
Dimension: pressure (P)
Expressed as Fundamental Units: kg/(m·s²)

Unit: joule (J)
Dimension: energy (E)/work (W)/heat (Q)
Expressed as Fundamental Units: kg·m²/s²

Unit: degree Celsius (°C)
Dimension: Celsius temperature T_c

The beauty of the SI is that we can express nearly every other unit of measurement in terms of these fundamental units. For example, we can express thermal energy in terms of kilograms, meters, and seconds.

$$\frac{\text{kg} \cdot \text{m}^2}{\text{s}^2}$$

But expressing a measurement in terms of several fundamental units is often cumbersome. They can become complicated and hard to remember. To avoid this problem, scientists have established other units that they have derived from combinations of the fundamental units. There are twenty-two **derived units** approved for use in the SI. For example, the unit of thermal energy mentioned above is called the *joule* (J).

$$1 \text{J} = 1 \frac{\text{kg} \cdot \text{m}^2}{\text{s}^2}$$

The box at left lists some of the commonly used derived SI units.

SI Unit Prefixes

The SI allows scientists to scale units up or down so that they are convenient to measure and express quantities of different sizes. To change the size of a unit, the SI adds prefixes representing powers of ten to the names of the base units. For example, the meter is the base unit of length, but a meter is not always the best size for a particular measurement. It is more manageable to use the *centimeter*, a unit that is one one-hundredth of a meter, to measure the length of your shoe. At the same time, to measure the distance from your house to the school, it is usually more convenient to use the *kilometer*, a unit that is 1000 times as long as a meter.

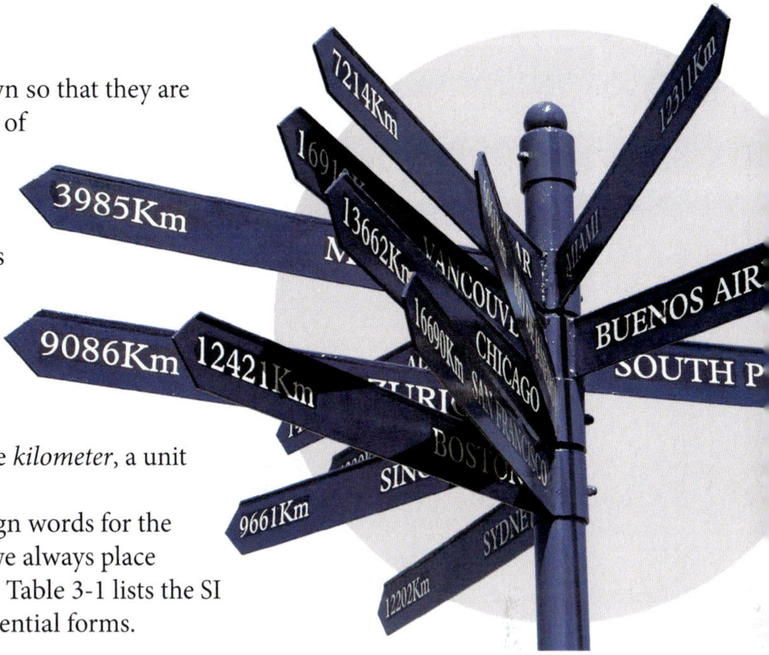

The prefixes used in the SI are based on foreign words for the related power of ten. As the word *prefix* implies, we always place prefixes in front of the base unit that they modify. Table 3-1 lists the SI prefixes with their symbols, meanings, and exponential forms.

TABLE 3-1 *SI Unit Prefixes*

Prefix	Prefix Symbol	Meaning	Factor	Exponential Form
peta-	P-	quadrillion	1 000 000 000 000 000	10^{15}
tera-	T-	trillion	1 000 000 000 000	10^{12}
giga-	G-	billion	1 000 000 000	10^{9}
mega-	M-	million	1 000 000	10^{6}
kilo-	k-	thousand	1000	10^{3}
hecto-	h-	hundred	100	10^{2}
deka-	da-	ten	10	10^{1}
—	(none)	base	1	10^{0}
deci-	d-	tenth	0.1	10^{-1}
centi-	c-	hundredth	0.01	10^{-2}
milli-	m-	thousandth	0.001	10^{-3}
micro-	μ-	millionth	0.000 001	10^{-6}
nano-	n-	billionth	0.000 000 001	10^{-9}
pico-	p-	trillionth	0.000 000 000 001	10^{-12}
femto-	f-	quadrillionth	0.000 000 000 000 001	10^{-15}

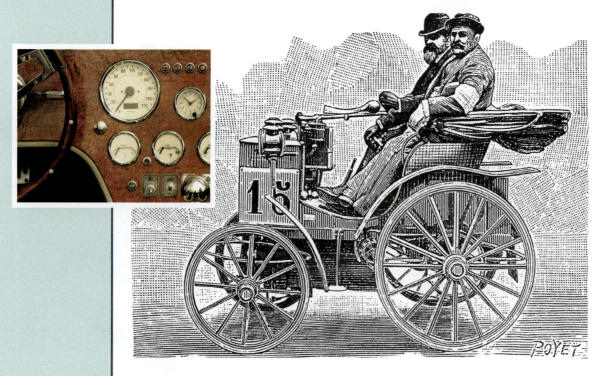

HOW IT WORKS

Speedometer

Some of you have your driver's license by now and are familiar with a vehicle's speedometer. But do you know how it measures and displays your speed?

Speedometers have been around since the early 1900s and were actually optional on early cars. But they quickly became standard equipment. Speedometers display the speed of a vehicle by measuring the rotational speed of the transmission. For many years, the usual design was a mechanical speedometer. This worked as a flexible shaft connected to the vehicle's transmission. At the end of the shaft was a permanent magnet. The rotating magnetic field of the spinning magnet would induce a current in a nearby aluminum disk. The induced current would produce another magnetic field. The interaction of the two fields would cause the aluminum disk to rotate. The disk was connected to a spring that would limit the disk's movement. The spring and magnetic forces were balanced and calibrated to the vehicle's speed. A pointer attached to the disk allowed the vehicle's indicated speed to be read from a dial.

Most speedometers today are electronic. Many of these systems still measure the speed of the rotating transmission. On cars with anti-lock braking systems (ABS), the system will often use the rotation rate of the wheels, a value which needs to be known for the ABS system to operate properly. In either case, the system consists of a magnetic input device that turns the rotation rate into an electronic signal. The signal is calibrated to vehicle speeds that are then displayed on the dashboard.

Speedometers have limitations just like all other measuring instruments. Some of these limitations are due to the design and construction of the system. Additional uncertainty can be introduced by aftermarket modifications to the vehicle. One common modification that can affect the accuracy of a speedometer is changing the wheel or tire size. Oversized tires travel over more ground per axle rotation, meaning that fewer axle rotations are needed to generate a given distance of vehicle travel. This produces indicated speeds on a speedometer that are lower than the vehicle's actual speed.

Unit Conversion

Many times we will know a measurement in a certain unit, but for some reason we need to express it in a different unit. In these cases, we use a mathematical process to change the units. In this process, called a **unit conversion**, we multiply our measurement by a **conversion factor**, which is a fraction that contains both the original unit and its equivalent value in a new unit. Mathematically, conversion factors are *always* equal to 1. Since multiplying by one doesn't change a quantity, unit conversion does not change the amount of the measurement, rather just the way that we are expressing it.

This process of using conversion factors is much like renaming fractions. If we are trying to add 1/2 to 1/4, we have to change the 1/2 into an equivalent fraction that has a four in the denominator. To do that, we multiply 1/2 by a fraction that is equal to 1, in this case, 2/2. Because 2/2 is equal to one, multiplying by it won't change the value of 1/2. But we have now changed the *form* of 1/2 to 2/4.

If we are trying to convert meters to millimeters, the conversion factor will contain the old and new units, m and mm. The equivalence expression is

$$1 \text{ m} = 1000 \text{ mm}.$$

Next, we create the conversion factor by dividing both sides of the equation by one one side or the other.

$$1 \text{ m} = 1000 \text{ mm} \qquad\qquad 1 \text{ m} = 1000 \text{ mm}$$

$$\frac{1 \text{ m}}{1000 \text{ mm}} = \frac{1000 \text{ mm}}{1000 \text{ mm}} \quad \text{or} \quad \frac{1 \text{ m}}{1 \text{ m}} = \frac{1000 \text{ mm}}{1 \text{ m}}$$

$$\frac{1 \text{ m}}{1000 \text{ mm}} = 1 \qquad\qquad 1 = \frac{1000 \text{ mm}}{1 \text{ m}}$$

The two final ratios are equivalent—they are both equal to 1. But which is the correct one to use? The good news is that the problem itself will let you know.

Let's look at a unit conversion between two units that you are very familiar with.

EXAMPLE 3-1: CONVERTING INCHES INTO FEET

Convert 55.6 in. into feet.

$$55.6 \text{ in.} = ? \text{ ft}$$

Solution

Since inches is in the numerator of the original measurement, inches will have to go in the denominator of the conversion factor so that they will cancel out when the problem is solved. Feet will go in the numerator.

$$55.6 \text{ in.} \left(\frac{\text{ft}}{\text{in.}}\right) = ? \text{ ft}$$

We know that 12 in. = 1 ft, so the numbers for the conversion factor are 12 and 1.

$$55.6 \text{ in.} \left(\frac{1 \text{ ft}}{12 \text{ in.}}\right) = ? \text{ ft}$$

Cancel the inches. Our units are now feet, which is what we are looking for.

$$55.6 \cancel{\text{in.}} \left(\frac{1 \text{ ft}}{12 \cancel{\text{in.}}}\right) = ? \text{ ft}$$

Evaluate.

$$\left(\frac{55.6 \text{ ft}}{12}\right) = 4.63 \text{ ft}$$

CASE STUDY

REDEFINING THE KILOGRAM

An important feature of the metric system of measurement was that the units were defined on the basis of physical objects. As the scientific community around the world adopted the modified metric system, they sought to redefine the units on the basis of physical *phenomena* and, if possible, physical *constants*. With the redefinition of the kilogram in 2019, that goal was achieved.

Originally, the kilogram was defined as the mass of "exactly" one liter of water. While this standard was reproducible anywhere in the world, it was impossible to reproduce with the precision desired. The kilogram was redefined in 1799 as the mass of a physical object made of platinum. In 1879, this platinum object was replaced with a platinum-iridium object as the standard for the kilogram. Replicas were produced to be used around the world. But over time the masses of the standard kilogram and the replicas changed ever so slightly. This prompted scientists to redefine the kilogram once again.

Today the kilogram is defined on the basis of the Planck constant, the speed of light, and the vibrational frequency of cesium-133. The redefinition provides an unchanging standard for the kilogram. But don't worry—day-to-day mass values have not changed.

Questions to Consider

1. What is a fundamental unit?
2. Why did scientists want standardized units?
3. Why is it better to define the SI fundamental units on the basis of physical constants instead of physical objects?
4. Which physical constant is the basis for the definition of the kilogram?

A technician holds a solid silicon sphere, which was used for development of the new definition of the kilogram.

Unit conversions can be simple or more complex. But know that complex unit conversions just need more conversion factors—however many we need to get from the units we start with to the units we want. Let's look at a few more, ranging from simple to complex.

EXAMPLE 3-2:
SIMPLE UNIT CONVERSION

Convert 132 547 cm to m.

Solution

$$132\,547 \text{ cm} \left(\frac{10^{-2} \text{ m}}{1 \text{ cm}} \right) = \left(\frac{1325.47 \text{ m}}{1} \right) = 1325.47 \text{ m}$$

Notice that in these examples we treat the units as if they were variables in math. We can add, subtract, multiply, and divide units as if they were variables, a process called *dimensional analysis*.

EXAMPLE 3-3:
COMPLEX UNIT CONVERSION

Convert 356 kg to µg

Solution

$$356 \text{ kg} \left(\frac{1000 \text{ g}}{1 \text{ kg}} \right) \left(\frac{1\,000\,000 \text{ µg}}{1 \text{ g}} \right)$$

$$= \left(\frac{356\,000\,000\,000 \text{ µg}}{1} \right)$$

$$= 3.56 \times 10^{11} \text{ µg}$$

Notice in Example 3-3 that using two conversion factors is convenient since we might not readily know the conversion factor from kilograms directly to micrograms. We converted kilograms to the base unit of grams and then from grams to micrograms. Also notice that we expressed the answer in scientific notation. That made for an abbreviated answer, but it also addressed another issue which we will discuss in Section 3.2. Let's look at one more example.

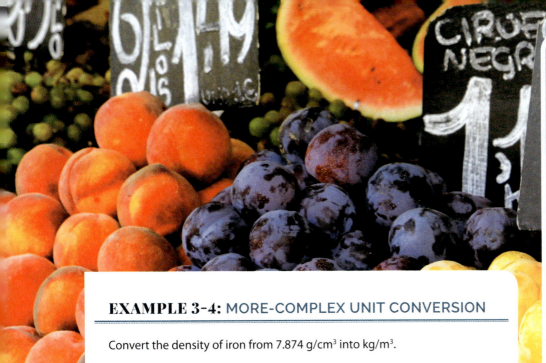

EXAMPLE 3-4: MORE-COMPLEX UNIT CONVERSION

Convert the density of iron from 7.874 g/cm³ into kg/m³.

Solution

$$7.874 \frac{\cancel{g}}{\cancel{cm^3}} \left(\frac{1 \text{ kg}}{1000 \cancel{g}}\right) \left(\frac{100 \cancel{cm}}{1 \text{ m}}\right)^3 = \left(\frac{7\,874\,000 \text{ kg}}{1000 \text{ m}^3}\right) = 7874 \text{ kg/m}^3$$

Notice in Example 3-4 that we needed three identical conversion factors to eliminate the cm³. We also placed cm in the numerator of the conversion factor because cm³ was in the denominator of the measurement. So you see that the problem tells you where to put the units.

3.1 SECTION REVIEW

1. What do we call the process by which scientists quantify their observations of their surroundings?

2. What is a dimension?

3. What measuring system is used by scientists around the world?

4. Explain how a measurement system is a model. Why is it important that scientists have a standardized system?

5. Explain one of the three reasons for the almost worldwide usage of the SI.

6. What is the SI fundamental unit for temperature, and on the basis of what physical constants is it defined?

7. Write an equation that states the relationship between the given units, then write a conversion factor that could be used to convert the first unit to the second. (*Example*: Millimeters and meters are related by the equation 1000 mm = 1 m. The conversion factor 1 m/1000 mm should be used.)

    a. grams and micrograms
    b. femtometers and meters
    c. liters and hectoliters
    d. gigahertz and hertz

8. Convert the following measurements.

    a. 3.8×10^{10} bytes to gigabytes
    b. 475 nm to meters
    c. 27.3 kL to milliliters
    d. 5.67 Mg to micrograms

Measurements in Chemistry

3.2 MEASUREMENTS

QUESTIONS
- » Why can't I measure something exactly?
- » Why do I need to be accurate and precise?
- » What affects the precision of measurements?
- » What are significant figures?

TERMS
accuracy • precision • significant figure

Is being accurate and precise the same thing?

The Limitations of Measurements

The Greek philosopher Protagoras once stated that "man is the measure of all things." According to that statement, man is the determiner of all reality. He determines right and wrong and good and evil rather than God.

If humans were correct in every life decision and absolutely accurate when it comes to things like measurement, there might be some reason to trust human abilities. But people aren't able to measure anything with complete accuracy. Since the Fall our abilities have been permanently affected. Scientists recognize that there is always some *uncertainty* involved with any measurement. There is always a difference between our measured values and the actual values of those measurements.

While we should always strive to measure as exactly as possible, it is impossible to measure the true value of any dimension. Scientists strive to be both as accurate and precise as possible.

Accurate but not precise

Accuracy

The **accuracy** of a measurement or instrument is a numerical evaluation of how close a measured value is to the expected or accepted value of the dimension being measured. For example, if the accepted value for the length of your textbook is 27.6 cm, and you measured the length as 27.7 cm, then you made a fairly accurate measurement. But what would you say if you measured it later with a different ruler and your measurement was 27.9 cm? Since the accepted length is 27.6 cm, there is some uncertainty in each of your measurements and more in the second than the first. Where did this uncertainty come from? The first possibility is that you made a procedural error. You can check for that by repeating your measurement. There is also the possibility that there is uncertainty in the instrument that you are using.

When scientists consider the uncertainty in an instrument, they refer to both random uncertainty and systematic uncertainty. Random uncertainty accounts for the slight variations when the same person makes repeated measurements with the same instrument. This uncertainty can cause the measurement to be greater or less than the actual measurement. But systematic uncertainty typically will skew the measurements in only one direction. Sources of systematic uncertainty could be a tape measure that has been stretched or bent or a balance that reads 0.2 g too much mass due to improper calibration. All sources of uncertainty guarantee that we can never measure exactly.

Scientists try to quantify uncertainty in measurements using various statistical techniques. In this course, you will use a relatively simple approach to uncertainty analysis that compares the measurement uncertainty with the size of the measurement, a quantity called *percent error*. This technique assumes that you know what the measurement should be. The formula for percent error is shown below.

$$\%_{error} = \left(\frac{value_{observed} - value_{accepted}}{value_{accepted}} \right) 100\%$$

This calculation can yield either positive or negative values depending on whether an observed value is too high or too low. In some applications, we are interested in only the magnitude of the percent error. In such cases, you will include absolute value signs around the numerator of the fraction. Written this way, the formula determines the *absolute percent error*.

$$absolute\ \%_{error} = \left(\frac{|value_{observed} - value_{accepted}|}{value_{accepted}} \right) 100\%$$

If the percent error is small, then the measurement is considered to be accurate; otherwise, it is inaccurate. Large and small is a subjective matter that depends on the specific application and the need for a particular degree of accuracy.

EXAMPLE 3-5: CALCULATING PERCENT ERROR

You measure the mass of the product of a chemical reaction to be 3.80 g. According to your theoretical calculations, you should have obtained 3.92 g. What is the percent error of your work?

Write what you know.

$value_{observed}$ = 3.80 g

$value_{expected}$ = 3.92 g

$\%_{error}$ = ?

Write the formula and solve for the unknown.

$$\%_{error} = \left(\frac{value_{observed} - value_{expected}}{value_{expected}} \right) 100\%$$

Plug in known values and evaluate.

$$\%_{error} = \left(\frac{3.80\ g - 3.92\ g}{3.92\ g} \right) 100\%$$

$$= \left(\frac{-0.12\ \cancel{g}}{3.92\ \cancel{g}} \right) 100\%$$

$$= -3.1\%$$

Your measurement was 3.1% below the expected value.

Measurements in Chemistry

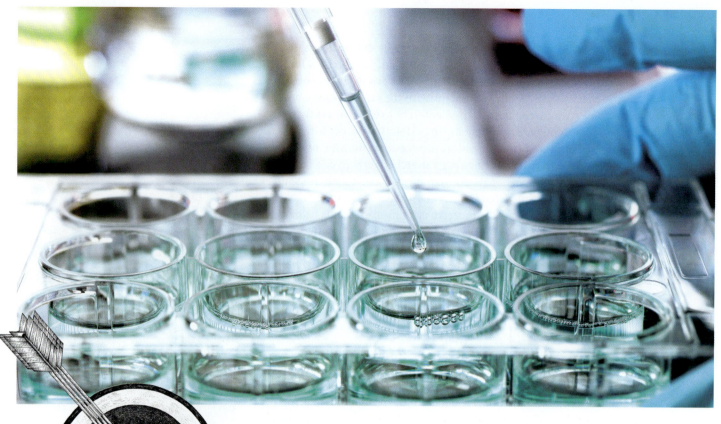

Precise but not accurate

Precision

Accuracy and **precision** are two terms that are often incorrectly used interchangeably. While accuracy indicates how close a measurement is to the actual value, precision is an evaluation of the exactness of a measurement or a measuring instrument. It indicates how repeatable a measurement is or how exactly one can make a measurement. Precision does not say how correct a measurement is.

Certain numbers can be known exactly. For example, you can count the number of beakers on a shelf. Counted numbers are exact. Some dimensions used in science are known exactly because they are defined. The speed of light, the second, and the meter are known exactly because scientists have defined them as a specific value.

But measurements do not have unlimited precision. Whenever we use an instrument to make a measurement, we translate the dimension that we measure into a number by using the scale on the instrument. The digits that we record in the measurement depend on how many of the large and small scale markings are known. These are the digits that we are certain about. But all measurements will include an estimated digit. When you use an analog (nondigital) metric instrument, you should estimate the measurement to the nearest tenth of the smallest whole decimal subdivision on the instrument's scale. For example, if a ruler is marked to the millimeter, then you should estimate the digit for the tenths of a millimeter place. If you are using a graduated cylinder that has 0.2 mL as its smallest marking, you can estimate to the tenths of a milliliter. You are limited to the tenths place because the smallest whole decimal mark is the milliliter. Instruments with finer scale graduations allow us to know more decimal places with certainty, and therefore our measurements are more precise because we know them more exactly.

TABLE 3-2 *Precision of Lab Instruments*

Instrument	Typical Smallest Marking	Precision Limit
Platform Balance	1 g	0.1 g
Triple-Beam Balance	0.1 g	0.01 g
Analytical Balance	0.001 g	0.0001 g
100 mL Graduated Cylinder	1 mL	0.1 mL
10 mL Graduated Cylinder	0.2 mL	0.1 mL
50 mL Burette	0.1 mL	0.01 mL
Thermometer	1 °C	0.1 °C

Does making accurate and precise measurements make a difference? Think back to the Chapter opener. As costly as that error was, it didn't put any lives at risk. But there have been situations where lives have been endangered due to errors regarding units and unit conversions. In 1983 an Air Canada flight ran out of fuel due to confusion between pounds of fuel and kilograms of fuel. Happily for the sixty-nine people on board, the pilots were able to safely land the plane at a closed Air Force base. Medical errors can put people's lives at risk too. Doctors prescribe medicine according to one's weight. Therefore, a patient's weight must be accurately known to calculate proper dosage. A nurse must then measure the proper dose before giving it to the patient. Errors in either of those measurements could be a matter of life or death. In one case, a patient had been prescribed 0.5 *grains* of a sedative. Instead, the patient was given 0.5 *grams*—more than 15 times the appropriate dose. The error was quickly discovered, and the patient made a full recovery. We must always strive to be both accurate and precise because lives may depend on it.

Integrity in Measurements

One purpose of this course is to train you to think like scientists as much as possible—to become student chemists. Scientists are acutely aware of their limitations, especially when it comes to measuring. For this reason, you need to be familiar with the uncertainty associated with measuring data.

Scientists try to report their quantitative results in the most accurate manner possible. To do this, they follow a set of rules for determining which digits of their answer are reliable or significant. Minimizing uncertainty in measurement is an important part of experiment design. Estimating the uncertainty of measurements is important when evaluating the validity of conclusions that are made on the basis of measured data.

Scientists must also report measurements correctly because it is a matter of integrity. Reporting measurements with more precision than was actually measured is dishonest. Reporting with too little precision is sloppy and calls into question the validity of one's work. In either case, scientists may lose their colleagues' respect and perhaps even their jobs. Similarly, you must make the correct reporting of measured data a matter of personal honesty in your work as a way to imitate Christ in your life. Scientists continually face the demand to make moral choices. Other choices may arise when a scientist is alone in his laboratory. He may be tempted to falsify data on an experiment. However good an excuse the scientist may think he has, God says, "A false balance is abomination to the Lord: but a just weight is his delight" (Prov. 11:1). Deception through falsifying measurements is clearly displeasing to God. Many Old Testament passages connect righteousness to right weights and measures, and just like physical measurements, all things are measured against a standard—the standard of the Bible.

Scientists observing Joseph Black demonstrating a steam power experiment. Today the peer review process holds scientists accountable.

Measurements in Chemistry

MINI LAB

ACCURACY AND PRECISION

Accuracy and precision in measuring depend on both the instrument used and how well it is used. This lab activity will demonstrate the effect of different instruments on the precision of measurements.

How do instruments affect accuracy and precision of measurements?

EQUIPMENT
- paper rulers (3)

1. Define *precision*.

Procedure

A Observe the three rulers.

2. Compare the three rulers.

B Copy the table below on a piece of paper.

C Measure the length and width of the rectangle at right with each of the rulers. Record the measurements in your table.

Conclusion

3. Which ruler is most precise? least precise?

4. If the actual length of the rectangle is 7.69 cm, with which ruler did you most accurately measure?

Going Further

5. Give an example in which a highly precise scale would be desired.

6. Give an example in which a low-precision scale would be acceptable.

Ruler	Smallest Marked Increment	Length (cm)	Width (cm)
1			
2			
3			

rectangle to measure

Significant Figures When Recording Data

Read the temperature indicated on the thermometer in the image at left. The major divisions are labeled every 10 °C, and the minor, unlabeled divisions mark every degree. The fluid column is past the twenty-two mark, but it doesn't reach twenty-three. You will have to estimate between 22 °C and 23 °C. Since the thermometer is calibrated to the ones place, we will estimate to the tenths place. It looks like 22.6 °C. We always measure to one-tenth of the smallest whole decimal increment. So even if the thermometer were indicating exactly 24 °C, you would still record the measurement to the tenths place: 24.0 °C. The uncertainty of the tenths digit is why the last digit is called the *estimated*, *uncertain*, or *least significant figure*. It would be improper to report more decimal places for this temperature.

You can see from the preceding discussion that in that temperature measurement, only three digits were significant. These are what we call the **significant figures** (SFs) of that measurement, that is, those that are known for certain plus one estimated digit. We estimated the digit in the tenths place because the ones place was the smallest subdivision on the instrument's scale. Therefore, the rightmost significant figure will always be the estimated digit. The purpose of significant figures is to indicate the precision of your measurement; doing so protects the integrity of the data.

SIGNIFICANT FIGURES IN RECORDED DATA

When scientists record their data, they record only the significant figures. When you work with another person's data, you must be able to quickly determine which digits in a measurement are significant, especially if you are going to perform calculations with them. Below are some easy-to-use rules that will maintain precision in solutions to chemistry problems.

SF Rule 1: Significant figures apply only to measured data. Some quantities, such as counted or pure numbers or defined values, are known exactly. These numbers are assumed to have an infinite number of significant figures.

  a. Significant figures do not apply to counted or pure numbers.

  Examples: 500, one pair, $\sqrt{3}$, π

  b. Significant figures do not apply to conversion factors.

  Examples: $\dfrac{60 \text{ s}}{1 \text{ min}}$, $\dfrac{1000 \text{ g}}{1 \text{ kg}}$, $\dfrac{1 \text{ °C}}{1 \text{ K}}$

SF Rule 2: All nonzero digits are significant figures.

25.4 mL (3 SFs)   13.78 g (4 SFs)

SF Rule 3: All zeros between nonzero digits are significant figures.

100.5 °C (4 SFs)   1.09 g (3 SFs)

SF Rule 4: All ending zeros to the right of the decimal point are significant figures.

20.0 s (3 SFs)   250.00 L (5 SFs)

0.075 kg (2 SFs)   0.0010 s (2 SFs)

You may ask, "What about measurements that end up as whole numbers?" For example, consider 2500 mL. The rules don't address a number like this because it is ambiguous. Whoever measured the volume as 2500 mL could have measured to two, three, or four significant digits, but there is no way for us to know which when the number is written this way. In cases like this, assume that the ending zeroes are not significant. But we can eliminate this kind of ambiguity by using scientific notation to correctly express measurements. Look at the three examples at right.

Standard Notation	Number of SFs	Scientific Notation
2500 mL	(2 SFs)	2.5×10^3 mL
2500 mL	(3 SFs)	2.50×10^3 mL
2500 mL	(4 SFs)	2.500×10^3 mL

You should see that using scientific notation eliminates all the confusion about significant figures in these cases.

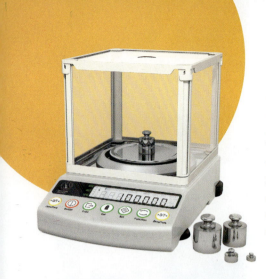

If you think about it, you may realize that significant figures not only tell us about the precision of measurements, but they also tell us about the instruments used to make those measurements. If we look at the measurement 35.47 cm, we recognize that there are four SFs. The digits three, five, and four are all certain, and the seven is uncertain. The measurement was taken precisely to the hundredths place. So what does that tell us about the instrument? For the scientist to estimate the last digit to one-tenth of the smallest marking, the smallest marking on the instrument must have been 0.1 cm. Let's practice determining the number of significant figures in measurements.

EXAMPLE 3-6: SIGNIFICANT FIGURES IN DATA

Determine the number of significant figures in the following quantities and the precision of each instrument's smallest scale increments.

- **a.** 9.370 kg
- **b.** 63 000 g
- **c.** 0.020 50 km
- **d.** 12 cookies
- **e.** 0.0001 s
- **f.** 4.50×10^3 mL

Solution

- **a.** 4 SFs (SF Rules 2 and 4); The scale was calibrated in 0.01 kg increments.
- **b.** indeterminate; It could be two, three, four, or five significant digits. We will assume that it has 2 SFs (SF Rule 2). The number should be expressed in scientific notation. The scale calibration depends on the significant digits.
- **c.** 4 SFs (SF Rules 2, 3, and 4); The scale was calibrated in 0.0001 km increments.
- **d.** This is an exact or counted number; therefore, significant figures are not applicable (SF Rule 1a).
- **e.** 1 SF (SF Rule 2); The scale was calibrated in 0.001 s increments.
- **f.** 3 SFs (SF Rules 2 and 4); The scale was calibrated in 100 mL increments.

3.2 SECTION REVIEW

1. Define the concept of *accuracy*.
2. What is uncertainty in a measurement?
3. Why can't I measure something exactly?
4. The density of oxygen gas is 1.429×10^{-3} g/cm³. Calculate the percent error of an experimentally determined density of 1.091×10^{-3} g/cm³.
5. In three separate trials and under the same conditions, you measure a gas's density to be 2.865 kg/m³, 2.852 kg/m³, and 2.860 kg/m³. The actual density of the gas is 3.214 kg/m³. Evaluate the precision and accuracy of your measurements.
6. What is the purpose of using significant figures in measurements? What is the benefit of using significant figures?
7. List the significant figures in each of the following quantities. Also, state the place value of the estimated digit.
    - **a.** 2.40 cm
    - **b.** 0.010 g
    - **c.** 200 km
    - **d.** 102 students
    - **e.** π
    - **f.** 3.00×10^4 mm
8. Write the following in scientific notation with the appropriate number of significant figures.
    - **a.** 0.0740 cm (3 SFs)
    - **b.** 15 200 g (3 SFs)
    - **c.** 230 000 000 km (4 SFs)
    - **d.** 0.000 000 581 4 ms (4 SFs)

3.3 PROBLEM SOLVING IN CHEMISTRY

Most scientists perform mathematical calculations in the normal course of their work. Math is an essential tool of science because creating scientific models often involves mathematical formulas. Those formulas are workable descriptions of natural processes. You will learn later that the order established by the Creator that governs the behavior of matter nearly always involves mathematical relationships.

Calculations with Measured Data

When using a calculator to compute the result of a scientific equation, you must be careful about which numbers you report in your answer. You might be inclined to record all the digits displayed in a calculator's answer. Let's compute the density of a sample of the mineral galena, a lead-bearing ore, to understand why we shouldn't do that. We measure the sample's volume to be 5.8 cm³ and its mass to be 44.03 g. We then calculate the density by dividing the mass by the volume.

$$\rho = \frac{m}{V} = \frac{44.03 \text{ g}}{5.8 \text{ cm}^3}$$

$$= 7.591\,379\,31 \, \frac{\text{g}}{\text{cm}^3}$$

The result on a calculator indicates that the density is precise to the hundred-millionths (10^{-8}) of a gram per cubic centimeter. But the scientist measured the mass to a hundredth (10^{-2}) of a gram, and the volume to a tenth (10^{-1}) of a cubic centimeter. Therefore, the calculated density is about a million times more precise than either the measurement of the mass or volume! In scientific calculations, the precision of a computed result should have approximately the same precision as the measured data used to calculate it. As with the original measurements, scientists must avoid adding or losing precision.

QUESTIONS
» Why do significant figures apply to calculations?
» How do I know how many significant figures to round my answer to?
» Why is an orderly process for problem solving important?
» How do I know whether my answer is reasonable?

TERMS
none

How do I solve problems in chemistry?

Measurements in Chemistry

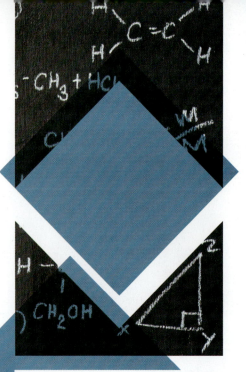

Below are some rules to follow for making calculations with measured data. The rules provided here are adequate for high-school subjects, but they do not always yield answers that would be valid in scientific research. Scientists doing research must determine the precision for each calculation on the basis of approved statistical methods. Those methods take into account the types of calculations and the sources and kinds of measured data.

RULES FOR ADDING AND SUBTRACTING

Math Rule 1: Measured data must be of the same dimension and have the same units before we can add or subtract them. You cannot add apples and oranges—or meters and grams!

Math Rule 2: The sum or difference of measured data cannot have greater precision than the least precise quantity in the measurements being added or subtracted. When adding or subtracting, you must round the result to the least precise decimal place of the estimated digits in the data.

EXAMPLE 3-7: MAINTAINING PRECISION WHEN ADDING

What is the perimeter of a piece of paper that is 11.53 cm long and 5.6 cm wide?

Write what you know.

$l = 11.53$ cm
$w = 5.6$ cm
$P = ?$

Write the formula and solve for the unknown.

$P = l + l + w + w$

Plug in known values and evaluate.

$P = 11.5\underline{3}$ cm $+ 11.5\underline{3}$ cm $+ 5.\underline{6}$ cm $+ 5.\underline{6}$ cm
$ = 34.\underline{26}$ cm

The underlined digits identify the estimated decimal place in each number. Note that the sum has two estimated digits, but only one is allowed. We round the sum to the estimated digit with the least precise decimal place.

The answer is 34.3 cm.

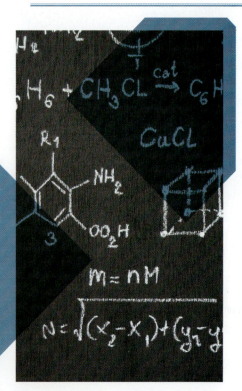

RULES FOR MULTIPLYING AND DIVIDING MEASURED DATA

Math Rule 3: Measured data can be of any dimension when multiplying or dividing. As a general rule, the units for the same dimension will be the same in multiplication and division problems, but there is no restriction on multiplying or dividing meters and centimeters. When multiplying and dividing, treat units as variables using the dimensional analysis process that was mentioned in Section 3.1.

Math Rule 4: The product or quotient of measured data cannot have more SFs than the quantity with the fewest SFs. Round results as necessary. Think back to the density calculation that we did on page 65. Because the volume had only 2 SFs, the density we calculated should have been rounded to 7.6 g/cm^3.

Math Rule 5: The product or quotient of a measurement and a counted number, conversion factor, or defined value has the same number of decimal places, or same precision, as the original measurement. In Example 3-7 above, we used the formula for perimeter shown below.

$$P = l + l + w + w$$

But many of you probably recognized that the formula could be written more simply as

$$P = 2l + 2w.$$

The answer using this formula comes out the same as the formula that we used before.

$$P = 2(11.53 \text{ cm}) + 2(5.6 \text{ cm})$$
$$= 23.0\underline{6} \text{ cm} + 11.\underline{2} \text{ cm}$$
$$= 34.3 \text{ cm}$$

Multiplication is just repetitive addition. Therefore, multiplying a measurement by a pure number should not contradict Math Rule 5. And since division is the inverse operation of multiplication, the same principle applies.

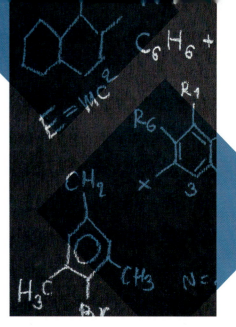

EXAMPLE 3-8: MAINTAINING PRECISION WHEN MULTIPLYING

What is the area of a piece of paper that is 11.53 cm long and 5.6 cm wide?

Write what you know.

$l = 11.53$ cm
$w = 5.6$ cm
$A = ?$

Write the formula and solve for the unknown.

$A = lw$

Plug in known values and evaluate.

$A = (11.53 \text{ cm})(5.6 \text{ cm})$
$= 64.568 \text{ cm}^2$

Since 5.6 cm has two SFs and 11.53 cm has four SFs, the answer is limited to 2 SFs.

The answer is 65 cm².

RULE FOR COMPOUND CALCULATIONS

Math Rule 6: Calculations involving multiple operations are called compound calculations. For solutions requiring more than one calculation, do not round after each step. Instead, keep track of the significant figures throughout the calculation as if you were going to round your intermediate answers. Round the final answer, taking into account the SFs of the intermediate answers.

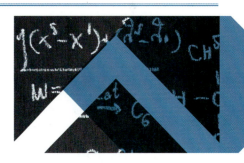

EXAMPLE 3-9: MAINTAINING PRECISION IN COMPOUND CALCULATIONS

The formula for the area of a trapezoid is

$$A = \tfrac{1}{2}(b_1 + b_2)h.$$

What is the area of a trapezoid with bases of 140.75 m and 83.0 m and a height of 125 m?

Write what you know.

$b_1 = 140.75$ m
$b_2 = 83.0$ m
$h = 125$ m
$A = ?$

Write the formula and solve for the unknown.

$A = \tfrac{1}{2}(b_1 + b_2)h$

Plug in known values and evaluate.

$A = \tfrac{1}{2}(140.7\underline{5} \text{ m} + 83.\underline{0} \text{ m})(12\underline{5} \text{ m})$
$= \tfrac{1}{2}(223.\underline{75} \text{ m})(12\underline{5} \text{ m})$
$= 13\,984.375 \text{ m}^2$

Since 223.75 should be rounded to four SFs and 125 cm has three SFs, the answer is limited to three SFs. Therefore, we would round our answer to 14 000 m². But you can't tell whether this answer has two, three, four, or five SFs. We will need to put the answer in scientific notation with three SFs.

The answer is 1.40×10^4 m².

Worldview and Problem Solving in Chemistry

Problem-solving skills are necessary for daily living. Buying a car, building a house, balancing a checkbook, and even baking cookies all require problem-solving skills. When we use these skills as part of fulfilling the Creation Mandate of Genesis 1:28, we discover more about the creation around us so that we may glorify God and serve our fellow human beings.

Chemistry provides opportunities to gain experience in problem solving. You can think of chemists as trained problem solvers. Chemists may be called on to identify unknown chemicals found in a river, to calculate the ingredients needed to make the most effective drug, or to create a plastic that can withstand extreme temperatures.

Problems in chemistry are solved by an orderly, reasoned approach with a correct application of the mathematical rules that you learned in the previous two sections. The following page describes a proven, systematic method for solving the chemistry problems that you will face.

Steps in Problem Solving

There are many different approaches to problem solving. Below is a framework for setting up most problems.

Step 1: Write what you know. Read the problem! This is the most important step, but it is too often overlooked. Make a list of variables and their numerical values, including their units.

Step 2: Write the formula and solve for the unknown. Once you have identified what you know and what your unknown is, you can determine which formula will help solve for the unknown. Formulas are mathematical models of natural phenomena. They relate quantities, whether variables or constants, in *literal equations*—those written so that most or all quantities are represented by letters rather than by numbers. Once you have identified the formula, write it down in its standard form. Use algebraic operations to solve for the unknown.

Step 3: Plug in known values and evaluate. Replace all the variables with known quantities that you have measured or been given. Cancel units to check that the result has the same units as the dimension for which you are solving (dimensional analysis).

a. Estimate and calculate. Estimate your answer by using round numbers and mental calculations. Estimating can help uncover calculator errors or mistakes in setting up the problem. Finally, calculate your answer.

b. Check. Students are often so eager to get an answer that they fail to check their work. Go back and reread the problem; did you answer the question that was asked? Is your answer close to the estimate (within a power of ten)? Check that your answer has appropriate units.

Step 4: Record your answer. Once you are satisfied that the problem was solved correctly, round the answer to the proper number of significant figures. Use scientific notation as needed.

EXAMPLE 3-10: SOLVING A PROBLEM IN CHEMISTRY

The mass measurements of two pieces of iron are 45.2 g and 67.95 g, and their volume measurements are 5.7 cm³ and 7.43 cm³. What is the density of iron on the basis of each sample? The accepted value for the density of iron is 7.874 g/cm³. What does this imply about our accuracy and precision?

Write what you know.

$m_1 = 45.2$ g
$m_2 = 67.95$ g
$V_1 = 5.7$ cm³
$V_2 = 7.43$ cm³
$\rho = ?$

Write the formula and solve for the unknown.

$\rho = \dfrac{m}{V}$

Plug in known values and evaluate.

Estimate: $45 \div 6 = 8$, $70 \div 7 = 10$

$\rho_1 = \dfrac{45.2 \text{ g}}{5.7 \text{ cm}^3}$

$= 7.929\,824\,561 \dfrac{\text{g}}{\text{cm}^3} = 7.9 \dfrac{\text{g}}{\text{cm}^3}$ (2 SFs)

$\rho_2 = \dfrac{67.95 \text{ g}}{7.43 \text{ cm}^3}$

$= 9.145\,356\,662$ g/cm³

$= 9.15$ g/cm³ (3 SFs)

Check: We were looking for density, which has the units g/cm³. Our answers are close to the estimate.

Record your answer. The two calculated densities for the iron samples are 7.9 g/cm³ and 9.15 g/cm³. The first measurements seem to be more accurate than the second, but the second measurements were more precise.

CAREERS

SERVING AS A CLIMATOLOGIST: *WEATHER WATCHERS*

If you have been paying attention to the news, you will recognize that there are some questions about climate change and global warming. Actually there are many questions on the issue. As scientists attempt to answer these questions, climatologists are making headlines. You could say that climatologists are a hot commodity today!

Wallace Smith Broecker was a climatologist who spent his career studying the earth's climate. He did extensive work on the carbon cycle, chemical oceanography, and energy movement by ocean currents. He was also one of the early scientists to bring the term *global warming* into the public arena.

Climatologists study weather conditions over long periods of time. They work in meteorology but are not concerned with the weather over a period of days, weeks, or months. They are interested in the weather over years and decades.

One of the biggest challenges for a meteorologist is the vast amount of data that must be modeled to make a prediction about what will happen over the next few days. For a climatologist, this task is exponentially more challenging because he works with years' and years' worth of data. The other challenge to the study of climate is that in some cases no data exists from the time period that he is researching or from many of the locations that he is interested in.

Christians are called to exercise good stewardship over God's creation. Part of that involves studying the earth and its systems. We have an obligation to understand our climate and the way that our actions affect it. Climate affects food production, which in turn affects our ability to fulfill the command to fill the earth. It also influences the frequency and intensity of storms. Christians have an obligation to understand the climate and changes to it if they are to glorify God and help others.

3.3 SECTION REVIEW

1. Why do significant figures apply to calculations?
2. What restrictions apply to units when adding or subtracting measurements?
3. Explain how we determine the number of allowed significant figures when multiplying or dividing measurements.
4. Summarize the steps to orderly problem solving when dealing with a mathematical chemistry problem.
5. Why are literal equations important in science? Give an example of a formula that you have used before.
6. Perform the following operations. Answers should include appropriate significant figures and, if needed, scientific notation.

   a. 4.020 m + 0.23 m + 40.2 m − 10.00 m

   b. 4.718 cm − 3.94 cm

   c. 1.5 L − 275 mL

   d. (3.0 g/mL)(2.54 mL)

   e. $\dfrac{8.95 \times 10^3 \text{ g}}{9.8 \times 10^2 \text{ cm}^3}$

   f. $\dfrac{71.6 \text{ g} + 128.32 \text{ g}}{7.81 \text{ mL} + 14.2 \text{ mL}}$

   g. 2(98.95 cm − 10.1 cm)(1.98 m − 100.2 cm)

03 CHAPTER REVIEW
Chapter Summary

3.1 MEASUREMENT SYSTEMS

- Measuring is an essential part of studying the universe. It involves comparing a measurable quantity, called a dimension, to a scale graduated in units appropriate to the dimension. A measurement is the product of a number and the dimensional unit.

- Measuring is also an essential part of glorifying God. God's command to exercise dominion (Gen. 1:28) requires us to work with matter, time, and space. Useful and accurate measurements are vital to that work.

- Scientists have developed a metric system of units called the Système International d'Unités (SI). The SI has seven fundamental units from which other metric units are derived.

- Metric units that are smaller or larger than a base metric unit can be created by using prefixes that stand for powers of ten. Thus, unit conversions between different-sized SI units are a simple matter compared with those in the English system.

TERMS
measurement
SI
fundamental unit
derived unit
unit conversion
conversion factor

3.2 MEASUREMENTS

TERMS
accuracy
precision
significant figure

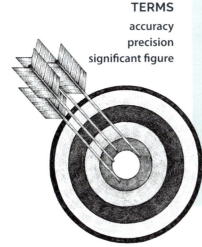

- Accuracy is a quantitative evaluation of how close a measurement is to the expected or accepted value of a dimension being measured.

- Precision is a quantitative evaluation of how exact and reproducible a measurement is and is indicated by the reported significant figures.

- Because measurements have uncertainties, scientists use significant figures to indicate the precision of measurements and the relative uncertainty of the data.

3.3 PROBLEM SOLVING IN CHEMISTRY

TERMS
none

- Simple rules determine the significant figures in mathematical results from calculations involving measured data.

- In problems involving addition and subtraction, results are limited by the least precise measurement (decimal place).

- In multiplication and division, results are limited by the measurement with the least number of significant figures.

Measurements in Chemistry 71

03 CHAPTER REVIEW

Chapter Review Questions

RECALLING FACTS

1. What is a measurement?
2. (True or False) A scientific measuring instrument requires some kind of graduated scale.
3. What are the two required components of every measurement?
4. What are the benefits of the SI?
5. What are the SI fundamental units for mass, length, and time? Is there an SI fundamental unit for volume?
6. What is the SI derived unit for pressure, and how is it expressed in SI fundamental units?
7. What is the unit, including its prefix for 10^{-6}, of the SI derived unit for energy?
8. Identify sources of uncertainty in measurements.
9. (True or False) Percent error is sometimes used to quantify the accuracy of a measurement or answer.
10. Define *precision*.
11. What affects the precision of measurements?
12. Which digits are the significant figures in a measurement?
13. What is the only reliable way to report the correct significant figures in a whole number measurement that contains trailing zeros?
14. What is the general rule for the precision of a calculated value when compared to the measured values that went into the calculation?
15. What determines the number of significant figures when adding or subtracting measurements?
16. What restrictions apply to units when multiplying or dividing measurements?
17. Describe a literal equation.

UNDERSTANDING CONCEPTS

18. What is the benefit of using one standardized measuring system in science?
19. Compare fundamental and derived units.
20. Which prefixed unit would be most appropriate to measure the following?
    a. the distance to the moon
    b. the length of a grain of rice
    c. the time for light to travel across the United States
    d. the mass of a blue whale
21. Create a graphic organizer with the terms *base unit, decimal system, derived unit, fundamental unit, metric system,* and *prefix*.
22. Explain how measurement is a form of modeling.
23. Explain the process for converting units.
24. Solve the following by using unit conversions.
    a. How many meters are in 738.4 nm?
    b. How many picobytes are in 0.003 55 Tb?
    c. A leap year has how many seconds in it?
    d. What is the density of antimony (6697 kg/L) in g/mL?
25. Evaluate the statement, "I can know the measurement of an object."

26. The density of mercury is 13.534 g/mL. What would be your percent error if you experimentally determined the density to be 13.000 g/mL?

27. Compare accuracy and precision.

28. During an experiment, you used a thermometer marked to 1 °C and obtained temperature readings of 45.2 °C, 44.9 °C, and 45.1 °C. The expected temperature was 45.0 °C. Evaluate the precision and accuracy of the measurements.

29. Why should a scientist be accurate and precise in his research?

30. Summarize the rules for determining significant figures in recorded data.

31. List in order which digits are significant in the following numbers:

  a. 230 mL **d.** 0.094 05 g/cm^3

  b. 0.005 40 s **e.** 700 jelly beans

  c. 1440 **f.** 1.05×10^6 cm

32. Why do scientists control the number of significant figures in calculations involving measurements?

33. Perform the following operations. Answers should include appropriate significant figures and scientific notation if needed.

  a. (6.98 cm − 2.8 cm)(1.7 cm^2)

  b. $\dfrac{\pi (2.1 \text{ m})^2 (121 \text{ m})}{6.83 \text{ s}}$

  c. 2(1.75 cm)

  d. $(4.08 \times 10^{-3} \text{ g}) - (7.2 \times 10^{-4} \text{ g})$

  e. $\dfrac{3.410 \times 10^{-2} \text{ cm}^3}{5.08 \times 10^{-3} \text{ cm}}$

  f. (86.00 cm)(3.228 cm)(0.37 cm)

  g. $\dfrac{796.13 \text{ cm} + 12.223 \text{ m}}{0.725 \text{ s} + 11.32 \text{ s}}$

  h. $(1.374 \times 10^{-2} \text{ m}^3) + (7.11 \times 10^3 \text{ cm}^3)$

34. When you consider Genesis 1:28, what do you think is the value of methodical problem solving?

35. After solving a problem, how can you know whether your answer is reasonable? Why is it important to know whether your answer is reasonable?

CRITICAL THINKING

36. Write the conversion factors for converting the following units.

  a. inches into kilometers

  b. centuries into seconds (assume 365.25 days per year)

37. Solve the following by using unit conversions.

  a. The earth's mass is estimated to be 5.9722×10^{24} kg. What is its mass in milligrams?

  b. A lightyear (ly), the distance traveled by light in one year, is equal to 9.4607×10^{15} m. The Milky Way has a diameter of 175 kly. What is the diameter of our galaxy in centimeters?

  c. The volume of the earth is approximately 1.083×10^{12} km^3. How many milliliters is this?

  d. A box is 0.28 m high, 0.0515 m wide, and 0.174 m long. How many liters of air does this box contain?

38. A particular forest covers 25 397 acres, and there is an average of 378 white oak trees on each acre. An average leaf of a white oak has a mass of 1.7 g, and each tree has an average of 4500 leaves on it. Assuming that all the trees drop all their leaves each year, how many kilograms of white oak tree leaves fall on the floor of that forest each year?

39. When making compound calculations, why do we delay rounding until the end of the calculation?

40. When making the calculation below, how many significant figures would you round to and why?

$$r = \sqrt[3]{\left(\dfrac{3}{4}\right)\left(\dfrac{92.75 \text{ m}^3}{\pi}\right)}$$

41. An experiment requires 45.6 g of isopropyl alcohol as a reagent. Instead of measuring the mass of the liquid on a balance, you decide to determine the mass by using an equivalent volume measured in a graduated cylinder. The density of isopropyl alcohol is 0.7851 g/mL. What volume of the alcohol do you need for the experiment?

42. Use a systematic approach to solve the following density problems. Answers should include the correct units and the allowed significant figures.

  a. Mercury, which is sometimes used in thermometers, has a density of 13.534 g/mL at room temperature. What volume of mercury contains 10.0 g?

  b. Copper has a density of 8.920 g/cm^3. What mass of copper will occupy 45 cm^3?

  c. A solid will float on any liquid that is denser than itself. The volume of a piece of calcite, which is often used in the production of chalk, is 12.5 cm^3, and its mass is 33.93 g. On which of the following liquids will the calcite float: acetone (ρ = 0.785 g/mL), water (ρ = 0.998 g/mL), methylene bromide (ρ = 2.497 g/mL), bromine (ρ = 3.119 g/mL), and mercury (ρ = 13.534 g/mL)?

03 CHAPTER REVIEW

ethics

ETHICAL MEDICAL TESTING

In Chapter 1 we outlined a process for making ethical decisions in chemistry. If you need to review, see the discussion on pages 8–9. To help you understand it better, we will outline the process by looking at the important issue of ethical medical testing.

The Strategy

Ethical issues can be complex and can therefore benefit from an orderly approach. The strategy outlined here is based on the ethics triad taught in Chapter 1 and is organized into five steps.

❶ What information can I get about this issue?

To formulate an appropriate and informed position, we need data and information, so we will need to do some research on a topic before we can evaluate how to ethically approach the issue. For this exercise, we can find the needed information in the example on the next page.

❷ What does the Bible say about this issue?

According to 2 Peter 1:3, Christians have all they need to live godly lives through the knowledge of God. We gain our knowledge of God through His Word. While the Bible addresses some issues directly, we can rely on its foundational principles to address all issues in life. Consider the biblical principles of the image of God in man, the Creation Mandate, and God's whole truth.

❸ What are the acceptable and unacceptable options?

As we address ethical issues, we should consider all possible options. There are some options that will come to mind immediately, others that are popular, and even some that are not initially obvious. We must test each of these options against the standard of the Bible (1 Thess. 5:21). Compare each option to the biblical teaching and principles that we considered in Step 2. Some of the options will fit with the biblical outcomes of human flourishing, a thriving creation, and God's glory, but others may not. Reject any options that are inconsistent with biblical principles and outcomes.

❹ What are the motivations of the acceptable options?

Besides considering principles that inform our decisions and consequences biblically, we also need to be asking how this option will help me and others grow. This is the third approach—considering biblical motivations. We should reject options that adversely affect our growth and be cautious of options that don't seem to help us grow in faith, hope, and love. Recall the biblical motivations of faith in God, hope in God's promises, and love for God and others.

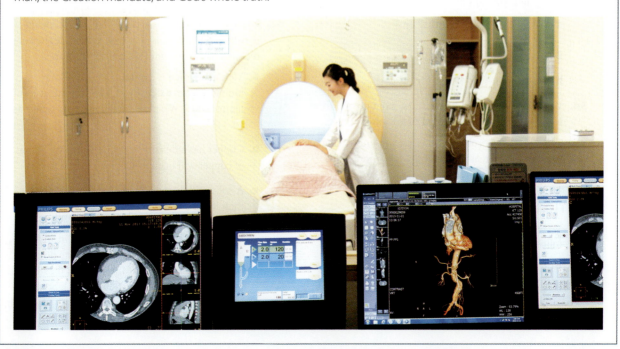

5 What action should I take?

Once we have assessed the different options according to the ethics triad, we can plan a course of action or form an opinion. We should always be ready to give justification for the action that we suggest on the basis of biblical analysis.

The Issue: Ethical Medical Testing

One of the many ethical issues in the field of chemistry is that of medical testing. Should testing be done? It is fairly obvious that for medical practice to advance there is a need for testing. Doctors need information on new procedures and new medicines. Are new medicines effective for all patients equally? Does the new medicine interact negatively with other medicines? How can we get the needed information in the most ethical manner?

1. What information can I get about this issue?

From 1932 until 1972, the US Public Health Service conducted a long-term study on the progression of syphilis, a sexually transmitted disease, in African-American men. The subjects of the study were never told that they had the disease. Even after the confirmation that penicillin was an effective treatment for the disease, the study was continued for an additional twenty-five years *without any of the participants being treated with penicillin*. A number of the men involved in the study died of the disease, and some transmitted the disease to their wives. Some children of the participants were infected at birth.

2. What does the Bible say about this issue?

All life is to be protected, but human life, the special creation of God, must be protected above all else. Biblical principles also tell us to love our neighbors. Loving our neighbors implies meeting the health and safety needs of those around us. Also, we glorify God when we learn more about His creation, especially about His special creation, human beings. These two principles intersect as new discoveries in medical procedures and medicines allow us to treat patients better.

3. What are the acceptable and unacceptable options, and what may be the consequences of each option?

One option is to forego all medical testing, which would mean that we would either stop introducing new procedures or medicines or we would introduce them without knowing their effectiveness. Another option is to test but with a participant-beware mentality. While this would ultimately benefit patients, it might be harmful to participants, which would result in a decrease in willing participants. A third option would be to do the research but to include what is called *advised consent*, which means that anyone participating in a study must be educated about the purpose, duration, risks, and benefits of the intended research along with any option for withdrawing from the study. Each participant would then have the option of consenting or not. In any instance that we opt for medical testing, that testing must be modified or stopped in the event that it is determined to have more risk for the participants than benefits gained from the research.

4. What are the motivations of the acceptable options?

Balancing the risks of research with the benefits of information gained from the research demonstrates the value we place both on the patients and the research participants. By getting advised consent from participants, we demonstrate their value as God's special creation. We must put the needs of human beings above our quest for scientific knowledge. By putting others first, we demonstrate our love for God and His creation.

5. What action should I take?

Medical testing is necessary to advance medical treatment. We must research new and better ways to help people deal with the results of the fall. We must balance the needs for research with the risks to the participants. We must have the advised consent of all participants. We must be willing to modify or even end a research project if the risks outweigh the benefits.

43. Why is Step 1 so important?
44. Which step(s) in this process explore(s) biblical principles?
45. Which step(s) in this process explore(s) biblical outcomes?
46. Which step(s) in this process explore(s) biblical motivations?
47. Why is Step 5 necessary?

Chapter 4

ATOMIC STRUCTURE

The year is 1955, and it is a year full of historic events. In the small town of Anaheim, California, the conversion of 160 acres of former orange and walnut groves into a new theme park—Disneyland—is completed. Popular TV shows like *I Love Lucy* fuel an explosion in the growth of television ownership, from 9% of American homes to 64% in only five years. In Montgomery, Alabama, an African-American woman named Rosa Parks refuses to give up her seat on a bus to a white passenger, sparking the equal rights movement. And at Penn State University, physicist Erwin Müller and his student, Kanwar Bahadur, become the first men to *see* atoms. Of course, they needed a special tool to see them—the field ion microscope—which the two scientists co-invented. Incredibly, scientists had discovered much about the structure of atoms long before they ever saw them. In this chapter we'll look at some of the models that scientists developed to explain the atomic world and how those models changed in light of new evidence.

4.1 Early Thoughts About Matter 77
4.2 The Development of Atomic Models 80
4.3 Useful Notations 85

4.1 EARLY THOUGHTS ABOUT MATTER

Investigating Atoms

Our ability to image atoms has improved since 1955. More recently, scientists have used a technique known as scanning tunneling microscopy (STM). But before the invention of such powerful instruments of observation, most of the evidence about atoms was collected indirectly. Scientists ingeniously used fragments of evidence acquired over nearly two centuries to piece together different theories called atomic models. These models are foundational to what we know about chemical and physical properties of matter, chemical bonding, and chemical reactions. As scientists continue to study matter, they continue to revise their models. The current model is itself open to further revision.

Christians are not troubled by science producing models rather than ultimate truth because they realize that the purpose of science is not to provide the ultimate answers for life. But this does not mean that Christians are unconcerned about research or about improving existing scientific models. Our God-given curiosity leads some of us to studying chemistry, and some who are so led will use that knowledge to manage His creation wisely. These scientists are interested in working with the most accurate models to better perform their work.

QUESTIONS
» Who conceived of the idea of atoms?
» What evidence led to the acceptance of the atomic model?
» What was Dalton's atomic theory?

TERMS
law of definite proportion • law of multiple proportions

How has our understanding of matter changed through history?

Atoms: From Philosophy to Science

The ancient Greeks debated about the nature of matter. Some of them, including Aristotle, held to a *continuous theory of matter*. They believed that matter could be continuously subdivided without end. This idea was an offshoot of the Greeks' purely mathematical view of the world, where dividing something in half always resulted in two new wholes that could be divided in half again. Other Greek philosophers disagreed, believing that matter was made of separate, discrete particles, an idea known as *atomism*. Democritus is usually credited with articulating this position. He said that matter contained definite particles and that it could *not* be divided infinitely without losing its properties. The Greek word that Democritus used to name those smallest particles, *atomos*, means "cannot be divided." Today we use the term *atom* for the smallest particle capable of chemical interactions.

Belief in philosophical atomism continued throughout the ancient and medieval eras, but it was not widespread. It wasn't until the seventeenth century that people started to use science to prove that matter is in fact composed of atoms or particles. Philosophical atomism was thus replaced by scientific atomism.

Democritus

▶ Democritus believed that atoms of solids had hooks holding them together, that atoms of liquids were slippery, and that atoms of salts and spices were sharp.

Atomic Structure **77**

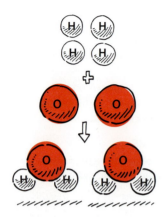

law of definite proportion

Many scientists during the seventeenth and eighteenth centuries grappled with the particle theory of matter. Famous men of science such as Isaac Newton, Robert Boyle, and John Dalton disagreed on whether matter is continuous. The first real experimental support for the existence of atoms was finally obtained in the late 1700s. But no one recognized the full significance of this data for several decades.

Eighteenth-century chemists observed that compounds always contained a set ratio for the masses of elements in the compound. This was true regardless of how much of the compound was analyzed or where it came from. For instance, decomposing a 9.0 g sample of water always produced 8.0 g of oxygen and 1.0 g of hydrogen, a ratio of 8:1. Decomposing an 18.0 g water sample produced 16.0 g of oxygen and 2.0 g of hydrogen—the same ratio as the smaller sample. Any sample of water always contained 8 g of oxygen for every 1 g of hydrogen.

Further experimentation with many compounds confirmed that the ratios of composition by mass were unique and constant for each compound. This discovery led to the **law of definite proportion**, formulated by French chemist Joseph Proust in 1794. This law states that every compound is formed of elements combined in specific ratios by mass that are unique for that compound.

Dalton's Atomic Model

FIRST EXPERIMENTAL MODEL

It wasn't until the early 1800s that an atomic model that was based on experimental evidence instead of philosophy was developed. This new model was framed by the work of John Dalton, an English schoolteacher. Chemists of Dalton's day knew that some elements could combine in different proportions to form different compounds.

John Dalton

For example, it was known that carbon could form two different oxides with oxygen. In one oxide, 100 g of carbon combines with every 133 g of oxygen. For the other oxide, every 100 g of carbon combines with 266 g of oxygen. Through a careful examination of these and other compounds, Dalton realized that whenever a fixed amount of one element could combine with different masses of a second element, those masses always occurred in a ratio that could be reduced to small, whole numbers. This relationship is now known as the **law of multiple proportions**.

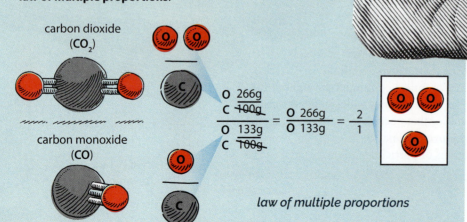

law of multiple proportions

John Dalton imagined atoms as solid spheres of different sizes and masses.

For example, the ratio of oxygen in the two carbon oxides described earlier (carbon monoxide and carbon dioxide) is 266 g to 133 g, or 2:1. By coupling this insight with what he knew from the law of definite proportion and the law of the conservation of matter, Dalton came to some remarkable conclusions.

1. All matter is made of atoms.

2. Atoms are indivisible; they cannot be created or destroyed.

3. The atoms of an element are all alike.

4. The atoms of one element are different from the atoms of all other elements, especially their masses.

5. Atoms combine chemically in small, whole number ratios.

Dalton's ideas about matter were based on two premises. First, combinations of atoms from different elements form compounds; second, atoms of different elements have different masses. Dalton assigned relative masses to atoms of various elements so that he could determine the masses of the atoms in relation to each other. Through painstaking analysis, Dalton and other chemists of his day determined that oxygen was much more massive than hydrogen and slightly more massive than carbon. He made the same calculation for other atoms as well.

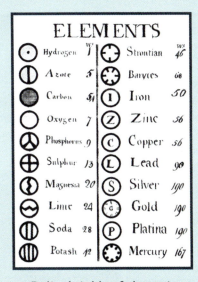

Dalton's table of elements

The exact values in Dalton's table were not accurate. In some cases, even the order of elements from smallest to largest was incorrect. Nevertheless, Dalton's accomplishments were notable, given the small amount of information available to him. His theory inspired new experimentation that brought new insights into the nature of matter. Within fifty years, many scientists had accepted the atomic theory, and chemistry advanced significantly. Dalton's model is still the foundation of present atomic theories.

4.1 SECTION REVIEW

1. State the main difference between atomism and the continuous theory of matter.

2. The law of definite proportion states that every compound has a definite composition by mass. What does that mean?

3. Compare the types of evidence used by the Greeks and later by John Dalton to support their respective models of the atom.

4. State the law of multiple proportions in your own words.

5. What two premises formed the basis for Dalton's atomic model?

6. Are the atomic models of Democritus and Dalton directly related? Why or why not?

4.2 THE DEVELOPMENT OF ATOMIC MODELS

QUESTION
» Why are there different atomic models?
» How did we learn that atoms could be divided?
» How were different subatomic particles discovered?

TERMS
electron • nucleus • proton • atomic number • neutron

What are atoms made of?

The invention of the electrochemical battery by Alessandro Volta in 1800—about the same time as Dalton's discovery—allowed scientists to study how matter behaves in the presence of an electric current. They found that a gas sealed under very low pressure in a glass tube can carry an electrical current between two electrodes. The glass tube was called a *gas discharge tube*.

Scientists noticed another strange phenomenon. When most of the gas molecules were removed by a vacuum pump, the current decreased. But when even more gas was removed, the current gradually began to increase, and the tube glowed with an eerie green light. Later, scientists determined that the color of the light depended on what gas was in the tube. Removing all gas molecules caused the glow to disappear, but current continued to flow. Obviously, something other than the gas carried the current. In a series of experiments during the late 1800s, physicists observed that current through the discharge tube came from the negative electrode, called the *cathode*. Emissions from the cathode were called *cathode rays*.

Thomson's Model
DISCOVERY OF THE ELECTRON

English physicist J. J. Thomson finally explained cathode rays. He noted that cathode rays travel in straight lines, unaffected by gravity. That phenomenon could be explained only if the cathode rays were waves, similar to light waves, or particles moving at incredibly fast speeds. He also observed that a magnet could deflect cathode rays. Thus, the rays could possibly be tiny charged particles.

When Thomson passed a cathode ray between two electrically charged plates, the ray bent toward the positively charged plate. Accordingly, Thomson concluded that the cathode-ray particles were negatively charged.

A cathode ray tube with a magnet deflecting rays.

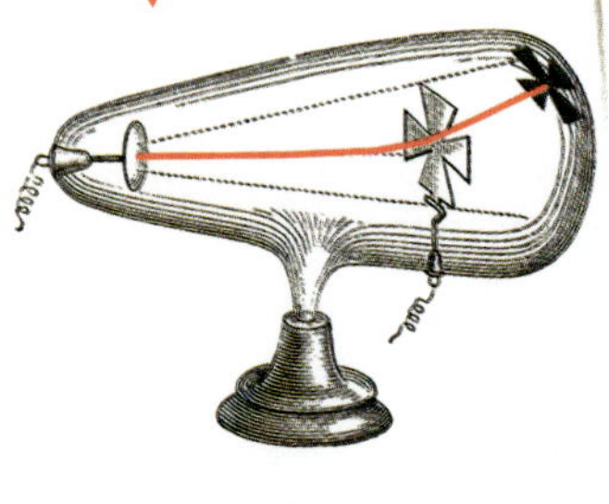

J. J. Thomson

Another scientist, George Johnstone Stoney, named these particles **electrons** (e⁻) in 1894. The *e* in the electron symbol indicates the fundamental charge, which was determined to be 1.602×10^{-19} coulombs.

In 1897 Thomson determined the charge-to-mass ratio (*e/m*) of the particle. He found that this ratio was surprisingly large. Compared with the charge, the mass of the electron was almost nothing (it was later determined to be 9.109×10^{-31} kg). He concluded that he was working with a particle fundamentally different from atoms and molecules, the ordinary particles of matter. Thomson concluded that every atom contained smaller, negatively charged particles.

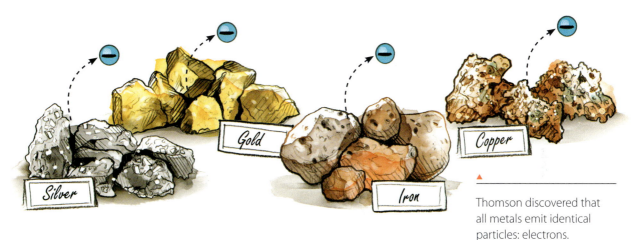

Thomson discovered that all metals emit identical particles: electrons.

Thomson further proved that the same type of particle was emitted from atoms of every element tested. Regardless of the gas in the tube or the type of cathode material, he obtained the same results. Gold, silver, iron, and copper all gave off the same particles.

Dalton's atomic model with its indivisible atoms could not explain how electrically charged particles could exist in an atom, why the atom was neutral, and how negative charges could leave an atom. A new, more workable model was needed to account for Thomson's observations.

In his model, negatively charged electrons are embedded in a positively charged substance that completely surrounds them. The charge on the positive material balances out the negative charges on the electrons so that the atom is neutral. Under certain conditions, electrons could be removed from the atom.

The proposed model looked something like an English plum pudding. The electrons resembled negatively charged "plums" in a positively charged "pudding." Thus, Thomson's model is often known as the *plum pudding model* of the atom. An analogy that you might be more familiar with is a lump of chocolate chip cookie dough—the chocolate chips represent the electrons, just like the plums in Thomson's plum pudding. Thomson's work earned him knighthood, as well as a Nobel Prize in Physics in 1906.

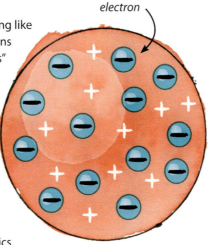

plum pudding model

Thomson's atomic model illustrates a fundamental aspect of the work of science: whenever current scientific models prove insufficient to explain new observations, they must be modified or else replaced by newer, more workable models. As you'll see in the next section, Thomson's model needed such modification almost immediately.

Rutherford's Model
DISCOVERY OF THE NUCLEUS

Wilhelm Röentgen's discovery of x-rays in 1895 triggered further experiments that led to the discovery of nuclear radiation. In the twenty years before World War I, the brightest minds in the scientific world pioneered the fascinating field of radioactive particles. One of these brilliant scientists was Ernest Rutherford, a native of New Zealand teaching at Cambridge University in England. His experiments with alpha particles, a form of radiation, led to new insights about the inner parts of the atom.

Ernest Rutherford

Rutherford discovered that alpha particles are positively charged ions with a mass equal to that of about 7300 electrons. These particles are emitted at very high speeds from some radioactive elements. Their large mass and high speed give alpha particles a great amount of energy. They also are strongly charged, having a 2+ electrical charge. If an alpha particle strikes a screen coated with zinc sulfide, its energy is converted into a brief flash of light that allows it to be observed.

In 1908, under his direction, one of Rutherford's assistants, Hans Geiger, and a student, Ernest Marsden, designed a series of experiments in which a beam of alpha particles was aimed at a sheet of thin gold foil (right) that was in turn surrounded by a curved zinc sulfide screen.

As the particles struck the foil, most of them went straight through, illuminating the screen in a direct line with the alpha particle source. A few, as expected, were slightly deflected and were detected by the flashes of light on the zinc sulfide screen at slight angles from the source.

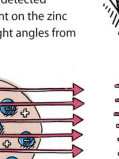

Thomson's atomic model: Showing alpha rays crossing straight through atom

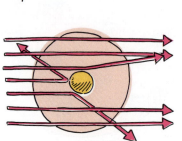

Rutherford's atomic model: Showing alpha rays reflecting off nucleus

Unexpectedly, about one in 20,000 particles was deflected at a large angle or even ricocheted back toward the source. Rutherford was astounded. He compared the deflected alpha particles to a cannon shell bouncing off tissue paper. Thomson's model could not explain these observations—already it was time for a new model!

Rutherford reasoned that atoms must be mostly empty space, otherwise most of the alpha particles should have been deflected. Because such a small percentage of the alpha particles were greatly deflected, there must be little likelihood that an alpha particle would collide with a central, dense region of an atom. Rutherford reasoned that this region, the **nucleus**, as he called it, must be very small. Subsequent calculations indicated that the diameter of a nucleus was only 1/100 000 the size of an entire atom. Rutherford's model is sometimes known as the nuclear model of the atom (right).

Rutherford's model of the atom had a tiny, positively charged, massive nucleus surrounded by electrons.

Rutherford's carbon atom: Protons and electrons, no neutrons

At first, Rutherford did not investigate the charge of the nucleus. Further experimentation, however, determined that the nucleus had to be positive. A decade after discovering the nucleus, Rutherford identified the positive particles in the nucleus, which he called **protons** (p^+, or p).

A proton has an electrical charge of $+1e$, exactly opposite the electron's charge of $-1e$. The mass of a proton is approximately 1836 times that of an electron, or 1.673×10^{-27} kg.

Shortly after the discovery of the proton, scientists recognized a relationship between the number of protons in the nuclei of atoms and their chemical and physical properties. The number of protons in the nucleus is called an atom's **atomic number**.

Completing Rutherford's Model
DISCOVERY OF THE NEUTRON

Rutherford and others realized that atoms have much more mass than the sum of the masses of protons and electrons. Rutherford suggested that neutral particles supplied the missing mass without changing an atom's charge. He thought that the particle might consist of a closely joined proton and electron. Because this hypothetical particle did in fact turn out to be electrically neutral, it eluded detection for many years.

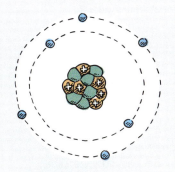

Chadwick's carbon atom: Protons, electrons, and neutrons

In 1932 James Chadwick observed neutral particles radiating from a sample of metal bombarded by alpha particles. Appropriately enough, these particles were named **neutrons** (n^0, or n). A neutron has a mass of 1.675×10^{-27} kg, 1839 times the mass of an electron—just slightly greater than the mass of a proton. Neutrons and protons make up the nucleus and contain almost all of the mass of an atom.

James Chadwick

MINILAB

INDIRECT OBSERVATION

What do Democritus, John Dalton, J. J. Thomson, and Ernest Rutherford have in common other than their mutual interest in atoms? None of them ever saw the thing that so interested them—and no one ever has, directly at least. Instead, these pioneers observed how atoms—matter—interacted with other matter. This was a type of indirect observation. In this activity, you'll practice this art of indirect observation.

How can we "observe" what we can't see?

Procedure

A Obtain a numbered box from your teacher. Record the number of your box on a piece of paper. The box contains a secured wooden block of unknown shape and a marble. Your job is to see whether you can figure out the shape of the block by rolling the marble around inside the box. You may not open or shake the box—only tilt it to roll the marble.

1. Why are your observations about the block a type of indirection observation?

Conclusion

2. Draw what you think the shape of your block is.

After you have drawn your block, your teacher will tell you what shape of block is in your box.

3. How close was your hypothesis compared to the actual shape?

Going Further

4. Suggest an additional test that might be performed on the box that, without opening it, could give additional information about the shape of the block inside.

EQUIPMENT
- observation boxes

4.2 SECTION REVIEW

1. What inventions permitted the investigation of the nature of atoms?

2. What four characteristics did Thomson and others determine about cathode rays that resulted in the discovery of the electron?

3. How did Thomson's discovery change the model of the atom?

4. After considering the evidence of the Geiger-Marsden experiments, how did Rutherford describe the atom, and what evidence was his model based on?

5. Why are there different atomic models?

6. Why was the discovery of the neutron difficult? Why was it important for the development of the atomic model?

7. Create a table that displays the mass, charge, and location within an atom for each of the three major subatomic particles.

8. Do you agree or disagree with the idea that Rutherford's atomic model was more workable than Thomson's model, even though Rutherford's model still had no accounting for neutrons? Defend your answer.

4.3 USEFUL NOTATIONS

It didn't take long for Rutherford's nuclear model to supplant Thomson's plum pudding model as the prevailing atomic model of the day. But the amazing pace of discovery about the atomic world suggested that Rutherford's model would soon be modified too. You'll learn more about that in the next chapter. For now we'll focus on the three basic subatomic particles and how chemists and physicists have developed notations to easily identify the numbers and arrangement of these particles within an atom. These symbols will help you understand the various properties of each element.

Atomic Number

We have already mentioned that the atomic number of an element is the number of protons in the nucleus. The atoms of each element contain a unique number of protons, which determines the identity of the element. If the atomic number of an atom changes, then its identity (element) also changes. It is very important to know the atomic number (Z) of an element. The size of an atom's positive nuclear charge determines the number of electrons surrounding the nucleus of an electrically balanced (neutral) atom. In electrically balanced atoms, the number of electrons is equal to the number of protons.

Mass Number and Isotopes

After Chadwick's discovery of neutrons, scientists could explain an observation that had puzzled them. When ionized atoms of a single gaseous element were "shot" between electrically charged plates, the electrostatic force deflected the atoms. But not all the atoms were deflected the same amount. Some seemed to be heavier and were deflected less, while others seemed to be lighter and were deflected more. Since these atoms all had the same number of protons, the differences in mass were attributed to differences in the number of neutrons in their nuclei. Repeated experiments confirmed this observation.

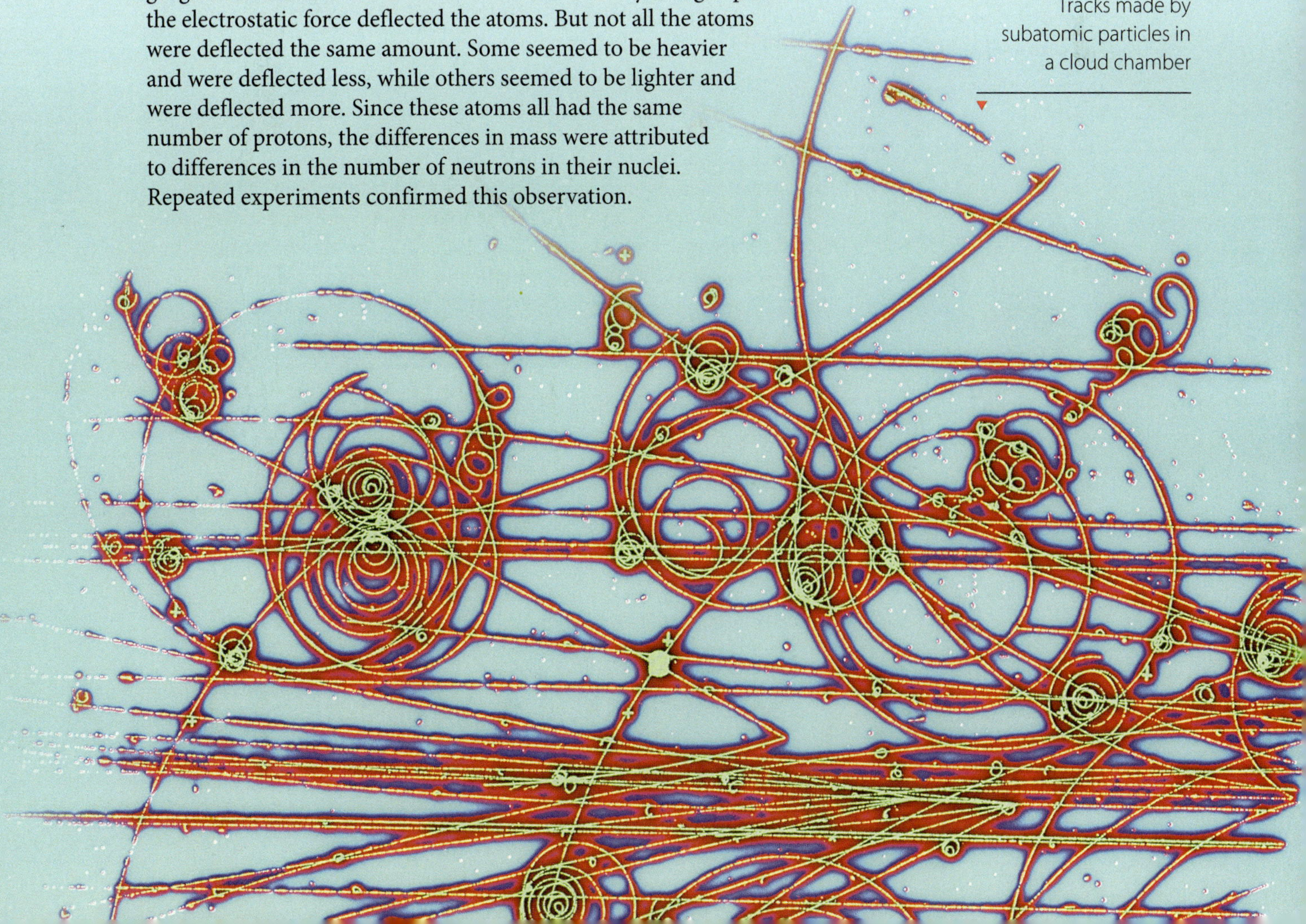

QUESTIONS
» How can I tell how many of each kind of subatomic particle are in an atom?
» Are mass number and atomic mass the same thing?
» How is atomic mass determined?

TERMS
isotope • mass number • isotope notation • unified atomic mass unit • atomic mass

Are all carbon atoms the same?

Tracks made by subatomic particles in a cloud chamber

carbon isotopes

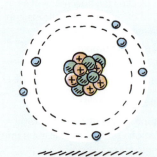

carbon-12

6 protons

6 neutrons

6 electrons

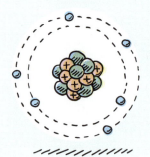

carbon-13

6 protons

7 neutrons

6 electrons

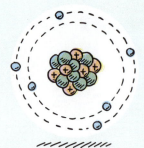

carbon-14

6 protons

8 neutrons

6 electrons

If the number of neutrons can vary from atom to atom in an element, then the atoms of a given element are not truly identical, contrary to what Dalton and other nineteenth-century scientists thought. Physicists and chemists call atoms with the same number of protons but different numbers of neutrons **isotopes**. Isotopes of an element generally have the same chemical properties. But certain physical properties that depend on mass, such as density, evaporation, and diffusion rate, vary between isotopes. In order to identify the isotopes of a given element, scientists established the **mass number** (A). The mass number is a whole number that is equal to the sum of the protons and neutrons in an atom.

$$A = Z + N$$

Isotopes of an element are named by writing the element's name followed by a hyphen and the isotope's mass number. For example, the most common isotope of oxygen ($Z = 8$) is oxygen-16. An atom of oxygen-16 contains 8 protons and 8 neutrons. The name of an isotope is sometimes shortened to just the element's symbol followed by a hyphen and the mass number. For example, oxygen-16 can be written O-16 and carbon-14 can be written C-14.

When chemists want to indicate the specific isotope in a chemical expression, they use a special notation called **isotope notation**. This notation is based on the element's symbol and includes the mass number as a leading superscript and the atomic number as a leading subscript. The generalized isotope notation is shown at right. The isotope notations for oxygen-16 and carbon-14 are respectively written as

$$^{16}_{8}O \text{ and } ^{14}_{6}C.$$

In some instances, the atomic numbers can be omitted from an isotope's notation since the element's identity is unique to its atomic number. If chemists are considering a group of isotopes of the same element, their notations will often be written with only the mass number superscripts. The reason in this case is that the atomic number for each isotope is the same and is implied by the element's symbol.

EXAMPLE 4-1 WRITING ISOTOPE NOTATION

Write the isotope notation for each of the following.

    a. sodium-24   **c.** lead-208

    b. hydrogen-2   **d.** uranium-235

Solution

    a. sodium-24 ($Z = 11$): $^{24}_{11}Na$

    b. hydrogen-2 ($Z = 1$): $^{2}_{1}H$

    c. lead-208 ($Z = 82$): $^{208}_{82}Pb$

    d. uranium-235 ($Z = 92$): $^{235}_{92}U$

EXAMPLE 4-2 COUNTING NEUTRONS IN ISOTOPES

hydrogen isotopes

We can easily determine the number of neutrons for a given isotope by subtracting the atomic number from the mass number. Determine the number of neutrons in the nucleus of each isotope listed in Example 4-1.

Solution
Use $N = A - Z$.

a. sodium-24: $N = 24 - 11 = 13$ neutrons

b. hydrogen-2: $N = 2 - 1 = 1$ neutron

c. lead-208: $N = 208 - 82 = 126$ neutrons

d. uranium-235: $N = 235 - 92 = 143$ neutrons

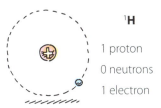

^{1}H
1 proton
0 neutrons
1 electron

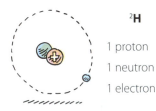

^{2}H
1 proton
1 neutron
1 electron

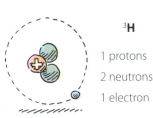

^{3}H
1 protons
2 neutrons
1 electron

In some instances, the identity of a particular isotope can have real consequences for its physical properties. For example, hydrogen has three known isotopes: ^{1}H; ^{2}H (or D), called *deuterium*; and ^{3}H (or T), called *tritium*. Notice that we have omitted the atomic numbers from the notations in this example, as described previously. The most important deuterium compound contains two atoms of the ^{2}H isotope and is written as D_2O, called "heavy" water, which in large quantities is poisonous to humans. Its only large-scale use is in nuclear reactors. Tritium is radioactive and highly unstable. It is an extremely toxic radiation source, even in small quantities. Tritium is used as a "trigger" in nuclear fusion bombs, known more commonly as hydrogen bombs. So for hydrogen, having an extra neutron or two makes a big difference!

CAREERS

SERVING AS A SCIENCE TEACHER:
ENCOURAGING INQUIRING MINDS

In 2018, over 15 *million* students attended high school in the United States. That number represents a huge need for teachers. In recent years, the United States has experienced a growing shortage of teachers, and science teachers are especially hard for schools to find.

Have you ever thought about what science teachers find especially rewarding? People engaged in the work of science, including science teachers, are usually driven by an intense curiosity about the world and how it works. Science teachers find satisfaction in helping others discover their own sense of wonder about God's creation. Christian science teachers have an additional motivation. Jeff Foster, a seventh-grade science teacher, says, "What motivates me as a science teacher is using the study of God's creation to provide exciting opportunities to help students take the next important steps in their relationship with the Creator."

Without a constant influx of new scientists, trained by science teachers like Jeff, the process of scientific inquiry would be greatly hampered.

Hydrogen
Element : H
Atomic Number (Z): 1
Atomic Mass (u) : 1.01

Sodium
Element : Na
Atomic Number (Z): 11
Atomic Mass (u) : 22.99

Atomic Masses of Some Common Elements

Carbon
Element : C
Atomic Number (Z): 6
Atomic Mass (u) : 12.01

Sulfur
Element : S
Atomic Number (Z): 16
Atomic Mass (u) : 32.06

Oxygen
Element : O
Atomic Number (Z): 8
Atomic Mass (u) : 16.00

Calcium
Element : Ca
Atomic Number (Z): 20
Atomic Mass (u) : 40.08

Atomic Mass

Once the chemists of Dalton's day started to analyze chemical compounds, they tried to determine the masses of individual atoms. They knew that water was made of hydrogen and oxygen, containing 8 g of oxygen for every 1 g of hydrogen. Because no one knew the real formula for water, Dalton assumed that there must be one atom of each. Thus, water's chemical formula would have been OH. So he assigned the smallest atom, hydrogen, a mass of 1 and oxygen a mass of 8. We now know that the real formula is H_2O, so oxygen's mass is really 16.

Chemists express these masses in daltons, or **unified atomic mass units** (Da, or u). You may see it abbreviated amu, but this is viewed less favorably. An atomic mass unit is currently defined as one-twelfth the mass of a carbon-12 atom, or as approximately $1.660\,54 \times 10^{-27}$ kg. It is about equal to the mass of a proton (1.0073 u) or a neutron (1.0087 u). The electron is far less massive (0.000 55 u).

Why are atomic masses not whole numbers? In nature, samples of even pure elements are mixtures of the element's isotopes, which have different mass numbers. Thus an **atomic mass** as printed in references such as the periodic table is the weighted average of the masses of the element's isotopes as they occur in nature. A weighted average is an average that gives more importance to certain values than to others. In the case of atomic mass, more common isotopes (higher percent abundance) are given more weight than less common isotopes. Thus, the listed atomic mass is closer to the mass of a typical atom than a simple average would be.

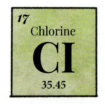

For example, a sample of chlorine is made of two chlorine isotopes, Cl-35 (34.969 u, 75.76% abundance) and Cl-37 (36.966 u, 24.24% abundance). The simple average mass of these isotopes is 35.968 u. But because Cl-35 is three times as common, it is much more likely that a randomly chosen atom would have the smaller mass. Thus, the weighted average atomic mass of a sample containing innumerable atoms will be closer to 35 u than to 37 u.

We calculate the average atomic mass of a chlorine atom by following the steps below.

1. Convert the percentages to decimals.

   75.76% = 0.7576

   24.24% = 0.2424

2. Multiply the mass of each isotope by its percent abundance to find the mass contributed by each isotope in the sample.

   a. mass of Cl-35 atoms in sample: 0.7576 × 34.969 u = 26.492 514 4 u

   b. mass of Cl-37 atoms in sample: 0.2424 × 36.966 u = 8.960 558 4 u

3. Add the results to get the weighted average mass of the sample.

   26.492 514 4 u + 8.960 558 4 u = 35.45 u

   The weighted average mass of a typical chlorine atom is 35.45 u.

Remember also that since your final answer is a *sum*, it can have no more decimal places than the least precise value being added. Though we know three decimal places in 8.961 u, we know only two decimal places in 26.49 u. Therefore, the sum of the two values can have only two decimal places.

EXAMPLE 4-3 COMPUTING A WEIGHTED AVERAGE ATOMIC MASS

Magnesium has three naturally occurring isotopes: Mg-24 (23.985 u) with an abundance of 78.99%, Mg-25 (24.986 u) with an abundance of 10.00%, and Mg-26 (25.983 u) with an abundance of 11.01%. Determine the average atomic mass of magnesium.

Solution

1. Convert the percentages to decimals.

    78.99% = 0.7899
    10.00% = 0.1000
    11.01% = 0.1101

2. Multiply the mass of each isotope by its percent abundance to find the mass contributed by each isotope in the sample.

    Mg-24: 0.7899 × 23.985 u = 18.945 751 5 u
    Mg-25: 0.1000 × 24.986 u = 2.4986 u
    Mg-26: 0.1101 × 25.983 u = 2.860 728 3 u

3. Add the results to get the weighted average mass of the sample.

    18.945 751 5 u + 2.4986 u + 2.860 728 3 u = 24.305 079 8 u

The weighted average mass of a typical magnesium atom is 24.31 u.

As you perform these kinds of calculations, remember to observe the rules for significant figures provided in Chapter 3. As expected, the weighted average is closest to the mass of the magnesium-24 isotope because it has the highest natural abundance.

4.3 SECTION REVIEW

1. Identify the symbol or notation that is used to represent the following information.

    a. atomic number

    b. mass number

    c. number of neutrons

    d. identification of different isotopes

2. Compare an element's mass number with its atomic mass.

3. Determine the number of protons, neutrons, and electrons in the neutral atoms of the following elements.

    a. silicon-28   b. calcium-44

4. How is the atomic mass unit currently defined?

5. Two naturally occurring isotopes of aluminum exist, Al-26 and Al-27. Aluminum's atomic mass is 26.98 u. What can you conclude about the relative abundance of each isotope?

6. Lithium is a mineral that exists in the form of two stable isotopes. Testing has shown that 7.59% of lithium exists as Li-6 (6.015 u), and the remaining 92.41% is Li-7 (7.016 u). Calculate lithium's atomic mass.

04 CHAPTER REVIEW
Chapter Summary

4.1 EARLY THOUGHTS ABOUT MATTER

- Democritus, a Greek philosopher, first proposed that matter was made of discrete particles rather than being infinitely divisible.

- Atomism was not investigated scientifically until the eighteenth century. Joseph Proust confirmed that every chemical compound had a unique and constant ratio of composition.

- Quantitative chemistry began with John Dalton's particle model of the atom. He suggested that atoms of different elements were unique, with different masses and other properties. They were also indivisible and indestructible.

TERMS
law of definite proportion
law of multiple proportions

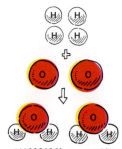

TERMS
electron
nucleus
proton
atomic number
neutron

4.2 THE DEVELOPMENT OF ATOMIC MODELS

- J. J. Thomson discovered electrons in the late nineteenth century after analyzing cathode rays in gas discharge tubes, establishing that atoms could be subdivided. His plum pudding model had negative electrons embedded in a positive substance.

- Using alpha particle radiation, Ernest Rutherford and his assistants discovered the atomic nucleus. Later he discovered that positively charged protons resided in the nucleus.

- Nearly three decades after the discovery of the proton, James Chadwick discovered and named the neutron. This accounted for the extra mass that Rutherford could not explain.

TERMS
isotope
mass number
isotope notation
unified atomic mass unit
atomic mass

4.3 USEFUL NOTATIONS

- The atoms of an element can differ according to the number of neutrons in their nuclei. These different forms of atoms are called isotopes.

- The mass number of an isotope is the sum of its protons and neutrons.

- Scientists identify isotopes with either isotope name or isotope notation.

- Atomic mass is the average of the masses of all the naturally occurring isotopes of an element. The average is weighted on the basis of the percent abundance of each isotope.

Chapter Review Questions

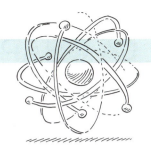

RECALLING FACTS

1. Who conceived of the idea of atoms?
2. What evidence led to the acceptance of the atomic model?
3. Briefly describe each of the following atomic models.
   a. Dalton's model
   b. plum pudding model
   c. nuclear model
4. How did we learn that atoms could be divided?
5. Describe the discoveries or advancements that made Dalton's model and the plum pudding model obsolete.
6. What is the charge in coulombs on an electron? on a proton?
7. (True or False) Protons and neutrons have the same mass.
8. How were different subatomic particles discovered?
9. Rank the three basic subatomic particles by mass from least massive to greatest.
10. Describe an isotope of an element. Give two ways that it can be represented.
11. What characteristic of an element does each of the following identify?
    a. atomic number
    b. mass number
    c. atomic mass

UNDERSTANDING CONCEPTS

12. Restate the law of definite proportion in your own words.
13. What characteristics of electrons contributed to their being the first subatomic particle to be detected? Explain.
14. Explain how changes to the atomic model made the model more workable.
15. Sketch each of the following atomic models.
    a. Dalton's atomic model
    b. plum pudding model
    c. Rutherford's model
16. Explain how atoms can be electrically neutral even though they contain charged particles.
17. Which aspects of Dalton's atomic model have proved to be incorrect? Explain.
18. How can you tell how many of each kind of subatomic particle are in a neutral atom when given an isotope's notation?

04 CHAPTER REVIEW

19. For each of the following isotopes, tell how many protons, neutrons, and electrons are present.

    a. $^{9}_{4}\text{Be}$

    b. $^{45}_{21}\text{Sc}$

    c. $^{127}_{53}\text{I}$

    d. $^{133}_{55}\text{Cs}$

20. Natural boron contains both B-10 and B-11 isotopes. Considering boron's listed atomic mass (10.81 u), which isotope do you think is more common? Explain.

21. Write isotope notation for each of the atoms whose composition is given below.

    a. fourteen protons, fourteen neutrons

    b. one proton, one neutron

    c. eighty protons, one hundred twenty neutrons

    d. twenty-six protons, thirty neutrons

    e. fifty-two protons, seventy-eight neutrons, fifty-two electrons

22. A naturally occurring sample of the element gallium (Ga) is a mixture of two isotopes. In a sample of gallium, 60.11% of the atoms have a mass of 68.9256 u and 39.89% of the atoms have a mass of 70.9247 u. Calculate the atomic mass of gallium.

CRITICAL THINKING

23. Evaluate the following statement: "The particle model of matter evolved continually from the time of Democritus around 400 BC until the time of Dalton in the early 1800s and on through the nineteenth and twentieth centuries." Do you agree or disagree? Explain.

24. Summarize the contributions to our understanding of atoms made by Dalton's, Thomson's, and Rutherford's atomic models using a single sentence for each.

25. A classmate tells you that only two naturally occurring isotopes of beryllium exist, Be-7 and Be-9, and that since beryllium's atomic mass is 9.01 u, we know that Be-9 must be the most common. Is your classmate correct? Explain.

26. Copper has two stable isotopes: Cu-63 (62.930 u) and Cu-65 (64.928 u). If the atomic mass of copper is 63.546 u, what is the percent abundance for each of the two isotopes?

27. The differences in physical properties caused by differences in mass between isotopes of an element are known as the *kinetic effect*. This effect is particularly apparent among the isotopes of hydrogen described on page 87. Why do you think this is so?

Use the Radium Girls ethics box below to answer Question 28.

28. Use the ethics triad from Chapter 1 and the strategy modeled for you in Chapter 3 to formulate in four paragraphs a Christian position on the topic of informing employees of potential workplace hazards. Be sure to address each leg of the triad: biblical principles, biblical outcomes, and biblical motivations. In the last paragraph, formulate your own position on the issue.

ethics

RADIUM GIRLS

During World War I, US soldiers needed to be able to tell time accurately—even in the dark. They could do so because the dials of their wristwatches were marked with glow-in-the-dark paint. Women working in the New Jersey factory of the United States Radium Corporation applied the luminescent paint to the watches. They were instructed to use their lips to maintain the pointed shape of their brushes. They were told that the paint was harmless. Because of the novelty of the paint, some women also painted their fingernails or faces with it.

Unhappily for these women, the glow of the paint was caused by radioactive uranium. United States Radium Corporation knew about the dangers associated with radiation exposure but withheld that information from its employees. Many of the women later endured severe medical conditions caused by working with the radioactive paint. In 1927, some of the "Radium Girls," as they came to be known, sued their former employer and settled out of court for damages. This and similar lawsuits filed against other companies eventually led to laws being implemented to improve workplace safety.

Atomic Structure

Chapter 5

ELECTRON ARRANGEMENT

The concept of atomism was widely accepted by 1897, the year in which J. J. Thomson discovered electrons. But imagine the awe and wonder that the scientific community must have felt as they discovered that ordinary matter was made of other subatomic particles as well. And the pace at which new particles were discovered was dizzying. Even before James Chadwick verified the existence of neutrons in 1932, other scientists had proposed the existence of *antimatter*, particles similar to ordinary matter but having opposite electric charges. The first of these, the positron, was discovered in the same year as the neutron. Other, equally fantastic particles of both matter and antimatter were soon discovered and given strange-sounding names such as muon (1937), kaon (1947), and neutrino (1956).

Perhaps even more amazing, it was eventually discovered that some particles of matter, such as protons, were themselves made of even smaller particles. The first of these *elementary particles* (other than the electron, which was subsequently classified as an elementary particle) was the quark, discovered in 1968. Particles composed of elementary particles were thereafter termed *composite particles*.

Since the discovery of the quark, other elementary and composite particles have been discovered. Current theories hypothesize the existence of many more. At facilities around the world, such as the Fermi National Accelerator Laboratory near Chicago, physicists continue to hunt for clues about the nature of matter and the most basic particles of which it is made.

5.1 Bohr Model *95*
5.2 Quantum-Mechanical Model *100*
5.3 Electron Configurations *106*

5.1 BOHR MODEL

Discovery of Electron Energy Levels: Bohr's Model

As you can see from the Chapter opener, Rutherford's nuclear model, which you learned about in Chapter 4, was still a very incomplete picture of atomic structure. Rutherford's experiments supplied much information about the nucleus of an atom, but what about the electrons? Many questions were still unanswered. How were the electrons arranged? Were they moving? If so, how? Niels Bohr, a young Danish physicist working in Rutherford's laboratory, took up the task of answering some of these nagging questions. He didn't understand why electrons did not simply fall into the nucleus. Physicists in the first decades of the twentieth century conjectured that some unknown force was holding electrons at a distance.

The science of *spectroscopy* (see How It Works on page 98) helped Bohr devise a model that describes the movement of electrons around a nucleus. You may recall from a physical science course that when white light passes through a prism, it is dispersed to form a continuous spectrum of colors. These visible colors are only a small part of a larger spectrum of all forms of electromagnetic radiation, ranging from radio waves to gamma rays. As scientists developed improved tools and techniques for spectroscopy, they discovered that the color spectrum produced by sunlight was not continuous—it contained gaps. Later, when scientists used spectroscopy to analyze the light produced by heated elements, they discovered that these also produced incomplete spectra in the form of a series of colored lines. The pattern of colored lines, called an **emission spectrum**, turned out to be unique for each element. The explanation of these bright lines in emission spectra eluded scientists for several decades.

QUESTIONS
» Why was Rutherford's nuclear model replaced?
» What is spectroscopy?
» How does the Bohr model explain the emission spectra of atoms?

TERMS
emission spectrum • Bohr model • principal energy level • ground state • excited state

Why are fireworks different colors?

The emission spectrum of hydrogen compared with the complete spectrum of visible light

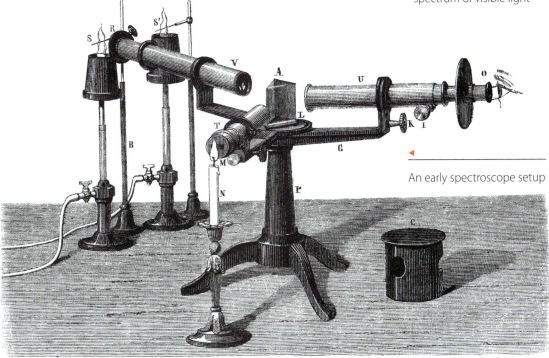

An early spectroscope setup

Electron Arrangement 95

Bohr's Model
DISCOVERY OF ELECTRON ENERGY LEVELS

In 1913 Danish physicist Niels Bohr devised a mathematical atomic model that explained the emission spectra. He first deduced that emission spectra are discrete frequencies of electromagnetic radiation given off by energized atoms.

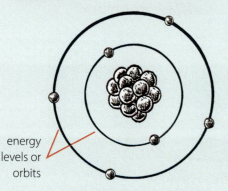

Neils Bohr

Bohr model of carbon

energy levels or orbits

How could energized atoms produce only these frequencies but not others? To answer that question, the **Bohr model** shown at left (sometimes referred to as the *planetary model*) suggested that electrons exist only in a series of distinct energy levels known as orbits at increasing distances from the nucleus. In his model, electrons normally orbit the nucleus in the lowest available energy level, much like planets orbit the sun in our solar system. But if the right amount of energy "excites" the electrons, they can absorb the exact amount of energy required to jump up to a higher energy level.

Once the electrons are at their higher energy levels, they quickly fall back to a lower energy level and shed their excess energy. This excess energy is radiated as electromagnetic energy.

The difference in potential energy between the higher and lower energy levels determines the frequency of the energy emitted. If the frequency of electromagnetic energy emitted falls in the range visible to human eyes, the color is observed as a line in an emission spectrum. Using his mathematical model, Bohr actually predicted the exact frequencies of the lines in hydrogen's emission spectrum.

Today Bohr's electron energy levels are called **principal energy levels** (n), or energy shells. The distance of an electron's orbit from the nucleus corresponds to a particular energy level. Electrons at greater distances from the nucleus possess higher potential energies. Any change in energy levels is distinct, with no stopping between levels. Physicists describe these energy levels as *quantized*, meaning that energy absorbed or emitted as electrons move between levels occurs only in certain amounts, or quantities. It's similar to comparing a flight of stairs to a ramp. The motions needed to ascend a ramp can be of any size, whether large, small, or any amount in between. But the motions needed to ascend stairs are determined by the fixed distances between steps; that is, the motions are quantized.

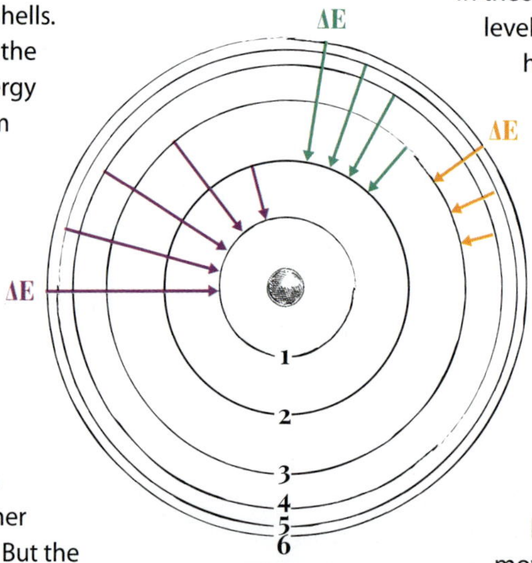

principal energy levels

ΔE

In theory, many principal energy levels exist depending on how energetic an atom is. To date atoms in their lowest energy state, or **ground state**, are known to contain only up to seven energy levels. That number of levels is sufficient to provide energy levels for all the electrons contained in the known elements. Atoms that have absorbed energy and moved some of their electrons to higher energy levels are said to be in an **excited state**.

96 Chapter 5

The figure below shows that outer energy levels are spaced so closely together that measuring the specific energy differences between them is very difficult. Therefore, electrons jumping between these levels emit low-energy radiation, such as invisible infrared light. On the other hand, electrons dropping from an outer energy level to the innermost levels release a large amount of energy, corresponding to the ultraviolet region of the electromagnetic spectrum.

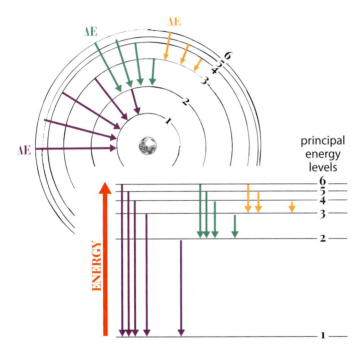

It is still common to see atoms depicted in the Bohr model style seen on page 96. The Bohr model is relatively easy to imagine and to illustrate, and it works well for describing some chemical phenomena, such as bonding. Bohr's model also works well for modeling elements with one electron, that is, neutral hydrogen and ionized helium. In fact, Bohr had developed his model by studying hydrogen.

But it had a serious shortcoming—it did not accurately predict the emission spectra of atoms with more than one electron. Although Bohr's model was a great accomplishment, he soon saw that modifications to his theory were necessary to make it more workable. Scientists of the early twentieth century, including Bohr himself, set out to accomplish this task.

WORLDVIEW INVESTIGATION

EXOPLANETS

If you believed everything you saw in the movies, you might think that life is common throughout our universe. In reality, Earth is the only place we know of where life exists. But some scientists are intently searching for life elsewhere in our solar system and beyond. But first they have to find planets where life could possibly exist. Planets outside our solar system are called *exoplanets*, and thousands of them have so far been found. But there's a big difference between just any exoplanet and one that could potentially harbor life. One of the key tools that scientists use to analyze far-off worlds is spectroscopy.

Introduction

As a writer for your school's science blog, you've been asked to look into cutting-edge applications of spectroscopy.

Task

Research the way(s) in which spectroscopy is currently being used to identify exoplanets that might harbor life. As part of your research, examine the motivations of scientists involved in exoplanet research and prepare a position statement from a biblical worldview.

Procedure

1. Do an internet search using the keywords "exoplanet spectroscopy." Find out how spectroscopy is being used to identify exoplanets that might harbor life.
2. Note how the worldviews of the websites you visit affect their view of exoplanetary research and the possibility of extraterrestrial life.
3. Plan your blog entry and collect any photos or videos that you might use to accompany it.
4. Create your blog entry and show it to another person for evaluation. Make any suggested revisions, if necessary.
5. Present your final blog entry to your class or family.

Conclusion

Secular exoplanetary research is usually driven by a naturalistic worldview that does not view life on Earth as being specially created. This worldview colors how secular scientists view both the possibility of extraterrestrial life and the question of whether and how we should look for it.

Electron Arrangement

Spectroscopy: Fingerprinting Atoms

Where do you think helium was first found? If you guessed that it was isolated from air, found in some mineral, or detected anywhere else on Earth, you missed the mark by some 150 million kilometers. Surprisingly, scientists discovered helium by looking at the sun. Scientists made this remarkable discovery in 1868 using one of the most diverse tools known to science: spectroscopy.

Spectroscopy is the study of how matter produces and interacts with electromagnetic radiation. When atoms are highly energized, they release light and other forms of radiation. In other instances, atoms absorb electromagnetic radiation. Because the atoms of every element have a unique arrangement of electrons, every element has its own characteristic pattern of energy interactions. The types of light that elements emit and absorb serve as their "fingerprints."

Atoms can absorb and emit all kinds of electromagnetic radiation. When an atom receives a small amount of thermal energy, electricity, or light of the proper wavelength, one or more of its electrons becomes more energetic, or "excited." It jumps from its original position to a higher energy level. Electrons in higher levels are extremely unstable, so within a fraction of a second they fall back to some lower energy level. When they fall, they give off a burst of electromagnetic energy whose frequency (or color) depends on the energy difference between the two levels.

Normally these individual colors of light cannot be observed because they are all mixed together. A prism or diffraction grating, however, can separate the colors of light so that they can be detected. When light from atoms in an incandescent material is analyzed, the distinct bands of an emission spectrum appear.

The simplest device for studying spectra, called the *prism spectroscope* (see image on page 95), was invented in the 1850s by Gustav Kirchhoff and Robert Bunsen at the University of Heidelberg. In these early devices, light from energized atoms was focused onto a prism, and the diffracted light was viewed through a telescope. Since then numerous improvements have been made. Instead of prisms, scientists now use diffraction gratings to separate the beams of light into their separate components. More sensitive optical detection systems have been developed, and it is now possible to study spectral lines in great detail. In many cases what was thought to be one line has turned out to be a compilation of many lines. Cameras have replaced the telescopes of the old spectroscope, and it is now possible to photograph an entire spectrum instantaneously. Armed with these newer tools and the basic theory of spectroscopy, scientists can quickly determine what elements are in a sample of water, in clay from an archaeological relic, or in the plasma of a distant star.

5.1 SECTION REVIEW

1. What is the most significant difference of the Bohr model of the atom compared with Rutherford's model?

2. Explain how emission spectra led Bohr to suggest the idea of energy levels.

3. What is the term used to describe the normal energy levels found in an unexcited atom?

4. What does it mean to say that an electron's energy within an atom is quantized?

5. Explain why scientists found it necessary to search for a replacement for the Bohr model.

6. The colors of fireworks are produced when chemical salts are burned in the presence of oxygen. On the basis of the material in this section, what conclusion(s) can you draw concerning these chemical salts and the colors they produce?

MINILAB

LIGHTS, SPECTROSCOPE, ACTION!

Spectroscopy equipment can be expensive, but you can make an inexpensive spectroscope using mostly household items. Try it!

Can I identify the materials used in different light sources?

Procedure

A On one end of the shoebox, draw a rectangle 1.9 cm tall and 3.8 cm wide. Cut the rectangle out with the utility knife or scissors. Then draw and cut out another rectangle of the same size on the other end of the box.

B Cut the index card in half and tape the two halves over one of the rectangular holes, leaving a 0.4 cm gap between them. This slit is where the light will enter your spectroscope.

C Cut out a piece of diffraction grating and *lightly* tape it over the other rectangular hole (you may have to re-tape it later). This hole is where you will view the spectrum that is produced by the light coming through the slit.

D Put the lid on your box and put a rubber band around it. Hold up your box to a light source and look through it. If the spectrum that you see appears to be oriented top-to-bottom, remove the diffraction grating and rotate it 90°.

Once the diffraction grating is taped on correctly, your spectroscope is ready to use.

E Darken the room as much as possible. Then turn on the first light source and observe it through your spectroscope.

F Observe and record the colors that you see.

G Repeat Steps E and F for additional light sources.

1. What differences did you observe for the spectra produced by different light sources?

Conclusion

2. What can you conclude about the materials that produce light in various light sources?

Going Further

3. Suggest a way to use your spectroscope observations to identify possible materials in the light sources that you observed.

EQUIPMENT

- utility knife or scissors
- diffraction grating
- shoebox with lid
- index card
- tape
- rubber band
- different light sources

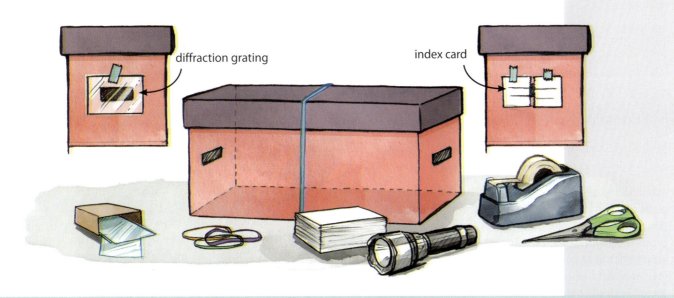

Electron Arrangement

QUESTIONS

» How is the quantum-mechanical model different from the Bohr model?

» What are orbitals?

» How does the development of the Bohr model demonstrate the modeling nature of science?

TERMS

quantum-mechanical model • photon • wave-particle duality • Heisenberg uncertainty principle • orbital • principal quantum number • energy sublevel • spin

Where are the electrons in atoms?

5.2 QUANTUM-MECHANICAL MODEL

Modern Physics and the Quantum-Mechanical Model

As physicists delved ever deeper into the subatomic structure of matter, it became evident that classical Newtonian mechanics could not describe what they were observing. Gradually, a new theory arose to describe matter on a submicroscopic scale—*quantum mechanics*. The atomic model that grew out of this new kind of physics came to be known as the **quantum-mechanical model**.

It started with a new way of thinking about light. For more than a century, scientists had known that light had properties of a wave—it could be refracted through a lens and it could be diffracted around a sharp edge. Light also could reflect off a flat surface like a particle. At the beginning of the twentieth century, Einstein's theories of matter and energy suggested that light consisted of massless particles called **photons**.

Louis de Broglie
DE BROGLIE'S HYPOTHESIS

During the 1920s a young French physicist named Louis de Broglie suggested that if waves could behave like particles, then particles could behave like waves. This concept is known as *de Broglie's hypothesis*. Whether an object acts as a particle or as a wave depends on how the object is observed.

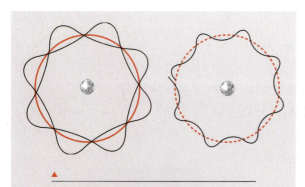

▲ De Broglie's electrons are described as waves that fit around the nucleus when there are whole numbers of waves possible in the distance that they travel. For example, an electron in the second principal energy level is exactly two waves long.

According to de Broglie's hypothesis, all forms of matter act like waves, including baseballs, pizzas, chemistry students, as well as smaller objects such as electrons and protons. De Broglie waves, or matter waves, increase in wavelength as the object's mass decreases and its speed increases. But such wavelengths are significant only for atomic and subatomic particles.

Louis de Broglie

The de Broglie wavelengths of large objects are so small as to be negligible, even at atomic sizes. On the other hand, for an electron, small whole numbers of de Broglie wavelengths can fit around the circumference of the electron's orbit in Bohr's model of the atom.

So, is an electron a particle, or is it a wave? It's both! The electron has a dual nature—it acts as both a particle and a wave, a condition scientists call **wave-particle duality**. This dual nature explains the movements of electrons that seem to defy the classical laws of motion described by Isaac Newton.

Scientists had discovered that light behaves both as a wave (left image) and as a particle (right image). But De Broglie noticed that electrons didn't always act like particles—they can also act like waves.

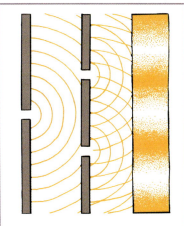

Young's double slit experiment: wave interference of light

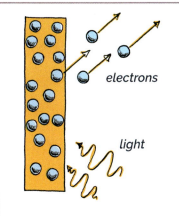

photoelectric effect: particle nature of light

Werner Heisenberg
UNCERTAINTY PRINCIPLE

Scientists were further perplexed when they tried to describe the location of electrons more precisely within the atom. In 1927 Werner Heisenberg determined that it is impossible to know both the energy or momentum and the exact position of an electron at the same time. This concept, known as the **Heisenberg uncertainty principle**, is a fundamental property of all submicroscopic systems, one that cannot be overcome or avoided. The reason for the uncertainty is due to the limits on our ability to observe atoms because of their extremely small size—smaller even than individual waves of visible light.

$$\underset{\text{uncertainty in momentum}}{\underset{\text{uncertainty in position}}{\Delta x \Delta p}} \geq \frac{h}{4\pi} = \frac{\hbar}{2}$$

All information about atomic matter can be obtained only by observing the effects of collisions between the matter and other particles or photons. But the collisions that provide information about matter simultaneously change the phenomenon that we are trying to measure. Therefore, the very act of observing properties of subatomic matter changes the properties that we are observing. This aspect of the quantum-mechanical model is a good example of the modeling nature of science, as opposed to finding truth—even with our best and most up-to-date scientific models, there are limits on the extent to which we can know our world.

Werner Heisenberg

The uncertainty principle gave scientists a whole new view of the atom. The figures at right show the regions of highest probability for the locations of electrons around an atom. The distance of highest probability corresponded very closely to the radius for Bohr's orbits in the simple atoms for which he had made predictions. Bohr's precise, planetlike orbits were replaced by **orbitals**, or three-dimensional regions of the most probable position for an electron.

1s sublevel

2s sublevel

An orbital is actually a four-dimensional map because it contains location information in three geometrical directions (*x*, *y*, and *z*), as well as the probability of finding the electron in time. Orbitals with low energies predict that the electron's average distance from the nucleus is small. Orbitals with higher energies predict that the electron's average distance from the nucleus is greater.

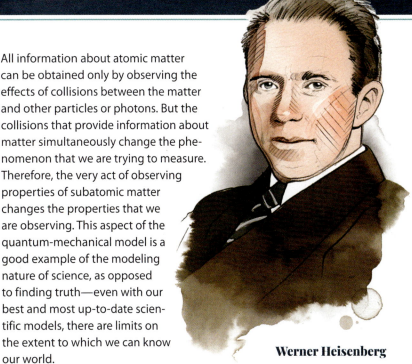

4f orbital

Electron Arrangement

Electron Addresses

As scientists gathered more evidence about the structure of electrons in the atom, they came to realize that every electron has its own energy-determined location within an atom. To identify each electron when the atom is in its ground state, scientists developed an address system that is based on the electron's quantum mechanical properties. This address system uses a series of four numbers called *quantum numbers*. Quantum numbers are solutions to the various wave equations that scientists use to describe the energy, momentum, and probable location of an electron. For this course, we need only examine a portion of this system.

PRINCIPAL QUANTUM NUMBER (n)

The **principal quantum number** (n) identifies the principal or main energy level of an electron. In essence, the principal quantum number indicates the average or most likely distance from the nucleus where an electron resides. The numbers are positive integers, beginning at 1, that is, $n = 1, 2, 3, \ldots, 7$.

The highest value of n for large atoms in the ground state is 7, although it can be much higher in excited atoms. These numbers correspond to the rows, or periods, on the periodic table, which you will learn more about in Chapter 6. All electrons in a given energy level have the same principal quantum number. The number of total electrons allowed in a principal energy level equals $2n^2$, where n is the principal quantum number.

SUBLEVELS AND ORBITALS

Scientists have developed models that describe the spatial arrangement of electrons. As the models are discussed in the following subsections, you will see the orderly way in which God designed atoms. These theoretical models have been extensively confirmed by analysis of emission spectra and other kinds of tests.

Unlike the Bohr model, the quantum-mechanical model subdivides all but the first principal energy level into **energy sublevels**, or *energy subshells*. These sublevels are further subdivided into orbitals, each of which may contain up to two electrons. The four energy sublevels have been assigned the letters *s, p, d,* and *f* respectively to avoid confusion when describing an atom's electron structure as it relates to chemical behavior. Because quantum theory predicts where an electron is *most likely* to be found, it is most accurate to call an illustration of a sublevel a *probability plot*.

The s Sublevel

The s sublevel is the simplest of all. A probability plot of an s sublevel has a spherical shape. Unlike the other sublevels, it contains only one orbital, so the orbital is often called the s orbital as well. Every principal energy level has an s sublevel that contains one s orbital. Remember that an orbital can hold a maximum of two electrons, though it may contain fewer. Each of the two possible electrons in a sublevel has an opposite **spin**—a form of angular momentum carried by electrons.

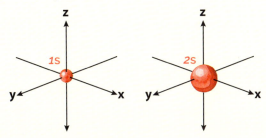

The p Sublevel

The three dumbbell-shaped orbitals of a p sublevel probability plot are arranged symmetrically around the origin of a three-dimensional x-y-z graph. One lies on the x-axis, one on the y-axis, and one on the z-axis. Because of this, p orbitals are sometimes designated p_x, p_y, and p_z. There are no p orbitals in the first principal energy level since it has only one s sublevel. Every neutral atom in its ground state with two or more principal energy levels has a p sublevel. Since each orbital can hold a maximum of two electrons, a p sublevel can hold a maximum of six electrons.

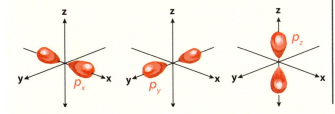

The d Sublevel

A d sublevel probability plot has a more complicated shape. The five orbitals in the d sublevel are oriented as shown at left. Atoms in their ground state containing three or more principal energy levels may have d sublevels. A d sublevel may contain up to ten electrons.

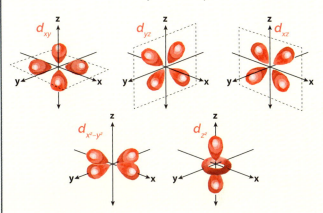

The f Sublevel

An f sublevel probability plot is even more complicated than the d sublevel plot. The figure below shows the seven possible f orbitals. Only ground-state atoms with four or more principal energy levels may have an f sublevel. An f sublevel may contain up to fourteen electrons.

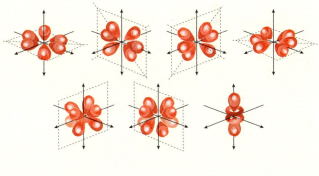

Putting Them All Together

You may have noticed that some orbitals overlap one another. They overlap because they are probability regions for the locations of electrons—regions where electrons *might* be found—not definite locations. When all of the orbitals for an atom are combined, a roughly spherical region is produced around the atom called the *electron cloud* (right). This cloud functions similarly to the hard spheres of Dalton's atomic model since the electron cloud of one atom tends to prevent other atoms from intruding into its space.

You may also have noticed that there are potentially a lot of principal energy levels and sublevels to keep track of. Scientists keep track of which sublevel is being discussed by combining the principal quantum number of the energy level with the letter of one of its sublevels. An electron in the 3p sublevel, for example, is located in one of the three p orbitals of the p sublevel of the third principal energy level. This notation will be used in later discussions.

Electron Capacities

There are limits on the number of electrons that can occupy the various principal energy levels, sublevels, and orbitals of the quantum mechanical atom. The first energy level of any atom contains one *s* sublevel with only one orbital, which can hold two electrons. Thus the total capacity of any first principal energy level is two electrons. Each second energy level has one *s* and one *p* sublevel. Since every *p* sublevel has three orbitals that can hold a combined maximum of six electrons, the total capacity of the second principal energy level is the sum of the two electrons from the *s* sublevel and the six electrons from the *p* sublevel—a total of eight electrons. The total capacities of the other principal energy levels can be determined by a similar procedure. Table 5-1 lists the total allowed electrons in each of the seven ground-state principal energy levels.

TABLE 5-1 *Electron Capacities of Energy Levels and Sublevels*

Principal Energy Level	Allowed Sublevels	Orbitals in Each Sublevel	Electron Capacity of Each Sublevel	Total Possible Electron Capacity
1	s	1	2	2
2	s	1	2	8
	p	3	6	
3	s	1	2	18
	p	3	6	
	d	5	10	
4	s	1	2	32
	p	3	6	
	d	5	10	
	f	7	14	
5	s	1	2	32
	p	3	6	
	d	5	10	
	f	7	14	
6	s	1	2	18
	p	3	6	
	d	5	10	
7	s	1	2	8
	p	3	6	

Relative Energies of Sublevels

In Bohr's model, as the energy level of an electron increases, the distance of the electron from the nucleus increases. In the quantum-mechanical model, this simpler view was replaced by a more complex relationship. Principal energy levels become larger in size as their energy level number increases, but their sublevels are also ranked by their relative energies.

Beginning with the innermost energy level, the 1s sublevel has the least energy. The probability plot for the 1s orbital is the closest to the nucleus of any orbital. The 2s and 2p sublevels, in the second principal energy level, have more energy; they are farther from the nucleus. Note that all three 2p orbitals have the same relative energy, and the electrons in these orbitals have more energy than those in the 2s sublevel. Farther out and higher in energy are the 3s and 3p sublevels.

When you consider Table 5-1, you might expect that the next higher energy sublevel after 3s and 3p should be 3d. However, as the number of electrons in an atom increases, their interactions become increasingly complex. As a result, the next energy sublevel to be filled in a ground-state atom is actually 4s. In Section 5.3, you will learn how the periodic table reflects this filling order, but there is also a simple diagram that can help you. The figure at right is sometimes called the *diagonal rule* because the parallel diagonal lines help you remember the filling order. If you start with the top arrow and work your way down, you can see the order in which the sublevels fill.

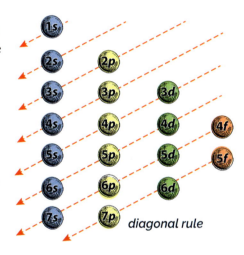

diagonal rule

5.2 SECTION REVIEW

1. Explain the significance of the discovery of wave-particle duality.

2. Why is it so difficult to describe both the energy and the position of an electron?

3. Describe the main differences between the Bohr and quantum-mechanical models in their conceptions of the electron.

4. How do the principal quantum numbers of the quantum-mechanical model relate to the Bohr model of the atom?

5. Give the electron capacity of each energy sublevel.

6. How does the quantum-mechanical model give evidence of God's orderly design in our universe?

7. Does science consist of a fixed body of knowledge? State a position on this question and defend it using information from this section.

QUESTIONS

» What determines the arrangement of electrons in atoms of different elements?

» How does the periodic table predict the arrangement of electrons in an atom?

» How can I indicate the arrangement of electrons in an atom?

» How can I tell which electrons in an atom are valence electrons?

TERMS

electron configuration • aufbau principle • orbital notation • Pauli exclusion principle • noble gas notation • Hund's rule • valence electron • electron dot notation • ion • cation • anion

5.3 ELECTRON CONFIGURATIONS

Electron Configuration and the Periodic Table

The modern periodic table, which you will learn more about in Chapter 6, is actually organized by **electron configuration**—the arrangement of electrons in an atom of an element in its ground state. Atomic numbers increase from left to right across the rows. Each row number corresponds to the highest principal energy level for the elements in that row. Certain regions, or blocks, of the table indicate sublevels in which electrons are located.

Why does it matter how the electrons are arranged?

Let's look at how this works. As you read the periodic table from left to right, electrons are added to orbitals in the order seen on the table.

The first electrons to be added to a new energy level are always in the *s* sublevel, which can hold only two electrons. Therefore, the first two elements of each row are in what we call the *s* block of the periodic table. The first row can contain only two elements because the first energy level can hold only two electrons. (Helium, however, is placed in the right-hand column because its energy level is full.) Similarly, the last six electrons to occupy the outermost energy level fill the *p* sublevels, so those elements occupy the *p* block of the table, which is six blocks wide.

PERIODIC TABLE — ELECTRON CONFIGURATION

The periodic table reveals the order in which electron sublevels are filled as the atomic number increases.

Legend:
- *s* sublevel
- *d* sublevel
- *p* sublevel
- *f* sublevel

106 Chapter 5

Notice the *d* energy sublevel elements in the middle of the periodic table on page 105. Notice that their energy level numbers are one number lower than the row in which they appear. This is because the *d* sublevels are higher energy states than the *s* sublevel in the next higher energy level. We call this the *d* block of the periodic table. The two separate rows at the bottom of the table contain the elements in which *f* sublevels are being filled. This is called the *f* block. The energy levels of the sublevels that these elements fill are offset by two compared with the row in which they appear. This is because *f* sublevels are the next higher energy sublevel after the *s* sublevels that are two energy levels higher. This sequence of filling orbitals is reflected in the diagonal rule structure that you saw on page 105.

We can use the preceding information to compare electrons in atoms of different elements. Take zinc and bromine for instance. Both of them are in the fourth row of the periodic table, which tells us that both elements have electrons in principal energy levels 1–4. However, zinc is located in row 3*d* of the *d* block, while bromine is in row 4*p* of the *p* block portion of the same row. This tells us that the highest-energy electrons of a zinc atom are located in its 3*d* energy sublevel, even though zinc has some electrons in energy level 4. Bromine, on the other hand, has its highest-energy electrons in its 4*p* sublevel. As described more fully below, "highest-energy electrons" are not necessarily in the principal energy level that has the highest value; there is overlap of sublevels from different main energy levels.

10	11	12	13	14	15	16	17	18
								2 Helium **He** 2
			5 Boron **B** 2, 3	6 Carbon **C** 2, 4	7 Nitrogen **N** 2, 5	8 Oxygen **O** 2, 6	9 Fluorine **F** 2, 7	10 Neon **Ne** 2, 8
			13 Aluminum **Al** 2, 8, 3	14 Silicon **Si** 2, 8, 4	15 Phosphorus **P** 2, 8, 5	16 Sulfur **S** 2, 8, 6	17 Chlorine **Cl** 2, 8, 7	18 Argon **Ar** 2, 8, 8
28 Nickel **Ni** 2, 8, 17, 1	29 Copper **Cu** 2, 8, 18, 1	30 Zinc **Zn** 2, 8, 18, 2	31 Gallium **Ga** 2, 8, 18, 3	32 Germanium **Ge** 2, 8, 18, 4	33 Arsenic **As** 2, 8, 18, 5	34 Selenium **Se** 2, 8, 18, 6	35 Bromine **Br** 2, 8, 18, 7	36 Krypton **Kr** 2, 8, 18, 8
46 Palladium **Pd** 2, 8, 18, 18	47 Silver **Ag** 2, 8, 18, 18, 1	48 Cadmium **Cd** 2, 8, 18, 18, 2	49 Indium **In** 2, 8, 18, 18, 3	50 Tin **Sn** 2, 8, 18, 18, 4	51 Antimony **Sb** 2, 8, 18, 18, 5	52 Tellurium **Te** 2, 8, 18, 18, 6	53 Iodine **I** 2, 8, 18, 18, 7	54 Xenon **Xe** 2, 8, 18, 18, 8
78 Platinum **Pt** 2, 8, 18, 32, 17, 1	79 Gold **Au** 2, 8, 18, 32, 18, 1	80 Mercury **Hg** 2, 8, 18, 32, 18, 2	81 Thallium **Tl** 2, 8, 18, 32, 18, 3	82 Lead **Pb** 2, 8, 18, 32, 18, 4	83 Bismuth **Bi** 2, 8, 18, 32, 18, 5	84 Polonium **Po** 2, 8, 18, 32, 18, 6	85 Astatine **At** 2, 8, 18, 32, 18, 7	86 Radon **Rn** 2, 8, 18, 32, 18, 8
110 Darmstadtium **Ds**	111 Roentgenium **Rg**	112 Copernicium **Cn**	113 Nihonium **Nh**	114 Flerovium **Fl**	115 Moscovium **Mc**	116 Livermorium **Lv**	117 Tennessine **Ts**	118 Oganesson **Og**

63 Europium **Eu** 2, 8, 18, 25, 8, 2	64 Gadolinium **Gd** 2, 8, 18, 25, 9, 2	65 Terbium **Tb** 2, 8, 18, 27, 8, 2	66 Dysprosium **Dy** 2, 8, 18, 28, 8, 2	67 Holmium **Ho** 2, 8, 18, 29, 8, 2	68 Erbium **Er** 2, 8, 18, 30, 8, 2	69 Thulium **Tm** 2, 8, 18, 31, 8, 2	70 Ytterbium **Yb** 2, 8, 18, 32, 8, 2	71 Lutetium **Lu** 2, 8, 18, 32, 9, 2
95 Americium **Am** 2, 8, 18, 32, 25, 8, 2	96 Curium **Cm** 2, 8, 18, 32, 25, 9, 2	97 Berkelium **Bk** 2, 8, 18, 32, 27, 8, 2	98 Californium **Cf** 2, 8, 18, 32, 28, 8, 2	99 Einsteinium **Es** 2, 8, 18, 32, 29, 8, 2	100 Fermium **Fm** 2, 8, 18, 32, 30, 8, 2	101 Mendelevium **Md** 2, 8, 18, 32, 31, 8, 2	102 Nobelium **No** 2, 8, 18, 32, 32, 8, 2	103 Lawrencium **Lr** 2, 8, 18, 32, 32, 8, 3

After learning the significance of the sublevel blocks in the periodic table, you can quickly determine the arrangement of electrons for any element. This method is the first of many ways in which the periodic table will be useful to you.

Writing Electron Configurations

You should now understand how electrons are arranged in a neutral atom in its ground state. In neutral atoms, the numbers of electrons and protons are exactly equal. Since the periodic table is arranged by increasing atomic number, and since the atomic number is equal to the number of protons in an atom's nucleus, we can use the atomic number to determine the total number of electrons in a neutral atom.

The **aufbau principle** states that electrons must fill the lowest available energy sublevels before any can be placed in higher-energy sublevels. In practice this means that we can determine the arrangement of electrons in an atom by adding electrons to an atom with a lower atomic number, that is, one with fewer electrons. Progressing from hydrogen through the higher-numbered elements, electrons add to the least energetic orbital possible. Generally, lower-energy orbitals must be filled before higher-energy orbitals.

Hydrogen ($Z = 1$) has one electron that resides in the $1s$ sublevel. Hydrogen's electron configuration, or arrangement of electrons, is written as $1s^1$. The coefficient 1 tells us that hydrogen's electron is in the first energy level. The letter s indicates that the electron is in the s sublevel of the first energy level. The superscript 1 tells how many electrons occupy that sublevel. In this case, there is one electron in the $1s$ sublevel, which is also the only orbital in the atom.

Physicists and chemists also use a special notation, called **orbital notation**, to illustrate the electron configuration of an atom. Orbital notation consists of horizontal lines representing the orbitals labeled by sublevel. Each electron occupying the orbital is written as an up or down arrow on the line. The opposite directions of the two arrows represent the opposite spin that each electron must have when they occupy the same orbital. The fact that two electrons in the same orbital must have opposite spins is one of the conclusions of the **Pauli exclusion principle**, first formulated by Wolfgang Pauli in 1925. The orbital notation for hydrogen is written as follows.

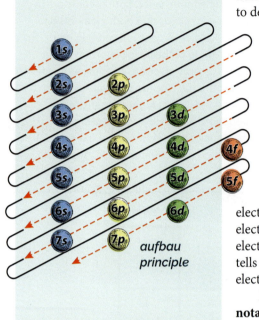

aufbau principle

Helium ($Z = 2$) has two electrons when neutral. The second electron joins the first to fill the $1s$ sublevel. Helium's electron configuration is thus $1s^2$. The $1s$ sublevel is now full and helium's notation is shown below.

Pauli's exclusion principle

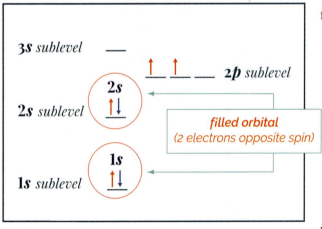

Lithium ($Z = 3$) has three electrons. Two electrons occupy the lower energy sublevel, $1s$. The third electron can reside only in the next higher energy sublevel, $2s$. Its electron configuration is written as $1s^2 2s^1$. Notice that lithium is on the second row of the periodic table. Similarly, the electron configuration of beryllium is $1s^2 2s^2$.

As shorthand for electron configurations, chemists often write the symbol of the element that fills the previous energy level in brackets before adding the next sublevels. Doing so shortens the notation and emphasizes the highest-energy electrons. Because helium fills the first energy level, lithium's abbreviated configuration can be written [He] $2s^1$. Because this last column of the periodic table is a group called the noble gases, this abbreviated form is sometimes called **noble gas notation**.

When determining the electron configuration of carbon ($Z = 6$), you have a choice in which blank to place the arrows. Its electron configuration is [He] $2s^2 2p^2$. But what about the orbital notation? Are the two electrons in the $2p$ sublevel in the same orbital (below left), or are they in different orbitals (below right)?

```
         2s      2p                    2s      2p
[He] :  ↑↓    ↑↓  __ __      [He] :   ↑↓    ↑  ↑  __
```

This question was answered by Friedrich Hund after he carefully studied the movements of electrons within the orbitals of a single sublevel. He determined that for orbitals of the same sublevel, the lowest energy state exists when the number of unpaired electrons with the same spin is maximized. In other words, **Hund's rule** states that as electrons fill a sublevel, all orbitals receive one electron with the same spin before electrons begin to pair up. Therefore, by Hund's rule, carbon's orbital notation is the second option. Notice also that there are three blanks for the p sublevel. These blanks stand for the three orbitals in the p sublevel: p_x, p_y, and p_z.

```
         2s      2p
[He] :  ↑↓    ↑   ↑   __
              px  py  pz
```

Review Table 5-2 to see the orbital notations and electron configurations of the next eight elements after helium.

Friedrich Hund

Hund's Rule

Place an electron of the same spin in each $2p$ orbital before putting a second electron in any $2p$ orbital.

↑ electron **UP** spin   ↓ electron **DOWN** spin

TABLE 5-2 *Electron Locations for Elements 3–10*

Atomic Number	Name	Orbital Notation 1s	2s	2p			Electron Configuration	Noble Gas Notation
3	lithium (Li)	↑↓	↑				$1s^2 2s^1$	[He] $2s^1$
4	beryllium (Be)	↑↓	↑↓				$1s^2 2s^2$	[He] $2s^2$
5	boron (B)	↑↓	↑↓	↑	__	__	$1s^2 2s^2 2p^1$	[He] $2s^2 2p^1$
6	carbon (C)	↑↓	↑↓	↑	↑	__	$1s^2 2s^2 2p^2$	[He] $2s^2 2p^2$
7	nitrogen (N)	↑↓	↑↓	↑	↑	↑	$1s^2 2s^2 2p^3$	[He] $2s^2 2p^3$
8	oxygen (O)	↑↓	↑↓	↑↓	↑	↑	$1s^2 2s^2 2p^4$	[He] $2s^2 2p^4$
9	fluorine (F)	↑↓	↑↓	↑↓	↑↓	↑	$1s^2 2s^2 2p^5$	[He] $2s^2 2p^5$
10	neon (Ne)	↑↓	↑↓	↑↓	↑↓	↑↓	$1s^2 2s^2 2p^6$	[He] $2s^2 2p^6$

Using the aufbau principle and Hund's rule, you can determine the electron configuration and orbital notation of any element.

EXAMPLE 5-1 WRITING ELECTRON CONFIGURATION

Give the ground-state electron configuration of manganese and its noble gas notation, then draw its orbital notation.

Solution
The order for filling the sublevels can be determined by the diagonal rule from Section 5.2 or by looking at the periodic table. We know that manganese is element number 25, so it has twenty-five electrons. It is in row four, which indicates that there are electrons in energy levels 1, 2, 3, and 4. Reading across the rows on the periodic table tells us that we are going to fill the 1s, 2s, 2p, 3s, 3p, 4s, and 3d sublevels in that order. Therefore, our electron configuration is

$$1s^2\,2s^2\,2p^6\,3s^2\,3p^6\,4s^2\,3d^5.$$

When writing the full form, always check to make sure that the sum of the superscripts equals the atomic number.

Since the electron configuration for argon is $1s^2\,2s^2\,2p^6\,3s^2\,3p^6$, we can substitute argon's symbol into this long expression to shorten it.

$$[Ar]\,4s^2 3d^5$$

All sublevels prior to the final one are filled, but the 3d sublevel is only partly filled. Since there are five electrons in the 3d sublevel and five orbitals in any d sublevel, each orbital is occupied by one electron; no electron pairing occurs (Hund's rule). The following is the orbital notation.

Mn : [Ar] 4s ↑↓   3d ↑ ↑ ↑ ↑ ↑

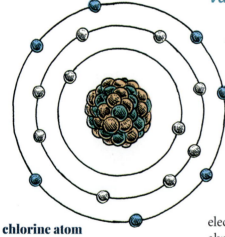

chlorine atom
(7 valence electrons)

Valence Electrons

Atoms can have many electrons, but those in the outermost energy level are the most important because they are the ones most likely to be involved in chemical bonding. They are also the ones that give elements many of their physical and chemical properties. These special electrons are called **valence electrons**.

Valence electrons are not necessarily the last electrons to fill an orbital. Rather, they are the electrons that have the highest principal quantum number, n. These are the electrons that occupy the highest-numbered s and p sublevels, even in atoms that have higher-energy electrons occupying inner d and f sublevels. Study the following example to see how this concept works.

As you will see is the case with zinc in Example 5-2, valence electrons can *never* be in the d or f orbitals because these orbitals are always filled after filling the s orbital of the higher energy level.

EXAMPLE 5-2 IDENTIFYING VALENCE ELECTRONS

How many valence electrons do neutral atoms of the following elements have?

   a. sulfur   **b.** zinc

Solution
a. The electron configuration for sulfur is $1s^2 2s^2 2p^6 3s^2 3p^4$. The outermost energy level—the one with the largest value for n—is energy level three, which contains two s and four p electrons. The total number of valence electrons for sulfur is six.

b. The electron configuration for zinc is $1s^2 2s^2 2p^6 3s^2 3p^6 4s^2 3d^{10}$. Even though the electron configuration obtained by using the periodic table shows that the 3d orbitals are occupied after the 4s sublevel, valence electrons are defined as those that occupy the outermost principal energy level—the fourth, in the case of zinc. Therefore, zinc has two valence electrons. This number of valence electrons is typical for most of the metals in the middle of the periodic table.

Electron Dot Notation

Sometimes scientists represent the valence electrons of an atom using **electron dot notation**. Dots that represent valence electrons are placed around the element's symbol. Because valence electrons involve only the outer *s* and *p* electrons (a maximum of eight electrons), no more than eight dots are used. The electron dot notations for the elements in the third row of the periodic table are shown in Table 5-3.

There is no standard way of arranging the dots. In this textbook, we will adopt the following convention. The first two dots, which represent the *s* valence electrons, are placed in a pair to the left of the element symbol, since the *s* block is on the left side of the periodic table. For the next element in atomic number sequence, the third dot is placed at the top of the symbol, the next to the right, and the next under the symbol. These three unpaired electrons occupy the three valence *p* orbitals according to Hund's rule. The remaining three electrons pair with the single electrons in the same order until the *p* sublevel is filled.

In later chapters, we will discuss bonding between atoms. In bonding, the positions of the *s* and *p* electron dots can be shifted around the element symbol to best illustrate bonds formed with adjacent atoms. But the number of valence electrons does not change.

TABLE 5-3 *Electron Dot Notation for Elements of Row 3*

Valence Electrons	Electron Dot Notation
1	•Na
2	:Mg
3	:Al•
4	:Si•
5	:P•
6	:S•
7	:Cl:
8	:Ar:

EXAMPLE 5-3 ELECTRON DOT NOTATION

Write the electron dot symbols for the atoms of the following elements.
- **a.** selenium
- **b.** boron

Solution

a. Selenium ($Z = 34$) has the noble gas notation [Ar] $3s^2 3p^4$ and therefore has six valence electrons.

$$:\!\overset{..}{\text{Se}}\!\cdot$$

b. Boron ($Z = 5$) has the electron configuration [He] $2s^2 2p^1$ and therefore has three valence electrons.

$$:\!\overset{\cdot}{\text{B}}$$

Ionized Atoms

Electrons move between energy levels within an atom when they gain or lose energy. Given enough energy, an electron can escape or be pulled from an atom entirely. Other atoms may acquire electrons to attain a more stable electron configuration. Losing or gaining an electron causes an imbalance in the number of protons and electrons, thus giving an atom a net electrical charge. An atom or group of atoms with a nonzero charge is an **ion**.

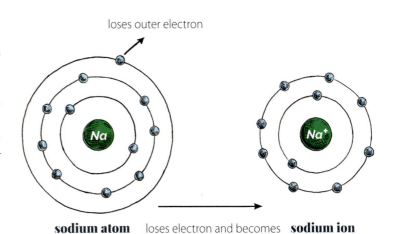

sodium atom loses electron and becomes sodium ion

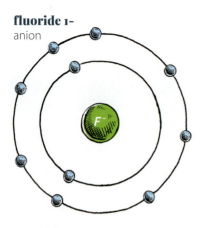

fluoride 1−
anion

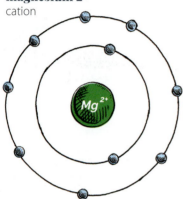

magnesium 2+
cation

Ions may have positive or negative charges. If an atom loses an electron, it will have a net positive charge because there will be one less negative charge compared with the positive nuclear charge due to the protons. Positive ions are called **cations**. If a neutral atom gains an electron, it will have a net negative charge when compared with the positive nuclear charges of the protons. Negative ions are called **anions**.

Electron dot notation can also be used to represent isolated ions. For example, if a neutral fluorine atom gains an electron, it acquires an electrical charge of 1−. We indicate this by placing a superscripted negative sign (the one is assumed) to the right of the symbol to represent ionized fluorine.

$$:\ddot{\text{F}}: + 1e^- \longrightarrow [:\ddot{\text{F}}:]^-$$

Similarly, a magnesium atom that loses two electrons will have a 2+ charge. Notice that instead of a "+2" charge we write "2+."

$$:\text{Mg} \longrightarrow [\text{Mg}]^{2+} + 2e^-$$

EXAMPLE 5-4 ELECTRON DOT NOTATION FOR IONS

Write the electron dot symbol for a chloride anion (Cl⁻).

Solution
Chlorine ($Z = 17$) has the electron configuration [Ne] $3s^2\,3p^5$ and therefore has seven valence electrons. As a 1− anion, it acquires an additional electron, for a total of eight valence electrons.

$$[:\ddot{\text{Cl}}:]^-$$

5.3 SECTION REVIEW

1. How did electron configurations determine the shape of the modern periodic table?

2. Draw a sketch of the periodic table showing where the different blocks (*s*, *p*, *d*, and *f*) are located.

3. How do the aufbau principle and Hund's rule guide us in determining the electron structure of an atom in its ground state?

4. What rule explains why the electron pair in any orbital must have opposite spins?

5. For each of the following atoms, (1) give the electron configuration, (2) give the noble gas notation, and (3) draw the orbital notation.
   a. silicon
   b. iron

6. From the following noble gas notations, give (1) the name of the element, (2) its number of valence electrons, and (3) the ground state energy sublevel(s) in which the element's valence electrons reside.
   a. [Kr] $5s^2\,4d^5$
   b. [Ar] $4s^2\,3d^{10}\,4p^3$

7. From the following orbital notations, give (1) the name of the element, (2) its number of valence electrons, and (3) the ground state energy sublevel(s) in which the element's valence electrons reside.
   a. [Ar]: ↑↓ ↑↓ ↑↓ ↑↓ ↑↓ ↑↓ ↑↓ ↑↓ ↑
           4s   3d   4p
   b. [Ar]: ↑↓ ↑
           4s 3d

8. Each of the following notations for magnesium is incorrect in some way. Explain why the notation is incorrect and then write the correct notation.
   a. [Ar]$3s^2$
   b. Mg: ↑↓ ↑↓ ↑↓ ↑↓ ↑↓ ↑ ↑
          1s 2s   2p   3s

05 CHAPTER REVIEW
Chapter Summary

5.1 BOHR MODEL

- Spectroscopy revealed that each element produced a unique emission spectrum.
- Niels Bohr created his atomic model to explain the emission spectra of elements.
- Bohr determined that electrons could have only certain amounts of energy (i.e., they were quantized) and that their energy corresponded to their distance from the nucleus.
- The energy levels described by Bohr are called principal energy levels.
- The Bohr model could predict only the emission spectra of atoms with single electrons such as hydrogen or ionized helium.

TERMS
emission spectrum
Bohr model
principal energy level
ground state
excited state

TERMS
quantum-mechanical model
photon
wave-particle duality
Heisenberg uncertainty principle
orbital
principal quantum number
energy sublevel
spin

5.2 QUANTUM-MECHANICAL MODEL

- Bohr and many others refined the model of the atom, producing the quantum-mechanical model of the atom in the first half of the twentieth century.
- Electrons and all matter behave both as particles and as waves, depending on how they are observed.
- Every electron in an atom has a unique set of four quantum numbers, including a principal quantum number (n).
- Each principal energy level consists of a number of energy sublevels comprised of one or more orbitals. Orbitals indicate probability regions where electrons might be found.
- As atomic number increases, electrons occupy those regions that increase in relative energy and average distance from the nucleus. In larger atoms, the higher principal energy levels overlap, and some inner sublevels are filled after outer sublevels.

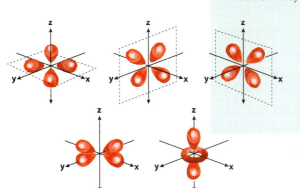

Electron Arrangement

05 CHAPTER REVIEW

Chapter Summary

TERMS
- electron configuration
- aufbau principle
- orbital notation
- Pauli exclusion principle
- noble gas notation
- Hund's rule
- valence electron
- electron dot notation
- ion
- cation
- anion

5.3 ELECTRON CONFIGURATIONS

- The arrangement of electrons in an atom can be illustrated in three ways: electron configuration, orbital notation, and electron dot notation.

- The aufbau principal states that as atomic number increases, atoms add electrons to the least energetic orbital possible.

- According to the Pauli exclusion principle, any given orbital can hold only two electrons, and these must have opposite spins.

- According to Hund's rule, as electrons are added to sublevels with more than one orbital, one electron with the same spin occupies each orbital before any orbital gets an electron with opposite spin.

- Valence electrons are those in the highest energy level of an atom and are the electrons most likely to be involved in chemical bonding.

- Atoms that gain or lose electrons through any process become ionized. Losing electrons produces positive ions called cations. Gaining electrons produces negative ions called anions.

Chapter Review Questions

RECALLING FACTS

1. Why did the Bohr model replace Rutherford's nuclear model?
2. How did observations of emission spectra lead to conclusions that the energy given off by excited atoms is quantized?
3. How many ground state principal energy levels are known to exist?
4. Briefly describe each of the following atomic models.
   a. Bohr model
   b. quantum-mechanical model
5. Why did scientists quickly replace the Bohr model?
6. If the Bohr model has been replaced, why do we see so many atoms still depicted using the Bohr model?
7. What unusual and unexpected aspect of electrons and, indeed, of all matter was discovered during the development of the quantum-mechanical model?
8. (True or False) The larger an object, the longer the wavelength of its de Broglie wave.
9. Briefly describe the contribution to the quantum-mechanical model made by each of the following physicists.
   a. Louis de Broglie   c. Wolfgang Pauli
   b. Friedrich Hund   d. Werner Heisenberg
10. How is the specific location of each electron in an atom identified?
11. Describe the arrangement of the orbitals in a p sublevel.
12. What is the electron capacity of the following?
    a. an orbital
    b. an s sublevel
    c. a p sublevel
    d. the fourth principal energy level
    e. the second principal energy level
13. How many orbitals are in
    a. an s sublevel?
    b. a d sublevel?
    c. the second principal energy level?
    d. the fourth principal energy level?
14. Define *electron configuration*.
15. State the aufbau principle.
16. What do the horizontal lines in an orbital notation represent? What do the arrows represent?
17. (True or False) The $3d$ sublevel electrons are at a higher energy than the $4s$ sublevel electrons that are in the next higher principal energy level.
18. What are valence electrons, and why are they important?

UNDERSTANDING CONCEPTS

19. Why does the emission spectrum of an element include specific wavelengths?
20. Compare the ground state and excited state.
21. One of your classmates says that matter can't act like waves because we never see a baseball moving in a wavelike fashion. How do you respond?
22. Create a graphic organizer using the terms *Bohr model, energy sublevel, orbit, orbital, principal energy level, principal quantum number, quantum-mechanical model*.
23. What is the main difference between electron configuration and orbital notation?
24. What exactly is excluded by the Pauli exclusion principle?
25. Explain the process of writing each of the following depictions of electrons.
    a. electron configuration
    b. orbital notation
    c. noble gas notation
26. Why are there up and down arrows in an orbital notation?
27. From the following notations, give (1) the name of the element, (2) its number of valence electrons, and (3) the ground state energy sublevel(s) in which the element's valence electrons reside.
    a. $1s^2 2s^2 2p^6 3s^2 3p^6 4s^2 3d^1$
    b.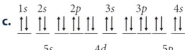
28. For each of the following atoms, (1) write out the ground-state electron configuration using noble gas notation, (2) write out the orbital notation, and (3) show the number of electrons in each principal energy level.
    a. oxygen b. potassium c. bromine
29. For each of the following atoms, (1) write out the ground-state electron configuration, (2) write out the noble gas notation, and (3) indicate how many valence electrons the element has.
    a. sulfur
    b. titanium
    c. barium

30. Of electron configuration, noble gas notation, orbital notation, and electron dot notation, which depiction gives us the most information?
31. Identify the element and how many valence electrons are represented in the following electron arrangement depictions.
    a. $1s^2 2s^2 2p^6 3s^2 3p^1$
    b. [Kr] $5s^2 4d^{10} 5p^5$
    c. 1s 2s 2p 3s 3p 4s
       ↑↓ ↑↓ ↑↓↑↓↑↓ ↑↓ ↑↓↑↓↑↓ ↑↓
    d. [Kr]: 5s ↑↓ 4d ↑↓↑↓↑↓↑↓↑↓ 5p ↑ __ __
    e. :S̈e·

32. Explain how anions and cations are formed.
33. Draw the electron dot notation for the following.
    a. sodium   c. cobalt
    b. nitrogen   d. tellurium

CRITICAL THINKING

34. Despite its current workability, why can't the quantum-mechanical model be considered the ultimate truth about atomic structure?
35. This chapter has only very briefly touched on the development of the quantum-mechanical model. Many scientists contributed to it other than those that have been named. Do a keyword search for "quantum-mechanical model" or "atomic theory" and identify other individuals who made important contributions to the quantum-mechanical model. Summarize your findings in a paragraph.
36. In each case below, indicate what is wrong with the electron depiction, including the rule that is violated.
    a. $1s^2 2s^2 2p^6 3s^2 3p^6 4s^2 3d^6 4p^2$
    b. Ni: 1s ↑↓ 2s ↑↓ 2p ↑↓↑↓↑↓ 3s ↑↓ 3p ↑↓↑↓↑↓ 4s ↑↓ 3d ↑↓↑ __ __
    c. F: 1s ↑↓ 2s ↑↓ 2p ↑↑ ↑↑ ↑↑
    d. :Z̈r

Chapter 6

PERIODIC TABLE & ELEMENTS

Throughout history, man has been interested in what matter is made of. The ancient Greeks thought that everything consisted of combinations of four elements—earth, water, air, and fire. Scientists searched for new elements. They isolated some elements from compounds, and even made new elements by smashing other elements together. Many scientists envision discovering some new element with amazing properties.

Interest in elements with unique properties is common in the world of fiction, especially science fiction. An internet search will quickly identify at least ninety fictional elements and their alloys with incredible properties. You may have heard of adamantium, vibranium, and unobtainium. Captain America's shield is made of vibranium, a rare metal element which has a low density but is very strong. Most amazing is its ability to absorb energy within the bonds between its atoms. Adamantium is an indestructible metal alloy. Once the components mix, the alloy must be molded within minutes. In solid form, it is a silver-gray metal similar to steel. Unobtanium is a room-temperature superconductor, which means that it can conduct electricity without any loss of energy.

While it may be fun to imagine discovering one of these fictional elements, scientists have worked for years to understand the 118 natural and manmade elements that are known today. Throughout this chapter, we will learn about these elements and their properties and how they are used to glorify God and help others.

6.1 The Periodic Table *117*
6.2 Periodic Trends *127*
6.3 Elements by Their Groups *133*

6.1 THE PERIODIC TABLE

One of the most recognizable tools of the chemist is the periodic table of the elements. This visual depiction of all the known elements provides a powerful description of how the elements are both similar to and different from one another. The **periodic table** organizes essential information about elements in a manageable format. We will learn throughout this chapter that the shape and arrangement of the table are closely related to the designed structure of the atoms themselves. But first, let's see how the step-by-step process of doing science resulted in the table's development.

Early Organizational Attempts

Alchemists in the Middle Ages suspected that some of the substances they worked with were pure, in the sense that they were not composed of other substances. But it was not until 1793 that the first list of elements was published. French chemist Antoine Lavoisier presented a list of thirty substances that he declared could not be broken down into simpler substances. He used the word *element* to describe such materials. Unknown to him, several of his elements were compounds of oxygen with other elements. Lavoisier had not yet been able to decompose those compounds.

QUESTIONS
» How did we get the periodic table?
» What can the periodic table tell us?
» Why are certain groups of elements together?
» Is there a reason for the arrangement of the elements?

TERMS
periodic table • periodic law • transuranium element • group • period • metal • nonmetal • metalloid

Why does the periodic table look the way it does?

Antoine Lavoisier

Shortly after John Dalton formulated his atomic theory in 1803, he produced his own list of elements. His list also contained many substances that were compounds, and he revised his list several times, eventually documenting as many as sixty elements. Dalton identified his elements and compounds with a complex system of symbols that was not well accepted by other scientists.

John Dalton

Periodic Table and Elements 117

Cl

Br

I

Johann Döbereiner

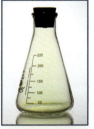

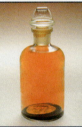

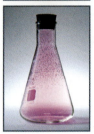

The Discovery of Element Periodicity

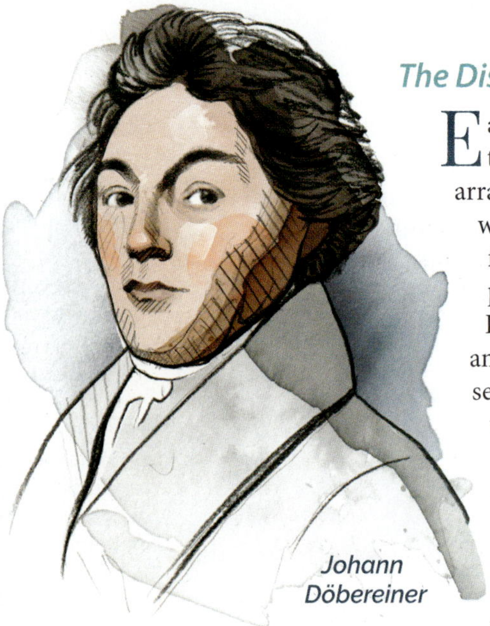

Early in the 1800s, chemists began to arrange the known elements systematically. When arranged by atomic mass, groups of elements with similar properties emerged. One of the first scientists to discover such common properties was German chemist Johann Döbereiner. In 1829 he announced that among the known elements, he had observed several groups, or *triads*, of three elements with similar properties. One of Döbereiner's triads was the group chlorine, bromine, and iodine. Each of those elements occurs as a gas with a distinctive color and has properties similar to the other two.

As scientists discovered more elements, Döbereiner's triads did not hold up. Elements with similar properties joined the triads to form tetrads and pentads. However, Döbereiner's practice of grouping elements with similar chemical properties was an important step in developing the concept of *periodicity* of the elements. Periodicity exists when some measurable property regularly repeats in a sequential list or time sequence.

John Newlands

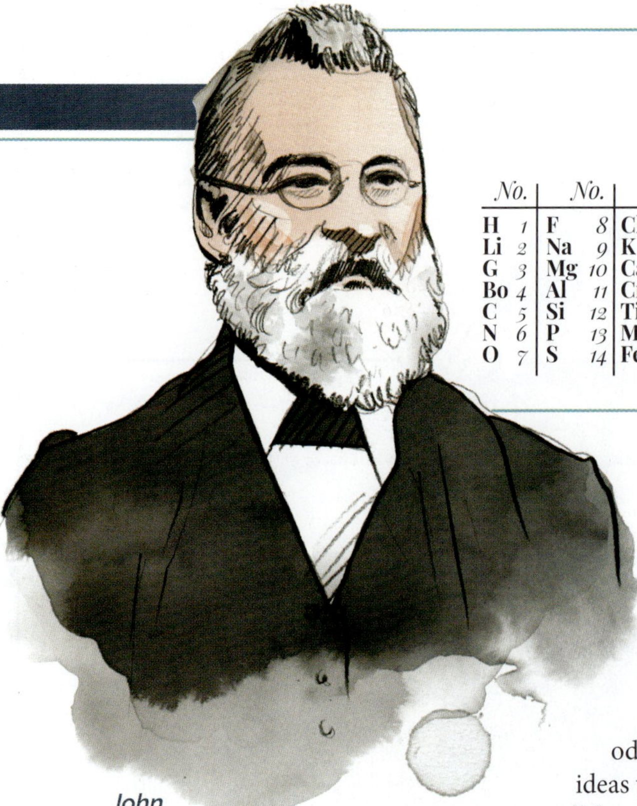

John Newlands
LAW OF OCTAVES

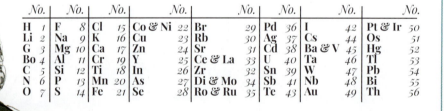

	No.		*No.*		*No.*		*No.*		*No.*		*No.*		*No.*		*No.*
H	*1*	**F**	*8*	**Cl**	*15*	**Co & Ni**	*22*	**Br**	*29*	**Pd**	*36*	**I**	*42*	**Pt & Ir**	*50*
Li	*2*	**Na**	*9*	**K**	*16*	**Cu**	*23*	**Rb**	*30*	**Ag**	*37*	**Cs**	*44*	**Os**	*51*
G	*3*	**Mg**	*10*	**Ca**	*17*	**Zn**	*24*	**Sr**	*31*	**Cd**	*38*	**Ba & V**	*45*	**Hg**	*52*
Bo	*4*	**Al**	*11*	**Cr**	*19*	**Y**	*25*	**Ce & La**	*33*	**U**	*40*	**Ta**	*46*	**Tl**	*53*
C	*5*	**Si**	*12*	**Ti**	*18*	**In**	*26*	**Zr**	*32*	**Sn**	*39*	**W**	*47*	**Pb**	*54*
N	*6*	**P**	*13*	**Mn**	*20*	**As**	*27*	**Di & Mo**	*34*	**Sb**	*41*	**Nb**	*48*	**Bi**	*55*
O	*7*	**S**	*14*	**Fe**	*21*	**Se**	*28*	**Ro & Ru**	*35*	**Te**	*43*	**Au**	*49*	**Th**	*56*

In 1864 British chemist John Newlands presented a classification scheme that, like Döbereiner's, organized elements by atomic mass. He noticed a pattern of elements with similar properties when he arranged them in eight groups of seven. Therefore, every eighth element had similar properties. Newlands had both musical and scientific training and he saw a correlation between the two fields. He labeled the periodicity in the behaviors of elements the *law of octaves*. His ideas were not well received, and most scientists ridiculed his parallel to music. But Newlands's idea that atomic mass and chemical properties might be related was correct. Today his law of octaves is recognized as an important step toward the formulation of what came to be known as the *periodic law*.

Mendeleev's Periodic Table

Credit for the development of the modern periodic table goes mostly to Russian chemist Dmitri Mendeleev. Like Döbereiner and Newlands, Mendeleev arranged elements by their atomic masses. Elements with similar chemical properties appeared in the same rows. But unlike earlier chemists, Mendeleev did not just organize the elements that were already known; his table left specific locations open for elements that he predicted would eventually be discovered.

Mendeleev noted that in many cases the atomic masses increased more than usual from those of the preceding elements. He reasoned that undiscovered elements belonged in the gaps, so he left blanks in his chart. Using information about the physical and chemical characteristics of nearby elements, he predicted the properties of the elements that would fit into the blanks. Not only did Medeleev predict where on the table these elements would fit, he also predicted their properties, such as mass, density, and melting point. After its publication in 1869, Mendeleev's table was dismissed by most chemists until one of the missing elements, gallium, was discovered in 1875. As amazing as it was to discover the missing elements, Mendeleev's predictions about their properties were also found to be accurate.

Another innovation of Mendeleev's table involved the elements now known as *transition metals*. These elements did not fit into the major groups of the chart, but they all had similar chemical and physical characteristics. Those similarities made them difficult to purify, so scientists often incorrectly identified the mixtures of two or more of these metals as elements. Mendeleev put them in the chart but did not let them interfere with the groupings of the other elements.

Mendeleev summarized his chart by formulating a periodic law, believing that the properties of elements vary periodically with their atomic mass. But even after scientists discovered most of Mendeleev's predicted elements, the table still had some problems. Arranging the elements in order of increasing atomic mass did not always produce a table that grouped similar elements. Because of the importance of the characteristics, scientists realized that the periodic table needed refining.

Dmitri Mendeleev

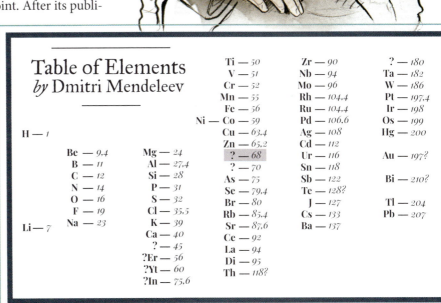

Mendeleev's table contained gaps for undiscovered elements. The highlighted element (atomic mass 68) on Mendeleev's table turned out to be the metal gallium, which melts in your hand.

The Modern Periodic Table

Even before protons were part of the atomic model, scientists were making advances that led to our current periodic table. Significantly, this shows that even when we do not completely understand something in our universe, we can still develop basic ideas about how it works.

English physicist Henry Moseley, one of Ernest Rutherford's students, suggested the solution to the apparent problem with Mendeleev's table. Even though Rutherford discouraged him from doing so at the time, Moseley started working in the relatively new field of the interactions of x-rays with matter. Moseley finally convinced Rutherford that this line of research would reveal information about the structure of the atom. He developed a technique for counting the protons in an atom's nucleus. In 1912 he found that if he arranged the elements in order of increasing atomic number, problems in the periodic table disappeared. Moseley's work led to a revision of Mendeleev's periodic law. The modern **periodic law** states that the properties of elements vary periodically with their *atomic numbers*.

Henry Moseley

Through the early twentieth century, scientists continued adding elements to the periodic table. As noted earlier, the transition elements were problems for chemists because of their similar properties. With the invention of the mass spectrograph, a device that uses electrical charges to sort particles by their mass, scientists finally were able to isolate these elements.

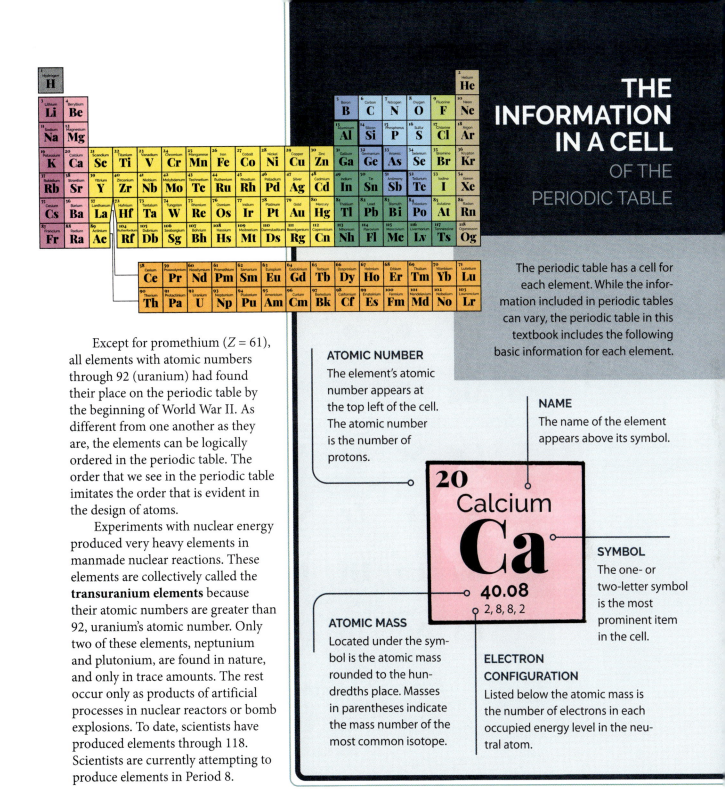

THE INFORMATION IN A CELL
OF THE PERIODIC TABLE

The periodic table has a cell for each element. While the information included in periodic tables can vary, the periodic table in this textbook includes the following basic information for each element.

ATOMIC NUMBER
The element's atomic number appears at the top left of the cell. The atomic number is the number of protons.

NAME
The name of the element appears above its symbol.

SYMBOL
The one- or two-letter symbol is the most prominent item in the cell.

ATOMIC MASS
Located under the symbol is the atomic mass rounded to the hundredths place. Masses in parentheses indicate the mass number of the most common isotope.

ELECTRON CONFIGURATION
Listed below the atomic mass is the number of electrons in each occupied energy level in the neutral atom.

Except for promethium ($Z = 61$), all elements with atomic numbers through 92 (uranium) had found their place on the periodic table by the beginning of World War II. As different from one another as they are, the elements can be logically ordered in the periodic table. The order that we see in the periodic table imitates the order that is evident in the design of atoms.

Experiments with nuclear energy produced very heavy elements in manmade nuclear reactions. These elements are collectively called the **transuranium elements** because their atomic numbers are greater than 92, uranium's atomic number. Only two of these elements, neptunium and plutonium, are found in nature, and only in trace amounts. The rest occur only as products of artificial processes in nuclear reactors or bomb explosions. To date, scientists have produced elements through 118. Scientists are currently attempting to produce elements in Period 8.

EXAMPLE 6-1 GETTING INFORMATION FROM THE PERIODIC TABLE

Use the periodic table to determine the following information for element number 19.
  a. name
  b. symbol
  c. atomic mass
  d. electron configuration

Solution
  a. potassium
  b. K
  c. 39.10 u
  d. 2, 8, 8, 1

Modern Periodic Table

Together the various cells form columns and rows in the periodic table. A column of elements represents a **group**, sometimes referred to as a *family*, of elements with similar physical and chemical properties due to their similar electron configurations. Horizontal rows of elements are **periods**, or *series*. The period number on the left side of the table indicates the highest principal energy level n that electrons occupy in ground-state atoms of those elements.

Period 2 begins with the element lithium. Each succeeding element has one more valence electron.

period (series)

	1	2		3	4	5	6	7	8	9
1	1 Hydrogen **H** 1.01									
2	3 Lithium **Li** 6.94	4 Beryllium **Be** 9.01								
3	11 Sodium **Na** 22.99	12 Magnesium **Mg** 24.31								
4	19 Potassium **K** 39.10	20 Calcium **Ca** 40.08		21 Scandium **Sc** 44.96	22 Titanium **Ti** 47.87	23 Vanadium **V** 50.94	24 Chromium **Cr** 52.00	25 Manganese **Mn** 54.94	26 Iron **Fe** 55.85	27 Cobalt **Co** 58.93
5	37 Rubidium **Rb** 85.47	38 Strontium **Sr** 87.62		39 Yttrium **Y** 88.91	40 Zirconium **Zr** 91.22	41 Niobium **Nb** 92.91	42 Molybdenum **Mo** 95.95	43 Technetium **Tc** (98)	44 Ruthenium **Ru** 101.07	45 Rhodium **Rh** 102.91
6	55 Cesium **Cs** 132.91	56 Barium **Ba** 137.33		57 Lanthanum **La** 138.91	72 Hafnium **Hf** 178.49	73 Tantalum **Ta** 180.95	74 Tungsten **W** 183.84	75 Rhenium **Re** 186.21	76 Osmium **Os** 190.23	77 Iridium **Ir** 192.22
7	87 Francium **Fr** (223)	88 Radium **Ra** (226)		89 Actinium **Ac** (227)	104 Rutherfordium **Rf** (267)	105 Dubnium **Db** (268)	106 Seaborgium **Sg** (269)	107 Bohrium **Bh** (270)	108 Hassium **Hs** (269)	109 Meitnerium **Mt** (278)

58 Cerium **Ce** 140.12	59 Praseodymium **Pr** 140.91	60 Neodymium **Nd** 144.24	61 Promethium **Pm** (145)	62 Samarium **Sm** 150.36
90 Thorium **Th** 232.04	91 Protactinium **Pa** 231.04	92 Uranium **U** 238.03	93 Neptunium **Np** (237)	94 Plutonium **Pu** (244)

group (family)

Group 1 starts with hydrogen with its one valence electron. Even though each new element in the group has another energy level, they each still have only one valence electron.

A group number (1–18) identifies each column of the periodic table. The numbering convention used in the periodic table in this textbook is the one used by IUPAC. You may see other group numbers shown on other periodic tables.

Under the IUPAC numbering scheme, the group numbers uniquely identify each of the groups, which is a tremendous advantage over other numbering systems. Elements in the two separate rows at the bottom of the table, the *inner transition metals*, are not in the numbered columns.

122   Chapter 6

Remember that regions of the periodic table represent energy sublevels of the atom according to the quantum-mechanical model of the atom (see Chapter 5).

Groups 1 and 2 represent the *s* sublevel. The *p* block is at the far right side of the periodic table (Groups 13–18). The *d* sublevels are represented by Groups 3–12.

At the very bottom of the periodic table are two separate rows that represent the *f* sublevel. This explains why the periodic table does not have a uniform rectangular shape.

10	11	12	13	14	15	16	17	18
								2 Helium **He** 4.00
			5 Boron **B** 10.81	6 Carbon **C** 12.01	7 Nitrogen **N** 14.01	8 Oxygen **O** 16.00	9 Fluorine **F** 19.00	10 Neon **Ne** 20.18
			13 Aluminum **Al** 26.98	14 Silicon **Si** 28.09	15 Phosphorus **P** 30.97	16 Sulfur **S** 32.06	17 Chlorine **Cl** 35.45	18 Argon **Ar** 39.95
28 Nickel **Ni** 58.69	29 Copper **Cu** 63.55	30 Zinc **Zn** 65.38	31 Gallium **Ga** 69.72	32 Germanium **Ge** 72.63	33 Arsenic **As** 74.92	34 Selenium **Se** 78.97	35 Bromine **Br** 79.90	36 Krypton **Kr** 83.80
46 Palladium **Pd** 106.42	47 Silver **Ag** 107.87	48 Cadmium **Cd** 112.41	49 Indium **In** 114.82	50 Tin **Sn** 118.71	51 Antimony **Sb** 121.76	52 Tellurium **Te** 127.60	53 Iodine **I** 126.90	54 Xenon **Xe** 131.29
78 Platinum **Pt** 195.08	79 Gold **Au** 196.97	80 Mercury **Hg** 200.59	81 Thallium **Tl** 204.38	82 Lead **Pb** 207.24	83 Bismuth **Bi** 208.98	84 Polonium **Po** (209)	85 Astatine **At** (210)	86 Radon **Rn** (222)
110 Darmstadtium **Ds** (281)	111 Roentgenium **Rg** (282)	112 Copernicium **Cn** (285)	113 Nihonium **Nh** (286)	114 Flerovium **Fl** (289)	115 Moscovium **Mc** (290)	116 Livermorium **Lv** (293)	117 Tennessine **Ts** (294)	118 Oganesson **Og** (294)

63 Europium **Eu** 151.96	64 Gadolinium **Gd** 157.25	65 Terbium **Tb** 158.93	66 Dysprosium **Dy** 162.50	67 Holmium **Ho** 164.93	68 Erbium **Er** 167.26	69 Thulium **Tm** 168.93	70 Ytterbium **Yb** 173.05	71 Lutetium **Lu** 174.97
95 Americium **Am** (243)	96 Curium **Cm** (247)	97 Berkelium **Bk** (247)	98 Californium **Cf** (251)	99 Einsteinium **Es** (252)	100 Fermium **Fm** (257)	101 Mendelevium **Md** (258)	102 Nobelium **No** (259)	103 Lawrencium **Lr** (266)

The periodic table is a visual model of the organization of all the known elements. In it we can see the principal energy levels of the atoms (periods). We can see the orbitals (*s*, *p*, *d*, and *f*) represented in the regions of the periodic table. We can see elements grouped by their electron configurations. While the elements could be shown in a simple rectangle, the modern periodic table models the intricate structure within the atoms. The design of the periodic table mimics the design that God has built into the atoms themselves.

EXAMPLE 6-2 GETTING AROUND THE PERIODIC TABLE

a. What element in Period 2 is a member of Group 15?

b. The element with atomic number 10 belongs to what period and group?

c. Element number 31 is in which group, period, and sublevel block?

Solution

a. nitrogen

b. Neon appears in the second row and the far right column. It is in Period 2 and Group 18.

c. Gallium is in Group 13, Period 4, and the *p* block.

Modern Periodic Table

The vast majority of elements—over 80%—are metals, and the degree to which elements exhibit metallic properties decreases from left to right across the periodic table.

The left side and the middle of the table contain metals.

METALS

Metals are usually solid, lustrous (shiny), malleable, and ductile, and they are good conductors of heat and electricity, though not all metals exhibit all these properties. For example, manganese is a brittle solid and mercury is a liquid.

LANTHANOID AND ACTINOID SERIES

Two rows of elements, called the *lanthanoid series* and the *actinoid series*, are found at the bottom of the table. These are *f*-block elements that include additional interior *f*-sublevel electrons with increasing atomic number. The lanthanoid series fits into the table immediately after lanthanum, and the actinoid series fits into the table after actinium.

The final arrangement of the table as seen above was based on the work of American physicist Glenn Seaborg, for whom element 106, seaborgium, is named. If we were to place the two series in their proper places within the periodic table, the table would change into an oversized, unmanageable shape.

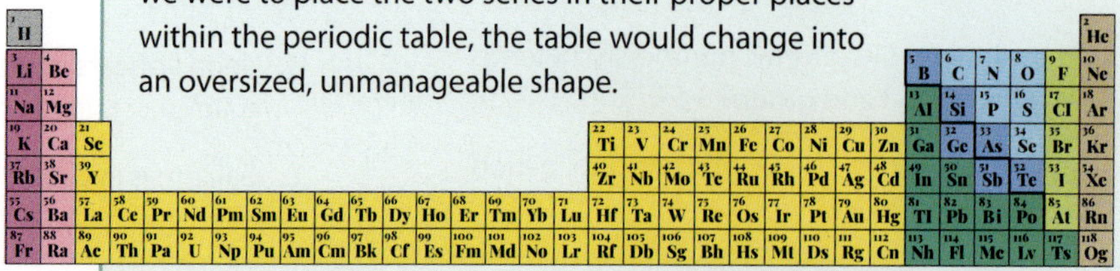

124 Chapter 6

NONMETALS

The right side of the table contains nonmetals. The heavy stairstep line toward the right side of the table marks the boundary between the metals and the nonmetals.

Nonmetals are generally gases or soft, crumbly solids. Some exceptions include bromine, which is a liquid, and diamond (a form of carbon), which is the hardest known natural substance. Nonmetals in their natural forms are generally poor conductors of thermal energy and electrical current.

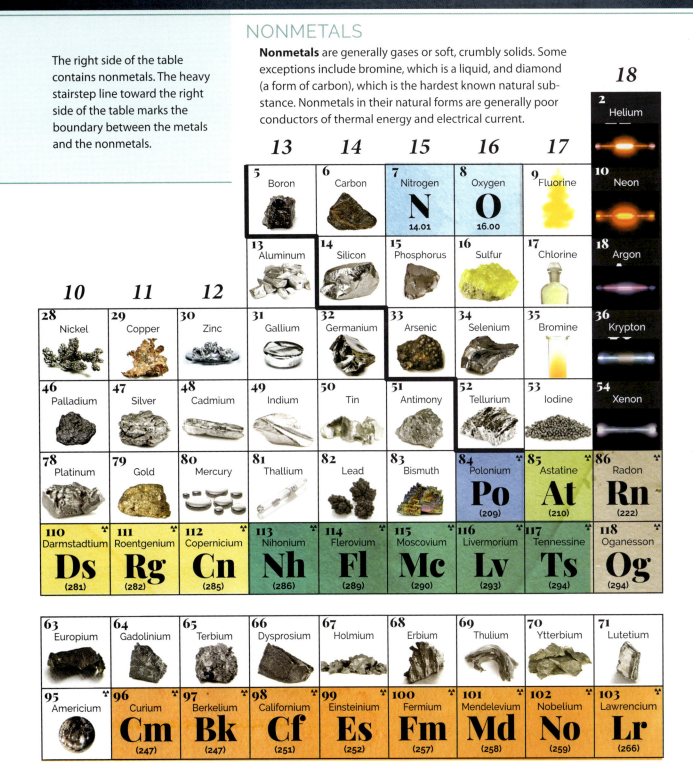

METALLOIDS

Along the stairstep line are the **metalloids**, or *semiconductors*, which have properties between metals and nonmetals. Metalloids have metallic luster but tend to be crumbly, brittle solids. They can conduct electricity better than nonmetals but not as well as metals. Since the metalloids mark the transition between metals and nonmetals, there is some debate as to which elements belong in this group.

Periodic Table and Elements **125**

WORLDVIEW INVESTIGATION

ELEMENT ORIGINS

The world began at some point—all scientists agree on this. But did it begin with a big bang or through the creative work of God? Naturalistic scientists work to develop theories that will explain observations that rely on purely naturalistic mechanisms. Some of these theories seem workable, meaning that they explain the observations fairly well. How do we interpret these theories in light of the Creation account of Genesis?

Task

You are working for a local television station. They have started to air a segment on science topics and have asked you to produce a 3–5 minute video on the origins of chemical elements. You want to do this showing both the secular and creationist viewpoints.

Procedure

1. Research the origin of elements by doing an internet search using the keywords "nucleosynthesis" and "origin of the elements."

2. To get a biblical perspective, go to the Answers in Genesis website and search for "nucleosynthesis" and "origin of the elements."

3. Plan and produce your video, citing your sources properly. Have a classmate review your video and provide feedback.

4. Complete your video and submit by the deadline.

Conclusion

Scientific *data* doesn't contradict the Bible, but scientific *theories* often do. Scientific theories are models that are influenced by worldview. Christians must seek to understand scientific theories with an open mind and then test them against the truth of the Bible. As with all models, theories may need to be modified or even replaced if they are not workable.

6.1 SECTION REVIEW

1. What is the purpose of the periodic table?

2. What common theme emerged as various scientists created lists and tables of elements?

3. State the modern periodic law.

4. Identify the name, period, group, and sublevel block for each of the following elements.
   a. the element with atomic number 17
   b. the element whose symbol is K
   c. the element with an atomic mass of 32.06 u

5. What do we call the vertical columns in the periodic table?

6. Use information from the periodic table to complete the following table.

Name	Symbol	Z	Period	Group
cadmium				
		56		
	Sn			
			5	2

7. In general, how do the electron configurations of the elements in a group compare?

8. Describe the locations of metals, nonmetals, and metalloids on the periodic table.

6.2 PERIODIC TRENDS

In the last section, you learned that the arrangement of the periodic table directly reflects the electron structure of atoms. Also, as atomic number increases from left to right across a period, the elements become less metallic until they are nonmetals at the right side of the table. In this section, we will examine various measurable properties that vary periodically with atomic number. These trends are important because they impact how elements interact with one another to form compounds. We will focus on how electron structure affects these properties.

QUESTIONS
» How do properties change as we move through the periodic table?
» Why do properties change as they do throughout the periodic table?
» Can I predict relative values of properties using a periodic table?

TERMS
atomic radius • ionization energy • electron affinity • electronegativity

Atomic Radius

The distance from the center of an atom's nucleus to its outermost electron is its **atomic radius**. Most atomic radii are measured using x-ray diffraction, in which scientists direct a beam of x-rays with very short wavelengths at a solid sample of an element. The resulting diffraction pattern of x-rays allows scientists to measure the distance between adjacent nuclei using simple geometry.

The extent of the electron cloud around an atom determines its radius. Electrons cannot travel too far from the nucleus because of the positive attraction of the protons. The electrons of adjacent atoms repel each other, holding them at a distance. Since the radius of an atom depends on its electron cloud, one property of an atom that you would expect to show periodic variation is its atomic radius.

What does fluorine's position on the periodic table tell us?

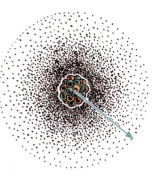

atomic radius

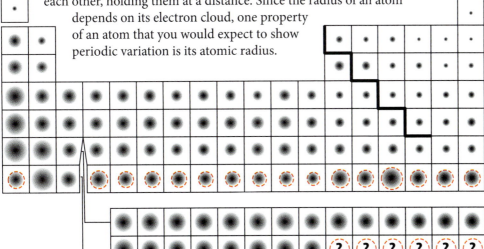

Periodicity of Atomic Radii
The dashed circles around some radii indicate that these radii are estimated or predicted values. The circles with a question mark are atoms with no radius data available.

In general, the radii of atoms decrease in size from left to right across a period. At first thought this pattern may surprise you since each successive atom in a period has an additional electron. As atomic number increases across a period, electrons are added to the *s* and *p* sublevels of the outermost shell or to the *d* or *f* sublevel of an interior shell. You would think that an extra electron would increase the size of its cloud. But from left to right across a period, the positive nuclear attraction for the negative valence electrons also increases disproportionately. Remember that each element also has one more *proton* than the previous element on the table. Each additional positive charge increases the pull on *all* the valence electrons. Thus, as the atomic number increases in a period, the outer-shell electrons are held progressively tighter, and the average distance from the nucleus to the outer-shell electrons decreases.

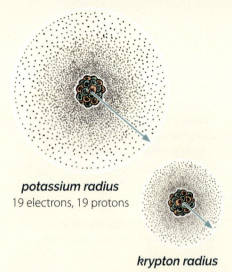

potassium radius
19 electrons, 19 protons

krypton radius
36 electrons, 36 protons

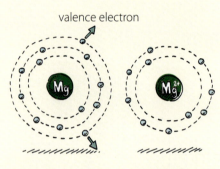

magnesium atom | **magnesium ion (cation)**

Magnesium has two loosely held valence electrons. When magnesium loses its two valence electrons, it also loses its outer energy level, which makes the magnesium ion much smaller than the magnesium atom.

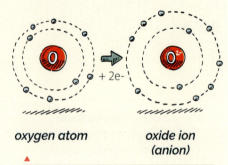

oxygen atom | **oxide ion (anion)**

Oxygen has six valence electrons and typically becomes an ion by gaining two additional valence electrons. With the addition of those electrons, the repulsive force between the valence electrons increases. This repulsion results in an anion that is larger than the original atom.

The shell where an electron is added also affects how the atom's radius changes with increasing atomic number. When electrons are added to the outermost shell, the negative charge of the inner electron shells is unchanged, so the change in nuclear charge (number of protons) affects the outermost electron shell the most. But if electrons are added to inner energy levels, such as the d or f sublevels, the outer electrons are shielded and the electric force on them is partially blocked by the inner electrons from the additional nuclear charge. Thus, the atomic radius is affected less in transition metals.

Atomic radius increases when moving down a group or column in the periodic table because another principal energy level (an electron shell) is added with each subsequent period. Higher energy generally corresponds to an electron with a greater average distance from the nucleus. Therefore, the atomic radius increases as energy levels are added.

Ionic Radius

We can extend our discussion of atomic radii to ions of single atoms. Remember, ions are atoms that have lost or gained electrons. Cations—positive ions—are smaller than their parent atoms because they have fewer electrons than the fixed number of protons. The resulting charge imbalance more strongly attracts the remaining electrons, making the electron cloud smaller. In many cations, the atom will lose its entire outer electron shell or principal energy level. This loss markedly reduces the ion's size as the excess nuclear charge draws in the electrons of the next lower energy level.

On the other hand, anions—ions that have resulted from a gain of electrons—are larger than their neutral atoms. The additional electrons are generally added to the outer, principal energy level, resulting in a larger cloud. The cloud is larger for two reasons. First, the excess negative charge is more effective in shielding the other valence electrons from the positive nuclear charge, so the nucleus holds the additional electron less strongly. Second, the added negative charge tends to repel the negative electrons in the same shell, forcing them apart and increasing the size of the shell.

We can observe the general trend for ionic radius by looking at the image below. The trend looks odd at first glance, as the ionic radius changes abruptly toward the right side of the periodic table. But notice that this abrupt change occurs at the transition from the metals to the nonmetals. Notice too that the trend in ionic radius matches the trend in atomic radius and is the same for metals (cations) and nonmetals (anions), decreasing across the period and increasing down the group. But the trend restarts with the nonmetals.

periodicity of ionic radii for common ionic forms

Ionization Energy

Some atoms lose their electrons easily, while others stubbornly hold on to theirs. Atoms that form compounds either lose or take electrons to some extent. The ease with which atoms either acquire or donate electrons is a very important property for predicting chemical reactivity and is related to chemical bond types.

To remove an electron, work must be done to overcome the attractive forces that hold the electron in the atom. The amount of energy required to remove an electron from an atom is called the **ionization energy**. For a single neutral atom in the gaseous state, the minimum energy required to remove the first electron from its outermost shell to make it a cation is called its first ionization energy. Ionization energy values generally increase from left to right within a period and decrease from top to bottom within a group. This trend is the opposite of the trend in atomic radii.

Ionization energies generally increase from left to right across a period because the strength with which protons in the nucleus attract the outer-shell electrons increases. It thus becomes more difficult to remove an electron that is closer to and more strongly attracted to an increasingly positive nucleus. First ionization energies decrease down a group for two reasons. First, the outer electrons are in increasingly higher energy levels and are thus farther away from the nucleus. The nuclear-charge attraction drops off quickly with distance. Second, the outer electrons are significantly shielded from the positive charge in the nucleus by the inner electron shells. This effect becomes more pronounced with more energy levels. You can see that ionization energy is clearly related to the atomic radius. A larger atomic radius means a smaller ionization energy, and a smaller atomic radius means a larger ionization energy.

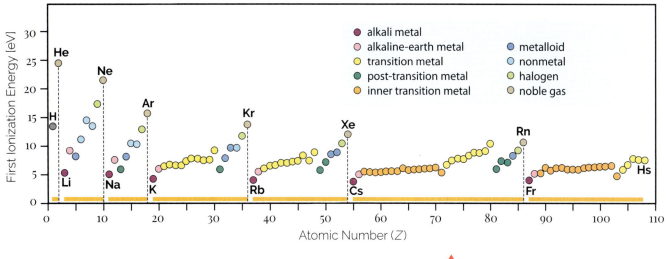

periodicity of first ionization energy

Notice that in each period the noble gases have the largest ionization energies because they have a full outer energy level and do not gain or give up electrons easily. Because of this, they rarely react with other elements to form compounds. On the other hand, metals have the smallest ionization energies. Metals with the lowest first ionization energies tend to be the most chemically reactive. Their low ionization energies explain why they react so energetically with nonmetals.

CASE STUDY

WHAT'S IN A NAME?

In the early days of the periodic table, scientists toiled over laboratory samples to isolate elements from compounds. They discovered many elements through these separation procedures. But today the periodic table is growing through a very different process. Nuclear physicists have learned how to synthesize new elements by smashing smaller atoms together. Very few of the collisions form new atoms and the new atoms decay in just fractions of a second, transforming into other elements.

As of 2020 the newest element is element 117, which scientists synthesized in 2010. Scientists from the Oak Ridge National Laboratory in Tennessee synthesized a sample of berkelium-249. Scientists at the Joint Institute for Nuclear Research in Russia bombarded the sample with calcium-48. The collisions produced atoms with 117 protons and either 176 or 177 neutrons. These atoms were radioactive with a half-life of less than 60 ms.

When scientists discover a new element, they publish their results in a scientific journal. If other researchers validate the finding, IUPAC assigns a temporary name and symbol to the element. This naming system assigns Latin or Greek roots for each of the digits of the element's atomic number. The element name consists of these three roots combined with an *-ium* ending. The symbol is composed of a single letter for each root (see table above right). So element 117 was named *ununseptium* (Uus) while it awaited its official name.

Digit	Root	Symbol
0	nil	n
1	un	u
2	b(i)	b
3	tr(i)	t
4	quad	q
5	pent	p
6	hex	h
7	sept	s
8	oct	o
9	en(n)	e

To get a permanent name and chemical symbol, the research must be verified. Only after verification will the IUPAC consider a naming proposal. The four most recent elements to receive official names are nihonium ($Z = 113$), moscovium ($Z = 115$), tennessine ($Z = 117$), and oganesson ($Z = 118$), all named in 2016.

Questions to Consider

1. How do the methods of discovering new elements in the mid-twentieth century differ with those of today?
2. Why does a new element get a temporary name and symbol?
3. Scientists are now trying to synthesize elements in Period 8. What would be the temporary name and symbol of the first element in that period (element 119)?
4. Why do you think IUPAC chose the names for elements 95 and above?

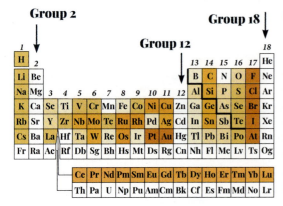

Electron affinity depends on the size of the atom and the fullness of the highest energy sublevels. Shading in the table indicates increasing electron affinity.

Electron Affinity

Electron affinity is the change in energy when an electron is added to a neutral atom to form an anion. In most cases, the atom releases energy when an electron is added. We can think of electron affinity as the opposite of ionization energy. Electron affinity measures how strongly an atom attracts additional electrons. This property is most affected by the fullness of an atom's highest energy sublevel. For incomplete sublevels, most elements will easily accommodate the additional electron, releasing energy. For elements in which the highest energy sublevels are full, such as Group 2 elements (*s* sublevel is already full), Group 12 elements (*d* sublevel is already full), and Group 18 elements (*p* sublevel is already full), the additional electron is repelled unless energy is expended to insert the electron into the next higher energy sublevel.

In general, electron affinity becomes larger from left to right along a period. As an energy level continues to fill, it has a stronger attraction for electrons. The trend from top to bottom within a group varies, depending on location in the table.

Electronegativity

As you have seen, the attraction of the nucleus for the surrounding electron cloud influences periodic trends. In the elements toward the right end of a period, the relatively stronger nuclear charge pulls the outer-shell electrons closer to the nucleus and decreases the size of the atom. Thus, it makes the removal of electrons more difficult (increases first ionization energy) and encourages the addition of electrons (increases electron affinity).

The measure of the attraction between the nucleus and shared valence electrons is called **electronegativity**. In most cases, scientists determine the electronegativity for atoms bonded *in molecules* that share electrons with other atoms. Thus, electronegativity is a measure of an atom's ability to attract and hold electrons in a molecule.

The values of element electronegativity were determined mathematically rather than experimentally. The first chemist to quantify electronegativity was the American chemist Linus Pauling. After examining the strengths of chemical bonds for many different compounds, he noted that fluorine held its electrons the strongest and cesium the weakest. Using a fairly simple mathematical relationship, he arbitrarily assigned fluorine an electronegativity of 4 Pauling units and computed the other elements' values relative to fluorine.

The table below shows the electronegativity values for the most common elements. Notice that there are values for only two of the noble gases. Other noble gases were not known to form compounds with any other elements in Pauling's time. Scientists later believed that they could compute electronegativities for all the elements using atomic properties distinct from bonding properties. They have now determined the electronegativity values for all the noble gases. These newer scales have been adjusted to agree with Pauling's electronegativity scale. Of the three periodic properties that measure the attraction between valence electrons and the nucleus of an atom, electronegativity has the widest use. It plays a central role in predicting how atoms chemically combine.

Linus Pauling

Electronegativities of the Elements
Shading in the table indicates increasing electronegativity values.

1	2	3	4	5	6	7	8	9	10	11	12	13	14	15	16	17	18
H 2.2																	He
Li 1.0	Be 1.6											B 2.0	C 2.6	N 3.0	O 3.4	F 4.0	Ne
Na 0.9	Mg 1.3											Al 1.6	Si 1.9	P 2.2	S 2.6	Cl 3.2	Ar
K 0.8	Ca 1.0	Sc 1.4	Ti 1.5	V 1.6	Cr 1.7	Mn 1.6	Fe 1.8	Co 1.9	Ni 1.9	Cu 1.9	Zn 1.7	Ga 1.8	Ge 2.0	As 2.2	Se 2.6	Br 3.0	Kr 3.0
Rb 0.8	Sr 1.0	Y 1.2	Zr 1.3	Nb 1.6	Mo 2.2	Tc 1.9	Ru 2.2	Rh 2.3	Pd 2.2	Ag 1.9	Cd 1.7	In 1.8	Sn 2.0	Sb 2.1	Te 2.1	I 2.7	Xe 2.6
Cs 0.8	Ba 0.9	La 1.1	Hf 1.3	Ta 1.5	W 2.4	Re 1.9	Os 2.2	Ir 2.2	Pt 2.3	Au 2.5	Hg 2.0	Tl 1.6	Pb 1.9	Bi 2.0	Po 2.0	At 2.2	Rn
Fr 0.8	Ra 0.9	Ac 1.1	Rf	Db	Sg	Bh	Hs	Mt	Ds	Rg	Cn	Nh	Fl	Mc	Lv	Ts	Og

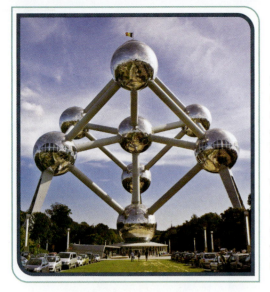

Much of science is about modeling the world around us. Recall that a workable model is one that explains most of the observations and makes accurate predictions. The periodic table is a model of the atomic structure of elements and the physical and chemical properties that result from that structure. From what you have seen in this chapter, you should recognize that the periodic table is a workable model. It explains much of what we understand about the structure of atoms. It has also demonstrated its predictive power over its 150-year history. It has accurately predicted the existence of once-unknown elements and has also predicted their chemical and physical properties. In upcoming chapters you will use the periodic table to predict the formulas of chemical compounds and the results of chemical reactions.

EXAMPLE 6-3: USING PERIODIC TRENDS TO PREDICT PROPERTIES

For each of the following pairs of elements, use a periodic table and your knowledge of periodic trends to compare the atomic and ionic radii, ionization energy, and electronegativity values for each pair.

  a. nitrogen and phosphorus

  b. rubidium and iodine

Solution

  a. Nitrogen atoms have smaller atomic and ionic radii than phosphorous atoms. This is due to phosphorus having an additional energy level. So even though phosphorous has more protons and electrons—implying a stronger electric pull—its valence electrons are shielded from the nucleus, resulting in a larger radius. Nitrogen's smaller radius means that it holds its valence electrons more tightly, resulting in a higher ionization energy and electronegativity.

  b. Iodine atoms have smaller atomic and ionic radii than rubidium atoms. This is due to iodine having more electrons and protons. The extra particles result in a stronger electric force, pulling the valence electrons closer to the nucleus. An iodine atom's smaller radius means that it holds its valence electrons tighter, resulting in higher ionization energy and electronegativity.

6.2 SECTION REVIEW

1. Describe the trends in atomic radius.
2. Why are anions larger than neutral atoms of the same element?
3. Explain why ionization energy changes as it does along rows and columns of the periodic table.
4. (True or False) Since work has to be done to remove an electron, ionization energy values are always positive.
5. Describe the trends in electron affinity.
6. Why is energy input required to add an electron to zinc?
7. Of the eighteen elements in Period 4, which
    a. has the largest atomic radius?
    b. has the smallest first ionization energy?
    c. has the largest electron affinity?
    d. has the lowest electronegativity?
    e. have the highest electronegativity?
    f. are electrical semiconductors?
    g. is one of the least reactive elements?
8. Periodic trends model the properties of atoms of different elements. Which of the properties in this section has the least consistency in its trends? What does this say about the workability of this model?

6.3 ELEMENTS BY THEIR GROUPS

The Importance of Descriptive Chemistry

Descriptive chemistry is the study of elements and the compounds they form. In this section, you will explore the physical and chemical properties of groups of elements. You will see how scientists have observed the periodicity of atomic properties in the laboratory. Elements in groups on the periodic table have similar electron configurations. Because of their similar configurations, they also have similar physical and chemical properties.

As you read, think about how descriptive chemistry can contribute to serving God and others. Technologies such as detergents, lasers, artificial joints, medicine, fuel cells, nuclear power, fertilizer, and even pest control can be beneficial to people and are the result of understanding elements and the compounds they form. Chemistry is one way to obey God's great command to love our neighbors.

QUESTIONS
» What are the groups of the periodic table?
» Why are elements in particular groups?
» What can we tell about an element on the basis of its chemical group?

TERMS
alkali metal • alkaline-earth metal • transition metal • inner transition metal • lanthanoid series • actinoid series • oxide • sulfide • halogen • noble gas

What do the elements in a group have in common?

MINILAB

DENSE, DENSER, DENSEST?

There are many physical and chemical properties that change in regular ways along the rows and columns of the periodic table. You have already learned about the trend in atomic radius. You can also quickly see the trend in atomic mass by looking at the periodic table. What do you think this implies about the trend in densities?

Can we predict density from information on the periodic table?

1. What is the periodic trend for atomic radius and atomic mass from left to right across a row?

2. On the basis of your answer to Question 1, what do you predict would be the trend in density?

Procedure

A Use an element handbook or an internet source to find the densities of scandium, manganese, and nickel.

3. Do the densities change as expected?

B On the basis of the densities that you looked up, predict the densities for titanium, vanadium, and copper.

Conclusion

4. How well were you able to predict the density values for these elements?

Going Further

C Look up the densities for germanium, selenium, and krypton.

5. Does the periodic trend continue?

6. Look at the periodic table on pages 122–23. How can what you observed in Step C tell us about the factors that influence the density of elements?

EQUIPMENT
• none

germanium

The Hydrogen Group
ELEMENT GROUP OF THE PERIODIC TABLE

Hydrogen, the lightest element and the most abundant in the entire universe, has an electron configuration similar to that of the Group 1 metals. However, hydrogen is a gas at room temperature, which is a property of only nonmetals. It can easily gain another electron or lose its one electron, giving it a 1– or 1+ charge. It is often considered a group by itself, which is why its cell in the periodic table is colored differently and separated from the rest of the elements.

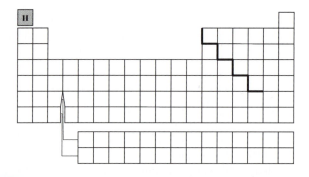

PHYSICAL PROPERTIES

Hydrogen is a colorless, odorless, and tasteless gas. It occurs naturally as a diatomic molecule (H_2). Because it has the least mass of all gases, its molecules move at high speeds and diffuse faster than other gases at the same temperature. Hydrogen molecules have little attraction for each other, so they remain in the gaseous state down to temperatures as low as –253 °C and solidify at –259 °C. But hydrogen will change into a metallic form at extreme pressures, that is, greater than 2.84×10^{11} Pa.

CHEMICAL PROPERTIES

Hydrogen has unique chemical properties. On the one hand, it has a single electron in its only occupied energy level, so it chemically reacts like other Group 1 elements. On the other hand, it is also only one electron shy of filling its only energy level, so it is chemically similar to Group 17 elements. Some periodic tables show it as a member of both groups on the periodic table.

At room temperature, hydrogen molecules are very stable. But at high temperatures and pressures, molecular hydrogen can easily split apart and combine, sometimes explosively, with other elements. A mixture of hydrogen and oxygen molecules sparked by a flame or electrical discharge burns to form water vapor. When properly controlled, combustion of liquid hydrogen with liquid oxygen as rocket fuel provides a powerful boost to spacecraft.

Hydrogen combines with many elements to form important compounds. It forms ammonia (NH_3) when it combines with nitrogen atoms. When combined with the elements of Group 17, hydrogen forms halides such as hydrogen chloride and hydrogen fluoride. These compounds can also be strong acids. Occasionally hydrogen combines with the reactive metals of Groups 1 and 2 to form ionic compounds. Also, hydrogen combines with carbon to form millions of compounds called *hydrocarbons*. We will study these important compounds when we survey organic chemistry (Chapter 20) and biochemistry (Chapter 21).

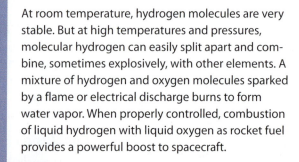

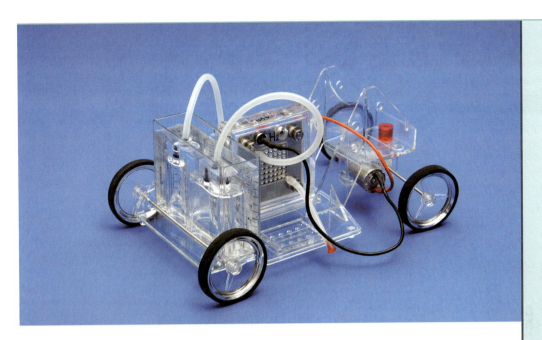

Hydrogen Fuel Cell

Gasoline burning vehicles are the standard mode of transportation for many people around the world. But as concerns increase over the supply of gasoline and the exhaust products of internal combustion engines, people are looking for alternatives, such as hybrid vehicles, electric vehicles, and hydrogen fuel cells. Of the three, hydrogen fuel cells have the advantage of producing only water and thermal energy as byproducts.

Hydrogen fuel cells consist of a cathode and an anode separated by an electrolyte solution. At the anode, hydrogen, the fuel, is converted into hydrogen ions and free electrons. The hydrogen ions move through the electrolyte directly to the cathode. The electrons move from the anode through the electric circuit, where they do work, and then continue on to the cathode. At the cathode, the electrons combine with the hydrogen ions and oxygen from the air to form water. Water and some thermal energy are released into the atmosphere.

Since a hydrogen fuel cell converts chemical energy directly into electrical energy, it is very efficient (about 55%). This efficiency is more than double that of an internal combustion engine and about equal to that of an electric vehicle. Hydrogen fuel cell exhaust consists solely of water and thermal energy, making hydrogen fuel cells better for the environment than both internal combustion engines and most electric vehicles (depending on the method used to generate the electrical energy). Additionally, a hydrogen fuel cell has no moving parts, which means that in theory it should be more reliable than an internal combustion engine. In the search for abundant clean energy, the hydrogen fuel cell is a promising candidate.

HOW IT WORKS

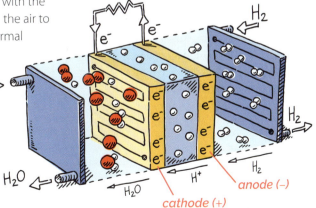

Group 1: Alkali Metals
ELEMENT GROUP OF THE PERIODIC TABLE

Group 1 elements, the **alkali metals,** are metals that are very chemically reactive. These elements have electron configurations with one valence electron, which they easily lose to form cations with a 1+ charge. Sodium is the most abundant alkali metal. It is found in stars and is the sixth most common element in the earth's crust.

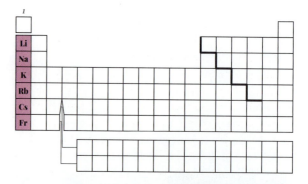

Lithium

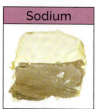
Sodium

Potassium

Rubidium

Cesium

87 Francium
Fr
(223)

Solid alkali metals are soft enough to be cut with a knife.

PHYSICAL PROPERTIES

Like most metals, alkali metals conduct heat and electricity well and have a bright metallic luster on freshly cut surfaces. But they have low densities, some less dense than water, and are very soft at room temperature.

CHEMICAL PROPERTIES

A solitary, loosely held electron in the outermost energy level makes alkali metals very reactive. An alkali metal donates its valence electron readily to attain a stable electron configuration like that of a noble gas. It generally reacts readily with water in a reaction that liberates hydrogen in the process, producing enough heat to ignite the hydrogen in an explosion. Because of their high reactivity, we find none of these elements in their pure metallic form. Their high reactivity also makes storage difficult; they must be stored so that they cannot react with water vapor or oxygen in the air.

sodium in water reaction

Group 2: Alkaline-Earth Metals
ELEMENT GROUP OF THE PERIODIC TABLE

The chemical term *earth* originally applied to metal-oxygen compounds, which are mostly geologic minerals that dissolve slightly in water. Some of these compounds were similar to compounds of alkali metals, so they were given the more specialized name **alkaline-earth metals**. Today the term applies to the metals in Group 2. The electron configuration of these elements includes two valence electrons, which they can lose easily. They form 2+ cations when they lose their valence electrons to obtain a noble gas electron configuration.

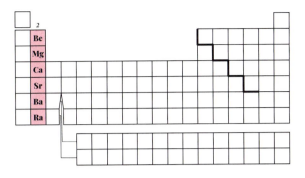

Beryllium

Magnesium

PHYSICAL PROPERTIES

Alkaline-earth metals are all solid at room temperature and have typical metallic properties. They are denser, harder, and have higher melting points than the alkali metals. In pure samples of alkaline-earth metals, freshly cut surfaces range from bright silver to white in color. The metals quickly oxidize to a dull gray or yellow color. Densities are slightly higher than those of the alkali metals, and hardness is much higher. All the alkaline-earth metals are malleable.

Calcium

Strontium

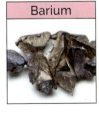
Barium

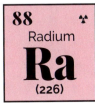
88 Radium
Ra
(226)

CHEMICAL PROPERTIES

Each neutral alkaline-earth metal atom has two electrons in its outermost *s* sublevel. An alkaline-earth metal typically donates its two electrons when it combines with a nonmetal element. Reactivity in this group increases from top to bottom in the column due to larger atoms holding their electrons less strongly. An example of this trend is how these elements react with water. Beryllium does not react with water, magnesium reacts with steam, and calcium reacts vigorously with even cold water.

calcium in water reaction

Periodic Table and Elements **137**

Groups 3-12: Transition Metals
ELEMENT GROUP OF THE PERIODIC TABLE

All the *d*-block elements on the periodic table belong to the group called **transition metals**. This name traces back to an earlier theory that the properties of these elements changed gradually but predictably into the properties of nonmetals. The term transition still retains some significance, however, as a label for the physical location of these elements in the center of the periodic table. This region of the periodic table contains some of the most commonly used metals, such as iron and copper, as well as the precious metals silver, gold, and platinum. The electron configurations of these elements include either one or two valence electrons. They can form cations with various charges.

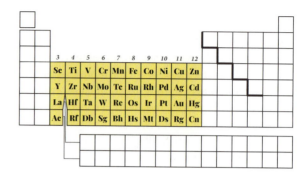

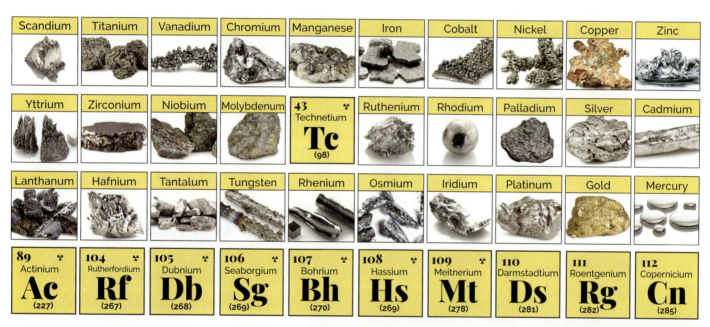

PHYSICAL PROPERTIES

Transition metals exhibit the qualities we typically associate with metals. Unlike the first two metal groups, most of these metals have high densities, are reasonably hard, and have considerable toughness or strength. They typically have a shiny luster on freshly cut surfaces, conduct heat and electricity well, and are ductile and malleable. All except mercury (below) are solids at room temperature.

CHEMICAL PROPERTIES

Because their highest-energy electrons occupy interior *d* sublevels, transition metals generally have similar chemical properties. But there are also many exceptions, which are caused by the specific quantum mechanical relationships between their electron structure and nuclear charges. Metals showing little chemical reactivity, such as gold, silver, and platinum, can resist corrosion for centuries. Other more reactive metals, like iron and copper, quickly corrode when exposed to moist air. Nearly two-thirds of all naturally occurring transition metals occur as native minerals—elements found as pure solids in nature.

138   Chapter 6

Inner Transition Metals

ELEMENT GROUP OF THE PERIODIC TABLE

We typically display the lanthanoid and actinoid series of elements below the main periodic table in two rows of fourteen cells. These are called series rather than periods because they are the *f* blocks of the sixth and seventh periods and belong between the *s* block and *d* block for those periods, respectively. These two series are called the **inner transition metals**.

LANTHANOID SERIES

The **lanthanoid series** extends from cerium through lutetium. These elements were once called the rare-earth elements because they were difficult to identify and even more difficult to purify from their ores. They occur in high concentrations in a considerable number of minerals and in low concentrations throughout the earth's crust.

ACTINOID SERIES

The **actinoid series** extends from thorium through lawrencium. We find only the first five elements of the actinoid series in nature. The rest have been formed only in man-made nuclear reactions. These elements consist of many radioactive isotopes that are produced artificially and are very unstable.

PHYSICAL PROPERTIES

The majority of the lanthanoid series elements are weakly attracted to magnetic fields because of their unpaired electrons. All the elements in the lanthanoid series occur naturally except promethium. Freshly cut surfaces of pure forms of these metals are bright and silvery.

ytterbium

CHEMICAL PROPERTIES

The highest-energy electrons of the inner transition metals occupy the 4*f* and 5*f* sublevels. Although the atoms of the elements lanthanum and actinium contain no 4*f* or 5*f* electrons themselves, their outer electron structures resemble those of the lanthanoid and actinoid elements closely.

The lanthanoid elements display great uniformity in their chemical behavior. Thus, they are difficult to purify from samples containing the other elements of the series.

Similar to the lanthanoid series, the actinoid series elements often give up most or all of their highest-energy electrons, including their 5*f* sublevel electrons, when combining with other elements. The actinoid series compounds are dangerous because they are radioactive.

Groups 13–16
ELEMENT GROUP OF THE PERIODIC TABLE

Groups 13–16 consist of combinations of post-transition metals, metalloids, and nonmetals. The post-transition metals, as their name implies, follow the transition metals on the periodic table. Post-transition metals include well-known elements such as aluminum, tin, and lead, as well as obscure elements such as thallium, indium, and gallium.

Metalloids, as we noted earlier, congregate around the stairstep line between the cells in Groups 13–16. Their characteristics—luster, hardness, conductivity, and chemical reactivity—lie somewhere between those of the metals and the nonmetals. Metalloids have a somewhat metallic luster, often showing a variety of colors beyond the typical silver or gray metallic color. But they tend to be more crystalline and brittle with lower ductility and malleability than most metals. They typically have fair electrical conductivity, but some are valuable as semiconductors.

Because of its high reactivity, we don't find aluminum as a native mineral. We typically find it bonded with oxygen atoms in aluminum ores, the most common of which is called *bauxite* (see facing page).

140   Chapter 6

Group 13: Boron Group

ELEMENT GROUP OF THE PERIODIC TABLE

Group 13 consists of five post-transition metals and a metalloid. There are significant differences between the chemical properties of boron, a metalloid, at the top of the column and the five metals below it. Boron has some metalloid properties, but the characteristics we associate with metalloids are more recognizable in other groups. Elements in this group have three valence electrons, which they can lose to form 3+ cations.

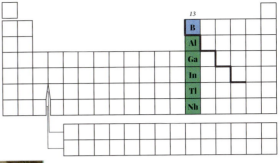

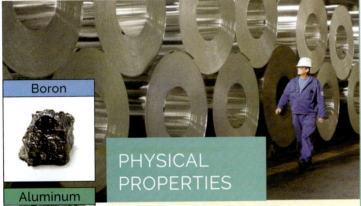

Boron

Aluminum

Gallium

Indium

Thallium

113
Nihonium
Nh
(286)

PHYSICAL PROPERTIES

The two most economically important elements of Group 13 are boron and aluminum. Boron, in its impure form, is a crumbly, black solid. It can transmit portions of the infrared spectrum (most materials absorb infrared), and it conducts electricity well but only at high temperatures.

Aluminum is a very important metal because it combines high strength with low density, especially in its alloys. Unlike the metalloid boron, aluminum has a silvery metallic luster, is easily machined, and is highly conductive. Aluminum is the most common metal in the earth's crust, found mostly in clays. It was initially difficult to refine, but once an inexpensive means of refining it was developed, the price of aluminum decreased significantly.

The other Group 13 metals are increasingly metallike as their atomic numbers increase. They have a silvery metallic luster, are conductive, and are malleable and ductile. Gallium's melting point is low enough that it will melt in your hand. Thallium is a soft metal and has other properties like those of alkaline-earth metals.

CHEMICAL PROPERTIES

Boron has the electron configuration of the elements in Group 13 but has several properties of silicon, which is in Group 14. Its small size allows the nucleus to hold electrons as though the nucleus had a greater positive charge than it does.

Like iron, aluminum corrodes in the presence of atmospheric oxygen. How then did aluminum get its reputation for being a durable, corrosion-resistant metal? The difference lies in the nature of the corrosion products. While many metal oxides are porous, aluminum oxide forms an impenetrable shield against further oxidation. Although fairly soft when pure, aluminum can be alloyed with many other elements to form many harder, more useful metals.

bauxite

Group 14: Carbon Group

ELEMENT GROUP OF THE PERIODIC TABLE

As in Group 13, a wide range of properties is present in Group 14. The only true nonmetal, carbon, is at the top of the column. Two metalloids, silicon, and germanium follow after carbon. Two common metals appear toward the bottom of the column—tin and lead. We know little about a third metal at the bottom, flerovium. It was identified by Russian physicists in 1999 and received its official name from IUPAC in 2012.

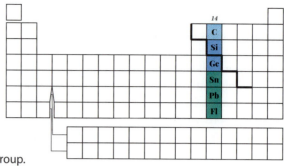

Carbon is more important than all the other elements in its group. Carbon compounds are the basis for life. Because of the sheer number of carbon-based biological compounds, there is an entire area of chemistry, called *biochemistry*, devoted to their study. There are also millions of other carbon compounds that we study in another area of chemistry called *organic chemistry*. You will be introduced to organic chemistry in Chapter 20 and biochemistry in Chapter 21. As noted in Chapter 4, we define the unified atomic mass unit on the basis of an isotope of carbon, carbon-12. All members of the carbon group have four valence electrons.

114
Flerovium
Fl
(289)

PHYSICAL PROPERTIES

Elements in Group 14 are solids at room temperature, though recent experiments indicate that flerovium might be a gas. Carbon itself takes several forms. It can occur as an amorphous solid that is very soft and dull black. In graphite, it is a soft, black solid that feels slippery because it is made of sheets of carbon atoms that slide easily across each other. In diamond, carbon is a clear-to-slightly-colored, extremely hard, crystalline solid. Scientists are also engineering a whole new generation of materials using carbon nanotubes.

The metalloids silicon and germanium are brittle solids with a metallic luster. The metals tin and lead are fairly common. Tin is a silvery white or gray metal that is malleable and somewhat ductile. Lead is a soft, gray metal that is highly malleable and ductile.

CHEMICAL PROPERTIES

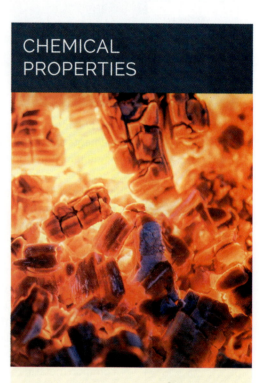

Carbon generally has a low chemical reactivity, but it combusts readily in oxygen to form carbon dioxide or carbon monoxide. Silicon does not react with air, water, or acids at low temperatures. Tin is relatively inactive and is often used to plate other metals to reduce corrosion. Lead is very inactive, especially in the presence of acids.

Group 15: Pnictogens
ELEMENT GROUP OF THE PERIODIC TABLE

Group 15 exhibits a dramatic range of properties from top to bottom. First is the element nitrogen, a gas, followed by four naturally occurring solids: phosphorus, arsenic, antimony, and bismuth. Nitrogen and phosphorus are nonmetals, arsenic and antimony are metalloids, and bismuth and moscovium are post-transition metals. Neutral forms of these elements have five valence electrons, but they have an unusual ability to transfer or acquire electrons as needed when combining with other elements.

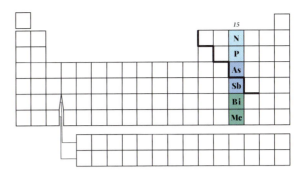

7 Nitrogen N 14.01

Phosphorus

Arsenic

Antimony

Bismuth

115 Moscovium Mc (290)

PHYSICAL PROPERTIES

Nitrogen normally exists as a diatomic molecule in the gaseous state. It has no taste, color, or odor, and it accounts for approximately 78% of the volume of the earth's atmosphere.

Phosphorus does not occur as a native element because it is too reactive. It exists in different polyatomic forms, depending on the number and arrangement of the atoms when it is synthesized. All are solid forms with a variety of colors. Some common forms are white (or yellow), red, and black (or violet) phosphorus. White phosphorus is phosphorescent and is the source for this term.

Liquid nitrogen can be used to make ice cream.

CHEMICAL PROPERTIES

Nitrogen as an elemental gas is essentially inert. The atoms in these molecules are tightly bonded together and are difficult to split. Only under relatively high temperatures can the element be forced to combine with other elements.

In contrast, white phosphorus is extremely reactive and will burn spontaneously in atmospheric oxygen. For this reason, it must be stored underwater and handled with tongs to avoid burns. It is poisonous, even in small amounts.

Arsenic readily bonds with nonmetals, can oxidize in moist air, and reacts with many metals. Antimony reacts to form compounds with hydrogen, oxygen, and halogens, forming many hydrides, hydroxides, oxides, and halides. Bismuth reacts with many acids and with water at high temperatures.

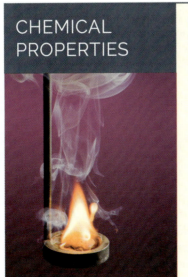

phosphorus burning

Periodic Table and Elements

Group 16: Chalcogens
ELEMENT GROUP OF THE PERIODIC TABLE

Group 16 includes several reactive and important elements. Oxygen is the most easily recognized nonmetal element in this group, although sulfur and selenium are also nonmetals. Tellurium is a highly reactive metalloid with few industrial uses. Polonium is either a post-transition metal or a metalloid. Livermorium, first identified by scientists at the Lawrence Livermore National Laboratory in 1999, is thought to be metallic. The elements in this group have six valence electrons. They often will gain two additional electrons, becoming 2– anions to achieve a stable, noble gas configuration.

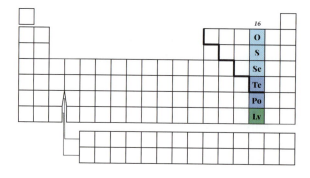

8 Oxygen O 16.00

Sulfur

Selenium

Tellurium

84 Polonium Po (209)

116 Livermorium Lv (293)

PHYSICAL PROPERTIES

Oxygen—a colorless, odorless, tasteless gas—forms about 21% of the earth's atmosphere. It is the most abundant element by mass in the earth's crust and is slightly soluble in water. Through God's design, enough oxygen dissolves in lakes, rivers, and oceans to sustain fish and aquatic plants. Atmospheric oxygen exists in two forms: a diatomic gas (O_2) and a triatomic gas called ozone (O_3). The design of Earth's atmosphere includes a naturally sustained layer of ozone high in the stratosphere that screens out most of the harmful forms of ultraviolet radiation from the sun.

Sulfur exists in a variety of forms depending on the arrangement of its atoms. Native sulfur is a brittle, yellow, crystalline solid. When heated to 115 °C, sulfur melts into a straw-colored liquid that can crystallize into another form. If the molten sulfur is quickly cooled by being poured into water, it forms amorphous globs with a plastic-like consistency. Eventually, the amorphous form transforms into the crystalline form.

Selenium has numerous forms. The most common is a brittle, gray, crystalline solid. The metalloid tellurium comes in both a black amorphous form and a brittle, silver-white, crystalline form.

Polonium is a highly radioactive, silver-colored metal.

molten sulfur

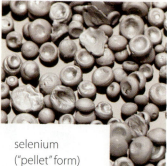

selenium ("pellet" form)

CHEMICAL PROPERTIES

While sulfur and oxygen do not have similar physical properties, they do have similar chemical properties. Oxygen is one of the most reactive elements, having an electronegativity second only to fluorine. Its strong attraction for electrons accounts for its ability to combine with nearly every other element to form compounds called **oxides**.

Sulfur is reactive at room temperature, but it does not match the reactivity of oxygen. Metals such as zinc, calcium, and iron combine with sulfur to form compounds called **sulfides**. Sulfur combines with nonmetals such as oxygen and Group 17 (halogen group) elements to form compounds such as sulfur dioxide (SO_2), sulfur dichloride (SCl_2), and sulfur dibromide (SBr_2).

Group 17: Halogens
ELEMENT GROUP OF THE PERIODIC TABLE

The **halogens** are probably the most chemically uniform group of elements other than the noble gases. They form salts—the name halogen means "salt-producing"—when they bond with reactive metals. They are so reactive that they are difficult to obtain and keep in their elemental forms. The first five halogens are nonmetals and the last, tennessine, discovered in 2010, is so new that we know little about it. The halogens have seven valence electrons, which means that their electron configuration is very close to that of a noble gas. But the fact that they have seven and not eight valence electrons makes them very reactive as they look to gain one more electron. When they gain an additional electron, they become a 1– anion.

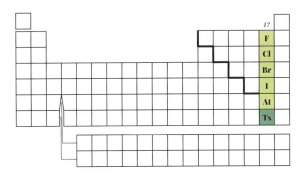

Fluorine

Chlorine

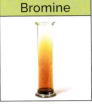
Bromine

Iodine

85
Astatine
At
(210)

117
Tennessine
Ts
(294)

PHYSICAL PROPERTIES

deposition of solid iodine

Halogens show a definite trend in their physical properties. As their atomic numbers increase, their densities, melting points, and boiling points increase, and their colors exhibit increasingly darker hues. Fluorine is a pale yellow gas with a low density, chlorine is a denser, greenish-yellow gas, bromine is a deep, reddish-brown liquid, and iodine is a grayish-black, crystalline solid.

CHEMICAL PROPERTIES

lead iodide precipitate

Halogens are particularly adept at taking electrons because they have high electronegativities, making them form compounds easily. Fluorine's electronegativity is greater than that of any other element, making it very difficult to separate from other elements. Identified in 1886, it was one of the last halogens to be discovered. Fluorine is extremely reactive and ignites when exposed to many substances, including water. The resulting chemical changes produced by reactions with halogens often take the form of spectacular releases of heat and light. When fluorine combines with metals, it often forms a protective layer of metallic fluoride that prevents all the metal from reacting.

As might be expected, the chemical properties of the other halogens are similar to those of fluorine, though less reactive. Fluorine, chlorine, bromine, and iodine are poisonous to humans in any but trace quantities.

Periodic Table and Elements **145**

Group 18: Noble Gases
ELEMENT GROUP OF THE PERIODIC TABLE

The Group 18 elements are called **noble gases** because they do not react with other elements except under unusual conditions of pressure and temperature. This low reactivity is a result of their electron configuration with eight valence electrons (two in the case of helium). Lord Rayleigh and Sir William Ramsay discovered one of these elements by carefully measuring the constituents of air. They noticed that nitrogen isolated from the atmosphere had more mass than nitrogen separated from ammonia. Suspecting that the atmospheric nitrogen contained some unknown substance, they separated the residual gas and found that it would not react with other elements. Because of its chemical sluggishness, this gas was called argon, meaning "the lazy one."

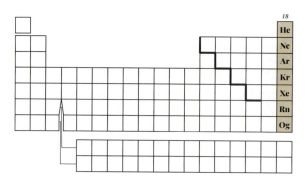

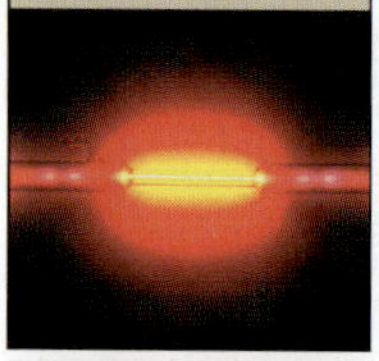

PHYSICAL PROPERTIES

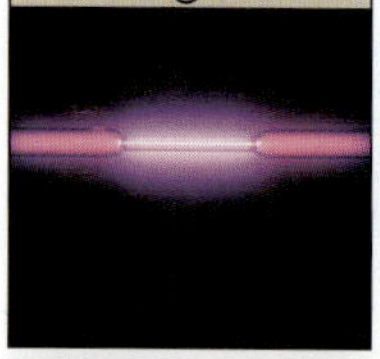

All noble gases are colorless, odorless, and tasteless. Extremely low boiling points and freezing points indicate that the individual atoms of these gases have little attraction for each other. Though highly radioactive, the newest noble gas to be discovered, oganesson, is expected to have similar physical properties.

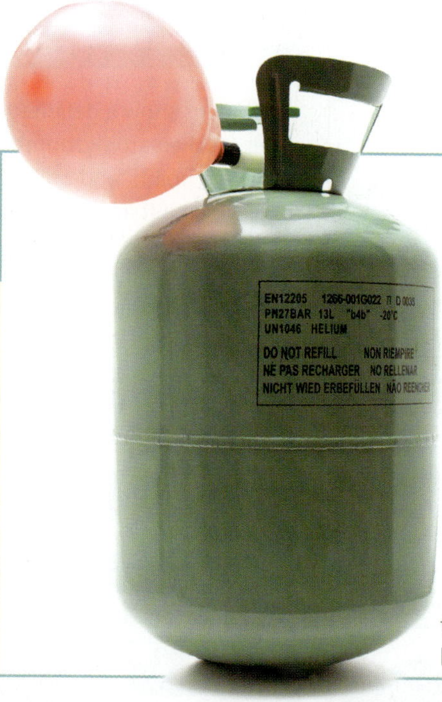

tank of helium

86 Radon
Rn
(222)

118 Oganesson
Og
(294)

CHEMICAL PROPERTIES

As their name implies, noble gases tend to avoid combining with other elements. After many attempts to force noble gases to show some chemical reactivity, scientists were finally able to force all the noble gases except neon and argon to combine with other elements, including other noble gases. The results: several short-lived, unstable compounds.

CAREERS

SERVING AS A CHEMICAL ENGINEER: *CHEMISTRY EVERYWHERE*

Almost everything you have touched today has been impacted by chemical engineers. As with all engineers, chemical engineers work to solve problems, specifically problems with chemicals. This career is directly involved with the command from Scripture to love others.

Chemical engineers must be well versed in the field of chemistry, but as engineers, they have to take an interdisciplinary approach to solving problems. Chemical engineering problems are solved by incorporating physics, biology, and mathematics with the principles of chemistry. Chemical engineers are involved in the production, testing, and transportation of just about everything that is manufactured and in safely getting those products to our homes. Chemical engineers work in food production, water treatment, electronics production, and textiles

industries. Some, like Yuki Abe (inset), have specialized in the field of biochemical engineering. Dr. Abe helps clients find the cure for cancer and rare diseases.

If you like the challenge of solving problems and also like to work in many fields of math and science, chemical engineering may be for you. You can serve others by solving problems that impact people's lives every day.

6.3 SECTION REVIEW

1. Identify the group name for each of the following groups.
   a. Group 1
   b. Group 3–12
   c. lanthanoid series
   d. Group 15
   e. Group 17
   f. Group 18

2. What characteristics set apart the alkali metals from the other metals?

3. Why were pure metals like gold, silver, and copper known in Old Testament times, but metals like sodium, aluminum, and potassium were not discovered until relatively recently?

4. What is one major chemical property of all the noble gases that makes them useful? Why does that property make them useful?

5. Identify each of the following elements by its group or series. A few elements are classified in more than one group. (For example, an element is always in a group but it may also be a post-transition metal or metalloid. Include both groups for these elements.)
   a. antimony
   b. tungsten
   c. argon
   d. iron
   e. bromine

6. Choose three of the periodic groups and describe their physical and chemical properties.

06 CHAPTER REVIEW

Chapter Summary

TERMS
periodic table
periodic law
transuranium element
group
period
metal
nonmetal
metalloid

6.1 THE PERIODIC TABLE

- Humans have attempted to classify matter into elemental materials since before the time of Christ.

- Russian chemist Dmitri Mendeleev developed the first ordered table of elements. He established the original periodic law when he ordered the elements by atomic mass.

- English physicist Henry Moseley developed the modern periodic law, which states that the properties of the elements vary periodically with their atomic numbers.

- The arrangement of the periodic table follows the structure of atoms and therefore predicts elemental properties.

- The periodic table is divided into regions by the elements' properties. The three types of elements are the metals, the nonmetals, and the metalloids.

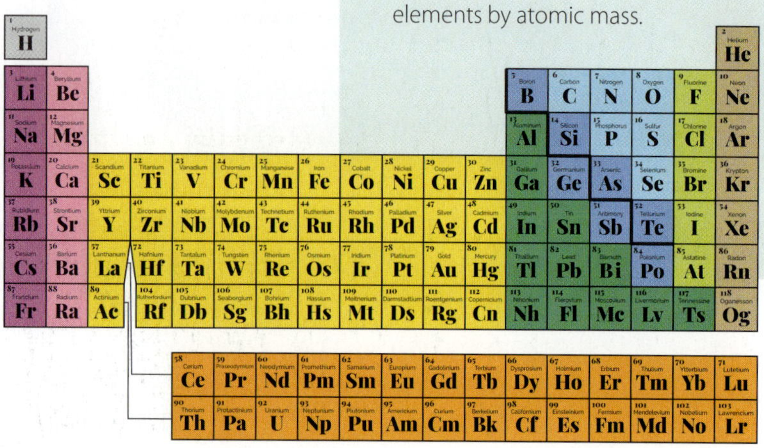

6.2 PERIODIC TRENDS

- We can use the periodic table to identify periodic trends in properties of elements. Those trends are attributable to the relative size of the nuclear charge compared with the size and arrangement of the electrons.

- Atomic radius and ionic radius generally decrease across a period and increases down a group.

- Electron affinity, first ionization energy, and electronegativity generally increase across a period and decrease down a group.

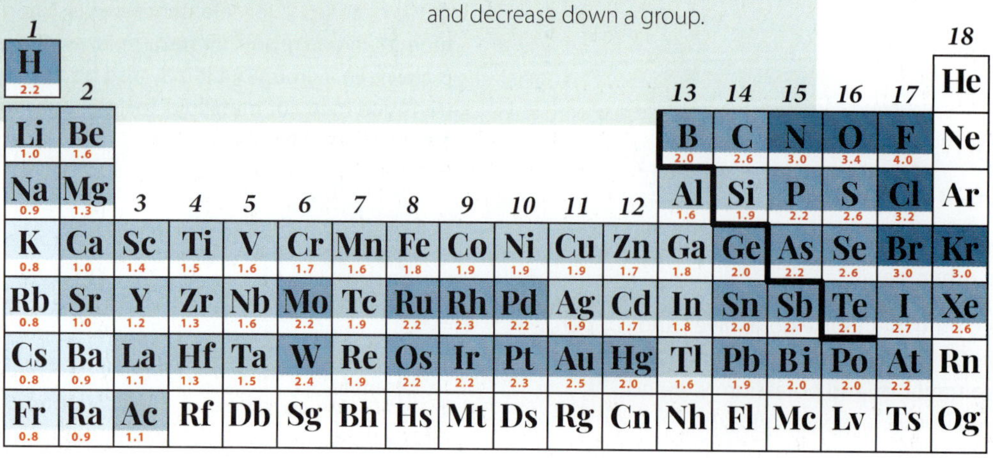

TERMS
atomic radius
ionization energy
electron affinity
electronegativity

TERMS
alkali metal
alkaline-earth metal
transition metal
inner transition metal
lanthanoid series
actinoid series
oxide
sulfide
halogen
noble gas

6.3 ELEMENTS BY THEIR GROUPS

- Hydrogen is often set apart on a periodic table because of its unique properties. It is the lightest and most plentiful element in the universe.

- The most "metallic" metals are on the left side of the periodic table. Metallic character is defined by how loosely an atom holds its valence electrons. Metals typically have a shiny metallic luster and they conduct heat and electricity well.

- The transition metals include the majority of the important structural metals and their alloys.

- The nonmetals are characterized by being gases, liquids, or crumbly solids at room temperature. They typically are poor conductors of heat and electricity.

- The metalloids exhibit intermediate properties of luster, rigidity, conductivity, color, and other features.

- The halogens are the most chemically reactive nonmetals because they require only one electron to attain a noble gas configuration.

- The noble gases are essentially inert. They have full outer electron shells, so they are very stable.

Chapter Review Questions

RECALLING FACTS

1. Why were Döbereiner's triads replaced?
2. What were the transition metals of Mendeleev's periodic table?
3. What convinced the scientific community that Mendeleev's periodic table worked? How does this illustrate one aspect of a good scientific model?
4. Why were several elements in odd places in Mendeleev's table? How was the problem corrected?
5. Give two names for horizontal rows in the periodic table.
6. What are groups?
7. (True or False) The shape and arrangement of the periodic table directly reflect the electron structure of the atoms composing the elements.
8. Identify the scientist who
   a. played the lead role in developing the structure of the modern periodic table.
   b. formulated the concept of triads.
   c. proposed that elemental properties vary in octaves.
   d. predicted the existence of several missing elements at the time of his work.
   e. ordered the periodic table by atomic number.
   f. devised the commonly used electronegativity scale.
9. How do the sizes of cations and anions compare with their corresponding neutral atoms?

Periodic Table and Elements **149**

06 CHAPTER REVIEW

10. Identify the term that matches each definition.
    a. the energy change associated with the formation of an anion
    b. a measure of an atom's size
    c. the energy required to remove an electron
    d. a measure of an atom's ability to attract or hold valence electrons in a molecule
11. Describe the trends in ionization energy.
12. Describe the trends in electronegativity.
13. Why are elements in groups?
14. In which group do we find elements that we traditionally think of as metals?
15. What is unusual about most of the inner transition metals in the actinoid series?
16. Which Group 14 element is prominent in chemistry as well as in other branches of science? Explain.
17. Which element
    a. is the most abundant in the earth's atmosphere by volume?
    b. is the most abundant in the earth's crust by mass?
18. Identify the group number for each element group.
    a. halogens
    b. boron group
    c. noble gases
    d. pnictogens
    e. chalcogens
    f. alkaline-earth metals
    g. transition metals
    h. alkali metals

UNDERSTANDING CONCEPTS

19. Summarize the development of the periodic table.
20. Describe two benefits of the periodic table.
21. What information can be found on the periodic table?
22. Name the element that appears at each of the following addresses in the periodic table.
    a. Period 5, Group 13
    b. Period 4, Group 2
    c. Period 2, Group 16
23. In general, how do the electron configurations of the elements in a row in the periodic table compare?
24. Why is the periodic table arranged as it is?
25. You are working on a project that needs an element that is a solid with high density and conductivity but that is also relatively nonreactive. Where on the periodic table would you look? Explain.
26. Explain why atomic radius changes as it does along rows and columns of the periodic table.
27. Why does electron affinity change as it does along rows of the periodic table?
28. Of the elements in Group 13, predict which has (a) the largest atomic radius, (b) the largest electronegativity, and (c) the smallest ionization energy. Explain your predictions.
29. If an element's first ionization energy is large, what can you predict about its atomic radius, electron affinity, and electronegativity?
30. Why are cations smaller than the neutral atoms of the same element?
31. Which of the stable alkaline-earth metals (Be, Mg, Ca, Sr, and Ba) fit in each of the following descriptions?
    a. has the largest atomic radius
    b. has the largest ionization energy
    c. has the largest ionic radius
    d. has the smallest electronegativity
32. Do you think that electronegativity is the main property that determines how strongly an isolated neutral atom holds on to its valence electrons? Defend your answer.
33. Periodic trends are models of the properties of atoms of different elements. How well do the trends in atomic radii predict the radii of atoms? What does this say about the workability of this model?
34. Explain why hydrogen is often considered a group by itself.
35. Of the elements Al, Au, Br, C, Ca, Cl, Cs, F, Fe, H, He, Hg, K, Mg, Na, O, Si, and W, which fit in each of the following descriptions?
    a. the principal metal found in human bones
    b. can exist as diamond, graphite, or charcoal
    c. gaseous at room temperature
    d. liquid at room temperature
    e. soft, highly reactive metal
    f. a constituent of bauxite
    g. most electronegative element
    h. a gas with a similar electron configuration as alkali metals

36. Identify each of the following elements by its group and by its period or series.
    a. cesium
    b. cerium
    c. uranium
    d. calcium
    e. lithium

CRITICAL THINKING

37. If element number 119 is discovered, where would it be located on the periodic table? What would you predict about its atomic structure?

38. If element number 119 is discovered, which group of elements would it be in? What would you predict about its physical and chemical properties?

39. Why do you think chlorine was isolated before fluorine?

Use the ethics box below to answer Question 40.

40. Use the ethical decision-making process outlined in Chapter 3 (pages 74–75) to explain how a Christian should respond to balancing the need for rare materials with the potential harm to the environment and people.

ethics

RARE-EARTH ELEMENTS AND RISKS

As you have learned in this chapter, there are many different elements with a variety of properties that are useful in many different ways. Some of these elements are abundant and easily obtainable while others are not. Many elements in the lanthanoid series are useful yet difficult to obtain in sufficient amounts. Scientists have developed a number of processes that could provide us with adequate supplies. But other scientists are concerned that the processes could expose people and the environment to hazardous levels of chemical compounds and even radiation.

Chapter 7

CHEMICAL BONDS

A world without plastic—is such a place even possible? Think about it—plastic is literally everywhere. In one form or another it's in your personal electronics, your family's car and indoor plumbing, the beverage cup you used at the restaurant last night, the toothbrush with which you brushed afterward, and an innumerable host of other products that people use every day. It's hard to believe that plastic has been around only since 1907 when the first synthetic plastic—Bakelite—was invented by Leo Baekeland in Yonkers, New York.

There's little argument that plastic has made our lives easier. It has many advantages over some of the more traditional materials that it has largely displaced, such as wood, glass, or ceramics. But plastic has a darker side too. The need for plastics drives some of the demand for fossil fuels from which many plastics are made. Most plastic degrades slowly and ends up in landfills, where it can persist for centuries. Plastic waste even winds up in food chains, gradually accumulating in the guts of birds and marine life. As Earth's population continues to grow, questions will need to be answered about how best to deal with plastic waste.

7.1 Bonding Basics *153*
7.2 Types of Bonds *158*
7.3 Properties of Compounds *167*

7.1 BONDING BASICS

Chemical Bonds and the Octet Rule

The plastics you read about in the Chapter opener are a group of chemical compounds classified as *polymers*—large molecules made of regularly repeated smaller subunits. The existence of polymers and other kinds of compounds is due to a tendency that most atoms have to form chemical bonds with other atoms. **Chemical bonds** are durable electrostatic attractions that form between atoms. Through this mechanism, God made it possible for the relatively small variety of elements to combine in seemingly endless ways, forming all of the various compounds of which both we and our world are made. In this section, we'll look at how and why atoms form chemical bonds.

You may recall that the second law of thermodynamics implies that natural systems tend toward a state of minimum energy. We can see this in everyday events. For example, balls roll downhill, leaves hang limp after the wind has calmed, and hot lava spontaneously cools. Similarly, an atom naturally minimizes how much energy it has. In Chapter 5, you learned that electrons can exist only at certain quanta of energy, and they are most stable in the lowest, or ground state, energy level. Under certain conditions, an atom becomes more compact because the average distance of the electrons from the nucleus decreases. With a smaller average distance, the electrons' potential energy relative to the nucleus is lowered, much like the gravitational potential energy of a person on a ladder is lowest when he is on the bottom rung closest to the ground. You also learned that when the entire outer *s* and *p* sublevels are full, atoms are smallest and thus the most stable. Noble gases are good examples of atoms with stable electron configurations.

Because of these inherent properties, atoms naturally link with other atoms to form chemical bonds. These bonds rearrange electrons to maximize stability. Bonding usually releases energy as the atoms' electron configurations attain lower energy states. This energy may be released as heat, light, or some other form of energy.

> A simplified representation of the consequences of bonding. Often the electron structures of bonded atoms are more stable and are at a lower energy state than those of unbonded atoms.

QUESTIONS
» Why do atoms bond?
» Are there different kinds of bonds?
» What makes a bond polar?
» Can I predict what kind of bond two atoms will make?

TERMS
chemical bond • octet rule • covalent bond • ionic bond • metallic bond • polarity • polar covalent bond

How do different atoms form bonds?

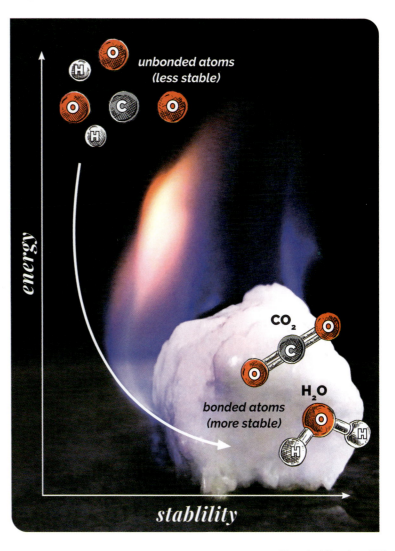

Chemical Bonds 153

The ease with which an atom can gain or lose electrons determines its chemical reactivity. If atoms can lose, gain, or share enough electrons to attain a noble-gas electron configuration by bonding, they will be more stable. In general, a full valence shell contains eight electrons. Thus, scientists have observed that by attaining a full valence shell, atoms maximize stability, a principle called the **octet rule**.

Octet rule
8 valence electrons for each atom

double bond structure

O=C=O

single bond structure

Octet rule
8 valence electrons for carbon atom, 2 for each hydrogen atom (an exception to the rule)

Note that the octet rule does not apply to transition and inner transition metals or to the first four or five elements in the periodic table. In the case of hydrogen, lithium, and beryllium, the nearest noble-gas configuration has only two electrons. For transition and inner transition metals, not only valence electrons but also high-energy electrons from deeper d and f sublevels may be removed or shared during bonding. Thus, a full valence octet in a transition metal cation may not affect the reactivity of these metals.

Many elements give up electrons to attain a full valence shell. Alkali and alkaline-earth metals tend to lose their one or two outer s sublevel electrons when they bond with other elements. Losing those electrons exposes the underlying noble-gas electron configuration from the previous period.

Elements rearrange the smallest number of electrons possible in bonding processes. Atoms with fewer than four valence electrons tend to lose electrons relatively easily, resulting in correspondingly low ionization energies. Atoms with more than four valence electrons tend to gain electrons, resulting in relatively high electron affinities.

EXAMPLE 7-1: ACQUIRING AN OCTET OF VALENCE ELECTRONS

State two ways in which (a) sodium atoms and (b) oxygen atoms could attain eight valence electrons. Which process is more likely for each element? Why?

Solution

a. Sodium, a Group 1 element with one valence electron, could attain eight valence electrons by gaining seven electrons. Or it could lose the one valence electron in its third energy level, leaving the second energy level with its eight electrons as the valence shell. It will likely lose the one electron since that electron is loosely held. It would be nearly impossible to force seven extra electrons into the atom because it has a low affinity for additional electrons.

b. Oxygen, a Group 16 element with six valence electrons, could gain two electrons to fill its valence shell, or it could lose the six valence electrons, uncovering the full shell of the first energy level. It is more likely to gain two electrons because it has a high electron affinity and a high first ionization energy, which would make removing electrons difficult.

TYPES OF CHEMICAL BONDS

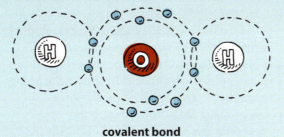

covalent bond
hydrogen + oxygen = water

Bond type is affected by how atoms attain a stable electron configuration. If two atoms both have high electronegativity values, neither can completely remove electrons from the other. But they can share electrons to complete an octet. Bonds formed by sharing electrons are called **covalent bonds**. These most often occur between nonmetals.

When atoms with very different electronegativity values combine, the atom with higher electronegativity takes electrons from the other atom. This electron transfer forms a cation and an anion. These ions are held together through attraction of opposite charges in an **ionic bond**. Typically, ionic bonds involve a metal bonded to a nonmetal.

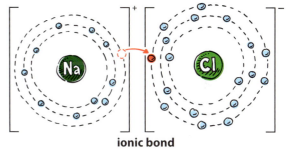

ionic bond
sodium + chlorine = sodium chloride

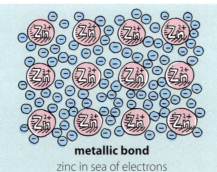

metallic bond
zinc in sea of electrons

Atoms with low electronegativity values usually have only a few loosely held valence electrons. These atoms may bond by sharing their easily lost electrons among many atoms. These electrons are mobile and are no longer associated with any specific nucleus. Metal atoms typically interact this way to form **metallic bonds** when in the solid state.

Polarity and Bond Character

Covalent and ionic bonds are affected by a property called **polarity**, the tendency of an object to form two localized regions of opposite character or charge. For example, the earth has a north magnetic pole and a south magnetic pole; it has magnetic polarity. A battery has two electrical poles where positive and negative charges collect; it is electrically polarized. Similarly, a bond can be described as polar if electrons are unequally shared between two atoms.

In the simplest case, let's look at diatomic hydrogen, shown at right. Each hydrogen atom has only a single valence electron. If the two atoms are brought close together, each will attract the other's electron to fill its valence energy level and achieve a more stable electron configuration. Since neither can take the electron from the other, they share the electrons equally. The electrons spend most of their time between the nuclei of the hydrogen atoms. No poles form because the hydrogen atoms are identical. A diatomic hydrogen molecule is held together by a completely nonpolar covalent bond in which the positive nuclei are attracted to the negative central region containing the bonding pair of electrons.

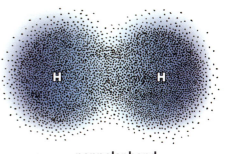

nonpolar bond
diatomic hydrogen

Chemical Bonds

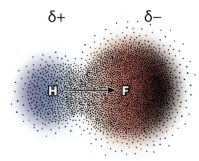

polar covalent bond

hydrogen (δ+ charge)
and fluorine (δ− charge)

Fluorine atoms each have three pairs of electrons and a single unpaired electron in their valence shells. They have identical, very strong electronegativities. Each needs just one more electron to have a full octet. As in hydrogen, electrons are shared equally, and, just as with hydrogen, the bond is nonpolar.

Consider what happens if a hydrogen atom and a fluorine atom bond, shown at left. The hydrogen has a single valence electron and the fluorine atom needs only a single electron to complete its valence octet. The fluorine atom has a greater electronegativity than the hydrogen atom. When they bond, the electron is pulled much closer to the fluorine and spends a greater amount of time around the fluorine nucleus than around the hydrogen nucleus. This unequal sharing of electrons between atoms of two different elements is called a **polar covalent bond**. Because of this unequal sharing of electrons, an uneven distribution of electric charge on the molecule as a whole is produced—on average, the fluorine end of the molecule will be negatively charged and the hydrogen end will be positively charged. These opposite partial charges can be indicated in illustrations by the characters δ− and δ+. Thus, both the hydrogen-fluorine bond and the resulting molecule are electrically polar. But as you'll see in Chapter 8, not all molecules that contain polar bonds are themselves polar.

Linus Pauling noticed that covalent bonds between different elements (e.g., hydrogen and fluorine) require more energy to break than the bonds between atoms of either pure element alone (e.g., hydrogen and hydrogen or fluorine and fluorine). Pauling explained these stronger bonds as an effect of polarity. Polar bonds are stronger because the opposite charges of the polarized atoms hold the atoms together. This effect was quantified by assigning electronegativity values to each element, depending on its ability to attract electrons to itself in a chemical bond (see page 131). Bonds range from nonpolar to highly polar depending on the difference between the atoms' electronegativities.

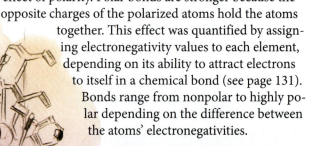
Linus Pauling

What happens if we attempt to bond two atoms with very different electronegativities?

Let's combine rubidium (electronegativity value of 0.8), an active metal, with fluorine (4.0), an active nonmetal. For a rubidium and fluorine bond, the electronegativity difference is 3.2.

The "shared" electrons spend nearly all their time around the fluorine atom and almost no time around the rubidium atom—the bond is essentially ionic in nature. But the electron transfer is not 100% complete because the electron still spends a small amount of time around the rubidium nucleus.

But the fluorine has essentially captured the valence electron from the rubidium atom, resulting in a positively charged rubidium cation and a negatively charged fluorine anion.

HOW IT WORKS

Recycled Plastic

One way to deal with plastic waste, of course, is to recycle it. These days, bins for collecting plastic to be recycled are everywhere—it's possible that your family may even have one at home. But what happens to plastic that's been targeted for recycling? One type of plastic that is commonly recycled is polyethylene terephthalate, or PET—the kind of plastic that is typically used to make plastic water bottles. Once the empty bottles have been separated from other kinds of plastic, they are sorted by color and then crushed, washed, dried, and shredded. The shredded particles, known as *flakes*, are processed to remove any remaining bits of paper and other residue. Afterward, the flakes can be spun into thread or yarn. These fibers are then used to make a variety of products, from jackets to shoes and hats.

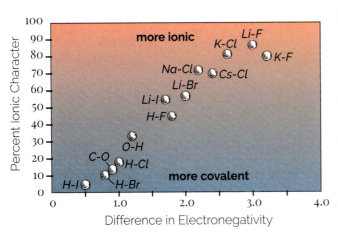

It is tempting to think of bonds as falling into one of the three categories so far described—either completely covalent, completely ionic, or completely metallic—as if each group has very distinct boundaries. But classifying bonds is not that straightforward. In reality, completely covalent bonds and highly ionic bonds exist at the opposite ends of a spectrum of bond polarities (see left). Bonds between *different* elements can never be 100% covalent, and as differences in electronegativities between bonding elements become greater, the bonds they form become increasingly ionic. But it is also impossible to have a completely ionic bond resulting from electron transfer because that would imply that the donor atom had no affinity at all for any donated electrons. A more accurate way to describe bonds is to designate them as either predominantly covalent or predominantly ionic. In Section 7.2 we will look more closely at the different types of chemical bonds.

7.1 SECTION REVIEW

1. Define *chemical bond*.
2. Generally, why do chemical bonds form between atoms?
3. Use the periodic table to determine the type of bond that would form between (a) cesium and sulfur, (b) chlorine and bromine, (c) magnesium and fluorine, and (d) silver and mercury.
4. State the octet rule. To what elements does it apply?
5. What are the three main bond types that can form between atoms?
6. Define *polarity*.
7. What chiefly determines the polarity of a bond?
8. (True or False) A bond between two atoms can be 100% covalent, but it can never be 100% ionic.

7.2 TYPES OF BONDS

QUESTIONS
» How do the various kinds of bonds form?
» How can I model bonds?

TERMS
Lewis structure • double bond • triple bond • formula unit • crystal lattice • polyatomic ion • electron sea model • delocalized electron

Covalent Bonding

Many familiar compounds consist of covalently bonded elements. Those elements have similar electronegativity values and nearly full valence electron shells. They are usually nonmetals and are often close to each other on the periodic table. When they bond, they each maintain influence over their valence electrons. To acquire a stable noble gas electron configuration, atoms share electrons to fill vacancies in their valence orbitals.

Diatomic oxygen (O_2) is an example of a completely covalent molecule. An isolated neutral oxygen atom has six valence electrons. Each oxygen atom needs two more electrons to attain an octet. Since both atoms in the bond have the same electronegativity, neither can completely pull the required electrons from the other to itself. Instead, both atoms share their unpaired electrons, filling their $2p$ orbitals and thus completing their valence shells. The two shared electrons that make up each pair are called a *bonding pair* in contrast to the other valence electron pairs that do not participate in bonding. These are called *lone pairs*, or *nonbonding pairs*, of electrons.

Electrons in bonding pairs have the highest probability of being found between the nuclei of the bonded atoms. There they form a dense region of negative charge. This region attracts the positive nuclei, holding the atoms together. This attraction is called *electrostatic force*—the force exerted between electrical charges. Opposite charges attract and like charges repel, according to the law of charges.

In the case of our two oxygen atoms, the positive nucleus of each atom is attracted to the negative electrons of the other atom, reducing its energy and increasing stability. But as the atoms draw closer, they reach a point at which the repulsion between their positively charged nuclei balances the attraction to the electron cloud.

What determines the bond type that forms between two atoms?

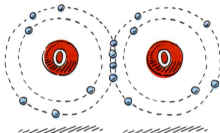
diatomic oxygen

Diatomic Elements

The atoms of hydrogen, nitrogen, oxygen, and the halogens (fluorine, chlorine, bromine, and iodine) are not stable by themselves. To exist in a pure form, they must occur as diatomic molecules. Their formulas are H_2, N_2, O_2, F_2, Cl_2, Br_2, and I_2.

Memorize this list as the "Hydrogen Seven" for future reference. On a periodic table you'll note that the cells for the six elements, in addition to hydrogen, form the shape of a number seven that starts at nitrogen (element number 7).

LEWIS STRUCTURES

American chemist Gilbert Lewis developed a useful notation that illustrates how atoms form molecules. **Lewis structures** are two-dimensional diagrams that show the bonds between different atoms.

Covalent bonds are symbolically represented in Lewis structures in several ways. Covalent bonds involve one or more pairs of electrons shared between two atoms. Each shared electron pair contributes to the valence-electron structure of *both* atoms involved in the bond.

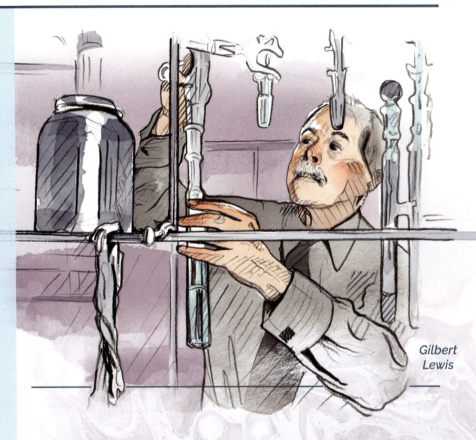

Gilbert Lewis

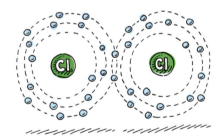

When illustrating the formation of a covalent bond, such as in diatomic chlorine, begin with the electron dot notation for each atom. Position the unpaired electrons between the bonding atoms. After bonding, the pair of dots lies between the bonded elements. To make this into a Lewis structure, replace each bonding pair of electrons with a dash to simplify the diagram. Each dash thus represents two electrons. The nonbonding electron pairs are retained unchanged.

$$:\!\ddot{\text{Cl}}\!\cdot \; + \; \cdot\!\ddot{\text{Cl}}\!: \; \rightarrow \; :\!\ddot{\text{Cl}}\!:\!\ddot{\text{Cl}}\!: \; \text{or} \; :\!\ddot{\text{Cl}}\!-\!\ddot{\text{Cl}}\!:$$

As we noted in the previous subsection, hydrogen can also form a diatomic molecule. But rather than completing an octet in this instance, hydrogen atoms share their single electrons to fill their 1s sublevel, which is their only shell.

$$\text{H}\cdot \; + \; \cdot\text{H} \; \rightarrow \; \text{H}\!:\!\text{H} \; \text{or} \; \text{H}-\text{H}$$

With each atom having filled its valence shell in this manner, they act much like chlorine or other Group 17 elements that are bonded together as diatomic molecules.

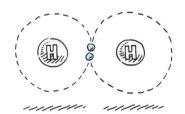

LEWIS STRUCTURES

Covalent bonds form not only between diatomic nonmetals but also between two different nonmetals. Probably the most familiar covalent compound is water, a compound of hydrogen and oxygen.

$$H\cdot + H\cdot + \cdot\ddot{\underset{\cdot\cdot}{O}}: \rightarrow H:\ddot{\underset{\cdot\cdot}{O}}: \quad \text{or} \quad H-\ddot{\underset{\cdot\cdot}{O}}:$$
$$\phantom{H\cdot + H\cdot + \cdot\ddot{O}: \rightarrow}\;H\phantom{:\ddot{O}:} \quad\quad \phantom{H-\ddot{O}:}H$$

A lone oxygen atom has two unpaired electrons in its valence p sublevel. It needs two electrons to complete its valence octet. Hydrogen atoms need only one electron to complete their valence shells. An oxygen atom and two hydrogen atoms can attain complete valence shells by sharing electrons. The central oxygen atom forms covalent bonds with each of the two hydrogen atoms.

Sometimes sharing one electron pair is not enough to fill an atom's valence octet. As you just read, oxygen must share two electron pairs to attain a noble-gas electron configuration. It can accomplish this by bonding with two other atoms, as in water. But it can also share two unpaired electrons with another oxygen atom to form a **double bond**, that is, a covalent bond consisting of two shared electron pairs.

$$:\ddot{O}\cdot + \cdot\ddot{O}: \rightarrow :\ddot{O}::\ddot{O}: \quad \text{or} \quad :\ddot{O}=\ddot{O}:$$

Triple bonds form in similar fashion when atoms share three pairs of electrons. Diatomic nitrogen molecules in the atmosphere consist of two nitrogen atoms bonded by a triple covalent bond.

$$:N\cdot + \cdot N: \rightarrow :N\!:\!\!:\!\!:N: \quad \text{or} \quad :N\equiv N:$$

Many organic compounds contain carbon atoms that form a triple bond with other atoms. One example is acetylene (ethyne), used for welding torches.

$$H:C\!:\!\!:\!\!:C:H \quad \text{or} \quad H-C\equiv C-H$$

The effect of having different numbers of shared electron pairs does more than just create different Lewis structures.

For example, two carbon atoms can form either a single, double, or triple bond between them, resulting in three different chemical compounds—ethane (C_2H_6), ethene (C_2H_4), and ethyne (C_2H_2). These three compounds have markedly different physical and chemical properties. Ethane's melting point, for example, is −182.8 °C, while ethyne's melting point is over 100 degrees higher at −80.8 °C.

The difference in properties is a result of each molecule's different internal structure. Bond strength, for example, increases from single bonds (weakest) to triple bonds (strongest). The length of these bonds also varies, with single bonds being the longest of the three bonds and triple bonds the shortest.

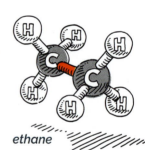
ethane

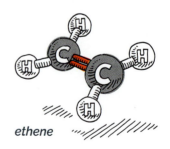
ethene

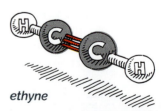
ethyne

Let's look at another example of how Lewis structures are written.

We'll construct the Lewis structure for formaldehyde (CH_2O), a chemical used as a preservative. You will see that Lewis structures reveal something about the shape of a molecule. And a molecule's shape affects its chemical properties.

Steps for Drawing Lewis Structures

1 Determine how many valence electrons are available by multiplying the number of atoms of each element by its number of valence electrons.

H: 2 × 1 e^- = 2 e^-

C: 1 × 4 e^- = 4 e^-

O: 1 × 6 e^- = 6 e^-

Total: 12 e^-

2 Determine the number of electrons needed to give each atom a full valence shell by multiplying the number of atoms of each element by the number of electrons in its full valence shell.

H: 2 × 2 e^- = 4 e^-

C: 1 × 8 e^- = 8 e^-

O: 1 × 8 e^- = 8 e^-

Total: 20 e^-

3 Find the difference between the electrons needed and electrons available. Divide this number by two to determine the number of bonds needed.

20 e^- − 12 e^- = 8 e^- ÷ 2 = 4 bonds

4 a. If carbon is in the compound, place it in the center; otherwise place the least electronegative element in the center.

C

b. Place the remaining element symbols for the compound around the central atom.

H   C   O
    H

5 Create as many single bonds as possible between outer atoms and the central atom.

H – C – O
    |
    H

6 Form any double or triple bonds to account for the number of bonds needed.

H – C = O
    |
    H

7 Place unbonded pairs of electrons around each atom as needed to give them full valence shells.

H – C = Ö:
    |
    H

8 Check your structure by confirming that the molecule has only the total number of electrons allowed in Step 1 (remember that each bond represents two electrons). Then check that each atom has a full valence shell, recalling that shared electrons count toward filling the valence shell of each atom making the bond.

12 electrons used (Step 1)
all valence shells filled

A Lewis structure can give you a general idea of a molecule's shape, but it doesn't depict the actual three-dimensional structure. It simply indicates the number and kinds of bonds that form between valence electrons of atoms. Predicting the actual geometry of a molecule is covered in Chapter 8.

EXAMPLE 7-2: DRAWING LEWIS STRUCTURES

Draw the Lewis structures for the following compounds.

a. trichloromethane ($CHCl_3$)

b. silicon dioxide (SiO_2)

Solution

a. Determine how many valence electrons are available and needed.

Available	Needed
C: $1 \times 4\ e^- = 4\ e^-$	C: $1 \times 8\ e^- = 8\ e^-$
H: $1 \times 1\ e^- = 1\ e^-$	H: $1 \times 2\ e^- = 2\ e^-$
Cl: $3 \times 7\ e^- = 21\ e^-$	Cl: $3 \times 8\ e^- = 24\ e^-$
Total: $26\ e^-$	Total: $34\ e^-$

$$34\ e^- - 26\ e^- = 8\ e^- \div 2 = 4 \text{ bonds}$$

Carbon will be the central atom. Position the three chlorine atoms and one hydrogen atom around the central carbon. Sketch in the single bonds and unshared electron pairs.

$$\begin{array}{c} \text{Cl} \\ \text{H C Cl} \\ \text{Cl} \end{array} \rightarrow \begin{array}{c} \text{Cl} \\ \text{H}-\text{C}-\text{Cl} \\ \text{Cl} \end{array} \rightarrow \begin{array}{c} :\!\ddot{\text{Cl}}\!: \\ \text{H}-\text{C}-\ddot{\text{Cl}}\!: \\ :\!\ddot{\text{Cl}}\!: \end{array}$$

Verify that each carbon and chlorine atom has an octet of electrons and that the hydrogen has a pair. The hydrogen may be placed on any side of the carbon atom. Then verify that twenty-six valence electrons are used.

b. Determine how many valence electrons are available.

Available	Needed
Si: $1 \times 4\ e^- = 4\ e^-$	Si: $1 \times 8\ e^- = 8\ e^-$
O: $2 \times 6\ e^- = 12\ e^-$	O: $2 \times 8\ e^- = 16\ e^-$
Total: $16\ e^-$	Total: $24\ e^-$

$$24\ e^- - 16\ e^- = 8\ e^- \div 2 = 4 \text{ bonds}$$

Silicon will be the central atom because it is less electronegative than oxygen. Position the two oxygen atoms on opposite sides of the silicon since the regions of shared electrons will repel each other. Create the single bonds between the silicon atom and the oxygen atoms.

$$\text{O}-\text{Si}-\text{O}$$

Note that this arrangement results in the silicon atom now having six valence electrons, and each oxygen atom is still short one electron to make a full octet. By forming a double bond between silicon and each of the oxygen atoms, however, all atoms can acquire full valence octets.

$$\text{O}=\text{Si}=\text{O} \rightarrow :\!\ddot{\text{O}}=\text{Si}=\ddot{\text{O}}\!:$$

Verify that sixteen valence electrons are used.

Ionic Bonding

When sodium, a soft, shiny, reactive metal, is placed in an atmosphere of poisonous chlorine gas, a violent chemical reaction takes place. The result is sodium chloride, an edible, white, crystalline compound commonly known as table salt. The explosive release of heat and light indicates that the resulting compound contains much less energy than the sum of the energies in the original elements. What happened?

Sodium atoms (electronegativity value 0.9) in Group 1 have a single valence electron; they need to lose just one electron to form an octet. Their first ionization energy is very small. Chlorine (3.2) in Group 17 has a very high electronegativity, needing one electron to complete its stable valence octet. When these two elements react, chlorine atoms essentially "steal" the loosely held electrons from the sodium atoms (see right). Both the sodium cations and the chlorine anions now have stable valence octets, and the excess energy is released as heat and light. The electronegativity difference (3.2 − 0.9 = 2.3) between these elements is large, as is typical of ionic bonds.

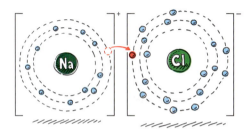

There are a few exceptions to the octet rule when it comes to ionic bonding. Lithium and chlorine atoms, for example, bond in a way that is similar to the way that sodium and chlorine atoms bond. But when lithium loses its outer electron, forming a 1+ lithium cation, it exposes its first energy level, which contains only two electrons. Even though this is not an octet, the level is full and lithium's stability increases.

The Structure of Ionic Compounds

Chemical compounds formed from metals and nonmetals are called *ionic compounds*. A sample of an ionic compound, such as a crystal of sodium chloride, consists of many bonded cations and anions held together by a strong electrostatic attraction. Ionic compounds are not made of distinct molecules like covalent compounds. Even though ionic compounds contain charged particles, the sum of all the ionic charges in a sample equals zero, so they are electrically neutral. The ratio of cations to anions needed to achieve this neutrality determines the chemical formula for the compound, called a **formula unit**. For example, let's use electron dot notation to examine the formation of table salt.

sodium chloride

Atoms before bonding:

Na• —1e⁻→ :Cl̈:

Ions after electron transfer:

[Na]⁺ [:C̈l̈:]⁻

Ratio: 1 Na⁺ : 1 Cl⁻, or NaCl

Notice that brackets are used to indicate ions, with the charge shown as a superscript outside the right-hand bracket.

From this example, we can see that the formula unit for table salt is NaCl. Although there are no molecules in an ionic compound, it is often convenient to work with its formula unit. We do not call a formula unit a molecule because it does not exist as a distinct particle, like a group of covalently bonded atoms.

Chemical Bonds

Now let's look at a more complicated example. To form calcium fluoride, calcium atoms lose electrons to strongly electronegative fluorine atoms (electronegativity difference is 3.0). Calcium atoms need to lose two electrons to attain an octet, but fluorine atoms need to gain only one electron. Therefore, every calcium atom in the compound gives its electrons to two fluorine atoms. As with sodium chloride, this combination can involve many ions, but the ratio is always two fluorine anions to every calcium cation.

Atoms before bonding:

$$Ca{:} \xrightarrow{2e^-} \begin{array}{c} {:}\ddot{F}{:} \\ {:}\ddot{F}{:} \end{array}$$

Ions after electron transfer:

$$\left[Ca\right]^{2+} \begin{array}{c} \left[{:}\ddot{F}{:}\right]^- \\ \left[{:}\ddot{F}{:}\right]^- \end{array}$$

Ratio: 1 Ca^{2+} : 2 F^-, or CaF_2

calcium fluoride

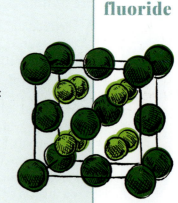

Ions in ionic compounds assemble themselves into a closely packed arrangement. The orderly, three-dimensional pattern that develops depends on the ratio and relative sizes of the anions and cations. The ions position themselves as close as possible to oppositely charged particles and as distant as possible from similarly charged particles. This orderly arrangement of ions is called a **crystal lattice**. We will discuss crystal lattices more in Chapter 13.

Polyatomic Ions

If certain elements with high electronegativity values, such as oxygen, are part of a polyatomic molecule, one or more of the missing valence electrons can sometimes be filled by nearby extra electrons instead of by sharing electrons with other atoms. In these cases, a covalently bonded anion forms, called a **polyatomic ion** (see below). Examining the Lewis structure of a polyatomic ion will show that every atom in the ion has an octet. The charge of the ion shows how many electrons were transferred to achieve those octets.

$$\text{atomic ion:} \quad {:}\ddot{Cl}{:} + 1e^- \longrightarrow \left[{:}\ddot{Cl}{:}\right]^- \text{ or } Cl^-$$

$$\text{polyatomic ion:} \quad {:}\ddot{O}{-}H + 1e^- \longrightarrow \left[{:}\ddot{O}{-}H\right]^- \text{ or } OH^-$$

Just as the chloride ion (Cl^-) has an extra electron, the hydroxide ion (OH^-) also has one more electron in addition to those supplied by the oxygen and hydrogen atoms.

Most polyatomic ions are anions, formed by acquiring additional electrons to make up their octets. In relatively rare cases, however, some form by losing electrons to acquire a positive charge, forming polyatomic cations. Some important examples of these cations include the hydronium (H_3O^+) and ammonium (NH_4^+) ions.

The Lewis structures of polyatomic ions can be drawn just like those of covalent compounds. The only difference is that when determining the electrons available, you will subtract the charge. Square brackets are placed around Lewis structures of the ions, and the charge is written as a superscript outside the brackets.

EXAMPLE 7-3: DRAWING POLYATOMIC ION LEWIS STRUCTURES

Draw the Lewis structure of the ammonium ion (NH_4^+).

Solution

Determine the electrons available. Note that the charge on the ion is 1+, so we will subtract (+1).

$$4 \times 1 + 1 \times 5 - (+1) = 8$$

Determine the electrons needed.

$$4 \times 2 + 1 \times 8 = 16$$

Determine the number of bonds.

$$16 - 8 = 8 \div 2 = 4$$

Assemble the four hydrogen atoms around the central nitrogen atom and form as many bonds as possible by sharing the eight available electrons with the central atom. Check that the octets and all eight available electrons are used.

$$\left[\begin{array}{c} H \\ | \\ H-N-H \\ | \\ H \end{array} \right]^+$$

Remember that the Lewis structure for an ion is placed in brackets and that the net charge of the ion is written as a superscript outside the right-hand bracket.

Polyatomic ions usually act as single, charged particles in chemical reactions and solutions, though they can be decomposed under some conditions. Their stable electron structures often allow them to survive chemical reactions to form compounds without splitting up. In Chapter 9 you will learn more about polyatomic ions and the compounds they form.

Metallic Bonding

The metallic bond forms entirely different kinds of materials. If they were limited to the octet rule, metal elements would have a problem forming bonds between themselves. Several properties of metals prevent them from easily losing or gaining electrons to acquire a noble-gas configuration. Most metal atoms need six or more electrons to fill their valence shells. Also, the transition and inner transition metals have gaping holes in their interior *d* and *f* sublevels, which must be filled before their valence shells can be filled. Since they have low electronegativity values, they cannot take electrons from other metal atoms. So what holds the atoms in a piece of metal together?

A different type of bond must form. The theory that explains how metal atoms bond together also explains why metals have such unique properties. This model is called the **electron-sea model**, or the *free-electron model*. According to this model, metals are an extensive crystal lattice of metal cations surrounded by and submerged in a "sea" of mobile electrons (see above right). Since metals have a low first ionization energy, their electrons are shared among adjacent metal atoms and lose their association with their parent atoms. Because of their mobility, these electrons are shared among all the atoms and are said to be *delocalized*. The negative charge of the **delocalized electrons** acts as a "glue" to hold the positively charged metal ions together.

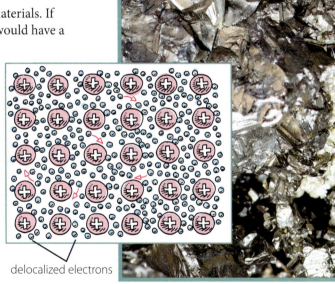
delocalized electrons

TABLE 7-1 *Bonds and Their Properties*

Bonding Elements	Valence Electrons	Type of Bond	Bond Polarity
nonmetal–nonmetal	tightly shared	covalent	low to medium
metal–nonmetal	transferred	ionic	high
metal–metal	widely shared	metallic	n/a

Chemical Bonds

Metallic bonds are possible because metals generally have low electronegativities and because the differences between their electronegativities are small. Metallic bonds do not occur when there are only a few atoms, though covalent and ionic bonds form with small numbers of atoms. The metallic bond can exist only for an arrangement of metal atoms in a crystal lattice.

EXAMPLE 7-4: PREDICTING BOND CHARACTER

Predict the predominant type of bond formed between atoms of each of the following pairs of elements:

a. nitrogen and oxygen
b. silver and copper
c. cesium and fluorine

Solution

a. Nitrogen and oxygen are both nonmetals with similar electronegativities. Their bond will be covalent.

b. Silver and copper are both metals, so the bond will be metallic.

c. Cesium is a metal and fluorine is a highly electronegative nonmetal. Because of the large difference between their electronegativities, the bond will be ionic.

7.2 SECTION REVIEW

1. Which kind(s) of element form(s) covalent bonds? Give two reasons for this.

2. What forces hold covalently bonded atoms together?

3. What number of valence electrons do most atoms (other than the first few elements, the transition metals, and inner transition metals) tend to have after forming bonds?

4. Which seven elements occur naturally as diatomic, covalent molecules?

5. Why do we not use the term *molecule* to refer to the product of an ionic bond?

6. What term describes the arrangement of particles in an ionic compound?

7. Predict the formula units of ionic compounds that form from the following atoms according to the number of valence electrons of each.

   a. Mg and Br
   b. Al and Cl
   c. Rb and I
   d. Sr and F
   e. Ca and O
   f. Al and O

8. State two ways that covalent polyatomic particles can become ions.

9. Draw the Lewis structures for each of the following.

   a. PO_4^{3-}
   b. $COBr_2$
   c. CN^-
   d. $SiHCl_3$

10. What kinds of elements normally participate in metallic bonding? State two factors that make metallic bonding possible.

11. Give one similarity and one difference between metals and solid ionic compounds.

Scanning electron micrograph of quartz

7.3 PROPERTIES OF COMPOUNDS

Properties of Covalent Compounds

Covalent compounds generally consist of distinct molecules. Various forces, called *intermolecular forces* (Chapter 13), attract one molecule to another, but these forces differ in strength from one compound to another. As the strength of intermolecular forces in covalent compounds increases, more energy is required to separate their molecules. At room temperature, covalent compounds exist as gases, liquids, or solids with relatively low melting points due to their generally weak intermolecular forces. Those that are solids commonly lack the density, hardness, and rigidity of metals and ionic substances. Covalent compounds may exist in a wide array of colors and lusters, and they generally are poor conductors of heat and electricity.

Some notable exceptions to these general characteristics are materials known as **network covalent substances**. Rather than forming individual molecules, the atoms in such substances are covalently bonded into a continuous three-dimensional network. The best example is diamond, the hardest natural substance known. A diamond consists of a three-dimensional crystalline array of carbon atoms, each bonded to four adjacent carbon atoms. Similarly, network covalent compounds called silicates form a wide variety of minerals in the earth's crust. Quartz is a common network covalent compound of silicon and oxygen. Network covalent materials often occur as relatively hard, brittle crystals with high melting points, glassy lusters, and unusual electrical properties.

> **QUESTIONS**
> » How do the properties of compounds differ according to bond type?
> » Why do some compounds conduct heat and electricity better than others?
> » How do intermolecular forces affect the melting point of a compound?
>
> **TERMS**
> network covalent substance • alloy

How do compounds with different bond types behave differently?

Properties of Ionic Compounds

Ionic bonds give the compounds that they form distinct properties. Since ions are held in place by strong electrostatic forces, the resulting compounds are dense, brittle, and hard solids. The melting points of many of these compounds are at least 800 °C, illustrating just how strong their bonds are. Their brittleness is also related to the forces between ions. If one row or layer of these ions is shifted even slightly, the misalignment creates repulsive forces, causing them to fracture. Because of their orderly structures, crystals of ionic compounds can usually be split, or cleaved, along a flat surface.

Solid ionic compounds are very poor conductors of electricity and heat since conductivity depends on the mobility of electric charges. In ionic compounds, all valence electrons are tightly bound, and the ions themselves that compose the crystal are fixed in position, making such compounds effective insulators. But if the compounds are dissolved or melted, the ions themselves are free to move, and so in dissolved or molten forms, ionic substances easily conduct electricity.

Properties of Metals and Alloys

The electron-sea theory explains many characteristics of metals and their **alloys**—mixtures consisting of a metal and one or more other elements. One such alloy is brass, a mixture of copper and zinc. In metals, delocalized electrons are free to move and carry electrical current and thermal energy. When a metal object is heated, its atoms vibrate faster. This energy is transferred to the delocalized electrons. They carry their energy from ion to ion, increasing atomic vibrations throughout the material. In this way, thermal energy moves quickly through a metallic object.

The luster, or shine, of metals is explained by the combination of the principles of quantum theory, spectroscopy, and free-electron theory. According to quantum theory, in any given metal the delocalized valence electrons move within a number of different orbitals. Metals in the *d* block on the periodic table have empty outer *p* sublevel orbitals and many empty orbitals in the *d* sublevel. All of this unused space allows electrons to move freely within a band of different energy sublevels of an individual atom or to nearby atoms. Therefore, delocalized electrons are able to absorb and instantly reemit nearly any wavelength of light shining on the metal as they jump between allowed valence energy levels and their ground state. When this factor is combined with the tendency of metallic crystalline structures to contain flat reflective surfaces, the result is the familiar shiny, metallic, and sometimes mirror-like luster of metals.

Delocalized electrons are also responsible for the malleability and ductility of most metals. Metal atoms are closely packed into vast three-dimensional crystalline structures held together by the "sea" of delocalized electrons. If a force such as a hammer blow displaces some of the atoms, they tend to move along planes within the metal firmly held in place by the "electron sea." Metals, unlike ionic compounds, do not cleave along flat surfaces. The shape of a metal object can be changed by hammering or stretching without significantly affecting its strength.

Brass is an alloy of copper and zinc.

A force can cause metal ions to move within the sea of delocalized electrons but still remain intact. This contributes to the malleability and ductility of metals.

Using Chemistry to Solve Problems

Plastics are covalently bonded compounds made of very long molecules. As you read in the Chapter opener, plastics have proved to be very useful, but they also pose a slew of difficult problems. One of the biggest questions of course is what to do with all the bottles, grocery bags, and food containers after we have finished using them. Many of these single-use items are simply thrown away and end up in landfills. Recycling might seem like the obvious answer, but at the time of this writing it is still less expensive in most cases for plastics producers to begin with new raw materials rather than to use recycled plastic.

Chemists aren't sitting idly on the sidelines as the war of public opinion over plastic is waged. Many chemists are using their knowledge of chemical bonding to engineer new kinds of plastics that remain practical and useful but pose less risk to both people and the environment. It's one more way in which chemistry is proving a powerful tool to help solve problems in our world.

WORLDVIEW INVESTIGATION

BIODEGRADABLE PLASTIC

One significant problem with conventional plastics is that they do not break down, or degrade, easily. This explains why plastic waste is so persistent in the environment. Even when conventional plastics do degrade, the products formed during decomposition are often themselves toxic. One possible solution to these problems that has been implemented to some extent and continues to be explored is *biodegradable plastic*.

Task

You are the assistant manager at a restaurant that provides single-use plastic sporks to customers, and part of your job is ordering supplies. Some customers have been requesting that the restaurant replace its regular plastic sporks with biodegradable sporks. Your task is to research the feasibility of the switch and make a recommendation to your manager.

Procedure

1. Research the current status of biodegradable plastics by doing an internet search using the keywords "biodegradable plastic."
2. As part of your presentation, include reasons for and against making the switch to biodegradable sporks, the pros and cons of each type of spork, and a cost-benefit analysis of the proposed switch.
3. Prepare a report of your findings and conclude it with a recommendation for your manager. Be sure to defend your recommendation on the basis of your findings. Remember to cite your sources.
4. Have a classmate review your article and give you feedback.
5. Complete your report and submit it.

Conclusion

Sometimes potential solutions to problems require extra costs to implement. The economic cost of such solutions is a necessary but often overlooked factor to consider when making decisions about how best to manage God's creation and its resources.

MINI LAB

PIE PAN PREDICTIONS

You know from everyday experience that some substances melt at lower temperatures than others. If you put a dab of butter in a cast iron skillet and heat it, the butter melts, but the skillet does not—which is a good thing! In this lab activity, you will examine whether there is a correlation between bond type and melting point.

Does bond type affect a substance's physical properties?

1. State a hypothesis regarding the order in which the following substances will melt: table salt (ionic compound), sugar (covalent compound), and paraffin wax (covalent compound).

Procedure

A Use a permanent marker to divide the bottom of the pie pan into three equal-sized wedges.

B Place a small sample of table salt in one of the wedges, sugar in another, and paraffin wax in the third.

C Place the pan on the hot plate. Turn on the hot plate and set it to HIGH. Observe the pan for five minutes, then turn the hot plate off. *Remember that the hot plate and pan will remain hot for some time.*

Conclusion

2. In what order did the three substances melt? Did this confirm your hypothesis?

3. On the basis of your results, what can you infer about the relationship between bond type and melting point?

Going Further

4. This activity actually tells you something about a third type of bond. What type of bond is it, where is it observed in the procedure, and what can you infer about that bond type's relationship to its melting point?

5. As you will learn in Chapter 8, molecules can be polar or nonpolar. A polar molecule, like a polar bond, has positive and negative regions, producing an electric force between molecules. On the basis of this information, which of the covalent substances—sugar or paraffin wax—would you predict is a polar molecule? Explain.

EQUIPMENT

- hot plate
- pie pan, 9 in. aluminum
- permanent marker
- table salt
- sugar
- paraffin wax

7.3 SECTION REVIEW

1. Why do covalent compounds typically have relatively low melting points?

2. What characteristic of network covalent substances makes them markedly different from other covalent compounds?

3. How does the structure of ionic compounds affect their ability to conduct electricity?

4. What is an alloy?

5. How does the electron configuration of metals affect their appearance?

07 CHAPTER REVIEW
Chapter Summary

TERMS
chemical bond
octet rule
covalent bond
ionic bond
metallic bond
polarity
polar covalent bond

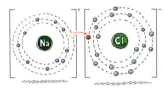

7.1 BONDING BASICS

- Atoms form chemical bonds to attain the low-energy, stable electron configurations of the noble gases. This principle is called the octet rule.

- There are three chemical bond types defined by the relationship of electrons to the bonded atoms. Covalent bonds share electrons between two atoms. Ionic bonds transfer electrons from one atom to another. Metallic bonds share electrons among many atoms.

- Differences in electronegativities determine bond polarity and character.

7.2 TYPES OF BONDS

- Covalent bonds occur between two nonmetals because they have similar electronegativity values.

- Covalent molecules can be conveniently represented by Lewis structures.

- Covalent molecules can contain single (one pair of shared electrons), double (two pairs of shared electrons), or triple (three pairs of shared electrons) bonds to complete the atoms' octets.

- Ionic bonds occur between a metal and a nonmetal because they have significantly different electronegativity values.

- Ionic compounds consist of crystal lattices. The ratio of cations to anions needed to cancel each other's charge determines the chemical formula for the compound.

- Covalently bonded polyatomic ions may form when groups of atoms need to lose or gain electrons to achieve their octets.

- Metallic bonds occur between metal atoms because they loosely hold their valence electrons. They can easily give these up, creating a sea of electrons that hold together the network of metal ions.

TERMS
Lewis structure
double bond
triple bond
formula unit
crystal lattice
polyatomic ion
electron sea model
delocalized electron

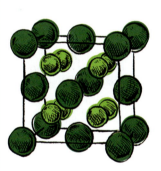

TERMS
network covalent substance
alloy

7.3 PROPERTIES OF COMPOUNDS

- Covalent compounds tend to be gases, liquids, soft solids, or brittle and crumbly solids (although there are notable exceptions). They tend to be poor conductors of heat and electricity.

- Ionic compounds generally form solid crystalline substances with high melting points. In their solid state they are poor conductors.

- Metals consist of many metal cations immersed in a "sea" of mobile, delocalized electrons, which explains their ductility, malleability, conductivity, and luster.

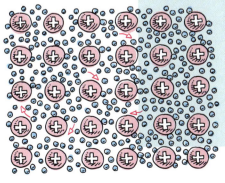

07 CHAPTER REVIEW

Chapter Review Questions

RECALLING FACTS

1. Does the octet rule always refer to a stable arrangement of eight valence electrons? Explain.
2. Why are there no completely ionic bonds?
3. Why are the diatomic elements not found in nature as single atoms?
4. What do Lewis structures depict?
5. Why does hydrogen act like a Group 17 element when forming covalent bonds?
6. Define *double bond*.
7. Summarize the steps for drawing Lewis structures.
8. What is a polyatomic ion?
9. Summarize the current model of metallic bonding.
10. How are the bonding electrons arranged in the following bonds?
    a. a nonpolar covalent bond
    b. a polar covalent bond
    c. a predominantly ionic bond
    d. a metallic bond
11. Why are metals good conductors of thermal energy and electricity?

UNDERSTANDING CONCEPTS

12. Explain why atoms tend to form chemical bonds.
13. Is it possible to have a completely covalent bond? Explain.
14. What makes a bond polar?
15. Create a graphic organizer using the terms *chemical bond*, *covalent bond*, *ionic bond*, *metallic bond*, *octet rule*, *polarity*, and *valence electron*.
16. Compare single, double, and triple covalent bonds.
17. Describe how atoms form covalent bonds.
18. Considering electronegativity alone, state the most probable way (losing, gaining, or sharing) in which the following atoms could attain a valence octet. Also state the number of electrons involved.
    a. K
    b. Ca
    c. Ga
    d. Ge
    e. As
    f. Se
    g. Br
    h. Kr
    i. C

19. Draw the Lewis structures for each of the following.
    a. H_2
    b. HCl
    c. CH_4
    d. CF_2Cl_2
    e. H_2S

20. The following covalent compounds each contain at least one double or triple bond. Draw their Lewis structures.
    a. H_2CO
    b. CS_2
    c. CO_2
    d. C_2H_4
    e. C_2H_2

21. Use electron dot notation to show the ionic bonding between lithium and nitrogen.
22. Your study partner says that a molecule of salt consists of one atom of sodium sharing a valence pair of electrons with one atom of chlorine. Explain the problem(s) with her description.
23. Draw the Lewis structures of the following polyatomic ions.
    a. HS^-
    b. SO_3^{2-}
24. The polyatomic ions in Question 23 follow the octet rule. Phosphate (PO_4^{3-}) does not. Draw a Lewis structure for the phosphate ion.
25. Sketch the bonding that occurs in a metal.
26. Which one of the three bond types is distinctly different from the other two in regard to the location of electrons? Explain why this is so.
27. Identify the types of atoms (metal or nonmetal) in the following compounds and then tell whether the compounds are predominantly ionic, covalent, or metallic.
    a. sodium chloride (NaCl)
    b. red brass (Sn-Cu-Zn alloy)
    c. magnesium bromide ($MgBr_2$)
    d. carbon dioxide (CO_2)
    e. stainless steel (Fe-Ni-Cr alloy)
28. Compare covalent, ionic, and metallic bonds, including the general arrangement of electrons in each type.
29. How do intermolecular forces affect the melting point of a compound?

30. Why do ionic solids typically exhibit cleavage?
31. Describe the general properties of compounds according to their predominant bond type.

CRITICAL THINKING

32. Why do you think that electronegativity differences can't precisely predict the nature of bonds between two atoms?
33. Examine the following electron dot notations for the atoms found in a molecule of sulfur dioxide. Propose a Lewis structure for sulfur dioxide that satisfies the octet rule for all three atoms.

$$:\ddot{S}\cdot \quad :\ddot{O}\cdot \quad :\ddot{O}\cdot$$

34. Why are double bonds stronger than single bonds?
35. Why are triple bonds shorter than single bonds?
36. Hypothesize some limitations of Lewis structures?

Use the Plastic ethics box below to answer Question 37.

37. Use the ethics triad from Chapter 1 and the strategy modeled for you in Chapter 3 to formulate a four-paragraph Christian position on the question of plastic waste. Be sure to address each leg of the triad: biblical principles, biblical outcomes, and biblical motivations. In the last paragraph, formulate your own position on the issue.

ethics

PLASTIC—WONDER PRODUCT OR DESTROYER OF WORLDS?

In an effort to help curb plastic wastes, California enacted a law in 2019 that prohibits restaurants from giving a customer a plastic straw unless one is requested. This may seem silly when you consider that plastic straws make up a very tiny fraction of plastic waste, but the problem of plastic waste is a real one and will need real answers. Statistics from the US Environmental Protection Agency show that in 2015, the United States produced over 34 million tons of plastic but recycled only a little over 3 million tons. A whopping 26 million tons wound up in landfills. As the world's population grows, demand for plastic will likely only increase—as will the amount of waste produced as a result.

Chapter 8

BOND THEORIES & MOLECULAR GEOMETRY

What are scientists looking for when they send orbiters, probes, and rovers to other planets? Often they are looking for water, especially liquid water. Why is liquid water so important to these scientists? They know that water is vital to the existence of living organisms. Secular scientists believe that if there is water, then there may be life.

What is it about water that makes it so important for living organisms? Water exists on Earth in all three states of matter. Water is unique in that in its solid form it is less dense and therefore floats in liquid water. Water can dissolve many different materials—it is considered a universal solvent.

The behavior of water, like that of all matter, is caused by the internal structure of its molecules. Throughout this chapter we will look at the current theories for how atoms bond to form molecules and what determines the shapes of those molecules. With this information, we will understand the amazing design of the water molecule and how it was specifically designed to support us.

8.1 Bond Theories *175*
8.2 Molecular Geometry *182*

8.1 BOND THEORIES

The Limits of Lewis Structures

As noted in the Chapter opener, water is an amazing substance. What gives water the unique properties that make it invaluable for supporting life? As you will repeatedly hear throughout this chapter, the composition and internal arrangement (structure) of matter determine its chemical and physical characteristics. So what is so special about the structure of water?

In Chapter 7 you learned to use Lewis structures to represent molecules whose atoms are held together by covalent bonds. Lewis structures clearly show the bonds between atoms as well as the unbonded electrons, but because they are simplified two-dimensional representations of three-dimensional molecules, they leave out valuable information about the compounds they represent. For instance, consider the Lewis structures for water.

Can you conclude anything about the three-dimensional shape of a water molecule from those structures? Because of the limitations of Lewis structures, it's hard to tell anything. For a more complete description of chemical bonds and the molecular shapes that they produce—especially their three-dimensional structures—we must use the quantum mechanical theory. We will use two models for bonds: the *valence bond theory* and the *molecular orbital theory*. In Section 8.2, we will learn about the three-dimensional shapes of molecules. We will see that it is water's internal structure that actually determines many of its physical and chemical properties.

Orbitals and Valence Bond Theory

As you learned in Section 5.2, electrons occupy various sublevels of principal energy levels. These sublevels, named *s*, *p*, *d*, and *f*, contain varying numbers of orbitals. We model these orbitals as spatial regions where electrons have a high probability of being found. Each orbital can hold a maximum of two electrons. An *s* sublevel has one orbital, a *p* sublevel has three, a *d* sublevel has five, and each *f* sublevel has seven. The *s* and *p* orbitals have relatively simple shapes, while the *d* and *f* orbitals are much more complex.

Covalent bonds involve two atoms sharing pairs of valence electrons with opposite spins. These shared electrons occupy partially filled valence orbital regions that overlap between bonded atoms. When the two bonding orbitals are superimposed, the overlapping space containing both electrons becomes available to both nuclei. Effectively, both atoms acquire another valence electron that fills the vacancy in that particular orbital. This concept is called the **valence bond theory**, just one of several models that describe the formation of covalent bonds. The valence bond theory is also called the *localized electron theory*. It is important to realize that no single theory of bonding accounts for everything that chemists observe about covalent bonds.

> **QUESTIONS**
> » Do Lewis structures model molecules well?
> » How do bonds form?
> » Are all covalent bonds the same?
> » Do the rules for bonding have exceptions?
>
> **TERMS**
> valence bond theory • sigma (σ) bond • pi (π) bond • resonance • molecular orbital theory

Where are the electrons in a chemical bond?

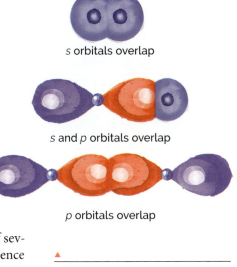

the valence bond theory

FORMING BONDS IN THE VALENCE BOND THEORY

According to the valence bond theory, covalent bonds form when partially filled orbitals of two atoms overlap. Which orbitals are involved and the way that they are aligned will determine whether the bond is a *sigma bond* or a *pi bond*. Let's look more closely at these two types of bonds.

SIGMA BONDS

A **sigma (σ) bond** forms when orbitals overlap along the bond axis that connects two nuclei, forming an end-to-end type of overlap. A region of high electron concentration forms on the *bond axis*. The hydrogen molecule (H_2) is the simplest diatomic molecule and forms when the spherical $1s$ orbital of each hydrogen atom overlaps and merges with the other to share a pair of electrons with opposite spins.

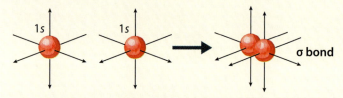

Valence *p* orbitals also form sigma bonds. Halogens, such as fluorine (F_2), form a single sigma bond when each atom's unfilled, dumbbell-shaped valence *p* orbital overlaps the other atom's orbital end to end.

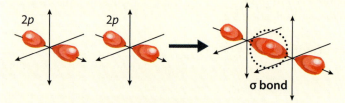

In some molecules, such as hydrogen fluoride (HF), the sigma bond forms when an *s* and a *p* orbital overlap end to end.

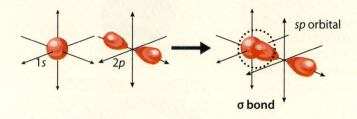

PI BONDS

Double and triple bonds form when more than one set of orbitals from the *p* sublevel overlap. One set always forms a sigma bond, overlapping end to end. The second bond is different because the overlap is side to side. The two lobes of an orbital extend toward the corresponding lobes of the other atom's orbital. The lobes overlap above and below the bond axis to form two regions of high electron concentration parallel to the bond axis. Even though the two separate regions overlap, they are both parts of one bonding orbital containing two electrons. Such a side-to-side orbital overlap is called a **pi (π) bond**. The double bond in ethene (C_2H_4) contains one sigma bond and one pi bond.

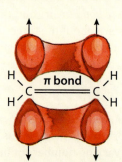

For triple bonds, as in molecular nitrogen, one of the bonds is a sigma bond formed by two lobes of two *p* orbitals overlapping end to end. The two remaining bonds are pi bonds. Pairs of *p* orbitals overlap above and below the bond axis and to either side of the bond axis to form these pi bonds.

Sigma bonds are the strongest type of covalent bond because the region of highest electron probability lies on the bond axis. This location maximizes the effectiveness of the electrostatic attraction between the electrons and the nuclei. Pi bonds are weaker than sigma bonds because the electrons are spread out over a greater volume of space around the bond axis. Also, the electrons in pi bonds are less attracted to the nuclei than in sigma bonds because the region of highest concentration is not along the bonding axis, and so the electrons pull on the nuclei at an angle. But since pi bonds almost always occur along with a sigma bond, the combination of a sigma and a pi bond together is stronger than either bond by itself. As would be expected, triple bonds are stronger than either double or single bonds.

Molecular Resonance

Multiple bonds introduce another limitation of Lewis structures. Consider the sulfur dioxide molecule (SO$_2$). We can draw it as shown.

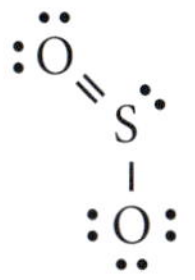

But there is no reason why the double bond could not exist between the sulfur and the other oxygen.

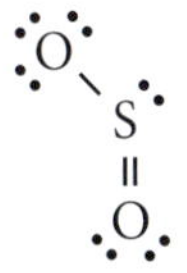

Because the double bond is stronger than the single bond, there should be a difference between the two ends of the molecule. But experimental evidence reveals that the molecule acts as if it is completely symmetrical. Therefore, the sulfur-oxygen bonds must be identical.

For some molecules, like SO$_2$, no single Lewis structure can completely describe the distribution of electrons. Such molecules exhibit **resonance**. A molecule displaying resonance does *not* exist in one of the configurations, nor does it oscillate between the possible configurations. Rather, the bonds take on an intermediate character because electrons are mobile and their wave functions can occupy different regions of a molecule at the same time. You can think of these bonds as being the average of the single and double bond possibilities. You may see these depicted as hybrid Lewis structures.

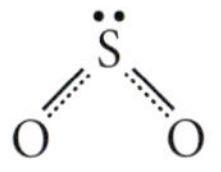

The dotted line represents the half bond. We don't show the unbonded electrons around each oxygen due to their varying numbers.

Resonance bonds are like a mixture of two primary colors. The new mixture does not oscillate between the two original colors but forms a new color. The sulfur dioxide molecule acts as though there are one and a half bonds between the central sulfur and each oxygen. One familiar example of a molecule exhibiting resonance is ozone (O_3).

$$:\ddot{\underset{..}{O}}-\overset{..}{O}=\ddot{O}: \longleftrightarrow :\ddot{O}=\overset{..}{O}-\ddot{\underset{..}{O}}:$$

Another is the nitrate ion (NO_3^-).

$$\left[\begin{array}{c}:\ddot{\underset{..}{O}}-\overset{..}{N}=\ddot{O}:\\|\\:\ddot{\underset{..}{O}}:\end{array}\right]^- \longleftrightarrow \left[\begin{array}{c}:\ddot{\underset{..}{O}}-\overset{..}{N}-\ddot{\underset{..}{O}}:\\||\\:\ddot{O}:\end{array}\right]^- \longleftrightarrow \left[\begin{array}{c}:\ddot{O}=\overset{..}{N}-\ddot{\underset{..}{O}}:\\|\\:\ddot{\underset{..}{O}}:\end{array}\right]^-$$

In each case, there are multiple "correct" Lewis structures, but in reality, there are no single and double bonds. The three bonds have been found experimentally to be identical to each other with a nature intermediate between typical single and double bonds. It has also been shown experimentally that the strength of the bonds is less than a double bond and stronger than a single bond. You could think of the bonds in nitrate as 1⅓ bonds.

When the Octet Rule
DOES NOT WORK

Up to this point, the goal in drawing Lewis structures has always been to have eight valence electrons for every atom except for atoms that are stable with the same configuration as helium. But some molecules form without an octet for each bonded atom. There are three situations when this can happen.

The Presence of an Odd Number of Valence Electrons

Some molecules have an odd number of valence electrons to share. Nitrogen dioxide (NO_2), a gaseous byproduct of combustion engines, is one such example. Nitrogen contributes five valence electrons, while the two oxygen atoms bring a total of twelve. There is no way to make seventeen electrons form octets, so one atom ends up with seven electrons.

$$:\ddot{O}=\dot{N}\diagdown \ddot{\underset{..}{O}}:$$

Substances with these unpaired electrons are called *free radicals*. They tend to be fairly reactive and when found in living organisms can cause damage.

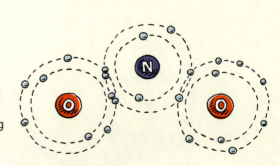

CAREERS

SERVING AS A PATENT ATTORNEY: PROTECTING THE GIVING OF PROPER CREDIT

Our country's Founding Fathers envisioned a nation marked by creative improvements in science and technology. The United States Constitution and federal law provide a legal framework for the protection of people's inventions. Chemists and chemical engineers need the help of trained patent counsel in navigating through these laws to achieve protection for their inventions.

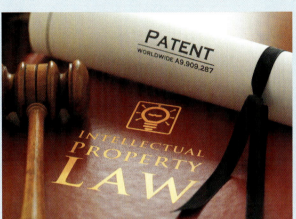

A patent attorney working on chemical patents must have an undergraduate degree, and often an advanced degree, in chemistry, chemical engineering or another scientific discipline; a law degree from a law school; permission to practice law within at least one state; and permission to practice before the US Patent and Trademark Office (USPTO). The attorney works with the scientist or engineer to define and describe the inventor's creation in a legal document called a patent application. The attorney then undertakes the process, often involving years, of applying to the USPTO for a patent. If obtained, the patent provides legal rights to the inventor.

Patent protection reflects the concept of ownership of property, a theme appearing with approval many times in the Bible. An individual should be able to enjoy the fruits of his labor. Patent attorneys like Mark Quatt (now retired) work to ensure that these inventions are legally secured. People use their gifts and talents to create things, and patent attorneys work to love and honor people by giving them proper credit.

Electron Deficiency

Elements such as boron are content with fewer than four pairs of valence electrons. Boron tends to form molecules in which it is stable with only six valence electrons. For example, boron has just three valence electrons, and fluorine already has seven of the eight electrons that it needs for stability. Boron forms three single bonds with the highly electronegative fluorine atoms.

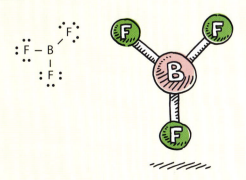

Hypervalent Molecules

In other molecules, the central atom has more than eight valence electrons. Sulfur hexafluoride (SF_6) is a common example. Each fluorine atom forms a single bond with the central atom, sulfur, which ends up with six pairs of electrons. Such situations occur only in Period 3 and higher. You may hear these molecules described as having *expanded octets*. Elements that have expanded octets have the ability to use *d*-sublevel electrons for bonding.

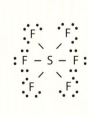

Bond Theories and Molecular Geometry

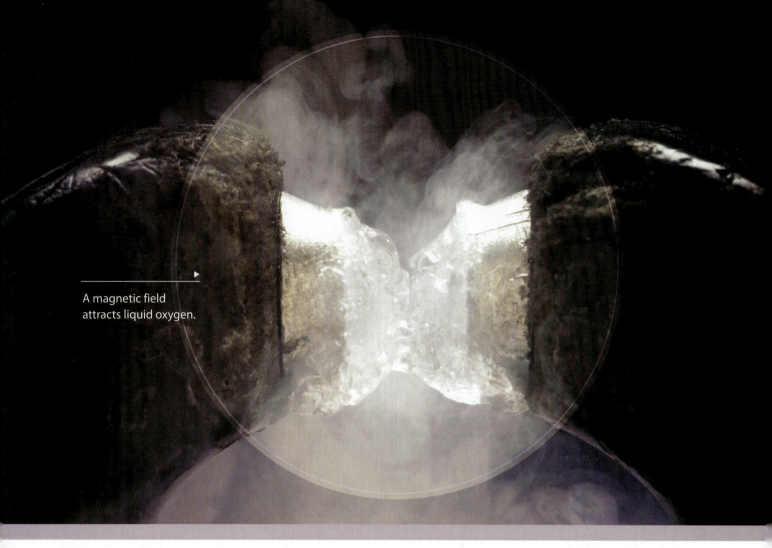

A magnetic field attracts liquid oxygen.

Molecular Orbital Theory

According to valence bond theory, diatomic oxygen should be colorless and nonmagnetic. When a group of scientists liquefied diatomic oxygen, they were surprised to find that it was pale blue and demonstrated magnetic properties. The molecular models at the time didn't predict either of these characteristics, so it was time to update the model.

The valence bond theory begins to explain why molecules have certain shapes, but it can't explain all observations. For example, the valence bond theory prohibits unpaired electrons that could account for unusual chemical and physical properties of certain compounds. As the scientists working with liquid oxygen discovered, valence bond theory does not correctly model the bonds between oxygen atoms.

A more recent model of bonding explains some of those observations. The **molecular orbital theory** suggests that the atomic orbitals of a molecule's atoms are replaced by totally new molecular orbitals when a molecule forms. Each molecule has a unique set of molecular orbitals equal to the sum of the atomic orbitals of the original atoms (see image at left). Some molecular orbitals encircle two, three, four, or even more atoms; often they encircle the entire molecule. The molecular orbital theory ranks the resulting orbitals in order of increasing energy. Electrons

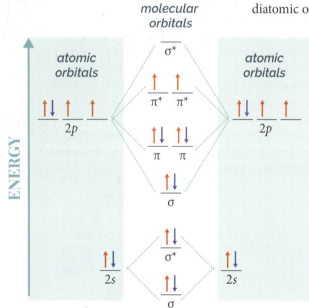

the molecular orbitals for O_2

from the original atoms fill low-energy molecular orbitals before they fill high-energy orbitals. The arrangement of electrons in these orbitals predict whether bonds will form; whether the bonds will be single, double, or triple; and whether the bonds will contain unpaired electrons. Scientists associate the presence of unpaired electrons in molecular orbitals with *paramagnetism*, the attraction of the molecule to a magnetic field. With this explanation of paramagnetism, molecular orbital theory correctly predicts the observed magnetic properties of diatomic oxygen.

In molecular orbital theory, there are both *bonding* (σ and π) and *antibonding* (σ^* and π^*) orbitals. When a molecule forms, the electrons in the atomic orbitals can either work together or interfere with each other because they behave like waves. When electron waves constructively interfere, they reinforce each other and bonding orbitals form. The location of the bonding orbitals is between the nuclei. The electrons in these orbitals stabilize the molecule because atoms can share them. They are lower in energy than the atomic orbitals from which they form, which is how chemical bonds store energy.

Whenever atomic orbitals combine, both bonding and antibonding orbitals form. If the electron waves destructively interfere with each other, an orbital forms on the outside of the molecule, far from the two nuclei. Antibonding orbitals are higher in energy than the atomic orbitals from which they form. The electrons in an antibonding orbital spend little time between the two nuclei, so these antibonding orbitals destabilize the molecule. Molecular orbital theory can thus be used to predict whether a bond will form between two atoms. If equal numbers of electrons inhabit bonding orbitals and antibonding orbitals, then no bond will form.

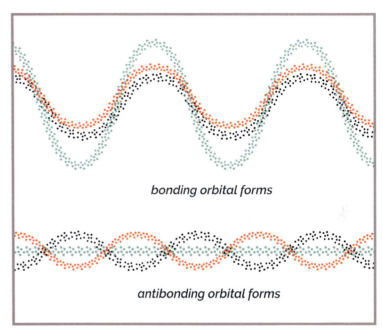

The electron waves at the top are constructively interfering to form a bonding orbital. When electron waves destructively interfere (bottom), an antibonding orbital forms.

8.1 SECTION REVIEW

1. How does the Lewis structure for water shown below demonstrate one of the key limitations of the Lewis structure as a model for molecules?

    H   H
     \\ /
     :Ö:
     ˙˙

2. According to the valence bond theory, how do covalent bonds form?

3. What type(s) of bonds make up a double bond?

4. Compare sigma and pi bonds.

5. Why can't Lewis structures accurately depict molecules that exhibit resonance?

6. What does it mean that an element is electron deficient? that it is hypervalent?

7. Describe the molecular orbital theory.

8. (True or False) According to the molecular orbital theory, bonding orbitals are higher in energy than the atomic orbitals from which they form.

8.2 MOLECULAR GEOMETRY

QUESTIONS
- What is VSEPR?
- How do electron configurations change for hybridized orbitals?
- How can I predict the shape of a molecule?
- How can a molecule have polar bonds but not be a polar molecule?

TERMS
valence shell electron pair repulsion (VSEPR) theory • tetrahedral molecule • trigonal pyramidal molecule • trigonal planar molecule • linear molecule • bent molecule • polar molecule • molecular dipole moment

VSEPR and Molecular Shape

One of the basic principles of electricity is that opposite charges attract and like charges repel, which means that electrons always repel, whether they are bonded or unbonded. Charge repulsion determines the shape of molecules held together by covalent bonds. The repulsion of valence shell electrons in a molecule determines the arrangement of covalent bonds and unbonded pairs of electrons around the central atom. Chemists developed the **valence shell electron pair repulsion (VSEPR) theory** to account for these observations.

VSEPR theory focuses on the locations of high electron densities surrounding the central atom in a molecule. The Lewis structure is used to determine the number of regions of high electron density. One of these regions can be either an unbonded pair of electrons or a single, double, or triple bond.

Why does the shape of a molecule matter?

MOLECULE SHAPES

In the VSEPR model, regions of high electron density repel each other until they are separated by the maximum distance possible. This repulsion will establish the *bond angle*. The number of atoms surrounding the central atom, in conjunction with the bond angle, will determine the shape of the molecule.

Let's look at a few of the most common molecular shapes according to VSEPR theory.

TETRAHEDRAL

Carbon tetrachloride (CCl_4), a once-popular dry-cleaning solvent, has four regions of electrons (four single bonds) around the carbon atom. How are the chlorine atoms arranged around the central carbon? The electrons in the single bonds repel each other. Geometrically, the chlorine atoms are farthest from each other when positioned at the vertices of a tetrahedron. The carbon atom is at the center of the tetrahedron. The shape of such a molecule is tetrahedral. A **tetrahedral molecule** contains four regions of electrons with four atoms bonded to the central atom. The angles between the bonds are approximately 109.5°.

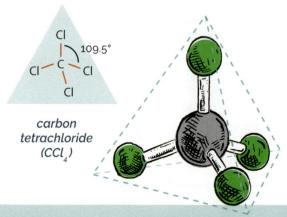

carbon tetrachloride (CCl_4)

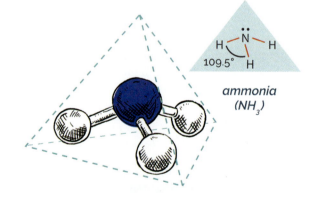

ammonia (NH_3)

TRIGONAL PYRAMIDAL

Ammonia (NH_3) also has four regions of electrons but has only three atoms bonded to the central atom. The three hydrogen atoms occupy three of the four vertices of the tetrahedron with a pair of unbonded electrons at the other vertex. While the expected bond angle is 109.5°, the unbonded pair of electrons produces a stronger repulsion force. This greater force decreases the bond angle slightly. This **trigonal pyramidal molecule**—shaped like a triangular pyramid—has four regions of electrons with three bonded atoms around the central atom.

TRIGONAL PLANAR

Formaldehyde (CH_2O) has three regions of electrons surrounding the central carbon atom—one double bond and two single bonds. The three regions of electrons produce bond angles of approximately 120°. It also has three atoms surrounding the central atom, resulting in a Y-shaped molecule, with all of the atoms in a single plane. This arrangement results in a **trigonal planar molecule**.

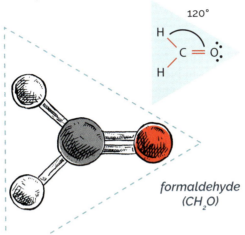

formaldehyde (CH_2O)

carbon dioxide (CO_2)

LINEAR

Carbon dioxide (CO_2) has two regions of electrons around the central carbon—two double bonds. A molecule with two regions of electrons around the central atom is a **linear molecule**, regardless of whether the regions are bonds or unbonded electrons. The bonds in a linear molecule are 180° apart and point in opposite directions. Diatomic molecules are linear.

BENT

Water (H_2O) molecules consist of an oxygen atom surrounded by four regions of electrons—two unbonded pairs and two single bonds—forming a **bent molecule**. The hydrogen atoms occupy two of the tetrahedral vertices. A line drawn from one end of this molecule to the other forms an angle, which is why this kind of molecule has a bent shape. Bond angles for bent molecules are typically either 109.5° (four regions of electrons) or 120° (three regions of electrons). The bond angle in water is 104.5° due to the extra repulsion from the unbonded electrons.

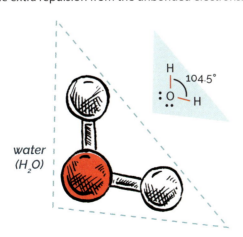

water (H_2O)

EXAMPLE 8-1:
PREDICTING MOLECULAR SHAPES

Predict the shapes of the following molecules.

a. $\overset{\cdot\cdot}{\underset{\cdot\cdot}{O}}\diagdown\overset{\cdot\cdot}{\underset{}{O}}\diagup\overset{\cdot\cdot}{\underset{\cdot\cdot}{O}}$

b. $H-C\equiv N:$

c. $:\overset{\cdot\cdot}{\underset{\cdot\cdot}{F}}-\overset{\cdot\cdot}{\underset{|}{N}}-\overset{\cdot\cdot}{\underset{\cdot\cdot}{F}}:$
     $\quad\quad\;\;:\overset{\cdot\cdot}{\underset{\cdot\cdot}{F}}$

Solution

a. There are three regions of electrons around the central oxygen. But since only two of those regions involve bonds with other atoms, the shape of a trioxygen (ozone) molecule is bent.

b. There are two regions of electrons around the central carbon in hydrogen cyanide. The three atoms lie in a straight line. This molecule's shape is linear.

c. Nitrogen trifluoride has four regions of electrons around the central nitrogen. Since only three atoms bond with the nitrogen atom, the molecule has a trigonal pyramidal shape.

Orbital Hybridization

Among the inadequacies of the valence bond theory is its explanation of how carbon bonds. In its ground state, carbon has two partially filled *p* orbitals and one full *s* orbital in its valence shell.

$$[\text{He}]: \underline{\uparrow\downarrow}\ \underline{\uparrow}\ \underline{\uparrow}\ \underline{}$$
$$\quad\quad\quad 2s \quad\quad 2p$$

According to valence bond theory, it would seem probable that carbon could fill these *p* orbitals by forming two covalent bonds. That explanation violates the octet rule because it leaves an entire *p* orbital empty. Also, carbon routinely forms four covalent bonds. Sometimes the four bonds are equivalent, as in methane (CH_4). A modification of bonding theory was necessary to account for these observations.

One explanation for the behavior of carbon suggests that one of the electrons in the 2s orbital moves, or is promoted, to the empty *p* orbital. Then carbon would have four unpaired electrons.

$$[\text{He}]: \underline{\uparrow\downarrow}\ \underline{\uparrow}\ \underline{\uparrow}\ \underline{} \xrightarrow{\text{promotion}} [\text{He}]: \underline{\uparrow}\ \underline{\uparrow}\ \underline{\uparrow}\ \underline{\uparrow}$$
$$\quad\quad 2s \quad\quad 2p \quad\quad\quad\quad\quad\quad\quad\quad 2s \quad\quad 2p$$

But the different energies of the *s* and *p* orbitals could not form identical bonds with the four hydrogen atoms. The theory that accounts for this observation states that the single *s* and three *p* orbitals combine to form four orbitals within a single valence energy level. Scientists call the process by which new kinds of orbitals with equal energies form from a combination of orbitals with different energies *orbital hybridization*. Orbital hybridization is common between nonmetals and even in metals involved in covalent bonding. In the case of carbon, one *s* and three *p* orbitals hybridize and form four new orbitals referred to as sp^3 hybrid orbitals. The new carbon abbreviated orbital notation is shown below.

$$[\text{He}]: \underline{\uparrow\downarrow}\ \underline{\uparrow}\ \underline{\uparrow}\ \underline{} \xrightarrow{\text{hybridization}} [\text{He}]: \underline{\uparrow}\ \underline{\uparrow}\ \underline{\uparrow}\ \underline{\uparrow}$$
$$\quad\quad 2s \quad\quad 2p \quad\quad\quad\quad\quad\quad\quad\quad\quad 2sp^3$$

Carbon's four hybridized sp^3 orbitals bond equally with hydrogen's *s* orbitals.

Why do atoms form hybrid orbitals in the first place? Promoting electrons to higher energy levels requires extra energy. But the main lobes of the individual hybrid orbitals are much larger than the lobes of *s* or *p* orbitals. They can point to and overlap the orbitals of other atoms more effectively, forming stronger bonds. The increase in stability from forming bonds more than offsets the energy required to create hybrid orbitals.

Other hybrids of *s* and *p* orbitals are possible. For example, in the covalent molecule boron trifluoride (BF_3), the central boron atom is bonded to three fluorine atoms, and each can share only one electron. Boron has three valence electrons—a pair in the 2s orbital and a single 2p orbital electron. To form three identical bonds with the three fluorine atoms, the boron atom must promote one of its *s* electrons to an empty *p* orbital and then hybridize its *s* and *p* orbitals to produce three equivalent hybrid orbitals. Since two *p* orbitals are involved, they are called sp^2 hybrid orbitals. The empty third *p* orbital that is "left over" (unhybridized) is not involved in bonding in this molecule.

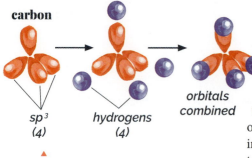

Notice that boron shares only six electrons by forming three sigma bonds with fluorine. This compound is a classic example of how a more stable electron configuration is attained through hybridization rather than by following the octet rule. The new boron orbital notation is shown below.

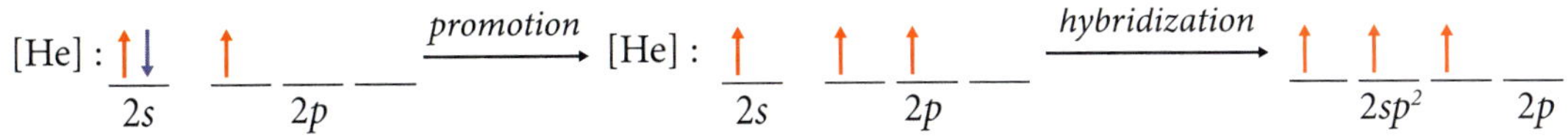

[He] : ↑↓ ↑ __ __ —*promotion*→ [He] : ↑ ↑ ↑ __ —*hybridization*→ ↑ ↑ ↑ __
    2s   2p   2s   2p   2sp^2 2p

In molecules with sp^3 orbitals, the four orbitals arrange themselves at the vertices of a regular tetrahedron because of mutual repulsion. The tetrahedral arrangement determines the locations of bonded atoms and lone pairs of electrons. Molecules that contain hybrid sp^3 orbitals can form tetrahedral, trigonal pyramidal, and bent molecules. The number of hybrid orbitals formed is always equal to the number of s and p orbitals containing unpaired electrons after promotion.

The same basic principles hold when determining the shapes of sp^2 and sp hybrid compounds. Table 8-1 provides additional examples for you to study. Remember that we determine the shape of a molecule by looking at the arrangement of atoms around the central atom, *not* by looking at the unbonded pairs.

TABLE 8-1 *Representative Molecular Shapes*

Regions of Electrons	Number of Bonds	Expected Bond Angle	Example	Lewis Structure	Spatial Arrangement	Geometry (actual bond angle)
4 sp^3 hybrid	4	109.5°	CH_2Br_2 dibromomethane	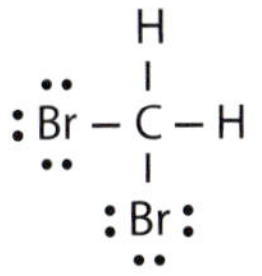	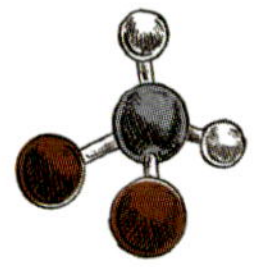	tetrahedral (108°–112°)
	3		PH_3 phosphane	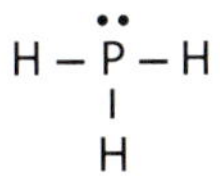	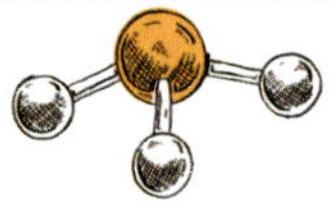	trigonal pyramidal (93.5°)
	2		H_2S hydrogen sulfide	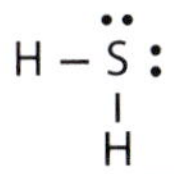		bent (92.1°)
	1	180°	HCl hydrogen chloride	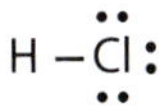		linear (180°)
3 sp^2 hybrid	3	120°	BI_3 triiodoborane	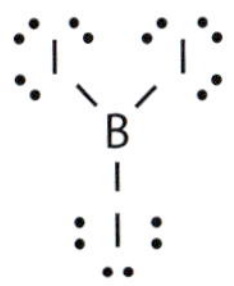	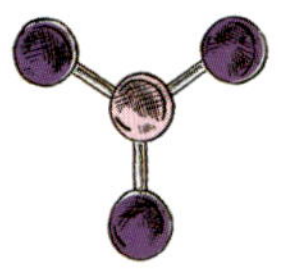	trigonal planar (120°)
	2		GeF_2 germanium difluoride (resonance)	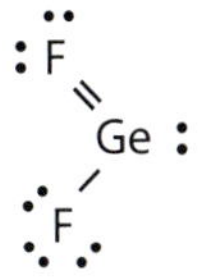		bent (94°)
2 sp hybrid	2	180°	BeF_2 beryllium fluoride	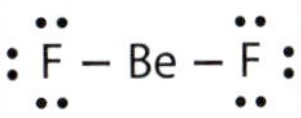		linear (180°)
	1		CO carbon monoxide	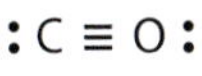 :C≡O:		linear (180°)

WORLDVIEW INVESTIGATION

REFRESHING WATER

Water is essential to life and covers over three-quarters of the earth's surface. In many locations around the world, the problem of keeping drinking water separate from wastewater is a life or death issue. Throughout history, scientists have attributed numerous epidemics to drinking water that was contaminated by wastewater. Many people are working on the issue of how to adequately treat wastewater. One novel approach is to have nature do much of the cleaning for us. In Northern California, one system uses ponds and marshes to clean wastewater.

Task

You are working for a missionary organization that is researching different options for dealing with the wastewater issue. Your task is to research wastewater treatment systems like the one in Arcata, California. You will write a two-page paper highlighting how the system works. You will be expected to include benefits and drawbacks to the system.

Procedure

1. Research wastewater treatment in general by doing a keyword search for "wastewater treatment."
2. Research the specific use of biological processes to treat wastewater by doing a keyword search for "Arcata wastewater treatment plant."
3. Write your paper and have a classmate review it and provide feedback.
4. Complete your paper and turn it in. Make sure that you have properly cited your sources.

Conclusion

In living life, we constantly use water and produce wastewater. To meet the needs of people around us, we have to think about how to properly treat our wastewater.

Dipole Moment: A Measure of Polarity

As you learned in Chapter 7, chemical bonds occur along a spectrum from ionic, through polar covalent, to purely covalent bonds. All bonds formed between atoms of different elements are polar to some extent. The magnitude of the bond polarity depends on the electronegativity difference between the two elements. The greater the difference in electronegativity, the more polar the bond will be. If the bonds within a molecule can be polar, can an entire molecule have a negative end and a positive end—what we could call a polar molecule? For simple molecules consisting of two atoms of different elements, polar bonds mean polar molecules. In other words, a molecule formed from a single polar bond will itself be polar—a **polar molecule**—with distinct positive and negative partial charges.

But if a molecule contains polar bonds, does that always mean that the molecule must be polar too? You might think so, but we must take into account the geometric symmetry of the molecule itself. As you learned earlier in this section, molecules can take on a variety of shapes. The polarities of individual bonds may be quite high, but if those bonds are arranged symmetrically, they balance each other. As a result, the molecule overall is nonpolar.

Consider the four molecules at right—two polar and two nonpolar. The red shading indicates a region of negative charge, blue shading a region of positive charge. In the cases of methane (CH_4) and carbon dioxide (CO_2), the bonds between the outer atoms and the central atom are polar but balanced. The molecular symmetry arranges the partial charges so that there are no "ends" or poles that have different partial charges. In the cases of water (H_2O) and ammonia (NH_3), the asymmetry of the bonds in conjunction with the unbonded electron pairs ensures that there are distinct regions of negative and positive charge on the molecular surfaces.

When scientists map the strength of charge around a molecule, an image emerges that bears little resemblance to the arrangement of the atoms. The most significant features are the locations of the positive and negative partial charges. For asymmetrical molecules, one end is predominantly negative and the other positive.

Because of the distribution of attached atoms on one side of the molecule, bent and trigonal pyramidal molecules have an asymmetrical electron density, resulting in molecules that are always polar. This rule is true even if all the outer atoms have the same electronegativity values. The basic symmetrical geometric shapes of molecules are tetrahedral, trigonal planar, and linear. Other less common shapes are also symmetrical. Symmetrical molecules are not polar if the outer atoms are the same, but even symmetrical molecules are polar if the outer atoms are different. Consider two linear molecules: carbon dioxide (CO_2) and carbonyl sulfide (COS). Carbon dioxide is nonpolar because the oxygen atoms balance each other. Carbonyl sulfide, on the other hand, is polar because the outer atoms, oxygen and sulfur, have different electronegativity values. Table 8-2 summarizes these guidelines.

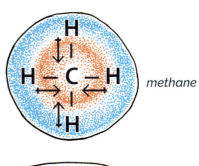
methane

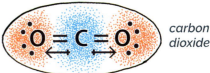
carbon dioxide

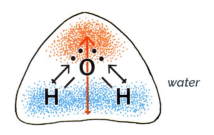
water

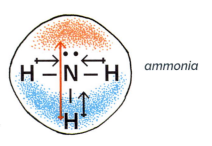
ammonia

TABLE 8-2 *When Molecules Are Polar*

Atoms Bonded to the Central Atom	Four Regions of Electrons	Three Regions of Electrons	Two Regions of Electrons
4	tetrahedral; polar if outer atoms are different	n/a	n/a
3	trigonal pyramidal; always polar	trigonal planar; polar if outer atoms are different	n/a
2	bent; always polar	bent; always polar	linear; polar if outer atoms are different
1	linear; polar if outer atoms are different	linear; polar if outer atoms are different	linear; polar if outer atoms are different

HOW IT WORKS

WATER STRIDERS

If you have ever visited a lake or pond, you have seen water striders—those insects that seem to skate across the surface of the water. You might wonder how they do that. The answer is in the design of both water molecules and the water striders themselves.

As you learned in this section, water is a bent molecule with polar bonds. This means that water is a polar molecule. The positive and negative ends of adjacent water molecules attract each other. At the surface of a body of water, the molecules are pulled down and to the sides, producing surface tension. The surface tension of water is very high compared with other liquids.

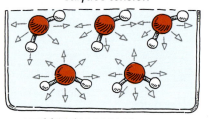
surface tension

The design of the water strider allows it to take advantage of the high surface tension of water. A strider has long legs that allow it to spread its weight over a large area of water. Its legs are also strong and flexible. Leg strength allows a strider to hold its body out of the water while the leg's flexibility enables the strider to keep its balance on the surface of the water. A strider gathers food with its front legs. The middle legs provide propulsion, and hind legs are used to steer and brake. This efficient design allows a strider to travel rapidly over the surface of the water.

The most amazing design feature of a water strider is that it has tiny hairs covering its body. And "covered" means that there are about a million hairs on every square *millimeter* of the strider! These hairs increase the contact area between a strider and the water's surface. The hairs also trap air, making a strider more buoyant than an insect without this hair.

Water striders are yet another example of the amazing creative work of God.

Scientists describe the magnitude and direction of molecular polarity with a quantity known as the **molecular dipole moment** (μ). It is the vector sum of all bond polarities that exist within the molecule. As a vector sum, the molecular dipole moment factors in both the magnitude and direction of the bond polarities for each bond in the molecule. The unit for dipole moments (bond and molecular) is the *debye* (D). This is comparable to the SI derived unit A·s·m (ampere-second-meters). The larger the value of μ, the more polar the bond or molecule will be. As with bond polarity, a crossed arrow is used to point from the more positive region of a molecule to its more negative region.

TABLE 8-3 *Dipole Moments of Several Substances*

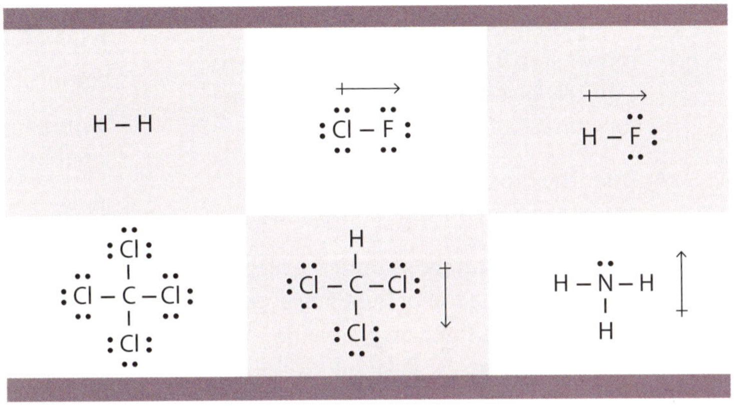

H–H	:Cl–F:	H–F:
:Cl–C(Cl)(Cl)–Cl: (CCl₄)	:Cl–C(H)(Cl)–Cl: (CHCl₃)	H–N(H)–H (NH₃)

EXAMPLE 8-2: DETERMINING MOLECULAR POLARITY

Determine whether the following molecules are polar. If they are, indicate the direction of their molecular dipole moments.

a. CO_2

b. H_2S

c. ClBr

Solution
Draw the molecule's Lewis structure and then predict its shape. If the molecule is asymmetrical, it must be polar. If the molecule is symmetrical, determine the polarity of the molecule by examining the distribution of charge around the central atom and calculating the electronegativity difference for each bond.

a.
$$\ddot{\underset{..}{O}} = C = \ddot{\underset{..}{O}}$$

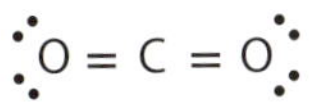

Because there are two regions of electrons around the central atom, CO_2 is linear—a symmetrical shape. Since both outer atoms have the same electronegativity, they are "pulling" on the carbon equally, resulting in a nonpolar molecule.

b.

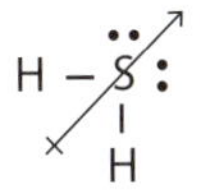

H–S̈:
 |
 H

Because it has four electron regions but only two outer atoms, the molecule's shape is bent. Since all bent shapes are polar, H_2S is polar. The electronegativity difference is 2.6 – 2.2 = 0.4.

c. ClBr must be linear since it is a diatomic molecule. The electronegativity difference between chlorine and bromine is 3.2 – 3.0 = 0.2. Therefore, electrons spend more time closer to the chlorine, resulting in the following dipole moment.

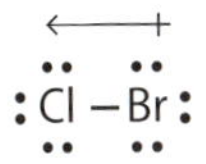

:Cl–Br:

Bond Theories and Molecular Geometry

MINI LAB

A PILE OF WATER

Some people don't like bugs, while others find them fascinating. Regardless of how you feel about bugs, you may have marveled at a water strider gliding across a pond or river (see How It Works box on page 188). A water strider's ability to stride across the water results from its design and the surface tension of water. Surface tension is a result of cohesion—the attractive forces between the liquid molecules. Water has a high surface tension because of its bent shape and polar bonds.

Can we pile water on a coin?

1. Is water a polar molecule? Explain. (*Hint*: A Lewis structure may be helpful.)

Procedure

A Place the dime tail-side up on the laboratory table.

B Using the pipette, carefully place a drop of water on the top surface of the dime.

2. Does the water spread out on the surface of the dime?

C Continue adding drops until the water spills off the top of the dime. Count the total drops that you were able to put on the dime.

Conclusion

3. What did you notice about the surface of the water on top of the dime before it spilled over?

D Completely dry the dime and then repeat step B using the other pipette to carefully place a drop of alcohol on the top surface of the dime.

4. Does the drop of alcohol act the same as water on the surface of the dime?

E Continue adding drops until the alcohol spills off the top of the dime. Count the total drops that you were able to put on the dime.

5. Of which substance were you able to add more drops? Why do you think that was true?

Going Further

6. How does the water strider demonstrate God's design in nature?

EQUIPMENT
- beakers, 50 mL (2)
- disposable pipettes (2)
- dime
- water, 25 mL
- isopropyl alcohol, 25 mL

190 Chapter 8

Water Molecules Designed for Usefulness

All of God's Creation is amazing. God designed the common water molecule elegantly for the benefit of man and other organisms. Throughout the chapter we have seen that God designed polar chemical bonds to form water molecules with a bent shape. This combination of design features produces a polar molecule with many useful properties.

The design of water produces high surface tension that is vital to a water strider's ability to live on the surface of water. The design of the water molecule also explains its unusual property of being denser as a liquid than it is as a solid. As a result, ice floats in water, which means that surface water freezes from the top down. Because of this, most bodies of water on Earth will not freeze solid, allowing aquatic animals to survive through winter.

Water is critical in the transport systems of plants. It can also dissolve many different materials—it's often called the universal solvent. This allows water to be the medium for cellular processes. Doctors use the polarity of water in numerous medical technologies, such as magnetic resonance imaging. And how many of us go through the day without using a microwave oven, which heats food rapidly by heating water molecules within food? In a microwave, heating occurs because of the polarity of water molecules.

God's design of molecules meets the needs of the creation and also provides a foundation for amazing conveniences.

Seeking the Perfect Bonding Model

The valence bond model and the molecular orbital model both have their places in chemistry. Since neither of them accounts for all features of molecular bonds, scientists use the one that best explains a particular situation. It would, of course, be better to have a single theory that explains every observation, and scientists continue to work toward such a solution. But we should not be troubled that science is incapable of exhaustively explaining the universe on the subatomic level. This should motivate scientists to continually improve their models and adjust their theories. God has not commissioned us to know everything about our world in its minutest detail. But He has commissioned us to rule over and manage the creation (Gen. 1:26–28). The model-making done by scientists is a valuable contribution to fulfilling this commission.

The limitations of scientists in discovering the details of how the chemical world works also remind us of the greatness of God. The unseen chemical world that functions within and all around us is in many ways still mysterious. But God designed the way that the chemical world works, and He understands every intricacy of a world too small for us to see.

8.2 SECTION REVIEW

1. What basic principle of force governs VSEPR theory? What property of molecules is significantly affected by that principle?

2. Explain what determines the bond angle and the shape of molecules in the VSEPR theory.

3. Draw the Lewis structure for each of the following and then predict its shape and ideal bond angle.
   a. HCN
   b. OF_2
   c. PO_4^{3-}

4. Discuss the difference between the tetrahedral, trigonal pyramidal, and trigonal planar molecular geometries.

5. How are sp^3 hybrid orbitals produced? Which group of elements would most likely bond using sp^3 hybrid orbitals?

6. Define *polar molecule*.

7. Evaluate each of the following Lewis structures. If it is polar, copy it and add the crossed arrow notation to show the direction of the dipole moment.

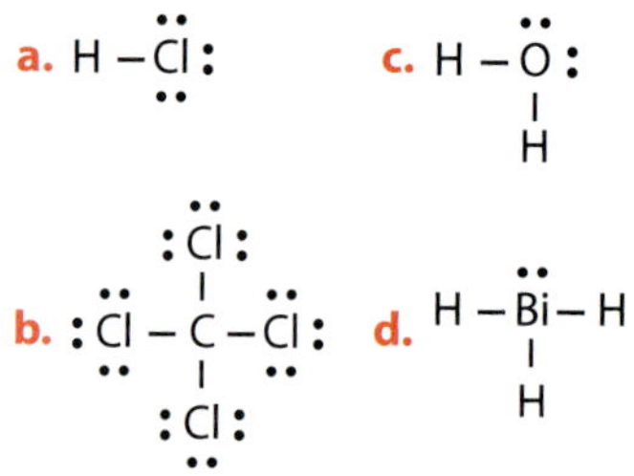

   a. H−Cl:

   b. :Cl−C(−Cl)(−Cl)−Cl:

   c. H−O−H

   d. H−Bi(−H)−H

8. How many polar bonds are found between atoms in each of the molecules below? Indicate whether the molecule is polar.
   a. C_2H_2
   b. CH_4
   c. CS_2
   d. CF_2Cl_2
   e. CO
   f. H_2S

08 CHAPTER REVIEW
Chapter Summary

8.1 BOND THEORIES

- The valence bond theory states that two elements form a bond when they share two unpaired electrons of opposite spin. The orbitals containing the shared electrons overlap.

- Bonding orbitals overlap in one of two ways. When orbitals overlap along their bonding axis, a sigma bond forms. When side-by-side orbitals overlap, a pi bond forms.

- Double bonds consist of a sigma bond and a pi bond. Triple bonds are made up of a sigma bond and two pi bonds.

- Molecular resonance occurs when a molecule has a combination of single and multiple bonds, and the location of the bonds can change. While we think of these molecules as having two types of bonds, in reality, they have identical bonds with an intermediate nature.

- Molecules typically bond to allow their atoms to form complete octets. But some molecules are stable without a full octet. This can occur if there is an odd number of valence electrons in the molecule, if the molecule is electron deficient, or if the molecule is hypervalent.

- The molecular orbital theory states that atomic orbitals are replaced by molecular orbitals when atoms combine to form a molecule.

TERMS
valence bond theory
sigma (σ) bond
pi (π) bond
resonance
molecular orbital theory

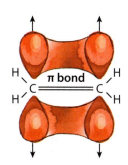

8.2 MOLECULAR GEOMETRY

- The valence shell electron pair repulsion (VSEPR) theory accounts for the arrangement of bonds and unbonded electrons in a molecule. Regions with a high density of valence electrons repel each other, forcing bonds and lone pairs of electrons to position themselves as far apart as possible around the central atom.

- The bond angles and the number of atoms attached to the central atom determine a molecule's shape. The number of regions of electrons around the central atom determines bond angles.

- Bond hybridization theory describes how orbitals with different energies can form new hybridized orbitals with the same energy.

- If a molecule has polar bonds, the vector sum of the bond dipole moments is equal to the molecular dipole moment. A molecule in which the bond dipole moments don't cancel out is a polar molecule.

TERMS
valence shell electron pair repulsion (VSEPR) theory
tetrahedral molecule
trigonal pyramidal molecule
trigonal planar molecule
linear molecule
bent molecule
polar molecule
molecular dipole moment

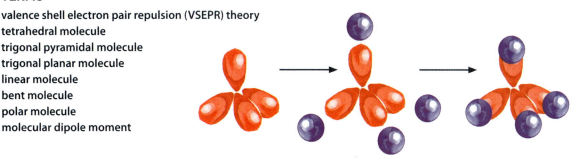

08 CHAPTER REVIEW

Chapter Review Questions

RECALLING FACTS

1. What information about molecules do Lewis structures represent well?
2. Describe how sigma bonds form.
3. Define *resonance*.
4. (True or False) The atmospheric pollutant sulfur dioxide (SO_2), a molecule exhibiting resonance, has single and double bonds that switch back and forth so rapidly that chemists cannot detect their state.
5. What two characteristics of liquid oxygen contradict predictions from the valence bond theory but are explained by the molecular orbital theory?
6. Define *VSEPR theory*.
7. How would we designate hybrid orbitals that combine an *s* orbital with two *p* orbitals?

UNDERSTANDING CONCEPTS

8. Evaluate the workability of Lewis structures to represent molecules.
9. Sketch a sigma bond, including the bond axis, between
   a. two *s* orbitals.
   b. an *s* orbital and a *p* orbital.
   c. two *p* orbitals.
10. Why do pi bonds result in two regions for electrons?
11. How do single and double bonds behave in a molecule that exhibits resonance?
12. Draw the two equivalent Lewis structures and the hybrid structure for NO_2^-.
13. Explain the three exceptions to the octet rule described in your textbook.
14. Describe the following molecule shapes and how they form according to VSEPR theory.
    a. linear
    b. trigonal planar
    c. bent
    d. trigonal pyramidal
    e. tetrahedral
15. Draw the Lewis structure for each of the following and then predict the shapes and ideal bond angle.
    a. ClO_2^-
    b. $COBr_2$
    c. PH_3
16. How do electron configurations change for hybridized orbitals?
17. Write the abbreviated orbital notation showing the hybridized orbitals for boron when it bonds with fluorine to form BF_3.
18. Why are bonds in most covalent compounds polar? How do polar bonds contribute to polar molecules?
19. Draw the Lewis structure for each of the following compounds. Determine whether its molecules are polar or nonpolar. If they are polar, indicate the direction of the dipole moment.
    a. CF_4
    b. NI_3
    c. SBr_2
20. Explain how a tetrahedral molecule could have a dipole moment.
21. How many polar bonds occur between atoms in each of the molecules below? Indicate whether the molecule is polar or nonpolar.
    a. H_2
    b. H_2CO
    c. HCl
    d. C_2H_4
    e. CO_2
22. Explain how the shape and structure of water confirm the Bible's teaching about God's care for creation.

CRITICAL THINKING

23. A classmate tells you that the textbook indicates that a sigma bond is stronger than a pi bond. Therefore a single bond must be stronger than a double bond since the double bond would be the average of the strong sigma bond and the weaker pi bond. Do you agree? Explain.
24. A classmate is looking at the Lewis structure below. She asks whether the triple bond forms from three sigma bonds. How do you reply?

$$:N \equiv N:$$

25. How does boron's position in the periodic table hint that it is stable when bonded to only three atoms?
26. What shape would you expect SF_6 to have?
27. Are the bonds in SF_6 polar? Would you expect it to be a polar molecule? Explain.
28. A classmate tells you that nonpolar molecules never have polar bonds, while all molecules with polar bonds are polar. Do you agree? Explain.

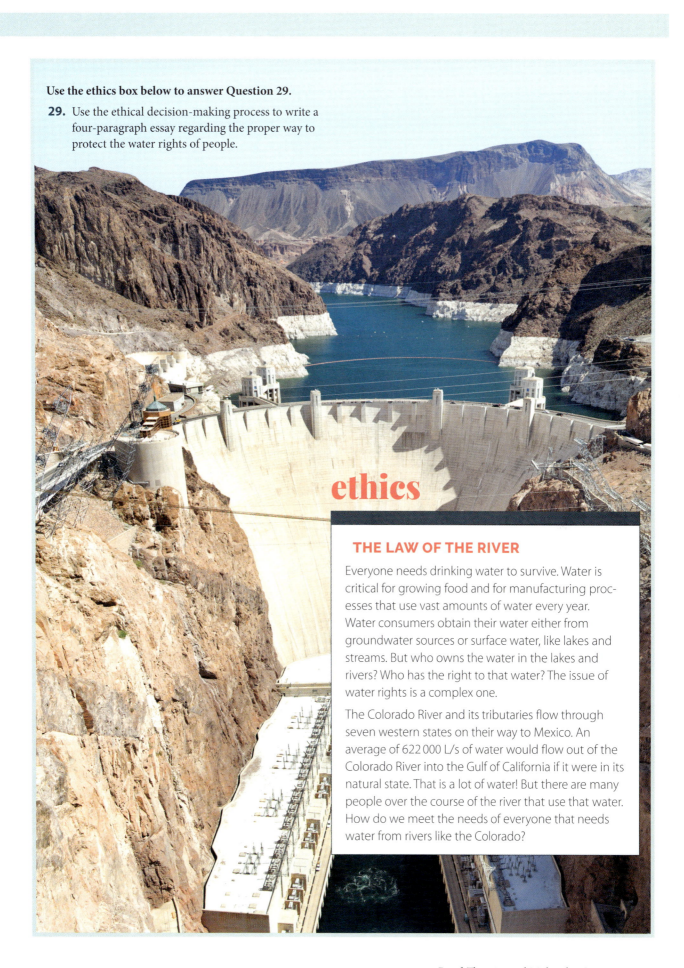

Use the ethics box below to answer Question 29.

29. Use the ethical decision-making process to write a four-paragraph essay regarding the proper way to protect the water rights of people.

ethics

THE LAW OF THE RIVER

Everyone needs drinking water to survive. Water is critical for growing food and for manufacturing processes that use vast amounts of water every year. Water consumers obtain their water either from groundwater sources or surface water, like lakes and streams. But who owns the water in the lakes and rivers? Who has the right to that water? The issue of water rights is a complex one.

The Colorado River and its tributaries flow through seven western states on their way to Mexico. An average of 622 000 L/s of water would flow out of the Colorado River into the Gulf of California if it were in its natural state. That is a lot of water! But there are many people over the course of the river that use that water. How do we meet the needs of everyone that needs water from rivers like the Colorado?

Chapter 9

CHEMICAL COMPOUNDS

Perhaps you've heard the old joke about the hazardous chemical known as *dihydrogen monoxide*. On average, this dangerous substance kills about ten people every day in the United States, yet it's commonly found in homes, businesses, and even in our air and soil. Of course, the reason why it is so common is that dihydrogen monoxide is the fancy chemical name for plain, everyday water.

Your fear of dihydrogen monoxide may be allayed now that you know that it's just water, but every day you are likely to run into or use chemical substances with equally exotic names. You probably applied some aluminum chlorohydrate ($Al_2Cl(OH)_5$) this morning and afterward may have eaten some pyrrolidine-2-carboxylic acid ($C_5H_9NO_2$), and washed it down with some 1,3,7-trimethylpurine-2,6-dione ($C_8H_{10}N_4O_2$). Don't panic, though—those are just the IUPAC names for the common ingredient in antiperspirants, an amino acid found in proteins, and caffeine.

Contrary to what you might think, chemists don't sit around dreaming up torturously hard names for high-school students to pronounce. Naming things is part of exercising dominion, and there's actually a standardized system in place for naming chemical substances. To the trained eye, chemical names give information about both the kind of substance being named and its structure.

9.1 Ionic Compounds *197*
9.2 Covalent Compounds *210*
9.3 Acids *213*

9.1 IONIC COMPOUNDS

Chemistry is much more than a study of the elements that compose matter. It also examines reactions that occur between elements and compounds. And like many human endeavors, chemistry has its own distinct **nomenclature**, or naming system. These are the words and phrases that mean something to those doing a particular kind of work, such as chemistry. To do chemistry effectively in this course, you'll need to learn some—but thankfully not all—of this unique terminology. You'll need to learn how to identify and write proper chemical formulas and name them. Then you can write balanced equations and classify them by reaction type. But don't worry—we won't start off with any tongue twisters like those in the Chapter opener.

Oxidation Numbers

In Chapter 7 we saw how electrons are lost, gained, or shared when bonds form. These bonds involve specific numbers of electrons, usually from the valence shells of the participating atoms. Most elements stabilize by forming bonds to attain a stable valence octet of electrons. Scientists have a "bookkeeping" system to keep track of electrons involved in bonding. The system uses oxidation numbers when writing formulas for chemical compounds. Although electrons are often shared to some extent, oxidation numbers assign each electron to a particular element within a compound.

Oxidation numbers, also called *oxidation states*, represent the number of electrons that an atom loses or gains when it bonds. A *negative* number means that the bonded atom has *gained* that many negatively charged electrons, and a *positive* number means that the atom has *lost* that many electrons. Low-electronegativity metals in compounds typically have positive oxidation numbers, while nonmetals with high electronegativities have negative oxidation numbers.

QUESTIONS
» How can I find the oxidation number for an element in a compound?
» What determines the formula for an ionic compound?
» How are ionic compounds named?
» Why do the names of some compounds contain roman numerals?

TERMS
nomenclature • oxidation number • binary compound • Stock system • oxyanion • hydrate • anhydrous compound

How do I write formulas for and name ionic compounds?

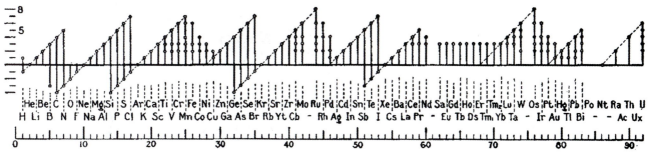

Fig. 1.

▲
A graph showing oxidation states for elements by Irving Langmuir, an early 1900s chemist

Oxidation numbers were originally assigned to elements in compounds after experimentally determining the composition of the compounds. After countless thousands of analyses, chemists developed rules that govern how oxidation numbers are assigned to elements. As with any system of scientific rules, which are in effect models, there are always exceptions. But these rules allow scientists to predict how an element typically combines with other elements. Let's have a look at these rules.

Rules for Oxidation Numbers

RULE 1: *THE FREE-ELEMENT RULE*

The oxidation number of atoms in their uncombined or pure form as free elements is 0. This includes those elements that exist naturally as diatomic elements, such as hydrogen and the halogens.

RULE 2: *THE ION RULE*

The oxidation number of a monatomic ion is equal to the charge of the ion. For example, when a bromine atom gains an electron to become a bromide anion, it has an oxidation number of −1.

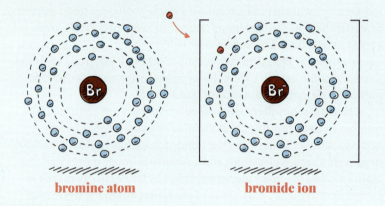

bromine atom **bromide ion**

RULE 3: *THE ZERO-SUM RULE*

The sum of the oxidation numbers of all the atoms in a compound must be 0. Compounds are not electrically charged.

Because the opposite charges in an ionic compound cancel out and because the oxidation numbers within ionic compounds are equal to the charges on the ions, oxidation numbers within ionic compounds must also cancel out, that is, add up to 0.

$$\overset{+1}{Na^+} + \overset{-1}{Cl^-} \longrightarrow \overset{0}{NaCl}$$

$$\overset{+2}{Mg^{2+}} + 2\overset{-1}{Cl^-} \longrightarrow \overset{0}{MgCl_2}$$

This rule applies to covalent compounds as well. Shared electrons reside closer to the more electronegative atom in a covalent bond. Therefore, the more electronegative atom is assigned a negative oxidation number on the basis of how many electrons it acquires by sharing. The less electronegative atom is assigned a positive oxidation number for the number of electrons that it "donates" by sharing with the more electronegative atom. Generally, the element with the highest electronegativity in a compound determines the oxidation numbers of the other elements in the compound.

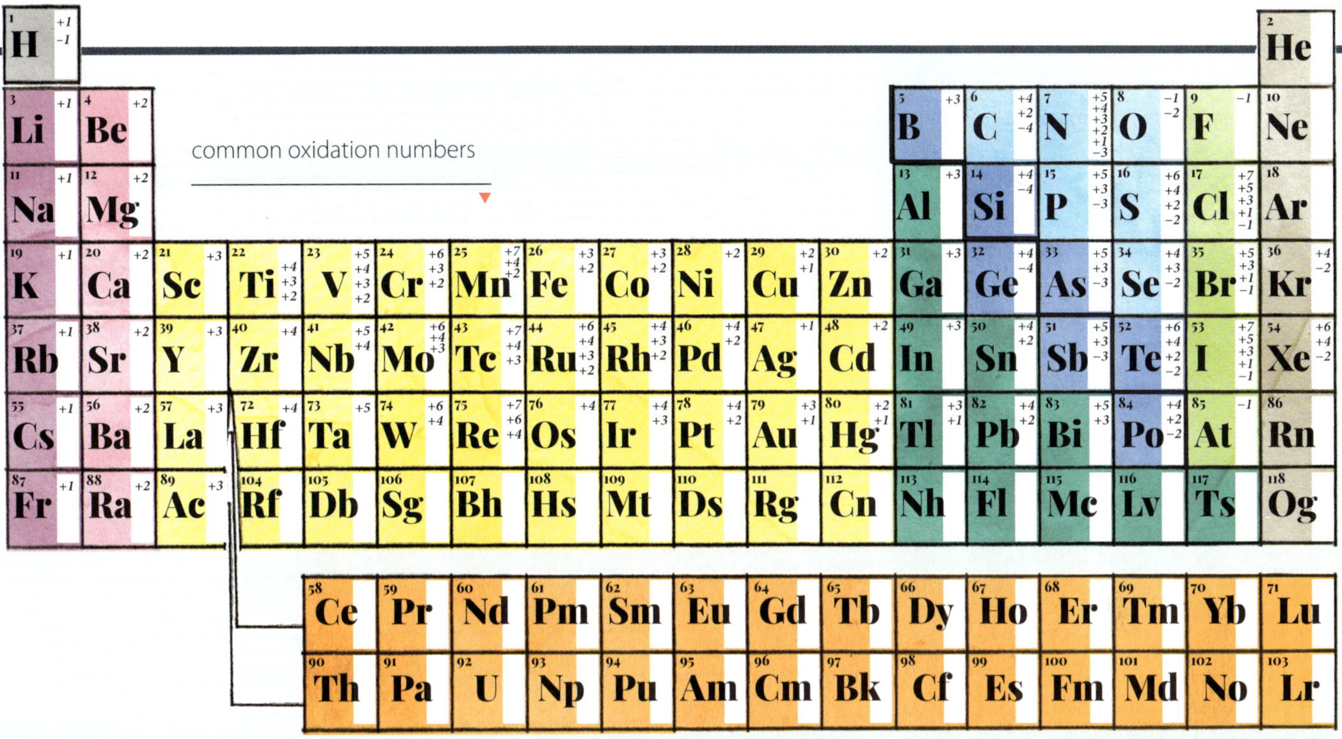

common oxidation numbers

RULE 4: THE SPECIFIC OXIDATION NUMBER RULE

Certain elements regularly have the same oxidation numbers in compounds. This makes writing chemical formulas much easier.

- **A.** Alkali metals (Group 1) always have a +1 oxidation number in compounds.

- **B.** Alkaline-earth metals (Group 2) always have a +2 oxidation number in compounds.

- **C.** Hydrogen usually has a +1 oxidation number when bonded to another nonmetal because the compound typically contains a more electronegative element. Hydrogen has a −1 oxidation number when bonded to metals to form ionic compounds called *metallic hydrides*.

- **D.** Oxygen usually has a −2 oxidation number except when bonded to fluorine. Because fluorine is the most electronegative element, oxygen has a positive oxidation number when combined with fluorine. Another exception is the peroxide ion, O_2^{2-}, in which each oxygen atom has a −1 oxidation number.

- **E.** Halogens (Group 17) have an oxidation number of −1 when bonded to metals. When halogens are bonded to other nonmetals, the element with the higher electronegativity is assigned the negative number. Fluorine always has an oxidation number of −1 because it is the most electronegative element.

Finding Oxidation Numbers

The few rules just listed determine the oxidation numbers for elements in multitudes of compounds. If one rule contradicts another, the rule listed first should be followed. If you can't figure out the oxidation number of an atom from these rules, you can use an algebraic equation to solve for it. Write an equation that is based on the fact that the sum of all the oxidation numbers in a compound must equal 0 (Rule 3).

Chemical Compounds

EXAMPLE 9-1: DETERMINING OXIDATION NUMBERS

Determine the oxidation number of each element in the following compounds.

a. Na_2O
b. H_2SO_4

Solution

a. The sum of the oxidation numbers of all the atoms in the compound must add up to 0 (Rule 3). The sodium atom always has a +1 oxidation number (Rule 4A), and the oxygen atom is normally −2 (Rule 4D). You can check these oxidation numbers using an algebraic equation.

2 × oxidation number for sodium + oxidation number for oxygen = 0

$$2(+1) + (-2) = 0$$

b. Because hydrogen is less electronegative than oxygen and sulfur, it has a +1 oxidation number (Rule 4C). Oxygen has an oxidation number of −2 (Rule 4D). Use an equation to determine the oxidation number of sulfur for this compound.

2 × oxidation number for hydrogen + oxidation number for sulfur + 4 × oxidation number for oxygen = 0

$$2(+1) + (\text{oxidation number for sulfur}) + 4(-2) = 0$$

oxidation number for sulfur = +6

Solving this equation, we determine that sulfur's oxidation number must be +6 (Rule 3). By checking the periodic table on page 199, we can confirm that +6 is a possible oxidation number for sulfur.

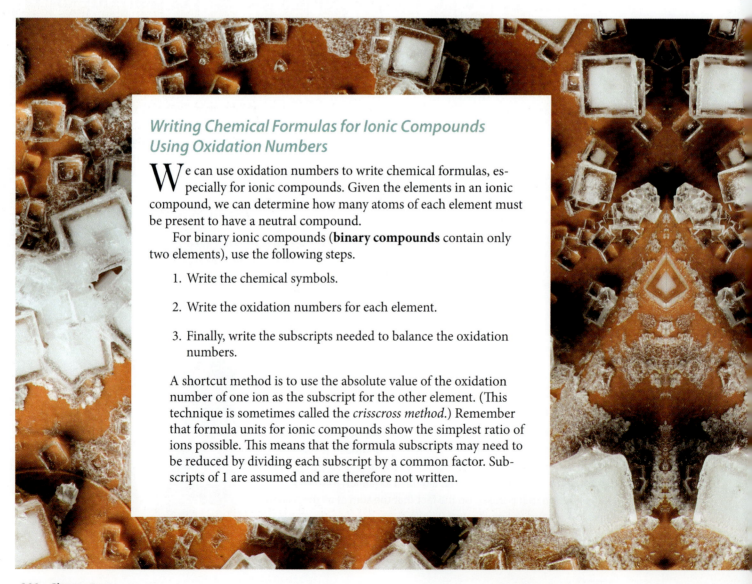

Writing Chemical Formulas for Ionic Compounds Using Oxidation Numbers

We can use oxidation numbers to write chemical formulas, especially for ionic compounds. Given the elements in an ionic compound, we can determine how many atoms of each element must be present to have a neutral compound.

For binary ionic compounds (**binary compounds** contain only two elements), use the following steps.

1. Write the chemical symbols.

2. Write the oxidation numbers for each element.

3. Finally, write the subscripts needed to balance the oxidation numbers.

A shortcut method is to use the absolute value of the oxidation number of one ion as the subscript for the other element. (This technique is sometimes called the *crisscross method*.) Remember that formula units for ionic compounds show the simplest ratio of ions possible. This means that the formula subscripts may need to be reduced by dividing each subscript by a common factor. Subscripts of 1 are assumed and are therefore not written.

EXAMPLE 9-2:
WRITING IONIC COMPOUND FORMULAS

Write the formula for each ionic compound:
  a. barium and iodine b. calcium and oxygen

Solution

a. Barium is an alkaline-earth metal with an oxidation number of +2 (Rule 4B). An iodine atom (a halogen) always has a −1 oxidation number when combined with a metal (Rule 4E). For the sum of the oxidation numbers to equal 0 (Rule 3), two iodine atoms must be included in the formula.

The crisscross method leads to this solution.

$$\overset{+2}{Ba} + \overset{-1}{I} \longrightarrow Ba_1 I_1 = Ba_1 I_2 = BaI_2$$

b. Calcium is also an alkaline-earth metal, so it has an oxidation number of +2 (Rule 4B). Oxygen always has a −2 oxidation number when combined with metals (Rule 4D).

$$\overset{+2}{Ca} + \overset{-2}{O} \longrightarrow Ca_2 O_2 = Ca_2O_2 = CaO$$

Though the first formula, Ca_2O_2, has equal negative and positive charges, it is incorrect because it is not in the simplest possible ratio.

One exception to the simplification step just shown is the peroxide ion, O_2^{2-}. For example, sodium peroxide has the formula Na_2O_2, not NaO, because the peroxide ion exists as a covalently bonded diatomic particle.

Oxidation Numbers and Polyatomic Ions

Polyatomic ions are groups of atoms that are covalently bonded and carry a charge. Since polyatomic ions are found in many compounds and reactions, they have their own oxidation number rule.

Rule 5: The oxidation numbers of all the atoms in a polyatomic ion add up to its charge.

For example, consider the hydroxide ion, OH^-. The oxygen atom has an oxidation number of −2, and that of the hydrogen atom is +1. The sum of these numbers is −1, which equals the charge on the ion. Polyatomic ions remain intact during most chemical reactions, so using the charge of the ion as you would use an oxidation number simplifies many problems.

EXAMPLE 9-3:
OXIDATION NUMBERS IN POLYATOMIC IONS

What is the oxidation number of the lead atom in $Pb(OH)_2$?

Solution
(oxidation number for lead) + 2(charge of hydroxide ion) = 0
(oxidation number for lead) + 2(−1) = 0
oxidation number for lead = +2

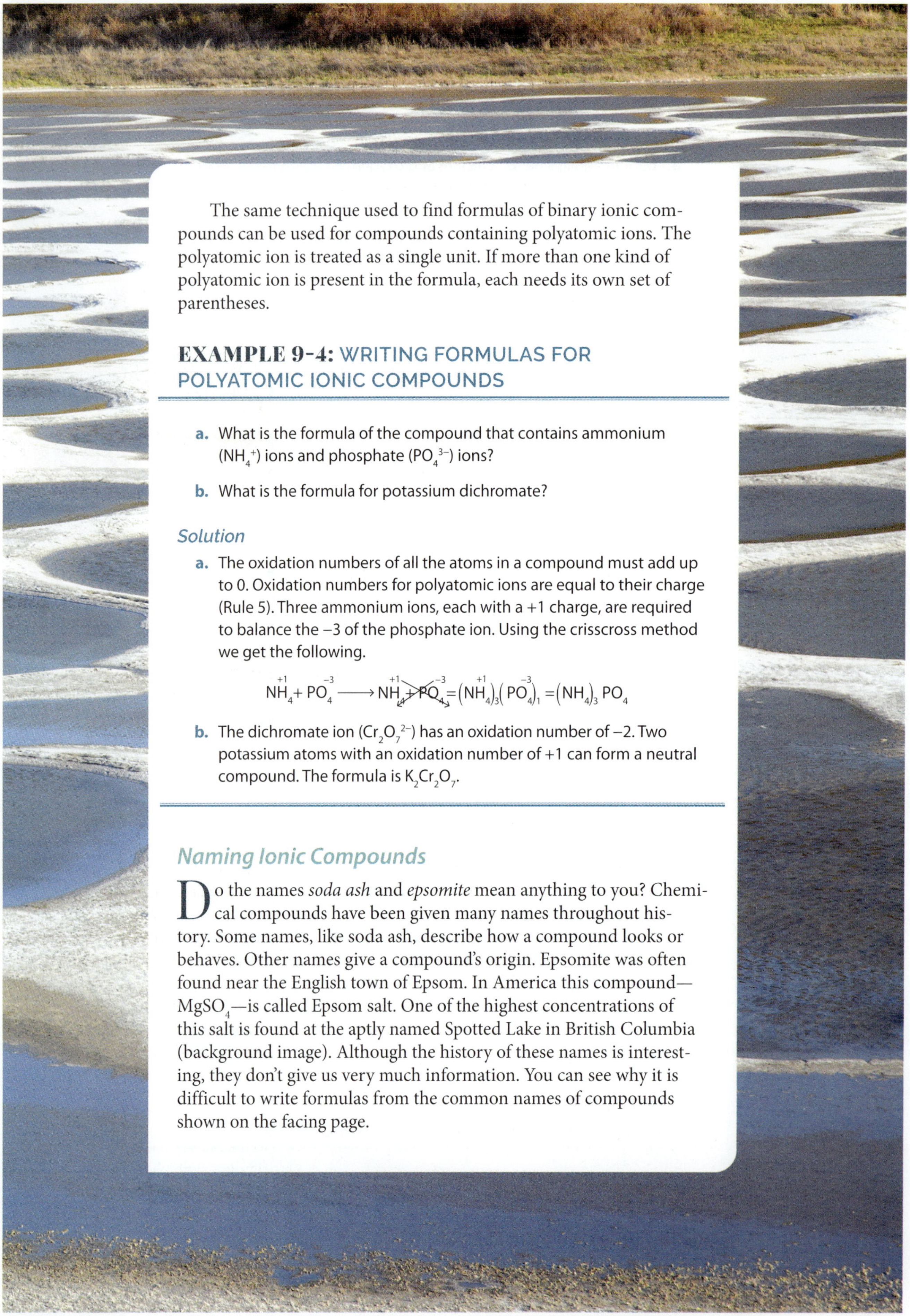

The same technique used to find formulas of binary ionic compounds can be used for compounds containing polyatomic ions. The polyatomic ion is treated as a single unit. If more than one kind of polyatomic ion is present in the formula, each needs its own set of parentheses.

EXAMPLE 9-4: WRITING FORMULAS FOR POLYATOMIC IONIC COMPOUNDS

a. What is the formula of the compound that contains ammonium (NH_4^+) ions and phosphate (PO_4^{3-}) ions?

b. What is the formula for potassium dichromate?

Solution

a. The oxidation numbers of all the atoms in a compound must add up to 0. Oxidation numbers for polyatomic ions are equal to their charge (Rule 5). Three ammonium ions, each with a +1 charge, are required to balance the −3 of the phosphate ion. Using the crisscross method we get the following.

$$\overset{+1}{NH_4} + \overset{-3}{PO_4} \longrightarrow \overset{+1}{NH_4} \underset{}{\searrow\!\!\!\!\nearrow} \overset{-3}{PO_4} = (\overset{+1}{NH_4})_3 (\overset{-3}{PO_4})_1 = (NH_4)_3 PO_4$$

b. The dichromate ion ($Cr_2O_7^{2-}$) has an oxidation number of −2. Two potassium atoms with an oxidation number of +1 can form a neutral compound. The formula is $K_2Cr_2O_7$.

Naming Ionic Compounds

Do the names *soda ash* and *epsomite* mean anything to you? Chemical compounds have been given many names throughout history. Some names, like soda ash, describe how a compound looks or behaves. Other names give a compound's origin. Epsomite was often found near the English town of Epsom. In America this compound—$MgSO_4$—is called Epsom salt. One of the highest concentrations of this salt is found at the aptly named Spotted Lake in British Columbia (background image). Although the history of these names is interesting, they don't give us very much information. You can see why it is difficult to write formulas from the common names of compounds shown on the facing page.

Common Names
OF SOME INDUSTRIAL CHEMICALS

As more and more compounds were discovered and synthesized, chemists realized that they could not continue to rely on memorized names. In the twentieth century, IUPAC developed a systematic way to name compounds that follows a standardized set of rules. IUPAC names of compounds are packed with information. From these names scientists gain an understanding of the elements present in a compound, the type of compound, its intermolecular attractions, and its general properties.

Common Name: slaked lime
IUPAC name: calcium hydroxide
Formula: $Ca(OH)_2$

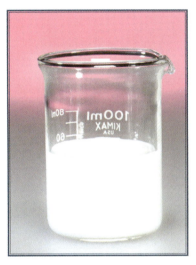

Common Name: milk of magnesia
IUPAC name: magnesium hydroxide
Formula: $Mg(OH)_2$

Common Name: soda ash
IUPAC name: sodium carbonate
Formula: Na_2CO_3

Common Name: lye
IUPAC name: sodium hydroxide
Formula: NaOH

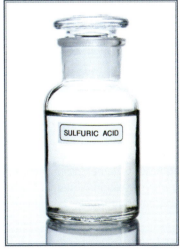

Common Name: oil of vitriol
IUPAC name: sulfuric acid
Formula: H_2SO_4

Today the term *soda ash* is used very little. Instead, chemists refer to this compound as *sodium carbonate*. The name *epsomite* has also given way to a more informative name, *magnesium sulfate*. Some common names like *water* and *ammonia* are still used today, but only when the compound is familiar. Let's look now at how some different kinds of compounds are named according to the IUPAC system. We'll start simple—with binary ionic compounds.

Chemical Compounds

NAMING BINARY IONIC COMPOUNDS

Binary ionic compounds, which are two-element compounds consisting of a metal and nonmetal, are named after the two ions involved. The cation is named first, followed by the anion. The positive ions (cations) use the same name as their parent atoms (sodium atoms form sodium ions). Negative ions (anions) take the *-ide* ending (e.g., chlorine atoms form chloride ions). Table 9-1 lists examples of *-ide* endings for nonmetals.

SOME COMMON IONIC COMPOUNDS

Compound: Al_2O_3
Name: aluminum oxide

Compound: MgO
Name: magnesium oxide

Compound: K_2S
Name: potassium sulfide

Compound: NaCl
Name: sodium chloride

TABLE 9-1 *Names of Nonmetallic Ions*

Element	Ion
antimony	antimonide
arsenic	arsenide
astatine	astatide
boron	boride
bromine	bromide
carbon	carbide
chlorine	chloride
fluorine	fluoride
hydrogen	hydride
iodine	iodide
nitrogen	nitride
oxygen	oxide
phosphorus	phosphide
selenium	selenide
silicon	silicide
sulfur	sulfide
tellurium	telluride

EXAMPLE 9-5:
NAMING BINARY IONIC COMPOUNDS

Name CaI_2.

Solution
Since CaI_2 is a compound of a metal and a nonmetal, it is a binary ionic compound. Its name, *calcium iodide*, is formed by listing the ions, beginning with the cation. Note that the name must end in -ide since it is a binary compound. *Iodine* is changed to *iodide*.

IONIC COMPOUNDS AND MULTIPLE OXIDATION NUMBERS

Many metals have more than one possible oxidation number, especially transition metals. Oxidation numbers for the transition metals are determined by how many of their electrons participate in chemical bonds. These different oxidation numbers pose a problem when

204   Chapter 9

it comes to naming ionic compounds. Take copper chloride, for example. Copper can form ionic bonds with chlorine in two different ratios, either a 1:1 copper-to-chlorine ratio or a 1:2 ratio. If someone says "copper chloride," which ratio do they mean? In instances like this, where the metal in an ionic compound can have more than one oxidation number, we make a slight addition to the naming system that we've seen so far. This modified nomenclature is called the **Stock system**, named after the German chemist Alfred Stock, and uses a roman numeral placed immediately, without a space, after the cation's name to show its oxidation number. This is sometimes called the *roman numeral system*. The chemical formula will determine the oxidation number and, therefore, the name of this type of ionic compound. Let's look at how this works.

Alfred Stock

EXAMPLE 9-6:
USING THE STOCK SYSTEM TO NAME IONIC COMPOUNDS

a. Name HgO.

b. Name Hg_2O.

Solution

a. Mercury is a metal that can have more than one oxidation number. The name of the anion is not affected by this and will still be called *oxide*. By consulting a periodic table, we see that oxygen's primary oxidation number is −2. We also see that mercury's primary oxidation number is +2, which means that mercury and oxygen can bond in a one-to-one ratio just as shown in the formula. Since mercury's oxidation number in this case is +2, we add the roman numeral "II" after the cation, enclosed by parentheses. The Stock system name is thus mercury(II) oxide.

b. From the periodic table we see that mercury can also have an oxidation number of +1. In this case, two mercury atoms can bond with a more electronegative oxygen atom, with the result being that all of the oxidation numbers add up to 0. The name for this compound is thus mercury(I) oxide because of mercury's +1 oxidation number in this compound.

Table 9-2 below gives examples of ionic compounds containing metals with more than one oxidation number. You will need to become familiar with which metals occur in more than one oxidation state so that you will know when to use the Stock system chemical names.

TABLE 9-2 *Stock System Compound Names*

Compound	Stock System Name
CuBr	copper(I) bromide
$CuBr_2$	copper(II) bromide
$FeCl_2$	iron(II) chloride
$FeCl_3$	iron(III) chloride
SnO	tin(II) oxide
SnO_2	tin(IV) oxide

Different cations of the same metal can combine with the same anion to form very different compounds.

WORLDVIEW INVESTIGATION

IUPAC

You're fresh out of college, and it's your first day working as a substitute teacher—no problem! Except that your first assignment is covering for Mrs. Santopietro's chemistry class for two days. On the first day, your notes say that you're supposed to go over IUPAC names for compounds. All goes well until a student raises his hand and asks, "What's IUPAC?" Great question—but you don't know the answer! So you tell the student, "I'll get back to you tomorrow on that."

Task

Find out who or what IUPAC is, what they do, and why they do it. Then create a brief presentation on the basis of your findings.

Procedure

1. Do an internet search using the keyword "IUPAC." Find out what you can about the history, purpose, structure, and work of the organization.
2. Note whether any of the websites you visit hints at a particular worldview expressed by IUPAC or its member organizations.
3. Plan your presentation and collect any photos or videos that you might use to accompany it.
4. Create your presentation and show it to another person for evaluation. Make any suggested revisions, if necessary.
5. Present your presentation to your class or family.

Conclusion

The work of science is a complex task that faces a huge potential for confusion if people use different terminology. Streamlining the process of dominion requires scientists to work together toward agreement on standardized nomenclature.

EXAMPLE 9-7: CONVERTING STOCK SYSTEM NAMES TO FORMULAS

We can also convert from Stock system names back to chemical formulas. For example, what is the formula for iron(III) oxide?

Solution

The roman numeral identifies the +3 oxidation number for iron. Oxide ions have a −2 oxidation number. A combination of two iron atoms for every three oxide ions results in an electrically neutral compound. You can find this ratio using the crisscross method described on page 200. The formula is Fe_2O_3.

NAMING POLYATOMIC IONIC COMPOUNDS

Tables 9-3 and 9-4 list the formulas and names of polyatomic ions. You should become familiar with these lists. Some generalizations can help make this table easier to learn. The only positive polyatomic ion on the list is the ammonium ion (NH_4^+). Anions that contain oxygen and one other element are called **oxyanions**. These ions often have two or more forms with different numbers of the same elements. For example, perchlorate, chlorate, chlorite, and hypochlorite all contain chlorine and oxygen in differing amounts. If there are only two forms of oxyanions, the form with more oxygen atoms ends in -ate and the form with fewer oxygen atoms ends in -ite. For example, SO_4^{2-} is the sulfate ion and SO_3^{2-} is the sulfite ion. For some polyatomic ions, especially those that contain halogens, there are more than two oxyanions. In such cases, the form with the most oxygen atoms has the prefix *per-*, and the form with the fewest oxygen atoms has the prefix *hypo-*. You can see how this works for the oxyanions of chlorine, bromine, and iodine in Table 9-3.

TABLE 9-3 *Common Oxyanions*

Oxidation Number	Greatest Number of Oxygens: per___ate	Base Number of Oxygens: ___ate	Fewer Number of Oxygens: ___ite	Fewest Number of Oxygens: hypo___ite
−1	perbromate (BrO_4^-)	bromate (BrO_3^-)	bromite (BrO_2^-)	hypobromite (BrO^-)
	perchlorate (ClO_4^-)	chlorate (ClO_3^-)	chlorite (ClO_2^-)	hypochlorite (ClO^-)
		hydrogen sulfate* (HSO_4^-)	hydrogen sulfite* (HSO_3^-)	
	periodate (IO_4^-)	iodate (IO_3^-)	iodite (IO_2^-)	hypoiodite (IO^-)
		nitrate (NO_3^-)	nitrite (NO_2^-)	
	permanganate (MnO_4^-)			
−2		carbonate (CO_3^{2-})		
		chromate (CrO_4^{2-})		
		dichromate ($Cr_2O_7^{2-}$)		
		oxalate ($C_2O_4^{2-}$)		
		sulfate (SO_4^{2-})	sulfite (SO_3^{2-})	
		thiosulfate ($S_2O_3^{2-}$)		
−3		arsenate (AsO_4^{3-})	arsenite (AsO_3^{3-})	
		borate (BO_3^{3-})		
		phosphate (PO_4^{3-})	phosphite (PO_3^{3-})	

*Hydrogen sulfate and hydrogen sulfite are not oxyanions, but they follow the same naming convention and so are included in Table 9-3.

TABLE 9-4 *Other Polyatomic Ions*

Name	Formula	Name	Formula
acetate	$C_2H_3O_2^-$	hydrogen sulfide	HS^-
amide	NH_2^-	hydroxide	OH^-
ammonium	NH_4^+	peroxide	O_2^{2-}
azide	N_3^-	silicate	SiO_4^{4-}
cyanate	OCN^-	tetraborate	$B_4O_7^{2-}$
cyanide	CN^-	thiocyanate	SCN^-
hydrogen carbonate	HCO_3^-		

Chemical Compounds

CAREERS

SERVING AS AN ANESTHESIOLOGIST: THE COMFORT OF NOT FEELING A THING

Chances are you have had some form of surgery. Chances are also excellent that you remember little, if anything, about the actual procedure. If so, you can thank an anesthesiologist for that particular blessing.

We can scarcely imagine the routine barbarity of surgery prior to the mid-nineteenth century. Obviously, having any part of your body cut open is extremely painful, even if the motivation for doing so is pure. Prior to the discovery of the analgesic (pain reducing) and amnestic (memory inhibiting) effects of diethyl ether ($C_4H_{10}O$) and chloroform ($CHCl_3$), prepping a patient for surgery was often little more than giving them a shot of rum and restraining them in some fashion. But since the discovery of anesthetics—chemical compounds that induce anesthesia—such practices are happily now a thing of the past.

An anesthesiologist is a specialized medical doctor who works with other doctors to provide care for patients before, during, and after an operation. A trained anesthesiologist must not only know how to sedate a patient but also how to sustain his vital functions during a procedure. Good people skills are a must as well since patients often experience anxiety before an operation. Working as an anesthesiologist, you can honor God's command to love others by helping them cope with what can be a traumatic experience.

TABLE 9-5
Compounds with Polyatomic Ions

Compound	Name
NH_4Cl	ammonium chloride
$Ba(ClO_3)_2$	barium chlorate
$Ca(ClO)_2$	calcium hypochlorite
$Mg(NO_2)_2$	magnesium nitrite
KOH	potassium hydroxide
Na_2SO_4	sodium sulfate

Once you know the names of polyatomic ions, you can name compounds that contain them. Simply name the cation and then the anion, just as with binary ionic compounds (but without modifying the ending of the anion's name if it is a polyatomic ion). Table 9-5 gives some examples of compounds containing polyatomic ions along withv their names. You'll notice that if more than one polyatomic ion is required to make a compound neutral, such as in barium chlorate, parentheses are placed around the ion and the subscript is placed outside the parentheses.

Some ion names in Tables 9-3 and 9-4 contain the word *hydrogen*. Continue to treat these as one single ion. That is, when they combine with metal ions, simply follow the same procedure as for other polyatomic ionic compounds—name the cation and the anion. For example, $NaHCO_3$ is called sodium hydrogen carbonate and $Mg(HS)_2$ is called magnesium hydrogen sulfide.

EXAMPLE 9-8:
NAMING POLYATOMIC IONIC COMPOUNDS

Name NH_4BrO_3.

Solution

The compound NH_4BrO_3 is ionic, so we simply name the ions. The name ammonium bromate is formed from the names of the two polyatomic ions included in the formula. Note that the bromate ion is polyatomic, so its name is *not* modified.

HYDRATES

Hydrates are compounds that hold a certain amount of water within their crystalline structures. This water is called the *water of hydration*. Water molecules combine with certain compounds in specific ratios determined by their crystalline structures. Formulas of these compounds indicate the presence and number of water molecules by a centered dot followed by the number of water molecules (e.g., $Na_2CO_3 \cdot 7H_2O$).

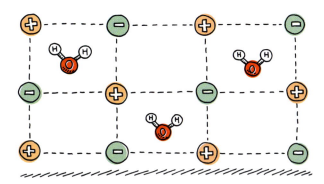

To name hydrates, the word *hydrate* preceded by a Greek prefix (Table 9-8) is added to the end of an ionic compound's name. In order to identify the dehydrated form of a compound as distinguished from its hydrates, the term *anhydrous* is used. An **anhydrous compound** has no water in its crystalline structure.

Table 9-6 lists sodium carbonate in its anhydrous form and several of its hydrate forms.

TABLE 9-6
Hydrate Compound Names

Formula	Name
Na_2CO_3	sodium carbonate (anhydrous)
$Na_2CO_3 \cdot H_2O$	sodium carbonate monohydrate
$Na_2CO_3 \cdot 7H_2O$	sodium carbonate heptahydrate
$Na_2CO_3 \cdot 10H_2O$	sodium carbonate decahydrate

9.1 SECTION REVIEW

1. What is the relationship between electrons and elements with negative oxidation numbers? between electrons and elements with positive oxidation numbers?

2. Determine the oxidation number of each element in $KClO_3$, potassium chlorate.

3. Write the formulas for the ionic compounds made by the following combinations of elements.
   a. aluminum and chlorine
   b. lithium and fluorine

4. Why are common names not always useful?

5. What is the formula of the compound that contains iron(II) and sulfate ions? How can you verify that the formula is correct?

6. What is the formula of the compound that contains hydrogen and phosphate ions?

7. Name the following compounds.
   a. Rb_2S
   b. $BaCl_2$
   c. $CaSO_4$
   d. FeC_2O_4

8. Name the following polyatomic ions.
   a. NH_4^+
   b. CN^-
   c. $C_2H_3O_2^-$
   d. NO_3^-
   e. MnO_4^-

9. Give the formula for the compound copper(II) sulfate pentahydrate.

QUESTIONS
» Why do names of covalent compounds use prefixes?
» How do I determine the name of a covalent compound on the basis of its formula?
» Can I determine the formula of a covalent compound on the basis of its name?

TERMS
Greek prefix system

How are the names of covalent compounds different from those for ionic compounds?

9.2 COVALENT COMPOUNDS

Nonmetals with Multiple Oxidation Numbers

Like many metals, some nonmetals can also have more than one oxidation number. In addition, a nonmetal can usually bond with another nonmetal in more ways than a metal ion could with the same nonmetal. This fact is reflected in the large number of different compounds that can be formed from the same two elements. For example, nitrogen can combine with oxygen in several different ratios, assuming nearly any oxidation number from +5 to −3 in the process. Table 9-7 shows possible compounds of nitrogen and oxygen. Each of these compounds is very different. For example, N_2O and NO_2 both make you go to sleep, but NO_2 can do so quickly and permanently! Because of this greater variety of potential compounds, chemists use a slightly different system for naming covalent compounds. The system is more specific about the kinds and numbers of atoms found in a covalently bonded molecule.

TABLE 9-7 *Compounds of Nitrogen and Oxygen*

Compound	Name	Oxidation Number for N	Properties
N_2O_5	dinitrogen pentoxide	+5	white crystalline solid, melting point of 41 °C
NO_2	nitrogen dioxide	+4	toxic brownish gas, used in manufacture of explosives
N_2O_3	dinitrogen trioxide	+3	pale blue liquid, decomposes at 3 °C
NO	nitrogen monoxide	+2	colorless gas, used to make nitric acid
N_2O	dinitrogen monoxide	+1	laughing gas, used as an anesthetic

Naming Binary Covalent Compounds

Let's first consider covalent compounds composed of only non-metals. Binary covalent compounds are not named using the Stock system. Instead, a **Greek prefix system** is used, in which the prefixes indicate how many atoms of each element are in a binary covalent compound (see Table 9-8).

There are a few simple rules to keep in mind when using this system. Have a look at these, and then we'll study an example of how the system is used.

- The prefix *mono-* is used only for the second element in the compound unless it's needed for emphasis or clarity.

- Generally, when the prefix system is used, any doubled vowels are omitted; for example, CO is named *carbon monoxide*, not *carbon mono-oxide*. There are a few exceptions to this rule, such as when the prefixes *di-* or *tri-* are used with *iodide*, as in nitrogen triiodide.

- The least electronegative element comes first, followed by the more electronegative element.

- The ending of the last element is changed to -ide.

- Unlike what we do in formula units, the atoms in covalent compounds are not reduced to the lowest whole-number ratio.

Deducing the formula of a binary covalent compound from its name is relatively simple since the number and kind of each element is implicit in the name. It's rather like doing the naming process in reverse. Take the name of silicon dioxide, for example. There is no prefix for silicon, so that means that there is only one silicon atom in the formula. The prefix *di-* in dioxide tells us that there are two atoms of oxygen in the formula. Thus, the formula is SiO_2.

TABLE 9-8 *Greek Prefixes*

Prefix	Number
mono-	1
di-	2
tri-	3
tetra-	4
penta-	5
hexa-	6
hepta-	7
octa-	8
nona-	9
deca-	10

EXAMPLE 9-9:
NAMING COVALENT COMPOUNDS

Name PCl_3.

Solution
PCl_3 is not ionic because all its elements are nonmetals. It should thus be named according to the Greek prefix system. There is only a single phosphorus atom, so the prefix *mono-* is not used. There are three chlorine atoms, signifying the need for the *tri-* prefix, and chlorine will also take the *–ide* ending. Thus, phosphorus trichloride is the accepted name.

EXAMPLE 9-10:
WRITING COVALENT COMPOUND FORMULAS

What is the molecular formula for tetraphosphorus decaoxide?

Solution
The prefix *tetra-* means "four," and *deca-* indicates "ten," so the formula is P_4O_{10}. Remember: Do not reduce the ratio of atoms in the formula for a covalent compound.

Chemical Compounds **211**

MINILAB

SAME STUFF, DIFFERENT NAME?

Dichloroethene is a covalent compound with the formula $C_2H_2Cl_2$. The problem is, it's not the *only* compound with that formula. How does this kind of situation affect the way that we do the work of chemistry?

Are complex-sounding chemical names really necessary?

Procedure

A Use a molecular modeling set to build a model of dichloroethene as shown (see below right). Make sure to include a double bond between the carbons.

B Examine the model that you have made and then answer the following questions.

1. Without changing the configuration of the central carbon atoms, are there any other ways to arrange the hydrogen and chlorine atoms on the carbons? If so, how many?

2. Sketch the Lewis structure(s) for the alternative arrangement(s) of dichloroethene along with a sketch of the original arrangement.

3. Do you think that the molecule(s) you sketched for Question 2 will have the same chemical properties? Explain.

C Create a naming convention for distinguishing between the versions of dichloroethene that you have modeled. Label your sketches in Question 2 with the names that you have created.

D Compare your sketches with those of your classmates.

4. Did everyone come up with the same naming convention?

Conclusion

5. How did this activity demonstrate the need for a standardized naming system?

EQUIPMENT
- molecular modeling set

9.2 SECTION REVIEW

1. Why is the Greek prefix system used in naming covalent compounds?

2. Name the following compounds.
   a. SeF_6
   b. NI_3
   c. P_2O_5
   d. CBr_4

3. Write the formulas for the following compounds.
   a. sulfur dioxide
   b. triphosphorus pentanitride
   c. disulfur dibromide
   d. diboron tetrafluoride

9.3 ACIDS

Binary Acids

Acids are aqueous solutions of covalent compounds whose formulas usually begin with *hydrogen*. Acids consisting of hydrogen and one other nonmetal are called **binary acids**. When they are not dissolved, these covalent compounds are named as outlined in Section 9.1. But if they are dissolved in water, that is, in an aqueous solution, they are named differently. For example, as a gas, HCl is called *hydrogen chloride*. When HCl is dissolved in water, however, it forms an acid called *hydrochloric acid*. In general, the names of binary acids include the prefix *hydro-*, the root name for the nonmetal with an *-ic* ending, and the word *acid*. Table 9-9 gives some examples of these name changes.

TABLE 9-9 *Binary Acid Names*

Formula	Common Name (gases)	Acid Name (hydro___ic acid)
HBr	hydrogen bromide	hydrobromic acid
HCl	hydrogen chloride	hydrochloric acid
H_2S	hydrogen sulfide	hydrosulfuric acid

Ternary Acids

Ternary acids contain more than two elements, namely hydrogen, oxygen, and at least one other nonmetal. The oxygen and other nonmetal(s) are often combined in a polyatomic ion. The names of ternary acids are derived from the anions in the acids. If the anion's name ends in -ate, the ending changes to -ic and the word *acid* is added. If the anion's name ends in -ite, the ending changes to -ous and the word *acid* is added. Table 9-10 lists several examples.

TABLE 9-10 *Ternary Acid Names*

Anion in Ternary Acid		Ternary Acid	
Formula	Name	Formula	Name
ClO_3^-	chlorate	$HClO_3$	chloric acid
ClO_2^-	chlorite	$HClO_2$	chlorous acid
ClO^-	hypochlorite	$HClO$	hypochlorous acid
NO_3^-	nitrate	HNO_3	nitric acid
NO_2^-	nitrite	HNO_2	nitrous acid
PO_4^{3-}	phosphate	H_3PO_4	phosphoric acid
SO_4^{2-}	sulfate	H_2SO_4	sulfuric acid
SO_3^{2-}	sulfite	H_2SO_3	sulfurous acid

QUESTIONS
» Are all acids the same?
» How are acids named?
» Do acid names give any hints about their structure?

TERMS
binary acid • ternary acid

Why is naming acids so complex?

Take note that the rules for using suffixes when naming acids are slightly different from those used for naming ions that you learned in Section 9.1. Sulfur, for example, produces *sulf*ide ions but its name is not similarly shortened when naming acids (e.g., *sulfur*ous acid rather than *sulf*ous acid). The root for the element phosphorus is treated similarly.

EXAMPLE 9-11: NAMING ACIDS

a. Name HI when dissolved in water.

b. Name $HBrO_4$ when dissolved in water.

Solution

a. Since the formula HI contains hydrogen and one nonmetal, iodine, it must represent a binary acid. The *–ine* ending of iodine is thus changed to *–ic*, and the complete name of the acid becomes *hydroiodic acid*.

b. The formula $HBrO_4$ starts with *hydrogen*, which identifies it as an acid. Since the compound contains a polyatomic ion, its name is derived from the name of the ion. The BrO_4^- ion is the perbromate ion. The *-ate* ending changes to *-ic*. The correct name is therefore *perbromic acid*.

EXAMPLE 9-12: DERIVING FORMULAS FROM ACID NAMES

a. What is the formula for hydrosulfuric acid?

b. What is the formula for sulfuric acid?

Solution

a. The name *hydrosulfuric acid* contains the root *hydro-*, which tells us that this is a binary acid. The other nonmetal is sulfur. Since sulfur's oxidation number as a monatomic ion is −2, one sulfur atom needs two hydrogen atoms to balance the charge. Therefore, the correct formula is H_2S.

b. The absence of a *hydro-* prefix alerts us that this is a ternary acid. The *–ic* suffix on the sulfur tells us that the polyatomic anion is sulfate (SO_4^{2-}), which has an oxidation number of −2. Therefore, one sulfate ion needs two hydrogen ions, making the formula for sulfuric acid H_2SO_4.

9.3 SECTION REVIEW

1. What is a binary acid?

2. What is the difference between hydrogen iodide and hydroiodic acid?

3. Name the following acids.

   a. HF
   b. HClO
   c. H_2Se
   d. H_2SO_3

4. Write the correct formula and name for the binary acid that forms from the ion F^-.

5. Write the correct formula and name for the ternary acid that forms from the ion BO_3^{3-}.

214 Chapter 9

09 CHAPTER REVIEW
Chapter Summary

TERMS
nomenclature
oxidation number
binary compound
Stock system
oxyanion
hydrate
anhydrous compound

9.1 IONIC COMPOUNDS

- Oxidation numbers help keep track of valence electrons so that we can write formulas correctly. A negative number indicates that an element has gained electrons in a compound, and a positive number indicates that an element has lost electrons in a compound.

- Many elements have more than one possible oxidation number.

- The sum of the oxidation numbers in a molecule is 0, and polyatomic ions have sums equal to their charge.

- Formulas for binary ionic compounds are determined by writing the symbols for the ions and then writing subscripts that will balance the oxidation numbers of the elements.

- Polyatomic ions participate in compounds as single charged units. Those that contain oxygen and one other compound are called oxyanions. Polyatomic anions retain their ending in compound names.

- Names for binary ionic compounds consist of the cation, followed by the anion with its ending changed to -ide.

- For metals that have more than one possible oxidation number, the Stock system uses a roman numeral after the element's name to indicate the cation's oxidation number.

- A hydrate is an ionic compound that has water molecules incorporated into its structure in a fixed ratio to the formula units. Greek prefixes are used to indicate the number of water molecules in a hydrate.

TERMS
Greek prefix system

9.2 COVALENT COMPOUNDS

- Because nonmetals can have many oxidation numbers, two nonmetals can bond in numerous ways. Therefore, Greek prefixes are used to indicate the number of each type of atom in a covalent molecule.

- For binary covalent compounds, the least electronegative element is named first. The prefix *mono-* is not used for the first element in the name.

- The name of the second element in a binary covalent compound takes the *–ide* ending.

TERMS
binary acid
ternary acid

9.3 ACIDS

- Binary acids consist of hydrogen and one other nonmetal. Ternary acids consist of hydrogen, oxygen, and one other nonmetal.

- Names for binary acids begin with the *hydro-* prefix, followed by the anion root modified with the *–ic* acid ending.

- Both binary and ternary acids have formulas beginning with *hydrogen*.

- Names for ternary acids begin with the anion name, modified with the appropriate suffix (*-ate* becomes *–ic* and *-ite* becomes *-ous*), and end with *acid*.

09 CHAPTER REVIEW

Chapter Review Questions

RECALLING FACTS

1. (True or False) An element's oxidation number is the same as its group number on the periodic table.
2. What is the oxidation number of elemental magnesium? Which oxidation number rule applies?
3. (True or False) The sum of all the oxidation numbers in a compound should equal 0.
4. What two oxidation numbers can hydrogen have? When would it have each of these?
5. What determines the subscripts in an ionic compound?
6. In what order do we name the ions in a binary ionic compound?
7. Explain the significance of the roman numerals used in the names of some ionic compounds.
8. Why are Greek prefixes used in covalent compounds?
9. What is the Greek prefix that indicates four atoms of an element? seven atoms? nine atoms?
10. What does the *hydro-* prefix indicate about an acid?
11. What is the difference between a binary and a ternary acid?

UNDERSTANDING CONCEPTS

12. Explain why
    a. fluorine's oxidation number is always negative.
    b. the oxidation numbers of alkali metals are always positive.
    c. elements such as P, N, and S have positive oxidation numbers in some compounds but negative oxidation numbers in others.
13. When the following pairs of atoms bond, which atom has a positive oxidation number?
    a. H, O
    b. Na, S
    c. N, S
    d. Mg, H
14. Give the compound name as well as the oxidation number of each atom in the following binary ionic compounds.
    a. LiCl
    b. Mg_3N_2
    c. CaO
    d. NaI
    e. Al_2S_3
    f. CuCl
15. Give formulas for the following ionic compounds.
    a. zinc chloride
    b. calcium phosphide
    c. potassium chloride
    d. barium chloride
    e. strontium oxide
    f. calcium chloride
16. Give the compound name as well as the oxidation number of each atom in the following polyatomic ionic compounds.
    a. $AgNO_3$
    b. NH_4NO_3
    c. $NaNO_2$
    d. $Zn_3(PO_4)_2$
    e. $(NH_4)_2S$
17. Give the formula for each of the following compounds containing polyatomic ions.
    a. ammonium bromate
    b. calcium carbonate
    c. barium phosphate
    d. aluminum acetate
    e. potassium permanganate
    f. barium chromate
18. Give the compound name (using the Stock system) as well as the oxidation number of each atom in the following compounds.
    a. $PbCl_2$
    b. HgS
    c. CoS
    d. $Fe(OH)_3$
    e. $PbCrO_4$
    f. $Sn(C_2H_3O_2)_2$
19. Give the formula for each of the following compounds.
    a. iron(III) oxide
    b. copper(I) hydroxide
    c. lead(IV) chromate
    d. lead(II) arsenate
    e. tin(IV) chloride
20. Name the following hydrates.
    a. $CaSO_4 \cdot 2H_2O$
    b. $MgSO_4 \cdot 7H_2O$
    c. $Na_2SO_4 \cdot 10H_2O$
    d. $NiSO_4 \cdot 6H_2O$
    e. $Na_2S_2O_3 \cdot 5H_2O$
    f. $FeSO_4 \cdot 7H_2O$

21. Give the formula for each of the following hydrates.
    a. iron(III) bromide hexahydrate
    b. barium chloride dihydrate
    c. lead(II) acetate decahydrate
    d. cobalt(II) chloride hexahydrate
    e. sodium tetraborate decahydrate
    f. magnesium carbonate pentahydrate

22. Give the compound name as well as the oxidation number of each atom in the following covalent compounds.
    a. N_2O_3
    b. I_2O_5
    c. P_4O_6
    d. S_2Cl_2
    e. PCl_3
    f. Cl_2O_7

23. Give the formula for each of the covalent compounds below.
    a. carbon disulfide
    b. sulfur dioxide
    c. boron trichloride
    d. phosphorus pentabromide
    e. dinitrogen pentasulfide
    f. dibromine monoxide

24. Which is a ternary acid, HCl or $HClO_3$? Defend your answer.

25. Name each of the following acids.
    a. HBr
    b. $HClO_4$
    c. H_2S
    d. H_3PO_4

26. What is the chemical formula for nitrous acid? Explain how you arrived at your answer.

27. Name the following compounds. Give both the chemical name when not dissolved in water and when in solution.
    a. HF
    b. HI

28. Give the formula for each of the following compounds.
    a. hydroselenic acid
    b. hydrotelluric acid

29. Identify the anion and give the ternary acid name for each of the following compounds.
    a. H_3AsO_4
    b. $HC_2H_3O_2$
    c. H_2CO_3
    d. $HMnO_4$
    e. $H_2C_2O_4$

30. Give the formula for each of the following compounds.
    a. periodic acid
    b. chromic acid
    c. phosphoric acid
    d. cyanic acid
    e. bromous acid

CRITICAL THINKING

31. Give the compound name and the oxidation number of each atom in $Zn(C_2H_3O_2)_2$.

32. Name the substances SO_3 and SO_3^{2-}, and explain the differences between them.

33. Explain why the subscripts in S_2Cl_2 are not reduced to their lowest whole numbers.

34. Why would it be a mistake to identify an acid as hydrochlorous acid?

Use the ethics box below to answer Question 35.

35. Write a four-paragraph response on the ethics of mandatory drug testing using the ethics triad model presented in Chapter 3.

ethics

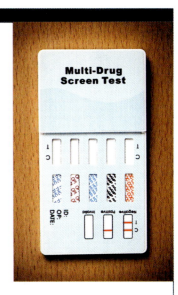

DRUG TESTING

The study of chemistry has led to the discovery of many useful drugs, from everyday aspirin to specialized drugs for treating cancer and other serious conditions. Yet because of our fallen nature, humans have used many drugs for less noble purposes. News headlines often include stories about the ravages of drug addiction, performance-enhancing drugs in sports, or seemingly endless debates about so-called "recreational drugs." What is typically not debated is that many of these controversial drugs impair cognitive and motor function—one's ability to think clearly and act normally. Because of this, many businesses have made mandatory drug testing a condition of employment. But some people argue that this is an invasion of privacy. How should Christians view mandatory drug testing?

Chapter 10

CHEMICAL REACTIONS & EQUATIONS

Alfred Nobel, the son of an inventor and engineer, developed an early fascination with explosives. The most commonly used explosive of the day was nitroglycerin, which is very powerful but also highly unstable. Merely striking it with a hammer can cause it to explode. Compelled by his brother's death in a nitroglycerin explosion, Nobel worked to make a more stable, safer form of nitroglycerin. In 1867, he mixed nitroglycerin with diatomaceous earth, developing dynamite.

Nobel went on to develop other explosives, eventually holding over 350 patents. He amassed a fortune in arms manufacturing, even though he was a pacifist. By the end of his life, he owned over ninety weapons factories.

Having been called the "Merchant of Death" by a French newspaper, Nobel was unhappy with his legacy. And so he used his fortune to establish awards for people working in the fields of physics, chemistry, medicine, and literature. A fifth prize is also now awarded to people or organizations that promote peace. Today we know these prizes as Nobel Prizes.

Chemistry does have the power to destroy, but it also has the power to heal. Throughout this chapter, we will investigate chemical reactions and their uses.

10.1 Chemical Equations *219*

10.2 Types of Reactions *228*

10.1 CHEMICAL EQUATIONS

Signs of Reactions

As you have learned in previous chapters, elements and compounds consist of tiny particles that are in constant motion. Because of their constant motion, these particles can collide with each other, and if the conditions are right, new substances can form. When elements and compounds collide and become a new substance, we say that a chemical change or chemical reaction has occurred. During a chemical reaction, the chemical bonds between atoms are broken, the atoms are rearranged, and new chemical bonds form. You will learn more about the mechanisms of chemical reactions in Chapter 16. Since we can't see individual atoms and how they interact in a chemical reaction, let's learn the observable signs that indicate that a reaction may have occurred.

The only way to be sure that a chemical reaction has occurred is to observe whether a new chemical substance has been formed. That being said, there are also some other forms of evidence that suggest that a reaction has taken place.

QUESTIONS
» What happens to the atoms in a chemical reaction?
» How can I tell whether a chemical reaction has occurred?
» How is a chemical equation a model of a chemical reaction?
» How do I write an equation for a chemical reaction?

TERMS
precipitate • chemical equation • reactant • product

What do chemical equations do for us?

Evidences of CHEMICAL REACTIONS

CHANGE IN ENERGY ▶
Energy is required to break existing chemical bonds, and energy is released when new bonds form. Those two energies are almost never equal. Therefore, chemical changes are almost always accompanied by a change in energy, often observed as a temperature change. The change could also be noticed as the production of light or sound, such as when iron reacts with potassium nitrate in sparklers.

PRODUCTION ▶ OF A GAS
Oftentimes chemical reactions will create a gas. This may be seen as bubbles in a liquid or as colored vapor. Chemical reactions inflate automobile airbags amazingly quickly by converting numerous solid substances into large amounts of nitrogen gas.

▼ FORMATION OF A PRECIPITATE
Not all substances can be dissolved in water. Therefore, when different solutions are mixed, the resulting substances may not be soluble. In such cases, the substance will come out of solution as a solid. A solid that comes out of solution due to a chemical or physical change is called a **precipitate**. Potassium iodide and lead(II) nitrate are soluble in water, but when mixed they react to form insoluble lead(II) iodide, a yellow solid.

▼ COLOR CHANGE
The chemical and physical properties of substances depend on their internal structure. When materials go through a chemical reaction, their physical properties change. One of the most noticeable changes is when the substances change color. We are all familiar with the reddish-brown color of copper, but some people don't realize that the green-hued Statue of Liberty is made of copper. Its familiar green color is a result of a chemical reaction forming a copper-containing compound.

While the occurrence of any one of these phenomena can be evidence of a chemical reaction, some of them can also occur with a physical change. The more forms of evidence that are present, the more certain we can be that a reaction has occurred.

Knowing that a reaction has occurred is just the starting point. What substances reacted? What substances did the reaction produce? How much of each substance was used or produced? Chemists pack the answers to these questions into an expression called a *chemical equation*.

MINILAB

CONSERVING ATOMS

In chemical reactions, the atoms in the substance that react are rearranged to form new substances. The law of conservation of matter tells us that atoms cannot be created or destroyed. Therefore, all the atoms that are in the substances going into the reaction must come out. Conversely, all the atoms that come out of a reaction must have gone into the reaction.

How can I model a chemical reaction?

1. What is the chemical formula for water? for hydrogen? for oxygen? Don't forget your diatomic molecules.

Procedure

A Using the molecular modeling kit, make a hydrogen molecule and an oxygen molecule.

B Now rearrange the atoms to form as many water molecules as you can. Remember that you have to use all the atoms that you start with.

2. How many water molecules could you make from the hydrogen and oxygen molecules that you started with?

3. Were you able to use all the atoms that you started with?

C Make the smallest possible number of additional molecules—hydrogen, oxygen, or both—to make additional water molecules with no atoms left over. Then rearrange the atoms to make water molecules.

4. What additional molecules did you need to make? How many did you make?

Conclusion

5. How many of each starting substance did you need? How many atoms of each element did this represent?

6. How many water molecules were you able to make? How many atoms of each element did this use?

Going Further

7. What could we call this process of accounting for all the atoms on each side of a chemical reaction?

EQUIPMENT

- molecular modeling kit

220 Chapter 10

The Information in Chemical Equations

A **chemical equation** is a model that uses chemical symbols, coefficients, and other symbols to represent a chemical reaction. A chemical equation is similar to the equations that you use in math. For instance, it has symbols and opposite sides that are equivalent. But in the place of an equal sign that equates two quantities, a chemical equation has an arrow that separates the chemical substances. To accurately describe reactions, chemical equations must meet four requirements.

First, chemical equations must identify all the substances involved in a reaction. For example, some water softeners remove calcium salts, such as calcium hydrogen carbonate, from hard water by adding calcium hydroxide. The two compounds react to form water and calcium carbonate, which settles out of the solution. The substances that take part in a reaction appear on the left-hand side of a chemical equation and are called **reactants**. Substances that are produced by a reaction appear on the right-hand side and are called **products**. In the equation below, calcium hydrogen carbonate and calcium hydroxide are reactants, and water and calcium carbonate are products. As mentioned above, an arrow separates the reactants from the products and shows the direction of the reaction.

Second, chemical equations must show the composition of substances involved in a reaction. Because formulas communicate more information than names, we use them in chemical equations.

$$Ca(HCO_3)_2 + Ca(OH)_2 \longrightarrow H_2O + CaCO_3$$

If we were to read the equation for this reaction, we would say, "Calcium hydrogen carbonate reacts with calcium hydroxide to produce, or yield, water and calcium carbonate."

But we are not quite finished with this equation. The third requirement of chemical equations is that they must account for all the atoms involved in a reaction. According to the law of conservation of matter, matter cannot be created or destroyed in chemical reactions. The atoms that go into a reaction must come out of the reaction because a chemical reaction is a rearranging of the atoms involved. Different compounds form, but the total number and types of atoms must remain constant.

As it now stands, our chemical equation does not obey the law of conservation of matter, as illustrated at right. Note the difference between the number of atoms on the reactant and product sides of the equation. If you total them, sixteen atoms in the reactants produce eight atoms in the products. Can sixteen atoms turn into eight atoms? Absolutely not! To be correct, the equation must show equal numbers of atoms both before and after the reaction.

Element	Atoms in Reactants	Atoms in Products
Ca	2	1
H	4	2
C	2	1
O	8	4

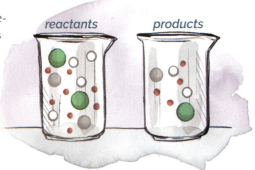

reactants   products

CASE STUDY

WASTE NOT, WANT NOT

Americans generate millions of tons of unusable hazardous waste every year. A substance is a hazardous waste if it can catch fire, react, corrode other materials, or cause health problems like cancer. We have many methods of handling and treating our hazardous waste.

The chemical 1,1-difluoroethane is very useful as a refrigerant and in aerosol applications. But it is also a health hazard, so we must properly treat it once we have used it. We treat 1,1-difluoroethane by incinerating it or by burning it in a furnace to produce electricity. But when 1,1-difluoroethane is burned, it reacts in the following manner, producing poisonous gases as a byproduct.

$$C_2H_4F_2 + O_2 \xrightarrow{\Delta} HF + CO + H_2O \text{ (unbalanced)}$$

These toxic products are easily treated and disposed of. Other methods of treating 1,1-difluoroethane are not as effective.

Questions to Consider

1. Name the toxic chemicals produced when we burn 1,1-difluoroethane. If this chemical produces toxic products when incinerated, why is it treated this way?
2. Give three biblical reasons why you ought to be concerned about caring for the environment.
3. What are some practical things that you can do to minimize the production of hazardous waste?

Chemical equations that follow the law of conservation of matter are called *balanced* chemical equations. Balancing an equation involves adjusting the number of atoms, ions, molecules, or formula units by placing numbers called *coefficients* in front of the formulas. Coefficients tell how many atoms, ions, molecules, or formula units are present. When you count atoms, multiply a coefficient by the number of atoms of every element in the formula that it precedes. If there is no coefficient, there is just one atom, ion, molecule, or formula unit.

Let's see how the equation for removing calcium salts from hard water looks when balanced. Note that the number of atoms on each side of the equation is now the same.

$$Ca(HCO_3)_2 + Ca(OH)_2 \longrightarrow 2H_2O + 2CaCO_3$$

We read our balanced equation, "One formula unit of calcium hydrogen carbonate reacts with one formula unit of calcium hydroxide to produce two water molecules and two calcium carbonate formula units." This method of balancing an equation by adjusting coefficients is called *balancing by inspection*.

Element	Atoms in Reactants	Atoms in Products
Ca	2	2
H	4	4
C	2	2
O	8	8

The final requirement of chemical equations is that the coefficients must be whole numbers in the smallest possible ratio. If coefficients were not whole numbers, they would be fractions, which would imply that we had a part of an atom or molecule. But if we had only a part of an atom or molecule, it would no longer be that substance. In the example above, half a water molecule isn't a water molecule anymore. The need for the smallest whole number ratio has to do with the fact that we are analyzing a reaction. We need to know what is required for a particular reaction to occur. In the above reaction we must have one formula unit of each reactant for the reaction to occur. While the reaction would occur with two of each, only one of each is required.

Balancing Equations by Inspection

As we just saw, chemical equations must be balanced so that equal numbers of each atom appear on both sides of the equation. Here are a few general guidelines to help you balance equations by inspection. Consider the reaction that occurs when nitrogen monoxide reacts with oxygen in the atmosphere to produce nitrogen dioxide.

1. Write formulas for all reactants and products. Make sure that all the formulas are correct. *Once you have written the correct chemical formulas, do not change them!*

$$NO + O_2 \longrightarrow NO_2$$

2. Check to see whether the equation is balanced. One technique is to use a table like the one on the right to keep track of the elements on the reactant and product sides of the equation. You can update the number of atoms as you adjust the coefficients.

Element	Atoms in Reactants	Atoms in Products
N	1	1
O	3	2

3. Adjust coefficients until there are equal numbers of atoms on both sides of the equation. When balancing equations by inspection, start with the most complicated molecules and save simple molecules until last. *Never change a compound's subscripts to balance an equation!* Doing so would change the identity of the substance.

In the example above, we need an additional oxygen atom on the product side of the equation. The only way to get another oxygen atom in the product is to add another NO_2 molecule. Changing the coefficient to 2 means that you now have two of every atom in the molecule.

$$NO + O_2 \longrightarrow 2NO_2$$

Element	Atoms in Reactants	Atoms in Products
N	1	2
O	3	4

This still leaves the equation unbalanced. In fact, doubling the amount of NO_2 molecules didn't even solve the problem with the unbalanced oxygen atoms. In addition to that issue, there are now two nitrogen atoms in the product but only one in the reactants. But we needn't panic. Putting a 2 in front of the NO in the reactants will produce equal numbers of every atom on both sides.

$$2NO + O_2 \longrightarrow 2NO_2$$

Element	Atoms in Reactants	Atoms in Products
N	2	2
O	4	4

Chemical Reactions and Equations

4. Check to be sure that coefficients are whole numbers in the simplest ratio possible.

If in balancing the above equation we arrive at

$$NO + \frac{1}{2}O_2 \longrightarrow NO_2,$$

we have a fractional coefficient. To rid a chemical equation of fractions, multiply all the coefficients by the least common denominator so that they become whole numbers.

$$2NO + 2\left(\frac{1}{2}O_2\right) \longrightarrow 2NO_2$$

$$2NO + O_2 \longrightarrow 2NO_2$$

Similarly, if we arrive at the equation

$$4NO + 2O_2 \longrightarrow 4NO_2,$$

it again is balanced, but the coefficients are not in the simplest possible ratio. Dividing each coefficient by the greatest common factor yields the lowest possible ratio.

$$\frac{4}{2}NO + \frac{2}{2}O_2 \longrightarrow \frac{4}{2}NO_2$$

$$2NO + O_2 \longrightarrow 2NO_2$$

Special Symbols in Equations

Special symbols pack additional information into equations. Double arrows between reactants and products show that a reaction happens in the reverse direction as well as the forward direction. Such reactions are called *reversible reactions*; you will learn more about these in Chapter 17.

$$3Fe + 4H_2O \rightleftharpoons Fe_3O_4 + 4H_2$$

Symbols in chemical equations can also indicate the physical states of reactants and products. Gases are indicated by (g). Liquids are represented by (l). If a substance is dissolved in water, (aq), meaning "aqueous," appears after the formula. Since acids always appear as dissolved in water, (aq) should be put after every acid. For example, H_2SO_4 is hydrogen sulfate, but $H_2SO_4\,(aq)$ is sulfuric acid. Solids are indicated by (s). See examples of these symbols in the following reaction.

$$3Fe(s) + 4H_2O(l) \rightleftharpoons Fe_3O_4(s) + 4H_2(g)$$

We indicate special reaction conditions with symbols above and below the arrow. A delta (Δ) above the arrow means that the reactants are heated. We can communicate other conditions such as pressure, light, or specific temperatures by placing those words above the arrow.

Catalysts are substances that change the rate of the reaction but do not undergo permanent changes themselves. When a catalyst is used, we write its chemical formula above the arrow. We can include these extra pieces of information when reading chemical equations aloud. For instance, the equation

$$2KClO_3(s) \xrightleftharpoons[]{Fe_2O_3, \Delta} 2KCl(s) + 3O_2(g)$$

can be read as, "solid potassium chlorate heated in the presence of iron(III) oxide yields solid potassium chloride and oxygen gas."

The symbols generally used in chemical equations appear in Table 10-1.

TABLE 10-1 *Symbols Used in Chemical Equations*

Symbol	Use
+	placed between reactants and between products
⟶	means "yields" or "produces"; separates reactants from products
⇌	used in place of a single arrow to indicate a reversible reaction
(g)	indicates a gaseous reactant or product
(s)	indicates a solid reactant or product
(l)	indicates a liquid reactant or product
(aq)	indicates that the reactant or product is in an aqueous solution (dissolved in water)
$\xrightleftharpoons[]{\Delta}$	indicates that heat must be supplied to reactants for the reaction to occur
$\xrightarrow{MnO_2}$	indicates that a catalyst must be supplied to reactants for the reaction to occur

EXAMPLE 10-1: WRITING A BALANCED EQUATION

Solid iron and gaseous chlorine react to form iron(III) chloride powder. Write and balance the chemical equation for this reaction.

Solution
First, write the formulas. Iron atoms can react as individual atoms, but gaseous chlorine is a diatomic element.

$$Fe(s) + Cl_2(g) \longrightarrow FeCl_3(s)$$

Second, count the atoms.

Element	Atoms in Reactants	Atoms in Products
Fe	1	1
Cl	2	3

Third, adjust the coefficients. Start with the element having the largest subscript in the most complex formula—the Cl in the $FeCl_3$. To balance an element with subscripts of 2 and 3, you will need to use the least common multiple, which in this case is 6.

$$Fe(s) + 3Cl_2(g) \longrightarrow 2FeCl_3(s)$$

Element	Atoms in Reactants	Atoms in Products
Fe	1	2
Cl	6	6

The chlorine atoms are balanced, but now the iron atoms are not. Twice as many iron atoms are needed on the reactant side.

$$2Fe(s) + 3Cl_2(g) \longrightarrow 2FeCl_3(s)$$

Element	Atoms in Reactants	Atoms in Products
Fe	2	2
Cl	6	6

Fourth, check the equation. The number of atoms is balanced, and all coefficients are whole numbers in the simplest ratio possible.

Chemical Reactions and Equations

EXAMPLE 10-2: WRITING A BALANCED EQUATION

Ethane gas (C_2H_6) reacts with oxygen gas, producing gaseous carbon dioxide and liquid water. Write a balanced equation for this reaction.

Solution
Remember that oxygen is a diatomic element.

$$C_2H_6(g) + O_2(g) \longrightarrow CO_2(g) + H_2O(l)$$

Element	Atoms in Reactants	Atoms in Products
C	2	1
H	6	2
O	2	3

Start with the most complex formula (C_2H_6). Leave any free elements (O_2) until last because they are easiest to balance. Balancing carbon and hydrogen atoms results in an equation that is still unbalanced.

$$C_2H_6(g) + O_2(g) \longrightarrow 2CO_2(g) + 3H_2O(l)$$

Element	Atoms in Reactants	Atoms in Products
C	2	2
H	6	6
O	2	7

Now seven oxygen atoms are on the product side. Since this oxygen molecule is diatomic, divide the number of atoms needed (7) by the number found in each oxygen molecule (2) to get 7/2 as a coefficient.

$$C_2H_6(g) + \frac{7}{2}O_2(g) \longrightarrow 2CO_2(g) + 3H_2O(l)$$

Element	Atoms in Reactants	Atoms in Products
C	2	2
H	6	6
O	7	7

The equation is now balanced, but since not all of the coefficients are whole numbers, the equation must be multiplied by 2 to produce a balanced, simplified equation.

$$2C_2H_6(g) + 7O_2(g) \longrightarrow 4CO_2(g) + 6H_2O(l)$$

Element	Atoms in Reactants	Atoms in Products
C	4	4
H	12	12
O	14	14

EXAMPLE 10-3: BALANCING AN EQUATION

Aqueous aluminum nitrate and aqueous sodium carbonate react to form a precipitate of aluminum carbonate and aqueous sodium nitrate. Write a balanced equation for this reaction.

Solution

$$Al(NO_3)_3(aq) + Na_2CO_3(aq) \longrightarrow Al_2(CO_3)_3(s) + NaNO_3(aq)$$

Since all the formulas are equally complex, start with any of them. We need twice as much aluminum on the reactant side. We also need three times as much carbon on the reactant side.

$$2Al(NO_3)_3(aq) + 3Na_2CO_3(aq) \longrightarrow Al_2(CO_3)_3(s) + NaNO_3(aq)$$

Now we need six times as much nitrogen and sodium on the product side.

$$2Al(NO_3)_3(aq) + 3Na_2CO_3(aq) \longrightarrow Al_2(CO_3)_3(s) + 6NaNO_3(aq)$$

Now the chemical equation is balanced.

$$2Al(NO_3)_3(aq) + 3Na_2CO_3(aq) \longrightarrow Al_2(CO_3)_3(s) + 6NaNO_3(aq)$$

Element	Atoms in Reactants	Atoms in Products
Al	~~1~~ 2	2
N	~~3~~ 6	~~1~~ 6
Na	~~2~~ 6	~~1~~ 6
C	~~1~~ 3	3
O	~~12~~ 27	~~12~~ 27

Notice that the two polyatomic ions remain the same throughout the reaction and that none of those elements occur elsewhere in the reaction. When that happens, you can keep track of the polyatomic ions (NO_3^{1-} and CO_3^{2-}) as units instead of individual nitrogen, carbon, and oxygen atoms.

Limitations of Balanced Equations

Although balanced equations give a lot of information about reactions and are vital to the study of chemistry, they do have several limitations. Remember that chemical equations are models of reactions and not the reactions themselves. Also, the fact that we can write an equation doesn't mean that the reaction occurs spontaneously. The equation below is balanced, but the reaction that it represents does not occur spontaneously. This reaction doesn't occur because silver is less reactive than sodium is. We will look more closely at the concept of reactivity in Section 10.2.

$$Ag + NaCl \longrightarrow Na + AgCl$$

Chemical equations also do not tell whether a reaction goes to completion. An equation shows us the reactants and products, but it doesn't show how a reaction occurs. Some reactions involve more than one step. Their equations do not show the intermediate steps of the reaction or the order in which those steps take place. For example, the chemical reaction represented by the equation

$$4C(s) + 6H_2(g) + O_2(g) \longrightarrow 2C_2H_5OH(l)$$

occurs as the culmination of three simpler intermediate reactions—

$$4C(s) + 4O_2(g) \longrightarrow 4CO_2(g)$$

$$6H_2(g) + 3O_2(g) \longrightarrow 6H_2O(l)$$

$$4CO_2(g) + 6H_2O(l) \longrightarrow 2C_2H_5OH(l) + 6O_2(g)$$

—but you certainly cannot tell that from the equation for the overall reaction.

10.1 SECTION REVIEW

1. What happens to the atoms in chemical reactions?

2. List several evidences that tell whether a chemical reaction has occurred.

3. What are three essential functions of a chemical equation?

4. Why can't subscripts be changed to balance chemical equations?

5. Explain what each red symbol means.

   $$2HgO(s) \xrightarrow{\Delta} 2Hg(l) + O_2(g)$$

6. Balance the following equations.

   a. $P_4(s) + S_8(s) \longrightarrow P_4S_3(s)$

   b. $KClO_3(s) \xrightarrow{\Delta} KCl(s) + O_2(g)$

   c. $AgNO_3(aq) + Cu(s) \longrightarrow Cu(NO_3)_2(aq) + Ag(s)$

   d. $H_3PO_4(aq) + Ba(OH)_2(aq) \longrightarrow Ba_3(PO_4)_2(s) + H_2O(l)$

   e. $NaHCO_3(s) \xrightarrow{\Delta} Na_2CO_3(s) + CO_2(g) + H_2O(g)$

7. Write a balanced chemical equation for each of the following word equations.

   a. Aqueous aluminum sulfate reacts with aqueous barium hydroxide, producing precipitates of aluminum hydroxide and barium sulfate.

   b. Solid aluminum metal reacts with aqueous sulfuric acid, producing hydrogen gas and aqueous aluminum sulfate.

   c. Hydrogen peroxide breaks into water and gaseous oxygen when exposed to bright sunlight.

   d. Sucrose, $C_{12}H_{22}O_{11}$ (table sugar), burns in oxygen, producing gaseous carbon dioxide and gaseous water.

   e. Aqueous calcium hydroxide reacts with gaseous sulfur trioxide, producing a precipitate of calcium sulfate and liquid water.

8. What are some benefits to using chemical equations?

QUESTIONS
» How can I tell the type of reaction from a chemical equation?
» How do I know whether a reaction will occur?

TERMS
synthesis reaction • decomposition reaction • combustion reaction • single-replacement reaction • double-replacement reaction • complete ionic equation • spectator ion • net ionic equation

Are all chemical reactions the same?

Iron and oxygen combine to form rust.

10.2 TYPES OF REACTIONS

Most reactions can be classified by how a chemical change takes place. Grouping similar reactions leads to generalizations that can help us better understand them. In this section, we will discuss synthesis, decomposition, combustion, single-replacement, and double-replacement reactions.

Synthesis Reactions

Synthesis reactions, also known as *combination reactions*, combine two or more substances into a single product. These reactions follow a general pattern and can be recognized by having a lone product.

$$A + B \longrightarrow AB$$

EXAMPLES

1. Metals combine with nonmetals, other than oxygen, to form compounds called salts.

$$Ca(s) + Cl_2(g) \longrightarrow CaCl_2(s)$$

2. Metals combine with oxygen to form metal oxides.

$$2Mg(s) + O_2(g) \longrightarrow 2MgO(s)$$

3. Nonmetals react with oxygen to form nonmetal oxides.

$$P_4(s) + 5O_2(g) \longrightarrow 2P_2O_5(s)$$

4. Water and metal oxides combine to form metal hydroxides.

$$H_2O(l) + CaO(s) \longrightarrow Ca(OH)_2(s)$$

5. Water and nonmetal oxides combine to form oxyacids.

$$H_2O(l) + SO_3(g) \longrightarrow H_2SO_4(aq)$$

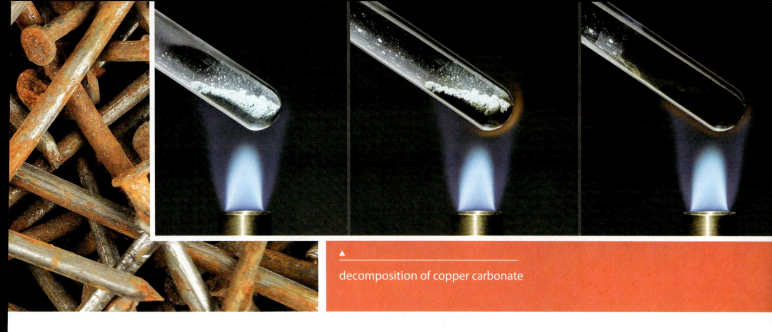

decomposition of copper carbonate

Decomposition Reactions

While synthesis reactions combine substances, **decomposition reactions** break single substances down into two or more simpler substances. Decomposition reactions are characterized by one substance becoming more than one.

$$AB \longrightarrow A + B$$

EXAMPLES

1. Oxygen can be driven out of some compounds.

   a. Metal oxides are usually very stable, but some compounds of less active metals decompose by heating them to high temperatures.

   $$2HgO(s) \xrightarrow{\Delta} 2Hg(l) + O_2(g)$$

   b. Metal chlorates, when heated, produce oxygen.

   $$2KClO_3(s) \xrightarrow{\Delta,\ MgO_2} 2KCl(s) + 3O_2(g)$$

   c. Water is decomposed by an electric current.

   $$2H_2O(l) \xrightarrow{elec.} 2H_2(g) + O_2(g)$$

2. Metal hydroxides release water vapor when heated.

   $$Mg(OH)_2(s) \xrightarrow{\Delta} MgO(s) + H_2O(g)$$

3. Metal carbonates release carbon dioxide when heated.

   $$CaCO_3(s) \xrightarrow{\Delta} CaO(s) + CO_2(g)$$

4. Some acids decompose into nonmetal oxides and water when heated.

   $$H_2CO_3(aq) \xrightarrow{\Delta} CO_2(g) + H_2O(l)$$

Chemical Reactions and Equations

HOW IT WORKS

Dynamite!

If you were to ask most people how to categorize the chemical reaction that occurs when dynamite explodes, most people would say that it is obviously a combustion reaction. But it is actually a decomposition reaction.

The explosive power of dynamite comes from nitroglycerin, which is a highly unstable liquid. To make nitroglycerin safer, and therefore useful, Alfred Nobel mixed nitroglycerin with clay material (see Chapter opener). He patented the resulting dynamite and made his fortune from it and other explosives.

When detonated, nitroglycerin rapidly decomposes according to the following chemical equation.

$$4C_3H_5N_3O_9(l) \longrightarrow 12CO(g) + 6N_2(g) + 10H_2O(g) + 7O_2(g)$$

In addition to the tremendous energy released when the reaction occurs, the reaction also produces a huge amount of gas, which expands. The rapidly expanding gas produces high temperatures and pressures on surrounding material. This combination makes dynamite useful in mining, demolition, and construction.

Combustion Reactions

In **combustion reactions**, a substance reacts with oxygen and releases energy in the form of heat and light. Burning candles and campfires are excellent examples of combustion reactions. In a combustion reaction, oxygen is *always* a reactant.

We have already seen some combustion reactions in this section. The synthesis reactions that form magnesium oxide and phosphorus pentoxide are also combustion reactions. Other common combustion reactions are the combustion of hydrogen gas to form water.

$$2H_2(g) + O_2(g) \longrightarrow 2H_2O(g)$$

Coal, a common fuel source in electrical power plants, is a fossil fuel that is mostly carbon. In a power plant, coal is burned to produce energy for electrical power production. The combustion of carbon is shown below.

$$C(s) + O_2(g) \longrightarrow CO_2(g)$$

Probably the most common form of combustion is the burning of hydrocarbons—compounds composed of hydrogen and carbon. (You will learn more about hydrocarbons in Chapter 20.) The octane in gasoline and the natural gases methane, butane, and propane are all hydrocarbons that are burned as fuels. All combustion reactions of hydrocarbons produce carbon dioxide gas and water vapor. The following is the reaction for the burning of methane.

$$CH_4(g) + 2O_2(g) \longrightarrow CO_2(g) + 2H_2O(l)$$

We must be careful when burning hydrocarbons; when insufficient oxygen is available, the reaction will be incomplete, and deadly carbon monoxide will be formed instead of carbon dioxide.

Single-Replacement Reactions

In **single-replacement reactions**, a more reactive element replaces a less reactive element in a compound. These kinds of reactions are also called *single-displacement reactions*. They occur frequently in solutions and involve less energy than synthesis, decomposition, or combustion reactions. Single-replacement reactions generally follow one of two forms depending on whether the cation or anion in a compound is replaced. In either case, you can identify a single-replacement reaction by noticing that one element is replacing another element in a compound.

Replacing the cation
$A + BX \longrightarrow AX + B$

Replacing the anion
$X + AY \longrightarrow AX + Y$

EXAMPLES

1. Atoms of reactive metals can replace less reactive cations in solutions and compounds. For instance, solid zinc will replace the copper ions in a copper(II) chloride solution, forcing the copper to precipitate out.

    $Zn(s) + CuCl_2(aq) \longrightarrow Cu(s) + ZnCl_2(aq)$

    Some metals will react with acids, replacing the hydrogen in the solution. The hydrogen bubbles off as hydrogen gas and the metal becomes part of a salt.

    $Mg(s) + 2HCl(aq) \longrightarrow MgCl_2(s) + H_2(g)$

    Very reactive metals react with water to produce hydrogen gas.

    $2Na(s) + 2H_2O(l) \longrightarrow 2NaOH(aq) + H_2(g)$

    In the equation below, writing water as HOH clearly shows how this is a replacement reaction.

    $2Na(s) + 2HOH(l) \longrightarrow 2NaOH(aq) + H_2(g)$

2. More reactive halogens can replace less reactive halogens that are in solution.

    When we bubble chlorine gas through a sodium bromide solution, the more reactive chlorine replaces the less reactive bromine. The bromine atoms escape as bromine gas.

    $Cl_2(g) + 2NaBr(aq) \longrightarrow 2NaCl(aq) + Br_2(g)$

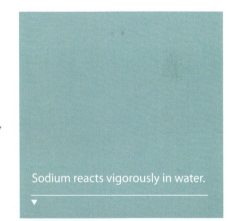

Sodium reacts vigorously in water.

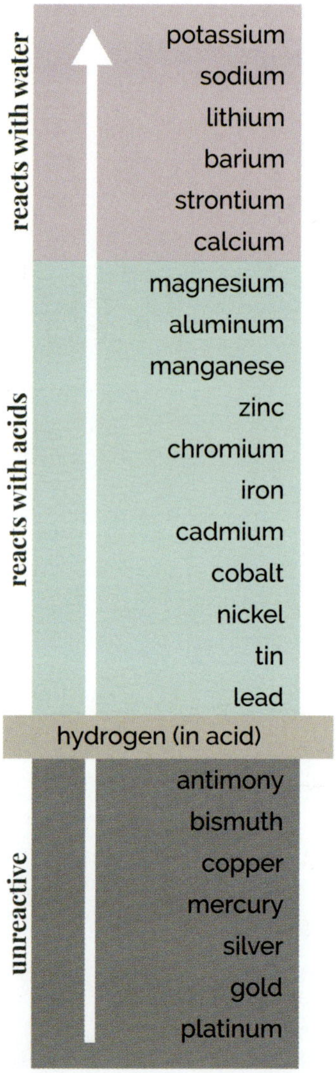

What determines whether a single-replacement reaction will occur? It depends on the reactivity of the elements that are involved. The reactivity of an element is related to its tendency to lose or gain electrons to form bonds. If unbonded atoms of an element are more likely to gain or lose electrons than the bonded atoms of another element, a reaction will probably take place. The more reactive element forces the less reactive element from the compound. The most reactive metal is potassium; the least reactive is platinum. Other metals can be arranged according to how reactive they are in a *reactivity series*, as shown at left.

Elements at the top of the list are the most reactive. Note that we include hydrogen even though it is not a metal. You can predict whether a reaction is probable from this series. For instance, will gold react with sodium chloride in solution? Gold is toward the bottom of the reactivity series. Its tendency to lose electrons and form ionic bonds is nowhere near that of sodium. Since no reaction will occur, a person can wear a gold ring when swimming in the ocean without worrying about losing a fortune!

$$Au(s) + NaCl(aq) \longrightarrow \text{no reaction}$$

With the aid of a reactivity series, it is also possible to predict which metals will react with acids. If a metal is more reactive than hydrogen, it can take the place of hydrogen. Because zinc is above hydrogen in the reactivity series, it will react with an acid such as dilute sulfuric acid.

$$Zn(s) + H_2SO_4(aq) \longrightarrow ZnSO_4(aq) + H_2(g)$$

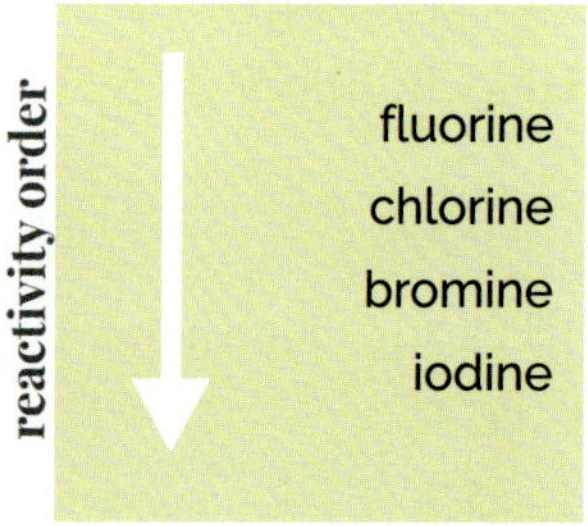

The halogens have a separate reactivity series, which follows the same order in which we find them on the periodic table. A halogen can replace another halogen that is below it in the series. Iodine, at the bottom of the series, cannot replace any of the other halogens because of its low reactivity.

$$Cl_2(g) + MgBr_2(aq) \longrightarrow$$
$$MgCl_2(aq) + Br_2(g)$$

Double-Replacement Reactions

In **double-replacement reactions**, components from two compounds switch places. They are also commonly called *double-displacement reactions*. You can recognize a double-replacement reaction by noticing that either the cations or the anions in two ionic compounds switch places. The reactions typically occur between compounds in aqueous solutions. Equations of double-replacement reactions have one of two general forms.

Cations switch
$$AX + BY \longrightarrow BX + AY$$

Anions switch
$$AX + BY \longrightarrow AY + BX$$

Whether or not a double-replacement reaction occurs depends on the products that might be formed. If a gas, water, or a precipitate forms, then the reaction occurs. Otherwise there is no reaction.

1. Production of a gas: A reaction between sodium cyanide and sulfuric acid will occur because hydrogen cyanide gas is one of the products.

$$2NaCN(aq) + H_2SO_4(aq) \longrightarrow Na_2SO_4(aq) + 2HCN(g)$$

2. Production of water: Mixing aqueous solutions of sodium hydroxide and hydrochloric acid results in a reaction occurring because water is produced.

$$NaOH(aq) + HCl(aq) \longrightarrow NaCl(aq) + H_2O(l)$$

3. Production of a precipitate: Mixing aqueous solutions of lead(II) nitrate and potassium chromate results in a reaction occurring. Solid lead(II) chromate, a brilliant yellow compound that is useful as a pigment in paint, precipitates out.

$$Pb(NO_3)_2(aq) + K_2CrO_4(aq) \longrightarrow PbCrO_4(s) + 2KNO_3(aq)$$

Sodium carbonate and lead(II) nitrate solutions react to form lead(II) carbonate precipitate.

CAREERS

SERVING AS AN EXPLOSIVE ORDNANCE DISPOSAL (EOD) TECHNICIAN:
THE BOMB SQUAD

This whole chapter has been about chemical reactions, but there are some people that train tirelessly to ensure that chemical reactions *don't* happen. Whether they are called to disarm unexploded munitions left from a conflict or to figure out a suspicious package, the bomb squad works to avoid reactions that could be catastrophic.

Bob Folger is a former explosive ordnance disposal expert for the US Navy. In retirement, he still works around the world disarming bombs and training the next generation to do the same. To be successful, an EOD technician needs to understand the chemical compounds that are found in the device. He must know what conditions could trigger an unintended ignition. He must also study different trigger methods. All this knowledge allows EOD specialists like Bob to render explosive devices safe for removal to a location where they can be disposed of.

Bob Folger studies hard to ensure that people around him are safe. EOD experts put themselves in harm's way because they recognize the value that God places on His image-bearers. Explosive ordnance disposal technicians show their love for God and others on a daily basis.

Lead(II) nitrate and potassium chromate solutions react to form lead(II) chromate precipitate.

It may be easy to determine whether a reaction will occur in the first two cases because it is easy to determine whether a gas or water will be produced. But how do you know whether a precipitate will form when two aqueous solutions are mixed? A precipitate forms when a substance comes out of solution, and a substance will come out of solution if it is insoluble. Table 10-2 lists some basic solubility rules to help you determine whether a precipitate forms. To use the rules, check the ions that make up each of the products. Read the table from top to bottom and exit the table as soon as one of the rules applies to the compound in question. If neither of the potential products "Forms Precipitate," no reaction occurs. If either of the products "Forms Precipitate," then the reaction does occur. Use Table 10-2 to determine the states [(s) or (aq)] of the products.

EXAMPLE 10-4:
DETERMINING WHETHER A PRECIPITATE WILL FORM

If the following aqueous solutions are mixed, determine whether the reaction will occur. If it will occur, write the balanced chemical equation and identify the precipitate.

a. calcium hydroxide mixed with sodium sulfate

b. potassium nitrate mixed with sodium chloride

Solution

a. Write the formulas for the reactants and then switch the cations or anions to determine the *potential* products.

$$Ca(OH)_2(aq) + Na_2SO_4(aq) \longrightarrow CaSO_4(?) + Na(OH)_2(?)$$

Check Table 10-2 for calcium sulfate and sodium hydroxide. The fourth line shows sulfates, which are soluble, except with six cations including calcium. Calcium sulfate is insoluble and forms a precipitate. So the reaction occurs; annotate calcium sulfate with an (s) for sodium hydroxide. The second line indicates that sodium compounds are soluble without any exceptions. Annotate sodium hydroxide with (aq). Balance the equation.

$$Ca(OH)_2(aq) + Na_2SO_4(aq) \longrightarrow CaSO_4(s) + 2Na(OH)(aq)$$

b. Write the formulas for the reactants and then switch the cations or anions to determine the *potential* products.

$$KNO_3(aq) + NaCl(aq) \longrightarrow KCl(?) + NaNO_3(?)$$

Check Table 10-2 for potassium chloride and sodium nitrate. The second line indicates that potassium compounds are soluble without any exceptions. Therefore, we would annotate its state of matter as (aq). Check the table again for sodium nitrate; the first line indicates that nitrates are always soluble. Therefore, sodium nitrate is also aqueous. Since no precipitate is formed, no reaction occurs.

$$KNO_3(aq) + NaCl(aq) \longrightarrow \cancel{KCl(?) + NaNO_3(?)} \quad \text{no reaction}$$

TABLE 10-2 *Solubility Rules*

Soluble (remains aqueous)	Exceptions (forms precipitate)
NO_3^-	
$Li^+, Na^+, K^+, Rb^+, NH_4^+$	
Cl^-, Br^-, I^-	$Cu^+, Ag^+, Hg_2^{2+}, Pb^{2+}$
SO_4^{2-}	$Ag^+, Ba^{2+}, Ca^{2+}, Sr^{2+}, Pb^{2+}, Hg_2^{2+}$

Insoluble (forms precipitate)	Exceptions (remains aqueous)
S^{2-}	$Mg^{2+}, Ba^{2+}, Ca^{2+}, Sr^{2+}$
CO_3^{2-}, PO_4^{3-}	Na^+, K^+, NH_4^+
OH^-	$Na^+, K^+, Ca^{2+}, Ba^{2+}$
O^{2-}	$Na^+, K^+, Ca^{2+}, Ba^{2+}, Sr^{2+}$

Ionic Equations

To better understand what is going on in double-replacement reactions, we use two different kinds of ionic equations. One shows all the particles present before and after the reaction; this is called a **complete ionic equation**. Ionic equations can be written only for reactions taking place in an aqueous solution. It is the aqueous environment that allows the ionization to occur.

Let's revisit the reaction of lead(II) nitrate solution with potassium chromate solution from the previous section. We know that the reaction occurs because of the formation of lead(II) chromate precipitate.

$$Pb(NO_3)_2(aq) + K_2CrO_4(aq) \longrightarrow PbCrO_4(s) + 2KNO_3(aq)$$

Let's turn this into a complete ionic equation by separating the soluble ionic compounds, indicated by (*aq*), into separate ions and leaving any insoluble ionic compounds, indicated by (*s*), combined.

$$Pb^{2+}(aq) + 2NO_3^-(aq) + 2K^+(aq) + CrO_4^{2-}(aq) \longrightarrow$$
$$PbCrO_4(s) + 2K^+(aq) + 2NO_3^-(aq)$$

In the complete ionic equations above, you will notice that some of the ions exist on both sides of the equation. These particles do not participate in the reaction. Because these ions are present but not participating, we call them **spectator ions**. In the example above, NO_3^- and K^+ are the spectator ions. Since they don't react with other ions, we can cancel them from both sides of the complete ionic equation, yielding a net ionic equation. **Net ionic equations** show only the ions that react, omitting any spectator ions.

$$Pb^{2+}(aq) + \cancel{2NO_3^-(aq)} + \cancel{2K^+(aq)} + CrO_4^{2-}(aq) \longrightarrow$$
$$PbCrO_4(s) + \cancel{2K^+(aq)} + \cancel{2NO_3^-(aq)}$$

Net ionic equation: $Pb^{2+}(aq) + CrO_4^{2-}(aq) \longrightarrow PbCrO_4(s)$

Many reactions between acids and bases that produce a salt and water (called *neutralization reactions*) are also double-replacement reactions. Notice the next example.

$$HCl(aq) + KOH(aq) \longrightarrow H_2O(l) + KCl(aq)$$

The complete ionic equation for this reaction follows.

$$H^+(aq) + Cl^-(aq) + K^+(aq) + OH^-(aq) \longrightarrow H_2O(l) + K^+(aq) + Cl^-(aq)$$

Cl^- and K^+ are spectator ions, so they can be omitted.

$$H^+(aq) + \cancel{Cl^-(aq)} + \cancel{K^+(aq)} + OH^-(aq) \longrightarrow H_2O(l) + \cancel{K^+(aq)} + \cancel{Cl^-(aq)}$$

The net ionic equation is

$$H^+(aq) + OH^-(aq) \longrightarrow H_2O(l).$$

Most neutralization reactions between acids and bases have this same form of net ionic equation. Double-replacement reactions usually reduce the number of ions in solution. Solid precipitates, such as lead(II) chromate, and the formation of the largely nonionizable water molecules in acid-base reactions (discussed in Chapter 18) reduce the number of ions.

10.2 SECTION REVIEW

1. What are the products of the combustion of any hydrocarbon?

2. Predict whether the following single-replacement reactions will occur. If they will, predict the products.

   a. $Ba(NO_3)_2(aq) + Ca(s) \longrightarrow$

   b. $BaCl_2(aq) + Br_2(l) \longrightarrow$

   c. $NaOH(aq) + K(s) \longrightarrow$

   d. Solid magnesium is added to an iron(III) chloride solution.

   e. Solid iron is added to aluminum sulfate solution.

3. Predict whether each of the following double-replacement reactions will occur. If the reaction will occur, write the balanced chemical equation and explain how you know that it will occur.

   a. Aqueous solutions of sodium hydroxide and copper(II) chloride are mixed.

   b. Aqueous magnesium nitrate solution and sodium sulfate are mixed.

   c. Aqueous solutions of barium hydroxide and hydrochloric acid are mixed.

4. Describe how you can recognize each of the five types of reactions outlined in this chapter.

5. Classify each of the following reactions.

   a. $8P_4(s) + 3S_8(s) \longrightarrow 8P_4S_3(s)$

   b. $2KClO_3(s) \xrightarrow{\Delta} 2KCl(s) + 3O_2(g)$

   c. $2AgNO_3(aq) + Cu(s) \longrightarrow$
      $Cu(NO_3)_2(aq) + 2Ag(s)$

   d. $4Fe(s) + 3O_2(g) \longrightarrow 2Fe_2O_3(s)$

   e. $2H_3PO_4(aq) + 3Ba(OH)_2(aq) \longrightarrow$
      $Ba_3(PO_4)_2(s) + 6H_2O(l)$

   f. $2NaHCO_3(s) \xrightarrow{\Delta}$
      $Na_2CO_3(s) + CO_2(g) + H_2O(g)$

   g. $C_3H_8(g) + 5O_2(g) \longrightarrow 3CO_2(g) + 4H_2O(g)$

6. What is a complete ionic equation?

7. Balanced chemical equations for two double-replacement reactions are given below. For each reaction, (1) write the complete ionic equation, (2) identify spectator ions, and (3) write the net ionic equation.

   a. $AgNO_3(aq) + HCl(aq) \longrightarrow AgCl(s) + HNO_3(aq)$

   b. $CaCl_2(aq) + Na_2CO_3(aq) \longrightarrow$
      $CaCO_3(s) + 2NaCl(aq)$

8. Solutions of lead(II) nitrate and sodium chromate are mixed and produce a precipitate of lead(II) chromate. (1) Write the balanced chemical equation, (2) give the complete ionic equation, (3) identify the spectator ions, and (4) write the net ionic equation.

10 CHAPTER REVIEW

Chapter Summary

TERMS
precipitate
chemical equation
reactant
product

10.1 CHEMICAL EQUATIONS

- Chemical reactions are a rearranging of the atoms in chemical substances to form new substances. Four signs that could indicate that a chemical reaction has occurred are a change in energy, the production of a gas, the formation of a precipitate, and a change of color.

- A balanced chemical equation shows the formulas and symbols for all substances involved in a chemical reaction and accounts for all atoms in both the reactants and the products to comply with the law of conservation of matter.

- Coefficients, subscripts, and special symbols provide more details on the quantities and conditions of a reaction.

- As a model of chemical reactions, chemical equations provide a lot of critical information but do not tell us everything about a reaction.

10.2 TYPES OF REACTIONS

- Five major categories of chemical reactions are synthesis, decomposition, combustion, single-replacement, and double-replacement.

- We can predict whether a single-replacement reaction will occur on the basis of the reactivity of the substances in the reaction. Double-replacement reactions will occur if at least one of the products is a precipitate, gas, or water.

- We can more accurately model what is going on in a double-replacement reaction with complete ionic and net ionic equations.

TERMS
synthesis reaction
decomposition reaction
combustion reaction
single-replacement reaction
double-replacement reaction
complete ionic equation
spectator ion
net ionic equation

Chapter Review Questions

RECALLING FACTS

1. What happens to chemical bonds in chemical reactions?
2. What is the only definitive way to know whether a chemical reaction has occurred?
3. Why is there typically a change in energy during a chemical reaction?
4. What is a precipitate?
5. Explain the steps to writing a balanced chemical equation.
6. Consider the reaction below.

    $HCl(aq) + H_2O(l) \rightleftharpoons Cl^-(aq) + H_3O^+(aq)$

    a. Why are two opposite arrows shown?
    b. Why is (aq) written after several of the substances?

Chemical Reactions and Equations

10 CHAPTER REVIEW

7. What determines whether a single-replacement reaction will occur?
8. Write the general form of the chemical equation for each type of reaction.
   a. synthesis
   b. decomposition
   c. single-replacement
   d. double-replacement
9. Define *spectator ion*.
10. What is a net ionic equation?

UNDERSTANDING CONCEPTS

11. Give one example each of the production of a gas by a physical change and a chemical change.
12. As two solutions are mixed in a beaker, the beaker gets warm and a white solid falls to the bottom. Did a chemical reaction occur? Explain.
13. Explain how a chemical equation is a model.
14. Balancing chemical equations satisfies which scientific law? Explain.
15. Refer to the case study on page 222 to balance the reaction for the burning of 1, 1-difluoroethane.
16. Balance each of the following equations. If the equation is already balanced, indicate that fact. Also, identify the type of reaction.
    a. $BaO_2(s) \xrightarrow{\Delta} BaO(s) + O_2(g)$
    b. $Li(s) + H_2O \xrightarrow{\Delta} LiOH(s) + H_2(g)$
    c. Acetylene gas (C_2H_2) burns in oxygen, producing carbon dioxide gas and water vapor.
    d. Sodium hydride forms when hydrogen gas is bubbled through molten sodium.
    e. $H_2SO_4(l) \xrightarrow{\Delta} H_2O(g) + SO_2(g) + O_2(g)$
    f. $NH_3(g) + HCl(g) \longrightarrow NH_4Cl(s)$
17. Balance the equations involved in the following processes. Indicate any that are already balanced. Also, identify the type of reaction.
    a. We use carbon black (pure carbon) in rubber tires and black ink. We produce it by breaking apart methane (CH_4).
    b. Ingested barium sulfate highlights the intestinal tract in x-ray pictures. We produce it by reacting solid barium with sulfuric acid. Hydrogen gas is also a product.
    c. Ammonia is produced commercially by combining nitrogen and hydrogen gases.
    d. Milk of magnesia is an aqueous suspension of magnesium hydroxide. When ingested, it reduces the amount of hydrochloric acid in the stomach by reacting with it to produce magnesium chloride and water.
    e. Copper(I) oxide forms when solid copper reacts with oxygen gas.
18. High-purity silicon is used to produce microcomputer chips. One process for producing chip-grade silicon entails three steps. Balance the equations involved in the following processes. Indicate any that are already balanced. Also, identify the type of reaction.
    a. The first step is to obtain impure silicon from molten sand (SiO_2).
    $$SiO_2(l) + C(s) \longrightarrow Si(l) + CO(g)$$
    b. The second step in the production of pure silicon is to produce silicon tetrachloride from the impure silicon.
    $$Si(s) + Cl_2(g) \xrightarrow{\Delta} SiCl_4(l)$$
    c. The last step in silicon production is to pass hot silicon tetrachloride vapor and hydrogen gas through a tube. Pure silicon deposits out of hydrogen chloride gas.
    $$SiCl_4(g) + H_2(g) \xrightarrow{\Delta} Si(s) + HCl(g)$$
19. What are some limitations to using a chemical equation?
20. Explain how the metal and halogen reactivity series are organized and what they tell us about a chemical reaction.
21. Predict whether the following single-replacement reactions are possible. If so, predict the products.
    a. $Ba(s) + 2H_2O(l) \longrightarrow$
    b. $I_2(aq) + MgCl_2(aq) \longrightarrow$
    c. Solid tin is added to a barium chloride solution.
    d. Fluorine gas is bubbled through hydrochloric acid.
22. Predict whether each of the following double-replacement reactions will occur. If so, write the balanced chemical equation and explain how you know that it will occur.
    a. Aqueous solutions of calcium nitrate and magnesium chloride are mixed.
    b. Aqueous solutions of sodium hydroxide and phosphoric acid are mixed.
    c. Solid iron(II) sulfide is mixed with hydrochloric acid. Hydrogen sulfide is a gas.

d. Aqueous solutions of sodium phosphate and magnesium nitrate are mixed.

$$3NaOH(aq) + H_3PO_4(aq) \longrightarrow Na_3PO_4(aq) + 3H_2O(l)$$

$$2Na_3PO_4(aq) + 3Mg(NO_3)_2(aq) \longrightarrow Mg_3(PO_4)_2(s) + 6NaNO_3(aq)$$

23. Balanced chemical equations for double-replacement reactions are given below. For each reaction (1) write the complete ionic equation, (2) identify spectator ions, and (3) write the net ionic equation.

    a. $H_2SO_4(aq) + 2KOH(aq) \longrightarrow$
       $K_2SO_4(aq) + 2H_2O(l)$

    b. $Mg(OH)_2(s) + 2HCl(aq) \longrightarrow$
       $MgCl_2(aq) + 2H_2O(l)$

    c. $Ba(NO_2)_2(aq) + Na_2SO_4(aq) \longrightarrow$
       $BaSO_4(s) + 2NaNO_2(aq)$

24. Hydrochloric acid reacts with solid aluminum hydroxide to form water and aluminum chloride solution. (1) Write the balanced chemical equation, (2) give the complete ionic equation, (3) identify the spectator ions, and (4) write the net ionic equation.

CRITICAL THINKING

25. A classmate states that when salt water is cooled and salt comes out of solution, it must be a chemical change because there is a temperature change (change in energy), a precipitate forms, and a color change occurs (clear liquid to a clear liquid and a white solid). Is she correct? Explain.

26. Another classmate writes the following balanced chemical equations for the given reactions. Is each equation correct? If not, explain the error and provide a properly balanced chemical equation.

    a. Aluminum solid reacts with oxygen to form aluminum oxide.

       $$2Al(s) + 3O_2(g) \longrightarrow 2AlO_3(s)$$

    b. $Ca(OH)_2(aq) + H_3PO_4(aq) \longrightarrow$
       $Ca_3(PO_4)_2(s) + H_2O(l)$

       $3Ca(OH)_2(aq) + 2H_3PO_4(aq) \longrightarrow$
       $Ca_3(PO_4)_2(s) + H_2O(l)$

    c. Aqueous solutions of iron(III) chloride and sodium hydroxide mix to form a precipitate of iron(III) hydroxide in sodium chloride solution.

       $$FeCl_3(aq) + 3NaOH(aq) \longrightarrow Fe(OH)_3(s) + 3NaCl(aq)$$

    d. Combustion of octane (C_8H_{18})

       $$C_8H_{18}(l) + \frac{25}{2}O_2(g) \longrightarrow 8CO_2(g) + 9H_2O(g)$$

27. Why do you think that decomposition reactions usually require added energy?

28. What is wrong with the following single-replacement reaction?

    $$Zn(s) + CuCl_2(aq) \longrightarrow Cl_2(g) + CuZn(s)$$

29. What properties of elements would you expect to have a significant impact on the reactivity of an element? Explain.

30. Where on the reactivity series (p. 232) would you expect to find the following elements? Explain.

    a. rubidium

    b. beryllium

    c. astatine

Use the ethics box below to answer Question 31.

31. Use the ethical-decision-making process modeled in Chapter 3 to prepare a four-paragraph response to the development of explosives from a biblical perspective.

ethics

EXPLOSIVES DEVELOPMENT

Alfred Nobel tried to develop a safer form of nitroglycerin to be used in mining operations. Much of Nobel's work and most of his companies were involved in weapons research and development, even though he was a pacifist. Today explosives are used in mining, construction, and controlled demolition, but explosives are also used in weapons for both conventional and unconventional warfare.

Chemical Reactions and Equations

Chapter 11

CHEMICAL CALCULATIONS

It's a cold night in the middle of winter, and somewhere a family is sleeping soundly. Suddenly, the stillness of the night air is shattered by a brash beeping sound. Awakened by the alarm, the family is able to move to a place of safety.

The family in the scenario described above wasn't awakened by a smoke detector. Rather, their alarm alerted  them to a dangerous buildup of poisonous carbon monoxide, a colorless, odorless, and tasteless gas. One source of it is the incomplete combustion of fuels such as the natural gas that is used to heat many homes. Every year, hundreds of people are killed in the United States by accidental carbon monoxide poisoning.

11.1 The Mole *241*

11.2 Stoichiometry *252*

11.3 Real-World Stoichiometry *259*

11.1 THE MOLE

Ideally, when the hydrocarbons in natural gas are burned, the only resulting products should be carbon dioxide and water. But like many other chemical reactions, this isn't always the case. Sometimes oxygen is the limiting factor—if there isn't enough available, the reaction cannot proceed to completion. Is there a way to determine how much oxygen is needed to combust a certain amount of natural gas? Or can we calculate how much gas will burn if only a limited amount of oxygen is available? It turns out that chemists have such a tool at their disposal—*stoichiometry*—which you will read about in Section 11.2. But before that we need to learn something about a certain peculiarly large number.

Avogadro's Number and the Mole

How many molecules do you think are in a test tube of water or a nugget of gold? It's a lot—sextillions, in fact. Yet these are the sizes of samples that chemists often work with. It is nearly impossible to work with single molecules or atoms in the laboratory. So how do we know how many particles are in a sample?

Because atoms are so small and are nearly impossible to count, scientists devised a unit called the mole to make counting atoms more practical. A **mole** (mol) is the amount of any substance that contains exactly $6.022\,140\,76 \times 10^{23}$ particles, though we will use 6.022×10^{23} in this textbook. This number is called **Avogadro's number** in honor of the Italian physicist Amedeo Avogadro. Even though it's a huge number, a mole is a defined quantity, much like a dozen is a defined quantity. A dozen donuts or a dozen eggs each consists of twelve items.

QUESTIONS
» What is a mole?
» How is the mole used in chemistry?
» Are there different kinds of chemical formulas?
» How can I determine a substance's percent composition?

TERMS
mole • Avogadro's number • molar mass • molecular formula • empirical formula • percent composition

How do we count atoms if we can't see them?

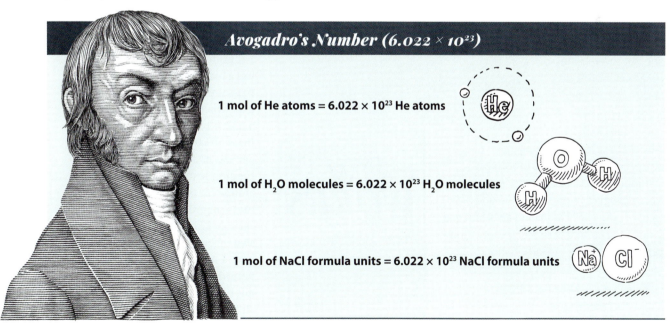

Avogadro's Number (6.022×10^{23})

1 mol of He atoms = 6.022×10^{23} He atoms

1 mol of H_2O molecules = 6.022×10^{23} H_2O molecules

1 mol of NaCl formula units = 6.022×10^{23} NaCl formula units

Because the mole contains such a large number of particles, it is used to describe only very small objects. No one has ever seen a mole of bricks or a mole of golf balls because no one has ever manufactured 602.2 sextillion of them! If one mole of marbles were spread out over the earth's surface, the earth would be covered over 2 km deep!

Chemical Calculations   **241**

EXAMPLE 11-1: USING AVOGADRO'S NUMBER

How many atoms are in a 4.5 mol sample of helium?

Write what you know.

4.5 mol He

Since one mole of any substance contains 6.022×10^{23} objects, this relationship can be used as a conversion factor to change 4.5 mol into the number of atoms.

Convert from moles of helium to atoms of helium.

$$4.5 \; \cancel{\text{mol He}} \left(\frac{6.022 \times 10^{23} \; \text{atoms He}}{1 \; \cancel{\text{mol He}}} \right) = 2.7 \times 10^{24} \; \text{atoms He}$$

Molar Mass

If you were to compare the mass of a dozen grapes with the mass of a dozen apples, you would find that the apples have considerably more mass. The number of objects is the same, but each apple has a much greater mass. Similarly, a mole of carbon atoms has a different mass than a mole of copper atoms. Can you guess which one contains more mass?

The mass of one mole of any pure substance is called its **molar mass**. These particles can be ions, atoms, molecules, or formula units. For example, a hydrogen atom has a mass of about 1.01 atomic mass units (u). Scientists have experimentally determined that 6.022×10^{23} hydrogen atoms have a mass of 1.01 g. Avogadro's number was assigned its value so that the atomic mass of an element (in atomic mass units) equals its molar mass (in grams per mole). A carbon-12 atom has a mass of 12 u, so 6.022×10^{23} carbon-12 atoms have a mass of 12 g. Thus, the periodic table can be used to find not only the mass of an individual atom, molecule, or formula unit but also the mass of one mole of these particles.

How can one mole of hydrogen atoms have a different mass than one mole of carbon atoms?

HOW IT WORKS

Carbon Monoxide Detector

Human ingenuity has produced not just one but several kinds of carbon monoxide detectors. Each kind works on the basis of one of carbon monoxide's chemical or physical properties. The kind of detector most commonly used in the United States is an electrochemical type. Inside this kind of detector is a fuel cell containing an electrolyte (a fluid that conducts electric current) and a pair of electrodes. As carbon monoxide seeps into the detector, it is oxidized by one electrode, producing carbon dioxide and some free electrons. The free electrons move to the other electrode, creating an electric current. The more carbon monoxide that enters the detector, the more electric current that is produced. If the current reaches a predetermined level, an audible alarm is triggered.

Remember that a mole, like a dozen, always contains a specific *number* of objects. The *mass* of each object determines the object's molar mass.

EXAMPLE 11-2: MASS AND MOLES

Calculate the mass of 0.500 mol of helium atoms.

Write what you know.

0.500 mol He

On average, a helium atom has a mass of 4.01 u, as shown on the periodic table. This number, expressed in grams, is also the mass of one mole of helium atoms. Thus, the molar mass of helium is 4.01 g.

Convert from moles of helium to mass of helium.

$$0.500 \text{ mol He} \left(\frac{4.01 \text{ g He}}{1 \text{ mol He}} \right) = 2.01 \text{ g He}$$

How many copper atoms are in a 4.00 g sample of pure copper wire? Answering such a question requires an additional step. We must first calculate the number of moles in 4.00 g of copper, then convert the number of moles to the number of particles. From the periodic table, we see that 1 mol of copper atoms is contained in 63.55 g, providing the necessary conversion factor. From there, the number of atoms can be calculated from the number of moles of copper. The complete calculation is shown below.

converts grams to moles

$$4.00 \text{ g Cu} \left(\frac{1 \text{ mol Cu}}{63.55 \text{ g Cu}} \right) \left(\frac{6.022 \times 10^{23} \text{ atoms Cu}}{1 \text{ mol Cu}} \right) = 3.79 \times 10^{22} \text{ atoms Cu}$$

converts moles to number of atoms

EXAMPLE 11-3: MASS AND ATOMS

How many atoms are in 33.3 mg of gold? (This amount is about the size of the period at the end of this sentence.)

Write what you know.

33.3 mg Au

Grams should first be changed to moles, and then the moles should be changed to atoms.

Convert from mass of gold to atoms of gold.

$$33.3 \text{ mg Au} \left(\frac{1 \text{ g Au}}{1000 \text{ mg Au}} \right) \left(\frac{1 \text{ mol Au}}{196.97 \text{ g Au}} \right) \left(\frac{6.022 \times 10^{23} \text{ atoms Au}}{1 \text{ mol Au}} \right) = 1.02 \times 10^{20} \text{ atoms Au}$$

Chemical Calculations

aluminum sulfate

Compounds and the Mole

Compounds contain two or more elements bonded together into molecules or formula units, so the atoms within them behave as one unit. The masses of molecules and formula units can thus be found simply by adding the masses of the atoms they contain. For example, a water molecule (H_2O) has a mass of

$$2(1.01 \text{ u}) + 1(16.00 \text{ u}) = 18.02 \text{ u},$$

where 1.01 u is the mass of a hydrogen atom and 16.00 u is the mass of an oxygen atom. A sodium chloride formula unit (NaCl) has a mass equal to the sum of one atom of sodium and one atom of chlorine.

$$1(22.99 \text{ u}) + 1(35.45 \text{ u}) = 58.44 \text{ u}$$

We use the molar mass of compounds and formula units just as we do with atoms. In any of these situations, we can easily express the molar mass in grams per mole (g/mol). As shown in Table 11-1, the numerical values for relative mass and molar mass are the same.

TABLE 11-1 *Masses of Compounds*

Object	Relative Mass	Molar Mass
H_2 molecule	2.02 u	2.02 g/mol
H_2O molecule	18.02 u	18.02 g/mol
NaCl formula unit	58.44 u	58.44 g/mol

EXAMPLE 11-4: CALCULATING MOLAR MASS

Find the molar mass of aluminum sulfate, $Al_2(SO_4)_3$.

Solution

Each formula unit contains two aluminum, three sulfur, and twelve oxygen atoms, so a mole of the compound $Al_2(SO_4)_3$ consists of 2 mol of aluminum atoms, 3 mol of sulfur atoms, and 12 mol of oxygen atoms. First, determine the mass that each element contributes to the compound.

Convert moles of elements into masses and then find the sum.

$$2 \text{ mol Al} \left(\frac{26.98 \text{ g Al}}{1 \text{ mol Al}} \right) = 53.96 \text{ g Al}$$

$$3 \text{ mol S} \left(\frac{32.06 \text{ g S}}{1 \text{ mol S}} \right) = 96.18 \text{ g S}$$

$$12 \text{ mol O} \left(\frac{16.00 \text{ g O}}{1 \text{ mol O}} \right) = 192.00 \text{ g O}$$

Then add the masses for each element together to get the molar mass for $Al_2(SO_4)_3$: 342.14 g/mol.

Types of Formulas

There are several ways to describe the composition of a chemical substance. *Structural formulas*, similar to the Lewis structures shown in Chapter 7, show the types of atoms involved, the exact composition of each molecule, and the arrangement of chemical bonds. These formulas are informative, but they can also be difficult to draw and often take up large amounts of space. The structural formula for water is simply H–O–H, but other structural formulas, such as the formula for a molecule of a fat (shown at right), can become quite complex.

Molecular formulas, like the familiar H_2O or CO_2, indicate the types and numbers of atoms found in a molecule, ion, or formula unit. These formulas are more convenient than structural formulas, but they do not give as much information about the arrangement of the atoms or locations and types of bonds found in a substance.

Empirical formulas tell what elements are present and give the simplest whole-number ratio of atoms in a compound. You have already seen that empirical formulas can be used to describe the composition of ionic compounds. When used for a molecular compound, an empirical formula might represent the actual molecular composition if the molecular formula contains a simple ratio, as in the case of H_2O. On the other hand, the empirical formula of a molecular compound may not represent the entire molecule. In Table 11-2, notice that ethene's molecular formula is C_2H_4. Every ethene molecule consists of carbon and hydrogen atoms in a 1:2 ratio. But ethene's molecular formula does not show the simplest whole-number ratio of atoms, so the empirical and molecular formulas for ethene are different.

All the compounds listed in Table 11-3 have the same empirical formula—CH_2O—but have very different molecular formulas. Notice though that the empirical formula and the molecular formula for formaldehyde are the same.

TABLE 11-2
Molecular and Empirical Formulas

Compound	Molecular Formula	Empirical Formula
water	H_2O	H_2O
ethene	C_2H_4	CH_2
hexane	C_6H_{14}	C_3H_7

TABLE 11-3 *Compounds with the Empirical Formula CH_2O*

Empirical Formula	Molecular Formula	Common Name
CH_2O	CH_2O	formaldehyde
CH_2O	$C_2H_4O_2$ (CH_3COOH)	acetic acid
CH_2O	$C_3H_6O_3$	L-lactic acid
CH_2O	$C_4H_8O_4$	D-threose
CH_2O	$C_5H_{10}O_5$	D-ribose
CH_2O	$C_6H_{12}O_6$	D-fructose
CH_2O	$C_7H_{14}O_7$	D-mannoheptulose

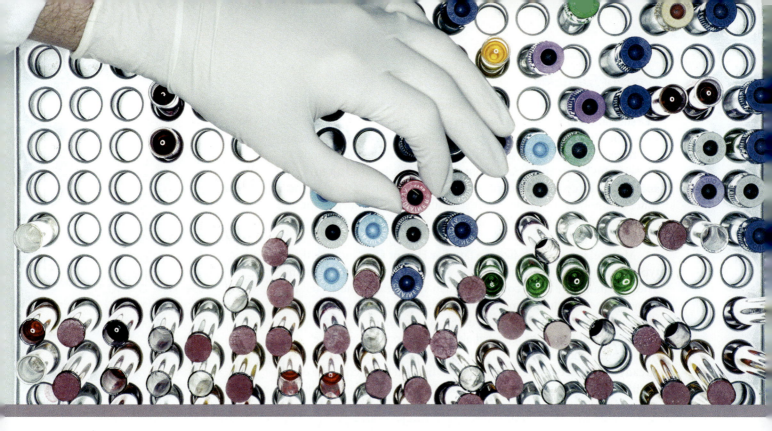

Percent Composition

Percent composition describes the mass composition of a compound by showing what percentage of its total mass comes from each element. A compound's percent composition often looks totally different from what its formula might suggest, as the circle graphs at left show. For example, though water has only one oxygen atom, this one atom contains the great majority of the compound's mass. The two hydrogen atoms contribute much less mass. Likewise, ethene has two hydrogen atoms for every carbon atom, but carbon is much more massive than hydrogen.

A laboratory analysis of a substance is usually expressed as a percent composition. It is essential to understand that percent literally means "per hundred." A general setup to calculate any percentage is (part/whole) × 100%. Suppose that a 60.00 g sample of water were decomposed into its elements and that 53.28 g of oxygen gas and 6.72 g of hydrogen gas were produced. The percent composition of water could be calculated from these results as follows.

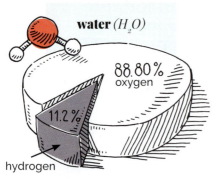

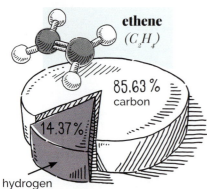

Hydrogen: There are 6.72 g of hydrogen in 60.00 g of water.

$$\left(\frac{part}{whole}\right)100\% = \left(\frac{6.72 \text{ g H}}{60.00 \text{ g H}_2\text{O}}\right)100\%$$

$$= 11.2\% \text{ hydrogen}$$

Oxygen: There are 53.28 g of oxygen in 60.00 g of water.

$$\left(\frac{part}{whole}\right)100\% = \left(\frac{53.28 \text{ g O}}{60.00 \text{ g H}_2\text{O}}\right)100\%$$

$$= 88.80\% \text{ oxygen}$$

When doing percent composition calculations, it's a good idea to check your work afterward by confirming that all the calculated percentages add up to 100% (allowing for rounding error).

EXAMPLE 11-5: PERCENT COMPOSITION

A laboratory analysis of a 30.00 g sample of aluminum sulfate, $Al_2(SO_4)_3$, showed that it contained 4.731 g of aluminum, 8.436 g of sulfur, and 16.833 g of oxygen. What is the percent composition of this compound?

Write what you know.

4.731 g Al, 8.433 g S, 16.835 g O

Convert masses of elements into percent of the compound.

$$Al: \left(\frac{4.731 \text{ g Al}}{30.00 \text{ g Al}_2(SO_4)_3} \right) 100\% = 15.77\% \text{ Al}$$

$$S: \left(\frac{8.433 \text{ g S}}{30.00 \text{ g Al}_2(SO_4)_3} \right) 100\% = 28.11\% \text{ S}$$

$$O: \left(\frac{16.835 \text{ g O}}{30.00 \text{ g Al}_2(SO_4)_3} \right) 100\% = 56.12\% \text{ O}$$

You can also calculate the percent composition of a compound directly from a formula.

EXAMPLE 11-6: PERCENT COMPOSITION DIRECTLY FROM A FORMULA

Find the percent composition of $Al_2(SO_4)_3$.

Solution

First, assume that you have a 1 mol sample of the compound. In this case, each mole of aluminum sulfate would contain the following.

Write what you know.

2 mol of Al (26.98 g/mol) = 53.96 g
3 mol of S (32.06 g/mol) = 96.18 g
12 mol of O (16.00 g/mol) = 192.00 g
Total = 342.14 g

Convert masses of elements into percent of the compound.

$$Al: \left(\frac{53.96 \text{ g Al}}{342.14 \text{ g Al}_2(SO_4)_3} \right) 100\% = 15.77\% \text{ Al}$$

$$S: \left(\frac{96.21 \text{ g S}}{342.14 \text{ g Al}_2(SO_4)_3} \right) 100\% = 28.11\% \text{ S}$$

$$O: \left(\frac{192.00 \text{ g O}}{342.14 \text{ g Al}_2(SO_4)_3} \right) 100\% = 56.117\% \text{ O}$$

Notice that the percentages are essentially the same using either method. This equivalence is due to the law of definite composition, which as you will recall explains that the ratio of elements in a compound is constant regardless of the amount of the compound.

EXAMPLE 11-7: FINDING COMPONENT MASS

How many grams of oxygen would a 65.00 g sample of silver nitrate ($AgNO_3$) contain?

Write what you know.

65.00 g $AgNO_3$

The problem can be solved by using the ratio of the molar masses of silver nitrate and the oxygen within silver nitrate to create a conversion factor.

Convert grams $AgNO_3$ into grams of oxygen.

$$65.00 \text{ g AgNO}_3 \left(\frac{48.00 \text{ g O}}{169.88 \text{ g AgNO}_3} \right) = 18.37 \text{ g O}$$

Calculations with Empirical Formulas

Empirical formulas contain the information necessary to calculate the percent composition of compounds since they preserve the ratio of atoms in a compound. To find percent composition, the mole ratio in the empirical formula must be converted to a mass ratio through a series of calculations.

Here's an example. The formula H_2O means that there are exactly 2 mol of hydrogen atoms for every 1 mol of oxygen atoms. One mole of water contains 2 mol of hydrogen atoms and 1 mol of oxygen atoms. Expressed in masses, 1 mol of water contains 2.02 g of hydrogen atoms and 16.00 g of oxygen atoms.

$$2 \text{ mol H} \left(\frac{1.01 \text{ g H}}{1 \text{ mol H}} \right) = 2.02 \text{ g H}$$

$$1 \text{ mol O} \left(\frac{16.00 \text{ g O}}{1 \text{ mol O}} \right) = 16.00 \text{ g O}$$

The total mass of one mole of water is 18.02 g. Now the percent composition of water can be found.

$$\left(\frac{2.02 \text{ g H}}{18.02 \text{ g H}_2\text{O}} \right) 100\% = 11.2\% \text{ H}$$

$$\left(\frac{16.00 \text{ g O}}{18.02 \text{ g H}_2\text{O}} \right) 100\% = 88.79\% \text{ O}$$

Analytical laboratories often do the reverse process, using the percent composition of an unknown compound determined experimentally to calculate its empirical formula. Suppose that a chemist is given 100.0 g of an unknown compound and is told to determine its empirical formula. After a careful analysis, the chemist concludes that 75.00 g (75.00%) of the sample's mass is carbon. The other 25.00 g (25.00%) comes from hydrogen. From this you might be tempted to think that since carbon contributes three times the mass that hydrogen does, the empirical formula must be C_3H. But since empirical formulas are based on mole ratios rather than mass ratios, the masses that the chemist determined next need to be converted to moles using molar mass.

$$75.00 \text{ g C} \left(\frac{1 \text{ mol C}}{12.01 \text{ g C}} \right) = 6.2447 \text{ mol C}$$

$$25.00 \text{ g H} \left(\frac{1 \text{ mol H}}{1.01 \text{ g H}} \right) = 24.752 \text{ mol H}$$

According to the above calculation, the empirical formula for this compound could be written as $C_{6.2447}H_{24.752}$. Although this formula is numerically accurate, it is not in its final form because empirical formulas must be written as whole-number ratios. Dividing both numbers by the smaller number gives the simplest form of the ratio and guarantees that one of the subscripts will be a 1.

$$\text{mol C : mol H} = \frac{6.2447}{6.2447} : \frac{24.752}{6.2447} = 1 : 3.96 \text{ or } 1 : 4$$

248 Chapter 11

Notice in this example that it is necessary to round slightly to get to a whole number. Experimental error or rounding during calculation can make the final results slightly off. Sometimes calculations produce ratios such as 1:1.5 or 1:1.33. These should not be rounded off. In general, if the decimal portion of the part of the ratio is greater than 0.1 and less than 0.9, you should not round off. The ratio 1:1.5 is equivalent to the whole-number ratio 2:3. The ratio 1:1.33 is equivalent to the whole-number ratio 3:4. Insight and practice are necessary to know when to round off and when to make another ratio. One method of obtaining a whole-number ratio, shown in Example 11-9 on the next page, is to express the ratio in fractions having a common denominator and then multiply both sides by the common denominator. Rounding is allowable in this case, and the empirical formula of the compound is thus C_1H_4, or CH_4. The flow chart below shows the steps used in this example to convert from percent composition to an empirical formula.

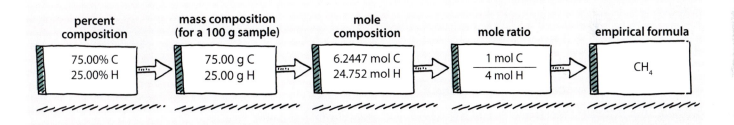

EXAMPLE 11-8:
CALCULATING EMPIRICAL FORMULAS FROM PERCENT COMPOSITION

A laboratory analysis of an unknown gas has shown that the gas is 72.55% oxygen and 27.45% carbon by mass. What is the empirical formula of the compound?

Write what you know.

72.55% O
27.45% C

Convert percentages of elements into masses of elements.

Though the percent composition of any quantity of a sample will be the same, if we choose the sample size to be exactly 100 g, the "%" unit can be changed to "g," and the calculation is simplified. So in a 100 g sample of the gas, there will be 72.55 g of oxygen and 27.45 g of carbon.

Convert masses into moles of each element.

$$72.55 \text{ g O}\left(\frac{1 \text{ mol O}}{16.00 \text{ g O}}\right) = 4.5343 \text{ mol O}$$

$$27.45 \text{ g C}\left(\frac{1 \text{ mol C}}{12.01 \text{ g C}}\right) = 2.2855 \text{ mol C}$$

Notice that in these problems the molar mass of each element is used, even for those such as oxygen that naturally exist as diatomic molecules.

Convert moles into a mole ratio.

From the calculation above it is seen that the mole ratio of oxygen to carbon (mol O : mol C) is 4.534 : 2.285. Next, the ratio is reduced to lowest terms.

$$\frac{4.5343}{2.2855} : \frac{2.2855}{2.2855} = 1.984 : 1.000$$

Determine the empirical formula.

Since rounding is permitted in this case, we see that for every 2 mol of oxygen, there is 1 mol of carbon. The empirical formula is therefore CO_2.

EXAMPLE 11-9: CALCULATING EMPIRICAL FORMULAS FROM MASS COMPOSITION

A 5.000 g sample of an unknown compound contains 1.844 g of nitrogen and 3.156 g of oxygen. Find the empirical formula.

Solution

The mass composition of the sample is already known, so we can bypass one step on the flow chart.

Write what you know.

1.844 g N
3.156 g O

Convert masses into moles for each element.

$$1.844 \text{ g N} \left(\frac{1 \text{ mol N}}{14.01 \text{ g N}} \right) = 0.131\underline{6}2 \text{ mol N}$$

$$3.156 \text{ g O} \left(\frac{1 \text{ mol O}}{16.00 \text{ g O}} \right) = 0.197\underline{2}5 \text{ mol O}$$

Determine the mole ratio.

Mol N : mol O = 0.1316 : 0.1973. Reduced to lowest terms, this is

$$\frac{0.131\,62}{0.131\,62} : \frac{0.197\,25}{0.131\,62} = 1.000 : 1.499.$$

The ratio should be 1 : 1.5 (N : O).

The ratio 1 : 1.5 is in its simplest form, but the numbers are not whole numbers. To put the ratio in whole numbers, express the numbers as fractions.

$$1 : 1.5 = \frac{2}{2} : \frac{3}{2}$$

Eliminate the fractions by multiplying both sides of the proportion by the common denominator.

$$2 \left(\frac{2}{2} \right) : \left(\frac{3}{2} \right) 2 = 2 : 3$$

Determine the mole ratio.

The empirical formula of the compound is N_2O_3.

Sometimes molecular formulas can be determined when the molar mass of the actual compound and its empirical formula are known. The molar mass of the compound is divided by the molar mass of the empirical formula. The resulting whole number is used as a multiplier for the subscripts in the empirical formula to produce the compound's molecular formula. Let's look at an example of this.

EXAMPLE 11-10:
CALCULATING MOLECULAR FORMULAS FROM PERCENT COMPOSITION

Caffeine, found in coffee, tea, and some carbonated beverages, is 5.170% hydrogen, 16.49% oxygen, 28.86% nitrogen, and 49.48% carbon by mass. The molar mass of caffeine is 194.22 g/mol. Find its molecular formula.

Write what you know.

5.170% H
16.49% O
28.86% N
49.48% C

Convert percentages into masses for each element.

In a 100 g sample of the compound, there will be 5.170 g of hydrogen, 16.49 g of oxygen, 28.86 g of nitrogen, and 49.48 g of carbon.

Convert masses into moles for each element.

$$5.170 \text{ g H} \left(\frac{1 \text{ mol H}}{1.01 \text{ g H}} \right) = 5.1188 \text{ mol H}$$

$$16.49 \text{ g O} \left(\frac{1 \text{ mol O}}{16.00 \text{ g O}} \right) = 1.0306 \text{ mol O}$$

$$28.86 \text{ g N} \left(\frac{1 \text{ mol N}}{14.01 \text{ g N}} \right) = 2.0599 \text{ mol N}$$

$$49.48 \text{ g C} \left(\frac{1 \text{ mol C}}{12.01 \text{ g C}} \right) = 4.1199 \text{ mol C}$$

Determine the mole ratio.

Finding the ratios of these elements is done by dividing all of their mole compositions by the lowest value, 1.031 mol. This gives us whole-number ratios (all the values allow for rounding).

$$H: \frac{5.1188 \text{ mol}}{1.0306 \text{ mol}} = 4.967$$

$$O: \frac{1.0306 \text{ mol}}{1.0306 \text{ mol}} = 1.000$$

$$N: \frac{2.0599 \text{ mol}}{1.0306 \text{ mol}} = 1.999$$

$$C: \frac{4.1199 \text{ mol}}{1.0306 \text{ mol}} = 3.998$$

Determine the empirical formula.

The values above give an empirical formula of $C_4H_5N_2O$, which has a molar mass of 97.11 g/mol.

Determine the molecular formula.

The molar mass of caffeine is divided by the molar mass of its empirical formula to produce a multiplier.

$$\frac{194.22 \text{ g/mol}}{97.11 \text{ g/mol}} = 2.000$$

The subscripts in the empirical formula, $C_4H_5N_2O$, are multiplied by this whole number to produce the molecular formula $C_8H_{10}N_4O_2$.

11.1 SECTION REVIEW

1. What is Avogadro's number, and what is its significance?

2. How many atoms are found in 6.00×10^2 g of iron?

3. How many moles of atoms are in 454 g of lead?

4. Find the molar mass of vitamin A ($C_{20}H_{30}O$).

5. If 4.7000×10^4 metric tons (t) of sulfuric acid are industrially produced in a year, how many moles are produced? (1 t = 1000 kg)

6. Give one benefit and one disadvantage of using structural formulas.

7. What is the difference between an empirical formula and a molecular formula?

8. Give the structural, molecular, and empirical formulas of hydrogen peroxide (also known as dihydrogen dioxide).

9. DDT, an insecticide banned in the United States for most uses, has the formula $C_{14}H_9Cl_5$. Find its percent composition.

10. Find the empirical and molecular formulas for acetone (molar mass 58.09 g/mol), used to make nail polish remover and other industrial products. Its composition is 62.02% carbon, 10.43% hydrogen, and 27.54% oxygen.

11. After reading the How It Works box (p. 242), explain why carbon monoxide detectors are not designed to go off immediately upon detecting the gas.

11.2 STOICHIOMETRY

QUESTIONS
» What is stoichiometry?
» Why are mole ratios so important?
» Can I predict how much of a chemical will take part in a chemical reaction?

TERMS
stoichiometry • mole ratio

How do I know how much reactant to use?

The United States produces about 10 million tons of phosphoric acid each year. Most of this is used to make fertilizers. One of the most common methods to produce this acid is to grind phosphate rock to a powder and mix it with sulfuric acid in water. The products include the intended phosphoric acid plus calcium sulfate dihydrate. A simplified equation for this reaction is shown below.

$$Ca_3(PO_4)_2(s) + 3H_2SO_4(aq) \; 6H_2O(l) \rightleftharpoons$$
$$2H_3PO_4(aq) + 3CaSO_4 \cdot 2H_2O\,(s)$$

Chemical engineers who manage this process must carefully track all the compounds involved. For instance, they need to know how much product is needed to supply the market and how much of each reactant is needed to make that amount of product. They can calculate quantities of reactants and products using the mole and mass relationships between them. **Stoichiometry** examines the mathematical relationships between the amounts of reactants and products in a chemical reaction.

Mole-to-Mole Conversions

Before we knuckle down to serious chemistry, let's look at how stoichiometry might be applied to a kitchen conundrum. Suppose you're getting ready to go to a party and you decide to bring some Rice Krispies® treats. You pull out your recipe.

Rice Krispies Treats (makes twenty-four 2 × 2 in. squares)

3 tbsp butter or margarine

1 10 oz pkg marshmallows (about 40)

6 cups Rice Krispies cereal

If you were really hungry, you could scale the recipe up to make more treats. You could also reduce the recipe if you needed to. As long as you preserve the ratios of the ingredients, you can adjust the recipe to suit your needs. For example, the ratio of cereal to marshmallows is 6 cups to 40 marshmallows. Suppose you had only 4.5 cups of cereal and sufficient amounts of the other ingredients. How many marshmallows would you need? You could use the recipe relationships and dimensional analysis as follows.

$$4.5 \text{ cups cereal} \left(\frac{40 \text{ marshmallows}}{6 \text{ cups cereal}} \right) = 30 \text{ marshmallows}$$

Further, you could determine how many treats you can make (assuming they were the same size) or how much margarine is needed. All of these can be determined from the relationships given in the recipe. For example,

$$4.5 \text{ cups cereal} \left(\frac{24 \text{ treats}}{6 \text{ cups cereal}} \right) = 18 \text{ treats.}$$

In a similar way, chemical equations show the ratios of substances involved in a reaction. Balanced chemical equations contain the numerical information necessary for detailed calculations. The coefficients in front of the substances show how many atoms, molecules, and moles are involved. For example, the equation

$$P_4(s) + 5O_2(s) \longrightarrow 2P_2O_5(s)$$

means that one P_4 molecule and five O_2 molecules form two P_2O_5 molecules. The coefficients can also represent the number of moles of a compound or element. This is key to stoichiometric calculations, since moles is a more practical or workable amount of matter than individual molecules! Coefficients give the numerical relationships between the substances in a reaction, like the ingredients in a recipe. This balanced equation shows that

1 mol P_4 reacts to form 2 mol P_2O_5 (a 1:2 ratio),

1 mol P_4 reacts with 5 mol O_2 (a 1:5 ratio), and

5 mol O_2 react to form 2 mol P_2O_5 (a 5:2 ratio).

Even if the original quantities were cut in half, the ratios would still hold true. In each case, the coefficients from the balanced equation give the ratios between the moles of one substance and the moles of another substance and are known as **mole ratios**. These mole ratios can be used as conversion factors when calculating how changes in the quantity of one substance affect the quantities of other substances in a chemical equation.

For example, suppose a chemical engineer wanted to produce 25.0 mol of diphosphorus pentoxide. How many moles of tetraphosphorus would he burn? From the balanced equation in the previous example, we can see that 1 mol of tetraphosphorus produces 2 mol of diphosphorus pentoxide. Using that information, he can set up the following unit conversion to solve the problem.

$$25.0 \text{ mol } P_2O_5 \left(\frac{1 \text{ mol } P_4}{2 \text{ mol } P_2O_5} \right) = 12.5 \text{ mol } P_4$$

The amount of oxygen required is calculated from either the 12.5 mol of phosphorus or from the 25.0 mol of diphosphorus pentoxide. Both methods use ratios formed from the coefficients in the balanced equation.

$$12.5 \text{ mol } P_4 \left(\frac{5 \text{ mol } O_2}{1 \text{ mol } P_4} \right) = 62.5 \text{ mol } O_2$$

or

$$25.0 \text{ mol } P_2O_5 \left(\frac{5 \text{ mol } O_2}{2 \text{ mol } P_2O_5} \right) = 62.5 \text{ mol } O_2$$

EXAMPLE 11-11: MOLE-TO-MOLE CONVERSIONS

If 25.0 mol of diphosphorus pentoxide reacts with water to form phosphoric acid, how many moles of water are required? The balanced equation of the reaction is shown below.

$$P_2O_5(s) + 3H_2O(l) \longrightarrow 2H_3PO_4(aq)$$

Write what you know and what you want to know.

25.0 mol P_2O_5
? mol H_2O

This problem amounts to changing from moles of one substance (P_2O_5) to moles of another substance (H_2O). Their relationship is found in the balanced equation, where the coefficients show that 3 mol of water are needed for every 1 mol of diphosphorus pentoxide.

Convert mol P_2O_5 to mol H_2O.

$$25.0 \text{ mol } P_2O_5 \left(\frac{3 \text{ mol } H_2O}{1 \text{ mol } P_2O_5} \right) = 75.0 \text{ mol } H_2O$$

Mass-to-Mole Conversions

Conversions between substances in a reaction must be done in moles. Why? Because the balanced equation—the recipe, so to speak—is based on moles, not mass. To convert from one substance to another, the amount of the substance must be in moles. So if we wanted to know, for instance, how many moles of product would be formed by reacting a certain mass of a reactant, we would first need to convert the grams of reactant to moles of reactant as an intermediate step.

Let's consider calculating the moles of diphosphorus pentoxide that would result from burning 1.55 kg of tetraphosphorus as described by the equation on page 253. Before the conversion factor

$$\frac{2 \text{ mol } P_2O_5}{1 \text{ mol } P_4}$$

can be used, the given mass of phosphorus in grams must be converted to moles. Remember from Section 11.1 that the molar mass of a substance is numerically equal to its atomic mass. The molar mass of P_4 is 4(30.97 g/mol), or 123.88 g/mol.

$$1.55 \text{ kg } P_4 \left(\frac{1000 \text{ g } P_4}{1 \text{ kg } P_4} \right) \left(\frac{1 \text{ mol } P_4}{123.88 \text{ g } P_4} \right) = 12.51 \text{ mol } P_4$$

The moles of phosphorus can now be converted to moles of diphosphorus pentoxide using the mole ratio obtained from the balanced equation.

$$12.51 \text{ mol } P_4 \left(\frac{2 \text{ mol } P_2O_5}{1 \text{ mol } P_4} \right) = 25.0 \text{ mol } P_2O_5$$

You can use dimensional analysis to combine both the grams-to-moles and the mole-to-mole conversion factors into one equation.

$$1.55 \text{ kg } P_4 \left(\frac{1000 \text{ g } P_4}{1 \text{ kg } P_4} \right) \left(\frac{1 \text{ mol } P_4}{123.88 \text{ g } P_4} \right) \left(\frac{2 \text{ mol } P_2O_5}{1 \text{ mol } P_4} \right) = 25.0 \text{ mol } P_2O_5$$

CAREERS

SERVING AS A CHEMICAL ABATEMENT SPECIALIST: *NOT YOUR AVERAGE CLEANUP CREW*

Asbestos, lead, mold, PCBs—what do these have in common? Each is a type of contaminant that might exist in a home, business, factory, or job site. Chemical agents like these can pose serious health risks to people, even if the contamination occurred many years in the past. The threat is often not readily apparent, either, since the problems usually exist in out-of-the-way places such as inside walls and ceilings, beneath floors, and even inside lamps, thermostats, and switches. Because of this, inspections take place whenever buildings are scheduled for demolition or renovation.

Chemical abatement specialists are specially trained technicians who identify chemical hazards and safely neutralize them. They use their knowledge of hazardous chemicals and the technologies needed to deal with them to help make safer working and living environments. As a chemical abatement specialist, you could demonstrate God's care for others by improving their lives through the wise remediation of chemical health risks.

EXAMPLE 11-12: MASS-TO-MOLE CONVERSIONS

How many moles of phosphoric acid can be formed from 3.550×10^3 g of diphosphorus pentoxide?

$$P_2O_5(s) + 3H_2O(l) \longrightarrow 2H_3PO_4(aq)$$

Write what you know and what you want to know.

3.550×10^3 g P_2O_5
? mol H_3PO_4

Note that the given value is in grams and the desired value is in moles. Since the conversion factors from the balanced equation are based on a mole ratio, the mass of diphosphorus pentoxide must first be converted to moles. Then the moles of diphosphorus pentoxide can be converted to moles of phosphoric acid according to the mole ratio from the equation above. These two steps should be done together as follows.

Convert g P_2O_5 to mol H_3PO_4.

$$3.550 \times 10^3 \text{ g } P_2O_5 \left(\frac{1 \text{ mol } P_2O_5}{141.94 \text{ g } P_2O_5} \right) \left(\frac{2 \text{ mol } H_3PO_4}{1 \text{ mol } P_2O_5} \right) = 50.02 \text{ mol } H_3PO_4$$

MINILAB

BLOWUP

Baking soda (sodium hydrogen carbonate, $NaHCO_3$) and vinegar (an aqueous solution of acetic acid, CH_3COOH) react together, producing carbon dioxide as one product. It's the ever-popular grade school science fair volcano reaction. Typically the reaction is started by pouring some baking soda into some vinegar. So if you wanted a bigger eruption, you could simply add more baking soda, right? Or no?

Do the reactants in a reaction always get used up?

1. If you were to do two separate volcano reactions but use five times as much baking soda in the second reaction as in the first, do you think that you would get five times as much carbon dioxide produced? Write your answer in the form of a hypothesis.

Procedure

A Fill each of the Erlenmeyer flasks with 200 mL of vinegar.

B Label the balloons 1–5.

C Use the laboratory balance and weighing paper to measure out 1.0 g of baking soda.

D Carefully add the baking soda to Balloon 1. Try to get all of the baking soda down into the large end of the balloon.

E Carefully stretch the mouth of the balloon over the top of one of the flasks without spilling any baking soda into the flask. Allow the balloon and baking soda to hang to one side.

F Repeat Steps C–E for Balloons 2–5, increasing the amount of baking soda by 1.0 g each time. It helps to keep track of the amounts by placing the same number of grams of baking soda into a balloon as the number marked on the balloon.

G Tip the baking soda from Balloon 1 into its flask and observe what happens to the balloon.

2. What happens to the balloon? What explains this result?

H Repeat Step G for Balloons 2–5 one at a time, taking time to observe what happens each time.

3. Describe what happens to Balloons 2–5 compared with Balloon 1.

4. Gently swirl the contents of the flask with Balloon 5 and describe what you see. What do you suppose this is?

Conclusion

5. Did your observations support your hypothesis in Question 1?

6. Suggest an explanation for the changing volume of gas produced in each flask.

EQUIPMENT

- laboratory balance
- Erlenmeyer flasks, 500 mL (5)
- weighing paper
- white vinegar, 1000 mL
- balloons (5)
- baking soda, 15 g
- goggles

Mass-to-Mass Conversions

When the mass of one substance in a reaction is known, the mass of a second substance can be calculated. This type of problem is the most practical one presented in this chapter because chemists frequently do this in their laboratories. Their balances measure in units of grams, not moles, so starting and ending with grams is the usual approach. To do similar calculations, you must be able to convert between mass and moles, so the preceding problems have prepared you for this task. For mass-to-mass conversions, the moles of both the given and desired substances must be determined before any unknown mass can be calculated. Let's see what this looks like.

EXAMPLE 11-13: MASS-TO-MASS CONVERSIONS

What mass of water will react with 3.550×10^3 g of diphosphorus pentoxide?

$$P_2O_5(s) + 3H_2O(l) \longrightarrow 2H_3PO_4(aq)$$

Write what you know and what you want to know.

3.550×10^3 g P_2O_5
? mass H_2O

Convert g P_2O_5 to g H_2O.

The mass of diphosphorus pentoxide must first be expressed as moles before the mole ratio conversion factor can be used. Once the number of moles of water is known, the mass of water can be calculated from its molar mass.

$$3.550 \times 10^3 \text{ g } P_2O_5 \left(\frac{1 \text{ mol } P_2O_5}{141.94 \text{ g } P_2O_5}\right)\left(\frac{3 \text{ mol } H_2O}{1 \text{ mol } P_2O_5}\right)\left(\frac{18.02 \text{ g } H_2O}{1 \text{ mol } H_2O}\right) = 1352 \text{ g } H_2O$$

Chemical Calculations

EXAMPLE 11-14: MASS-TO-MASS CONVERSIONS

How many grams of sodium chloride decompose to yield 27 g of chlorine gas?

Solution

First, write a complete balanced equation.

$$2NaCl\,(l) \xrightarrow{\Delta} 2Na\,(l) + Cl_2\,(g)$$

Next, convert from grams of chlorine gas to moles of chlorine gas, then to moles of sodium chloride, and then to grams of sodium chloride.

$$27\text{ g }Cl_2 \left(\frac{1\text{ mol }Cl_2}{70.90\text{ g }Cl_2}\right)\left(\frac{2\text{ mol NaCl}}{1\text{ mol }Cl_2}\right)\left(\frac{58.44\text{ g NaCl}}{1\text{ mol NaCl}}\right) = 45\text{ g NaCl}$$

The number of atoms or molecules involved in a reaction can also be calculated from known molar quantities. Suppose that Example 11-13 above asked, "How many water molecules will react with 3.550×10^3 g of diphosphorus pentoxide?" The solution to the problem is the same until the step in which the number of moles of water is converted to the mass of water. The definition of a mole can be used to convert the number of moles of water to the number of molecules of water.

$$3.550 \times 10^3\text{ g }P_2O_5 \left(\frac{1\text{ mol }P_2O_5}{141.94\text{ g }P_2O_5}\right)\left(\frac{3\text{ mol }H_2O}{1\text{ mol }P_2O_5}\right)\left(\frac{6.022 \times 10^{23}\text{ molecules }H_2O}{1\text{ mol }H_2O}\right)$$

$$= 4.518 \times 10^{25}\text{ molecules }H_2O$$

11.2 SECTION REVIEW

1. State a general outline for performing stoichiometric calculations.

2. Explain the process by which you would determine the mass of a reactant that is needed in order to produce a specific mass of product for a specified chemical reaction.

3. Draw a flow chart that shows the conversions between masses, moles, and particles of a substance. Additionally show the stoichiometric conversion between two substances.

Use the equation for iron rusting shown below to answer Questions 4–7.

$$4Fe(s) + 3O_2(g) \longrightarrow 2Fe_2O_3(s)$$

4. How many moles of iron(III) oxide—rust—can be produced from 3.2 mol of iron metal?

5. How many moles of oxygen will be consumed by the rusting process in Question 4?

6. What mass of iron metal is needed to form 100.0 g of rust?

7. How many molecules of oxygen will be needed to react with 50.0 g of iron metal?

8. About 60% of the sulfuric acid produced worldwide is used to produce phosphoric acid for fertilizers. The equation for one reaction used for such production is shown below.

$$Ca_5F(PO_4)_3(s) + 5H_2SO_4(aq) + 10H_2O(l) \xrightarrow{\Delta}$$
$$5CaSO_4 \cdot 2H_2O\,(s) + HF(aq) + 3H_3PO_4(aq)$$

About 1.6×10^4 metric tons of phosphoric acid are produced every day in the United States. How many metric tons of sulfuric acid are required?

11.3 REAL-WORLD STOICHIOMETRY

Limiting Reactants

Up to this point in our study of chemistry, we have assumed that chemical reactions all start with measured amounts of reactants and end with specific amounts of products. In fact, nature is rarely so tidy. Even in a laboratory, where conditions are controlled and reactants are carefully measured, at least one reactant is usually used up before the others, preventing more products from forming. The reactant that is completely used up is called the **limiting reactant**. Obviously, such conditions will leave at least some of one other reactant, and possibly several, in its original unchanged state. Any leftover substances are called **excess reactants**.

It's kind of like this: imagine that you are having a cookout for your sports team, and fourteen people say that they will eat only cheeseburgers. You have plenty of hamburgers and buns, but only twelve slices of cheese. In this case, the cheese is the limiting "reactant" because it limits how many cheeseburgers can be produced. The extra buns and hamburger patties are excess reactants.

Why can't we actually obtain a theoretical yield?

QUESTIONS
» What is a limiting reactant?
» What is the difference between theoretical and actual yield?

TERMS
limiting reactant • excess reactant • theoretical yield • actual yield • percent yield

Some examples will show how this works.

EXAMPLE 11-15: LIMITING REACTANT

Lithium hydroxide canisters used on spacecraft capture exhaled carbon dioxide gas and convert it to solid lithium carbonate and water. If a set of canisters contains 5.75 kg of LiOH, and each of the six crew members exhales 21 mol CO_2/day, will the lithium hydroxide be the limiting reactant for that day?

Write what you know and what you want to know.

5.75 kg LiOH
(6 astronauts)(21 mol CO_2/astronaut) = 126 mol CO_2
? mol CO_2 reacted

Write a complete balanced equation.

$$2LiOH(s) + CO_2(g) \longrightarrow Li_2CO_3(s) + H_2O(l)$$

Convert kg LiOH to mol LiOH, then from mol LiOH to mol CO_2.

$$5.75 \text{ kg LiOH} \left(\frac{1000 \text{ g LiOH}}{1 \text{ kg LiOH}}\right)\left(\frac{1 \text{ mol LiOH}}{23.95 \text{ g LiOH}}\right)\left(\frac{1 \text{ mol } CO_2}{2 \text{ mol LiOH}}\right) = 120 \text{ mol } CO_2$$

The LiOH in the canisters can eliminate 120 mol CO_2, but the astronauts are going to exhale 126 mol CO_2. This means that there is not enough LiOH to eliminate all the CO_2. Therefore the LiOH is the limiting reactant.

EXAMPLE 11-16: EXCESS REACTANT

In Example 11-15, how many grams of excess reactant remain after the reaction takes place?

Write what you know and what you want to know.

CO_2 remaining = CO_2 exhaled – CO_2 eliminated
= 126 mol CO_2 – 120 mol CO_2
= 6 mol CO_2

Convert mol CO_2 to g CO_2.

$$6 \text{ mol } CO_2 \left(\frac{44.01 \text{ g } CO_2}{1 \text{ mol } CO_2}\right) = 264.1 \text{ g } CO_2$$

You have 264.1 g of excess CO_2 reactant.

Percent Yield

Even if a scientist carefully measures out the proper number of grams of reactants on the basis of their mole ratios, there is a high probability that he will find less mass in his products than expected. In the problems found in this chapter, it has been assumed that all of the reactant would be used up and the reaction would go to completion. You have been calculating the **theoretical yields** of reactions—the maximum amount of product that could be produced from a given amount of reactant.

There are many reasons why an experimental yield will probably not match the theoretical yield of a reaction. In many reactions, there are competing side reactions that take place that are not figured into the calculations. Some reactions will stop prior to completion even though there are reactants remaining. Other reactions are reversible and reach an *equilibrium position* rather than going to completion. Often it's simply a matter of being unable to recover all of the product that is actually produced. For example, the products may need to be purified first, and some material may be lost in the process. Some dissolved products may cling to the inside of glassware or liquids may evaporate. The measured amount of product at the end of a reaction is called the **actual yield**. The actual yield is normally expressed in grams of the product.

Chemists often need to know how efficient a reaction is, especially in industry, where costs must be carefully controlled. For industry, a more important figure than theoretical or actual yield is the percent yield. **Percent yield** is the percentage of the theoretical yield that is actually produced and isolated. It is calculated by taking a ratio of the actual yield to the theoretical yield and multiplying by 100%.

EXAMPLE 11-17: PERCENT YIELD

Aluminum hydroxide is found in many antacid tablets because it reacts with hydrochloric acid from the stomach to form aluminum chloride and water.

a. If 15.00 g of aluminum hydroxide is found in a single tablet, determine the theoretical yield of aluminum chloride that would form if we assume that there was an excess amount of hydrochloric acid.

b. If 23.00 g of aluminum chloride was the actual measured amount of product, calculate the percent yield.

Write what you know and what you want to know.

$\qquad$ 15.00 g Al(OH)$_3$

$\qquad$? g AlCl$_3$

a. First, write the complete balanced equation.

$$Al(OH)_3(s) + 3HCl(aq) \longrightarrow AlCl_3(aq) + 3H_2O(l)$$

Convert g Al(OH)$_3$ to g AlCl$_3$.

$$15.00 \text{ g Al(OH)}_3 \left(\frac{1 \text{ mol Al(OH)}_3}{78.01 \text{ g Al(OH)}_3} \right) \left(\frac{1 \text{ mol AlCl}_3}{1 \text{ mol Al(OH)}_3} \right) \left(\frac{133.33 \text{ g AlCl}_3}{1 \text{ mol AlCl}_3} \right)$$

$$= 25.637 \text{ g AlCl}_3$$

The theoretical yield of this reaction is then 25.64 g of AlCl$_3$.

b. The percent yield is calculated by dividing the actual yield by the theoretical yield and multiplying by 100%.

$$percent_{yield} = \left(\frac{yield_{actual}}{yield_{theoretical}} \right) 100\%$$

$$= \left(\frac{23.00 \text{ g AlCl}_3}{25.637 \text{ g AlCl}_3} \right) 100\% = 89.71\%$$

11.3 SECTION REVIEW

1. Explain what limiting and excess reactants are.

2. Nitrogen and hydrogen react to form ammonia. The equation is shown below.

$$N_2(g) + 3H_2(g) \longrightarrow 2NH_3(g)$$

   Assume that 50.00 g each of N_2 and H_2 are reacted together. Answer the following questions.

   a. Will all of each reactant be used up? If not, which reactant is limiting?

   b. How many grams of ammonia will be produced?

   c. How many moles of any excess reactant will be left over?

3. What is the difference between theoretical and actual yield?

4. Use the reaction for rusting iron shown below to answer the following.

$$4Fe(s) + 3O_2(g) \longrightarrow 2Fe_2O_3(s)$$

   a. If 200.0 g of Fe reacts fully with O_2, what is the theoretical yield of Fe_2O_3?

   b. If the actual yield is 252.7 g, what is the percent yield?

Use the case study below to answer Questions 5–7.

5. If sulfuric acid is so dangerous, why do humans continue to produce it? (*Hint*: Do an internet search to find the uses of sulfuric acid.)

6. Since sulfuric acid is toxic and polluting, should Christians oppose its creation and use? Why or why not?

7. Does the world's need for sulfuric acid present an opportunity for Christian service? Explain.

CASE STUDY

SULFURIC ACID

Sulfuric acid is potentially very dangerous to both humans and the environment. It is highly corrosive and can cause third-degree burns on contact with skin. If it gets into the air, it can irritate eye, nose, throat, and lung tissue. It has the potential to contribute to acid rain. But despite sulfuric acid's reactivity and the risk that it presents to human health, chemical plants around the world produce it in massive quantities. In fact, it's the most-produced industrial chemical by volume in the world, and demand for it rises every year. Current production has been estimated at 270 million metric tons per year.

11 CHAPTER REVIEW

Chapter Summary

TERMS
mole
Avogadro's number
molar mass
molecular formula
empirical formula
percent composition

11.1 THE MOLE

- Chemists have devised a unit called the mole to simplify calculations related to chemical reactions. A mole is defined as the amount of substance that contains exactly
  $6.02214076 \times 10^{23}$
  (Avogadro's number) particles.

- The molar mass of a substance is the mass that contains one mole of its particles.

- The atomic masses found on the periodic table can be equated with the unit grams/mole when calculating the molar mass of a pure substance.

- Structural formulas, molecular formulas, and empirical formulas emphasize different characteristics of molecules. These different formulas are useful in different situations.

- An empirical formula gives the simplest whole-number ratio of atoms in a compound. In some cases, it is identical to the molecular formula.

- Percent composition is used to express the ratio of the mass of each element in a compound to the entire mass of the compound.

11.2 STOICHIOMETRY

TERMS
stoichiometry
mole ratio

- Stoichiometry deals with the mathematical relationships between the amounts of reactants and products in a chemical reaction.

- Stoichiometric calculations rely on mole ratios taken from balanced chemical reactions to determine masses of reactants and products.

TERMS
limiting reactant
excess reactant
theoretical yield
actual yield
percent yield

11.3 REAL-WORLD STOICHIOMETRY

- In most real chemical reactions one of the reactants is completely consumed while others remain. When there is no more of this limiting reactant, the reaction stops generating further products. Unused reactants are called excess reactants.

- Stoichiometric calculations are models of the actual chemical reaction and therefore are used to calculate theoretical yields. In reality, actual yields are typically less than theoretical yields. The ratio of the two is called the percent yield.

Chemical Calculations

11 CHAPTER REVIEW

Chapter Review Questions

RECALLING FACTS

1. Define *mole*.
2. Describe the relationship between Avogadro's number and a mole of a substance.
3. Define *molar mass*.
4. (1) How is the molar mass of an element found? (2) How is the molar mass of a compound found?
5. When determining a compound's empirical formula, if your calculated mole amounts are not equal to whole numbers, what must you do?
6. What term describes the calculations used to relate the amounts of reactants and products in a chemical reaction?
7. What are mole ratios and how are they obtained?
8. Explain why excess reactants often remain in a chemical reaction.
9. Give at least two reasons why the actual yield of a chemical reaction is typically less than the calculated theoretical yield.

UNDERSTANDING CONCEPTS

10. Give the molar mass for hydrogen, scandium, arsenic, iodine, and uranium.
11. How many atoms are in the following?
    a. 12.01 g C
    b. 16.00 g O
    c. 1.01 g H
12. How many moles
    a. of Fe are in 37.0 g Fe?
    b. of Kr are in 4.58×10^{20} atoms Kr?
    c. of $NaIO_3$ are in 3.25×10^{26} formula units $NaIO_3$?
13. Calculate the molar mass of the following compounds.
    a. N_2O_3
    b. Mg_3N_2
    c. $AgNO_3$
14. How many
    a. Fe atoms are in 0.256 mol Fe?
    b. Kr atoms are in 3.87 g Kr?
    c. $AlCl_3$ formula units are in 6.17 mol $AlCl_3$?
    d. $NaIO_3$ formula units are in 8.58 g $NaIO_3$?
15. What is the mass in grams of
    a. 6.58 mol Fe?
    b. 8.58×10^{28} atoms Kr?
    c. 1.05 mol $AlCl_3$?
    d. 3.17×10^{18} formula units $NaIO_3$?

16. Can a compound's empirical formula ever be the same as its molecular formula? Explain.

17. Iron(II) sulfate ($FeSO_4$) is a therapeutic agent for iron-deficiency anemia. It is administered orally as iron(II) sulfate heptahydrate. If 0.300 g $FeSO_4 \cdot 7H_2O$ contain 0.0603 g of iron, 0.0346 g of sulfur, 0.190 g of oxygen, and 0.0151 g of hydrogen, what is the percent composition of $FeSO_4 \cdot 7H_2O$?

18. Do the percentages that you calculated in Question 17 add up to 100%? Explain.

19. What mass of vitamin E contains 0.500 g C, given that vitamin E is 80.8% carbon by mass?

20. Epsom salt used as a laxative consists of $MgSO_4 \cdot 7H_2O$. How many grams of Mg are present in 0.0250 mol $MgSO_4 \cdot 7H_2O$, which is 9.86% Mg?

21. Limestone, which is foundational to cement, consists of calcium carbonate ($CaCO_3$). What is the percent composition of $CaCO_3$?

22. Aspirin is the common name for acetylsalicylic acid. If 100.00 g of aspirin contains 60.00 g of carbon, 4.480 g of hydrogen, and 35.53 g of oxygen, what is its empirical formula?

23. Lidocaine is a widely used local anesthetic. A laboratory analysis of lidocaine reveals that a 5.000 g sample of lidocaine contains 3.588 g of carbon, 0.473 g of hydrogen, 0.598 g of nitrogen, and 0.342 g of oxygen.

    a. What is the percent composition of lidocaine?

    b. What is the empirical formula of lidocaine?

24. Why is it important to use only balanced chemical equations when using stoichiometry?

25. Chlorine was the first of the halogens to be isolated. C. W. Scheele carried out the following reaction in 1774.

    $$NaCl(aq) + 2H_2SO_4(aq) + MnO_2(s) \longrightarrow 2Na_2SO_4(aq) + MnCl_2(aq) + 2H_2O(l) + Cl_2(g)$$

    a. If you start with 1.00 g NaCl and an excess of the other reactants, how many grams of Cl_2 will be produced?

    b. If you start with 1.00 g of H_2SO_4 and an excess of the other reactants, how many grams of Cl_2 will be produced?

    c. If you start with 1.00 g of MnO_2 and an excess of the other reactants, how many grams of Cl_2 will be produced?

    d. How many grams of NaCl must react to produce 1.00 g of Cl_2?

    e. How many grams of H_2SO_4 must react to produce 1.00 g of Cl_2?

    f. How many grams of MnO_2 must react to produce 1.00 g of Cl_2?

    g. How many grams of Na_2SO_4 will be produced along with 1.00 g of Cl_2?

    h. How many grams of $MnCl_2$ will be produced along with 1.00 g of Cl_2?

    i. How many grams of H_2O will be produced along with 1.00 g of Cl_2?

    j. Show that the law of mass conservation is upheld in the production of 1.00 g of Cl_2.

11 CHAPTER REVIEW

26. Hematite [iron(III) oxide] is converted to molten iron in a blast furnace and is then poured into molds. The balanced equation is shown below.

$$Fe_2O_3(s) + 3CO(g) \longrightarrow 2Fe(l) + 3CO_2(g)$$

  a. How many moles of Fe_2O_3 must react to produce 262 mol of Fe?

  b. How many moles of CO_2 are produced by the reaction of 64.0 mol of CO?

  c. How many moles of Fe_2O_3 must react to produce 765 kg of Fe?

  d. How many grams of CO_2 will be produced when 40.0 mol of CO react?

  e. What mass (in kg) of Fe is produced when 299 kg of Fe_2O_3 react?

27. Aqueous solutions of lead(II) nitrate and potassium iodide react to produce a lead(II) iodide precipitate. The balanced chemical equation is shown below.

$$Pb(NO_3)_2(aq) + 2KI(aq) \longrightarrow PbI_2(s) + 2KNO_3(aq)$$

Assume that 175.0 g of $Pb(NO_3)_2$ and 75.0 g of KI are reacted. Answer the following questions.

  a. Which compound is the limiting reactant?

  b. How many grams of lead(II) iodide should be produced?

  c. How many grams of the excess reactant will remain?

  d. If the reaction yields only 93.2 g PbI_2, what is the percent yield?

28. Bleach is an aqueous solution of sodium hypochlorite. A bleach solution may be prepared when chlorine gas is bubbled through aqueous sodium hydroxide.

$$2NaOH(aq) + Cl_2(g) \longrightarrow NaCl(aq) + NaOCl(aq) + H_2O(l)$$

  a. How many moles of NaOCl will be produced from 5.73 mol of Cl_2?

  b. How many moles of NaOH must react to produce 13.7 mol of NaCl?

  c. How many grams of H_2O will be produced if 0.750 mol of NaOH react?

  d. How many grams of Cl_2 must react to produce 6.70 mol of NaCl?

  e. What mass of NaOCl will be produced when 65.5 g of NaOH react?

  f. How many grams of NaOH will react with 37.5 g of Cl_2?

  g. What is the theoretical yield of NaCl in this reaction if 100.0 g of NaOH react?

  h. If your actual yield is 68.45 g of NaCl, what is the percent yield?

CRITICAL THINKING

29. Defend the use of stoichiometry, even though the theoretical yields that it predicts are not actually achieved in real life.

30. Consider the reaction that produces bleach.

$$2NaOH(aq) + Cl_2(g) \longrightarrow NaCl(aq) + NaOCl(aq) + H_2O(l)$$

For a particular application, you must produce 38.5g of sodium hypochlorite in the laboratory. If you know that the percent yield for the reaction is 94.4%, how many grams of NaOH must you start with?

Use the ethics box below to answer Question 31.

31. Write a four-paragraph response on the ethics of legally requiring the installation of carbon monoxide detectors. Use the ethics triad model presented in Chapter 3.

ethics

MANDATORY DETECTORS

Perhaps you have experienced a less threatening version of the story in the Chapter opener. It's the middle of the night, and you're sound asleep. Suddenly you are awakened by a shrill chirp. You try to go back to sleep, but sixty seconds later the chirp sounds again—and again, and again. It's the low-battery warning on one of the carbon monoxide detectors in your house, and it won't stop until someone gets out of bed to deal with the problem. These days, various detectors, along with their annoyances such as low-battery warnings and their need to be replaced every five to ten years, are usually required by building codes. That means that they're required by law. Can such a requirement be justified on a biblical basis?

Chapter 12

GASES

Chances are, you or someone you know has been in a car accident. Though some accidents cause only minor inconveniences, others cause major injuries and even death. Every year there are over three million car accidents in the United States, injuring about three million people and killing more than 30,000. This works out to one death every sixteen minutes. These realities are concerning for Christians because the Bible teaches that human life is valuable. Is there some way to minimize injuries and especially deaths from car accidents?

In response to statistics like these, scientists and engineers have been looking for ways to protect drivers and passengers. Engineers have incorporated antilock braking systems, collision avoidance warning systems, and crumple zones as standard design features of automobiles. One of the innovations that has had the greatest effect on crash statistics is the airbag. This device uses properties of gases to protect the occupants of a vehicle in an accident. You will learn throughout this chapter about gases, their behaviors, and how they participate in chemical reactions.

12.1 Properties of Gases *269*
12.2 Gas Laws *274*
12.3 Gas Stoichiometry *283*

12.1 PROPERTIES OF GASES

Kinetic-Molecular Description of Gases

Have you ever let a helium balloon go, watching it rise into the sky and drift with the wind? It seems like it will never stop rising until it reaches space! At some point, however, the balloon does stop rising. Either the helium no longer has a density low enough to make the balloon continue to rise, or the balloon pops when it is stretched beyond its capacity due to the difference between internal and atmospheric pressures. The kinetic-molecular theory, introduced in Chapter 2, explains the behavior of gases, such as helium in a balloon, on the basis of the motion of particles. The theory was developed to explain the observations of physical properties of gases obtained from many experiments. This theory is summarized in the following five statements.

Why do gases behave as they do?

QUESTIONS
» How do scientists describe gas behavior?
» What causes gas pressure?
» How are temperature, pressure, and volume related in gases?

TERMS
diffusion • effusion • Graham's law of effusion • fluid • compressibility • expansibility • pressure

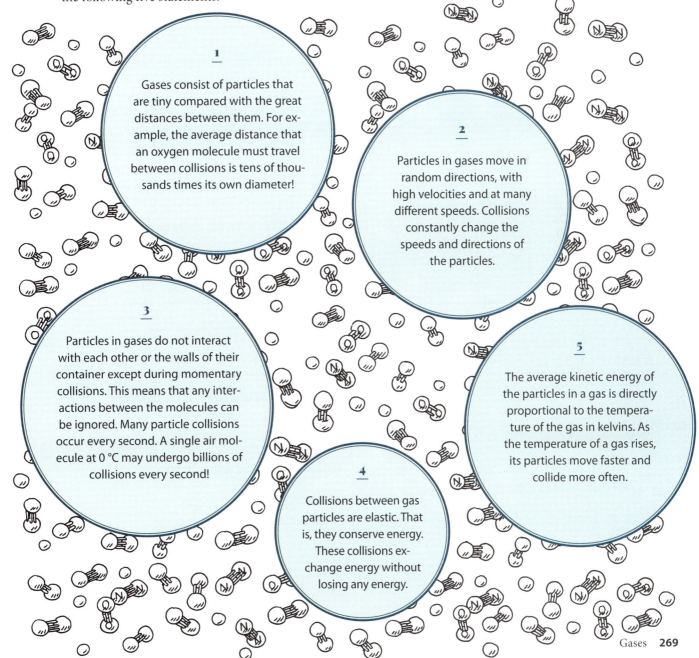

1 Gases consist of particles that are tiny compared with the great distances between them. For example, the average distance that an oxygen molecule must travel between collisions is tens of thousands times its own diameter!

2 Particles in gases move in random directions, with high velocities and at many different speeds. Collisions constantly change the speeds and directions of the particles.

3 Particles in gases do not interact with each other or the walls of their container except during momentary collisions. This means that any interactions between the molecules can be ignored. Many particle collisions occur every second. A single air molecule at 0 °C may undergo billions of collisions every second!

4 Collisions between gas particles are elastic. That is, they conserve energy. These collisions exchange energy without losing any energy.

5 The average kinetic energy of the particles in a gas is directly proportional to the temperature of the gas in kelvins. As the temperature of a gas rises, its particles move faster and collide more often.

Gases **269**

The Physical Properties of Gases

The kinetic-molecular theory is a very workable model for explaining the physical properties of gases that we observe.

DIFFUSION AND EFFUSION

When a gas enters a vacuum (a space containing no matter) or mixes with another gas, it spreads out to uniformly fill the entire volume over time. This process is called **diffusion**. Diffusion occurs because gas molecules are in constant motion. **Effusion** is a related process in which gas particles pass through a tiny opening into an evacuated chamber or space. Both of these processes are directly related to the speed and constant motion of gas particles.

In 1846 British chemist Thomas Graham experimented with various gases. In his studies he allowed gases at the same temperature to effuse into a vacuum. He discovered that the rate of effusion was inversely related to a gas's molar mass. Mathematically, **Graham's law of effusion**, also known as *Graham's law of diffusion*, states that the rate of effusion for a gas is inversely proportional to the square root of its molar mass. This relationship is true because at the same temperature, two different gases have the same kinetic energy. Therefore, the lighter of the two gases moves faster. Because lighter gas particles move faster, they spread more quickly. This law provides a way to experimentally compare molar masses.

Thomas Graham

$$\text{rate of effusion} \propto \frac{1}{\sqrt{\text{molar mass}}}$$

($\propto$ indicates "is proportional to")

This same law allows us to calculate the ratio between effusion rates of two different gases using the following formula.

$$\frac{\text{rate of effusion}_{gas\ 1}}{\text{rate of effusion}_{gas\ 2}} = \frac{\sqrt{\text{molar mass}_{gas\ 2}}}{\sqrt{\text{molar mass}_{gas\ 1}}}$$

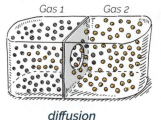
diffusion

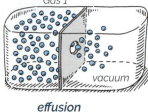
effusion

EXAMPLE 12-1: CALCULATING AN EFFUSION RATE RATIO

Calculate the ratio of effusion rates between nitrogen (N_2) and argon (Ar).

Write what you know.

$\text{molar mass}_{N_2} = 28.02$ g/mol

$\text{molar mass}_{Ar} = 39.95$ g/mol

$\dfrac{\text{rate of effusion}_{N_2}}{\text{rate of effusion}_{Ar}} = ?$

Write the formula and solve for the unknown.

$$\frac{\text{rate of effusion}_{N_2}}{\text{rate of effusion}_{Ar}} = \frac{\sqrt{\text{molar mass}_{Ar}}}{\sqrt{\text{molar mass}_{N_2}}}$$

Plug in known values and evaluate.

$$\frac{\text{rate of effusion}_{N_2}}{\text{rate of effusion}_{Ar}} = \frac{\sqrt{39.95\ \frac{\cancel{g}}{\cancel{mol}}}}{\sqrt{28.02\ \frac{\cancel{g}}{\cancel{mol}}}} = 1.194$$

This result indicates that the lighter nitrogen gas, which moves fast, will effuse 1.194 times faster than the argon gas.

FLUIDITY

Gas particles are able to move past one another like particles in liquids. Both liquids and gases can be poured, so both are considered **fluids**—they have the ability to flow, enabling them to take the shape of their container.

COMPRESSIBILITY AND EXPANSIBILITY

Though liquids do not easily compress, gases have the ability to change their volume to fit their containers. High pressures can squeeze gases into smaller volumes, a property that is called **compressibility**. When gases experience lower pressure, they quickly expand to fill the available space. This ability to spontaneously fill the available space is called **expansibility**. Gases will expand until they are constrained by an external force, thus they always fill their containers. Empty spaces between constantly moving gas molecules make these properties possible.

How Gases Cause Pressure

Gas molecules collide with each other billions of times each second. They also collide with things like trees, buildings, people, and the walls of their containers. Though individual collisions are not very forceful, they add up to produce a significant force. **Pressure** is the average force exerted per unit of area when particles collide against a boundary. It is the observable result of billions of molecular collisions. The total pressure is dependent on the number of collisions that occur and how energetic those collisions are.

We experience gas pressure every moment of our lives due to the pressure exerted on us as the gas in the atmosphere collides with our bodies. These collisions result in a normal atmospheric pressure at sea level of 101 325 pascals (Pa). The pascal, which is named for French physicist Blaise Pascal, is the SI unit of pressure. A *pascal* is defined as the pressure produced by one newton (the SI unit of force) acting on an area of one square meter: $1 \text{ Pa} = 1 \text{ N/m}^2$. A pascal is a small amount of pressure, so for many applications scientists will use kilopascals (kPa).

Pressure, Volume, and Temperature

A gas's volume depends not only on how many gas molecules are present but also on its temperature and pressure. According to the kinetic-molecular theory, a gas at a given temperature contains particles that are moving in random directions at many different speeds. These particles collide with the walls of their container, exerting a pressure on the walls. The average kinetic energy of these particles determines the gas's temperature. In a nonrigid vessel, the pressure exerted on the container will determine the volume of the container.

The graph at right shows how gas particles behave at different temperatures. At 0 °C, many hydrogen molecules move at speeds near 1500 m/s—some slower and some faster. The average speed of 1500 m/s causes the temperature to be 0 °C. At 500 °C, hydrogen molecules move faster, making the average speed greater. Let's look at three different scenarios to see how a gas's pressure, temperature, and volume are all related.

Helium gas is compressed into a canister for storage and shipping, but it quickly expands when released to inflate balloons.

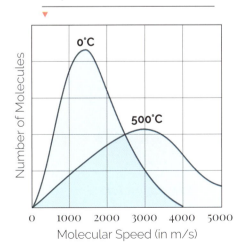

Notice how the average speed of hydrogen molecules shifts with temperature increase.

Why Do Pressure, Temperature, and Volume Change?

In each case, parentheses indicate which statement of the kinetic-molecular theory is referenced (see page 269).

Case 1:
PRESSURE AND VOLUME

What happens to the pressure of a gas if we change its volume? As volume goes down and temperature is held constant, pressure increases. We can do this with a piston in a cylinder: the movable piston moves inward, decreasing the volume. By decreasing the volume, the particles are forced closer together (kinetic-molecular theory Statement 1, or S1), increasing the number of collisions with the walls (S2, S3). With more collisions, pressure on the walls also increases (S4). Additionally, there is a reduction in the surface area of the container. If we were to increase the volume at a constant temperature, pressure would decrease due to fewer collisions occurring. This is an inverse relationship.

Internal Combustion Engine

compression
fuel mixture is compressed

power
burning fuel expands

spark plug fires

Steam Engine

piston

high *temperature*
high *volume*
pressure *remains constant*

low *temperature*
low *volume*
pressure *remains constant*

drop in temperature

Case 2:
TEMPERATURE AND VOLUME

You left your basketball out overnight. You pick it up and try to dribble it and notice that it doesn't bounce like it did yesterday afternoon. What happened? As the air in the ball cooled, the gas particles moved slower (S5). The slower-moving particles collided with the interior of the ball less often (S3) and with less energy (S4, S5). Because the pressure remained constant, the ball contracted and so now has less volume. As gas temperature decreased with the pressure held constant, the volume of the gas also decreased.

This is a direct relationship. If you were to let the ball warm up again, the volume would also increase and the ball would bounce just like it did yesterday.

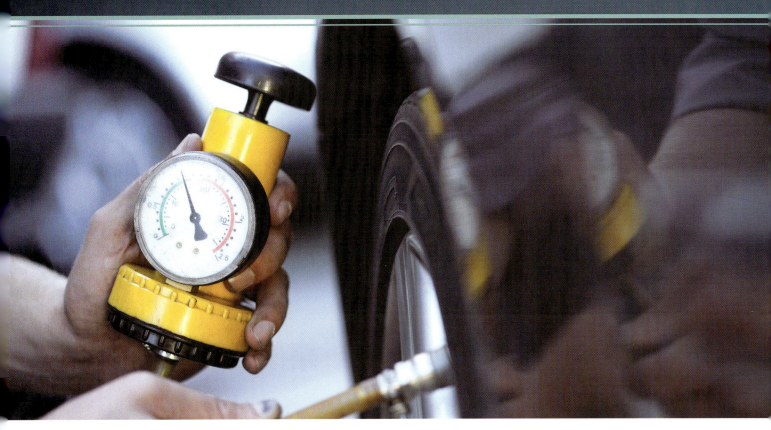

Case 3:
TEMPERATURE AND PRESSURE

The owner's manual of your car tells you the proper *cold* tire pressure. Why does the temperature matter for having the proper pressure? As the temperature of a gas increases, its particles have a higher average kinetic energy (S5). The particles move faster, collide more often (S3), and strike with more force (S4). As force increases, pressure increases because of additional and more forceful collisions. Therefore, in a fixed volume, as in your tires, pressure will increase as temperature increases. The cold inflation pressure ensures that you won't exceed the limits of the tires as they warm up. The converse is also true, meaning that you will likely have to add air to your tires when the weather turns cold. This relationship is also direct.

The relationships between pressure, temperature, and volume shown in these three cases can be summarized with the following proportionalities.

$$P \propto \frac{1}{V} \text{ (constant } T\text{)}$$

$$T \propto V \text{ (constant } P\text{)}$$

$$P \propto T \text{ (constant } V\text{)}$$

12.1 SECTION REVIEW

1. List the five statements of the kinetic-molecular theory of gases.

2. Define *diffusion*.

3. If samples of oxygen and carbon dioxide are at the same temperature and are released into a large room, which sample will diffuse faster? Explain.

4. What causes gas pressure?

5. Use the statements of the kinetic-molecular theory to explain why pressure in a cylinder increases if you compress the gas in the cylinder to one-fifth its original volume.

6. You have a sample of gas in a 1 L balloon. Use the kinetic-molecular theory to explain what will happen to the volume of the balloon if the temperature of the gas is doubled at a constant pressure.

7. Aerosol cans display a label warning against incinerating them. According to the kinetic-molecular theory, why shouldn't an aerosol can be incinerated?

QUESTIONS

» How can I predict the behavior of gases?

» How can I calculate changes in the pressure, temperature, and volume of a gas?

» Why do the gas laws work?

» How can I solve gas problems for mixtures of gases?

TERMS

standard temperature and pressure (STP) • Boyle's law • Charles's law • Gay-Lussac's law • combined gas law • Dalton's law of partial pressures

12.2 GAS LAWS

Standard Conditions

So far we have qualitatively described how gases behave. However, we can make these descriptions quantitative by using a series of gas laws that are based on the kinetic-molecular theory. These gas laws can be used to calculate volumes, pressures, and temperatures of gases at various conditions.

Why do balloons stop getting bigger?

Because the volume of a gas can change with temperature and pressure, reporting a volume without specifying these conditions is meaningless. Scientists have defined a standard temperature and pressure to be used when comparing volumes of gases. **Standard temperature and pressure (STP)** are 0 °C and 101.325 kPa.

Boyle's Law: Pressures and Volumes

Anglo-Irish physicist Robert Boyle was the first to study the relationship between the pressure and volume of a sample of gas. He summed up this relationship mathematically in what is called **Boyle's law**. The law states that the volume of a gas is inversely related to its pressure if the temperature is held constant (Case 1). In equation form, this law is

$$PV = k,$$

where P is pressure, V is volume, and k is a constant.

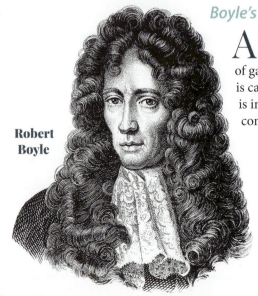

Robert Boyle

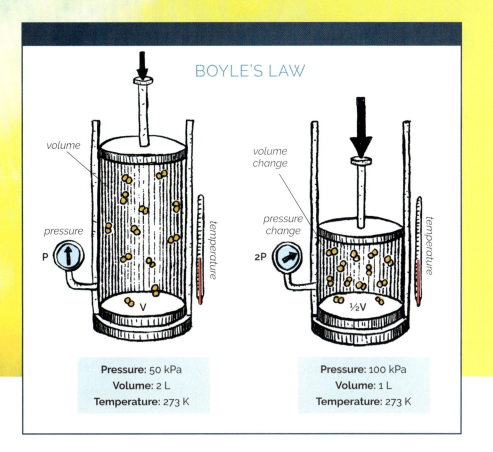

Think back to the cylinder in Case 1 on page 272. As the volume decreased, the particles were pushed together, increasing the number of collisions on a smaller surface area. So we see that, if the temperature is held constant, as the volume of a sample of gas goes down, pressure increases. The product of pressure and volume remains constant for the sample of gas. The equation below is a form of Boyle's law that can be used to solve problems when pressures and volumes of gases change.

$$P_1V_1 = k = P_2V_2$$

EXAMPLE 12-2: USING BOYLE'S LAW

Gas in a cylinder occupies 455 mL at 103.35 kPa. How much space does the gas occupy if we increase pressure to 113.48 kPa at a constant temperature?

Write what you know.

$V_1 = 455$ mL
$P_1 = 103.35$ kPa
$P_2 = 113.48$ kPa
$V_2 = ?$

Write the formula and solve for the unknown.

$$P_1V_1 = P_2V_2$$

$$\frac{P_1V_1}{P_2} = \frac{\cancel{P_2}V_2}{\cancel{P_2}}$$

$$V_2 = \frac{P_1V_1}{P_2}$$

Plug in known values and evaluate.

$$V_2 = \frac{(103.35 \text{ k}\cancel{\text{Pa}})(455 \text{ mL})}{113.48 \text{ k}\cancel{\text{Pa}}}$$

$$= 414 \text{ mL}$$

The increased pressure would compress the gas to 414 mL.

MINI LAB

CHANGING VOLUME

If you have ever watched a hot air balloon, you know that it rises as a burner heats the air in the balloon. If the burner is turned off, the balloon will begin to sink. We know that things float or sink depending on whether they are less dense or more dense than the fluid in which they are submerged. So a balloon sinks because it becomes more dense than the air around it. How does this happen?

How does a change in temperature of a gas affect its volume?

1. Hypothesize why you think a hot air balloon sinks in the air around it.

Procedure

A Fill the bucket half full with ice.

B Add water until the bucket is three-quarters full.

C Inflate the balloon and tie it closed.

2. Why does a party balloon expand as you blow it up? Why doesn't it continue inflating after you stop blowing into it?

D Using the marker, draw a line around the circumference of the balloon.

E Using the string, measure the circumference of the balloon. Mark the circumference on the string.

3. What do you think will happen to the balloon if it is submerged in the ice water?

F Stir the water and then submerge the balloon.

G After ten minutes, retrieve the balloon and measure the circumference again.

Conclusion

4. What happened to the circumference?

5. What happened to the volume?

Going Further

6. Why does a hot air balloon sink when the burner below it is off?

EQUIPMENT
- bucket
- ice
- balloon
- marker
- string
- spoon

Charles's Law: Temperatures and Volumes

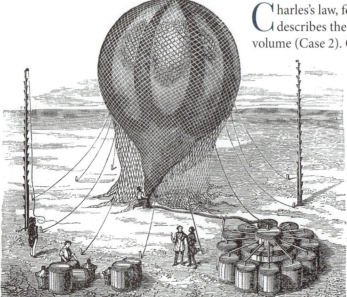

Charles's law, formulated by French physicist Jacques Charles, describes the relationship between a gas sample's temperature and volume (Case 2). Careful measurements of the volume of a gas at different temperatures and constant pressure yield data like that listed in Table 12-1. We can see from this table that temperature and volume are directly proportional when pressure is constant.

TABLE 12-1
Effect of Temperature on Volume

Temperature (K)	Volume (mL)
400	1000
200	500
100	250

276 Chapter 12

Graphing these values with temperature on the *x*-axis and volume on the *y*-axis shows that the volume would approach zero at a temperature of 0 K. This extrapolation of gas volumes led to the development of the Kelvin temperature scale. On this scale, the lowest temperature that is theoretically possible (−273 °C) is labeled 0 K—absolute zero.

Charles's law states that when the pressure of a sample of a gas is held constant, the temperature in kelvins and the volume are directly related. Mathematically, this law is stated as

$$\frac{V}{T} = k,$$

where *T* is temperature in kelvins, *V* is volume, and *k* is a constant.

Remember the basketball (Case 2 on page 272) that sat out overnight? As the temperature decreased, the particles' kinetic energy decreased (S5). This meant that the particles collided with the inside of the ball less often (S3) and the collisions were less forceful (S4). Since the volume of the ball was changeable, the ball's volume also decreased to maintain a constant pressure. So when a sample of gas is held at a constant pressure, no matter what changes occur, the ratio between volume and temperature in kelvins will remain the same.

$$\frac{V_1}{T_1} = k = \frac{V_2}{T_2}$$

Remember that the temperature has to be in kelvins for Charles's law to work properly, though the volume can be measured in any unit. The reason for this is that the Kelvin scale is designed on the basis of thermodynamic temperature. Also, if you were to not use an absolute temperature scale, you could end up with negative volumes.

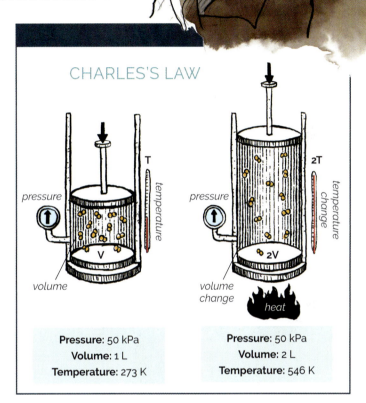

Jacques Charles

CHARLES'S LAW

Pressure: 50 kPa
Volume: 1 L
Temperature: 273 K

Pressure: 50 kPa
Volume: 2 L
Temperature: 546 K

EXAMPLE 12-3: USING CHARLES'S LAW

A sample of gas occupies 3.75 L at 25 °C. Assuming that pressure is held constant, how much space will it occupy at standard temperature?

Write what you know.

$V_1 = 3.75$ L
$T_1 = 25$ °C $= 25 + 273.15 = 298.15$ K
$T_2 = 0$ °C $= 273$ K
$V_2 = ?$

Write the formula and solve for the unknown.

$$\frac{V_1}{T_1} = \frac{V_2}{T_2}$$

$$\left(\frac{V_1}{T_1}\right)T_2 = \left(\frac{V_2}{\cancel{T_2}}\right)\cancel{T_2}$$

$$V_2 = \left(\frac{V_1}{T_1}\right)T_2$$

Plug in known values and evaluate.

$$V_2 = \left(\frac{3.75 \text{ L}}{298.15 \text{ K}}\right) 273 \text{ K}$$

$$= 3.43 \text{ L}$$

As the temperature decreased to standard temperature, the volume decreased to 3.43 L.

Gay-Lussac's Law: Temperatures and Pressures

The pressure in a car's tires increases as tires heat up on a lengthy trip. This is an example of **Gay-Lussac's law**, named in honor of French chemist Joseph Louis Gay-Lussac, which states that pressure is directly proportional to temperature in kelvins for a fixed mass of gas held at a constant volume (Case 3 on page 273). The law is expressed mathematically as

$$\frac{P}{T} = k,$$

where P is pressure, T is temperature in kelvins, and k is a constant.

As mentioned in Case 3, the pressure in the tires increased as temperature increased. Friction with the road raises the temperature of the air inside the tires. Gas molecules move faster (S5), collide more often (S3), and transfer more force (S4) to the inner walls of the tires. Since the volume is essentially fixed, pressure increases. As long as volume is held constant, Gay-Lussac's law holds true for many different pressures and temperatures. Again, remember that the temperature must be in kelvins.

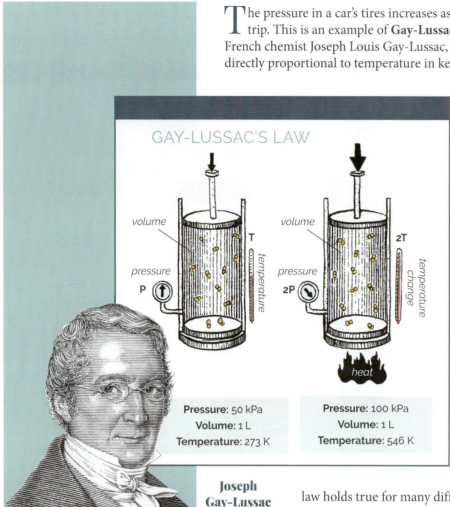

Joseph Gay-Lussac

$$\frac{P_1}{T_1} = k = \frac{P_2}{T_2}$$

EXAMPLE 12-4: USING GAY-LUSSAC'S LAW

Before Molly began a car trip, she measured the air pressure in her car tires and found that it was 221 kPa at 18 °C. After two hours of driving, she found that the pressure had increased to 248 kPa. What was the new temperature of the air in her tires?

Write what you know.

$P_1 = 221$ kPa
$T_1 = 18$ °C $= 18 + 273.15 = 291.15$ K
$P_2 = 248$ kPa
$T_2 = ?$

Write the formula and solve for the unknown.

$$\frac{P_1}{T_1} = \frac{P_2}{T_2}$$

$$\left(\frac{P_1}{T_1}\right) T_2 = \left(\frac{P_2}{T_2}\right) T_2$$

$$\left(\frac{P_1}{T_1}\right) T_2 \left(\frac{T_1}{P_1}\right) = P_2 \left(\frac{T_1}{P_1}\right)$$

$$T_2 = P_2 \left(\frac{T_1}{P_1}\right)$$

Plug in known values and evaluate.

$T_2 = 248 \text{ kPa} \left(\dfrac{291.15 \text{ K}}{221 \text{ kPa}}\right)$

$= 326.72$ K

$= 326.72 - 273.15 = 54$ °C

The air in her tires is now 54 °C.

The Combined Gas Law

The gas laws that we have discussed so far have applied to situations in which only two variables can change. But what happens when pressure, volume, and temperature all change? The three laws described so far can be combined to form a single equation called the **combined gas law**, which states that the pressure, volume, and temperature of a sample of gas vary jointly.

The formula for the combined gas law is

$$\frac{PV}{T} = k,$$

where P is pressure, V is volume, T is temperature in kelvins, and k is a constant.

$$\frac{P_1 V_1}{T_1} = k = \frac{P_2 V_2}{T_2}$$

The advantage of the combined gas law is that it uses one equation to solve problems in which pressure, volume, and temperature all change. The three previous gas laws can be derived from the combined gas law. For example, if temperature is constant in the above equation ($T_1 = T_2$) the two temperatures cancel from the combined gas law to produce Boyle's law.

EXAMPLE 12-5: USING THE COMBINED GAS LAW

A sample of gas has a volume of 3.6 L at 1.06×10^5 Pa and $-15\ °C$. What will its volume be at STP?

Write what you know.

$V_1 = 3.6\ \text{L}$
$P_1 = 1.06 \times 10^5\ \text{Pa}$
$T_1 = -15\ °C = 258.15\ \text{K}$
$P_2 = 1.01 \times 10^5\ \text{Pa}$
$T_2 = 0\ °C = 273.15\ \text{K}$
$V_2 = ?$

Write the formula and solve for the unknown.

$$\frac{P_1 V_1}{T_1} = \frac{P_2 V_2}{T_2}$$

$$\left(\frac{P_1 V_1}{T_1}\right)\left(\frac{T_2}{P_2}\right) = \left(\frac{\cancel{P_2} V_2}{\cancel{T_2}}\right)\left(\frac{\cancel{T_2}}{\cancel{P_2}}\right)$$

$$V_2 = \frac{P_1 V_1 T_2}{T_1 P_2}$$

Plug in known values and evaluate.

$$V_2 = \frac{(1.06 \times 10^5\ \cancel{\text{Pa}})(3.6\ \text{L})(273.15\ \cancel{\text{K}})}{(1.01 \times 10^5\ \cancel{\text{Pa}})(258.15\ \cancel{\text{K}})}$$

$$= 4.0\ \text{L}$$

By adjusting the temperature and pressure to STP, the volume expands to 4.0 L.

Dalton's Law of Partial Pressures: Mixtures of Gases

It is rare to find a naturally occurring pure gas. Even gases produced in laboratories contain some impurities. These mixtures complicate gas law calculations. English chemist John Dalton formulated a law to describe how gaseous mixtures behave, called **Dalton's law of partial pressures**. This law states that the total pressure of a mixture of gases equals the sum of the pressures of each individual gas, called the *partial pressures*. The general formula for Dalton's law of partial pressures is

$$P_{total} = \Sigma P_i,$$

where P_{total} is the total gas pressure and ΣP_i (the uppercase Greek letter *sigma* combined with P_i) is the sum of the partial pressures.

Suppose that 1 L of oxygen is added to 1 L of nitrogen, both gases at STP, in a 1 L container. As long as the molecules do not chemically react, they behave independently of each other. So oxygen molecules collide with the container and exert a pressure of 101.325 kPa, just as they did before they were mixed with the nitrogen molecules. The nitrogen gas particles also collide with the container, exerting an additional 101.325 kPa of pressure, so the total pressure from the two gases is 202.650 kPa. This happens because there are twice as many particles colliding with the sides of the container.

As another example, a sample of air contains 78% nitrogen, 21% oxygen, 1% argon, and traces of other gases. These percentages are based on volume. If the air mixture is at normal atmospheric pressure, 101.325 kPa, we find that 78% of the pressure (79.0335 kPa) comes from nitrogen, 21% of the pressure (21.2783 kPa) comes from oxygen, and 1% of the pressure (1.013 25 kPa) comes from argon. The sum of all the partial pressures equals the total pressure of the mixture.

John Dalton

CASE STUDY

WHEN OXYGEN IS BAD

God created the air in our atmosphere to meet the needs of all His created plants, animals, and humans. It is the perfect mix of nitrogen, oxygen, and other gases (1%). We cannot survive without oxygen. But there are times when nitrogen and oxygen can be dangerous to us. Deep-sea diving is one of those instances.

Two hazards of deep diving are nitrogen narcosis and oxygen toxicity, both of which are caused by breathing nitrogen or oxygen under the high-pressure conditions associated with deep diving (up to 3000 kPa, almost thirty times normal pressure). Nitrogen narcosis is a condition in which nitrogen has a narcotic effect on the human body. Initially noticeable as a feeling of euphoria, left unchecked it can progress to cognitive impairment, hallucinations, and even death. Oxygen toxicity occurs when too much oxygen is dissolved in the blood and is caused by breathing at higher than normal partial pressures. This condition affects the central nervous system, the eyes, and can also lead to death.

To combat these two risks, technical divers will use a special blend of compressed air for deep dives. The most common mix consists of oxygen, nitrogen, and helium. The exact proportions of each gas depend on the depth to which the diver will descend. By adding helium to the mix, the diver is able to reduce the partial pressure of oxygen while also reducing the narcotic effects of the nitrogen.

Questions to Consider

1. How does the nitrogen in the atmosphere demonstrate God's care for his creation?
2. What gas condition changes during deep diving, contributing to nitrogen narcosis and oxygen toxicity?
3. What does the term *partial pressure* mean?

In the laboratory, chemists often collect a sample of a gas by trapping it at the top of a water-filled container. The gas bubbles up through the water, collects at the top, and forces water out the bottom. This technique is called *collecting over water* or *collection by water displacement*. The gas is initially pure, but as it bubbles through the water, some water evaporates and mixes with the gas being collected, forming a mixture called a wet gas. Accurate measurements of the gas cannot be made when it is mixed with water vapor. The total pressure, which equals the atmospheric pressure, is made up of the partial pressures of the gas and water vapor. To find the pressure due to the dry gas, the pressure from the water vapor must be subtracted from the atmospheric pressure. The pressure from the water vapor depends on its temperature. Table 12-2 lists vapor pressures of water at various temperatures.

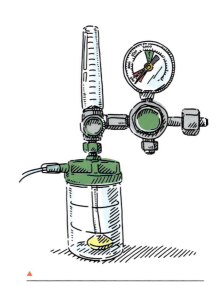

An oxygen humidifier. Ambulance equipment humidifies oxygen by bubbling it through water. Without this precaution, a patient's mouth, throat, and lungs would become dehydrated when receiving oxygen.

EXAMPLE 12-6: ADJUSTING FOR VAPOR PRESSURE

A 46 mL sample of O_2 gas is collected over water at 25 °C when the atmospheric pressure is 102 kPa. What volume of pure oxygen is this at STP?

Write what you know.

$V_1 = 46$ mL
$T_1 = 25\ °C = 298.15$ K
$P_{total} = 102$ kPa
$T_2 = 273.15$ K
$P_2 = 101.325$ kPa
P_1 (sample of O_2) = ?
V_2 (sample of O_2) = ?

To use the combined gas law, we must find the pressure of pure oxygen. Use Table 12-2 to find the pressure of water vapor at 25 °C.

$P_{H_2O} = 3.1690$ kPa
$P_{total} = P_1 + P_{H_2O}$
$P_1 = P_{total} - P_{H_2O}$
$= 10\underline{2}$ kPa $- 3.1690$ kPa
$= 9\underline{8}.831$ kPa

Now we have what we need to use the combined gas equation. Notice that Celsius temperatures must be converted.

Write the formula and solve for the unknown.

$$\frac{P_1 V_1}{T_1} = \frac{P_2 V_2}{T_2}$$

$$\left(\frac{P_1 V_1}{T_1}\right)\left(\frac{T_2}{P_2}\right) = \left(\frac{\cancel{P_2} V_2}{\cancel{T_2}}\right)\left(\frac{\cancel{T_2}}{\cancel{P_2}}\right)$$

$$V_2 = \frac{P_1 V_1 T_2}{T_1 P_2}$$

Plug in known values and evaluate.

$$V_2 = \frac{(9\underline{8}.831\ \text{kPa})(4\underline{6}\ \text{mL})(273.15\ \text{K})}{(298.15\ \text{K})(101.325\ \text{kPa})}$$
$= 41$ mL

Once we factor out the water vapor, we will have 41 mL of O_2, the dry gas.

TABLE 12-2 *Water Vapor Pressure at Various Temperatures*

Temperature (°C)	Vapor Pressure (kPa)
0.0	0.6113
5.0	0.8726
10.0	1.2281
15.0	1.7056
20.0	2.3388
25.0	3.1690
30.0	4.2455
35.0	5.6267
40.0	7.3814
45.0	9.5898
50.0	12.3440
55.0	15.7520
60.0	19.9320
65.0	25.0220
70.0	31.1760
75.0	38.5630
80.0	47.3730
85.0	57.8150
90.0	70.1170
95.0	84.5290
100.0	101.3200

12.2 SECTION REVIEW

1. Why have scientists established a standard temperature and pressure (STP)?

2. Explain Boyle's law on the basis of the kinetic-molecular theory of gases.

3. State Charles's law qualitatively and mathematically.

4. Describe how the Kelvin temperature scale was formulated from gas behavior.

5. Identify the gas law needed to answer the following questions and then use it to answer the question.

    a. A sample of gas at 1.19×10^5 Pa occupies 4.15 L. If the gas is allowed to expand to 11.26 L and the temperature remains constant, what will the pressure be?

    b. A sample of nitrogen gas occupies 137.4 mL at 19.7 °C. What volume will the gas occupy at 46.2 °C if pressure remains constant?

    c. A sample of gas in a 0.537 L container at 164.653 kPa and 16.5 °C undergoes a temperature change so that the pressure is changed to 41.997 kPa. If the volume has remained constant, what is the new temperature?

    d. If a sample of a gas at 122 kPa and 22.0 °C occupies 0.350 L, what pressure will the sample of gas exert on a 0.050 L container at 25.0 °C?

    e. A sample of gas occupies 3.51 L at 9.54×10^5 Pa and 20 °C. If the conditions change to 6.17×10^5 Pa at 20 °C, what will the volume be?

    f. A particular sample of pantothenic acid, a B vitamin, gives off 72.6 mL of nitrogen at 23 °C and 106 kPa. What is the volume of the nitrogen at STP?

    g. An experiment calls for 5.54 L of sulfur dioxide (SO_2) at 0.00 °C and 101 325 Pa. What will the volume of this gas be at 31.15 °C if pressure remains constant?

    h. A 1.25 L sample of a gas is at 107 kPa at 128 °C. If the volume remains constant, what will the gas's pressure be if it cools to 98 °C?

6. Use your knowledge of the gas laws to predict what will happen to the

    a. pressure in a can of spray paint when the gas inside is heated and no gas escapes.

    b. volume of a balloon as it is warmed when pressure is constant.

    c. volume of an inflated tire tube when it is immersed in cold water and pressure stays the same.

    d. pressure in a cylinder when the volume is increased by moving a piston while the temperature is constant.

    e. pressure as a sample of gas is allowed to expand and cool at the same time.

7. State Dalton's law qualitatively and mathematically.

8. A sample of 43.9 mL of hydrogen is collected by water displacement at a temperature of 20 °C. The barometer reads 101 185 Pa. What is the volume of the hydrogen at STP?

12.3 GAS STOICHIOMETRY

You have seen throughout this textbook that chemists are interested in how much product is produced when reactions occur. How much oxygen and hydrogen do you think would be produced when one mole of water decomposes? The answer depends on how you are measuring that quantity. If you compare the masses of oxygen and hydrogen produced, you would find that eight times more oxygen (16 g) is produced than hydrogen (2 g). But if you were to compare the volumes of the gases produced, then you would discover that twice as much hydrogen is produced as oxygen. Why is this so?

Gases in Reactions

Recall that stoichiometry is the mathematical process for predicting the amount of each product produced in a chemical reaction. In addition to his studies of pressure and temperature, Gay-Lussac studied the chemical reactions of gases. In particular, he measured and compared the volumes of gases that reacted with each other. One reaction that he studied was between hydrogen and chlorine to form hydrogen chloride gas. He found that if the gases had identical pressures and temperatures, 1 L of hydrogen combined with 1 L of chlorine to form 2 L of hydrogen chloride gas. When he studied the reaction between hydrogen and oxygen to form water, he found different ratios for their combining volumes. Another chemist, investigating the reaction between nitrogen and hydrogen to form ammonia, found yet another set of volume ratios.

In 1808, Gay-Lussac formulated the **law of combining volumes**, which states that under equivalent conditions, the volumes of reacting gases and their gaseous products are expressed in ratios of small whole numbers. Although he did not know it at the time, the ratios of these small whole numbers were the ratios of moles of reactants to moles of products. Notice in Table 12-3 that the volume ratios are identical to the mole ratios from the balanced chemical equations.

This law, along with the law of definite proportion, eventually led chemists to understand the relationship between atoms, molecules, and compounds.

QUESTIONS
» Do the volumes of gases in reactions just add together?
» How well do our models represent real gases?
» How can I determine the number of particles in a sample of gas if I know the pressure, volume, and temperature?
» Do reactions between gases work the same as other reactions?
» Can we predict amounts of chemicals in reactions with gases?

TERMS
law of combining volumes • Avogadro's law • standard molar volume • ideal gas • ideal gas law

How do gases actually behave?

TABLE 12-3
Comparing Volume Ratios and Mole Ratios of Chemical Reactions with Gases

Volume Ratio	Mole Ratio
1 L $H_2(g)$ + 1 L $Cl_2(g)$ ⟶ 2 L $HCl(g)$	1 mol $H_2(g)$ + 1 mol $Cl_2(g)$ ⟶ 2 mol $HCl(g)$
2 L $H_2(g)$ + 1 L $O_2(g)$ ⟶ 2 L $H_2O(g)$	2 mol $H_2(g)$ + 1 mol $O_2(g)$ ⟶ 2 mol $H_2O(g)$
1 L $N_2(g)$ + 3 L $H_2(g)$ ⟶ 2 L $NH_3(g)$	1 mol $N_2(g)$ + 3 mol $H_2(g)$ ⟶ 2 mol $NH_3(g)$

Volcanic eruptions involve chemical reactions as well as changes in the temperature, pressure, and volume of gases.

Molar Volume

The Molar Volume of a Gas

Gay-Lussac's law of combining volumes led Amedeo Avogadro to hypothesize that under equivalent conditions, equal volumes of gases contain the same number of particles. A series of experiments supported his hypothesis, and additional work led to the development of **Avogadro's law**, which states that the volume of a gas, maintained at a constant temperature and pressure, is directly proportional to the number of moles of the gas. In equation form, this law is

$$\frac{V}{n} = k,$$

where V is volume, n is the number of particles in moles, and k is a constant.

Every equivalent volume of gas in a reaction contains the same number of particles if the gases are at the same temperature and pressure.

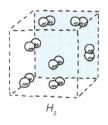
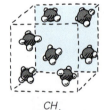
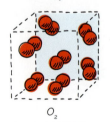

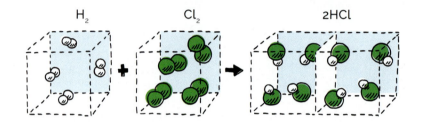

When hydrogen and chlorine react, 1 L of hydrogen contains the same number of molecules as 1 L of chlorine. One hydrogen molecule reacts with one chlorine molecule to form two hydrogen chloride molecules. Since twice as many hydrogen chloride molecules are produced, 2 L of gas results.

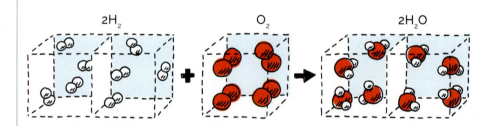

In the formation of water from hydrogen and oxygen, 2 L of hydrogen contain twice as many molecules as 1 L of oxygen. The 2 L of water vapor produced contain the same number of molecules as the 2 L of hydrogen and twice the number of molecules as the 1 L of oxygen.

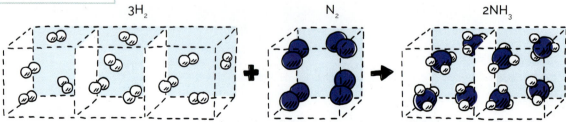

Likewise, the volumes of gases involved in the formation of ammonia (NH_3) show the relative number of molecules.

Avogadro's ideas eventually helped establish the mole as a unit of measure. Experiments also determined how many particles were present in a given volume. At STP, a volume of 22.4 L contains 6.022×10^{23} particles, or 1 mol, of a gas. For this reason, 22.4 L is called the **standard molar volume** of a gas. No matter what the gas is, 1 mol at STP occupies 22.4 L. The amount of space that a mole of gas will occupy—its molar volume—at other than STP will vary with temperature and pressure, but every gas will have the same molar volume at that same temperature and pressure.

The molar volume of a gas allows us to convert between moles of a gas and its volume as long as the gas is at STP. If the gas is not at STP, we can still determine the number of moles present by adjusting the volume of the gas by using the combined gas law that we studied in Section 12.2.

EXAMPLE 12-7:
CALCULATING GAS VOLUME FROM MOLES (STP)

What volume will 2.67 mol of ammonia occupy at STP?

Write what you know.

n = 2.67 mol
STP

Convert from moles to volume using molar volume.

$$2.67 \text{ mol NH}_3 \left(\frac{22.4 \text{ L NH}_3}{1 \text{ mol NH}_3} \right) = 59.8 \text{ L NH}_3$$

As you can imagine, most of the time a sample of gas will be at conditions other than STP. We could still use the molar volume of a gas to convert between volume and moles, but we would have to use the combined gas law (Section 12.2) to first determine what the volume of the gas would be at STP. There is an easier way to find the moles at nonstandard conditions, and that is to use the ideal gas law, which you will learn about later in this section.

Ideal Gases

Did you know that gases do not exactly obey Boyle's law for all pressures and temperatures? For that matter, Charles's law falls short too. Gas laws do not govern how gases behave—they merely describe. Boyle and Charles provided accurate descriptions of the volume, pressure, and temperature of gases, but the predictive ability of these laws depends on conditions. The gas laws that you have learned about are models—workable explanations or descriptions of the world around us.

Because the kinetic-molecular theory's description of gases is simple, it cannot exactly describe the behavior of most gases. An **ideal gas** is a gas whose behavior is perfectly predicted by the kinetic-molecular theory. Though no real-world gases are ideal, some come close under certain conditions. The kinetic-molecular theory makes several assumptions that are not always accurate. Let's reexamine two of the postulates of the kinetic-molecular theory.

1. Gas particles are tiny compared with the great distances between them.
2. Particles do not interact with each other or the walls of their container except during momentary collisions.

The kinetic-molecular theory assumes that particles, either atoms or molecules, are masses that take up no volume. Since gas particles, especially larger ones, take up at least some space, this postulate of the kinetic-molecular theory does not match the characteristics of real gases. The assumption that gas particles are separated by great distances is valid for gases at high temperatures and low pressures. However, high pressures or low temperatures can force gases to approach the point of *condensation*. Under these conditions, gas molecules slow down considerably and the spaces between them shrink. In these conditions, real gases behave quite differently than ideal gases. When a gas is near condensation, these interactions have an even greater impact on the particles. Consequently, the gas particles begin to coalesce, which results in pressures and volumes that are smaller than expected.

In general, real gases behave most like the ideal predicted by the kinetic-molecular theory when they have a low molar mass, are under low pressure (< 101.325 kPa), and are at a relatively high temperature (> 273 K). This is great news since we spend much of our time using gases at these high-temperature, low-pressure conditions, which means that the gas laws work well in our daily lives.

The molar volume of all gases is 22.4 L, right? Well, not exactly. The volume quantity 22.4 L is actually for ideal gases and is a useful approximation for real gases. When gaseous particles are large, molar volumes decline slightly because intermolecular forces become significant. Table 12-4 shows the molar volumes for some common gases.

TABLE 12-4

Values for Some Common Gases at STP

Gas	Gas Density (g/L)	Molar Mass (g/mol)	Standard Molar Volume (L)
H_2	0.0899	2.016	22.428
He	0.1785	4.003	22.426
Ne	0.9002	20.18	22.425
N_2	1.251	28.02	22.404
O_2	1.429	32.00	22.394
Ar	1.784	39.95	22.393
CO_2	1.977	44.01	22.256
NH_3	0.7710	17.03	22.094
Cl_2	3.214	70.90	22.063

The Ideal Gas Law

Boyle's law states that the volume of a gas is inversely proportional to its pressure when temperature remains constant. Charles's law states that volume is directly proportional to the Kelvin temperature when pressure is constant. Avogadro's law states that the volume is directly proportional to the number of moles of gas in the sample. These three laws can be written as shown below.

$$V \propto \frac{1}{P}$$

$$V \propto T$$

$$V \propto n$$

When they are combined, they give us the proportions below.

$$V \propto \frac{nT}{P} \text{ or } PV \propto nT$$

Using mathematics, we can change a proportion into an equation by including a constant. By including R, the *universal gas constant*, we change the proportionality into the **ideal gas law**, which relates the pressure, temperature, volume, and the number of moles in any sample of a gas. The ideal gas law allows us to solve for any of these quantities for a gas if we know the other three. The formula for the ideal gas law is

$$PV = nRT,$$

where P is the pressure, V is the volume, n is the number of particles in moles, and T is the temperature in kelvins. R is the universal gas constant. Rearranging the ideal gas law allows us to solve for the universal gas constant.

$$R = \frac{PV}{nT}$$

The gas constant's specific value depends on the units used for pressure, temperature, and volume (see Table 12-5). Since 1 mol of gas at STP (273.15 K and 101.325 kPa) occupies approximately 22.4 L, these values can be substituted into the equation. For example,

$$R = \frac{(101.325 \text{ kPa})(22.4 \text{ L})}{(1 \text{ mol})(273.15 \text{ K})}$$

$$= 8.31 \frac{\text{kPa} \cdot \text{L}}{\text{mol} \cdot \text{K}}$$

We can easily convert the value of the gas constant if working in other units.

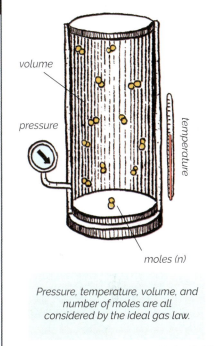

Pressure, temperature, volume, and number of moles are all considered by the ideal gas law.

TABLE 12-5
Universal Gas Constant Values

Pressure Unit	Volume Unit	R Value
kPa	L	$8.31 \frac{\text{kPa} \cdot \text{L}}{\text{mol} \cdot \text{K}}$
Pa	L	$8.31 \times 10^3 \frac{\text{Pa} \cdot \text{L}}{\text{mol} \cdot \text{K}}$
Pa	m³	$8.31 \frac{\text{Pa} \cdot \text{m}^3}{\text{mol} \cdot \text{K}}$

EXAMPLE 12-8: CALCULATING MOLES OF A GAS

How many moles of a gas are present in a 3.32 L sample at 154 kPa and 29 °C?

Write what you know.

$V = 3.32$ L
$P = 154$ kPa
$T = 29\ °C = 302.15$ K
$R = 8.31\ \dfrac{\text{kPa} \cdot \text{L}}{\text{mol} \cdot \text{K}}$ (from Table 12-5)
$n = ?$

Write the formula and solve for the unknown.

$$PV = nRT$$

$$\dfrac{PV}{RT} = \dfrac{nRT}{RT}$$

$$n = \dfrac{PV}{RT}$$

Plug in known values and evaluate.

$$n = \dfrac{(154\ \text{kPa})(3.32\ \text{L})}{(8.31\ \dfrac{\text{kPa} \cdot \text{L}}{\text{mol} \cdot \text{K}})(302.15\ \text{K})}$$

$$= 0.204\ \text{mol}$$

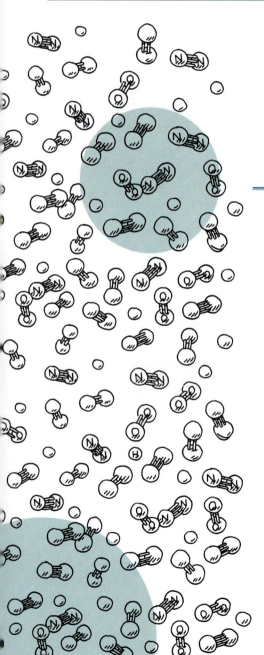

If chemists measure the mass, pressure, volume, and temperature of an unknown gas, they can determine its molar mass. They first use the ideal gas law to find the number of moles in the sample. They then form a ratio of the mass of the sample to the number of moles in the sample. This ratio is reduced to produce the molar mass of the gas.

EXAMPLE 12-9: CALCULATING MOLAR MASS OF A GAS MASS AT NONSTANDARD CONDITIONS

A 0.481 g sample of gas occupies 4.04×10^{-4} m³ at 9.71×10^4 Pa and 297 K. What is the molar mass of this unknown gas?

Write what you know.

$m = 0.481$ g
$V = 4.04 \times 10^{-4}$ m³
$P = 9.71 \times 10^4$ Pa
$T = 297$ K
$R = 8.31\ \dfrac{\text{Pa} \cdot \text{m}^3}{\text{mol} \cdot \text{K}}$ (from Table 12-5)
$n = ?$
molar mass $= ?$

Write the formula and solve for the unknown.

$n = \dfrac{PV}{RT}$ (see Example 12-8)

Plug in known values and evaluate.

$$n = \dfrac{(9.71 \times 10^4\ \text{Pa})(4.04 \times 10^{-4}\ \text{m}^3)}{(8.31\ \dfrac{\text{Pa} \cdot \text{m}^3}{\text{mol} \cdot \text{K}})(297\ \text{K})}$$

$$= 0.01589\ \text{mol}$$

Now, by dividing the mass by the number of moles we can determine the molar mass.

$$\text{molar mass} = \dfrac{0.481\ \text{g}}{0.01589\ \text{mol}}$$

$$= 30.3\ \text{g/mol}$$

Stoichiometry Involving Gases

In Chapter 11 you learned that the coefficients in balanced chemical equations give the molar ratios between reactants and products. The number of particles and their masses can be calculated once the number of moles is known. Throughout this section you have learned that by using the ideal gas law, we can convert between volume and moles of gases at any temperature and pressure. This allows us to accomplish stoichiometric calculations when given the volumes of gases in the reaction. Let's look at an example of a gas stoichiometry problem.

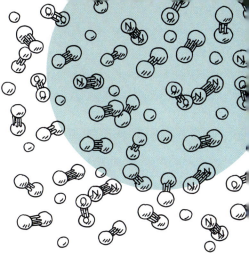

EXAMPLE 12-10: CALCULATING THE VOLUME OF A GASEOUS PRODUCT

When 78.5 g of calcium react with water to form calcium hydroxide and hydrogen gas, what volume of hydrogen gas will be produced at 315 K and 115 511 Pa?

Write what you know.

$$Ca(s) + 2H_2O(l) \longrightarrow Ca(OH)_2(aq) + H_2(g)$$
$m = 78.5$ g Ca
$P = 115\,511$ Pa
$T = 315$ K
$n_{H_2} = ?$
$V_{H_2} = ?$

Convert g Ca into mol H_2.

$$78.5 \, \cancel{g\,Ca} \left(\frac{1 \, \cancel{mol\,Ca}}{40.08 \, \cancel{g\,Ca}} \right) \left(\frac{1 \text{ mol } H_2}{1 \, \cancel{mol\,Ca}} \right)$$
$= 1.958\,582\,834$ mol H_2

Write what you know.

$n_{H_2} = 1.958$ mol H_2
$8.31 \times 10^3 \, \frac{Pa \cdot L}{mol \cdot K}$ (from Table 12-5)
$V_{H_2} = ?$

Write the formula and solve for the unknown.

$$PV = nRT$$
$$\frac{\cancel{P}V}{\cancel{P}} = \frac{nRT}{P}$$
$$V = \frac{nRT}{P}$$

Plug in known values and evaluate.

$$V = \frac{(1.958 \text{ mol } H_2)(8.31 \times 10^3 \, \frac{Pa \cdot L}{mol \cdot K})(315 \text{ K})}{(115\,511 \text{ Pa})}$$
$= 44.4$ L H_2

EXAMPLE 12-11: CALCULATING THE MASS OF PRODUCT FROM THE VOLUME OF A REACTANT

A volume of 3.58 L of oxygen gas at 180.3 kPa and 35 °C is reacted with hydrogen. How many grams of water will be produced?

Write what you know.

$$O_2(g) + 2H_2(g) \longrightarrow 2H_2O(l)$$
$V_{O_2} = 3.58$ L
$P = 180.3$ kPa
$T = 35$ °C $= 308.15$ K
$R = 8.31 \, \frac{kPa \cdot L}{mol \cdot K}$
$n_{O_2} = ?$
$m_{H_2O} = ?$

Write the formula and solve for the unknown.

$n = \frac{PV}{RT}$ (see Example 12-8)

Plug in known values and evaluate.

$$n = \frac{(180.3 \text{ kPa})(3.58 \text{ L } O_2)}{(8.31 \, \frac{kPa \cdot L}{mol \cdot K})(308.15 \text{ K})}$$
$= 0.2520$ mol O_2

Convert mol O_2 into g H_2O.

$$0.2520 \, \cancel{mol\,O_2} \left(\frac{2 \, \cancel{mol\,H_2O}}{1 \, \cancel{mol\,O_2}} \right) \left(\frac{18.02 \text{ g } H_2O}{1 \, \cancel{mol\,H_2O}} \right)$$
$= 9.08$ g H_2O

The Chapter opener discussed car crashes. Scientists have discovered how they can use chemistry and specifically chemical reactions that produce a gas to save lives during the milliseconds in which a car crash happens. Through the use of airbags, thousands of lives have been saved.

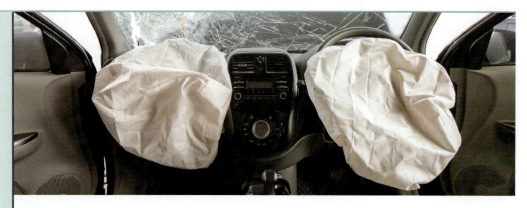

How It Works

Airbags

Vehicles often come to a sudden stop during a car accident. While this is not good for the car, it is potentially deadly for the occupants. An airbag applies less force *over a relatively longer* time period compared with the greater force that the steering wheel would apply over a very short time. Airbags also spread the force over a larger area on the body, reducing the severity of any injuries. All cars made since 1998 are equipped with front airbags, and with good reason. How does chemistry make this possible?

Airbags were first considered for aircraft in the 1940s. By the 1970s, scientists began planning for their use in cars. The original designers tried using compressed gases to inflate airbags. This approach was rejected due to issues with storing the compressed gas, maintaining consistency in inflation rates, and most importantly dealing with the limited inflation rate. To be effective, an airbag must inflate fast enough to make a difference. Since it takes only about fifty milliseconds after vehicle impact for a driver to strike the steering wheel, airbags must be designed to fully inflate before an occupant reaches them—faster than the blink of an eye.

An automobile airbag is folded and stored in the steering wheel or dashboard. When the airbag's control system senses a significant crash, it triggers a series of chemical reactions. In the primary reaction, sodium azide (NaN_3), a chemical also used in solid rockets, rapidly decomposes to form nitrogen gas and sodium metal.

$$2NaN_3(s) \longrightarrow 2Na(s) + 3N_2(g)$$

Since the sodium metal is hazardous, another reaction combines the sodium metal with potassium nitrate to form metal oxides and more nitrogen gas.

$$10Na(s) + 2KNO_3(s) \longrightarrow K_2O(s) + 5Na_2O(s) + N_2(g)$$

The nitrogen fills the airbag so quickly that it moves outward at about 89 m/s (about 200 mph)!

A third reaction with silicon dioxide consumes the hazardous metal oxides. Almost instantaneously, the airbag begins to deflate to provide softer cushioning before impact. After only two seconds, the airbag has completely deflated.

Scientists and engineers spend much of their time seeking ways to help others. Those who have worked on inventing and improving the airbag have saved many lives.

The Densities of Gases

The density of a gas, as with all types of matter, is defined as mass per unit of volume. Compared with solids and liquids, gases have very low mass per unit of volume. This is not surprising since gases have so much empty space between their particles. Most gases have densities about one-thousandth the densities of their solid or liquid states.

Gas density is measured in grams per liter. Some people assume incorrectly that all gases have at least similar densities. Helium (0.1786 g/L at STP) has such a low density that it is used to lift weather balloons. Air at STP has a density of 1.2922 g/L. Some gases, such as nitrogen dioxide (1.880 g/L), are so dense that they tend to sink in air and roll along the ground.

Gases have different densities because of the variations in their molar masses. Since 1 mol of any gas occupies the same volume at any combination of temperature and pressure, the molar mass determines the density of the gas. Gases with low molar masses have low densities. Gases with higher molar masses have greater densities. In other words, the density and molar mass of a gas are directly related.

$$\rho_{gas} = \frac{molar\ mass}{molar\ volume}$$

This equation is a powerful tool for calculating the density of a gas if its molar mass is known.

EXAMPLE 12-12: CALCULATING GAS DENSITY

What is the density of hydrogen gas at STP?

Write what you know.

molar mass H_2 = 2.02 g/mol
standard molar volume = 22.4 L/mol.

Write the formula and solve for the unknown.

$$\rho_{gas} = \frac{molar\ mass}{molar\ volume}$$

Plug in known values and evaluate.

$$\rho_{H_2\ at\ STP} = \frac{2.02\ \frac{g}{mol}}{22.4\ \frac{L}{mol}}$$

$$= 0.0902\ g/L$$

WORLDVIEW INVESTIGATION

GREENHOUSE GASES

If you pay any attention to the news, you are aware that there is a lot of concern about greenhouse gases and global warming. Many people don't know what greenhouse gases are or how they affect the earth. And there is much confusion about the greenhouse effect and its relationship to global warming. Should we be trying to fix the greenhouse effect and global warming?

Task

Since you are a member of your school's science club, the editor of the school newspaper has asked you to write an article about greenhouse gases from a biblical worldview. She specifically wants you to address the following five questions. (1) What are greenhouse gases? (2) How do they affect the earth? (3) What biblical principles relate to greenhouse gases and climate change? (4) Are greenhouse gases and climate change significant issues? (5) How do biblical principles inform your position? While you have heard a lot about greenhouse gases and global warming on the news, you realize that you know very little about the subject since you have never researched the issue yourself.

Procedure

1. Do appropriate research on the topic. To get started, do an internet search using keywords such as "greenhouse effect" or "greenhouse gases." You could even search for "Are the greenhouse effect and global warming the same thing?"
2. Write your article, being careful to properly cite your sources.
3. Have a classmate read your article and provide feedback.
4. Edit your article and submit it before press time.

Conclusion

Greenhouse gases, the greenhouse effect, and global warming are issues that humans will continue to deal with. We have an obligation to be good stewards of the world that God created. We must wisely decide on appropriate action. But to make appropriate decisions, we must be informed on the issue.

Analytical chemists can use this relationship to find the molar mass of an unknown gas. They can measure its density and then solve for its molar mass.

While understanding the densities of gases may seem like a novelty, the reality is that the behavior of gases on the basis of their densities can be a life or death matter. In 1986 many cattle (inset) and over 1700 people lost their lives when a large amount of carbon dioxide was released from deep within Lake Nyos in Cameroon. Because of its high density, the carbon dioxide gas remained at ground level and flowed through a river valley. Scientists investigated the event and determined that Lake Nyos experienced what is known as a *limnic eruption*—dissolved carbon dioxide being suddenly released from a lake. The tragedy of Lake Nyos led scientists to study other lakes with limnic activity to prevent a recurrence of this disaster. Their focus is on larger, more populated areas, such as the region surrounding Lake Kivu in East Central Africa. This is another example of scientists using their knowledge and skill to protect God's creation.

Lake Nyos, Cameroon

12.3 SECTION REVIEW

1. (True or False) Under equivalent conditions, equal volumes of gases contain the same number of particles.

2. How many moles of nitrogen gas occupy 14.7 L at STP?

3. When is the behavior of real gases most like ideal ones? Explain.

4. State the ideal gas law qualitatively and mathematically.

5. When do we use the combined gas law? the ideal gas law?

6. Use the ideal gas law to obtain the answers to the following questions.

    a. How many moles of oxygen will occupy 1.75 L at −118 °C and 5043 kPa?

    b. What is the temperature of 0.257 mol of oxygen occupying 6.78 L at 8.67×10^4 Pa?

    c. What is the pressure of 25.6 g of chlorine occupying 15.6 L at 218.6 °C?

    d. How much space will 0.514 mol of fluorine occupy at 58.5 °C and 1.076×10^5 Pa?

7. A technician produces hydrogen gas by reacting zinc with sulfuric acid. The equation for this single-replacement reaction is shown below.

    $$2Zn(s) + H_2SO_4(aq) \longrightarrow Zn_2SO_4(aq) + H_2(g)$$

    a. If the technician completely reacts 2.85 g Zn with excess acid, what volume of hydrogen gas will result at 98.29 kPa and 55.0 °C?

    b. When she finishes the procedure, she actually has 0.702 L of gas. Why did this occur?

8. How many grams of sodium azide are needed to produce enough nitrogen gas to fill a 57.6 L airbag at 98.8 kPa and 22 °C, assuming that only the nitrogen from the equation below is needed to fill it?

    $$2\,NaN_3(s) \longrightarrow 2\,Na(s) + 3\,N_2(g)$$

9. An unknown gas has a density of 0.714 g/L at STP. What is its molar mass?

12 CHAPTER REVIEW

Chapter Summary

TERMS
diffusion
effusion
Graham's law of effusion
fluid
compressibility
expansibility
pressure

12.1 PROPERTIES OF GASES

- The kinetic-molecular theory describes the properties of gases on the basis of particle movement.

- Key properties of gases include low density, diffusion, effusion, fluidity, compressibility, and expansibility.

- Pressure results from the collision of molecules against a boundary.

- The pressure, temperature, and volume of a sample of gas are all interrelated, and those relationships can be explained by the kinetic-molecular theory.

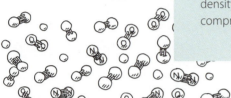

12.2 GAS LAWS

- Standard temperature and pressure (STP) is defined as 0 °C and 101.325 kPa.

- Boyle's law states that when temperature is constant, the volume and pressure of a gas are inversely proportional to each other.

- Charles's law states that when pressure is constant, there is a direct relationship between the Kelvin temperature of a gas and its volume.

- Gay-Lussac's law states that when volume is constant, the pressure and Kelvin temperature of a gas are directly proportional.

- The combined gas law allows calculations when the pressure, volume, and temperature of a gas all change.

- Dalton's law of partial pressures states that the total pressure of a mixture of gases equals the sum of the partial pressures.

TERMS
standard temperature and pressure (STP)
Boyle's law
Charles's law
Gay-Lussac's law
combined gas law
Dalton's law of partial pressures

TERMS
law of combining volumes
Avogadro's law
standard molar volume
ideal gas
ideal gas law

12.3 GAS STOICHIOMETRY

- Gay-Lussac's law of combining volumes states that the volumes of reacting gases and their gaseous products are expressed in ratios of small whole numbers. This principle allows the amounts of gases in moles to be calculated with stoichiometry.

- Avogadro's law states that the volume of a gas, maintained at a constant temperature and pressure, is directly proportional to the number of moles of the gas.

- At STP, 1 mol of any gas occupies approximately 22.4 L.

- The density of a gas can be determined by dividing its molar mass by its molar volume.

- The ideal gas law relates the pressure, volume, temperature, and quantity of gases.

- Scientists use the gas laws along with stoichiometric calculations to predict quantities of gases involved in chemical reactions.

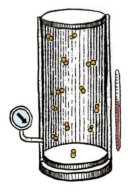

Chapter Review Questions

RECALLING FACTS
1. Define *pressure*.
2. Define *standard temperature and pressure* (STP).
3. State Boyle's law qualitatively and mathematically.
4. State the combined gas law qualitatively and mathematically.
5. Why is gas that is collected over water not pure?
6. State the law of combining volumes.
7. State Avogadro's law qualitatively and mathematically.
8. Define *standard molar volume*.
9. Define *ideal gas*.

UNDERSTANDING CONCEPTS
10. Use the kinetic-molecular theory of gases to explain why
    a. air has a low density.
    b. on a day with no wind, you can smell a dead skunk that is 500 yards away.
    c. a large amount of air can be pumped into a small bicycle tire.
    d. the temperature of a sample of gas doesn't change even though the collisions between particles transfer kinetic energy.
    e. we can compare the average kinetic energy in samples of gases by comparing their temperatures.
11. Calculate the ratio of the effusion rates between nitrogen and argon gases.
12. Why is the carbon dioxide gas that carbonates fountain drinks and powers paintball guns stored in sturdy canisters?
13. You have a sample of gas in a 1 L bottle. What will happen to the pressure if the gas is allowed to expand into a 2 L bottle at a constant temperature?
14. Explain why an inflated balloon becomes smaller when placed in a freezer.
15. Why does the pressure of a gas at constant volume decrease with decreased temperature?
16. When dealing with gases, why is a standard value for temperature and pressure important?
17. Why do we have to use the Kelvin scale when working with the gas laws?
18. Explain Gay-Lussac's law on the basis of the kinetic-molecular theory of gases.
19. Solve each of the following questions using the appropriate gas law. Also identify the gas law that you used each time.
    a. At 22.19 °C, a chemical reaction produces 74.30 mL of oxygen gas. How much space will the oxygen gas occupy at 54.37 °C if pressure remains constant?
    b. A sample of gas at 112.7 kPa occupies 997.4 mL. What will the pressure be if the gas is compressed to 253.9 mL at a constant temperature?
    c. A sample of nitrogen at 16 °C and 100.3 kPa has a volume of 2.60 L. What is the volume of the nitrogen at STP?

12 CHAPTER REVIEW

  d. A 73.4 mL sample of gas is collected at 3.54×10^5 Pa and 114.6 °C. If the gas is allowed to expand to 269 mL and cooled to 35.9 °C, what is the new gas pressure?

  e. A 95.1 mL sample of gas is collected at 124.8 kPa. How much space will the gas occupy if pressure is increased to 130.5 kPa and the temperature remains constant?

  f. A 13.20 mL sample of hydrogen sulfide is at 1.875×10^5 Pa at 55.6 °C. At what temperature will the pressure drop to 1.475×10^5 Pa if the volume is held constant?

  g. A sample of gas occupies 111 mL at 32.4 °C. What temperature is necessary to maintain a constant pressure if the gas is moved to a 41.3 mL container?

  h. A 19.8 L sample of helium has a pressure of 1.142×10^5 Pa at 53.8 °C. If the volume is held constant, what temperature will result in 9.86×10^4 Pa of pressure?

20. A sample of 117.2 mL of chlorine is collected by water displacement at a temperature of 85.0 °C. The barometer reads 105.695 kPa. What is the volume of the chlorine at STP?

21. Explain Avogadro's law on the basis of the kinetic-molecular theory of gases.

22. If the behavior of real gases cannot be exactly predicted by gas laws, why are they still useful?

23. Answer the following questions using the ideal gas laws.

   a. A canister holds 553 mL of krypton at STP. How many moles of Kr are present?

   b. A weather balloon contains 0.465 mol He at STP. How much space will that gas occupy?

   c. A 1.78 g sample of argon fill a 1.46 L tank at a pressure of 100.1 kPa. What is the temperature of the gas?

   d. A tank of oxygen gas contains 89.0 g of O_2. If the volume of the tank is 17.54 L, what is the pressure (in Pa) of the O_2 if the temperature is 21.4 °C?

24. When nitroglycerine (227.1 g/mol) explodes, N_2, CO_2, H_2O, and O_2 gases are released initially. Assume that the gases from the explosion are allowed to change to STP without reacting further.

$$4\,C_3H_5N_3O_9(s) \longrightarrow 6\,N_2(g) + 12\,CO_2(g) + 10\,H_2O(g) + O_2(g)$$

   a. If 16.7 mol of nitroglycerine reacts, how many liters of nitrogen are produced?

   b. What is the total volume of gas produced when 1.000 kg of nitroglycerine reacts?

25. A reaction between ammonia and oxygen is the first step in the preparation of nitric acid (HNO_3) on a commercial scale. The reactants are mixed at STP. The products are produced at 1058 °C and at 98.7 kPa.

$$4NH_3(g) + 5O_2(g) \xrightarrow{\text{catalyst}} 4NO(g) + 6H_2O(g)$$

   a. What volume of nitrogen monoxide (NO) is produced in the reaction vessel by the reaction of 0.500 mol O_2?

   b. What mass of water is produced by the reaction of 15.0 L of ammonia at STP?

   c. How many liters of oxygen must react to produce 33.5 L of nitrogen monoxide?

26. What is the density of chlorine gas (Cl_2) at STP?

CRITICAL THINKING

27. You set up an experiment comparing the diffusion rates of an unknown elemental gas with nitrogen gas. If the nitrogen diffused 1.591 times faster than the unknown gas, what is the molar mass of the unknown gas? Identify your unknown gas.

28. The descriptions of changes in pressure, temperature, and volume in Section 12.1 always stipulate that one of the properties is held constant. Since it seems unlikely that you can maintain one of these conditions constant in real-world situations, why do you think there is no description of these changing simultaneously?

29. Derive Charles's law from the combined gas law.

30. A sample of 75.6 mL of oxygen is collected by water displacement at a temperature of 48.1 °C. The barometer reads 116.5 kPa. What is the volume of the oxygen at STP?

31. Why are we taught to crawl out of a burning building?

32. Starting from the ideal gas law, explain Charles's law.

Use the ethics box to answer Question 33.

33. Write a four-paragraph response to the issue of airbag safety on the basis of a biblical worldview.

ethics

DEADLY SAFETY DEVICE?

The Chapter opener highlighted the dangers of automobile collisions. Scientists developed airbags to help protect occupants during a collision, and they have been instrumental in saving many thousands of lives in their five decades of use. Airbags inflate at about 200 mph in order to deploy quickly enough to protect vehicle occupants. Sadly, airbags have actually been the cause of a number of deaths. Because of this, some vehicle owners have considered deactivating the airbags in their cars. As a Christian, how do you balance the protective capabilities of airbags with their potential hazard? Is it appropriate to have your airbag disabled?

Chapter 13

SOLIDS & LIQUIDS

There are over 100,000 people in the United States on waiting lists for organ transplants. Those who make it to the top qualify for a transplant. More than 28,000 people receive organ transplants each year in the United States. That's 77 people every day who get a second chance at life.

After doctors remove an organ from a donor, surgeons have a brief window of time to transplant it into the person who needs it before the organ dies. A liver or pancreas can survive up to twelve hours, and kidneys can survive for twenty-four hours. But hearts and

lungs must be transplanted within four to six hours. To last for even a few hours, organs need special storage. If a recipient is hundreds of miles away, every minute could make a difference. How can doctors preserve organs so that they stay alive long enough to bring new hope for the people who need them?

13.1 Intermolecular Forces *299*
13.2 Solids *304*
13.3 Liquids *312*

13.1 INTERMOLECULAR FORCES

Defining Intermolecular Forces

As you will see later in this chapter, medical professionals use their knowledge of states of matter to preserve donated organs during transport. The key to understanding the differences between one state of matter and another is to look at how particles of matter behave in the presence of other particles. Molecules interact with each other—some more than others. The electrostatic attractions between molecules are called **intermolecular forces**, also known as *van der Waals forces*. Recall from Chapter 7 that electrostatic attractions are the forces exerted between electrical charges—opposite charges attract and like charges repel. When these forces occur between different molecules, they are called intermolecular forces. Intermolecular forces are much weaker than the bonds that hold molecules together. On a scale from 0 to 100, the strength of bonds within a molecule would measure 100 while the strength of forces between molecules would be only 0.001 to 15. Nonetheless, these intermolecular forces influence the physical properties of many substances. They explain why, at room temperature, hydrogen is a gas, water is a liquid, and sugar is a solid.

Intermolecular forces are classified into three groups, each with a different strength. Each force is the result of attraction between regions of charge on some molecules and regions of opposite charge on other molecules.

QUESTIONS
» Are molecules attracted to each other in the same manner as atoms attract other atoms?
» Are all interactions between molecules equally strong?
» How can I tell which intermolecular forces are at work?
» How do intermolecular forces affect a substance's physical properties?

TERMS
intermolecular force • dipole-dipole force • hydrogen bond • London dispersion force

Why is oxygen a gas, water a liquid, and iron a solid at room temperature?

Dipole-Dipole Forces

Consider polar molecules, which have regions of unevenly distributed electrical charge. The positive areas of one molecule attract the negative areas of other molecules. The greater the polarity, the stronger the attraction between the two molecules. These polar molecules align themselves so that the positive end of one molecule is near the negative end of another molecule. This attraction is called a **dipole-dipole force**. The dipole-dipole forces of polar molecules are similar to the electrostatic attraction of ions in crystals of ionic compounds, but they are much weaker. They are strong enough, however, to have a significant influence on the physical properties of many compounds.

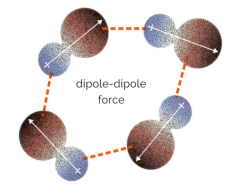

dipole-dipole force

Solids & Liquids **299**

Hydrogen Bonds

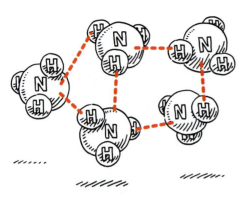

Hydrogen bonds are a special type of dipole-dipole interaction in which one of the participating atoms is hydrogen and the other element is either nitrogen, oxygen, or fluorine. Whenever hydrogen is bonded to one of these very electronegative elements, the shared electrons shift away from the hydrogen atom because of the greater electronegativity of the other element. The hydrogen nucleus, essentially a proton, is left exposed. Any negatively charged regions of other molecules containing nitrogen, oxygen, or fluorine attract the exposed proton. Since hydrogen is so small, interacting molecules can get extremely close to it. The combination of high polarity and close proximity produces the strongest of the intermolecular forces.

HYDROGEN BONDS IN WATER

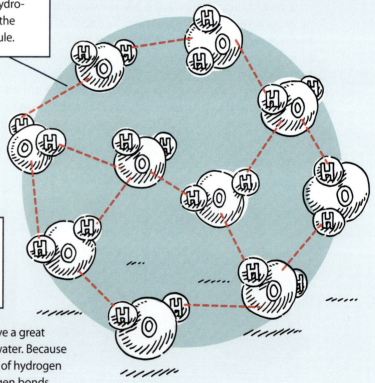

Hydrogen bonds form between the hydrogen atom of one water molecule and the oxygen atom of another water molecule.

They are not bonds in the same sense that ionic or covalent bonds are bonds; they are simply interactions between molecules. Like other intermolecular forces, hydrogen bonds are not as strong as the covalent bonds in water molecules. While the energy of covalent bonds in water is about 286 000 J/mol, the energy of hydrogen bonds in water is only about 20 000 J/mol.

The average hydrogen bond length between water molecules is 0.177 nm, compared with 0.099 nm for the O-H covalent bond. The longer length is indicative of a weaker attraction.

Despite the apparent weakness of these bonds, they have a great effect on the physical properties of substances such as water. Because there is a significant difference in the electronegativities of hydrogen and oxygen, water molecules are quite polar. The hydrogen bonds result in water remaining a liquid at room temperature.

TABLE 13-1 *Hydrogen Compounds Compared*

Compound	Molecular Mass (u)	Melting Point (°C)	Boiling Point (°C)
H_2Te	129.6	−49.0	−2.0
H_2Se	80.98	−65.7	−41.3
H_2S	34.09	−85.5	−59.6
H_2O	18.02	0.0	100.0

Table 13-1 compares water with other similar covalent compounds that contain hydrogen. Although water has a molecular mass that is much lower than that of the other compounds, hydrogen bonds raise its boiling and melting points.

Water is essential for life. Is it merely coincidental that atmospheric conditions on this planet allow water to exist in the liquid state? Hardly. The Lord providentially engineered the sun's thermal output, the dimensions of the solar system, the earth's distance from the sun, the earth's cloud cover, and the hydrogen bonding between water molecules to allow water to be a liquid. Water vapor contributes to the greenhouse effect that moderates our temperatures and promotes life. Other planets and moons do not have appreciable amounts of water. Those that do have water have surface temperatures and pressures that will not allow it to exist as a liquid. God says that this creation testifies to the fact that He is the Creator God and that He is to be worshiped as the Sovereign of the world (Isa. 45:18).

London Dispersion Forces

Even nonpolar substances exist as either solids or liquids at room temperature. Why is this? Dipole-dipole interactions and hydrogen bonds are not the only intermolecular forces. There is a third type of intermolecular force called the **London dispersion force**, also called a *dispersion force* or *London force* after the physicist Fritz London. Nonpolar molecules do not have permanent regions of electrical charge, yet they still attract each other. On average, electrons in a nonpolar bond spend an equal amount of time around each atom. Yet at any point, these electrons can momentarily concentrate at one end of the molecule. These concentrations occur when the negative electrons of one molecule repel the electrons of a nearby molecule. The concentrations form temporary dipoles. Although the imbalance of charges occurs for just an instant, this is long enough to attract an oppositely charged region of a neighboring molecule. Thus, dispersion forces result from the random, unequal dispersion of electrons around the atoms in a molecule. Dispersion forces occur between all molecules, but they are the only intermolecular force acting on nonpolar molecules. Table 13-2 ranks the relative strengths of dispersion forces compared with other attractive forces, including bonds and intermolecular forces.

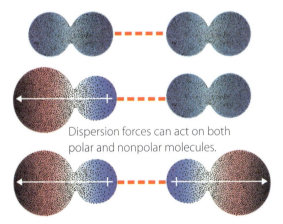

Dispersion forces can act on both polar and nonpolar molecules.

TABLE 13-2 *Relative Strengths of Attractive Forces*

1	ionic bond
2	metallic bond
3	covalent bond
4	hydrogen bond
5	dipole-dipole force
6	dispersion force

Solids & Liquids

Larger molecules have more electrons and thus are more likely to participate in dispersion forces at any given time. This tendency affects their physical properties. For example, consider the hydrocarbons. Methane, which has one carbon, is a gas at room temperature. Octane, with a chain of eight carbons, is a free-flowing liquid at room temperature. Octadecane, an 18-carbon molecule, is a solid at room temperature. All of these nonpolar compounds exhibit dispersion forces. Their differences in state depend on the number of carbon atoms and their associated electrons in each molecule.

EXAMPLE 13-1: INTERMOLECULAR FORCES

List the types of intermolecular forces that affect the following compounds.

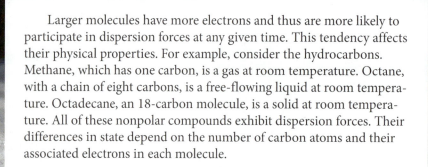

Solution

a. CO_2 is nonpolar, so only dispersion forces act.

b. HF is polar, and it has the necessary elements for hydrogen bonds to form. Thus, dipole-dipole forces, hydrogen bonds, and dispersion forces are present.

c. BrF molecules are polar, so dipole-dipole and dispersion forces are present.

EXAMPLE 13-2: BOND COMPARISON

On the basis of the bonds within each molecule and the forces between each molecule, predict which compound in each pair has a higher boiling point.

a. KF or BrF   b. Cl_2 or ICl

Solution

a. KF is an ionic compound that has extremely strong ionic bonds between its units. Its boiling point (1505 °C) is higher than that of the covalently bonded BrF (20 °C), which has only weaker dipole-dipole and dispersion forces between its units.

b. Both Cl_2 and ICl are covalently bonded. ICl molecules are slightly polar, while Cl_2 molecules are not. Since ICl molecules have dipole-dipole interactions as well as dispersion forces, their boiling point (97.4 °C) is higher than that of Cl_2 (−34.6 °C).

Just as ionic, covalent, and metallic bonds greatly affect the physical properties of the compounds that they form, intermolecular forces also affect these physical properties. The facing page summarizes bonds and intermolecular forces, the compounds they affect, and their influence on the physical properties of compounds.

TYPES OF BONDS AND INTERMOLECULAR FORCES

STRUCTURAL UNITS	INTERACTION BETWEEN UNITS		PROPERTIES
ions	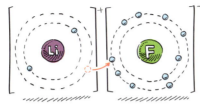	ionic bonds	very high melting/boiling points; usually soluble in polar solvents; electrical conductors when molten or dissolved
cations, mobile electrons (metals)	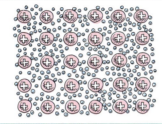	metallic bonds	high melting/boiling points; insoluble; electrical conductors in all states
polar molecules	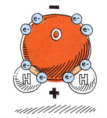	possibly hydrogen bonds, dipole-dipole forces, dispersion forces	melting/boiling points slightly higher than those of nonpolar molecules; usually soluble in polar solvents; nonconductors of electricity
nonpolar molecules	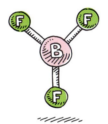	dispersion forces	low melting/boiling points; molecules soluble only in nonpolar solvents; nonconductors of heat and electricity

13.1 SECTION REVIEW

1. What are intermolecular forces?
2. How are the three types of intermolecular forces similar? How are they different?
3. How do you explain dispersion forces in nonpolar substances?
4. What types of intermolecular forces would you expect to find between molecules of H_2O? of BrCl? of hexane (C_6H_{14})?
5. Wax is nonpolar, yet it exists as a solid at room temperature. How can this be explained in terms of intermolecular forces?
6. Which substance, pure oxygen (O_2) or ozone (O_3), would you expect to have a higher boiling point? Explain.

13.2 SOLIDS

A Kinetic Description of Solids

QUESTIONS
» Why do solids have the properties they do?
» Do all solids have similar structures?
» How do I interpret a heating curve?
» What happens at the particle level during a phase change?
» What factors affect the formation of crystals?

TERMS
crystalline solid • amorphous solid • melting point • unit cell • allotrope • lattice energy

The strength of bonds and intermolecular forces in a solid affects the internal movements of its own particles. According to the kinetic-molecular theory of matter, particles in solids are always moving—or more precisely, vibrating. At low temperatures, these particles barely vibrate. As temperatures approach a solid's melting point, the particles have more kinetic energy and the vibrations get faster and bigger. Strong intermolecular forces stifle these vibrations, while weak forces allow the particles to move more freely. For example, table salt exists as a solid up to the scorching temperature of 801°C because attractions between the Na^+ ions and the Cl^- ions are extremely strong. Nonpolar diatomic bromine molecules, with only weak dispersion forces, can move more freely. Consequently, bromine melts at −7 °C, allowing it to exist as a liquid at room temperature.

Why can't I write with a diamond?

Intermolecular forces affect other properties besides a substance's boiling and melting points. Because solid particles have little motion, they are not fluids, have a fixed shape and volume, and have low rates of diffusion. If a silver coin and a copper coin were clamped together for several years, a few atoms would diffuse between the metals, but the rate would be imperceptible compared with the diffusion rate of the same atoms if they were in the liquid phase.

Solid benzene sinks in liquid benzene (left). Solid water floats in liquid water (right).

Density of Solids

Because particles in solids are held closely together, they usually have high densities. Since solids are so dense and have fixed shapes and definite volumes, they resist compression. Atoms, molecules, or ions would have to be deformed for solid matter to compress significantly. A substance is typically about 10% denser as a solid than as a liquid. Consequently, when liquid and solid forms of the same substance are together, the solid form almost always sinks to the bottom.

One notable exception to this trend is water. Down to a temperature of 4 °C, water becomes increasingly dense, but as it cools below this temperature, hydrogen bonds cause water molecules to orient themselves in a manner that spreads them farther apart from each other, forming an open hexagonal lattice. The increased space between the molecules causes ice to be less dense than water. This lower density is why ice cubes float in a glass of water. More importantly, open bodies of water such as lakes or ponds freeze from the top down. As ice forms on the surface, the water beneath is insulated. If the water is deep enough, it may never freeze solid during winter, allowing aquatic organisms to overwinter safely.

Crystalline and Amorphous Solids

Solids can occur in many shapes and forms. Some solids have naturally orderly shapes, forming regular three-dimensional patterns with distinct edges and sharp angles. When they are shattered, smaller shapes form with similar edges and angles. These solids are called **crystalline solids**. Ionic and metallic solids are usually crystalline. Other solids have no distinct shape or underlying pattern. When they split or shatter, irregular fragments result. Such solids are called **amorphous solids**. Covalent compounds can be crystalline or amorphous.

The differences between crystalline and amorphous solids result from the particular structures of the solids. The particles in crystalline solids—such as salt, sugar, and monoclinic sulfur—are arranged in well-defined, orderly, three-dimensional patterns. Atoms, ions, or molecules are stacked row upon row, column by column. The patterns, when repeated many times, result in crystals with regular shapes.

Particles in amorphous solids—such as rubber, asphalt, paraffin, or amorphous sulfur—are not arranged in any particular pattern. Their random, disordered structure results in globular shapes. Some amorphous solids are called *supercooled liquids*, or glasses. If a liquid can be cooled fast enough, its particles may not have time to form a preferred crystalline pattern. These molecules are locked in random positions. Volcanic obsidian is produced in this fashion.

The internal structure of amethyst is reflected in the external shape of its crystals. Obsidian is amorphous.

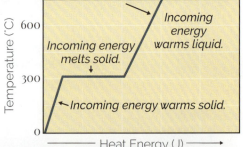

Melting and Freezing

Melting and freezing are changes between the solid and liquid states of a substance. Melting is the transition from a solid to a liquid. The temperature at which this process takes place is called the **melting point**. Freezing is the reverse process. For pure substances, the freezing point is the same as the melting point. Freezing points are not always cold temperatures. For example, iron freezes at 1538°C! When a lump of lead is placed in a ladle and heated, the atoms begin to vibrate more vigorously. Temperature, a measure of the vibrations of lead particles, rises (see graph at left). When the melting point of the lead is reached, the temperature stabilizes. At this point, all atoms are at the brink of liquefaction. Additional heat overcomes the attractive forces between lead atoms and melts the solid. Not until all the lead is melted will the temperature resume its upward climb.

The melting points of crystalline and amorphous solids are affected by the arrangements of their particles. Crystalline substances such as ice or lead have distinct melting points. Heating-curve graphs for crystalline substances show clear plateaus that correspond to sharp melting points. An entire sample melts at a clearly defined temperature because all particles are held by nearly identical forces. Amorphous solids, on the other hand, do not have sharp melting points. Their particles are in random positions at different distances from each other. Since the force strength varies with distance, not all of the particles are held together with identical force. As the temperature rises, only some intermolecular forces are overcome at each specific temperature. Amorphous solids gradually soften as more intermolecular forces are broken. The heating curve for an amorphous substance thus lacks the stairstep pattern seen in the heating curve for lead. Instead, the plateaus are replaced by a gradually ascending line.

Analytical chemists use melting points to determine the identity and purity of compounds. Many pure substances are crystalline, with their melting points listed in reference tables. Chemists can analyze the purity of a compound by observing its melting-point range. A narrow range indicates a pure sample, while a wide range indicates that impurities are present.

Sublimation and Deposition

Many caterers use dry ice (solid carbon dioxide) to refrigerate their foods. When dry ice warms, it changes directly into a gas instead of melting into a liquid as ice does. This state change bypasses the liquid phase and alleviates the problems of messy puddles. Sublimation is the direct change in state from the solid to the gaseous state. Sublimation occurs because individual molecules leave the surface of a solid and become gaseous. For example, the smell of naphthalene or 1,4-dichlorobenzene mothballs comes from the individual molecules that have sublimed. Sublimation causes snow to slowly disappear, even when the temperature remains below freezing. Substances sublime readily if they have many molecules that can easily leave their surfaces.

dry ice

Deposition occurs when a substance changes directly from a gas to a solid—the reverse of sublimation. When frost forms on your windshield or lawn, deposition has taken place. Deposition of water vapor high above the earth forms snowflakes.

HOW IT WORKS

Cryogenics

Imagine a world where you can use a banana to hammer nails into wood. In this world, a rubber bouncy ball shatters when it hits the ground. Believe it or not, this bizarre world actually exists—it is the world of cryogenics.

Cryogenics is the science of the supercold. The word *cryogenics* comes from the Greek word *kruos*, meaning "frost." Cryogenics does not deal with the moderately cold temperatures of your kitchen freezer, but with the intensely cold world of −150 °C and below. Applications of this fascinating area of science are found in such wide-ranging fields as space exploration, food preservation, and medicine.

Aerospace engineers use supercooled gases extensively in the space industry. Liquid hydrogen and oxygen are used together as a powerful fuel. Life support systems and space refrigeration systems also use cryogenic conditions.

Another application of cryogenics is rapidly freezing food. Some meats, pastries, and vegetables are frozen rapidly by being passed through cryogenic tunnels. As the food passes through frigid gases, its surface freezes rapidly. Cold temperatures then gradually spread to the core for an even freeze.

Liquefied gases have another interesting connection to food. Breeders can keep the eggs, sperm, and even embryos of livestock frozen in liquid nitrogen for years before being thawed and implanted into female animals. Some prize bulls that have been dead for years are still fathering calves through this technology.

Surgeons have used cryosurgery—the use of a freezing probe in place of a scalpel—with good success in removing warts, tonsils, and cataracts. Cryosurgery repairs and treats patients with less pain and hemorrhaging. Doctors have replaced many operations that previously required hospitalization with procedures that they can perform in the office.

As you can see, cryogenics deals with a wide range of applications. Scientists who specialize in applying an understanding of cryogenics to meeting the needs of people are called *cryogenic engineers*. It's a *cool* job that you might enjoy pursuing as a means to glorifying God and serving others.

Solids & Liquids 307

Crystalline Structures

The particles in crystals are arranged in orderly, repeating patterns. These patterns vary, as shown by the wide variety of beautiful shapes found in natural crystals. The beauty and predictability of naturally occurring crystals is yet another of God's fingerprints on our world. Several factors influence how the particles are arranged. Like electrical charges repel to maximize the distance between particles. At the same time, opposite charges seek to minimize that distance. These forces interact to produce the most stable crystal arrangement possible.

A crystal's three-dimensional pattern, or crystal lattice, depends on the number and kinds of particles involved, their relative sizes, and the charges of the ions. Scientists have found and classified seven basic classes of crystals—cubic, tetragonal, orthorhombic, monoclinic, triclinic, hexagonal, and trigonal.

Scientists mentally divide the natural structures of crystals into building blocks called **unit cells** that contain the fundamental patterns of the lattices. Unit cells usually consist of a specific number of ions. These portions of crystals can be compared to the basic pattern in a piece of wallpaper. The wallpaper pattern can be repeated indefinitely to cover large walls. Unit cells, when repeated many times in three dimensions, make up crystals.

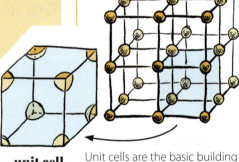

unit cell Unit cells are the basic building blocks for any type of crystal.

Some of the seven basic classes of crystals can be slightly modified by adding particles to the faces or interiors.

A body-centered crystal not only has particles at each of the corners, but it also has one in the center of the crystal.

A face-centered crystal has particles on the corners and on each face but no particle in the center of the crystal.

Some, but not all, of the seven basic crystal types have one or both of these modifications. These variations expand the seven basic classes into fourteen different lattices.

Because crystals are three-dimensional arrangements of unit cells, they often assume the same shape as their unit cells. Salt crystals, for example, are tiny cubes. When they are formed, their units build upon themselves to form larger cubes with approximately the same number of units along each edge. A sodium chloride unit cell, which contains four Na^+ ions and four Cl^- ions, has a length of 0.56 nm. A salt crystal 0.5 mm along each edge has nearly one million unit cells along each edge and approximately 10^{18} unit cells in its total structure.

But many crystals found in nature do not have the same shape as their constituent unit cells. Because of varying temperature, pressure, and other environmental factors, unit cells may stack in such a way that a different external structure forms. The "steps" on the surfaces of the crystals are only a fraction of a nanometer wide, so they are not visible to the eye. The crystal faces appear to be smooth even though they are ragged at the atomic level. Despite the fact that all the known minerals have one of the basic crystal structures, mineral crystals can exhibit a great variety of external forms.

POLYMORPHS AND ALLOTROPES

Some substances can form more than one type of crystal lattice. Such elements and compounds are said to be polymorphous. When Ca^{2+} and CO_3^{2-} ions crystallize at a low temperature, they fall into a rhombohedral lattice. Mineralogists call this form of calcium carbonate *calcite*. When these same ions crystallize at a high temperature, they orient themselves in an orthorhombic lattice and form a substance called *aragonite*, which is also calcium carbonate.

Pure elements that are polymorphous are called *allotropic elements*, and their different forms are called **allotropes**. Sulfur, phosphorus, arsenic, and carbon are a few allotropic elements. You are probably familiar with some of carbon's allotropes. One of them, diamond, is Earth's hardest naturally occurring substance. This hardness is the result of the internal arrangement of diamond's carbon atoms and the strong covalent bonds between them. Most diamonds are used in industrial cutting tools, but high-grade diamonds are considered valuable gemstones. Another carbon allotrope, graphite, consists of sheets of carbon rings stacked layer on layer. The sheets are held together only by dispersion forces. This makes graphite extremely soft—it easily flakes if rubbed against a sheet of paper. It's the "lead" in ordinary pencils. In fact, graphite's name is derived from the Greek word meaning "to write."

BINDING FORCES IN CRYSTALS

Crystals are shaped by electrical forces. Oppositely charged regions attract, and similarly charged regions repel. The balance between attractions and repulsions determines how tightly the particles in the crystal are bound. If attractions just barely overcome the repulsions, the crystal will have weak bonds. If attractions are much greater than repulsions, the crystal will have strong binding forces.

calcite

aragonite

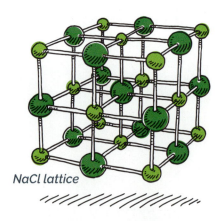

NaCl lattice

LATTICE ENERGY

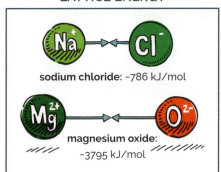

sodium chloride: −786 kJ/mol

magnesium oxide: −3795 kJ/mol

The **lattice energy** of a crystal is the energy released when gaseous particles form crystals. It is expressed as a negative number because heat is released during crystallization, leaving less energy in the products than in the reactants. Lattice energy is equal in magnitude to the energy necessary to pull a crystal apart. It shows the difference in energy between gas particles and those trapped in a crystal.

Binding forces must be overcome whenever a crystal is melted or dissolved. When a crystal melts, thermal energy is used to overcome binding forces. When a crystal dissolves, attractions between the particles of the crystal and the molecules of the solvent are able to break apart the crystal structure. For example, when table salt dissolves in water, its crystals absorb thermal energy from the water to overcome the lattice energy and break apart the ions. If a large amount of salt is added to a glass of water, you can feel the glass cool as the salt dissolves, using energy from its environment to break the crystal lattice.

What makes a crystal strong? In other words, what determines its lattice energy? The magnitude of the electrical charges in a crystal affects its stability. For example, sodium chloride, with a lattice energy of −786 kJ/mol, involves a sodium ion with a charge of 1+ and a chloride ion of 1−. But magnesium oxide, with a lattice energy of −3795 kJ/mol, involves a magnesium ion with a charge of 2+ and an oxide ion of 2−. Highly charged particles, like those in ionic compounds, interact strongly to produce strong crystals. The size of the particles also affects the binding forces. Small particles can be more tightly bound than large particles. A third factor—the geometric structure of the crystal—also affects the binding forces because both the structure and the size of the particles involved affect the distance between ions.

TYPES OF CRYSTALLINE SOLIDS

Crystalline substances can be classified by the types of particles they contain and how they are held together. This method of classification relies on an understanding of the types of bonds that we discussed in Chapter 7.

Atomic Crystals

These solids form only when noble gases freeze. The particles in these solids are individual atoms since noble gases do not typically form molecules. As you might expect, they are soft and are poor conductors of electricity.

Type: atomic
Unit Particles: atoms
Examples: Group 18 elements

Covalent Molecular Crystals

Molecular solids, such as table sugar, are made of covalently bonded atoms held together by any combination of dipole-dipole forces, hydrogen bonds, and dispersion forces. The intermolecular forces that attract the molecules to each other are weaker than the forces binding atoms together within each molecule. Thus, they have relatively low melting points. Their low melting points explain why molecular compounds are often gases or liquids at room temperature.

Type: covalent molecular
Unit Particles: molecules
Examples: H_2O (snowflake), CO_2 (dry ice), $C_{12}H_{22}O_{11}$ (table sugar)

Ionic Crystals

Ionic solids are made of a repeating network of ions defined by a unit cell. The ions may be monatomic or polyatomic. Remember from our earlier study that ionic compounds typically form between a metal and a nonmetal. The strong bonds between the alternating positive and negative ions in these crystals cause them to be strong and have high melting points. Many also fracture easily.

Covalent Network Solids

Some nonmetallic elements, such as carbon, are able to bond covalently to form a large network of atoms, as occurs with diamond. The particles in this and other covalent network solids are individual atoms bound together by covalent bonds. These crystals function more like giant molecules. The strength in these three-dimensional arrangements contributes to the generally high melting points of covalent network solids.

Type: Covalent network
Unit Particles: covalently bonded atoms
Examples: diamond (carbon atoms)

Type: ionic
Unit Particles: ions
Examples: NaCl, K_2SO_4

Type: metallic
Unit Particles: metal ions surrounded by mobile electrons
Examples: almost all metals

Metallic Crystals

Positive metal ions are surrounded by a sea of valence electrons in metallic solids. Because metals vary greatly in the strength of the metallic bonds that form between the metal ions and mobile electrons, metals have a wide range of physical properties. For example, mercury is a liquid at room temperature, while many others remain solid past 1000 °C. Most metals are lustrous and conduct electricity well because of their free electrons

13.2 SECTION REVIEW

1. How does the kinetic-molecular theory explain the properties of solids?
2. Explain the difference between crystalline and amorphous solids.
3. On a warming curve, what is indicated by the sloped portions? by the plateaued portions?
4. Draw a warming curve to represent a crystalline solid and one for an amorphous solid. Is there a difference between the two curves? If so, why?
5. How is a unit cell like a formula unit?
6. Explain why the lattice energy and the energy to break apart crystals of a substance differ in sign but are equal in magnitude.
7. How are covalent network solids similar to ionic solids?
8. Explain why covalent network crystals have relatively high melting points, while covalent molecular crystals, as a rule, melt at much lower temperatures.

13.3 LIQUIDS

If liquids were not so common, we would consider them amazing. Why does water form spherical drops? Why are bubbles round? Why does rubbing alcohol make your skin feel cold? What makes the surface of a liquid in a test tube curve? How can some insects skate across the surface of a pond? Although many properties of liquids are well known, they are not necessarily well understood.

A Kinetic Description of Liquids

Particles in liquids are held together by intermolecular forces that balance out their kinetic energy. The particles in a liquid substance move but with less energy than those same particles have as a gas. Because of the strength of their intermolecular forces, liquids are held closer together compared with gas particles having the same energy (temperature). Yet the attractive forces holding the particles together do not totally dominate, as in a solid. There is enough freedom for liquid particles to roll and slide over, around, and past each other. Liquid particles are not fixed in any one position, so liquids, like gases, are fluids that can flow and be poured.

Density of Liquids

Liquids have densities that are much greater than those of gases since there is little space between the particles of liquids. Because of this closeness, liquids cannot significantly expand or compress like gases can. People rely on this property every time they apply the brakes in their car. The pressure that is applied on the liquid through the brake pedal is transferred throughout the fluid, causing the brake pads to clamp down. Since the particles are already close together, liquids can transfer incredible pressures of even the most powerful hydraulic systems without being compressed much. The particles themselves would have to be crushed before a liquid could be appreciably compressed.

QUESTIONS
» Why do liquids have the properties they do?
» Aren't evaporation and boiling the same thing?
» How does energy move through evaporation?
» How do I interpret a phase diagram?

TERMS
cohesion • adhesion • surface tension • surfactant • viscosity • meniscus • evaporation • boiling • boiling point • distillation • phase diagram • triple point • critical temperature • critical pressure

What makes water special?

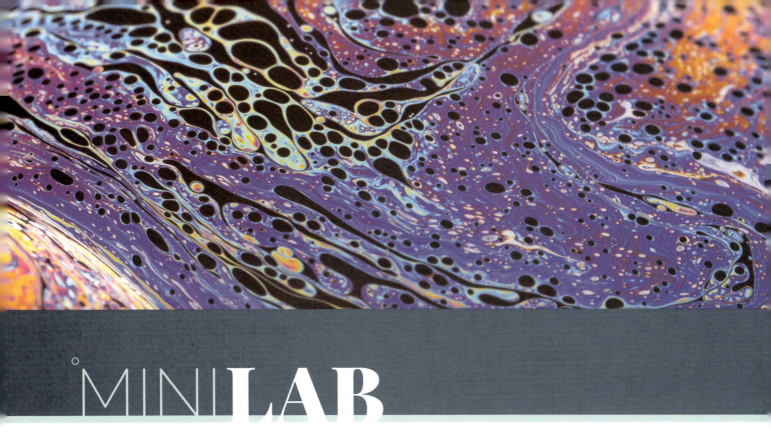

MINILAB

THROUGH THE VOID

If you took a course in physical science, you may recall that certain chemicals, called indicators, will change color in the presence of an acid or base. Some indicators are liquids. Typically they are used by adding the indicator drop-wise into the solution being tested. Bromothymol blue (BTB), for example, is a greenish liquid that will change to yellow when added to an acid. But what if the two liquids don't touch? Will the indicator still work?

Can liquid particles move through empty space?

1. Write a hypothesis about whether BTB will change color when near an acid but not in direct contact with it.

Procedure

A. Place the petri dish on a blank sheet of paper for easier observation. Remove the cover and place one drop of vinegar and one drop of BTB in the dish, about 3 cm apart. Do not mix the drops or otherwise allow them to come in contact.

B. Replace the lid on the petri dish and observe.

2. Describe what happens in the dish.

Conclusion

3. Suggest an explanation for what you observed.

Going Further

4. The creative title of this lab activity suggests that there is "nothing" in the petri dish other than the two liquids. Is that true? Explain.

EQUIPMENT

- petri dish with cover
- vinegar
- bromothymol blue (BTB)

Effects of Intermolecular Attractions

The properties of liquids are greatly affected by intermolecular forces. The attraction between the particles of a liquid is called **cohesion**. The attraction between the particles of a liquid and the particles of other materials is called **adhesion**. Cohesion and adhesion together cause many of the amazing properties of liquids that we observe.

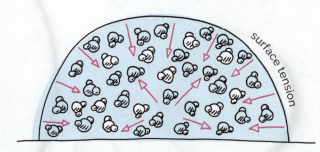

COHESION

Remember the water striders (How It Works on page 188)? You may recall that their ability to glide across the water's surface is partly due to a property of water itself. **Surface tension**, the elastic "skin" that forms on the surface of water and other liquids, is caused by the imbalance of intermolecular forces at the liquid's surface. A particle within a volume of liquid has neighbors on all sides. Each neighbor attracts interior particles equally in all directions. But a surface particle does not have other particles on one side. All forces exerted on surface particles are directed toward the interior of the liquid, and directionality of the forces binds surface particles together. If the force exerted by an object on the surface of a liquid is less than the liquid's surface tension, the object will be unable to penetrate the surface of the liquid. This phenomenon allows water striders and other insects to move across the surface of the water without sinking.

Surface tension affects the shapes of liquids. Raindrops are spherical because surface tension pulls water into the shape with the least surface area per unit of volume. Mercury, which has more surface tension than water, forms drops that are even more spherical than water drops.

Surface tension affects more than just the shape of raindrops. Have you ever tried to use water to get oil or grease off your hands with no soap? You need soap, or a detergent, to break down the grime. These both act as surface-active agents, or **surfactants**, meaning that they break down the normal surface tension of water by interfering with hydrogen bonds. This allows normally insoluble grease, oil, and other nonpolar liquids to dissolve. A soap molecule is a long molecule with a polar end and a nonpolar end. The polar end attracts water through dipole-dipole forces, while the nonpolar end attracts dirt, grease, and oil, allowing the water and the other substances to mix.

The attraction of liquid particles for one another affects their ability to flow. Thick and gooey liquids consist of particles that have a high cohesion. Liquids that flow easily consist of particles that have a low cohesion. The degree of cohesion and the shape of a fluid's particles affect a property called **viscosity**—a liquid's ability to resist flowing. The stronger the attractions between particles, the more viscous the

Surface tension causes drops of mercury lying on a flat surface to curve under, around their edges.

fluid will be. Molasses is viscous, so it can be expected to have some strong attractions between its particles. Thinner, less viscous fluids like water and gasoline pour well and spread out quickly because they have weak intermolecular attractions. Temperature affects viscosity. At cold temperatures, intermolecular forces are more effective on slow-moving particles. That is why cold syrup flows more slowly than hot syrup.

ADHESION

The same intermolecular forces that cause surface tension affect how a liquid interacts with the particles in its container. Have you ever observed the curved surface of a liquid in a container with a narrow neck? This curved surface, called a **meniscus**, results from intermolecular attractions within the liquid and between the liquid and the container. When attractions between the container and the particles of the liquid exceed the attractions between the liquid particles, the liquid climbs the walls of the container, forming a concave meniscus. In narrow glass tubes called *capillary tubes*, this effect becomes greatly exaggerated. Water and other liquids rise up narrow capillary tubes easily by *capillary action*. Mercury, on the other hand, has little attraction for glass but has strong internal cohesive forces. These cohesive forces are so strong that the surface molecules are pulled away from the glass as the molecules try to form a sphere, producing a convex meniscus. Thus, mercury does not exhibit capillary action, as shown below.

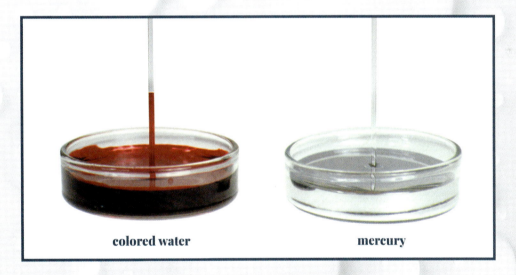

colored water　　　　　　　　　　　　**mercury**

Other fluid properties that liquids exhibit are *diffusibility* and *absorption*. A drop of food coloring spreads, or diffuses, throughout a glass of water. Spilled milk seeps into, or is absorbed by, a paper towel. Both cases are evidence that liquids consist of moving particles in unfixed positions. However, liquids diffuse in and are absorbed by other materials only when they are similar to one another. For instance, water wets a cotton cloth, but it does not wet wax candles. Water molecules are polar, so they readily adhere to sugar, cotton, skin, and other materials that have polar regions. Nonpolar materials, such as a freshly waxed car, cause water to bead up and roll away. On the other hand, nonpolar liquids wet nonpolar surfaces but not polar surfaces. In each case, intermolecular attractions determine how the liquid behaves.

Vaporization

Liquid particles do not all move at the same speed. Many move at similar speeds, but because particles are colliding, their speeds are not completely uniform. Molecules with above-average speeds sometimes break away from the liquid phase. This process, whether it occurs above or below the boiling point, is called *vaporization*. Vaporization can occur at any temperature, but it happens much more readily at higher temperatures. Condensation is the reverse of vaporization—it is the formation of a liquid from its gaseous state.

EVAPORATION

When vaporization occurs in a non-boiling liquid, it is known as **evaporation**—a relatively slow form of vaporization that occurs when liquid particles at the surface obtain enough energy to leave the liquid state. Evaporation is a cooling process. For example, you might shiver right after a shower or a swim because of the water removing thermal energy from your skin as it evaporates. During evaporation, liquid particles with the most kinetic energy leave the liquid state and enter the gaseous state. Remaining particles have a lower average kinetic energy and therefore a lower temperature.

Intermolecular attractions within a liquid affect the rate of evaporation. Strong attractions restrain particles from evaporating, while weak attractions allow quick evaporation. If temperature is increased, particles will move faster, overcoming intermolecular forces and evaporating more easily.

Particles that evaporate can also reenter liquids. Suppose that a liquid evaporates in a partially filled, closed jar. At first, the vapor particles enter the space above the liquid. During their random movements, these particles can bounce back to the surface of the liquid. As more particles evaporate, the space becomes crowded, making it more likely that some particles will return to the liquid. The pressure inside the container is determined by the number of vapor particles. The more particles that are evaporated, the greater the pressure, and vice versa. The pressure that is exerted by the evaporated particles over a liquid is called *vapor pressure*.

When enough particles have evaporated and the vapor pressure is high enough, the number of particles that go back into the liquid equals the number of particles that evaporate. At this point, the level of the liquid will remain the same. The volume of the liquid stays constant because the opposite process, condensation, nullifies the effect of evaporation. This situation, called a *dynamic equilibrium*, occurs when the two processes of condensation and evaporation balance each other so that no net effect can be observed.

BOILING

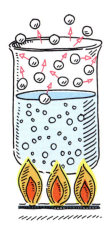

Boiling is a rapid change between the liquid and gaseous states caused by the application of heat or a dramatic decrease in pressure. Vapor bubbles form inside the liquid and rise to the surface. Vapor also escapes from a liquid's surface, as in evaporation.

Boiling is a relatively rapid form of vaporization that occurs when the vapor pressure of a liquid equals or exceeds the atmospheric pressure. Normally, atmospheric pressure is greater than vapor pressure at room temperature. The pressure prevents liquids from boiling. Table 13-3 shows that the vapor pressure of ethanol at 25 °C is only 9.07 kPa. The normal atmospheric pressure of 101.325 kPa can easily suppress this weak tendency to boil. But when the alcohol is heated to 78.4 °C, ethanol's vapor pressure equals normal atmospheric pressure and boiling occurs.

Vapor Pressure and Boiling Point

While most people equate boiling with particular temperatures, a liquid's temperature is not the sole deciding factor for when it boils. Climbers at the summit of Mount Everest would find that water boils at 70 °C, not at the expected 100 °C. At an elevation of 8850 m, the atmospheric pressure is only about 30 kPa, not the 101 kPa at sea level. Boiling occurs more readily because the vapor pressure of the liquid exceeds the lower atmospheric pressure at a lower temperature. When the vapor pressure reaches 30 kPa, the water will boil. Similarly, if atmospheric pressure is higher, a higher temperature is required for a liquid to boil, as is the case in hospital autoclaves or pressure cookers.

Because boiling points change with pressure, the **boiling point** of a liquid is defined as the temperature at which the vapor pressure equals the applied pressure. The *normal boiling point* is the temperature at which the vapor pressure equals 101.325 kPa. As you might expect, temperature and intermolecular forces affect vapor pressure just as they do evaporation. As the temperature in a liquid increases, its vapor pressure increases. The vapor pressure is less for a substance with strong intermolecular forces than for a substance with weaker intermolecular forces.

We can use the calculated vapor pressures of various substances to compare the rate at which they will evaporate. Suppose that mercury, water, and diethyl ether are spilled in a room at 25 °C. From Table 13-3 we can see that the mercury will exert 0.000 247 kPa of vapor pressure; water, 3.17 kPa; and diethyl ether, 72.1 kPa. These numbers show that diethyl ether will evaporate quickly, water will take longer to evaporate, and mercury will take decades to evaporate.

TABLE 13-3 *Vapor Pressures (in kPa)*

Substance	Temperature (°C)			
	0	25	50	100
Mercury	0.000 024 7	0.000 247	0.001 70	0.0364
Water	0.611	3.17	12.3	101
Ethanol	1.63	9.07	31.2	222
Methanol	3.16	16.3	53.9	—
Diethyl Ether	24.7	72.1	162	648

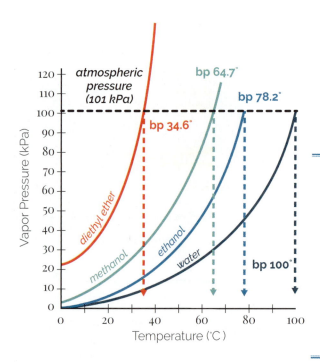

The vapor pressure of a substance at a specific atmospheric pressure can also be used to predict when a substance will boil. And we can use vapor pressure curves, such as the one at left, to answer questions related to boiling points. Let's look at how this works.

EXAMPLE 13-3:
CALCULATING THE BOILING POINT

What is the boiling point of water that is subjected to an atmospheric pressure of 66.7 kPa?

Solution

Using the vapor pressure curve, draw a horizontal line from 66.7 kPa on the vertical axis to the curve that represents water's vapor pressure. From that intersection, drop a vertical line to the axis representing the temperature. It will intersect at approximately 89 °C. At this temperature, the vapor pressure of water will match the given atmospheric pressure, and the water will boil.

Distilling Liquids

The processes of vaporization and condensation can be used to separate mixtures. This technique is called **distillation**. This is one way in which salt water can be purified, and it is also used to separate combinations of liquids into pure samples called *fractions*.

In the laboratory, the mixture to be separated is placed into a distillation flask and heated. The temperature rises steadily until it reaches the boiling point of the liquid that boils at the lowest temperature. The vapor of this liquid enters a condenser and flows over its water-cooled glass walls. The vapor soon condenses and drips into a collecting flask. Once all of the first liquid has been vaporized, the temperature can rise until the boiling point of another substance is reached. The process is repeated, and the other liquid is collected in another flask. This technique works well as long as the substances in the mixture have distinctly different boiling points and do not strongly interact. If distillation is precisely controlled, substances with boiling points only a few degrees apart can be separated.

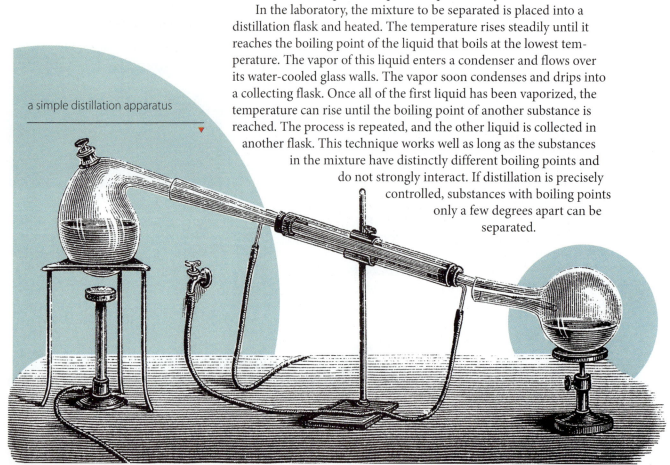

a simple distillation apparatus

Phase Diagrams

A **phase diagram** is a graphical way to represent the relationship between the temperature, pressure, and physical state of a substance. At right is a phase diagram of water. Note that all the points along the liquid-solid curve (AB) are melting points, all the points along the liquid-gas curve (AC) are boiling points, and all the points along the solid-gas curve (AD) are sublimation points.

Notice that the line of boiling points, the line of melting points, and the line of sublimation points intersect at point A. At this specific pressure and temperature (0.01 °C, 0.6 kPa), water can exist simultaneously in all three states. This specific set of conditions for a substance is called its **triple point**.

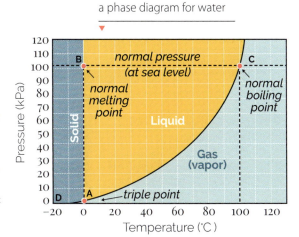

a phase diagram for water

EXAMPLE 13-4: DETERMINING THE STATE OF MATTER

Using the phase diagram for water, determine the state of water at the following conditions:

a. 60 °C, 53.3 kPa

b. −10 °C, 133 kPa

c. 80 °C, 26.7 kPa

Solution

a. The point that corresponds to 60 °C and 53.3 kPa is between the line of melting points and the line of boiling points. This region represents the conditions at which water is in the liquid state.

b. The point corresponding to −10 °C and 133 kPa is to the left of the line of melting points. Water is a solid under these conditions.

c. The point corresponding to 80 °C and 26.7 kPa lies beyond the line of boiling points. Water will exist as a gas under these conditions.

Critical Temperatures and Pressures

Oxygen, hydrogen, nitrogen, and helium, though gases at room temperature, can be liquids under the proper conditions. Low temperatures and high pressures can condense and even solidify any gas. Low temperatures slow down the molecules; high pressures pack them together.

Scientists have found that high pressures alone cannot liquefy gases. Their temperatures must be lowered past a certain point. This value, called the **critical temperature** (T_c), is the highest temperature at which a gas can be liquefied. Each gas has its own characteristic critical temperature.

Hydrogen gas at 35 K cannot be liquefied, even at tremendous pressures. The molecules are moving too quickly. If the temperature is lowered to 33 K, it can be squeezed into a liquid if enough pressure is applied. Gases that have critical temperatures above room temperature can be liquefied at room temperature by pressure alone. Other gases require a combination of refrigeration and compression.

The pressure that is required to liquefy a gas at its critical temperature is called the **critical pressure** (P_c). Hydrogen gas at its critical temperature can be liquefied under a pressure of 1297 kPa. If the gas is colder than its critical temperature, less pressure is required to liquefy it.

TABLE 13-4

Critical Temperatures and Pressures

Substance	T_c (K)	P_c (kPa)
hydrogen (H_2)	33.1	1298
nitrogen (N_2)	126	3395
oxygen (O_2)	155	5043
carbon dioxide (CO_2)	304	7387
ammonia (NH_3)	406	11 277
chlorine (Cl_2)	417	7991
water (H_2O)	647	22 064

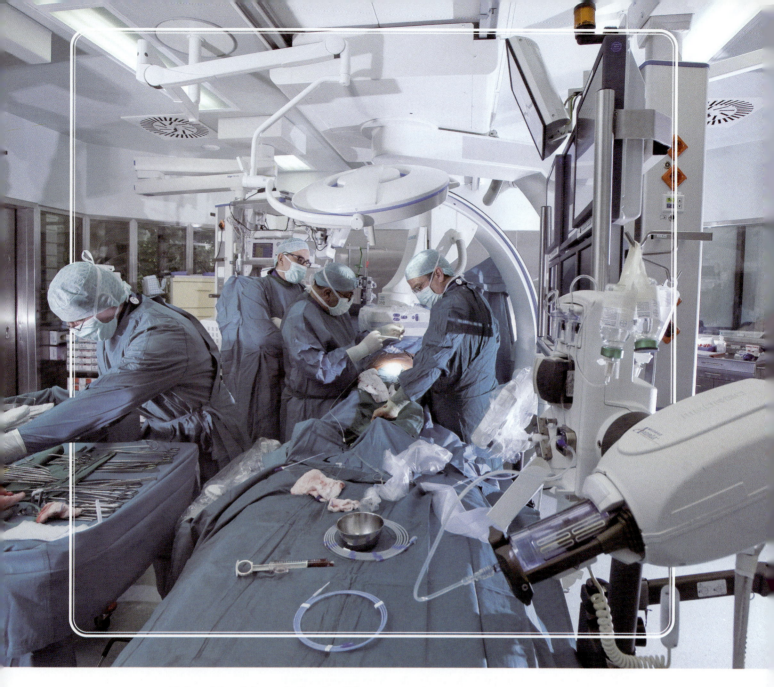

Using Liquids to Solve Problems

So how do doctors apply what you've learned in this chapter to help tackle the issue of organ donation described in the Chapter opener? People have been trying to transplant organs for thousands of years, mostly without success. Hundreds of years before Christ, Indian doctors performed skin grafts to treat burns. But it wasn't until the twentieth century that doctors developed safe and effective ways to transplant other organs to patients who needed them.

New types of transplants emerged during almost every decade between 1950 and 2000. The first successful kidney and bone marrow transplants were performed in the 1950s. The 1960s heralded the first successful pancreas and heart transplants. The 1980s saw the first successful lung and heart-lung transplants. And in the 1990s, cord-blood stem cell transplants and pancreatic islet transplants were successful. But how do all these organs and tissues stay alive in the transition from donor to recipient?

Currently, surgeons preserve most transplant organs by chilling them in the absence of blood flow, resulting in a condition called *cold ischemia*. Without blood flow, an organ will quickly die, so cooling is used to slow cell death and keep the organ viable during transport, at least for a little while. The organ is placed in a sterile container with special preservative solutions—liquids—surrounding it. This container is then placed in a beverage cooler filled with another icy solution designed to cool but not freeze the organ.

Kidneys get a different treatment. Doctors use a machine called a *pulsatile perfusion device* to continuously pump preservative solution through the kidney, keeping it viable for a longer period. Kidneys preserved in this fashion also survive transplantation better and are more likely to continue functioning after being transplanted into a recipient. Of course, the preserving liquid is different from the blood that normally pumps through the organ. As doctors continue to invent better ways to preserve organs, they use the properties of liquids to give new hope to those who are suffering. This is a way to use chemistry to show love to people.

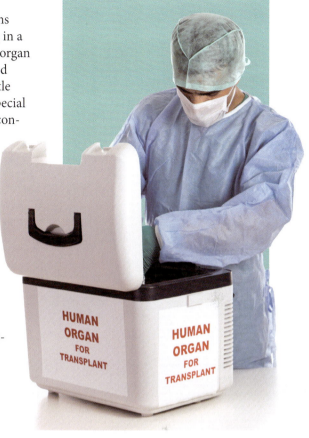

13.3 SECTION REVIEW

1. How does the kinetic-molecular theory explain the liquid phase of a substance?

2. What would happen to a paper clip floating on the surface of water if some detergent were added? Explain.

3. Explain the difference between cohesion and adhesion. Give an example of each.

4. Compare evaporation and boiling.

5. Why are the published values of the boiling points of substances always given for 101.325 kPa?

6. Describe what happens to the temperature of water as it changes phase from a solid to a gas.

Refer to the phase diagram for ethanol at right to answer Questions 7–8.

7. a. In what state is ethanol at 300 °C and 3000 kPa?

   b. In what state is ethanol at −145 °C and 1500 kPa?

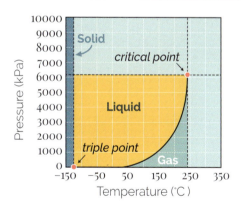

c. At what combination of temperature and pressure can ethanol exist in all three phases of matter?

8. If ethanol is held at a constant temperature of 200 °C while the atmospheric pressure is increased from 900 kPa to 5250 kPa, will it undergo a phase change? If so, what kind?

13 CHAPTER REVIEW

Chapter Summary

TERMS
intermolecular force
dipole-dipole force
hydrogen bond
London dispersion force

13.1 INTERMOLECULAR FORCES

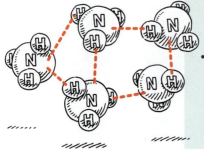

- Intermolecular forces are the electrostatic forces due to temporarily or permanently charged regions on particles. These forces influence the physical properties of substances.

- Hydrogen bonds, dipole-dipole forces, and dispersion forces are the three main types of intermolecular forces in order from strongest to weakest.

- The intermolecular forces present in a substance can be predicted on the basis of its composition and structure.

- The physical properties of a substance can be predicted on the basis of its intermolecular forces.

13.2 SOLIDS

- The properties of solids can be described and explained by the kinetic-molecular theory.

- Amorphous solids have no preferred arrangement of particles, whereas crystalline solids form regular three-dimensional patterns called crystal lattices. These can be described using the concept of unit cells.

- Melting is the change from the solid to the liquid state; freezing is the reverse process.

- Heating curves illustrate how adding thermal energy affects the temperature and state of matter.

- Phase changes can be explained on the basis of the interplay between kinetic-molecular theory and intermolecular forces. During phase changes, the temperature of a substance remains constant until the change is completed.

- Sublimation is the direct change from the solid to the gaseous state; deposition is the reverse process.

- The lattice energy of a crystal is reflected in its melting point and solubility and is partly determined by its particles' electric charge and size.

- Allotropes are different crystal lattice structures formed by an element.

TERMS
crystalline solid
amorphous solid
melting point
unit cell
allotrope
lattice energy

13.3 LIQUIDS

- As is true for solids, the properties of liquids can be described and explained by the kinetic-molecular theory.

- Cohesion, adhesion, surface tension, capillary action, and viscosity are all dependent on intermolecular forces.

- Vaporization is the change from the liquid to the gaseous state; condensation is the reverse process.

- There are two types of vaporization. Evaporation is vaporization any time the liquid phase is present and particles of the liquid get enough energy to overcome the intermolecular forces. Boiling is vaporization when the vapor pressure of a liquid equals the atmospheric pressure.

- Distillation is a technique used to separate mixtures by vaporizing and condensing the components, or fractions, on the basis of their boiling points.

- Phase diagrams show the temperature and pressure conditions under which a substance is solid, liquid, or gaseous.

TERMS
cohesion
adhesion
surface tension
surfactant
viscosity
meniscus
evaporation
boiling
boiling point
distillation
phase diagram
triple point
critical temperature
critical pressure

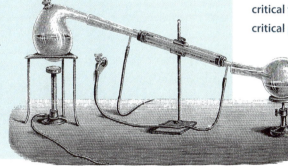

Chapter Review Questions

RECALLING FACTS

1. What is a hydrogen bond?
2. Arrange the attractive forces (bonds and intermolecular forces) within substances in order of increasing strength.
3. What makes water different from other substances in regard to the density of its solid and liquid states?
4. What is the major difference between crystalline and amorphous solids?
5. Name three factors that determine the structure of a crystal.
6. Name three factors that determine the strength of a crystal.
7. What causes surface tension?
8. Why do raindrops not assume triangular or cubic shapes?
9. Why does the surface of water in a glass test tube curve upward at the edges?
10. How does perspiration cool the skin on a person's body?
11. What is the essential difference between boiling and evaporation?

UNDERSTANDING CONCEPTS

12. A study buddy claims that the bonds between atoms within a molecule are stronger than the forces that exist between molecules. Is he correct? Explain.
13. Explain how you would determine which type(s) of intermolecular forces exist within a substance.
14. Predict the types of intermolecular forces that may act between the molecules in the following substances.

    a. CO_2 c. HCl
    b. NH_3 d. C_3H_8

15. Assume that the particles in table salt (NaCl) vibrate just as forcefully as the particles in lead (Pb). Explain why table salt remains a solid at 500 °C while lead exists as a liquid.
16. When thermal energy is being removed from a liquid, why does the temperature of the liquid at its freezing point remain constant until all the liquid freezes?

13 CHAPTER REVIEW

17. Using your understanding of intermolecular forces, determine which substance in each of the pairs should have the higher boiling point. Give your reasoning.

   a. NF_3, NH_3
   b. NaCl, HCl
   c. CF_4, CHF_3
   d. Cl_2, C_2H_5Cl

18. In terms of attractive forces and kinetic energies of particles, explain what happens during the following phase changes. *Example*: In melting, the particles gain enough kinetic energy to overcome the attractive forces that hold them in fixed positions.

   a. boiling
   b. evaporation
   c. freezing
   d. sublimation

19. Use the kinetic-molecular theory to explain each observation given.

   a. Wax melts near the flame of a burning candle.
   b. Liquid water may be converted into ice cubes in a freezer.
   c. Ginger ale flows to match the shape of a glass.
   d. Water gradually evaporates from a swimming pool.
   e. Water vapor condenses inside house windows on cold days.
   f. Snow gradually disappears, even when the temperature remains below freezing.
   g. Solids and liquids cannot be compressed as much as gases.

20. A white powder contains tiny cube-shaped grains and melts at a temperature between 141.6 °C and 142.2 °C. Is this solid more likely to be a crystalline solid or an amorphous solid?

21. Automobile engines are designed to be lubricated with motor oils of specific viscosities. These are rated from the lowest viscosity of 5W up to the highest viscosity of 70W. What can you conclude about the intermolecular attractions in an oil rated 10W?

22. Water rolls off a duck's back but thoroughly wets a head of human hair. What do these observations reveal about the chemical nature of these two surfaces?

23. After a jar of liquid has been sealed, the level of the liquid decreases slightly because of evaporation. After a slight decrease, the level of the liquid ceases to change. Why?

24. Crude oil is purified or refined through a process called *fractional distillation*. In this process the different components of oil (e.g., lubricating oil, grease, gasoline, and natural gas) are separated by distillation. What do you conclude about the physical properties of the different components?

25. Water in a truck's radiator can get hotter than 100 °C when the radiator is sealed tightly. How is it possible for water to exist as a liquid at temperatures above its normal boiling point?

CRITICAL THINKING

26. In this chapter, hydrogen bonds are described as a force between molecules, but might there be conditions under which it could also exist as a force *within* a molecule? Explain.

27. Predict which member in each of the following pairs of crystals has the stronger binding forces. Explain the reasons for your prediction.

   a. NaCl, I_2
   b. KBr, NaBr
   c. CaI_2, KI

28. Refer to the diagram below to determine

   a. the boiling point of ethanol when it is at normal atmospheric pressure.
   b. the boiling point of methanol at 96.0 kPa.
   c. the atmospheric pressure at which diethyl ether boils at 20 °C.

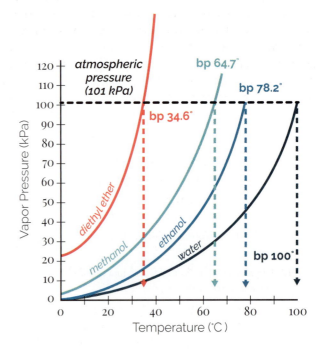

29. Refer to the phase diagram of water below to determine the state of water at the given conditions.

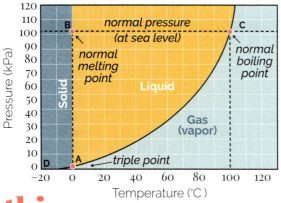

a. 80 °C, 40.0 kPa

b. 100 °C, 120.0 kPa

c. −10 °C, 66.7 kPa

30. A scientist claims to have a cooling apparatus kept at −100 °C by liquid nitrogen. Is this possible? Why or why not?

Use the ethics box below to answer Question 31.

31. Use the ethics process from pages 74–75 to write a four-paragraph assessment of the notion that cryonics offers hope as science advances our understanding of medicine.

ethics

CRYONICS

As you read in Section 13.2, applications of cryogenics are very useful, but there are a few dubious uses. *Cryonics*, one of the more macabre uses of cryogenics, is the process of deep freezing and storing a person who has been declared legally dead. People hope that a preserved person can be brought back to life at some time in the future.

Why would someone choose to do this? Some people hope that they can be stored until a cure is found for their incurable diseases. They are looking for healing for themselves or for their loved ones. People recognize that things are not the way they should be. Or there is the hope for a future reunion with loved ones, as with baseball great Ted Williams, probably the most famous person to undergo cryonics. Our desire for healing and life shows that we long for the world to be the way that God created it. However, there is currently no possibility of reviving a person preserved by cryonics since, as far as science has been able to determine, the biological processes of death are irreversible. Despite these facts, some people still put all their hope in science.

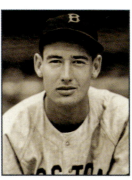
Ted Williams

What if advances in scientific technology did someday produce a way to bring these people back to life? Even if this happened, that person would still ultimately die, and their most fundamental problem would not be cured. People are still sinners, and death—both physical and spiritual—is part of God's penalty for sin. Adam and Eve's sin brought disease and death into the world, and as their children, we follow them in sin. But God's solution is far better than the faint hope of cryonics. All the promises of God for healing and eternal life find fulfillment in Jesus. He alone can save us from the deadly effects of sin and restore us to the abundant life that God wants us to enjoy. Whoever believes in Jesus will "not perish, but have everlasting life" (John 3:16).

Chapter 14

SOLUTIONS

Water covers about 70% of the earth's surface, yet many consider water scarcity to be one of the world's greatest health issues. Over 1.1 billion people do not have an adequate source of clean water. That's more than 14% of the world's population.

You may be wondering if there is so much water on the earth, why do so many people struggle to find water to drink? For some, there simply is no source of fresh water nearby. (The vast majority of Earth's water is salt water.) Even when we have fresh water, pollution is a big issue. Between water's salt content and contamination, much of the planet's available water is not potable. Therefore, many peo-

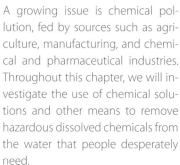

ple suffer from dehydration, parasites, cholera, dysentery, and other conditions.

A growing issue is chemical pollution, fed by sources such as agriculture, manufacturing, and chemical and pharmaceutical industries. Throughout this chapter, we will investigate the use of chemical solutions and other means to remove hazardous dissolved chemicals from the water that people desperately need.

14.1 The Dissolving Process *327*
14.2 Measures of Concentration *338*
14.3 Colligative Properties *343*
14.4 Suspensions and Colloids *349*

14.1 THE DISSOLVING PROCESS

Defining Solutions

We use solutions and their properties when we add antifreeze to a car's radiator, spread salt on an icy sidewalk, wear sterling silver, disinfect a cut, or clean windows. Many chemical reactions take place in solution. People breathe a solution called air, and most of the human body is a water-based solution. The study of chemistry is saturated with solutions!

Shampoos, soft drinks, and perfumes are mixtures of many ingredients. Their ingredients do not clump together, separate from each other, or fall to the bottom of a container. Their components cannot be filtered out because they are very small, typically less than 1 nm. These products are homogeneous mixtures since their compositions are uniform. They are not compounds because they are physically combined, not chemically bonded together.

Why is it so difficult to dissolve sugar in my iced tea?

The substances mentioned above are all **solutions**—homogeneous mixtures of variable composition in a single phase. The most abundant substance in a solution is called the **solvent**. In shampoo and soft drinks, the solvent is water; in perfume, the solvent is often an alcohol. Dissolved substances are called **solutes**. Solutes dissolved in a solvent make a solution.

QUESTIONS
» What are the parts of a solution?
» Are there different types of solutions?
» Why doesn't everything dissolve?
» How can I get sugar to dissolve more quickly in my tea?
» How can I get more sugar to dissolve in my tea?

TERMS
solution • solvent • solute • miscibility • solvation • dissociation • ionization • enthalpy of solution • solubility • unsaturated solution • saturated solution • supersaturated solution • Henry's law

Properties of Solutions

- Solutes cannot be filtered out of a solution.
- Solutes do not settle out of a solution.
- Solutions are uniform.
- Solutions are not chemically combined.
- Solutions can vary in their concentrations.

TYPES OF Solutions

miscible
seawater

LIQUID SOLUTIONS

Most of the solutions that we use are liquids because the solvent is a liquid. Liquid solvents can dissolve solids, other liquids, and even gases. Sweetened tea and seawater are common examples of solutions made by dissolving solid solutes in a liquid solvent.

A common liquid-liquid solution is rubbing alcohol. Rubbing alcohol is a mixture of 30% water and 70% iso-propyl alcohol. Both the solute and solvent are liquids. But not all liquid-liquid mixtures form a solution. We all know that oil and water don't mix. Scientists use the term **miscibility** to describe whether two liquids will mix to form a solution. When two liquids, like alcohol and water, mix to form a homogeneous mixture, they are said to be *miscible*. But liquids that separate after mixing, like oil and water, are said to be *immiscible*.

immiscible
oil and water

SOLID SOLUTIONS

Solids can also act as solvents. Liquids, gases, and even other solids can dissolve in solid solvents. The most common solid-solid solutions are metal alloys. Brass, a copper and zinc alloy, is a blend of the two metals in a uniform mixture. Gold and mercury can form a solid-liquid solution that in the past was used in the process for mining gold. A solution containing mercury is called an *amalgam*. Hydrogen gas can dissolve in palladium metal and produce a solid-gas solution. Scientists have used this process to purify hydrogen by allowing it to dissolve into the metal and then forcing it out with heat.

gold mercury almalgam

pewter: *an alloy made with tin, lead, and copper*

GASEOUS SOLUTIONS

Gases can also act as solvents. Although some scientists may argue the point, there is only one type of permanent gas solution at ordinary pressures—gaseous solvents dissolving gaseous solutes.

Gases cannot dissolve liquids or solids because the gaseous solvent can't separate liquids and solids into individual particles. The collection of liquid or solid particles is too large to give the mixture a single phase, so the mixture is not considered a solution. Scientists categorize these mixtures as *colloids* or *suspensions* (see Section 14.4). An example of a gaseous solution is the air we breathe. It is a solution of oxygen, carbon dioxide, argon, and trace amounts of other gases dissolved in nitrogen.

SCUBA tanks contain gaseous solutions.

TABLE 14-1 *Solutions*

		Solvents		
		Gas	Liquid	Solid
Solutes	Gas	gas-in-gas (air)	gas-in-liquid (soft drink)	gas-in-solid (H_2 in Pd)
	Liquid	does not exist	liquid-in-liquid (vinegar)	liquid-in-solid (amalgam)
	Solid	does not exist	solid-in-liquid (salt water)	solid-in-solid (alloy)

CASE STUDY

PHARMACEUTICAL POLLUTION

When most of us think of pharmaceutical products, we think of medicines that make us well. But there is a side to pharmaceuticals that is not so beneficial. In recent years, environmental scientists have identified a problem—pharmaceutical pollution in wastewater and ultimately in our water supply.

Most of these dissolved chemicals enter water systems from human waste, which we already know how to remove. Other chemicals enter systems through the disposal of unused chemicals. A small percentage enter the water supply from manufacturing. Today we know the chemicals that are present, but in many cases we need to develop a process for removing them at water treatment facilities.

Questions to Consider

1. Since these chemicals are in solution, why might filtering not eliminate them?
2. What process might work for removing these smaller solute particles?
3. How does the process that you mentioned in Question 2 remove pollution from water?

The Dissolving Process

So how does a solute dissolve in a solvent? For example, how does sugar dissolve in coffee? The dissolving mechanisms for each type of solution are different. To dissolve a substance in a liquid solution, the intermolecular forces within both the solute and solvent must be overcome.

DISSOLVING FOR LIQUID SOLUTIONS

Solvation is the dissolving process in solid-in-liquid solutions. In solvation, the particles of the solvent surround and separate the ions or molecules of the solute. Solvation in water is called *hydration*. The more general term, solvation, describes the dissolving of a solid compound in any liquid solvent. If the solute is an ionic compound, solvent molecules break up the crystal into ions in a process called **dissociation**. In some cases, covalent compounds form ions when they dissolve in solution, a process known as **ionization**.

A Sugar Solution

Dissolving sugar in water provides a good example of how this process works. In a sugar-water solution, the solvent molecules (water) separate the solute particles (sugar molecules) from each other. Before this happens, energy is needed to overcome the attractions between solvent particles. The positive regions of the water molecules attract the negative regions of the sugar molecules.

Negative regions of water molecules attract positive regions of sugar molecules in the same manner. Hydrogen bonds form and dispersion forces take effect. The water molecules then pull sugar molecules away from their neighbors. After solvation, diffusion carries the dissolved particles throughout the solution.

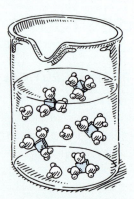

solvent (water)

solute (sugar)

solvent particles removing solute particles

solvent particles surrounding (solvating) solute

solute dissolved

Thermal energy is necessary to overcome the intermolecular forces between the solvent particles and between the solute particles. Therefore, the first steps of solvation are endothermic, absorbing thermal energy from the surroundings. After the separation, solute and solvent particles mix and begin to attract each other. This mixing produces a more stable arrangement of particles, releasing energy. This part of the solution process is exothermic. The net energy change during the dissolving process is the **enthalpy of solution** ($\Delta H_{solution}$), also known as the *heat of solution*. This value, specific to each solute, is expressed in kilojoules per mole (kJ/mol). The concept of enthalpy is discussed more in Chapter 15.

DISSOLVING FOR SOLID AND GASEOUS SOLUTIONS

In solid and gaseous solutions, solute and solvent particles mix predominantly by the process of diffusion. In gaseous solutions, molecules mix by random, constant motion. For solid solutions, mixing typically takes place while the components are in liquid form, allowing diffusion to take place.

MINILAB

OFF TO THE RACES

Many people find an ice-cold glass of tea refreshing. While we may differ on how sweet tea is supposed to be, many of us like some sugar in our tea.

How can we increase the rate of dissolving?

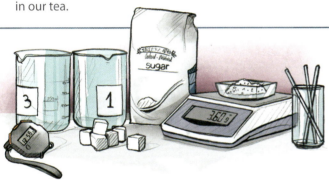

1. Name two conditions that you think might affect the rate at which sugar will dissolve in water.

2. In relationship to dissolving, what is different between the granulated sugar and the sugar cube?

Procedure

A Copy the table below onto a piece of paper.

B Label the four beakers 1–4.

C Use the laboratory balance and three of the weighing dishes to measure three 3.60 g samples of granulated sugar. Place the sugar cube in the fourth dish.

D Add 100 mL of room temperature water to Beakers 1–3 and 100 mL of hot water to Beaker 4.

E Start the stopwatch as you simultaneously add the granulated sugar to Beakers 1, 2, and 4 and the sugar cube to Beaker 3. Stir Beaker 2.

F Note the time when each sample of sugar finishes dissolving.

Conclusion

3. In which beaker did the sugar dissolve fastest?

4. From your observations, summarize the effect that different conditions had on dissolving the sugar.

Going Further

5. How can we most quickly dissolve sugar in a glass of tea?

EQUIPMENT

- laboratory balance
- beakers, 250 mL (4)
- weighing dishes (4)
- stopwatch
- stirring rod
- marker
- granulated sugar, 11 g
- sugar cube
- room temperature water, 300 mL
- hot water, 100 mL

Beaker	Conditions			Time (s)
	Water	Sugar	Other	
1	room temperature	granulated	none	
2	room temperature	granulated	stirred	
3	room temperature	cube	none	
4	hot	granulated	none	

Nonpolar hexane (clear liquid on left) and water (tinted with bromine) are immiscible and remain in distinct phases. Over time the bromine migrates to the hexane (top layer, right) because it is more soluble in hexane than in water.

Solvent Selectivity

Why do isopropyl alcohol and water mix but not oil and water? What determines which substances can mix or dissolve in others? The governing principle in solutions is *like dissolves like*. Polar solvents dissolve polar solutes, and nonpolar solvents dissolve nonpolar solutes.

Ionic compounds and polar covalent molecules have regions of electrical charge. These regions of charge can interact with neighboring ions or with other polar molecules. Isopropyl alcohol (C_3H_8O), acetic acid ($C_2H_4O_2$), ammonia (NH_3), and hydrogen chloride (HCl) are all polar substances. They dissolve in water because of dipole-dipole forces. Electrostatic forces also separate ions, as in a saltwater solution. Positive sodium ions are strongly attracted to the negative pole of water molecules, while negatively charged chloride ions are attracted to the positive pole. When a solute can dissolve in a solvent, it is said to be *soluble* in that solvent. **Solubility** is the maximum amount of a solute that can dissolve in a specific solvent under specific conditions, such as temperature and pressure. Typically chemists report solubility as the mass of solute that will dissolve in 100 mL of solvent.

Nonpolar substances such as hexane, pentane, and petroleum ether cannot mix with polar substances. Consequently, they are pushed out of polar solvents, forming distinct layers in the container. If you were to try to mix water and hexane, they would remain in two layers.

EXAMPLE 14-1: COMPARISONS OF SOLUBILITY

Considering the polarity of methanol, hexane, and aluminum chloride, determine which of the following substances can dissolve in water.

methanol   hexane   aluminum chloride

$$H-\underset{\underset{H}{|}}{\overset{\overset{H}{|}}{C}}-OH \qquad H-\underset{\underset{H}{|}}{\overset{\overset{H}{|}}{C}}-\underset{\underset{H}{|}}{\overset{\overset{H}{|}}{C}}-\underset{\underset{H}{|}}{\overset{\overset{H}{|}}{C}}-\underset{\underset{H}{|}}{\overset{\overset{H}{|}}{C}}-\underset{\underset{H}{|}}{\overset{\overset{H}{|}}{C}}-\underset{\underset{H}{|}}{\overset{\overset{H}{|}}{C}}-H \qquad [Al^{3+}][Cl^-]_3$$

Solution
Polar water molecules can dissolve polar methanol molecules and ionic aluminum chloride, but they cannot dissolve nonpolar hexane molecules.

Soluble versus insoluble: NaCl in water (left) and calcium carbonate in water (right)

Like dissolves like is a general rule, but there are several exceptions. Calcium chloride ($CaCl_2$) dissolves in water, but calcium carbonate ($CaCO_3$) does not. Both compounds are ionic, so why aren't both soluble in the polar solvent water? The strength of ionic attractions within compounds accounts for differences in the solubility of different solutes. Dissociation must overcome the lattice energy of the crystalline structures. Some ions so strongly attract each other that a specific solvent cannot break them apart; the solute is *insoluble* in that solvent. Table 14-2 describes the solubility of many ions and compounds. Salts are considered soluble if at least 1 g of solute will dissolve in 100 g of H_2O. They are insoluble if less than 0.1 g of solute will dissolve in 100 g of H_2O. Scientists identify solutes with solubility values between these values as slightly soluble. Substances such as calcium carbonate (chalk), though considered insoluble in water, do have limited solubility. Almost no substances have zero solubility.

TABLE 14-2 *Solubility Rules*

	Soluble (remains aqueous)	Exceptions (forms precipitate)
1	NO_3^-	
2	Li^+, Na^+, K^+, Rb^+, NH_4^+	
3	Cl^-, Br^-, I^-	Cu^+, Ag^+, Hg_2^{2+}, Pb^{2+}
4	SO_4^{2-}	Ag^+, Ba^{2+}, Ca^{2+}, Sr^{2+}, Pb^{2+}, Hg_2^{2+}
	Insoluble (forms precipitate)	**Exceptions (remains aqueous)**
5	S^{2-}	Mg^{2+}, Ba^{2+}, Ca^{2+}, Sr^{2+}
6	CO_3^{2-}, PO_4^{3-}	Na^+, K^+, NH_4^+
7	OH^-	Na^+, K^+, Ca^{2+}, Ba^{2+}
8	O^{2-}	Na^+, K^+, Ca^{2+}, Ba^{2+}, Sr^{2+}

EXAMPLE 14-2: SALT SOLUBILITY

Using Table 14-2, determine whether the following salts will be soluble in water.
  a. KCl
  b. $PbSO_4$
  c. BeO
  d. MgS

Solution
  a. soluble; Line 2 indicates that potassium compounds are soluble.
  b. insoluble; Line 4 indicates that sulfate compounds are soluble, but Pb^{2+} is an exception.
  c. insoluble; Line 8 indicates that oxides are insoluble.
  d. soluble; Line 5 indicates that sulfides are insoluble, but Mg^{2+} is an exception.

Solution Equilibria

Consider dissolving table salt in a beaker of water. Initially, during the dissociation process, the solution has not yet reached its maximum solubility. Since the solution contains less than the maximum amount of solute that it could hold at the current conditions, we consider this an **unsaturated solution**. The thermal energy required for the dissociation of salt is greater than the thermal energy given off in hydration. Energy is therefore absorbed in the solution process, and the solution becomes slightly cooler.

$$NaCl(s) + \text{heat} \longrightarrow Na^+(aq) + Cl^-(aq)$$

If we continue to add more and more salt, eventually no more salt appears to dissolve. The temperature and the amount of salt at the bottom of the glass remain constant. When a solution contains the maximum amount of solute at a given temperature, we call it a **saturated solution**. But this condition does not stop the dissolving process. The water molecules are still moving, dissociating ions. The amount of salt at the bottom of the glass remains constant because the reverse process, precipitation, also occurs.

$$Na^+(aq) + Cl^-(aq) \longrightarrow NaCl(s) + \text{heat}$$

When both processes occur at the same rate, no noticeable changes occur. This condition is called a *dynamic equilibrium*. We discuss equilibrium reactions more in Chapter 17.

$$NaCl(s) + \text{heat} \rightleftharpoons Na^+(aq) + Cl^-(aq)$$

▲ As sodium chloride dissociates, negative portions of water molecules attach to the Na⁺ ions. Positive portions of water molecules interact with Cl⁻ ions.

As strange as it may seem, it is possible to dissolve more than the maximum amount of solute—we call it a **supersaturated solution**. For example, the solubility of sodium thiosulfate ($Na_2S_2O_3$) in water is about 73 g/100 mL at room temperature. But if we raise the temperature of the solution to 100 °C, the solubility increases to 231 g/100 mL; thus, more sodium thiosulfate can dissolve. If the solution is then slowly and carefully cooled back to room temperature, the excess sodium thiosulfate will not crystallize out. Instead, 231 g of sodium thiosulfate remain in 100 mL of water—more than three times the expected amount! Supersaturated solutions are not in equilibrium—they are unstable. A scratch on the inner surface of the container, the addition of a small crystal of sodium thiosulfate, or even a sudden disturbance can cause the excess solute to crystallize rapidly, returning the solution to its saturated level.

334 Chapter 14

Equilibria can also exist for liquid-gas solutions. Carbonated drinks are solutions of carbon dioxide. When you open a bottle of carbonated beverage, carbon dioxide gas escapes with a fizz. If you replace the cap, the escaping gas collects above the liquid in the bottle. As the partial pressure of carbon dioxide increases, the rate at which carbon dioxide escapes from the solution decreases. Eventually, the rate at which carbon dioxide reenters the solution will match the rate at which it leaves the solution. Because the two processes oppose each other, the drink does not lose all its carbonation right away.

FACTORS THAT AFFECT THE
Rate of Solution

The rate at which a solute dissolves depends on its solubility and the frequency of collisions with solvent particles. The more collisions there are between the solute and solvent particles, the faster the solute dissolves. Temperature, stirring, and the amount of surface area of the solute exposed to the solvent all affect the frequency of these collisions.

The temperature of a solution affects the number of molecular collisions. Consider a glass of warm, unsweetened tea. Which should be put into the tea first—sugar or ice? If the sugar is to dissolve quickly, you should add it first. The faster moving particles in warm tea will dissolve sugar faster than the slower moving particles in iced tea.

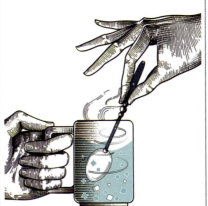

Stirring a solvent increases the number of collisions. A moving spoon brings the solvent molecules into contact with the sugar at the bottom of the glass more rapidly.

The number of collisions also depends on the amount of surface area exposed to solvent action, which explains why granulated sugar will dissolve more quickly than a sugar cube will. In a sugar cube, the bulk of the cube shields most of the solute from the solvent. In granulated sugar, the solvent has access to many more sugar molecules. Increased surface area speeds up the rate of dissolving.

FACTORS THAT AFFECT
Solubility

TEMPERATURE

Solid and Liquid Solutes

Will the solubility of a solute increase or decrease as temperature increases? In general, higher temperatures help more of a solid and liquid dissolve by increasing the space between the solvent particles. Increased temperature also increases solubility because most solids and liquids require thermal energy to dissolve. Suppose that a saturated solution of potassium chlorate ($KClO_3$) is at equilibrium.

$$KClO_3(s) + heat \rightleftharpoons K^+(aq) + ClO_3^-(aq)$$

Adding thermal energy is equivalent to adding more of a reactant. This energy favors the forward reaction, causing more of the potassium chlorate to dissolve until the solution reaches a new equilibrium position.

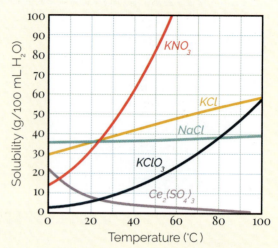

The graph above shows the solubility of different substances at temperatures from 0 °C to 100 °C. We can see on the graph the solubility curve for $KClO_3$, which shows how much $KClO_3$ can be dissolved in 100 mL of water.

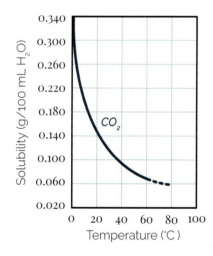

Gaseous Solutes

However, when gases dissolve, the opposite usually occurs—thermal energy is released.

$$CO_2(g) + H_2O(l) \rightleftharpoons H_2CO_3(aq) + heat$$

We could, therefore, think of thermal energy as a product of the process. Adding thermal energy is equivalent to adding more product to the reaction vessel. This additional energy drives the reverse reaction and causes carbon dioxide to escape from the solution until it establishes a new equilibrium position. Note on the graph at left how the solubility of carbon dioxide changes with temperature.

EXAMPLE 14-3: SOLUBILITY AND TEMPERATURE CHANGE

Use the solubility curves above to answer the following questions.

a. What is the solubility of KNO_3 in 100 mL of water at 50 °C?

b. At what temperature is the solubility of CO_2 0.160 g in 100 mL of water?

Solution

a. Enter the graph on the x-axis at 50 °C, move vertically until you intersect the KNO_3 curve, and then move horizontally until you reach the y-axis. The solubility of KNO_3 is approximately 83 g.

b. Enter the graph on the y-axis at 0.160 g, move horizontally until you intersect the CO_2 curve, and then move vertically until you reach the x-axis. The temperature is approximately 18 °C.

EXAMPLE 14-4: SOLUBILITY AND TEMPERATURE CHANGE

Use the solubility curves on the previous page to answer the following questions about the solubility of various compounds in water at different temperatures.

a. Which compound is the most soluble at 0 °C?

b. Which compound is the most soluble at 60 °C?

c. Which compounds show decreasing solubility as the temperature increases?

Solution

a. NaCl

b. KNO_3

c. $Ce_2(SO_4)_3$ and CO_2

Pressure

While pressure had little effect on the solubility of solid and liquid solutes, pressure drastically affects the solubility of gases in liquids. **Henry's law** states that the solubility of a gas is directly proportional to the partial pressure of the gas above the solution. We use the formula

$$\frac{S_1}{P_1} = k,$$

where S is the solubility of a gas at different partial pressures (P).

This law is vividly illustrated whenever we open a bottle of soda. Opening the bottle reduces the pressure from 385 kPa to atmospheric pressure, and the solubility of carbon dioxide in the beverage decreases, allowing the excess gas to escape. The escape of a gas from a liquid-gas solution is called *effervescence*.

14.1 SECTION REVIEW

1. Give an example of a liquid solution, a solid solution, and a gaseous solution.

2. Predict whether each of the following pairs of liquids is miscible or immiscible.
    a. olive oil and avocado oil
    b. gasoline and water
    c. iced tea and salt water
    d. olive oil and tea

3. Why do you think that water is often called the universal solvent? What makes it such a good solvent?

4. Define *solubility*.

5. Compare unsaturated, saturated, and supersaturated solutions.

6. Using the kinetic-molecular theory, explain each of the following observations.
    a. Sugar dissolves faster in hot coffee than it does in cold coffee.
    b. Finely ground salt dissolves faster than large chunks of salt do.
    c. Solutes dissolve faster when stirred.

7. Use the solubility curves on page 336 to answer the following questions.
    a. What is the solubility of KCl at 35 °C?
    b. At what temperature is 30 g the solubility of $KClO_3$?
    c. Consider the curve for KNO_3. What locations on the graph represent supersaturated, saturated, and unsaturated solutions?

8. Will a carbonated drink go flat faster if it is heated? Why?

14.2 MEASURES OF CONCENTRATION

QUESTIONS
» How can we describe the amount of dissolved solute?
» How can we calculate the amount of solute needed to make a solution?

TERMS
concentration • mass fraction • volume fraction • amount concentration • molality

How much sugar is really in my soda?

Defining Concentration

What is the best way to describe how much solute is in a solution? Terms such as *diluted* and *concentrated* refer to relative amounts of solute in a solvent. These qualitative descriptions are not sufficient to precisely describe the amount of solute in a solution. Quantitative measurements of solutes are much more useful in chemistry. **Concentration** is a measure of the amount of a solute in a certain volume or mass of either the solvent alone or the entire solution.

The most common ways to express concentration are mass fraction, volume fraction, amount concentration, and molality. Each has its advantages and unique uses. Let's look at each of these calculations.

Mass Fraction

One common method of expressing concentrations compares the mass of the solute with the mass of the solution. Known as **mass fraction**, this method reports the concentration as a decimal value or as a percent. We use the formula

$$w_{solute} = \left(\frac{m_{solute}}{m_{solution}}\right) 100\%,$$

where w_{solute} is the mass fraction, m_{solute} is the mass of the solute, and $m_{solution}$ is the mass of the solution. For example, oceanographers typically measure the salinity, or salt concentration, of seawater using the mass fraction. Typically, ocean water has 3.5 g of dissolved salts in every 100 g of solution.

$$w_{salt} = \left(\frac{3.5 \text{ g}}{100 \text{ g}}\right) 100\%$$

$$= 3.5\%$$

Although ecologists often use this measure of concentration, other measures are more common in chemical applications.

Volume Fraction

Another common method of expressing concentrations relates the volume of the solute to the volume of the solution. **Volume fractions** often appear on the ingredient labels of many household products, especially liquid-liquid solutions. Volume fraction is reported as a decimal value or as a percent. We use the formula

$$\varphi_{solute} = \left(\frac{V_{solute}}{V_{solution}}\right) 100\%,$$

where φ_{solute} is the volume fraction, V_{solute} is the volume of the solute, and $V_{solution}$ is the volume of the solution. Vinegar is an example of a solution whose concentration is reported as a volume fraction, as it consists of approximately 5.0% acetic acid by volume. That means that acetic acid contributes 5 mL for every 100 mL of solution. While you may see this measure of concentration used for some foods and medicines, chemists use it infrequently.

Amount Concentration

The most common expression of concentration in chemistry is called **amount concentration**, or *molar concentration*, and compares the moles of solute with the volume of the solution. You will often hear this measurement called *molarity*. The formula for amount concentration is

$$c_{solute} = \frac{n_{solute}}{V_{solution}},$$

where c_{solute} is the amount concentration in moles per liter, n_{solute} is the number of moles of the solute, and $V_{solution}$ is the volume of the solution in liters. For example, a solution containing 6 mol of NaCl per liter of solution has an amount concentration of

$$\frac{6 \text{ mol NaCl}}{1 \text{ L water}} = \frac{6 \text{ mol}}{\text{L}} \text{ NaCl}.$$

You may hear this referred to as a "six molar sodium chloride solution," and it is often written as "6 M NaCl solution."

EXAMPLE 14-5: CALCULATING AMOUNT CONCENTRATION

What is the amount concentration of a solution that contains 3.40 mol of solute in 245 mL of solution?

Solution

Write what you know.

$$n = 3.40 \text{ mol}$$
$$V = 245 \text{ mL} = 0.245 \text{ L}$$
$$c_{solute} = ?$$

Write the formula and solve for the unknown.

$$c_{solute} = \frac{n_{solute}}{V_{solution}}$$

Plug in known values and evaluate.

$$c_{solute} = \frac{3.40 \text{ mol}}{0.245 \text{ L}}$$
$$= 13.9 \frac{\text{mol}}{\text{L}}$$

The amount concentration of the solution is 13.9 mol/L.

EXAMPLE 14-6: CALCULATING MOLES FROM AMOUNT CONCENTRATION

You have 0.615 L of a 0.50 M HCl solution. How many moles of HCl does this solution contain?

Solution

Write what you know.

$$V_{solution} = 0.615 \text{ L}$$
$$c_{HCl} = 0.50 \text{ mol/L}$$
$$n_{solute} = ?$$

Write the formula and solve for the unknown.

$$c_{HCl} = \frac{n_{solute}}{V_{solution}}$$

$$c_{HCl}V_{solution} = \frac{n_{solute}}{\cancel{V_{solution}}}\cancel{V_{solution}}$$

$$n_{solute} = c_{HCl}V_{solution}$$

Plug in known values and evaluate.

$$n_{solute} = \left(0.50 \frac{\text{mol}}{\cancel{\text{L}}}\right)(0.615 \cancel{\text{L}})$$
$$= 0.31 \text{ mol}$$

The solution contains 0.31 mol HCl.

EXAMPLE 14-7: CALCULATING VOLUME FROM AMOUNT CONCENTRATION

A chemical reaction requires 0.180 mol of silver nitrate ($AgNO_3$). How many milliliters of a 0.800 M solution should be added to the reaction vessel to provide this amount?

Solution

Write what you know.

$$n_{AgNO_3} = 0.180 \text{ mol}$$
$$c_{AgNO_3} = 0.800 \text{ mol/L}$$
$$V_{solution} = ?$$

Write the formula and solve for the unknown.

$$c_{AgNO_3} = \frac{n_{AgNO_3}}{V_{solution}}$$

$$c_{AgNO_3}V_{solution} = \frac{n_{AgNO_3}}{\cancel{V_{solution}}}\cancel{V_{solution}}$$

$$\frac{\cancel{c_{AgNO_3}}V_{solution}}{\cancel{c_{AgNO_3}}} = \frac{n_{AgNO_3}}{c_{AgNO_3}}$$

$$V_{solution} = \frac{n_{AgNO_3}}{c_{AgNO_3}}$$

Plug in known values and evaluate.

$$V_{solution} = \frac{0.180 \cancel{\text{mol}}}{0.800 \frac{\cancel{\text{mol}}}{\text{L}}}$$

$$= 0.225 \cancel{\text{L}}\left(\frac{1000 \text{ mL}}{1 \cancel{\text{L}}}\right) = 225 \text{ mL}$$

You will need to add 225 mL of the solution to the reaction vessel.

DILUTING SOLUTIONS

To prepare a solution with a specific amount concentration, chemists add the solute to a small amount of solvent and then dilute the solution up to its required volume. Special pieces of glassware called *volumetric flasks* are carefully calibrated to yield exact amounts of solutions.

We sometimes also have a need to make a solution with a particular amount concentration from a stock solution with a high amount concentration. Let's look at an example of this type of calculation. We can solve many dilution problems with the equation

$$c_1 V_1 = c_2 V_2,$$

where c is the amount concentration and V is the volume.

Volumetric flasks help chemists prepare solutions accurately.

EXAMPLE 14-8: CALCULATING VOLUME DURING DILUTION

You need 565 mL of 1.54 M HCl. In the stock room, you have a stock solution of 12 M HCl. What volumes of stock solution and water should you mix to produce the required solution?

Solution
Start by determining the volume of stock solution that will provide the necessary amount of solute.

Write what you know.

$V_{solution} = 565$ mL
$c_{solution} = 1.54$ mol/L
$c_{stock\ solution} = 12$ mol/L
$V_{stock\ solution} = ?$

Write the formula and solve for the unknown.

$$c_{solution} V_{solution} = c_{stock\ solution} V_{stock\ solution}$$

$$\frac{c_{solution} V_{solution}}{c_{stock\ solution}} = \frac{c_{stock\ solution} V_{stock\ solution}}{c_{stock\ solution}}$$

$$V_{stock\ solution} = \frac{c_{solution} V_{solution}}{c_{stock\ solution}}$$

Plug in known values and evaluate.

$$V_{stock\ solution} = \frac{\left(1.54 \frac{mol}{L}\right)(565\ mL)}{12 \frac{mol}{L}}$$

$$= 72.5\ mL$$

Now find the volume of water required to make 565 mL of solution.

Write the formula and solve for the unknown.

$$V_{solution} = V_{water} + V_{stock\ solution}$$

$$V_{water} = V_{solution} - V_{stock\ solution}$$

Plug in known values and evaluate.

$$V_{water} = 565\ mL - 72.5\ mL = 492.5\ mL$$

To make the necessary solution, you will need 72.5 mL of 12 M HCl and 492.5 mL H_2O.

Molality

A fourth measure of a solution's concentration is called **molality**, which is the number of moles of solute per kilogram of solvent. We use the formula

$$b_{solute} = \frac{n_{solute}}{m_{solvent}},$$

wheres b_{solute} is the molality of the solution in moles per kilogram, n_{solute} is the number of moles of the solute, and $m_{solvent}$ is the mass of the solvent in kilograms. For example, a solution containing 6 mol of HCl per kilogram of solvent has a molality of 6 mol/kg. You may see this written "6 m HCl solution," which is read as "six molal hydrochloric acid."

Molality has the advantage that it is not affected by changes in temperature and pressure as amount concentration is. You will see one application that uses molality when you study colligative properties in Section 14.3.

Solutions 341

SEM of sodium nitrate

EXAMPLE 14-9: MOLALITY CALCULATION

What is the molality of a solution composed of 31.2 g of sodium nitrate ($NaNO_3$) dissolved in 415 g of water?

Solution

Write what you know.

$m_{NaNO_3} = 31.2$ g $NaNO_3$
$m_{H_2O} = 415$ g $H_2O = 0.415$ kg H_2O
$n_{NaNO_3} = ?$
$b_{NaNO_3} = ?$

Convert mass sodium nitrate into moles.

$$31.2 \text{ g NaNO}_3 \left(\frac{1 \text{ mol NaNO}_3}{85.00 \text{ g NaNO}_3} \right) = 0.367 \text{ mol NaNO}_3$$

Write the formula and solve for the unknown.

$$b_{solute} = \frac{n_{solute}}{m_{solvent}}$$

Plug in known values and evaluate.

$$b_{NaNO_3} = \frac{0.367 \text{ mol NaNO}_3}{0.415 \text{ kg}}$$
$$= 0.884 \frac{\text{mol}}{\text{kg}} \text{ NaNO}_3$$

Your solution has a molality of 0.884 mol/kg $NaNO_3$.

14.2 SECTION REVIEW

1. Define *concentration*.

2. What is the mass fraction of a solution made by dissolving 6.3 g NaCl in 125.0 g H_2O?

3. A scientist is working with 460.0 mL of vinegar solution with an 8.5% volume fraction. How much vinegar does the solution contain?

4. What does amount concentration compare?

5. Solve the following solution concentration problems.

   a. What is the amount concentration if we dissolve 38.97 g of sodium nitrate ($NaNO_3$) in water to make 475 mL of solution?

   b. How many moles of sodium hydroxide (NaOH) need to be dissolved in 725 mL of solution to produce a 0.85 M NaOH solution?

   c. What mass of solute should you measure out to prepare 0.550 L of 0.761 M K_2SO_4 solution?

   d. What volume of solution will produce a 1.750 M KNO_3 solution using 47.23 g KNO_3?

6. A scientist needs to dilute a 9.75 M HCl stock solution so that he has 2.54 L of 1.75 M HCl solution. How much stock solution does he need to add to water to make his new solution?

7. Solve the following solution concentration problems.

   a. A scientist dissolves 4.51 moles of sodium chloride (NaCl) in 2.57 kg of water. What is the molality of the solution?

   b. A radiator contains a mixture of 3.25 kg of ethylene glycol ($C_2H_6O_2$) in 7.75 kg of water. Calculate the molality of this solution.

   c. While making jelly, a confectioner boils a sugar solution, removing water, until it is a 2.74 molal sucrose ($C_{12}H_{22}O_{11}$) solution with 4.55 kg $C_{12}H_{22}O_{11}$ in it. How much water remains?

   d. How many grams of potassium iodide are needed to make a 2.77 m KI solution using 1.75 kg of water?

14.3 COLLIGATIVE PROPERTIES

As solutes are introduced to solutions, those solutions begin to behave differently than their pure solvents do. Vapor pressure, boiling point, freezing point, and osmotic pressure all change as the number of particles in the solution increases. **Colligative properties** are those properties of solvents that change as the number of solute particles in the solution change.

QUESTIONS
» Do dissolved particles change the physical properties of solvents?
» Why do they put rock salt in ice cream makers?
» What is osmosis?

TERMS
colligative property • vapor pressure depression • boiling point elevation • freezing point depression • osmosis • osmotic pressure

Vapor Pressure Depression

How can the same substance be both an antifreeze and a coolant?

At the surface of a solution, solute particles fill positions that are normally occupied by solvent molecules. If the solute is less likely to become a gas—that is, it is less *volatile*—fewer of the more volatile solvent particles will have access to the surface. Since fewer solvent particles have a chance to evaporate, the vapor pressure of the solution is lower than the vapor pressure of the pure solvent. This lowering of vapor pressure due to the presence of solute particles in a solution is called **vapor pressure depression**.

If we add a mole of sugar to a liter of water, the vapor pressure of the solution will be lower than that of pure water. If we add a mole of sodium chloride to a liter of water, the vapor pressure will be even lower. This is because sodium chloride, an ionic compound, dissociates into Na^+ and Cl^- ions, while covalent sugar remains as an intact molecule when dissolved. The larger number of particles in the salt solution more effectively lowers the vapor pressure. If we dissolve aluminum chloride ($AlCl_3$) in a liter of water, four moles of ions for every one mole of aluminum chloride will be released, further decreasing the vapor pressure.

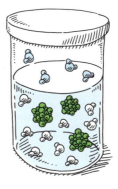 

The presence of the solute particles in the solution (left) causes a lower vapor pressure than in the pure solvent (right).

Boiling Point Elevation

Recall that boiling occurs when the vapor pressure of a liquid is equal to or exceeds the atmospheric pressure. Since the presence of solute particles lowers the vapor pressure, the vapor pressure at the expected boiling point will be lower than atmospheric pressure. Therefore, the solution will have to be heated more to raise the vapor pressure to atmospheric pressure. We call this **boiling point elevation** (ΔT_b)—the phenomenon that occurs when solute particles raise the boiling point of a solution.

A 1 molal (1 mol sugar/1 kg water) solution of sugar in water boils at 100.512 °C—an increase of 0.512 °C. The change in boiling temperature is the difference between the new boiling point and the original boiling point.

$$\Delta T_b = T_2 - T_1$$

For water, 1 mol of particles—molecules, atoms, or ions—in 1 kg of water elevates the boiling point of water 0.512 °C. This value is called the *molal boiling point constant* (K_b) for water. Other solvents have their own unique molal boiling point constants, as shown in Table 14-3.

As we dissolve more particles in a solution, the boiling point elevation increases. We calculate this value using the equation

$$\Delta T_b = i K_b b_{solute},$$

where ΔT_b is the change in boiling point in degrees Celsius, K_b is the molal boiling point constant for the solvent, and b_{solute} is the molality of the solution. The i in the formula is the van't Hoff factor, which accounts for the number of particles dissolved for each molecule or formula of the solute. In this textbook, we will assume that all molecules remain together and all ions dissociate completely. Therefore, for sugar, $i = 1$; for NaCl, $i = 2$; and for AgCl$_3$, $i = 4$.

A concentrated sugar solution boils at a temperature well above 100 °C.

▼

TABLE 14-3 *Boiling Point Elevations of Solvents*

Solvent	Normal Boiling Point (°C)	Molal Boiling Point Constant (K_b) (°C·kg/mol)
acetic acid	117.9	3.07
acetone	56.2	1.71
benzene	80.15	2.53
carbon tetrachloride	76.50	5.03
ethanol	78.26	1.22
ether	34.42	2.02
phenol	181.8	3.56
water	100.0	0.512

Let's look at an example of calculating boiling point elevation in a solution with a molecular solute.

EXAMPLE 14-10: CALCULATING BOILING POINT WITH A MOLECULAR SOLUTE

To make maple syrup, people boil maple sap (sugar water mixture) until the solution concentration is that of 1.34 kg of sugar dissolved in 0.500 kg of water. (Assume that the sugar is all sucrose, $C_{12}H_{22}O_{11}$). What is the final boiling point of this mixture?

Write what you know.

$m_{solute} = 1.34$ kg $C_{12}H_{22}O_{11}$
$i = 1$ (Sucrose is a molecular substance.)
$m_{solvent} = 0.500$ kg
$K_b = 0.512$ °C·kg/mol
$T_1 = 100.0$ °C
$n_{solute} = ?$
$b_{solute} = ?$
$\Delta T_b = ?$
$T_2 = ?$

Convert mass of sucrose to moles of sucrose.

$$1340 \text{ g } C_{12}H_{22}O_{11} \left(\frac{1 \text{ mol } C_{12}H_{22}O_{11}}{342.34 \text{ g } C_{12}H_{22}O_{11}} \right) = 3.914 \text{ mol } C_{12}H_{22}O_{11}$$

Determine the molality of the syrup.

Write the formula and solve for the unknown.

$$b_{solute} = \frac{n_{solute}}{m_{solvent}}$$

Plug in known values and evaluate.

$$b_{solute} = \frac{3.914 \text{ mol } C_{12}H_{22}O_{11}}{0.500 \text{ kg}}$$

$$= 7.828 \frac{\text{mol}}{\text{kg}} C_{12}H_{22}O_{11}$$

Determine the boiling point elevation.

Write the formula and solve for the unknown.

$$\Delta T_b = i K_b b_{solute}$$

Plug in known values and evaluate.

$$\Delta T_b = (1)\left(0.512 \frac{°C \cdot kg}{mol}\right)\left(7.828 \frac{mol}{kg}\right)$$

$$= 4.007 \text{ °C}$$

Determine the new boiling point.

Write the formula and solve for the unknown.

$$\Delta T_b = T_2 - T_1$$
$$T_2 = T_1 + \Delta T_b$$

Plug in known values and evaluate.

$$T_2 = 100.0 \text{ °C} + 4.007 \text{ °C}$$
$$= 104.0 \text{ °C}$$

The final temperature is 104.0 °C

The calculation of boiling point elevation in a solution with an ionic solute is identical except for the van't Hoff factor. For an ionic solute, we have to identify the number of ions that each formula unit contributes.

EXAMPLE 14-11: CALCULATING MOLALITY FROM BOILING POINT ELEVATION

What is the molality of a solution of NaCl in water if the solution has a boiling point that is 0.400 °C higher than water?

Write what you know.

$\Delta T_b = 0.400$ °C
$i = 2$ (NaCl is an ionic substance.)
$K_b = 0.512$ °C·kg/mol
$b_{solute} = ?$

Write the formula and solve for the unknown.

$\Delta T_b = i K_b b_{solute}$

$$\frac{\Delta T_b}{i K_b} = \frac{\cancel{i K_b} b_{solute}}{\cancel{i K_b}}$$

$$b_{solute} = \frac{\Delta T_b}{i K_b}$$

Plug in known values and evaluate.

$$b_{solute} = \frac{0.400 \cancel{°C}}{(2)\left(0.512 \frac{\cancel{°C} \cdot kg}{mol}\right)}$$

$$= 0.391 \frac{mol}{kg} \text{ NaCl}$$

This is a 0.391 m NaCl solution.

Solutions 345

TABLE 14-4 *Freezing Point Depressions of Solvents*

Solvent	Normal Freezing Point (°C)	Molal Freezing Point Constant (K_f) (°C·kg/mol)
acetic acid	16.6	−3.90
benzene	5.48	−5.12
ether	−116.3	−1.79
phenol	40.9	−7.40
water	0.00	−1.86

Freezing Point Depression

While dissolving solute in a solution will raise the boiling point, the same solute particles will cause solutions to freeze at temperatures lower than that of their pure solvents. A 1 molal solution of sugar in water freezes at −1.86 °C, which is 1.86 °C lower than the normal freezing point of water. This change in freezing point is called the **freezing point depression** (ΔT_f). This quantity is calculated similarly to boiling point elevation.

$$\Delta T_f = T_2 - T_1$$

As with changes to the boiling point, each solvent has its own *molal freezing point constant* (K_f), which allows us to calculate freezing point depression with the equation

$$\Delta T_f = iK_f b_{solute},$$

where ΔT_f is the change in freezing point in degrees Celsius, i is the van't Hoff factor, K_f is the molal freezing point constant for the solvent, and b_{solute} is the molality of the solution.

One application of freezing point depression is the addition of antifreeze to a car's engine cooling system. Antifreeze is a 25% (by volume) solution of ethylene glycol that depresses the freezing point by about 12 °C. As a dissolved solute, it also raises the boiling point. So the antifreeze in a car's radiator prevents the water in the radiator from freezing in the winter and also prevents it from vaporizing in the summer, which is why it is also considered a coolant.

Another cold weather application of freezing point depression occurs in regions that deal with snow and ice on roads. The salts sodium chloride and calcium chloride are often used to lower the freezing point of water and prevent ice from forming on roads. Calcium chloride ($CaCl_2$) lowers the freezing point of water more than sodium chloride (NaCl) because it dissociates into more ions than sodium chloride does. A mole of calcium chloride dissociates into three, rather than two, moles of ions. The effective concentration of the solution is tripled, rather than doubled, so there is a greater depression in the freezing point. This temperature depression prevents the water from freezing until the water is cooled to lower temperatures. Solid calcium chloride also has the added benefit of actively absorbing water from the air to dissolve itself, thus increasing the rate at which it dissolves and lowers the freezing point.

You might think that calcium chloride would be the natural choice for deicing all roads, but there is a catch. Sodium chloride in its raw form as rock salt is much cheaper than calcium chloride and is often effective enough to get the job done. In locations with heavy precipitation or extremely low temperatures, calcium chloride may be used alone or in combination with sodium chloride.

EXAMPLE 14-12: FREEZING POINT DEPRESSION FOR IONIC COMPOUNDS

Compare the freezing points of aqueous solutions of 3.00 m $CaCl_2$ and 3.00 m NaCl.

Write what you know.

b_{CaCl_2} = 3.00 mol/L $CaCl_2$
b_{NaCl} = 3.00 mol/L NaCl
i_{CaCl_2} = 3
i_{NaCl} = 2
K_f = −1.86 °C · kg/mol (water from Table 14-4)
T_1 = 0.00 °C (water from Table 14-4)
$\Delta T_{b\ CaCl_2}$ = ?
$T_{2\ CaCl_2}$ = ?
$\Delta T_{b\ NaCl}$ = ?
$T_{2\ NaCl}$ = ?

Determine the boiling point elevation.

Write the formula and solve for the unknown.

$\Delta T_b = iK_b b_{solute}$

Plug in known values and evaluate.

$\Delta T_{b\ CaCl_2} = (3)\left(-1.86\ \frac{°C \cdot kg}{mol}\right)\left(3.00\ \frac{mol}{kg}\right)$
$= -16.\underline{7}4\ °C$

$\Delta T_{b\ NaCl} = (2)\left(-1.86\ \frac{°C \cdot kg}{mol}\right)\left(3.00\ \frac{mol}{kg}\right)$
$= -11.\underline{1}6\ °C$

Determine the new boiling point.

Write the formula and solve for the unknown.

$\Delta T_b = T_2 - T_1$
$T_2 = T_1 + \Delta T_b$

Plug in known values and evaluate.

$T_{2\ CaCl_2} = 0.00\ °C + (-16.\underline{7}4\ °C) = -16.7\ °C$
$T_{2\ NaCl} = 0.00\ °C + (-11.\underline{1}6\ °C) = -11.2\ °C$

The calcium chloride will lower the freezing point to −16.7 °C while the sodium chloride will lower it only to −11.2 °C.

Molecular substances will also lower freezing points. The calculation of freezing point when using a molecular solute is identical to the above example except that molecular solutes will have a van't Hoff factor of 1. We have this scenario when adding antifreeze to a car's cooling system.

Increased Osmotic Pressure

Peel a fresh potato and slice it into strips. Add these immediately to a strong sugar solution. What happens? The semipermeable membrane that surrounds each cell in the potato pieces allows the small water molecules within to pass through it and enter the sugar solution. But the membrane prevents the larger sugar molecules from entering into the cells. Only a few water molecules leave the sugar solution to enter the cells. Many more enter the solution from the water within the cells. As a result, the potato slices will shrink as they lose water. If we wait about thirty minutes and then remove the potato slices, pat them dry, and put them on a balance, we can determine how much water they have lost. The change in mass is due to **osmosis**—the movement of a solvent across a semipermeable membrane due to differences in the solution concentration on each side of the membrane. You may remember osmosis from biology, where living organisms move water in and out of cells through osmosis. In the example of the potato, the solvent is water and the semipermeable membrane is the cell membrane. Osmosis occurs as the water molecules attempt to equalize the concentration on each side of the membrane.

The **osmotic pressure** of a solution is the amount of pressure required to prevent osmosis from occurring. The more solute particles there are in a solution, the greater the osmotic pressure. The osmotic pressure of a solution is a colligative property because it depends on the number of particles in solution.

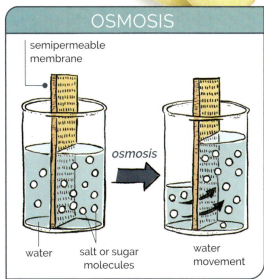

Reverse Osmosis

Osmosis is the movement of a solvent across a semipermeable membrane due to differences in solution concentration. In this section, you learned that the presence of solute particles increases the osmotic pressure—a colligative property. Scientists have discovered a way to use osmosis to address the issue of clean water around the world.

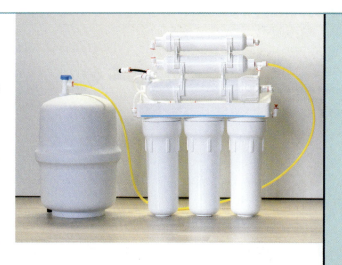

HOW IT WORKS

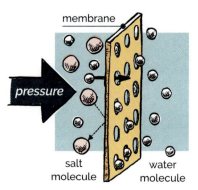

The process to purify water is called *reverse osmosis* (RO). In RO, the solvent is still moving across the membrane, but it is moving *away* from the high concentration solution rather than *toward* it—the opposite, or reverse direction, of what normally happens. In an RO system, the untreated water enters the system under high pressure. This pressure must be higher than the osmotic pressure, and the higher the pressure, the more effective the system. Under this high pressure, the solvent—pure water—leaves the solution and is collected on the "clean" side of the membrane. Depending on the intended use of the water, there may be additional sanitizing steps needed.

RO has many applications, with the most common being desalination to obtain drinking water from seawater. Other applications include drinking water purification, food production, military uses, and aquariums. RO has the advantage of having low energy costs compared with purification processes that involve high temperatures. Current research in RO technology is looking into using nanotechnology in the membrane.

14.3 SECTION REVIEW

1. What is a colligative property?
2. Explain how the presence of solute particles lowers the vapor pressure.
3. If we add one mole each of KCl, NaCl, Na_3PO_4, and $PbSO_4$ to equal amounts of water, which substance causes the greatest decrease in vapor pressure? Explain.
4. What is the boiling point of an aqueous solution of 75.5 g of glucose ($C_6H_{12}O_6$) dissolved in 150 g of water?
5. What is the freezing point of an aqueous solution of 75.5 g of glucose ($C_6H_{12}O_6$) dissolved in 150 g of water?
6. Which measure of concentration do we use in boiling point elevation and freezing point depression calculations? Explain.
7. What is osmosis?
8. Explain why
    a. red blood cells that are placed in pure water absorb water until they burst but do not absorb water when surrounded by blood plasma.
    b. red blood cells shrink when they are in a very concentrated salt solution.

14.4 SUSPENSIONS AND COLLOIDS

Up to this point, we have spent our time discussing solutions—homogeneous mixtures with a single phase. What about heterogeneous mixtures, which have more than one phase? Such mixtures can be further categorized on the basis of their particle sizes and are called either *suspensions* or *colloids*.

How are eggs and Jell-O® related?

QUESTIONS
» What is a colloid?
» What is the difference between solutions, suspensions, and colloids?
» What common substances are colloids?

TERMS
suspension • colloid • dispersed phase • dispersing medium • Tyndall effect • Brownian motion

Suspensions

Solutions are homogeneous mixtures because the dissolved particles are so small that we cannot see them. These particles are typically smaller than 1 nm. **Suspensions**, on the other hand, contain larger particles—greater than 1 μm. These particles are so large that they will eventually settle out of a mixture. This means that a suspension is a heterogeneous mixture. Two common suspensions are aerosols, which consist of a liquid suspended in a gas, and dust—solid particles suspended in a gas. If you have ever dusted the furniture, you know that eventually more dust will settle out of the air.

If solutions have dissolved particles of less than a nanometer and suspensions contain particles larger than a micrometer, what do we call a mixture with particles between a nanometer and a micrometer?

Colloids

A **colloid** is a heterogeneous mixture that contains intermediate-sized particles dispersed in a medium. The particles are sometimes called the **dispersed phase**; the medium is called the **dispersing medium**. A very familiar colloid is clouds, which consist of water droplets (dispersed phase) in the air (dispersing medium). The size of particles in a colloid cannot exceed 1 μm since particles any larger quickly settle out from their medium. Intermolecular collisions buffet particles smaller than 1 μm with sufficient force to counteract the constant tug of gravity. Since colloids are different from solutions, terms such as *solute*, *dissolved*, and *solvent* do not apply.

TABLE 14-5 *Particle Size Comparison*

Mixture	Particle Size
solution	< 1 nm
colloid	> 1 nm and < 1 μm
suspension	> 1 μm

TABLE 14-6 *Types of Colloids*

Dispersing Medium	Dispersed Phase	Common Name	Examples
Gas	gas		does not exist
	liquid	liquid aerosol	fog, clouds, mist
	solid	solid aerosol	smoke, dust in the air
Liquid	gas	foam	shaving cream, whipped cream
	liquid	emulsion	hand cream, mayonnaise (oil dispersed in water)
	solid	sol	paint, blood
Solid	gas	solid foam	plastic foam, marshmallows
	liquid	gel	gelatin, jelly
	solid	solid sol	cranberry glass (glass with dispersed metal)

In a colloid, both the dispersing medium and dispersed phase can be in any of eight possible combinations of states. Mixtures of gases with gases are the exception—they are always solutions. Table 14-6 summarizes the possible combinations of states for colloids.

PROPERTIES OF Colloids

In 1869, an Irish physicist named John Tyndall demonstrated that the particles in colloids were large enough to scatter light waves. This **Tyndall effect** is often used to distinguish colloids from solutions. A beam of light will pass through a solution without being seen because dissolved particles are too small to disturb light waves. The outline of a light beam will not show up in air because air is a solution. But when a beam of light goes through a colloid, the beam's outline shows up distinctly. The perfect illustration of the Tyndall effect is seen when automobile headlights shine through the air into a patch of fog (a colloid).

Tyndall effect. Can you tell which cylinder contains a colloid and which one contains a solution?

John Tyndall

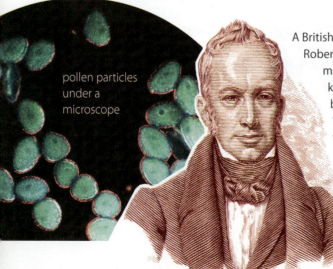

pollen particles under a microscope

A British botanist named Robert Brown discovered major proof for the kinetic-molecular theory by observing a colloid. In 1827, he watched pollen particles dispersed in water. He saw that the pollen particles moved slightly as if many small collisions were jostling them. This type of movement, called **Brownian motion**, results from colloidal particles colliding with fluid particles. This observation led scientists to accept the particle model of matter.

Robert Brown

Solving Problems with Solution Properties

In the Chapter opener, we discussed the challenge of providing clean drinking water to people around the world. Even if we address the issue of the proximity of people to water, we still haven't solved the problem of impurities in the water. Between bacteria, viruses, pollutants, and minerals in the water, we have a lot of work to do to provide clean drinking water to people. The good news is that we can address many of these issues with the chemistry concepts that we have discussed in this chapter.

Practical water treatment involves three steps. First, pathogens are killed by chlorinating the water with chlorine dioxide or a similar compound, forming a dilute solution. Then other compounds are added to clump impurities together, causing *coagulation*. This step, also called *flocculation*, clumps the impurities. This transitions the water from a colloid to a suspension. The suspended particles settle out over time. To speed the process, the water can also be passed through a filtration system. Finally, the water passes through a filter under pressure to quickly remove impurities. Even after all these processes, some toxins remain in the water as they are too small to be removed. As a result, people in some locations still have to boil their water before using it.

CAREERS

SERVING AS AN ENVIRONMENTAL SCIENTIST: *WATER WATCHER*

The Chapter opener highlights the challenge of providing drinking water to people around the world. While proximity to water is an issue, a much bigger problem are the pollutants that are in the water that may be available. This is a matter of particular interest for many environmental scientists.

Environmental science takes a multidisciplinary approach to solving problems. Environmental scientists need to incorporate knowledge and skills applicable to biology, chemistry, physics, and many other fields. Dr. Christopher Higgins from the Colorado School of Mines is an environmental engineer. Much of his work centers on what happens to organic compounds in wastewater and reclaimed water. He researches how those contaminants interact with biological systems.

Environmental science is an excellent career for someone interested in many fields of science. It is also a career in which you have tremendous opportunities to help others.

14.4 SECTION REVIEW

1. What terms used for colloids correspond to the terms *solvent* and *solute* used for solutions?

2. Compare solutions, colloids, and suspensions.

3. Give an example of a colloid for each of the following combinations of dispersing medium and dispersed phase.
    a. liquid; gas
    b. gas; liquid
    c. solid; solid

4. How can we determine whether a substance is a solution or a colloid? Why does this method of determination work?

5. What is the relationship between Brownian motion and colloids?

6. Why can't contaminants like salt be filtered out of contaminated water?

14 CHAPTER REVIEW

Chapter Summary

TERMS
solution
solvent
solute
miscibility
solvation
dissociation
ionization
enthalpy of solution
solubility
unsaturated solution
saturated solution
supersaturated solution
Henry's law

14.1 THE DISSOLVING PROCESS

- Solutions are homogeneous mixtures of variable composition in a single phase. The majority substance is the solvent, and it contains one or more solutes dissolved in it.

- Substances able to dissolve in a given solvent are soluble. Mutually soluble liquids are miscible.

- Solvation is the dissolving process of solids in liquids. First, solvent and solute particles have to be separated from like particles, an endothermic process. Next, the solvent and solute particles attract each other and release energy, an exothermic process. The enthalpy of solution is the change in energy.

- Solubility is the maximum amount of a solute that can dissolve in a specific solvent under specific conditions, such as temperature and pressure.

- A solution is saturated when it can dissolve no more solute at its current temperature and pressure.

- An increase in the temperature of a liquid solvent generally increases the solubility of solid solutes but decreases the solubility of gas solutes.

- An increase in pressure has little effect on the solubility of solids and liquids but increases the solubility of gases.

14.2 MEASURES OF CONCENTRATION

- Solution concentration is a comparison of the amount of solute with the amount of solution or solvent.

- The two most useful measures of concentration are amount concentration (moles of solute per liter of solution) and molality (moles of solute per kilogram of solvent).

TERMS
concentration
mass fraction
volume fraction
amount concentration
molality

14.3 COLLIGATIVE PROPERTIES

TERMS
colligative property
vapor pressure depression
boiling point elevation
freezing point depression
osmosis
osmotic pressure

- Colligative properties are physical properties of a solution that are related to the number of particles in solution.

- Colligative properties include vapor pressure depression, boiling point elevation, freezing point depression, and osmotic pressure.

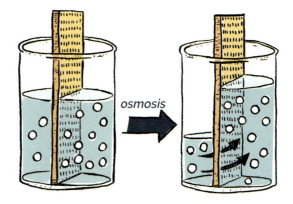

14.4 SUSPENSIONS AND COLLOIDS

TERMS
suspension
colloid
dispersed phase
dispersing medium
Tyndall effect
Brownian motion

- The main factor that determines whether substances will form a suspension, colloid, or solution is particle size.

- Suspensions have large particles that will eventually settle out.

- Colloids have particles of intermediate size—too small to settle out but too large to be considered dissolved.

- The Tyndall effect is the ability of colloidal particles to scatter light.

Chapter Review Questions

RECALLING FACTS

1. Define *solution*.
2. List the five properties of solutions.
3. What is the formula for calculating mass fraction?
4. What does volume fraction compare?
5. How does molality differ from the other measurements of concentration described in this chapter? How is this beneficial?
6. Which is the most common measure of concentration used in chemistry? Which is the least common?
7. How does a solute affect the colligative properties of a solution?
8. Define *freezing point depression*.
9. What is a suspension?
10. Define *colloid*.
11. What determines whether a mixture is a solution, colloid, or suspension?
12. Give an example of a colloid for each of the following combinations of dispersing medium and dispersed phase.
    a. liquid; liquid
    b. gas; solid
    c. solid; gas
13. What is the Tyndall effect? How is it used?

14 CHAPTER REVIEW

UNDERSTANDING CONCEPTS

14. Identify the solvent and solute(s) in each of the following solutions.

    a. salt water

    b. carbonated water

    c. air

    d. carbon steel (1% Mn, 0.9% C, 98.1% Fe)

15. How are dissociation and ionization different?

16. Explain changes in energy when a substance is dissolving.

17. Explain why water will dissolve lithium chloride but not carbon tetrachloride (CCl_4).

18. Create a graphic organizer for solutions. At a minimum, use the following terms: *concentrated solution, concentration, dilute solution, heterogeneous mixture, homogeneous mixture, mixture, saturated solution, solubility, solute, solution, solvent, supersaturated solution, unsaturated solution.*

19. Use the solubility curves on page 336 to answer the following questions.

    a. At what temperature will the solubility of KNO_3 be 65 g?

    b. What is the solubility of $KClO_3$ at 35 °C?

    c. If you dissolve 40 g of KCl in 100 mL of water at 50 °C, will the solution be supersaturated, saturated, or unsaturated? Explain.

20. A solute with a mass of 135.5 g needs to be dissolved in water to make a solution with a 3.55% mass fraction. How much water is needed?

21. A lab technician is mixing 1.754 L of solvent to make a solution with a 5.593% volume fraction. How much solute is needed?

22. An aqueous solution of anhydrous sodium tetraborate ($Na_2B_4O_7$) is sometimes used to fireproof wood.

    a. What is the amount concentration of 2.50 L of a solution that contains 1.85 mol $Na_2B_4O_7$?

    b. What is the amount concentration of 45.0 L of a solution that contains 6.78 kg $Na_2B_4O_7$?

    c. How many moles of $Na_2B_4O_7$ are in 600.0 mL of a 1.57 M $Na_2B_4O_7$ solution?

    d. A chemist needs 50.85 g $Na_2B_4O_7$ for a reaction. How many milliliters of a 1.870 M solution does she need?

23. For an experiment, you need 1.250 L of 0.300 M H_2SO_4 solution. If your stock solution is 17.4 M H_2SO_4, how much stock solution and water do you need to prepare the solution?

24. Disodium phosphorofluoridate (Na_2PO_3F) is a common source of fluoride in some modern toothpastes.

    a. What is the molality of a solution containing 2.75 mol Na_2PO_3F in 5.09 kg H_2O?

    b. How many moles of Na_2PO_3F would be contained in 125.0 g of H_2O if the solution were 0.0198 molal?

    c. What mass of disodium phosphorofluoridate is needed to make a 0.0223 m Na_2PO_3F solution with 500.0 kg of water?

25. From each of the following sets, choose the solution that has the lowest vapor pressure. Explain.

    a. 1.8 m CH_3OH b. 0.5 m Na_3PO_4

       0.7 m CH_3OH 0.5 m $MgCl_2$

       2.9 m CH_3OH 0.5 m NaCl

       0.2 m CH_3OH 0.5 m $Al_2(SO_4)_3$

26. How does the lowering of the vapor pressure due to dissolved particles explain the elevation of the boiling point in a solution?

27. A chef adds salt to the water when cooking eggs because she believes that the eggs will cook faster.

    a. Is this an accurate statement? Explain.

    b. What is the boiling point of the salt water if she dissolves 10.0 g NaCl in 1.00 L of pure water?

    c. How many grams of NaCl must be added to 1.00 L of water to raise its boiling point to 105.0 °C?

    d. In light of the answers to Questions 27b and 27c, do you think that adding salt to water is a convenient method for decreasing the cooking time of eggs?

28. Automobile coolant additive consists primarily of ethylene glycol ($C_2H_6O_2$). Manufacturers also add small amounts of dye and anticorrosion substances.

    a. The coldest temperature of the year in Augusta, Maine, is expected to be −26.0 °C. What must the molality of a solution of water (solvent) and ethylene glycol (solute) be for the radiator fluid to prevent the water from freezing?

    b. How many grams of ethylene glycol must be added to 4.00 L of water to attain this molality?

c. How many liters of antifreeze is this? Assume that the density of antifreeze is 1.113 g/mL and ignore the fact that there are small amounts of other substances present.

d. Why is it important to have a solution of coolant in a car's radiator during the summer?

e. What is the boiling point for this solution?

29. Describe two ways that colligative properties are used to make life better.

30. What is the only combination of phases that cannot form a colloid? Why?

31. Why do you think that water has the potential to be so contaminated?

CRITICAL THINKING

32. The Chapter opener focuses on the scarcity of drinking water for many around the world. Some Christians may minimize these kinds of concerns. How would you respond to the following statement?

"The poor people of Africa and Asia don't need drinking water nearly as much as they need the gospel. Christians should not put time and resources into water purification. They should put all their efforts into sending missionaries who can preach the Bible and plant churches."

33. The chief engineer at a soft drink bottling plant wants to increase the amount of carbonation in soda. What can he do to increase the amount of gas dissolved in soda?

34. Suppose that all the water on our planet had to be replaced by another liquid. List the physical and chemical properties that the new compound would need.

35. A 9.168 M aqueous solution of H_2SO_4 has a density of 1.4987 g/mL at 20 °C. How many grams of H_2SO_4 are in 50.00 g of this solution?

36. You mix 2.45 L of 6.51 M HCl solution with 4.30 L of 3.75 M HCl solution. What is the molarity of your new mixture?

37. An aqueous solution of an unknown molecular compound boils at 101.00 °C.

   a. What is the molality of the solution?

   b. What is the freezing point of the solution?

   c. What would the molar mass of the unknown compound be if we added 7.80 g of solute to 100.0 g of water to make the solution?

Use the ethics box below to answer Question 38.

38. Use the process for ethical decision making modeled in Chapter 3 to write a four-paragraph response on how to balance treatment cost, water usage, and discharge pollution.

ethics

WASTEWATER MANAGEMENT

You have probably noticed that water treatment is an important topic. The method used to treat wastewater is a multistep process, with each step adding cost. Most municipal water treatment plants balance the cost of treatment with the intended use of the water discharged from the plant. The water treatment plants in some cities release water with high levels of bacteria because the system discharges water to a stream or to the ocean. They reason that they are releasing the water to the environment, so there's no need to clean it any more than necessary. The downside to such thinking is that the body of water may become unusable for recreational uses like fishing or swimming.

Chapter 15

THERMO-CHEMISTRY

Order is seen everywhere in our universe. On this point everyone can agree. Even natural processes governed by physical laws can produce surprising beauty—snow crystals, ripples in a sand dune, or the turbulence in a whirlpool. But beyond these purely physical phenomena, there is something else in our world that we should not expect to find if it were created by naturalistic processes alone—information. Within every living thing, nucleic acids store unimaginable amounts of information. The mathematics of probability tells us that the chance of all this information incrementally assembling itself is essentially zero. An equally huge unlikelihood must be accounted for when considering naturalistic theories of origins, namely the origin of information itself. A theory of chemical evolution by natural means requires that self-replicating molecules arise at some point in time, by chance alone, from nonliving chemicals. Is such an event even possible? And does the study of chemistry shed any light on the question?

15.1 Thermodynamics and Phase Changes *357*

15.2 Thermodynamics and Chemical Changes *365*

15.3 Reaction Tendency *371*

15.1 THERMODYNAMICS AND PHASE CHANGES

The formation of a strand of DNA requires many chemical reactions. Can such reactions produce ever-increasing order in living things? In Chapter 2, we introduced the laws of thermodynamics. Those laws describe energy changes in matter. In summary, they state that matter and energy can be neither created nor destroyed but can be converted from one form to another. During these conversions, some energy becomes unusable. **Thermochemistry**, the branch of science that studies the transfer of energy during phase or chemical changes, is an important application of these laws. As you will see, thermochemistry has something to say about the order that we see in our universe, including in DNA molecules.

Measuring Heat and Temperature

In many investigations scientists are interested in thermal energy, the total kinetic energy of the particles within a substance. But we can't directly measure thermal energy, so scientists rely on measurements of two different properties: temperature and heat. Remember that temperature, measured in degrees Celsius or kelvins, expresses the average kinetic energy in the particles of a sample. Heat, on the other hand, measures the amount of thermal energy transferred from one substance to another. Heat is typically expressed in joules (J). The joule is the standard SI unit of work and can be used to measure any kind of energy.

Because it is impossible to determine the absolute amount of thermal energy in a substance, chemists measure the change in energy that occurs during a chemical reaction. This measurement is more useful to them anyway. Scientists have defined a property of matter called enthalpy to help them quantify the change in energy. **Enthalpy** (H) is the thermal energy content of a system at a constant pressure. For the most part, chemical reactions in nature or in the laboratory occur at a constant pressure.

Changes in enthalpy can be measured in a reaction vessel called a calorimeter. A **calorimeter** is an insulated container, similar to a thermos, in which a thermometer detects the temperature change that occurs during a chemical reaction. In one common kind of calorimeter, a chemical reaction occurs in a sealed reaction chamber submerged in a measured quantity of water. The change in the water temperature indirectly indicates how much energy the reaction has released or absorbed.

In this calorimeter, the enthalpy of the reaction is captured by the water and calculated from the temperature change.

QUESTIONS
» What is thermochemistry?
» Is there a difference between thermal energy and heat?
» Do all phase changes require the same amount of added energy?

TERMS
thermochemistry • enthalpy • calorimeter • sensible heat • latent heat • molar enthalpy of fusion • molar enthalpy of vaporization • specific heat

Does the temperature of water change as it freezes?

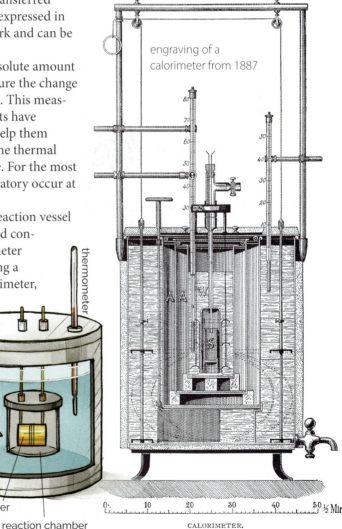

engraving of a calorimeter from 1887

stirrer
igniter
reaction chamber
thermometer

CALORIMETER.

Thermochemistry 357

ENTHALPY
of Phase Changes

The thermal energy transferred between substances can be divided into two categories. **Sensible heat** is the transfer of thermal energy that produces a temperature change in a substance. We can sense, or detect, this type of heat by a change in temperature.

A second category, called **latent heat**, is the thermal energy that produces a phase change when transferred to a substance. No temperature change is observed. Examine the heating curve below to see where sensible and latent heat affect water.

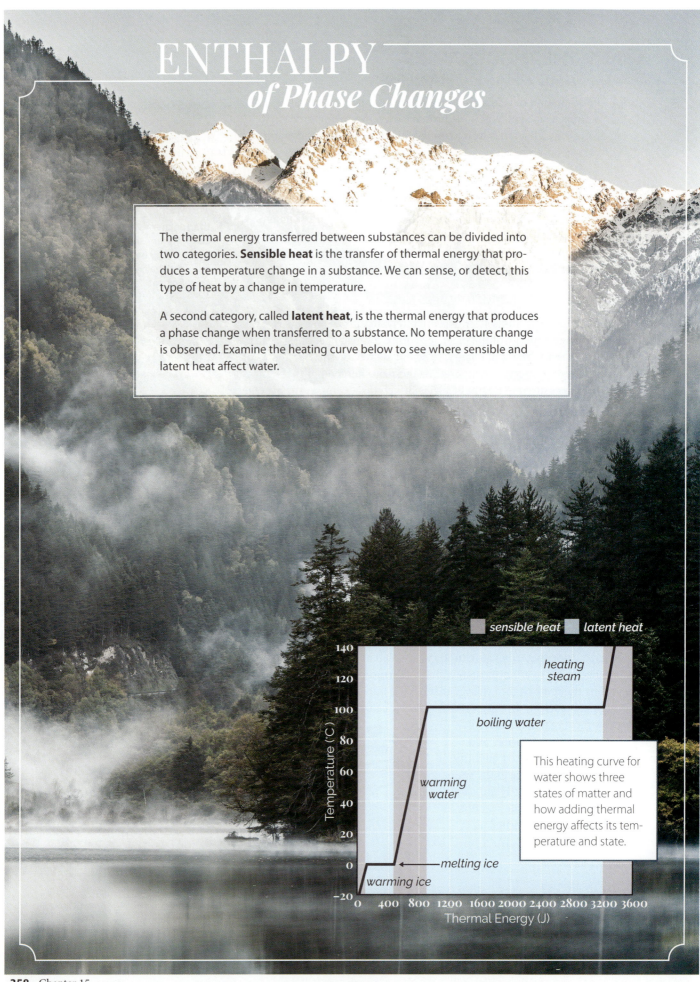

This heating curve for water shows three states of matter and how adding thermal energy affects its temperature and state.

Changing ice into liquid water involves both sensible and latent heat. Ice, at a temperature that is below its melting point, can be heated until it reaches a temperature of 0 °C (sensible heat). Any thermal energy added at this point does not raise the temperature. Instead, it serves to supply the energy needed to overcome the intermolecular forces (latent heat). Not until the last piece of ice has melted will the temperature of the water rise above 0 °C.

TABLE 15-1
Molar Enthalpies of Fusion and Vaporization

Substance	Formula	ΔH_{fus} (kJ/mol)	ΔH_{vap} (kJ/mol)
ammonia	NH_3	5.66	23.3
ethanol	C_2H_5OH	4.94	38.6
methanol	CH_3OH	3.17	35.2
water	H_2O	6.01	40.7

The amount of energy required to convert a mole of a liquid at its boiling point to its vapor at the same temperature is called the *heat of vaporization*, or the **molar enthalpy of vaporization** (ΔH_{vap}). The reverse of the enthalpy of vaporization is called the *enthalpy of condensation*; this quantity is equal in magnitude but opposite in sign to the enthalpy of vaporization.

A significant amount of energy must be added to a solid at its melting point to overcome intermolecular forces, and that amount differs for each substance. The thermal energy required is called the *heat of fusion*, or the **molar enthalpy of fusion** (ΔH_{fus}). It is defined as the quantity of energy required to melt one mole of a solid to a liquid with no temperature change. The same amount of energy is released when one mole of the liquid freezes.

Enthalpies of vaporization differ widely because the strength of intermolecular attractions in different liquids varies. For instance, it takes 40.7 kJ to vaporize 1 mol of water at 100 °C and 101.325 kPa, while a mole of ammonia at its boiling point can be vaporized by adding 23.3 kJ of energy.

EXAMPLE 15-1: CALCULATING LATENT HEAT

How much energy would be required to melt 3.54 kg of ice at 0 °C to water without any temperature change?

Write what you know.

$m = 3.54$ kg
$\Delta H_{fus} = 6.01$ kJ/mol

Convert mass of water to energy required.

$$3.54 \text{ kg } H_2O \left(\frac{1000 \text{ g } H_2O}{1 \text{ kg } H_2O} \right) \left(\frac{1 \text{ mol } H_2O}{18.02 \text{ g } H_2O} \right) \left(\frac{6.01 \text{ kJ}}{1 \text{ mol } H_2O} \right) = 1.18 \times 10^3 \text{ kJ}$$

You would need to add 1.18×10^3 kJ of energy to melt the ice.

Specific Heat

Have you ever noticed that an aluminum pan on a stove heats up faster than the water inside it? Why does this happen? You might explain this by observing that the pan receives the heat before the water does, but there is a better explanation.

Not all substances heat at the same rate or change temperature by the same amount under the same conditions. The amount of thermal energy that raises 1 g of water by 1 °C will raise the temperature of 1 g of aluminum by approximately 5 °C. Given the same amount of energy, the aluminum will have a greater temperature change. The **specific heat** (c_{sp}) of a substance is the amount of thermal energy required to raise the temperature of 1 g of the substance by 1 °C. The usual unit for this quantity is joules per gram kelvin (J/g · K). Degrees Celsius may be substituted for kelvins since the degree sizes are equal. For example, 1 g of water can be raised 1 °C with the input of 4.18 J of energy. Thus, the specific heat of water is 4.18 J/g ·°C.

Specific heats of some common substances are listed in Table 15-2. Substances with high specific heats require large amounts of energy for a given temperature change and will change temperature only slowly. Substances with low specific heats require comparatively small amounts of energy for a given temperature change and will change temperature rapidly under the same conditions. The thermal energy input can be found with the equation

$$Q = c_{sp} m \Delta T,$$

where Q is the thermal energy in joules, c_{sp} is its specific heat in joules per gram degree Celsius, m is the mass of the substance in grams, and ΔT is its change in temperature in degrees Celsius.

TABLE 15-2 *Specific Heats*

Substance	Specific Heat (J/g • °C)
lead	0.13
mercury	0.14
silver	0.23
copper	0.38
iron	0.45
chlorine	0.48
carbon (diamond)	0.51
aluminum	0.90
oxygen	0.92
benzene	1.73
steam	2.00
acetic acid	2.05
ice	2.11
ethanol	2.53
water	4.18

Rates of cooling are also related to specific heat. Substances with high specific heat values retain heat for longer periods than those with low specific heats. For example, the dry crust of a meat turnover may be slightly warm while the filling is still hot enough to scorch your tongue. Like enthalpies of vaporization, specific heats serve as rough indicators of intermolecular attractions. Particles that are strongly attracted to each other require extra thermal energy to increase their kinetic energies. Using the equation above for a cooling process, that is, releasing heat, would produce a negative change in temperature and a negative value for Q.

The heating curve of a substance reveals its relative specific heat. If temperature rises quickly with the addition of thermal energy, the specific heat must be low. In such a case, a small amount of heat results in a large change in temperature. A substance that has a gentle slope on its heating curve has a high specific heat because added energy does not change the temperature very much.

EXAMPLE 15-2: CALCULATING THERMAL ENERGY

A 28.35 g sample of silver is heated from 15.2 °C to 85.6 °C. How many joules were added to the sample?

Write what you know.

$c_{sp\ silver}$ = 0.23 J/g · °C (from Table 15-2)
m = 28.35 g
T_i = 15.2 °C
T_f = 85.6 °C
Q = ?

Write the formula and solve for the unknown.

$$Q = c_{sp}m\Delta T$$
$$= c_{sp}m(T_f - T_i)$$

Plug in known values and evaluate.

$$Q = \left(\frac{0.23\ J}{g \cdot °C}\right)(28.35\ g)(85.6\ °C - 15.2\ °C)$$
$$= \left(\frac{0.23\ J}{°C}\right)(28.35)(70.4\ °C)$$
$$= 459\ J$$

The given change in temperature requires the addition of 459 J of energy.

Thermochemistry

EXAMPLE 15-3: CALCULATING TEMPERATURE CHANGE

If 942 J of thermal energy are added to 50.0 g of water at 25.1 °C, what will its final temperature be?

Write what you know.

$Q = 942$ J
$c_{sp\ water} = 4.18$ J/g · °C (from Table 15-2)
$m = 50.0$ g
$T_i = 25.1$ °C
$T_f = ?$

Write the formula and solve for the unknown.

$Q = c_{sp} m\, \Delta T$
$ = c_{sp} m\, (T_f - T_i)$

$$\frac{Q}{c_{sp}m} = \frac{\cancel{c_{sp}m}\,(T_f - T_i)}{\cancel{c_{sp}m}}$$

$$T_f \cancel{-T_i + T_i} = \frac{Q}{c_{sp}m} + T_i$$

$$T_f = \frac{Q}{c_{sp}m} + T_i$$

Plug in known values and evaluate.

$$T_f = \frac{942\ \cancel{J}}{(4.18\ \cancel{J}/\cancel{g}\cdot°C)(50.0\ \cancel{g})} + 25.1\ °C$$
$ = 4.5\underline{0}7\ °C + 25.1\ °C$
$ = 29.6\ °C$

The final temperature is 29.6 °C.

A Complex Thermodynamic Problem

The last two subsections have described measurements of energy changes in substances. These include molar enthalpy of fusion, molar enthalpy of vaporization, and specific heat. How can this information be used?

Suppose you try your hand at frying ice cubes. You start with an ice cube with a mass of 5.00 g at a temperature of 0 °C. If that ice cube is heated until it is completely converted to steam at 100 °C, how many joules of energy must be added?

Remember, latent heat is thermal energy added to make a phase change with no change in temperature, and sensible heat is thermal energy added with a noticeable temperature change.

In this example, ice is being converted to water and then to steam. The first step is to gather all the needed values from Tables 15-1 and 15-2.

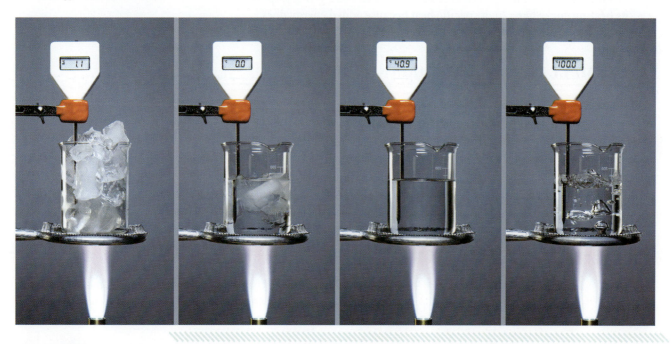

Use the heating curve at right to guide your calculations. You will need to calculate the energy needed to melt the ice (Segment 1), heat the liquid water (Segment 2), and finally, vaporize the water (Segment 3).

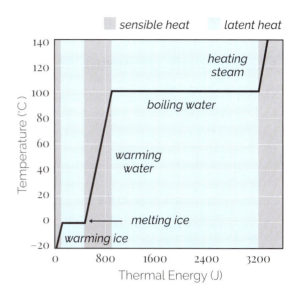

EXAMPLE 15-4:
CALCULATING ENERGY IN A COMPLEX THERMODYNAMIC PROBLEM

Write what you know.

$c_{sp\,water}$ = 4.18 J/g · °C (from Table 15-2)
m = 5.00 g
ΔH_{fus} = 6.01 kJ/mol (from Table 15-1)
ΔH_{vap} = 40.7 kJ/mol (from Table 15-1)
T_i = 0.0 °C
T_f = 100.0 °C
Q = ?

Segment 1 (melting ice)
Convert mass of ice to energy required to melt.

$$Q_1 = 5.00 \text{ g H}_2\text{O} \left(\frac{1 \text{ mol H}_2\text{O}}{18.02 \text{ g H}_2\text{O}} \right) \left(\frac{6.01 \text{ kJ}}{1 \text{ mol H}_2\text{O}} \right) = 1.6\underline{6}7 \text{ kJ}$$

Segment 2 (heating water)
Write the formula and solve for the unknown.

$Q = c_{sp} m \, \Delta T$
$Q = c_{sp} m \, (T_f - T_i)$

Plug in known values and evaluate.

$$Q_2 = \left(\frac{4.18 \text{ J}}{\text{g} \cdot °\text{C}} \right) (5.00 \text{ g}) (100.0 °\text{C} - 0.0 °\text{C})$$
$$= \left(\frac{4.18 \text{ J}}{°\text{C}} \right) (5.00 \text{ g}) (100.0 °\text{C})$$
$$= 2090 \text{ J} = 2.0\underline{9}0 \text{ kJ}$$

Segment 3 (boiling water)
Convert mass of water to energy required to boil.

$$Q_3 = 5.00 \text{ g H}_2\text{O} \left(\frac{1 \text{ mol H}_2\text{O}}{18.02 \text{ g H}_2\text{O}} \right) \left(\frac{40.7 \text{ kJ}}{1 \text{ mol H}_2\text{O}} \right) = 11.2\underline{9} \text{ kJ}$$

Total Energy

$Q_{total} = Q_1 + Q_2 + Q_3 = 1.6\underline{6}7 \text{ kJ} + 2.0\underline{9}0 \text{ kJ} + 11.2\underline{9} \text{ kJ}$
$= 15.0 \text{ kJ}$

It will take 15.0 kJ of energy to convert 5.00 g of ice at 0.0 °C to steam at 100.0 °C.

15.1 SECTION REVIEW

1. Define *thermochemistry*.

2. What is the difference between thermal energy, heat, and temperature? What units are used to measure each?

3. What does a scientist use a calorimeter for?

4. Compare sensible heat and latent heat.

5. Explain how sensible heat and latent heat interact to change a substance from a solid to a gas.

6. How does the specific heat of a substance relate to its ability to change temperature?

7. Define *molar enthalpy of fusion*, *molar enthalpy of vaporization*, and *specific heat*.

8. How much thermal energy is needed to accomplish the change in each case below?

    a. melt 355 g of ethanol at its melting point

    b. vaporize 515 g of methanol at its boiling point

9. How much heat energy is lost by a 53.6 g piece of copper when it cools from 95.7 °C to 22.5 °C?

MINI LAB

COMPARING THERMAL ENERGY TRANSFER

In days gone by, some people used to keep their feet warm at night by placing a pan of hot coals from a fire under the covers at the foot of the bed. Sounds risky, but it did demonstrate that thermal energy can be transferred from one object (a burning coal) through another object (the pan) to a third object (the cold feet). In this lab activity, you will use a similar process to test the heat-transferring abilities of some metal cubes.

Do equal volumes of metal contain the same amount of thermal energy?

1. Do you think that cubes of two different metals with equal volumes will transfer the same amount of thermal energy? State your answer in the form of a hypothesis.

Procedure

A. Fill each of the beakers with 300 mL of water. Place one beaker on the hot plate and record the starting temperature of the water in the other two beakers.

B. Use the tongs to place the aluminum and nickel cubes into the beaker on the hot plate. Set the hot plate on high and bring the water to a boil. Leave the cubes in the boiling water for 3 min.

2. What do we know about the temperatures of the two cubes at this point? Explain.

C. Turn off the hot plate. Carefully use tongs to remove one of the blocks and place it into one of the room-temperature beakers. Place the remaining block in the other beaker.

D. Monitor the temperature in the beakers with the cubes until each temperature no longer changes. Record the final temperature for each beaker. Subtract the initial temperature from the final temperature to determine the temperature change in each beaker.

3. Which cube caused a greater temperature change?

Conclusion

4. Suggest an explanation for the result that you observed.

Going Further

5. How might what you observed be important for a materials engineer to consider?

EQUIPMENT

- hot plate
- beaker, 400 mL (3)
- thermometers (2)
- tongs
- density cubes, aluminum and nickel (one of each)
- goggles

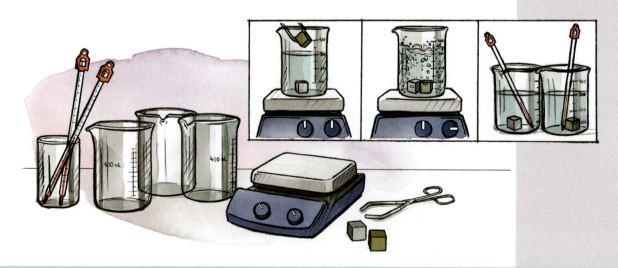

15.2 THERMODYNAMICS AND CHEMICAL CHANGES

Enthalpy (Heat) of Reaction

The concept of enthalpy is especially valuable to scientists studying the energy changes of chemical reactions. They call the change in enthalpy that occurs during a reaction the **enthalpy (heat) of reaction** (ΔH_r). The enthalpy change of a reaction is the difference between the enthalpy of the products and the enthalpy of the reactants.

$$\Delta H_r = H_{products} - H_{reactants}$$

The formation of water from its two elements illustrates this process. When 1 mol of hydrogen gas is ignited in the presence of 0.5 mol of oxygen gas, an explosive reaction forms 1 mol of water. Careful measurements reveal that 285.8 kJ of thermal energy are released in the explosion.

$$H_2(g) + 0.5 O_2(g) \longrightarrow H_2O(l) \quad \Delta H_r = -285.8 \text{ kJ}$$

Because the coefficients in this balanced equation represent moles of substances and not particles, it is permissible to use fractions in order to show the formation of a single mole of a product. The above equation shows the formation of 1 mol of water. Since there is a direct relationship between enthalpy and molar quantities, the standard balanced equation shown below is used. Note that 2 mol of water is produced.

$$2H_2(g) + O_2(g) \longrightarrow 2H_2O(l) \quad \Delta H_r = -571.6 \text{ kJ}$$

A chemical equation such as this one that shows the reactants, products, and amount of energy that is released or absorbed as heat is called a **thermochemical equation**. These equations show the physical state of each substance because the physical states affect how ΔH for the reaction is calculated.

QUESTIONS
» How can I know whether a reaction is exothermic or endothermic?
» How can I determine the energy change in a multistep reaction?

TERMS
enthalpy of reaction • thermochemical equation • standard molar enthalpy of formation • standard molar enthalpy of combustion • Hess's law

Are chemical reactions ever cold?

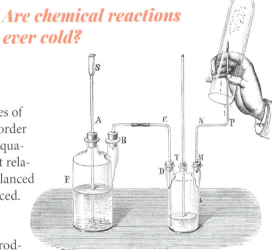

▲

Hydrogen combustion. The experiment shown above is demonstrating the production of water from hydrogen and oxygen.

Thermochemistry

Heat of Reaction

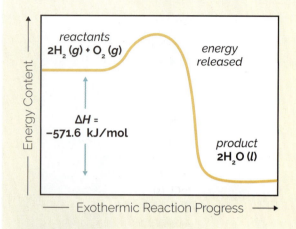

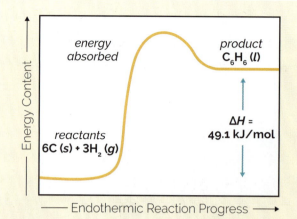

Exothermic

The products in an exothermic reaction have a lower enthalpy than the reactants. This means that energy is released during the reaction and the enthalpy of reaction (ΔH_r) is a negative value. Therefore, all exothermic reactions have negative values for ΔH_r. The graph above illustrates the energy changes in exothermic reactions. The products have a lower enthalpy than the reactants, and the difference is released as energy.

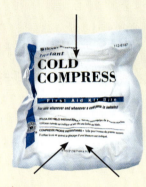

Endothermic reactions absorb energy, resulting in a greater amount of energy in the products.

Endothermic

Products of endothermic reactions have a higher enthalpy than the reactants because energy has been absorbed. An endothermic reaction has a positive value for ΔH_r. The graph above illustrates the difference in energies between the reactants and the products of an endothermic reaction.

The endothermic formation of one mole of benzene has the following thermochemical equation.

$$6C(s) + 3H_2(g) \longrightarrow C_6H_6(l)$$

$$\Delta H_r = 49.1 \text{ kJ}$$

If the ΔH_r is known, the ΔH_r of the reverse reaction can be easily calculated. The ΔH_r of the reverse reaction has the same magnitude but the opposite sign of the ΔH_r of the forward reaction.

For example, the reaction that formed a mole of liquid water had a ΔH_r of -285.8 kJ (the negative sign indicates that energy is given off). A reaction that decomposes the same amount of water has a ΔH_r of 285.8 kJ (energy absorbed).

Enthalpy (Heat) of Formation

Chemists could measure the change in enthalpy for thousands of known reactions, but measuring so many reactions would be difficult and the data would fill several volumes of books. Therefore, a quick technique for calculating changes in enthalpy was devised. Enthalpy in a substance varies with temperature and pressure, so these conditions must be specified. In thermodynamics, the *standard state* is defined as 25 °C (298 K, or roughly room temperature) and 101.325 kPa of pressure (normal atmospheric pressure at sea level).

iron burning in chlorine gas

The **standard molar enthalpy of formation** is defined as the enthalpy change to produce 1 mol of a compound in its standard state from its elements in their standard states. We use the notation $\Delta H°_f$; the degree symbol signifies that this ΔH refers to standard conditions. For example, the standard state of water is liquid; the standard state of gold is solid. Table 15-3 lists several standard molar enthalpies of formation. Appendix H contains a more extensive listing. A quick look at the appendix will show that most values for $\Delta H°_f$ are negative. Thus the synthesis reactions that produce most of the common compounds are exothermic. Compounds with large negative numbers release large amounts of energy when they form. Molecules such as Cl_2, H_2, and O_2 have no enthalpies of formation because these substances are the reference points with which the enthalpies of compounds are compared. This is true for every element in its naturally occurring pure form.

TABLE 15-3 *Standard Molar Enthalpies of Formation*

Compound	$\Delta H°_f$ (kJ/mol)
Fe_2O_3 (s)	−824.2
NH_4NO_3 (s)	−365.6
NO (g)	91.3
NO_2 (g)	33.2
$PbCl_2$ (s)	−359.4

EXAMPLE 15-5: CALCULATING $\Delta H°_F$

What is the $\Delta H°_f$ for $N_2 + 2O_2 \longrightarrow 2NO_2$?

Solution
Table 15-3 shows that the standard molar enthalpy of formation of nitrogen dioxide is 33.2 kJ/mol. The given reaction forms 2 mol of nitrogen dioxide.

$$2 \text{ mol NO}_2 \left(\frac{33.2 \text{ kJ}}{1 \text{ mol NO}_2} \right) = 66.4 \text{ kJ}$$

Enthalpy (Heat) of Combustion

Combustion reactions produce energy in the form of heat and light. They all involve a reaction with oxygen. The most significant combustion reactions occur with hydrocarbons—organic compounds containing carbon and hydrogen. Carbon dioxide and water are the typical products. The **standard molar enthalpy of combustion** ($\Delta H°_c$) is defined as the thermal energy released by the complete burning of one mole of a substance at standard conditions. For example, propane, the gas used for most outdoor grills, is useful because of the heat released during its combustion. The thermochemical equation for the complete combustion of propane is shown below.

$$C_3H_8(g) + 5O_2(g) \longrightarrow 3CO_2(g) + 4H_2O(l)$$

$$\Delta H°_c = -2219.2 \text{ kJ/mol}$$

A list of values for the enthalpy of combustion for some other common compounds is found in Table 15-4. All the values are negative because these reactions are all exothermic. As a rule, the larger the molecule is, the more energy is released.

TABLE 15-4 *Standard Molar Enthalpies of Combustion*

Substance	Formula	$\Delta H°_c$ (kJ/mol)
carbon monoxide (g)	CO	−283.4
graphite (s)	C	−393.5
hydrogen (g)	H_2	−285.8
methane (g)	CH_4	−890.8
propane (g)	C_3H_8	−2219.2
sucrose (s)	$C_{12}H_{22}O_{11}$	−5647

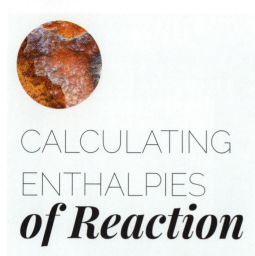

CALCULATING ENTHALPIES *of Reaction*

A chemical reaction can be viewed as two chemical processes. In the first step, each reactant breaks down to its elements in their standard states. In the second step, the elements in their standard states recombine to form the products. Remember that the formation and breakdown of a particular substance are opposite processes and must therefore have opposite signs. If a reaction is endothermic in one direction, it will be exothermic in the other.

> The following reaction between lead(II) oxide and carbon monoxide will illustrate this two-process approach.

$$PbO(s) + CO(g) \longrightarrow Pb(s) + CO_2(g) \quad \Delta H°_r = -65.7 \text{ kJ}$$

1 The first step in the reaction is the breakdown of lead(II) oxide and carbon monoxide.

The $\Delta H°_f$ for lead(II) oxide is -217.3 kJ/mol (Appendix H). The breakdown of lead(II) oxide, which is the reverse reaction, requires 217.3 kJ.

$$PbO(s) \underset{-217.3 \text{ kJ}}{\overset{217.3 \text{ kJ}}{\rightleftharpoons}} Pb(s) + \tfrac{1}{2}O_2(g)$$

The $\Delta H°_f$ for carbon monoxide is -110.5 kJ/mol, so the breakdown of carbon monoxide to its elements requires 110.5 kJ/mol.

$$CO(g) \underset{-110.5 \text{ kJ}}{\overset{110.5 \text{ kJ}}{\rightleftharpoons}} C(s) + \tfrac{1}{2}O_2(g)$$

The $\Delta H°$ for the first part of the reaction is obtained by adding together the two quantities of energy required to break the compounds apart. Their sum is 327.8 kJ.

2 The second part of the reaction is the formation of products.

The $\Delta H°$ for each of these changes can be easily found in the table listing values of $\Delta H°_f$. Since lead is an element in its standard state, its $\Delta H°_f$ is zero. The $\Delta H°_f$ for the formation of 1 mol of carbon dioxide from its elements at standard conditions is -393.5 kJ.

$$C(s) + O_2(g) \underset{393.5 \text{ kJ}}{\overset{-393.5 \text{ kJ}}{\rightleftharpoons}} CO_2(g)$$

The total $\Delta H°$ for the formation of products is therefore -393.5 kJ.

By adding the enthalpy for the breakdown of reactants to the enthalpy for the formation of products, we get the $\Delta H°_r$ for the reaction, which is -65.7 kJ. This result shows that the reaction is exothermic.

In this problem, calculating the $\Delta H°_r$ of the whole reaction involved combining the values for $\Delta H°_r$ in two partial reactions.

Hess's law states that the enthalpy change of a reaction equals the sum of the enthalpy changes for each step of the process.

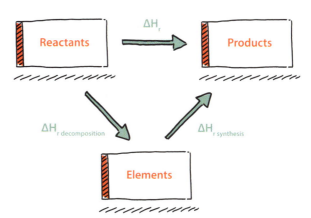

The example below shows how one reaction could be thought of as the sum of two other reactions. The first reaction breaks apart the reactants and the second puts them back together in a different arrangement.

$$PbO + CO \longrightarrow Pb + O_2 + C \quad \Delta H° = 327.8 \text{ kJ}$$

$$Pb + O_2 + C \longrightarrow Pb + CO_2 \quad \Delta H° = -393.5 \text{ kJ}$$

$$PbO + CO \longrightarrow Pb + CO_2 \quad \Delta H° = -65.7 \text{ kJ}$$

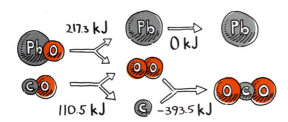

The enthalpy change of the net reaction equals the sum of the enthalpy changes for the two steps. Notice how the common reactants and products cancel each other. The mathematical version of this statement is shown below.

$$\Delta H°_{reaction} = \Delta H°_{reactant\ decomposition} + \Delta H°_{product\ synthesis}$$

Because the $\Delta H°_f$ values in the table refer to the formation of compounds from elements, the values for reactant-decomposition enthalpies will be opposite in sign to those listed. Consequently, the equation can be changed to a form that uses the data in Appendix H without changing the signs. The symbol Σ in this equation is the Greek letter *sigma*. It signifies a summation of numbers.

$$\Delta H°_{reaction} = \Sigma \Delta H°_{f(products)} - \Sigma \Delta H°_{f(reactants)}$$

> *There are two general principles for combining thermochemical equations.*
>
> 1. When using Hess's law, you may use any valid balanced equation, but remember to change the sign on the $\Delta H°$ if you reverse the equation.
>
> 2. Multiply the coefficients of a balanced equation to get the desired equation when the coefficients are added together, but remember to also multiply the $\Delta H°$ by the same number.

EXAMPLE 15-6: CALCULATING $\Delta H°$

Estimate the $\Delta H°_r$ for the combustion of methane using standard enthalpies of formation.

$$CH_4(g) + 2O_2(g) \longrightarrow CO_2(g) + 2H_2O(l)$$

Write what you know.

The equation requires the use of the following enthalpies of formation from Appendix H.

$\Delta H°_f (CO_2) = -393.5$ kJ/mol
$\Delta H°_f$ (liquid H_2O) $= -285.8$ kJ/mol
$\Delta H°_f (CH_4) = -79.6$ kJ/mol
$\Delta H°_f (O_2) = 0$ kJ/mol

A total of 2 mol of oxygen and water participate in the reaction. This is included when calculating the enthalpy change of the entire reaction.

Write the formula and solve for the unknown.

$$\Delta H° = \Sigma \Delta H°_{f\,(products)} - \Sigma \Delta H°_{f\,(reactants)}$$

$$\Delta H° = [\Delta H°_f (CO_2) + \Delta H°_f (H_2O)] - [\Delta H°_f (CH_4) + \Delta H°_f (O_2)]$$

Plug in known values and evaluate.

$$\Delta H_r = \left[1\text{ mol CO}_2\left(\frac{-393.5\text{ kJ}}{1\text{ mol CO}_2}\right) + 2\text{ mol H}_2\text{O}\left(\frac{-285.8\text{ kJ}}{1\text{ mol H}_2\text{O}}\right)\right] - \left[1\text{ mol CH}_4\left(\frac{-79.6\text{ kJ}}{1\text{ mol CH}_4}\right) + 2\text{ mol O}_2\left(\frac{0\text{ kJ}}{1\text{ mol O}_2}\right)\right]$$

$= (-393.5 \text{ kJ} + -571.6 \text{ kJ}) - (-79.6 \text{ kJ} + 0 \text{ kJ})$

$= -885.5$ kJ

This reaction is highly exothermic.

15.2 SECTION REVIEW

1. Define *endothermic reaction* and *exothermic reaction*.

2. Does an exothermic reaction have a negative or positive change in enthalpy? Explain.

3. Define *enthalpy of reaction*, *standard molar enthalpy of formation*, and *standard molar enthalpy of combustion*.

4. What is the difference between standard temperature and pressure (STP) and standard state?

5. How can we determine the energy in a multistep reaction?

6. Calculate $\Delta H°$ for each of the following reactions and tell whether each reaction is endothermic or exothermic.
   a. $N_2(g) + 3H_2(g) \longrightarrow 2NH_3(g)$
   b. $Ca(OH)_2(s) \longrightarrow CaO(s) + H_2O(g)$

7. Some instant ice packs contain ammonium chloride (NH_4Cl), which dissolves to produce two ions in a process similar to that involving ammonium nitrate. Given that the ammonium ion has a $\Delta H°_f$ of -132.5 kJ/mol and that the $\Delta H°_f$ of the chloride ion is -167.2 kJ/mol, calculate the enthalpy of solution. Remember from Chapter 14 that the enthalpy of solution is the net energy change during the dissolving process.

8. Compare the cooling efficiency of ice packs that use ammonium chloride with those that use ammonium nitrate, assuming that each reaction involves the same volume of water and molar quantity of salt. The equation for dissolving ammonium nitrate is shown below.

$$NH_4NO_3(s) \longrightarrow NH_4^+(aq) + NO_3^-(aq)$$

$$\Delta H_{soln} = 28.1 \text{ kJ/mol}$$

9. A key component in gasoline is octane (C_8H_{18}). If $\Delta H_f = -249.7$ kJ/mol, calculate the enthalpy of combustion for octane.

15.3 REACTION TENDENCY

Chemical compounds store energy. As long as bonds are not changed, the energy that they contain cannot be released. For chemical substances to react, there must be a transfer of energy, and this change in energy is more observable. Chemical bonds can be broken through *vibration*. In some cases, this vibration may be indirectly caused by the *rotation* or *translation* (lateral motion) of the molecule. An outside source of energy, such as a laser, can cause the vibration. The atoms can be vibrated so strongly that the attraction holding them together is overcome.

QUESTIONS
» What is entropy?
» What determines the likelihood of a reaction occurring?
» How do I determine the change in entropy of a system?
» Can I predict the favorability of a reaction?

TERMS
entropy • Gibbs free energy

Is dynamite dangerous?

Siege mortars, like this one intended for use in the Crimean War, rely on bond formation to release large amounts of energy.

Chemical Bonds and Enthalpy

In Section 15.2, you calculated the enthalpies of a reaction by using enthalpies of formation. Another method of analyzing the enthalpy changes of chemical reactions is to focus on the bonds between pairs of atoms. The strength of the bond between any two atoms depends on the elements that are bonding and on the length of the bond. Because bonded atoms are more stable than unbonded atoms, energy is always required to break bonds—that is, it is an endothermic process. Conversely, the formation of bonds always releases energy and is thus exothermic. The stronger the bond is, the more energy is released.

These relationships can help explain why some reactions give off energy and others require energy. Strong, stable bonds require large amounts of energy to break them. Strong bonds also give off large amounts of energy when they form. Weak bonds can be broken with small amounts of energy. Predictably, they give off small amounts of energy when they form.

A reaction that breaks strong bonds and forms weak bonds requires energy. More energy is required to break the strong bonds than is released when the weak bonds form. As a result, reactions that produce compounds with weaker, less stable bonds are endothermic.

$$\text{stronger bonds} \longrightarrow \text{weaker bonds}; \Delta H_r > 0 \text{ (endothermic)}$$

A reaction that forms stable, low-energy compounds from compounds with high-energy and weak bonds releases energy. The amount of energy that must be used to break the weak bonds is small compared with all the energy that is given off when the strong bonds form.

$$\text{weaker bonds} \longrightarrow \text{stronger bonds}; \Delta H_r < 0 \text{ (exothermic)}$$

In most cases, exothermic processes are thermodynamically favorable and endothermic processes are not. But this rule does not always hold true; a negative ΔH does not always indicate that a process is favorable.

Entropy and Reaction Tendency

high distribution of energy

low distribution of energy

We are surrounded by processes that occur *spontaneously*, yet many of those processes absorb rather than give off energy. In other words, they are endothermic. One example is the release of compressed gas into the atmosphere. As gas under high pressure is released, it gets colder. You may have noticed this phenomenon when dispensing whipped cream from a can. In Chapter 14 you learned that the formation of many solutions causes the temperature of the solvent to decrease. Ice cubes absorb thermal energy from their surroundings to melt into water. All of these are endothermic processes. Clearly, something in addition to enthalpy influences spontaneity.

Each of these examples shows us that processes tend to proceed naturally in a way that spreads out energy. The *second law of thermodynamics* relates this tendency to a very large system when it states that natural processes tend to decrease the usable energy in the universe. The energy of gas molecules is less concentrated after being released from the pressure of a can. The energy of solute particles is dispersed after solvation.

Entropy (S) is a measure of the dispersal of energy in a system. In natural processes, energy tends to spread out, not concentrate. A hot pan cooling off on a stovetop is an example of entropy in action. Gases, with their free-flying particles, have high entropies. Crystalline solids have particles arranged in definite, repeating, organized systems with energy stored in bonds; therefore, they possess low entropies and high energy density.

Because entropy is related to the kinetic energy of a substance, entropy increases as temperature increases. As the temperature becomes greater, the increased kinetic energy gives the particles greater motion. But it's not always directly proportional. For example, when ice (a crystalline solid) melts, entropy is increased because attractions in the crystal that store energy are broken and energy is dispersed. But the temperature does not change. Because the amount of entropy in a substance varies with temperature as well as pressure, these conditions must be specified. Table 15-5 lists the entropies per mole of several common substances at standard conditions (again indicated by the degree symbol). Although enthalpy values for pure elements are not listed in the extensive table in Appendix H, entropies are included because all substances have some entropy.

TABLE 15-5 *Standard Molar Entropies*

Substance	Formula	$S°$ (J/mol·K)
ammonia (*g*)	NH_3	192.8
ammonium chloride (*s*)	NH_4Cl	94.6
hydrogen bromide (*g*)	HBr	198.7
hydrogen chloride (*g*)	HCl	186.9

Calculating Entropy Changes

From the data in Appendix H, the entropy changes of many reactions can be calculated. An entropy change is the difference between the sum of the entropies of the products and the sum of the entropies of the reactants, similar to how the enthalpy of a reaction is calculated.

ΔS = total entropy in products − total entropy in reactants

$$\Delta S = \Sigma S_{products} - \Sigma S_{reactants}$$

A reaction with a positive ΔS increases entropy and spreads out energy. The products have more entropy than the reactants. A negative ΔS signifies a decrease in entropy and a concentration of energy in that part of the universe, though it results in a net loss of usable energy in the universe.

EXAMPLE 15-7: CALCULATING $\Delta S°$

Does the reaction $NH_3\,(g) + HCl\,(g) \longrightarrow NH_4Cl\,(s)$ produce an increase or decrease in entropy at standard conditions?

Write what you know.

Using entropies from Appendix H:
  $S°\,(NH_3) = 192.8$ J/mol·K
  $S°\,(HCl) = 186.9$ J/mol·K
  $S°\,(NH_4Cl) = 94.6$ J/mol·K

Write the formula and solve for the unknown.

$\Delta S° = \Sigma S°_{products} - \Sigma S°_{reactants}$

Plug in known values and evaluate.

$\Delta S° = (1 \cancel{\text{mol } NH_4Cl}) \left(\dfrac{94.6 \text{ J}}{\cancel{\text{mol } NH_4Cl} \cdot K} \right) - \left[(1 \cancel{\text{mol } NH_3}) \left(\dfrac{192.8 \text{ J}}{\cancel{\text{mol } NH_3} \cdot K} \right) + (1 \cancel{\text{mol } HCl}) \left(\dfrac{186.9 \text{ J}}{\cancel{\text{mol } HCl} \cdot K} \right) \right]$

= 94.6 J/K − 379.7 J/K
= − 285.1 J/K

The amount of entropy has decreased.

While the entropy of a specific system may increase or decrease, all processes, whether spontaneous or nonspontaneous, increase the total entropy of the universe. A positive $\Delta S°$ means a reaction is more favorable. A reaction with a negative $\Delta S°$ decreases entropy, and thus is less favorable.

The 70 m wide Darvaza gas crater in Turkmenistan burns methane gas that seeps from the ground.

CASE STUDY

ENTROPY AND LIFE

You've just finished a long, hard soccer game. The next thing on your mind is a cold drink, so you go to the vending machine. All you have is a $20 bill, so your shoulders sag in dejection when you see that the machine accepts only $1, $5, and $10 bills. Sure, you have enough money to get ten bottles, but your money isn't in a usable form.

This situation is similar to the way that your body gets and uses energy. When you eat food, your body is getting a great source of energy. But that energy is similar to your $20—there's plenty of it, but it's not in a usable form. If your body released the energy stored in fats, sugars, and other molecules in that food all at once, great damage would result. Your body's cells must break down food into a usable form. This is where entropy comes in to save the day.

Sugars from the digestive system are carried by the circulatory system to cells all over your body. Once these molecules reach body cells, the work of energy conversion begins. The energy-rich sugar molecules must be converted into a more usable form of energy for the cell. This conversion requires a process called *aerobic cellular respiration*. Aerobic cellular respiration requires oxygen—hence the word *aerobic*—which you obtain from breathing. Through a complicated series of steps, aerobic cellular respiration takes energy from the sugar molecules and stores that energy in less-energy-rich molecules of adenosine triphosphate—ATP. This process is less than 50% efficient, so over half the energy dissipates into forms that don't fuel the body. Most of the "lost" energy is given off as thermal energy and is used to maintain body temperature, so it is actually being put to good use. The secret to ATP's success lies in its three phosphate groups. The bonds holding each of the phosphates in place are high-energy bonds. By breaking the bonds, energy is released, and your body can use that energy for work.

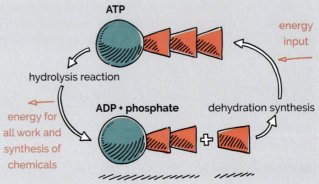

The body's supply of energy is maintained by recycling ATP. When a phosphate is broken off from ATP and energy is released, the ATP molecule is transformed into ADP—adenosine diphosphate. Through another type of reaction, a phosphate can be added back onto ADP, once again creating ATP. In just one of your muscle cells at work during that soccer game, ten million ATP molecules were used and recycled every second! Your body requires a constant source of ATP, and God's design and provision of energy for us is amazing. And it all demonstrates the second law of thermodynamics and entropy at work.

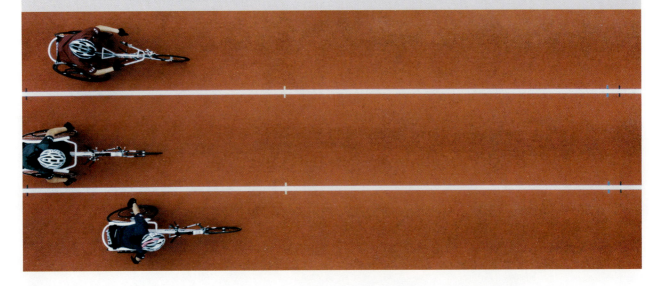

FREE-ENERGY CHANGE

Two tendencies drive reactions. The first is the tendency to decrease enthalpy (exothermic), and the second is the tendency to increase entropy. Reactions can be either exothermic or endothermic, and they can either increase or decrease entropy.

Four possible enthalpy/entropy combinations exist.

1. exothermic and increasing entropy ($-\Delta H$, $+\Delta S$)
2. exothermic and decreasing entropy ($-\Delta H$, $-\Delta S$)
3. endothermic and increasing entropy ($+\Delta H$, $+\Delta S$)
4. endothermic and decreasing entropy ($+\Delta H$, $-\Delta S$)

Reactions are driven by the combined effects of change in enthalpy (ΔH) and change in entropy (ΔS). A single expression that relates ΔH and ΔS is needed to determine how they combine.

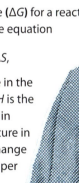

In the 1800s, American physicist J. Willard Gibbs (right) formulated a single criterion of favorability. He combined the change in enthalpy and the change in entropy into a single term called *free energy*, which is sometimes called **Gibbs free energy** (G) in his honor.

The free-energy change (ΔG) for a reaction can be calculated by the equation

$$\Delta G = \Delta H - T\Delta S,$$

where ΔG is the change in the free energy in joules, ΔH is the change in the enthalpy in joules, T is the temperature in kelvins, and ΔS is the change in the entropy in joules per kelvin.

A negative ΔG indicates a decline in free energy and signifies that a reaction may occur spontaneously (or is favorable). A positive ΔG indicates a net increase in free energy and signifies that a reaction is not favorable—nonspontaneous. The free-energy change can be negative under several conditions. A negative ΔH and a positive ΔS always contribute to a negative ΔG. When the two tendencies oppose each other, the temperature at which the reaction takes place determines the sign of ΔG. Table 15-6 shows the conditions at which ΔG is negative and the reaction is spontaneous.

TABLE 15-6 *When ΔG Is Negative*

	ΔH	ΔS	ΔG
Case 1	−	+	$-\Delta G$ at all T
Case 2	−	−	$-\Delta G$ at low T
Case 3	+	+	$-\Delta G$ at high T
Case 4	+	−	$+\Delta G$ at all T

Calculating Free-Energy Change

The ΔG of a reaction can be calculated in two different ways. First, we can calculate ΔG from the values for ΔH, ΔS, and T by substituting them into the free-energy equation. For example, for the reaction below, the free-energy change can be calculated from its enthalpy and entropy changes.

$$Mg(OH)_2\,(s) \longrightarrow MgO\,(s) + H_2O\,(l) \text{ at 298 K}$$

$$\Delta H° = (-601.6 \text{ kJ} - 285.8 \text{ kJ}) - (-924.5 \text{ kJ})$$

$$= 37.1 \text{ kJ}$$

$$\Delta S° = (27.0 \text{ J/K} + 70.0 \text{ J/K}) - (63.2 \text{ J/K})$$

$$= 33.8 \text{ J/K}$$

Note that the two values are not expressed in similar units. Converting the ΔS value from 33.8 J/K to 0.0038 kJ/K can remedy the problem. We calculate ΔG at standard conditions (298 K).

$$\Delta G° = \Delta H° - T\Delta S°$$

$$= 37.1 \text{ kJ} - 298 \text{ K}\left(0.0338\,\frac{\text{kJ}}{\text{K}}\right)$$

$$= 27.0 \text{ kJ}$$

Calculating $\Delta G°$ by the second method, where $\Delta G°$ values from Appendix H are used, gives a result very close to the first value.

$$\Delta G° = \left[(1\,\overline{\text{mol MgO}})\left(\frac{-569.3 \text{ kJ}}{\overline{\text{mol MgO}}}\right) + (1\,\overline{\text{mol H}_2\text{O}\,(l)})\frac{-237.1 \text{ kJ}}{\overline{\text{mol H}_2\text{O}\,(l)}}\right) \right] -$$

$$(1\,\overline{\text{mol Mg(OH)}_2})\left(\frac{-833.5 \text{ kJ}}{\overline{\text{mol Mg(OH)}_2}}\right)$$

$$= (-569.3 \text{ kJ} - 237.1 \text{ kJ}) - (-833.5 \text{ kJ})$$

$$= 27.1 \text{ kJ}$$

Since the ΔG is positive, the reaction is nonspontaneous at 298 K. At a higher temperature, the positive ΔS has a greater effect. This effect could be seen by estimating ΔG at 1000 K using the values for $\Delta H°_f$ and $S°$.

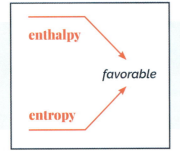

CASE 1: DECREASE IN ENTHALPY, INCREASE IN ENTROPY

Case 1 shows that exothermic reactions that increase entropy ($-\Delta H$, $+\Delta S$) are always favorable. They release energy and increase entropy. ΔG for such reactions is always negative.

EXAMPLE 15-8: CALCULATING ΔG, CASE 1

Calculate the Gibbs free-energy change for the following reaction.

$$2Ag\ (s) + S\ (s) \longrightarrow Ag_2S\ (s) \text{ at } 298\ K$$

Write what you know.

$\Delta H°\ (Ag_2S) = -32.6$ kJ
$S°\ (Ag_2S) = 144.0$ J/mol·K
$S°\ (Ag) = 42.6$ J/mol·K
$S°\ (S) = 32.1$ J/mol·K
$T = 298$ K

Write the formula and solve for the unknown.

$\Delta H° = \Sigma H°_{products} - \Sigma H°_{reactants}$
$\Delta S° = \Sigma S°_{products} - \Sigma S°_{reactants}$

Plug in known values and evaluate.

$\Delta H° = (-32.6\ kJ) - (-0.0\ kJ - 0.0\ kJ)$
$= -32.6$ kJ

$\Delta S° = (1\ \text{mol Ag}_2S)\left(\dfrac{144.0\ J}{\text{mol Ag}_2S \cdot K}\right) -$
$\left[(2\ \text{mol Ag})\left(\dfrac{42.6\ J}{\text{mol Ag}\cdot K}\right) + (1\ \text{mol S})\left(\dfrac{32.1\ J}{\text{mol S}\cdot K}\right)\right]$

$= 144.0\ \dfrac{J}{K} - \left(85.2\ \dfrac{J}{K} + 32.1\ \dfrac{J}{K}\right)$

$= 26.7\ \dfrac{J}{K} = 0.0267\ \dfrac{kJ}{K}$

Write the formula and solve for the unknown.

$\Delta G° = \Delta H° - T\Delta S°$

Plug in known values and evaluate.

$\Delta G° = -32.6\ kJ - 298\ K\left(0.0267\ \dfrac{kJ}{K}\right)$

$= -40.6$ kJ

The reaction is spontaneous.

CASE 2: DECREASE IN ENTHALPY, DECREASE IN ENTROPY

Case 2 shows that exothermic reactions that decrease entropy ($-\Delta H$, $-\Delta S$) may or may not be favorable. The two tendencies oppose each other. In cases like this, the temperature determines whether the reaction is favorable. The entropy change hinders the reaction less at low temperatures than it does at high temperatures, so ΔG is negative at low temperatures.

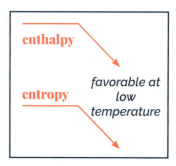

EXAMPLE 15-9: CALCULATING ΔG, CASE 2

Calculate the Gibbs free-energy change for the following reaction.

$$2\ Fe\ (s) + \dfrac{3}{2}O_2\ (g) \longrightarrow Fe_2O_3\ (s) \text{ at } 298\ K$$

Write what you know.

$\Delta H°\ (Fe_2O_3) = -824.2$ kJ
$S°\ (Fe_2O_3) = 87.4$ J/mol·K
$S°\ (Fe) = 27.3$/mol·K
$S°\ (O_2) = 205.2$ J/mol·K
$T = 298$ K

Write the formula and solve for the unknown.

$\Delta H° = \Sigma H°_{products} - \Sigma H°_{reactants}$
$\Delta S° = \Sigma S°_{products} - \Sigma S°_{reactants}$

Plug in known values and evaluate.

$\Delta H° = (-824.2\ kJ) - (-0.0\ kJ - 0.0\ kJ)$
$= -824.2$ kJ

$\Delta S° = (1\ \text{mol Fe}_2O_3)\left(\dfrac{87.4\ J}{\text{mol Fe}_2O_3\cdot K}\right) -$
$\left[(2\ \text{mol Fe})\left(\dfrac{27.3\ J}{\text{mol Fe}\cdot K}\right) + \left(\dfrac{3}{2}\ \text{mol O}_2\right)\left(\dfrac{205.2\ J}{\text{mol O}_2\cdot K}\right)\right]$

$= 87.4\ \dfrac{J}{K} - \left(54.6\ \dfrac{J}{K} + 307.8\ \dfrac{J}{K}\right)$

$= -275\ \dfrac{J}{K} = -0.275\ \dfrac{kJ}{K}$

Write the formula and solve for the unknown.

$\Delta G° = \Delta H° - T\Delta S°$

Plug in known values and evaluate.

$\Delta G° = -824.2\ kJ - 298\ K\left(-0.275\ \dfrac{kJ}{K}\right)$

$= -742.3$ kJ

The reaction is spontaneous.

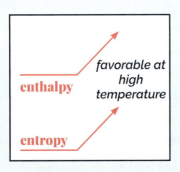

CASE 3: INCREASE IN ENTHALPY, INCREASE IN ENTROPY

Case 3 represents endothermic reactions that increase entropy (+ΔH, +ΔS). These reactions can be favorable if the entropy change is greater than the enthalpy change. High temperatures magnify the effect of the entropy change and make ΔG negative. As a result, these reactions can proceed. At lower temperatures, however, they have positive ΔG values, making them unfavorable.

EXAMPLE 15-10: CALCULATING ΔG, CASE 3

Calculate the Gibbs free-energy change for the following reaction.

$$2C\ (s) + H_2\ (g) \longrightarrow C_2H_2\ (g) \text{ at } 298\ K$$

Write what you know.

$\Delta H°\ (C_2H_2) = 227.4$ kJ
$S°\ (C_2H_2) = 200.9$ J/mol·K
$S°\ (C) = 5.7$/mol·K
$S°\ (H_2) = 130.7$ J/mol·K
$T = 298$ K

Write the formula and solve for the unknown.

$\Delta H° = \Sigma H°_{products} - \Sigma H°_{reactants}$
$\Delta S° = \Sigma S°_{products} - \Sigma S°_{reactants}$

Plug in known values and evaluate.

$\Delta H° = (227.4\ \text{kJ}) - (-0.0\ \text{kJ} - 0.0\ \text{kJ})$
$\quad\ \ = 227.4$ kJ

$\Delta S° = (1\ \text{mol}\ C_2H_2)\left(\dfrac{200.9\ \text{J}}{\text{mol}\ C_2H_2 \cdot K}\right) - \left[(2\ \text{mol}\ C)\left(\dfrac{5.7\ \text{J}}{\text{mol}\ C \cdot K}\right) + (1\ \text{mol}\ H_2)\left(\dfrac{130.7\ \text{J}}{\text{mol}\ H_2 \cdot K}\right)\right]$

$\quad\ \ = 200.9\ \dfrac{J}{K} - \left(11.4\ \dfrac{J}{K} + 130.7\ \dfrac{J}{K}\right)$

$\quad\ \ = 58.8\ \dfrac{J}{K} = 0.0588\ \dfrac{kJ}{K}$

Write the formula and solve for the unknown.

$\Delta G° = \Delta H° - T\Delta S°$

Plug in known values and evaluate.

$\Delta G° = 227.4\ \text{kJ} - 298\ K\left(0.0588\ \dfrac{kJ}{K}\right)$
$\quad\ \ = 209.9$ kJ

The reaction is nonspontaneous.

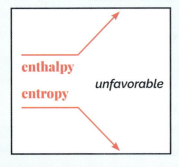

CASE 4: INCREASE IN ENTHALPY, DECREASE IN ENTROPY

Case 4 shows that endothermic reactions that decrease entropy (+ΔH, −ΔS) do not occur naturally. To proceed spontaneously, these reactions would have to concentrate energy by adding it to the chemical substances. Reactions that exhibit positive enthalpy changes can be forced if other types of energy are used. Otherwise, they are unfavorable at any temperature.

EXAMPLE 15-11: CALCULATING ΔG, CASE 4

Calculate the Gibbs free-energy change for the following reaction.

$$2C\,(s) + 2H_2\,(g) \longrightarrow C_2H_4\,(g) \text{ at } 298\text{ K}$$

Write what you know.

$\Delta H°\,(C_2H_4) = 52.4$ kJ
$S°\,(C_2H_4) = 219.3$ J/mol · K
$S°\,(C) = 5.7$/mol · K
$S°\,(H_2) = 130.7$ J/mol · K
$T = 298$ K

Write the formula and solve for the unknown.

$\Delta H° = \Sigma H°_{products} - \Sigma H°_{reactants}$
$\Delta S° = \Sigma S°_{products} - \Sigma S°_{reactants}$

Plug in known values and evaluate.

$\Delta H° = (52.4$ kJ$) - (-0.0$ kJ $- 0.0$ kJ$)$
$= 52.4$ kJ

$\Delta S° = (1 \text{ mol } C_2H_4)\left(\dfrac{219.3 \text{ J}}{\text{mol } C_2H_4 \cdot \text{K}}\right) -$
$\left[(2 \text{ mol } C)\left(\dfrac{5.7 \text{ J}}{\text{mol } C \cdot \text{K}}\right) + (2 \text{ mol } H_2)\left(\dfrac{130.7 \text{ J}}{\text{mol } H_2 \cdot \text{K}}\right)\right]$

$= 219.3\,\dfrac{\text{J}}{\text{K}} - \left(11.4\,\dfrac{\text{J}}{\text{K}} + 261.4\,\dfrac{\text{J}}{\text{K}}\right)$

$= -53.5\,\dfrac{\text{J}}{\text{K}} = -0.0535\,\dfrac{\text{kJ}}{\text{K}}$

Write the formula and solve for the unknown.

$\Delta G° = \Delta H° - T\Delta S°$

Plug in known values and evaluate.

$\Delta G° = 52.4$ kJ $- 298$ K$\left(-0.0535\,\dfrac{\text{kJ}}{\text{K}}\right)$
$= 68.3$ kJ

The reaction is nonspontaneous.

EXAMPLE 15-12: CALCULATING ΔG AT DIFFERENT TEMPERATURES

Determine whether the reaction between ammonia and hydrogen chloride is probable at 298 K and at 985 K according to the values shown in Appendix H.

$$NH_3(g) + HCl(g) \longrightarrow NH_4Cl(s)$$

Write what you know.

$H°\,(NH_3) = -45.9$ kJ/mol
$H°\,(HCl) = -92.3$ kJ/mol
$H°\,(NH_4Cl) = -314.4$ kJ/mol
$S°\,(NH_3) = 192.8$ J/mol · K
$S°\,(HCl) = 186.9$ J/mol · K
$S°\,(NH_4Cl) = 94.6$ J/mol · K
$T_1 = 298$ K
$T_2 = 985$ K

Write the formula and solve for the unknown.

$\Delta H° = \Sigma H°_{products} - \Sigma H°_{reactants}$
$\Delta S° = \Sigma S°_{products} - \Sigma S°_{reactants}$

Plug in known values and evaluate.

$\Delta H° = (1 \text{ mol } NH_4Cl)\left(\dfrac{-314.4 \text{ kJ}}{\text{mol } NH_4Cl}\right) -$
$\left[(1 \text{ mol } NH_3)\left(\dfrac{-45.9 \text{ kJ}}{\text{mol } NH_3}\right) + (1 \text{ mol } HCl)\left(\dfrac{-92.3 \text{ kJ}}{\text{mol } HCl}\right)\right]$

$= -314.4$ kJ $- (-138.2$ kJ$)$
$= -176.2$ kJ

$\Delta S° = (1 \text{ mol } NH_4Cl)\left(\dfrac{94.6 \text{ J}}{\text{mol } NH_4Cl \cdot \text{K}}\right) -$
$\left[(1 \text{ mol } NH_3)\left(\dfrac{192.8 \text{ J}}{\text{mol } NH_3 \cdot \text{K}}\right) +\right.$
$\left.(1 \text{ mol } HCl)\left(\dfrac{186.9 \text{ J}}{\text{mol } HCl \cdot \text{K}}\right)\right]$

$= 94.6$ J/K $- 379.7$ J/K
$= -285.1$ J/K $= -0.2851$ kJ/K

Write the formula and solve for the unknown

$\Delta G° = \Delta H° - T\Delta S°$

Plug in known values and evaluate. ($T_1 = 298$ K)

$\Delta G° = -176.2$ kJ $- 298$ K$\left(-0.2851\,\dfrac{\text{kJ}}{\text{K}}\right)$
$= -91.2$ kJ

Since $\Delta G°$ is negative, the reaction is probable at 298 K, making it spontaneous.

Even though the values of ΔH and ΔS at 985 K are not known, the value of ΔG can be estimated by using the 298 K figures in the free-energy equation and solving for ΔG.

Plug in known values and evaluate. ($T_2 = 985$ K)

$\Delta G = -176.2$ kJ $- 985$ K$\left(-0.2851\,\dfrac{\text{kJ}}{\text{K}}\right)$
$= 104.6$ kJ

The positive ΔG shows that the reaction is unfavorable at 985 K, making it nonspontaneous.

WORLDVIEW INVESTIGATION

HEAT DEATH

A cup of hot coffee cools off and a sound gets quieter as you get farther from its source. Some people have called entropy "time's arrow" because we naturally assume that a process continues in a direction that results in energy being spread out. And as energy spreads out, it becomes less useful for doing work. That's not a problem for a system as small as a cup of coffee—we can easily brew a new cup. But what about larger systems, such as our universe? If the naturalistic worldview were true, what consequences would the concept of entropy have for our universe?

Task

You're watching a TV documentary on the solar system. As often happens in such films, the narrator describes how the universe arose during the big bang, which you have heard of before. He also mentions that the solar system will die a "heat death." What does that mean?

Procedure

1. Research the concept of heat death by doing an internet search using the keywords "heat death" and "big freeze."

2. To get a biblical perspective, go to the Creation Ministries International website and search for "heat death."

3. Write a summary of your findings, including the concept of heat death, the evidence that it is based on, and a Christian response to the idea on the basis of what the Bible teaches about Earth's future state. Have a classmate review your summary and give you feedback.

4. Complete your summary and submit it by the deadline.

Conclusion

We are often exposed to popular ideas about the fate of our universe, many of which are voiced from a naturalistic perspective. Such narratives offer little hope for those who put stock in them. In contrast, the Bible offers an encouraging description of both the future state of believers and the new earth in which they will dwell.

Worldview Conflict in Thermodynamics

Energy radiates from the sun. Warmth from a fire in a fireplace spreads throughout the room. Natural processes in the universe tend to spread energy out, not concentrate it. In other words, it's very likely that in any natural process, spreading energy will happen. Entropy is a measure of the extent to which energy is spread out. The entropy of most objects decreases as they lose energy and increases as they gain energy. For example, the entropy of water vapor decreases as it deposits to form a snowflake. But the energy lost by the water vapor dissipates into the cooler environment and produces an even larger increase in its entropy. The net result is an increase of entropy in the universe.

So how is the formation of snowflakes different from the formation of the very first molecules that supposedly kick-started the evolution of life? Can we simply pump energy into a system to produce order and complexity? The molecules of life—amino acids, proteins, and DNA—are very energy dense and have low entropy compared with the smaller molecules from which they are formed. This means that according to the second law of thermodynamics, it is unlikely that these molecules would spontaneously form in any warm ocean of chemicals on an early earth. Adding energy from the sun wouldn't solve this problem either; there is no initial "machine" or process to convert that energy into a useful form to manufacture the molecules necessary for life. Life comes only from life.

Living things are energy converters. For example, plants live by converting energy from the sun into a form they can use. But this concentration of energy in the plant also produces a loss of usable energy somewhere else in the universe. In fact, only about 1%–2% of the energy that a plant receives from the sun is converted into sugars that the plant can use for food. Spontaneous chemical processes can produce only a limited degree of order and no degree of complexity and information. Living systems have an organization and complexity that spontaneous processes in nonliving systems cannot reproduce.

But entropy is not the enemy of complexity, nor is disorder. The second law shows how our world operates in ways that are predictable and reliable. God uses entropy to sustain the world and protect life through a helpful flow of energy, like energy from the sun or energy from our food. We can also see how God has provided for life by giving the universe a reservoir of usable energy. In the laws of thermodynamics, we can see the goodness and faithfulness of God at work.

15.3 SECTION REVIEW

1. A classmate tells you that since most chemical reactions are exothermic, breaking bonds must release energy. How do you respond?

2. What is entropy?

3. State whether each of the following is a case of increasing or decreasing entropy and give your reasoning for each answer.
   a. A car rolls to a stop.
   b. Ice forms on a car windshield overnight.
   c. Sugar dissolves in a cup of tea.
   d. A book falls from your desk.

4. Arrange the following examples in order of increasing entropy. (Assume that entropy usually increases with temperature.)
   a. 1 mol of H_2O (g) at 125 °C
   b. 1 mol of H_2O (g) at 140 °C
   c. 1 mol of H_2O (s)
   d. 1 mol of H_2O (l)

5. Calculate the standard entropy change ($\Delta S°$) for each reaction. On the basis of the entropy change, indicate whether the reaction tends to proceed naturally.
   a. $2KClO_3(s) \longrightarrow 2KCl(s) + 3O_2(g)$
   b. $SnO_2(s) + 2H_2(g) \longrightarrow Sn(s) + 2H_2O(l)$

6. Calculate the change in free energy ($\Delta G°$) for each reaction from Question 5 at 298 K, and tell whether the reactions are spontaneous or nonspontaneous at this temperature.

7. Which combinations of ΔH and ΔS signs (+ and −) always result in a reaction that is favorable? never result in a reaction that is favorable?

8. Under what conditions will a reaction with positive ΔH and ΔS values result in a reaction that is favorable? Explain.

15 CHAPTER REVIEW

Chapter Summary

15.1 THERMODYNAMICS AND PHASE CHANGES

- Thermochemistry studies the transfer of energy during chemical reactions or phase changes.

- Thermal energy is the total kinetic energy of the particles in a substance. Temperature measures the average kinetic energy of the particles in a substance. Heat is the transfer of thermal energy between substances.

- The thermal energy of a system at constant pressure is enthalpy.

- Latent heat is the transfer of thermal energy that causes a phase change in a substance, while sensible heat is the transfer of thermal energy that causes a temperature change in the substance.

- The molar enthalpy of fusion is the thermal energy needed to melt 1 mol of a solid at its melting point with no temperature change. The thermal energy required to convert 1 mol of a liquid at its boiling point to its vapor at the same temperature is its molar enthalpy of vaporization.

- The specific heat of a substance is the amount of thermal energy required to raise the temperature of 1 g of the substance 1 °C.

TERMS
thermochemistry
enthalpy
calorimeter
sensible heat
latent heat
molar enthalpy of fusion
molar enthalpy of vaporization
specific heat

15.2 THERMODYNAMICS AND CHEMICAL CHANGES

- When scientists study the energy changes in a reaction, they measure the enthalpy of reaction, which is defined as the total enthalpy of the products minus the total enthalpy of the reactants.

- The standard molar enthalpy of formation is the change in enthalpy that occurs when 1 mol of a compound is formed in its standard state from its elements in their standard states.

- Standard state refers to the state of matter that a substance has at 298 K and 101.325 kPa.

- The standard molar enthalpy of combustion is defined as the energy released by the complete burning of 1 mol of a substance at standard conditions.

- Hess's law states that the enthalpy change of a reaction equals the sum of the enthalpy changes for each step of the process. That law can be used to calculate the enthalpy of reaction.

TERMS
enthalpy of reaction
thermochemical equation
standard molar enthalpy of formation
standard molar enthalpy of combustion
Hess's law

15.3 REACTION TENDENCY

- Entropy (S) is the tendency of natural processes to spread out energy in the universe. The change in entropy (ΔS) is the entropy in the products minus the entropy in the reactants ($\Delta S = \Sigma S_{products} - \Sigma S_{reactants}$).

- Reactions that increase entropy (i.e., they have a positive ΔS) tend to be favorable, while reactions that decrease entropy (they have a negative ΔS) tend to be unfavorable.

- Gibbs free energy is helpful in determining whether overall reactions are spontaneous (favorable) or not. The free-energy change (ΔG) for a reaction can be calculated by the equation $\Delta G = \Delta H - T\Delta S$, where T is the temperature in kelvins.

TERMS

entropy
Gibbs free energy

Chapter Review Questions

RECALLING FACTS

1. What is enthalpy?
2. Why must a calorimeter be insulated?
3. Why is it necessary to designate the state of matter of each substance in a thermochemical equation?
4. Does a compound with a negative molar enthalpy of formation value result from an endothermic or exothermic reaction?
5. Why are most of the standard molar enthalpies of formation ($\Delta H°_f$) negative?
6. What law allows you to combine the steps of a chemical reaction to compute change in entropy?
7. Which kind of reaction is more thermodynamically favorable: exothermic or endothermic?
8. What two tendencies influence all chemical reactions?
9. Which of the following situations are possible according to the laws of thermodynamics?
    a. insect larvae spontaneously forming in rotting meat
    b. the human body converting the energy in food to other forms of energy
    c. the energy and matter in the universe coming into being from nothing without any intervention from God
    d. the invention of an automobile engine that is 100% efficient
10. Which combination(s) of ΔH and ΔS values will result in favorability depending on temperature?

15 CHAPTER REVIEW

UNDERSTANDING CONCEPTS

11. What would happen to the temperature and physical state of 1 mol of liquid diethyl ether at its standard boiling point if its molar enthalpy of vaporization (26.52 kJ/mol) were added?

12. Will soup take more or less time to cool than an equal volume of toast? Explain.

13. From the specific heat table (Table 15-2), what general conclusions can you draw about the relationship between specific heat values and the three common states of matter?

14. If 2586 J of energy are added to 128 g of benzene at 30.0 °C, what will its final temperature be?

15. A 47.35 g mass of copper is heated from 32.7 °C to 85.3 °C. How many joules were added to accomplish this?

16. Suppose you have an ice cube with a mass of 8.00 g at a temperature of 0.0 °C. How many joules of heat must be added to completely convert the ice cube to steam at 100.0 °C?

17. What is the enthalpy of formation ($\Delta H°_f$) for the following equation?

    $2\,C\,(s) + O_2\,(g) \longrightarrow 2CO\,(g)$

18. For each of the following reactions, find the enthalpy of reaction and tell whether the reaction is endothermic or exothermic.

    a. $Hg\,(l) + S\,(s) \longrightarrow HgS\,(s)$

    b. $HgS\,(s) \longrightarrow Hg\,(l) + S\,(s)$

    c. $2\,C\,(s) + 2H_2\,(g) \longrightarrow C_2H_4\,(g)$

    d. $C_2H_4\,(g) \longrightarrow 2\,C\,(s) + 2H_2\,(g)$

For Questions 19–23, refer to the following reactions.

   a. $4NH_3\,(g) + SO_2\,(g) \longrightarrow 4NO\,(g) + 6H_2O\,(g)$

   b. $3NO_2\,(g) + H_2O\,(l) \longrightarrow 2HNO_3\,(l) + NO\,(g)$

   c. $2NH_4NO_3\,(s) \longrightarrow 2N_2\,(g) + O_2\,(g) + 4H_2O\,(g)$

19. (UC) Calculate the standard change in enthalpy ($\Delta H°_r$) for each reaction, and tell whether each reaction is endothermic or exothermic.

20. Calculate the standard entropy change ($\Delta S°$) for each reaction.

21. Calculate the change in free energy ($\Delta G°$) for each reaction at 298 K, and tell whether each reaction is favorable (spontaneous) or unfavorable (nonspontaneous) at this temperature.

22. Estimate ΔG for each reaction at 985 K and tell whether each reaction is favorable (spontaneous) or unfavorable (nonspontaneous) at this temperature.

23. Estimate the temperature (if any) at which each reaction changes from being spontaneous to nonspontaneous (the temperature at which $\Delta G = 0$).

24. Under what conditions will a reaction with negative ΔH and ΔS values result in a reaction that is favorable? Explain.

CRITICAL THINKING

25. What is wrong with the statement, "As a substance freezes, it absorbs energy equal to its enthalpy of fusion"?

26. A 79.8 g piece of iron at 98.7 °C is placed in 215.0 g of water at 22.7 °C in a calorimeter. What is the final temperature of the water and iron?

27. One method of manufacturing sulfuric acid involves the four-step process shown below. Given the chemical equations for each step, determine the enthalpy of reaction for the overall reaction.

$$2S(s) + 3O_2(g) + 2H_2O(l) \longrightarrow 2H_2SO_4(l)$$

1. $S(s) + O_2(g) \longrightarrow SO_2(g) \quad \Delta H^\circ_r = -296.8 \text{ kJ}$

2. $SO_3(g) \longrightarrow SO_2(g) + \frac{1}{2}O_2(g) \quad \Delta H^\circ_r = 49.5 \text{ kJ}$

3. $H_2SO_4(g) \longrightarrow SO_3(g) + H_2O(l) \quad \Delta H^\circ_r = 101.1 \text{ kJ}$

4. $H_2SO_4(g) \longrightarrow H_2SO_4(l) \quad \Delta H^\circ_r = -68.9 \text{ kJ}$

28. How is developing an ice pack that uses enthalpy to help heal injuries an example of obeying two important commands from God?

Chapter 16

CHEMICAL KINETICS

For years people have understood that smoking is harmful. As people sought ways to quit smoking, the e-cigarette was introduced. Today some see smoking e-cigarettes—or vaping—as a "safe," high-tech alternative to smoking. The interest in these devices has been enormous, especially with teenagers. But is vaping really a safe alternative?

This chapter looks at chemical reactions from the viewpoint of the precise mechanisms for reactions and the speed at which they occur. These issues are of critical importance in view of reactions that process chemicals within our bodies.

16.1 Reaction Rates *387*
16.2 Reaction Mechanisms *396*

16.1 REACTION RATES

Kinetics

What is the difference between the decomposition of nitroglycerine in dynamite and the decomposition of food? For one, there is a huge difference in the amount of energy released, as you learned when studying thermodynamics in Chapter 15. But another part of the difference is how fast the reaction takes place. The field of chemistry called *kinetics* answers questions about the speed at which reactions occur and the way in which they get started. While thermodynamics answers the basic question, "Can it react?," kinetics answers the more specific questions, "Will it react?" and "How fast will it react?" **Kinetics** is the study of the rates of reactions and the steps by which they occur.

How is a fire different from an explosion?

The fact that a reaction is thermodynamically favorable does not mean that it will proceed automatically. Some reactions proceed at an extremely slow rate; others are rapid. Some need a push to get started; others proceed on their own.

QUESTIONS
» How can I tell how quickly a reaction will occur?
» Why don't all favorable reactions occur?
» How do reactions actually occur?
» What affects reaction rates?

TERMS
kinetics • collision theory • activation energy • activated complex • reaction rate • catalyst • homogeneous catalyst • heterogeneous catalyst • enzyme • inhibitor

FAVORABLE REACTIONS

As an example, consider the following three reactions. Each of them has a negative Gibbs free energy (ΔG), so they are all favorable (spontaneous) under the proper conditions of temperature and pressure.

the oxidation of a diamond

ΔG = –395.4 kJ

This reaction occurs, but not at a significant rate.

the burning of methane

ΔG = –882.0 kJ

Combustion of methane will not start unless it is given an energetic push (from a lighted match, for example) and is very rapid.

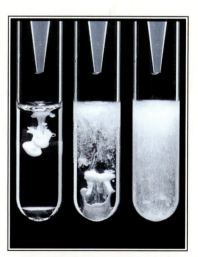

the mixing and the neutralization of $Ba(OH)_2$ and H_2SO_4

ΔG = –286.1 kJ

This third reaction proceeds as soon as the reactants are mixed together and occurs quickly.

Something besides thermodynamics must be used to explain why reactions have such a wide variety of rates. That something is kinetics.

Energy Diagrams

Thermodynamics relates only to the starting and ending points of a reaction and is therefore path independent. In chemistry, thermodynamics is concerned only with the change in energy from point A to point B. On the other hand, kinetics is path-dependent. Kinetics addresses how a reaction proceeds from point A to point B. Are the energy changes during a reaction simple and direct, or must the reactants gain some energy before the reaction can proceed? What steps are involved? Kinetics seeks to determine what happens between the starting point and end point of a reaction.

To understand the difference between thermodynamics and kinetics, consider a shopper strolling through a multilevel mall. She may change levels and go in and out of stores. Thermodynamics is concerned only with what level she entered and exited the mall, that is, with her beginning and ending energies. But kinetics would be interested in the specific path she took through the mall—which stores she entered and for how long, along with when and where she changed levels.

It can be helpful to graph the energy changes in a reaction on an energy diagram (see left). The horizontal scale represents the progress of the reaction being considered, and the vertical scale represents energy. The highest point on the energy diagram corresponds to the energy needed by the reactants to form products.

Collision Theory

Recall that, according to the kinetic-molecular theory, particles are in constant motion in random directions at speeds related to their temperature. With all that motion, collisions are bound to happen, and colliding particles may react. Suppose a hypothetical gas molecule (X_2) can combine with a different hypothetical gas molecule (Y_2). A sufficiently energetic collision will disrupt the electron orbitals of these molecules. For a moment, the four atoms form a single high-energy group; then the group separates into two new molecules (2XY) according to the equation below.

$$X_2 + Y_2 \longrightarrow 2XY$$

The **collision theory** of reactions states three conditions that must be met for any reaction to occur. First, the particles must collide in order to react, though a collision alone does not guarantee a reaction. The particles must also collide in the proper orientation necessary for the rearrangement of atoms and electrons. Finally, the collision must be energetic enough to form products. The theory can be summarized by saying that reactions follow only *effective collisions*—those that are sufficiently energetic and properly oriented.

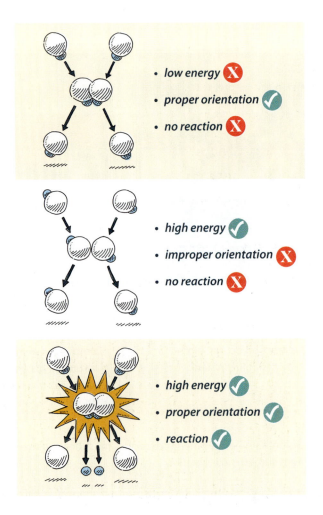

The collision theory explains why reactions occur at faster or slower rates, depending on reaction conditions. Any factor that increases the number of effective collisions increases the rate of a reaction. Therefore, the rate at which a chemical reaction proceeds is directly related to the frequency of effective collisions. Fast reactions occur when the three criteria are easily met: (1) collisions with (2) proper orientation and (3) sufficient energy. But if meeting even one criterion is difficult, the reaction will occur more slowly.

MINI LAB

CHANGING REACTION RATES

We all know that when there is little traffic on the highway, the probability of collision is low. As the amount of traffic increases, the chances of two cars colliding is greater. Do chemical reactions work the same way? Does increasing the concentration of the reactants affect the number of collisions and therefore the reaction rate?

How does concentration affect reaction rate?

1. Write a hypothesis regarding the effect of concentration on the rate of reaction.
2. Justify why you think concentration will affect the rate as you indicated in your hypothesis.

Procedure

A. Copy the table below onto a piece of paper.

B. Label the test tubes 1–4.

C. Fill the test tubes with vinegar and water according to the table.

3. What is being compared in Tubes 1 and 2?

4. What is being compared in Tubes 2, 3, and 4?

D. Add the indicated strip of metal to each test tube and record your observations.

Conclusion

5. What was the effect of using different metals? Why do you think this was true?

6. What was the effect of changing the concentration of the vinegar?

EQUIPMENT
- test tubes (4)
- test tube rack
- graduated cylinder, 25 mL
- vinegar, 55 mL
- water, 25 mL
- magnesium strips, 2 cm (3)
- copper strip, 2 cm
- marker

Test Tube	Vinegar (mL)	Water (mL)	Metal	Reaction
1	20	0	Cu	
2	20	0	Mg	
3	10	10	Mg	
4	5	15	Mg	

Chemical Kinetics

Activation Energy and the Activated Complex

Consider the formation of liquid water from hydrogen and oxygen gases.

$$2H_2(g) + O_2(g) \longrightarrow 2H_2O(l)$$

If hydrogen gas is mixed with oxygen gas in a tank, no water will be produced, even though the standard enthalpy of formation ($\Delta H°_f$) is −285.8 kJ/mol and the free-energy change (ΔG) is −237.1 kJ/mol, both strong indicators of a favorable reaction. Why don't these molecules immediately react to form water?

Even though a specific reaction may potentially be spontaneous because of a −ΔG value, it will not proceed unless the reactants collide with enough energy. How much energy is enough? The energy needed to jump-start a reaction, called the **activation energy** (E_a), is the minimum amount of kinetic energy that must be possessed by colliding molecules before they can react. Activation energy is like the first hill on a roller coaster. Many processes just don't happen (which is a good thing) because the activation energy hill is too high to overcome. As a rule, the lower the activation energy is, the faster the reaction will occur. Each reaction has a specific activation energy. Activation energy prevents the occurrence of many reactions that are actually spontaneous.

Although the reaction for the formation of water is exothermic, the initial breaking of bonds within the hydrogen and oxygen reactants is endothermic. Breaking bonds is always an endothermic process. The forming of new bonds releases energy. This released energy can provide the activation energy for more collisions. In the formation of water, the energy released when water molecules form is more than adequate to allow the reaction to continue without the further introduction of energy.

All reactions can be pictured as these two steps—the breaking of bonds and the formation of new ones. Although it is convenient to think of reactions taking place as two sequential steps, that view is somewhat misleading. It is not necessary, for instance, for all the bonds to break before new bonds start to form. The two processes often occur simultaneously.

As these two processes occur, colliding reactants can form an **activated complex,** an intermediate substance that forms as the reactants transition into the products. It appears briefly as old bonds are in the process of breaking and new bonds are forming. Its high-energy content makes it extremely unstable, so it is very short-lived (a few femtoseconds; 1 fs = 10^{-15} s). Relating this concept

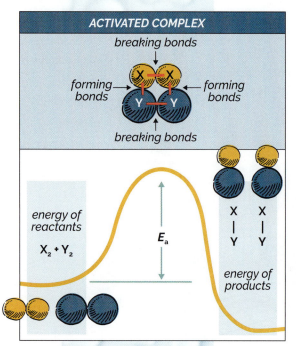

◄ An activated complex is a transitional structure early in a chemical reaction.

to the energy diagrams discussed previously, we see that the activation energy can also be defined in terms of the activated complex. It can be thought of as the minimum amount of energy needed to convert the reactants into the activated complex.

In conclusion, activation energy plays a crucial role in determining how fast a reaction will proceed under given temperature and pressure conditions. A large activation energy can prevent an otherwise favorable reaction from proceeding quickly or even at all.

Rates of Reactions

During a reaction, the concentrations of reactants and products change constantly. At the beginning of a reaction, the reactants are at their highest concentrations. As the reaction progresses, product concentrations increase as the concentrations of reactants decrease by being consumed. Consider the reaction between iodine chloride and hydrogen.

$$2ICl(g) + H_2(g) \longrightarrow I_2(g) + 2HCl(g)$$

The graph below shows how the concentrations of the substances in this reaction vary with time. Notice that the reactants are being consumed as the products form. Also note that the changes are most rapid early in the reaction and then gradually slow.

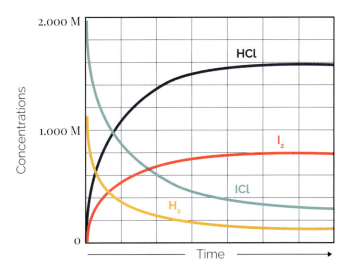

Reaction rate is a measure of how quickly reactants change into products. It can describe how fast the reactants disappear or how fast the products appear. Although reaction rates may be measured in several different units, they usually tell how fast concentrations change with time. Units such as amount concentration per second or moles per liter per hour are common.

Which reaction is most likely to proceed rapidly?

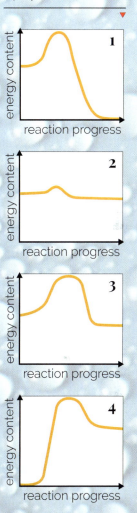

Remember that reactions depend on collisions that are properly oriented and have enough energy. Anything that increases the frequency, energy, and efficiency of these collisions should increase the rate of a reaction. Five of the most important factors affecting reaction rates are considered on this page spread.

Factors That Affect
REACTION RATES

Nature of Reactants

The most obvious factor that controls the rate of a reaction is the chemical nature of the reactants. Logically enough, reactive substances react quickly. Phosphorus, in its white form, spontaneously bursts into flames when exposed to air. It is so reactive that it must be stored under water. At the other extreme, the noble gases rarely or never react.

white phosphorus in water

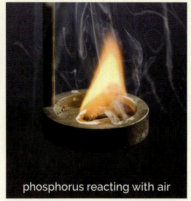
phosphorus reacting with air

Another variable is the particular combination of elements. Hydrogen and chlorine, under certain conditions, react vigorously; under those same conditions, hydrogen and nitrogen will react only weakly. The three test tubes shown below contain three metals in hydrochloric acid. Magnesium reacts the most, zinc reacts mildly, and copper doesn't react at all. This is related to the activity series that we saw in Chapter 10.

magnesium zinc copper

Concentration

Since changing the conditions of a reaction so that more collisions occur increases the reaction rate, there should be a definite connection between concentration and reaction rate. Notice in the two test tubes below that copper barely reacts in 1 M nitric acid, but reacts vigorously in 17 M nitric acid.

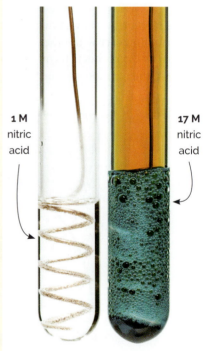
1 M nitric acid · 17 M nitric acid

Since there are seventeen times more hydrogen ions in the concentrated solution, there are many more collisions that can cause reactions with the copper. The effect of concentration on the reaction rate varies from reaction to reaction. Thus, each reaction must be studied experimentally.

Temperature

A rule of thumb in kinetics is that reaction rates double for every 10 °C rise in temperature. Higher temperatures increase reaction rates because they increase the kinetic energy of the reactants. According to the kinetic-molecular theory, increased kinetic energy results in faster-moving particles, which collide more often. These collisions will generally have more energy, meaning that they are more likely to react. Higher temperatures increase the number of effective collisions, increasing the reaction rate. Notice the two glow sticks shown above. The one intensely emitting light is in hot water, while the dimmer one is in ice water.

Presence of Catalysts

A **catalyst** is a substance that changes a reaction's rate without being consumed by the reaction. It is present during the reaction, but it is neither a reactant nor a product. A catalyst provides an alternative *reaction mechanism* from reactants to products that has a lower activation energy. A catalyst does not affect the concentration of products at the conclusion of a reaction, only the rate at which these products are formed. The catalyst also does not affect the enthalpy change (ΔH) of a reaction. Neither will it enable a thermodynamically impossible reaction to occur.

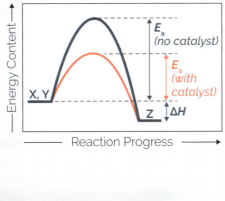

catalyst: powdered potassium iodide

3% hydrogen peroxide solution

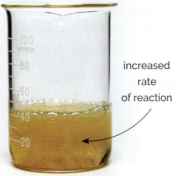

increased rate of reaction

Surface Area

In a mixture of gases or in a solute dissolved in a solvent, frequent collisions allow reactions, once started, to progress rapidly. But if the reactants are in two different phases, the reaction rate is dependent on how much contact occurs between the substances at their interface. Increasing the amount of exposed surface area allows more collisions to occur. If a substance is broken up into pieces rather than being in one piece, more surface area is exposed and the reaction will occur faster due to an increase in collision rate. For example, we all know that sugar has plenty of energy that is typically released slowly by digestion. But if sugar is in the form of fine dust with a greatly increased surface area, it can be deadly. The image above shows the results of a sugar explosion at a manufacturing facility.

MORE ABOUT *Catalysts*

How do catalysts work? One theory states that catalysts hold reactants in just the right positions for favorable collisions. For example, a mixture of hydrogen and oxygen gas does not react to form water at room temperature. But if a catalyst of powdered platinum is first introduced, the platinum causes an explosion as its surface becomes covered with adsorbed oxygen. *Adsorption* refers to the collection of one substance on the surface of another. It is different from absorption, which involves one substance entering into another. The platinum atoms stretch and weaken the bonds of the O_2 molecules as they are held in place, lowering the activation energy required for the reaction. The oxygen atoms then react rapidly with the hydrogen molecules to form water.

adsorption
the collection of one substance on the surface of another

absorption
the entering of one substance into another

SEVERAL TYPES OF CATALYSTS

A **homogeneous catalyst** is in the same phase as the reactants or in solution with a reactant. Homogeneous catalysts combine with one of the reactants to form an intermediate compound that will react more readily with the other reactant or reactants. A catalyst that is in a phase different than the reactants is a **heterogeneous catalyst**. Heterogeneous catalysts are materials capable of adsorbing molecules onto their surfaces. Platinum, nickel, palladium, and other powdered metals and metalloids are examples of heterogeneous catalysts.

One distinct class of catalysts comprises the naturally occurring biological substances known as **enzymes**. Enzymes are responsible for most of the essential biochemical reactions. More than a thousand enzymes have been identified, and each one is specific to a chemical reaction occurring within a living organism. The presence of an enzyme typically causes a reaction to occur millions of times faster than when the reaction is uncatalyzed. Without enzymes, these same processes would occur too slowly to be of any use to an organism. Enzymes also aid living things by providing a lower activation energy and thus a lower temperature at which chemical reactions can occur. The temperature that would be required for most reactions to proceed without an enzyme would cook cells and result in tissue damage or death for an organism.

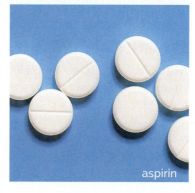

aspirin

Some catalysts, such as those that promote spoiling of food, are undesirable. An **inhibitor** is a substance used to reduce a catalyst's undesirable effects. Inhibitors work by bonding to a catalyst to decrease the catalyzed reaction rate or even completely stop the process. Like adsorption, this effect is the result of holding molecules in a certain position, but in this case the effect is to prevent effective collisions. Many important food preservatives act as inhibitors. Aspirin is another valuable inhibitor that works by blocking the action of two different enzymes that cause inflammation and pain.

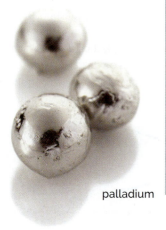

palladium

CASE STUDY

SPONTANEOUS COMBUSTION

It was a hot June day in 1933 in North Arlington, New Jersey. Swimmers seeking relief from the oppressive heat swarmed a beach along the bank of the Passaic River. Nearby was a storehouse full of pyroxylin—a highly combustible material used in lacquers, plastics, and artificial leather.

At 9:12 p.m., lingering swimmers noticed fire coming from the roof of the storehouse. With a puff, both ends of the building blew out, showering the crowd with bricks. Burning celluloid fell on everyone, and a vast flame swept over the surrounding area. Another blast occurred, and flames engulfed the beach as well as nine nearby buildings. Before it was over, ten people died and 180 were injured. The burning of the Atlantic Pyroxylin Waste Company remains a sobering yet predictable example of spontaneous combustion.

Spontaneous combustion starts without a flame or spark. Fires begun this way ignite because of the role that thermal energy plays in speeding up reaction rates. The Atlantic Pyroxylin Waste Company stored pyroxylin scraps swept up from factory floors. These scraps often had machine oil on them. The heat necessary for the fire to start came from the reaction between this oil and atmospheric oxygen.

Typically, a spontaneous combustion reaction doesn't generate great amounts of heat because it occurs slowly and the energy disperses. But if thermal energy cannot escape, it accumulates and the temperature increases. As the temperature increases, the reaction occurs faster, producing more heat. If unchecked, the reaction can generate temperatures above the auto-ignition point of the surrounding material.

Spontaneous combustion can occur in places other than factories. It can happen in soft coal, freshly cut bales of hay, and even in workshops, where oily and paint-saturated rags pose a threat. Preventing spontaneous combustion fires is a matter of controlling the rates of chemical reactions. Caution should be taken with combustible materials in powdered forms. The large amount of surface area allows reactions to proceed more quickly. More importantly, combustible material should be stored in a well-ventilated place where heat can be removed as fast as it is produced. Such actions will prevent heat from accumulating in a place where it can increase the rate of reaction.

Questions to Consider

1. Why does increased temperature increase a reaction rate?
2. What other factor seems to significantly impact spontaneous combustion? Explain.

16.1 SECTION REVIEW

1. Define *kinetics*.
2. Contrast thermodynamics and kinetics by indicating which question or questions each can answer.
3. Explain what an effective collision is according to the collision theory.
4. Give two reasons why a collision between two reactive molecules might not result in a reaction.
5. Why is the initial stage of almost all reactions endothermic?
6. Write the chemical equation for the reaction shown in the graph at right.
7. What are the five variables that can affect the rate of reaction?
8. How does a catalyst speed up a reaction rate?

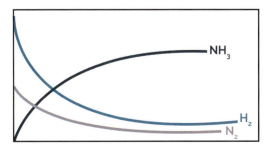

QUESTIONS

» What are the steps in a chemical reaction?
» Are there substances formed in reactions that are not products?
» What determines the rate of a reaction?
» How can I write a rate law?

TERMS

reaction mechanism • elementary step • rate-determining step • complex reaction • intermediate • reaction order • rate law • specific rate constant

16.2 REACTION MECHANISMS

Mechanisms

Chemists can easily identify the reactants of a chemical process. They also routinely analyze the products. But other molecules that form and then re-form during reactions are much more difficult to pinpoint. For chemists, observing reactions is like studying a factory's processes from outside the building. They can see the raw materials going in and the finished products coming out, but the individual steps involved in forming the products are more difficult to understand. The series of steps that make up a reaction is called a **reaction mechanism**.

How does water form when hydrogen is burned?

Understanding Reaction Mechanisms

To understand reaction mechanisms, consider the decomposition of NO_2.

- How does this reaction take place?
- Do two molecules collide to form an activated complex that splits into nitrogen monoxide and oxygen molecules?
- Do the three atoms in nitrogen dioxide all split apart and then rearrange themselves?
- Does an atom of oxygen separate from a molecule of nitrogen dioxide and then combine with another free oxygen atom?
- Or does something totally different occur?

Theoretically, any one of these proposed mechanisms could produce nitrogen monoxide and oxygen from nitrogen dioxide.

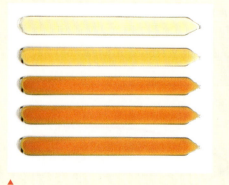

These vials show the color change as a reaction progresses to equilibrium.

Often, a reaction mechanism is made of two or more steps, called **elementary steps**. A possible reaction mechanism for the decomposition of nitrogen dioxide looks like the following.

Elementary Step 1: $\quad 2NO_2(g) \longrightarrow NO_3(g) + NO(g)$

Elementary Step 2: $\quad NO_3(g) \longrightarrow NO(g) + O_2(g)$

The rate at which the reaction occurs is dependent on the rates of the elementary steps. The reaction can proceed only as fast as the slowest elementary step, which is known as the **rate-determining step**. To understand this limiting factor, compare the reaction mechanism with an assembly line. Suppose that one job on an assembly line requires a worker to perform a task that is much more difficult than that of any other worker. The product can be produced only as fast as the worker with the most difficult task can complete his job. Similarly, a reaction can proceed only as fast as the rate-determining step. In the example of the decomposition of NO_2, the first step is the slowest, so it is the rate-determining step.

Elementary Step 1: $\quad 2NO_2(g) \longrightarrow NO_3(g) + NO(g) \quad$ slow

Elementary Step 2: $\quad NO_3(g) \longrightarrow NO(g) + O_2(g) \quad$ fast

Together, the individual elementary steps make up a **complex reaction**, or *net reaction*. During a complex reaction, there are substances that are formed in one step and consumed in the next. These substances are called **intermediates**. Intermediates do not appear in the chemical equation for the net reaction. So we can cancel the intermediates out of our elementary steps and add the two steps together to get the net equation.

Elementary Step 1: $\quad 2NO_2(g) \longrightarrow \cancel{NO_3(g)} + NO(g)$

Elementary Step 2: $\quad \cancel{NO_3(g)} \longrightarrow NO(g) + O_2(g)$

Net Equation: $\quad 2NO_2(g) \longrightarrow 2NO(g) + O_2(g)$

As another example, consider the reaction between hydrogen and iodine gases, one of the first to be studied in the field of kinetics.

$$H_2(g) + I_2(g) \longrightarrow 2HI(g)$$

Several reaction mechanisms for the production of hydrogen iodide (HI) are possible. In the illustrations below, the phases have been omitted to make it easier to follow the steps. However, all the reactants and products of this reaction are gases. Since they are all in the same phase, the reaction is a *homogeneous reaction*.

We cancel out the iodine atoms since they are intermediates and add the two elementary steps to obtain the net equation.

Elementary Step 1: $I_2 \longrightarrow \cancel{2I}$

Elementary Step 2: $\cancel{2I} + H_2 \longrightarrow 2HI$

Net Equation: $I_2 + H_2 \longrightarrow 2HI$

A second possible reaction mechanism has three steps. And again, by canceling the intermediates, we can add the steps to get the net equation.

Elementary Step 1: $I_2 \longrightarrow \cancel{2I}$

Elementary Step 2: $\cancel{I} + H_2 \longrightarrow \cancel{H_2I}$

Elementary Step 3: $\cancel{H_2I} + \cancel{I} \longrightarrow 2HI$

Net Equation: $I_2 + H_2 \longrightarrow 2HI$

Note that this reaction mechanism has an additional intermediate—H_2I.

Reaction mechanisms are temperature dependent; therefore, temperature is the deciding factor in determining which of the possible mechanisms will actually occur.

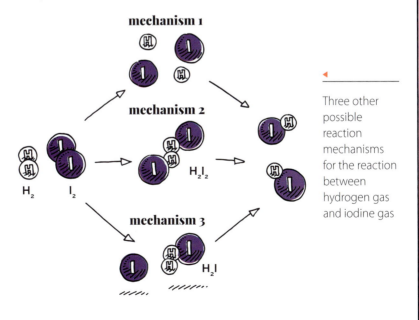

Three other possible reaction mechanisms for the reaction between hydrogen gas and iodine gas

Chemical Kinetics

Rate Laws and Reaction Orders

As stated before, the concentrations of reactants influence reaction rates. But the effect of concentration varies from reaction to reaction. For example, the rate of this reaction

$$2NO + O_2 \longrightarrow 2NO_2$$

has been experimentally determined. The amount concentration of reactants is signified by square brackets. The expression [NO] represents the molar concentration of nitrogen monoxide. It has been experimentally determined that doubling the concentration of nitrogen monoxide quadruples the reaction rate. Tripling the concentration makes the rate nine times greater. This means that the reaction rate is directly related to the square of the concentration of nitrogen monoxide gas. Experimental data also indicates that the rate of this reaction is directly proportional to the concentration of oxygen gas, $[O_2]$. Whatever factor the concentration of oxygen is changed by, the reaction rate changes by the same factor. The **reaction order** for a particular reactant indicates how the rate of a reaction is affected by that specific reactant's concentration. Because the reaction rate is directly proportional to the concentration of oxygen, the concentration of oxygen is linearly related and has an exponent of 1. The nitrogen monoxide concentration is related by a power of two, accounting for the square relationship. The reaction is second order for nitrogen monoxide and first order for oxygen. Remember that the symbol $\propto$ means "is proportional to."

$$rate \propto [NO]^2 \qquad rate \propto [O^2]$$

This expression, called a rate law, can be converted to an equation by the insertion of a numerical constant k. A **rate law** is an equation that mathematically describes how fast a reaction occurs. Each reaction has its own experimentally determined **specific rate constant** (k).

$$rate = k[NO]^2 [O^2]$$

The general form of a rate law is

$$rate = k[A]^m[B]^n,$$

where k is the specific rate constant, [A] and [B] are the molar concentration of the reactants, and m and n are the reaction orders for [A] and [B], respectively.

Table 16-1 gives several examples of rate laws. It is important to note that rate laws must be determined experimentally. If every chemical reaction occurred in a single step—an elementary reaction, as shown in its balanced chemical equation with no intermediates—it would be accurate to use the coefficients from the reactants as exponents for those substances. Single-step reactions are rare, however, so this is seldom an accurate method for writing rate laws. Remember that the exponents in the rate law are experimentally determined. They do not come from the stoichiometric coefficients in the chemical reaction.

In some cases, not all reactants appear in the rate law for the reaction. Changing the concentration of these reactants has no impact at all on the reaction rate. For those reactants, the reaction is zero order. The reactant would therefore not be included in the rate law. Can you find an example of such a reaction in Table 16-1?

Balanced chemical equations show only the reactants and the products they form, not the steps that the reactants take to form those products. Reactions often have several possible mechanisms, and an equation alone cannot reveal which mechanism occurred. If there are several possible reaction mechanisms, the rate law can sometimes be used to eliminate one or more of the possibilities. A possible mechanism can be eliminated if the experimentally determined rate law does not match the theoretical rate law derived from the proposed mechanism.

TABLE 16-1 *Rate Laws for Several Reactions*

Reaction	Rate =
$2H_2 + 2NO \longrightarrow N_2 + 2H_2O$	$k[H_2][NO]^2$
$2NO + Br_2 \longrightarrow 2NOBr$	$k[NO]^2[Br_2]$
$NO + O_3 \longrightarrow NO_2 + O_2$	$k[O_3][NO]$
$2NO + O_2 \longrightarrow 2NO_2$	$k[NO]^2[O_2]$
$C_4H_9Br + 2H_2O \longrightarrow C_4H_9OH + Br^- + H_3O^+$	$k[C_4H_9Br]$

The reaction between nitrogen dioxide and carbon monoxide at low temperatures has the following rate law.

$$NO_2(g) + CO(g) \longrightarrow NO(g) + CO_2(g); \text{ rate} = k[NO_2]^2$$

According to the rate law, changing the concentration of carbon monoxide [CO] has no effect on the rate of the reaction. Therefore, the concentration of carbon monoxide does not appear in the rate law. If this reaction is broken apart into its two most likely elementary steps as shown below, an examination of them will show why this is true.

Elementary Step 1: $NO_2 + NO_2 \longrightarrow NO_3 + NO$
Elementary Step 2: $NO_3 + CO \longrightarrow NO_2 + CO_2$
Net Equation: $NO_2 + CO \longrightarrow NO + CO_2$

carbon dioxide   nitrogen dioxide

Scientists have experimentally determined that the first step, in which the molecules collide, occurs at a slower rate than the second step. Consequently, the first step is the rate-determining step. In reactions with a single step, the rate law would then be based on the original balanced equation. Because carbon monoxide does not appear in the rate-determining step, the reaction rate does not depend on its concentration. *Remember that rate laws are always based on the rate-determining step.*

The exponent 2 on the $[NO_2]^2$ term of the rate law indicates that the concentration of nitrogen dioxide plays the crucial role in determining how fast the reaction occurs. The reaction is second order in NO_2. If the concentration of nitrogen dioxide is tripled, the rate is nine times faster. This relationship occurs because the coefficient 3 is also raised to the second power, $(3[NO_2])^2$.

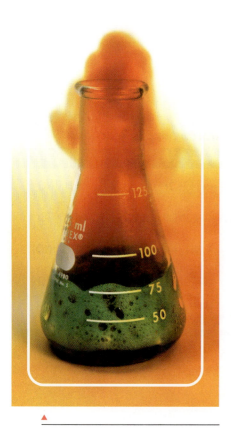

Concentrated nitric acid reacts with copper, generating nitrogen dioxide gas.

EXAMPLE 16-1: FORMULATING A RATE LAW

Nitrogen monoxide gas and chlorine gas combine as shown in the following equation.

$$2NO + Cl_2 \longrightarrow 2NOCl$$

A laboratory scientist collects the following data on three trials of this reaction. State the reaction order for each reactant and determine an appropriate rate law.

REACTION DATA

Trial	Initial [NO] (M)	Initial [Cl$_2$] (M)	Initial Rate (mol/L min)
1	0.50	0.50	0.0190
2	1.00	0.50	0.0760
3	0.50	1.00	0.0380

Solution
Notice that when the concentration of NO was doubled from Trial 1 to Trial 2, the reaction rate increased fourfold (0.0760/0.0190 = 4). Therefore, the reaction is second order in NO.

When the concentration of Cl_2 was doubled from Trial 1 to Trial 3, the reaction rate also doubled. The reaction is first order in Cl_2. In light of this,

$$rate = k[NO]^2[Cl_2].$$

EXAMPLE 16-2: INTERPRETING A RATE LAW

For the rate law $rate = k[A][B]^3$, what will happen to the rate of the reaction

  a. if the initial concentration of A is doubled?

  b. if the initial concentration of A is cut in half?

  c. if the initial concentration of B is doubled?

Solution
  a. The reaction rate will double.

  b. The reaction rate will be cut in half.

  c. The reaction rate will be eight times faster.

CAREERS

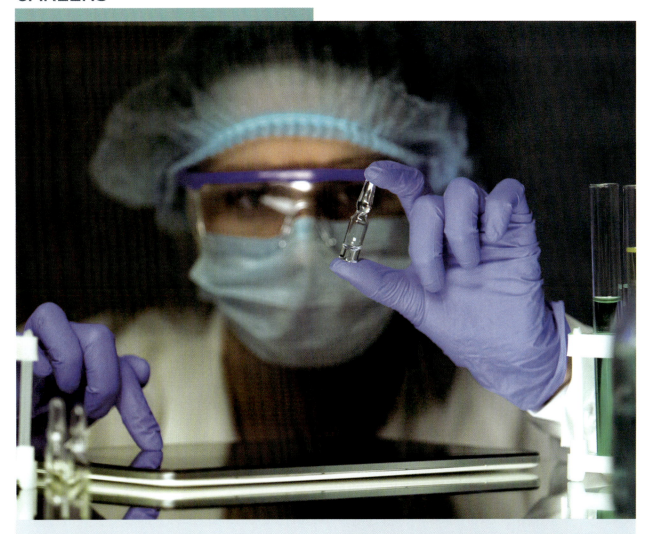

SERVING AS A PHARMACOLOGIST:
LIFESAVING CHEMISTRY

Daniella is a toddler from St. Petersburg, Russia. She was in the hospital recovering from surgery for a life-threatening condition when she contracted HIV during a blood transfusion. Now she will fight a lifelong battle with the disease—and she is not alone. Worldwide, 37.9 million people have contracted HIV, and another 1.7 million join them every year. Pharmacologists around the world are working to develop new drug regimens for diseases like HIV.

A cure for HIV starts in the laboratory of a pharmacologist—a scientist who investigates chemicals for their possible medicinal value. Once they have identified a potential medicine, pharmacologists plan and conduct experiments, testing to see how these chemicals will interact with biological processes in living things. As a drug goes through the approval process for use as a medication, a pharmacologist will test the drug in the laboratory, on animals, and later on people. When testing a medicine for conditions such as HIV, which requires lifelong medication, a pharmacologist must assess the long-term safety of the chemical. Pharmacologists need to be able to solve problems, work with sophisticated technology, collaborate with other scientists, and communicate with and lead others.

The work of pharmacologists helps suffering people. Relieving human suffering is one reason why Jesus came. Pharmacologists help people who are suffering with pain and disease, following Christ's example of shining as lights in a dark world through good works (Matt. 5:16).

Kinetics in the Real World

Pharmacokinetics is the study of how the human body processes medication. Scientists who specialize in this field study how the absorption, distribution, metabolism, and excretion of a drug can extend its beneficial effects. Absorption refers to how the body takes a drug into the bloodstream from its point of entry. Then in the distribution process, the body circulates the drug; the process varies with the drug since different parts of the body are affected differently by a given drug. The body then metabolizes the drug, that is, breaks it down into different components. The liver is the organ that is chiefly responsible for metabolizing drugs. Finally, the drug is eliminated from the body, usually by means of the kidneys. Sometimes drugs simply build up in the body.

The metabolism process of a drug is crucial. The rate at which this process takes place depends on the quantity of the drug and the rate at which the liver can metabolize it. Blood flow and the number of enzymes available to process the drug also affect how fast it is metabolized. The metabolism process, a series of elementary steps, determines how fast the drug takes action to relieve symptoms. The intermediates in the metabolism process, called *metabolites*, may contribute to the side effects of a drug.

Pharmacokinetics can help people with chronic illnesses. For people who need long-term relief of certain symptoms, slowing down the body's processing of medications can extend the duration of their relief from pain, seizures, asthma, or allergies. Time-release medications, which keep a more even concentration of medication in the body, can produce this effect. Sometimes derivations of medications can lessen the side effects and still provide relief from chronic illnesses.

Sustained-Release Medication

In the use of pharmaceutical chemicals, the goal is to maintain the amount of the substance in the body at optimal levels. When we take medicines, the substance goes through a number of physical and chemical processes. The substance often has to be dissolved to be available for transport throughout the body. It then goes through a number of chemical reactions with other substances. Over time, the chemical is processed by the body and usually eliminated. So what happens to the level of the drug in our system?

Upon dissolving, the concentration of the chemical usually increases to a maximum value and then decreases over time depending on the chemical reaction rates and metabolic rate of the individual. Doctors will calculate the dosing and administration schedule to maintain the drug close to the optimal level. But taking medication often can be a hassle and sometimes results in patients missing doses. Pharmacologists have been working on new medications that can alleviate many of these issues.

Pharmacologists often work on existing medicines to make their administration more effective. One advancement has been in the field of sustained-release medications. These are medications that typically contain a higher dose of the medicine but are formulated to release the chemicals over time. This allows doctors to reduce the number of times a medicine is taken and ensures that the chemical remains at the target concentrations longer.

There are a number of approaches that are used to control the release rate of these medicines. In some cases, the medicine is contained within a slow dissolving material. Other times it is enclosed by an insoluble material through which the medicine can diffuse. Some even use osmosis to move water into a capsule enclosing the drug, pushing it out of the capsule and into the body. Some of the most amazing technology actually releases the medicine in response to conditions in the body (e.g., pH levels).

In all these approaches, the pharmacologist is working to improve the lives of people around him. He strives to develop new ways to help doctors treat people in ways that allow them to live productive lives.

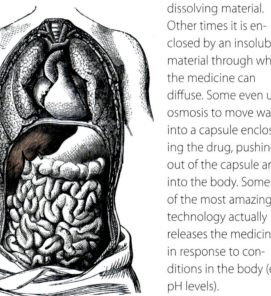

HOW IT WORKS

16.2 SECTION REVIEW

1. What term refers to the steps that make up a reaction mechanism?

2. Why do intermediates not appear in a net equation?

3. There is usually one step in a multi-step reaction that is rate-determining. How does the rate-determining step affect the overall rate of the reaction?

4. Use the reaction mechanism below containing the two elementary steps to answer the following questions.

   1: $2NO_2 \longrightarrow NO_3 + NO$ (slow)
   2: $NO_3 + CO \longrightarrow NO_2 + CO_2$ (fast)

   a. What is the intermediate?
   b. Which step is the rate-determining step?
   c. Write the net equation.

5. What is the net equation the mechanism of the complex reaction shown below?

   Elementary Step 1: $Cl + O_3 \longrightarrow O_2 + ClO$
   Elementary Step 2: $O_3 \longrightarrow O_2 + O$
   Elementary Step 3: $ClO + O \longrightarrow Cl + O_2$

6. What is reaction order?

7. The reaction $2NO(g) + 2H_2(g) \longrightarrow N_2(g) + 2H_2O(l)$ occurs in two steps, which could be either

   A: Elementary Step 1: $2NO + H_2 \longrightarrow N_2 + H_2O_2$
      Elementary Step 2: $H_2O_2 + H_2 \longrightarrow 2H_2O$

   or

   B: Elementary Step 1: $2NO + H_2 \longrightarrow N_2O + H_2O$
      Elementary Step 2: $N_2O + H_2 \longrightarrow N_2 + H_2O$

   a. What are processes A and B called?
   b. How can scientists decide whether A or B is correct?
   c. Suppose a scientist experimentally determines that both gases influence the reaction rate. Which step is the rate-determining step for this reaction? Can the scientist determine whether process A or B is the correct one from the information given? Explain.

8. A scientist is studying the following reaction.

   $2NO + 2H_2 \longrightarrow N_2 + 2H_2O$

   She collects the following data on three trials of this reaction. State the reaction order for each reactant and determine an appropriate rate law. Explain.

   REACTION DATA

Trial	Initial [NO] (M)	Initial [H_2] (M)	Initial Rate (mol/L min)
1	0.20	0.20	3.76×10^{-3}
2	0.40	0.20	1.50×10^{-2}
3	0.20	0.40	7.52×10^{-3}

16 CHAPTER REVIEW

Chapter Summary

TERMS
kinetics
collision theory
activation energy
activated complex
reaction rate
catalyst
homogeneous catalyst
heterogeneous catalyst
enzyme
inhibitor

16.1 REACTION RATES

- Kinetics is the study of the rates of reactions and the steps by which they occur.

- Energy diagrams show the changes in energy through a reaction, including the activation energy and enthalpy changes.

- According to the collision theory, chemical reactions are caused by collisions that are properly oriented and energetic enough to form products.

- The activation energy (E_a) is the minimum amount of kinetic energy that must be possessed by colliding particles before they can react.

- The activated complex is a transitional structure between the reactants and the products. It is unstable and short-lived.

- Reaction rates tell how quickly reactants change into products. They are affected by the nature, concentration, temperature, and surface area of the reactants, along with the presence of catalysts.

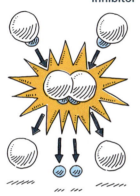

TERMS
reaction mechanism
elementary step
rate-determining step
complex reaction
intermediate
reaction order
rate law
specific rate constant

16.2 REACTION MECHANISMS

- The series of steps that make up a reaction is called a reaction mechanism.

- Complex reactions have at least two elementary steps and form intermediate substances that do not appear in the equation for the net reaction.

- A rate law is an equation that mathematically describes how fast a reaction occurs. These equations take the form

  $$rate = k[A]^m[B]^n,$$

  where k is the specific rate constant, [A] and [B] are the molar concentrations of the reactants, and m and n are the reaction orders for [A] and [B], respectively.

16 CHAPTER REVIEW

Chapter Review Questions

RECALLING FACTS

1. What questions do chemical thermodynamics and kinetics answer?
2. What term do we use to refer to the energy needed for a reaction to occur according to the collision theory?
3. What is an activated complex?
4. Define *reaction rate*.
5. A consumer product is available that, when applied to apple slices, prevents the enzymatic process that normally causes an apple slice to react with atmospheric oxygen and turn brown. Chemically speaking, what term describes this antibrowning agent?
6. What is the difference between a homogeneous catalyst and a homogeneous reaction?
7. What is a reaction mechanism?
8. What is an intermediate?
9. In reactions that have several possible reaction mechanisms, what factor determines the path that the reaction will follow?
10. In addition to the concentration of one or more of the reactants, what is found in all rate laws?

UNDERSTANDING CONCEPTS

11. How is the study of kinetics different from the study of thermodynamics?
12. When the reactants for a reaction with a negative ΔG are mixed, no reaction appears to occur. Suggest an explanation for the nonreaction.
13. According to the collision theory:
    a. Why does an increased temperature increase the reaction rate?
    b. Why does a greater concentration of reactants often increase the reaction rate?
    c. Why do powders react more quickly than a single mass of reactant?
    d. How does the presence of catalysts increase the reaction rate?
14. Draw an energy diagram for both an endothermic and an exothermic reaction. Include the following labels in your diagram: reactants, products, activated complex, activation energy, and ΔH.
15. The two curves in the energy diagram below represent the reaction shown below occurring under two different sets of reaction conditions. Assume that the two graphs are to scale with each other.

    $$2KClO_3(s) \longrightarrow 2KCl(s) + 3O_2(g)$$

    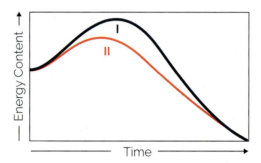

    a. Which set of conditions requires the lower activation energy? Justify your answer.
    b. Which reaction pathway would more likely include a catalyst? Justify your answer.
    c. Which reaction pathway requires higher temperatures? Justify your answer.

16. Of the three energy diagrams shown below, which is the most thermodynamically favorable? Which is the most likely to occur quickly? Explain.

    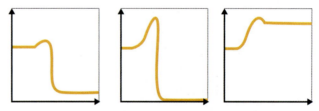

17. What happens to the concentration of the reactants and products in a chemical reaction? Explain.
18. Sugar needs temperatures much higher than 98.6 °F in order to burn. Yet sugar can be "burned" in your cells at this temperature. What is responsible for the ability of your body to burn sugar at this low temperature? Draw energy diagrams that illustrate the difference between the two situations.
19. Create a graphic organizer for kinetics using the terms *activated complex, activation energy, collision theory, complex reaction, elementary step, intermediate, kinetics, rate-determining step, reaction mechanism,* and *reaction rate*.
20. Why do activated complexes and intermediates not appear in net equations?

21. Use the elementary steps for the reaction mechanism below to answer the following questions.

   1: $2NO \longrightarrow N_2O_2$ (fast)
   2: $N_2O_2 + H_2 \longrightarrow N_2O + H_2O$ (slow)
   3: $N_2O + H_2 \longrightarrow N_2 + H_2O$ (fast)

   a. What are the intermediates?
   b. Which step is the rate-determining step?
   c. Write the net equation.

22. Why is it generally impossible to use only a balanced chemical equation to predict a rate law for a particular reaction?

23. A scientist postulates two possible rate laws for a reaction as shown below.

   A: $rate = k[NO_2][F_2]$ or B: $rate = k[NO_2]^2[F_2]$

   a. How can you tell which rate law is correct?
   b. What is the reaction order of F_2 in each rate law?
   c. What is the reaction order of NO_2 in each rate law?
   d. In which of the rate laws would a change in the concentration of NO_2 have a greater influence? Explain.
   e. How would doubling the concentration of F_2 in Rate Law A affect the reaction rate?
   f. From this information, can you tell whether the reaction is a complex reaction or a single-step reaction? Explain.

24. Acetone and bromine combine as shown in the equation below.

   $C_3H_6O + Br_2 \longrightarrow C_3H_5OBr + HBr$

   A scientist collects the following data on three trials of this reaction. State the reaction order for each reactant and determine an appropriate rate law.

   REACTION DATA

Trial	Initial $[C_3H_6O]$ (M)	Initial $[Br_2]$ (M)	Initial Rate (mol/L·min)
1	0.10	0.10	1.64×10^{-5}
2	0.10	0.20	1.64×10^{-5}
3	0.20	0.10	3.28×10^{-5}

CRITICAL THINKING

25. Why will a reaction involving the collision of three molecules proceed more slowly than one involving the collision of two molecules, all other things being equal?

26. Propose a possible two-step mechanism for the decomposition of HgO according to the equation below.

   $2HgO(s) \longrightarrow 2Hg(l) + O_2(g)$

Use the ethics box below to answer Question 27.

27. Research the issue of medical marijuana and write a four-paragraph response to the question of how a Christian should respond to a push within his state to legalize marijuana for medical uses.

ethics

MEDICAL MARIJUANA

Drug laws vary around the world. Even within countries there can be tremendous variation in the laws that exist and how they should be enforced. Within the United States, the federal government maintains a ban on the use of marijuana, while several states have legalized it for medical and even recreational use. Each state has its own program regarding regulation, sale, and enforcement. While some states recognize medical marijuana for patients from outside their state, others do not.

Chapter 17

CHEMICAL EQUILIBRIUM

In the early 1900s, most of the world's fertilizers for growing crops came from Chile's Atacama Desert. There, extensive beds of minerals were rich in sodium nitrate. But this resource was being quickly depleted, and there was growing concern that the world would soon be facing a food crisis.

Then German chemist Fritz Haber and his colleague Karl Bosch developed a chemical process that converted nitrogen in the air into fertilizers for crops. Their innovation, known as the Haber process, helped to dramatically increase farm productivity.

17.1 Equilibrium *409*
17.2 Le Châtelier's Principle *416*
17.3 Solution Equilibrium *424*

17.1 EQUILIBRIUM

What Is Chemical Equilibrium?

When you think about chemical reactions, like the Haber process, you might assume that most of them continue in the forward direction to completion and then stop, and indeed some do. These are known as *irreversible reactions*. But most natural processes are **reversible reactions**—reactions that occur in both the forward and reverse directions. As products form, a portion of those products simultaneously reverts to reactants. The Haber process is based on a reversible reaction of hydrogen and nitrogen. Left on its own, the reaction would quickly reach a balance between forward and reverse reactions. The Haber process gets around this reality by favoring the forward reaction at a high temperature and continuously removing the product as it forms, reducing the likelihood of the reverse reaction occurring.

Can chemical reactions go in the reverse direction?

QUESTIONS
» What does it mean that a reversible reaction is complete?
» Is a reversible reaction ever really finished?
» How can I derive the equation for an equilibrium constant from a balanced chemical equation?
» How can I calculate an equilibrium constant?

TERMS
reversible reaction • chemical equilibrium • equilibrium position • equilibrium constant • law of chemical equilbrium

REVERSIBLE REACTIONS SHARE THE FOLLOWING CHARACTERISTICS:

- Both reactants and products exist together as a mixture.

- The ratio in which reactants and products are present in the reaction vessel depends on the relative speeds of the forward and reverse reactions.

- Changing the rate of either the forward or the reverse reaction changes the ratio of products and reactants in the reaction vessel.

1928 photo of a chemical and pharmaceutical production facility located along the Rhine River in Ludwigshafen, Germany

Chemical equilibrium results when the forward and reverse reactions proceed at the same rate. The high temperature and product removal of the Haber process ensure that this point is never reached. Like the kinetic-molecular and atomic theories, chemical equilibrium is one of the big ideas of chemistry and explains many observations. While the study of thermodynamics determines whether a reaction can occur and kinetics determines how fast it will occur, the study of equilibrium determines how far a reaction will proceed toward completion.

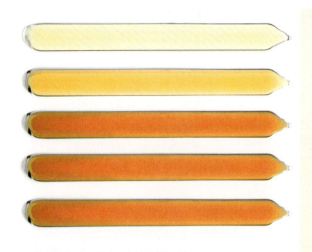

DYNAMIC EQUILIBRIUM

When you think of the term *equilibrium*, you probably picture a balanced scale or a never-ending game of tug-of-war. Yet these examples illustrate only one type of equilibrium—the static equilibrium of balanced physical forces. In chemistry, equilibrium refers to balanced changes, not balanced forces. Chemical equilibrium exists when two opposing reactions occur simultaneously at the same rate, balancing each other. Unlike the examples of balanced forces, chemical equilibria are dynamic. The particles involved constantly move and react. Particle activity continues even though no macroscopic changes seem to be happening.

Consider N_2O_4. This colorless gas can be sealed in a tube and cooled. When it warms, it changes into the brown gas NO_2. The heat energy excites the N_2O_4 particles so that they can break apart to form NO_2.

$$N_2O_4(g) \longrightarrow 2NO_2(g)$$

But not all the NO_2 molecules stay as they are. Some of them recombine to form more N_2O_4. The color changes until the rates of the forward and reverse reactions match each other and the two gases are in equilibrium.

$$2NO_2(g) \longrightarrow N_2O_4(g)$$

At varying temperatures, the gas in the tube is different colors (see above) depending on the proportions of the two gases.

calcium carbonate decomposing

Here's another example. As molecules of calcium carbonate ($CaCO_3$) decompose, they produce calcium oxide (CaO) and carbon dioxide (CO_2). As calcium oxide and carbon dioxide are allowed to accumulate, the chance that any particle of one will collide with a particle of the other increases. Some of these collisions result in the formation of the original reactant. As more calcium oxide and carbon dioxide accumulate, more calcium carbonate is re-formed.

$$CaCO_3(s) \longrightarrow CaO(s) + CO_2(g)$$

$$CaO(s) + CO_2(g) \longrightarrow CaCO_3(s)$$

Eventually the two reactions proceed at the same rate. To an observer it appears that the amounts of calcium carbonate, calcium oxide, and carbon dioxide remain constant; yet at the particle level, the two opposing processes continue. The half-head arrows in the chemical equation below show that both the forward and reverse reactions are occurring at the same rate.

$$CaCO_3(s) \rightleftharpoons CaO(s) + CO_2(g)$$

Physical changes can also be in equilibrium, such as evaporation and dissolving. For example, water molecules in a sealed jar continue to evaporate even after the air is completely saturated. Since the air is saturated, some of the vapor molecules condense at the same rate as liquid molecules evaporate.

$$H_2O(l) \rightleftharpoons H_2O(g)$$

Dissolving processes also reach dynamic equilibria at their saturation points. Particles dissolve and precipitate at equal but opposing rates.

$$NaCl(s) \rightleftharpoons Na^+(aq) + Cl^-(aq)$$

As you can see, chemical equilibria are vitally important. Many industrial processes increase their productivity by manipulating reversible reactions. But how can equilibrium reactions be controlled?

GETTING TO Equilibrium

To get a better understanding of equilibrium, consider the reaction of hydrogen gas with iodine gas to form another gas, hydrogen iodide. The graphs at right show the changing reaction rates and concentrations as this reaction progresses.

$$H_2(g) + I_2(g) \longrightarrow 2HI(g)$$

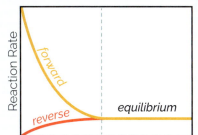
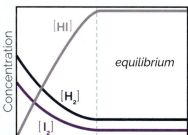

Initially the reaction vessel contains only hydrogen and iodine. The reaction can occur only in the forward direction, so the forward reaction rate is high and the reverse reaction rate is zero. As the reaction proceeds, the concentrations of hydrogen and iodine start to decrease as they are being converted to hydrogen iodide, whose concentration is thus increasing.

As hydrogen iodide is formed, some of it can decompose back into hydrogen and iodine and the equation looks slightly different.

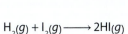

Notice that there are arrows pointing in both directions, though the forward reaction arrow is longer than the reverse reaction arrow. That's because prior to equilibrium, more of the forward reaction is happening. But as the reaction proceeds, the forward rate decreases, while the reverse reaction rate increases.

Ultimately, the forward and reverse reaction rates will be equal—the reaction reaches equilibrium. To the observer the reaction appears to have stopped. In reality, the forward and reverse reactions continue, but the concentration of each reactant and product remains constant. Understand, though, that the concentration of reactants does not necessarily equal the concentration of products when equilibrium has been reached. Each chemical substance achieves its own constant concentration depending on conditions. Again, the graphs show that at equilibrium the reaction rates are equal and the concentrations are constant, but the concentrations are not necessarily equal to each other. The equation is now updated to show both the forward and reverse reaction arrows equal in length.

$$H_2(g) + I_2(g) \rightleftharpoons 2HI(g)$$

Table 17-1 shows the concentrations of hydrogen gas, iodine gas, and hydrogen iodide. The concentrations of the substances at equilibrium is called the **equilibrium position** and is dependent on the conditions at equilibrium. In Section 17.2 we will look at how changing conditions shift the equilibrium position.

TABLE 17-1
Equilibrium Concentrations of H_2, I_2, and HI

	H_2 (g)	I_2 (g)	HI (g)
Beginning Concentration (c)	0.0275	0.0175	0
Change in Concentration (c)	−0.0148	−0.0148	+0.0296
Equilibrium Concentration (c)	0.0127	0.0027	0.0296

Chemical Equilibrium

Equilibrium Constants

Chemists often describe a reaction in equilibrium using a calculated value called an *equilibrium constant*.

Consider a reaction with the generalized equation

$$aA + bB \rightleftharpoons cC + dD,$$

where A, B, C, and D are chemical substances with coefficients *a*, *b*, *c*, and *d*. The **equilibrium constant** (K_{eq}) is a constant for a reversible chemical reaction at a particular temperature and has the following form.

$$K_{eq} = \frac{[C]^c [D]^d}{[A]^a [B]^b}$$

An equilibrium constant is a numerical expression of the ratio of the product equilibrium concentrations to the reactant equilibrium concentrations. If the value of K_{eq} is greater than 1, the forward reaction predominates and the concentrations of products (in the numerator) are greater than the concentrations of the reactants (in the denominator). If the value of K_{eq} is less than 1, the reverse reaction predominates and the concentration of products is small compared with the concentration of reactants.

ammonia

It is possible to determine the equilibrium constant for a reversible reaction because of the **law of chemical equilibrium**. This law states that at a specific temperature, a chemical system may reach a point at which the ratio of the concentration of the products to the reactants is constant.

CONSTANTS FOR HOMOGENEOUS EQUILIBRIA

In the equation $H_2(g) + I_2(g) \rightleftharpoons 2HI(g)$, the relationship for the equilibrium constant is written

$$K_{eq} = \frac{[HI]^2}{[H_2][I_2]}.$$

This reaction is in *homogeneous equilibrium* because all the reactants and products are in the same state.

EXAMPLE 17-1: FORMULATING HOMOGENEOUS EQUILIBRIUM CONSTANTS

Write the equilibrium constant for the formation of ammonia from its elements. The reaction is $3H_2(g) + N_2(g) \rightleftharpoons 2NH_3(g)$.

Solution

$$K_{eq} = \frac{[NH_3]^2}{[H_2]^3[N_2]}$$

CONSTANTS FOR HETEROGENEOUS EQUILIBRIA

Not all reactions that reach equilibrium are limited to substances in the same state. Some involve substances in different states. These can reach a *heterogeneous equilibrium*, that is, one in which not all the substances are in the same physical state.

Writing equilibrium constants for heterogeneous equilibria requires an additional bit of knowledge. Pure solids and liquids don't have concentrations, so their "concentrations" for the purpose of calculating equilibrium constants are both defined as 1. The presence of a solid or liquid in a reaction thus does not affect the value of an equilibrium constant. But gases and substances in aqueous solutions are included. To see how this works in practice, consider the equilibrium between solid zinc and hydrochloric acid to produce zinc chloride and hydrogen gas.

$$Zn(s) + 2HCl(aq) \rightleftharpoons ZnCl_2(aq) + H_2(g)$$

If we include the value for solid zinc, the equilibrium constant is calculated as follows.

$$K_{eq} = \frac{[ZnCl_2][H_2]}{[Zn][HCl]^2}$$

$$= \frac{[ZnCl_2][H_2]}{1\,[HCl]^2}$$

$$= \frac{[ZnCl_2][H_2]}{[HCl]^2}$$

As you can see, the value of 1 for solid zinc has no effect on the calculation and can thus be omitted entirely.

The sublimation of solid iodine crystals into iodine gas in a closed system also reaches a heterogeneous equilibrium.

$$I_2(s) \rightleftharpoons I_2(g) \quad K_{eq} = [I_2]$$

Again, the equilibrium is not dependent on the amount of solid iodine that might be present, so it is not included in the expression.

When acetic acid molecules mix with water, some of them release hydrogen ions to the water to form hydronium ions, H_3O^+. Since only an insignificant amount of water is used up in the reaction, the concentration of water molecules is defined as 1 and should be left out of the equilibrium constant.

$$HC_2H_3O_2(aq) + H_2O(l) \rightleftharpoons C_2H_3O_2^-(aq) + H_3O^+(aq)$$

$$K_{eq} = \frac{[C_2H_3O_2^-][H_3O^+]}{[HC_2H_3O_2]}$$

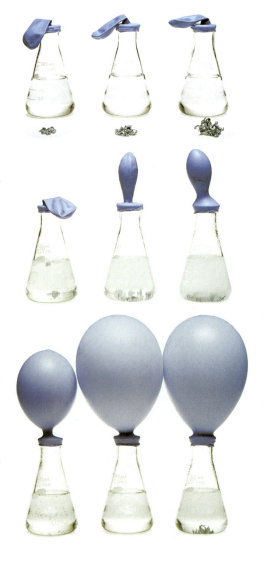

Zinc reacting with HCl to produce zinc chloride and hydrogen gas

EXAMPLE 17-2: FORMULATING HETEROGENEOUS EQUILIBRIUM CONSTANTS

Write the equilibrium constant for the reaction of sodium hydroxide and zinc chloride.

$$2NaOH(aq) + ZnCl_2(aq) \rightleftharpoons 2NaCl(aq) + Zn(OH)_2(s)$$

Solution

Since zinc hydroxide is a solid, its concentration is 1 in the equation for the equilibrium constant.

$$K_{eq} = \frac{[NaCl]^2}{[NaOH]^2[ZnCl_2]}$$

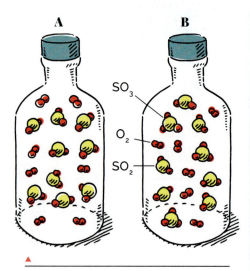

Flask A contains the reactant molecules and Flask B holds molecules of the reaction at equilibrium. Notice that all the reactants are still present with the product but in a different ratio.

DETERMINING THE VALUE OF K_{EQ}

The numerical value of an equilibrium constant must be determined experimentally. Suppose that 5 mol of sulfur dioxide and 4 mol of oxygen are placed in a 1 L container and allowed to react to form sulfur trioxide. At equilibrium, the concentrations are measured. It is found that 1.2 mol of sulfur trioxide, 3.8 mol of sulfur dioxide, and 3.4 mol of oxygen are present.

$$2\,SO_2(g) + O_2(g) \rightleftharpoons 2\,SO_3(g)$$

$$K_{eq} = \frac{[SO_3]^2}{[SO_2]^2[O_2]}$$

$$= \frac{(1.2)^2}{(3.8)^2(3.4)}$$

$$= 0.029$$

The equilibrium constant 0.029 describes the ratio between products and reactants at specific conditions.

EXAMPLE 17-3: CALCULATING THE VALUE FOR K_{EQ}

Calculate the numerical value of K_{eq} for the equilibrium involved in the formation of ammonia from nitrogen and hydrogen.

$$N_2(g) + 3H_2(g) \rightleftharpoons 2NH_3(g)$$

When the mixture of the gases reaches equilibrium, $[N_2] = 0.045$ M, $[H_2] = 0.055$ M, and $[NH_3] = 52$ M.

Write what you know.

$[N_2] = 0.045$ M
$[H_2] = 0.055$ M
$[NH_3] = 52$ M

Write the formula and solve for the unknown.

$$K_{eq} = \frac{[NH_3]^2}{[N_2][H_2]^3}$$

Plug in known values and evaluate.

$$K_{eq} = \frac{(52)^2}{(0.045)(0.055)^3}$$

$$= 3.6 \times 10^8$$

The equilibrium constant is 3.6×10^8.

SIGNIFICANCE OF THE VALUE OF K_{EQ}

Recall that the value of the equilibrium constant can tell you much about the equilibrium reaction. If the value of K_{eq} is much greater than 1, the reaction proceeds almost to completion as written, and the equilibrium lies far to the right. If the value of K_{eq} is much less than 1, the reaction does not proceed very far at all before reaching equilibrium, and the equilibrium lies to the left. In other words, the larger the value of K_{eq}, the more product-favored the reaction is; the lower the value of K_{eq}, the more reactant-favored the reaction is. If the value of K_{eq} lies between these two extremes, then significant concentrations of both reactants and products exist at equilibrium. In Example 17-3, the forward reaction is highly favored, resulting in almost exclusively products.

MINILAB

MIX, CHANGE, REPEAT

How can you tell whether a chemical reaction has taken place? Recall from Chapter 10 that there are different kinds of evidence that suggest that a reaction has occurred. But do they indicate whether a reaction is *finished* (i.e., reached equilibrium)?

Is it always possible to tell whether a reaction has reached equilibrium?

Procedure

Do an internet search using the keywords "Briggs-Rauscher reaction" to find videos that demonstrate the reaction. Select one to view.

1. Describe the first evidence that you observe that a reaction may have taken place.

2. Is there evidence that additional chemical changes occur? If so, describe them.

Conclusion

3. Explain what happens during the Briggs-Rauscher reaction on the basis of what you have learned in this chapter.

Going Further

4. Does the Briggs-Rauscher reaction ever reach equilibrium?

5. What is the term that chemists use to describe reactions like the Briggs-Rauscher reaction?

EQUIPMENT
- none

17.1 SECTION REVIEW

1. Why are most chemical reactions described as being reversible?

2. Under what conditions will a reversible reaction proceed to completion?

3. Explain why you don't run out of reactants if the forward reaction is still occurring at equilibrium.

4. Assuming that you start with only reactants, explain what happens to the reaction rates of forward and reverse reactions and the concentrations of products and reactants from the start of a reaction until equilibrium is reached.

5. Consider the following equation.

    $$CaF_2(s) \rightleftharpoons Ca^{2+}(aq) + 2F^-(aq)$$

    a. Write the equilibrium constant expression for the process.

    b. Is this an example of homogeneous or heterogeneous equilibrium?

6. Write the equilibrium constant expression for the following reaction.

    $$Na_2CO_3(aq) + CaCl_2(aq) \rightleftharpoons CaCO_3(s) + 2NaCl(aq)$$

7. Calculate the equilibrium constant for the chemical equation for the reaction of solid magnesium with hydrochloric acid to form magnesium chloride solution and hydrogen gas. The equilibrium position is [HCl] = 0.051 M, [MgCl$_2$] = 0.513 M, and [H$_2$] = 0.411 M.

8. Consider the reaction
    $$2NO(g) + O_2(g) \rightleftharpoons 2NO_2(g).$$

    a. What is the value of the equilibrium constant for this reaction if the equilibrium concentrations are [NO] = 0.897 M, [O$_2$] = 0.254 M, and [NO$_2$] = 0.0329 M?

    b. If, under the same conditions, [NO] = 1.13 M and [O$_2$] = 0.512 M, what will be the concentration of NO$_2$ in a reaction vessel?

Chemical Equilibrium

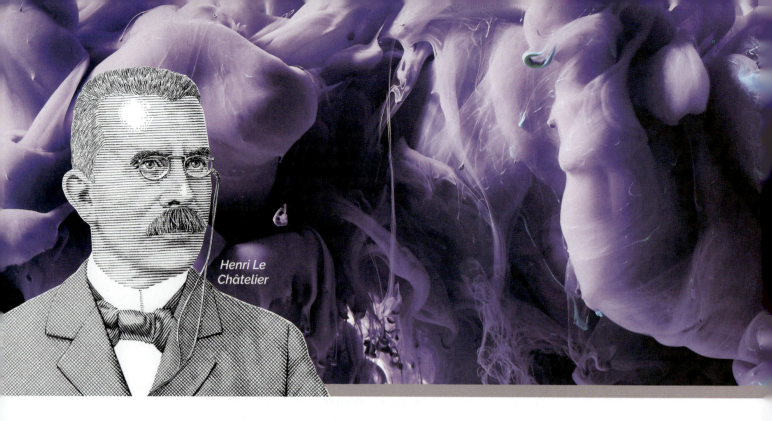

Henri Le Châtelier

17.2 LE CHÂTELIER'S PRINCIPLE

QUESTIONS
» How do different kinds of stresses affect chemical reactions?
» Can I predict how a chemical reaction will shift in response to a stress?

TERMS
Le Châtelier's principle

How can I produce more product without adding more of the reactants?

We have discussed how the concentrations of reactants and products change before they reach dynamic equilibrium. But we've also mentioned that conditions can impact the equilibrium position. How can concentrations be changed once equilibrium has been reached? Is it possible to shift an equilibrium from favoring one side of an equation to favoring the other?

In 1884, a French chemist named Henri Le Châtelier questioned how chemical systems in equilibrium would react if the conditions were changed. His findings, called **Le Châtelier's principle** in his honor, state that when a stress is placed on a reversible process, it will proceed in the direction that relieves the stress. We can put stress on a chemical system in equilibrium by changing the substance concentration, pressure, or temperature. Let's examine how each of these stresses influence equilibrium reactions.

The Effect of Concentration

Look back at Table 17-1 on page 411. What do you think would happen if the initial concentrations of H_2 and I_2 were doubled? More HI would be present at equilibrium. What would happen if HI were present initially? The reverse reaction would produce some H_2 and I_2. Changes in concentration definitely affect chemical equilibria, though not the value of K_{eq}. Consider the following equation for a reaction in equilibrium.

$$2SO_2(g) + O_2(g) \rightleftharpoons 2SO_3(g)$$

At one set of conditions, the K_{eq} has been measured to be 0.029.

$$K_{eq} = \frac{[SO_3]^2}{[SO_2]^2[O_2]} = 0.029$$

Suppose that more of the reactant sulfur dioxide were added to the container. You learned in Chapter 14 that increasing the concentration of a substance increases the rate of the reaction involving it because there are more collisions per unit of time. Thus, the system would be temporarily disturbed from equilibrium. Le Châtelier's principle predicts that the forward reaction would increase in order to reduce the extra sulfur dioxide—to reduce the stress of the additional sulfur dioxide. This reaction would not only increase the concentration of sulfur trioxide but also decrease the concentration of oxygen because it will have to react with the sulfur dioxide to form more sulfur trioxide. Despite all the changes, the equilibrium constant would remain unchanged once the rates again became equal. In other words, after adding the additional sulfur dioxide, the system would establish a new equilibrium position, again with a K_{eq} of 0.029.

If sulfur trioxide were added, the rate of the reverse reaction would increase; the effect would appear to be the consumption of some of the added substance because of it being converted to sulfur dioxide and oxygen. If some sulfur trioxide were removed, the reverse reaction involving it would decrease; in effect, the forward reaction would work to restore some of the lost amount because it would have a greater rate. Of course, the concentrations of sulfur dioxide and oxygen would decrease when the forward reaction predominated. This action would keep the value of the equilibrium constant the same once equilibrium was reestablished.

Remember that the concentrations of solids and liquids do not affect the K_{eq}, which means that adding or removing substances in those states will not place a stress on the system.

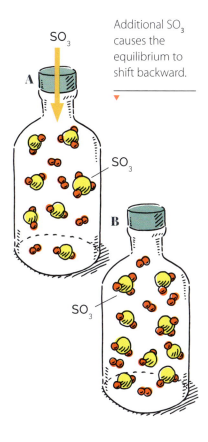

Additional SO_3 causes the equilibrium to shift backward.

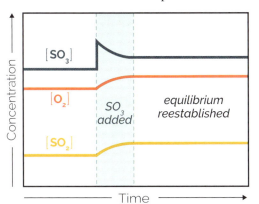

Chemical Equilibrium

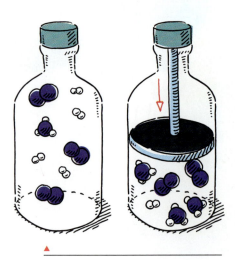

Another example: In the equation $N_2 + 3H_2 \rightleftharpoons 2NH_3$, additional pressure shifts the equilibrium forward.

The Effect of Pressure

Equilibrium reactions involving solids, liquids, and aqueous solutions are not significantly affected by a change in pressure because their particles are held more tightly, but reactions involving gases may be affected by such changes. Recall that 1 mol of any gaseous substance occupies 22.4 L at STP. In the equation

$$2SO_2(g) + O_2(g) \rightleftharpoons 2SO_3(g),$$

2 mol of sulfur trioxide are produced for every 3 mol of gaseous reactants. Since there are fewer gas molecules in the product than in the reactants, the forward reaction decreases the pressure.

But the value of K_{eq} for a reaction cannot be affected by changes in pressure alone. Suppose that the volume of a container was reduced, increasing the pressure inside it. This has the effect of increasing the *concentrations* of all the gases; thus, we can use Le Châtelier's principle to predict the effect. The reaction involving the larger number of molecules—the one whose concentrations are affected more—will increase more than the one with fewer, and the reaction will shift in the direction that produces fewer molecules. In this case, the forward reaction increases because it will reduce the total number of particles and thus counteract the increase in pressure. On the other hand, if the pressure on the reaction at equilibrium were decreased, the opposite would be true. The reverse reaction would have a greater rate and would act to fill the void—more sulfur dioxide and oxygen would be produced. In either case, the changes in reaction rates work to keep K_{eq} for the reaction constant.

Some gaseous reactions are not affected by pressure changes because there are equal numbers of gas particles in the reactants and products. For example, the reaction

$$H_2(g) + I_2(g) \rightleftharpoons 2HI(g)$$

has 2 mol of gas in the products for every 2 mol of gas in the reactants. The pressure of the entire system does not change with either the forward or the reverse reaction. Increased or reduced pressure on this equilibrium will not affect the direction of the reaction. Any reaction that has the same number of moles of *gas* product as it has *gas* reactant will be unaffected by pressure changes that result from volume changes.

EXAMPLE 17-4:
PREDICTING THE EFFECT OF PRESSURE

To produce more carbon dioxide in the following reaction, should you increase or decrease the pressure? Explain.

$$O_2(g) + 2CO(g) \rightleftharpoons 2CO_2(g)$$

Solution
The balanced equation shows that 3 mol of gaseous reactants combine to form 2 mol of gaseous products. High pressure causes the reaction that produces the smallest volume to predominate. Thus, increasing the pressure will produce more carbon dioxide gas.

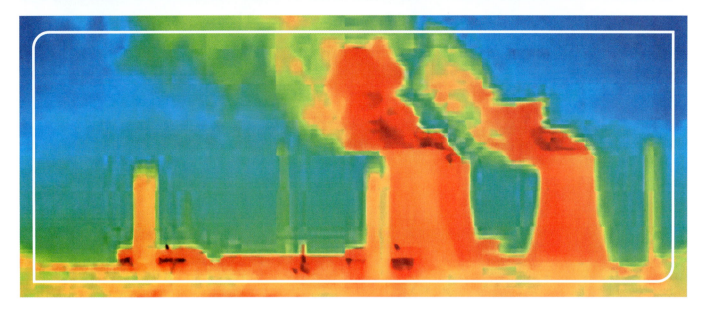

The Effect of Temperature

Recall that the equilibrium constant is a constant at a particular temperature. Therefore, temperature changes *do* change the value of K_{eq}. This is because thermal energy may be thought of as a reactant or product in chemical reactions. The exothermic reaction between sulfur dioxide and oxygen produces sulfur trioxide and thermal energy.

$$2SO_2(g) + O_2(g) \rightleftharpoons 2SO_3(g) + heat$$

If the temperature of this system is increased by adding thermal energy, the reaction will shift to the left, increasing the concentration of the reactants. Since the reactants are in the denominator, the K_{eq} value will decrease.

What would happen if you were to cool the system? If you think that the reaction would shift to the right, you are correct. This would produce more product, and since products are in the numerator of the equilibrium constant, the K_{eq} value would increase.

Le Châtelier's principle predicts how temperature affects each of these equilibria. Since thermal energy acts as a reactant in an endothermic reaction, increasing the temperature favors the forward reaction. The equilibrium of hydrogen, iodine, and hydrogen iodide—

$$H_2(g) + I_2(g) + heat \rightleftharpoons 2HI(g)$$

—will shift to produce more hydrogen iodide if heat is added to the reaction.

EXAMPLE 17-5:
PREDICTING THE EFFECT OF TEMPERATURE

The formation of ammonia is exothermic. Which reaction will be favored by increasing the temperature? What is the effect on the value of K_{eq}?

$$N_2(g) + 3H_2(g) \rightleftharpoons 2NH_3(g) + heat$$

Solution
Since thermal energy can be considered a product, additional heat encourages the reverse reaction and hinders the forward reaction. The value of K_{eq} decreases.

Additional heat shifts the equilibrium to the left in the equation $N_2(g) + 3H_2(g) \rightleftharpoons 2NH_3(g) + heat$.

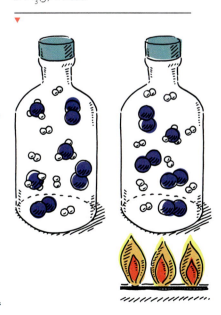

The Effect of Catalysts

As discussed in Chapter 16, catalysts are substances that change the rate of a reaction without undergoing any permanent change themselves. Since catalysts affect the rate of reaction, many people think that adding more catalyst will change the equilibrium position. However, a catalyst affects both the forward and reverse reactions, causing equilibrium to be reached sooner than if no catalyst were added. And though it causes equilibrium to be reached more quickly, the addition of a catalyst has no effect on either equilibrium concentrations or the value of K_{eq}.

Table 17-2 summarizes the effects on equilibrium reactions and K_{eq} by the addition of catalysts and by changes in concentration, pressure, and temperature.

Table 17-2

Effects of Stress on Equilibrium Concentrations and K_{eq}

Stress	Shift	Effect	Value of K_{eq}
add catalyst	none	none	no change
Increase Concentration			
of reactant	forward/toward right	more products form	no change
of product	reverse/toward left	more reactants form	no change
Decrease Concentration			
of reactant	reverse/toward left	more reactants form	no change
of product	forward/toward right	more products form	no change
Change Pressure			
increase	toward side with fewer moles of gas		no change
decrease	toward side with more moles of gas		no change
Increase Temperature			
exothermic	reverse/toward left	more reactants form	decreases
endothermic	forward/toward right	more products form	increases
Decrease Temperature			
exothermic	forward/toward right	more products form	increases
endothermic	reverse/toward left	more reactants form	decreases

The Haber Process

Huge quantities of ammonia (NH_3) are used every year in fertilizers, cleaning compounds, and explosives. Until World War I, industrial countries obtained nitrogen by importing saltpeter mined from Chilean *caliche* (a mineral partly composed of KNO_3 and $NaNO_3$). All countries had access to nitrogen (N_2) in the atmosphere, but no one knew how to liberate the atoms from the stable molecules. As an alternative to the expense of importing saltpeter, countries sought to form ammonia from atmospheric nitrogen.

Finding the appropriate reaction was no problem. The reaction

$$N_2(g) + 3H_2(g) \rightleftharpoons 2NH_3(g) + heat$$

seemed perfect. The relatively large equilibrium constant of this reaction (6.4×10^2 at 25 °C) indicated that the production of ammonia was the favored direction of the reaction. The problem was that nitrogen, hydrogen, and ammonia took a very long time to reach equilibrium. In 1909, Fritz Haber found a way to manipulate the conditions to overcome this difficulty. Today the Haber process provides ammonia for use in numerous industries.

The Haber process uses the following technique. Because 2 mol of gas are formed from 4 mol of gas, the products take up less space than the reactants. Le Châtelier's principle predicts that high pressures favor the production of ammonia. Because the forward reaction is exothermic, it proceeds best at low temperatures. Thus, engineers needed a process that used high pressure and low temperatures to make ammonia manufacturing easier.

Fritz Haber

In the Haber process, the reaction takes place under high pressures, at high temperatures, and with a catalyst. However, high temperatures significantly shift the equilibrium away from the formation of ammonia (at 500 °C, $K_{eq} = 1.5 \times 10^{-5}$). To compensate, the Haber process manipulates other variables to increase the production of ammonia. A catalyst (mostly Fe_3O_4 with traces of K_2O and Al_2O_3) helps to speed up the reaction. To further increase the yield, ammonia is removed by liquefaction as it forms. This continual removal keeps the forward reaction going.

HOW IT WORKS

A schematic of the ammonia synthesis loop of the Haber process

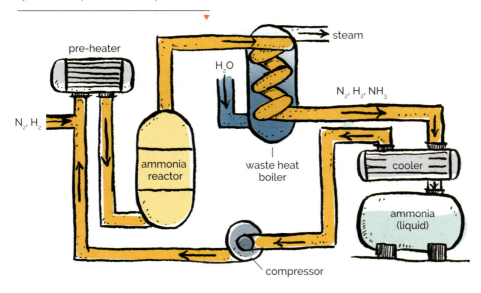

Equilibria and Industry

Chemical industries supply society with such diverse products as medicines, pesticides, fertilizers, paints, textiles, detergents, cosmetics, plastics, fuels, and building materials. Because the reactions used to make those products are usually reversible, the laws of equilibrium chemistry govern the processes. Methods of describing and controlling equilibria become indispensable tools in the fast, efficient, and economical production of consumer goods. The industrial production of esters illustrates this.

The characteristic flavors and odors of fruits are the results of naturally occurring compounds called *esters*. These compounds are parts of foods (left), cosmetics, and medicines. Because it is inefficient and expensive to extract esters from fruit, manufacturers of flavorings form esters artificially by reacting alcohols and organic acids.

$$\text{alcohol} + \text{organic acid} \rightleftharpoons \text{ester} + \text{water}$$

WORLDVIEW INVESTIGATION

ETHANOL

Introduction

Ethanol is an alcohol that forms naturally from the fermentation of the glucose found in fruits, vegetables, and grains. During this process, yeast breaks down glucose to form ethanol and carbon dioxide.

$$C_6H_{12}O_6 \rightleftharpoons 2C_2H_5OH + 2CO_2$$

Several factors, such as the presence of oxygen, affect the speed, progress, and direction of the reaction. This is the same process that makes bread rise. The ethanol in bread dough is decomposed by the high heat of baking, but of course the ethanol in fermented beverages is still there when they are consumed. Once in the body, ethanol has an effect on the equilibrium of the body's chemical processes.

But the modern use of ethanol isn't limited to beverages. Ethanol is used in industry for feedstock, antiseptics, paints, and personal care products like deodorant and shampoo. Ethanol is also used as a vehicle fuel. So are ethanol and the chemical processes used to make it good or evil?

Task

A chemical company plans to open a new ethanol plant in your town. Is this something that you as a Christian can or should support? Formulate a position paper on the new plant. You can kick-start your research by thinking about and answering the following questions.

Should scientists continue to find new ways to use ethanol?

Should a Christian use ethanol?

What should determine how a chemical process like fermentation is used?

How can we use the chemistry of ethanol for God's glory?

Procedure

1. Plan your position paper on the basis of your research. Cite any sources you used and note whether they display any evidence of a particular worldview regarding the topic.

2. Write your position paper and show it to another person for evaluation. Make any suggested revisions, if necessary.

3. Present your position paper to your class or family.

Conclusion

The question of whether to use a particular advancement of science is sometimes cut-and-dried for a Christian—but not always. Many times, the discoveries and products of scientific inquiry have both beneficial and detrimental applications. Christians must always be willing to be led by the Spirit and ready to apply biblical principles to questions regarding the usage of new technologies.

The general reaction works well, but it is very easily reversed. Neither the forward nor the reverse reaction dominates. Thus, efficient production of esters is difficult. Instead of going to completion, the reaction converts only a part of the reactants into the products. Rather than accept limited yields, manufacturers produce more ester per batch of reactants by forcing the forward reaction to prevail.

Manufacturers can shift the equilibrium by continually adding a reactant, removing a product, or both. If either the acid or the alcohol is inexpensive, huge quantities can be added to drive the equilibrium forward. Sometimes manufacturers can easily remove a product by boiling it off as it forms. Regardless of the technique that is used, ester production is increased.

Worldview Conflict and Equilibrium

Haber's use of equilibrium chemistry to efficiently produce ammonia on a large scale was perhaps the most significant scientific breakthrough of his century. It created a boom in agriculture and made food more plentiful and affordable. Haber's work is an example of using the earth's resources to help God's image bearers.

But as we saw with the work of Alfred Nobel in Chapter 10, chemistry can be a key to unlocking potential for both good *and* evil. The same ammonia-making process that served worldwide agriculture was also used to help create German munitions during World War I. Haber's work as director of the Kaiser Wilhelm Institute of Physical Chemistry and Electrochemistry in Berlin was also instrumental in developing gas warfare. He personally supervised the first use of poison gas at Ypres, Belgium, in 1915.

So how do we know whether we should turn the key of chemistry to unlock its potential? How can we judge whether it will bring good or evil? Chemistry itself is not the answer. Scripture gives us a moral compass and a worldview to see these issues as God sees them. We must apply the power of God's Word to the tool of science for His glory and mankind's greatest good.

A soldier wearing a gas mask in World War I

17.2 SECTION REVIEW

1. Give a definition of Le Châtelier's principle in your own words.

2. Summarize how changes to the concentration of chemical substances affect a system in equilibrium.

3. Explain why changing the temperature of a system in equilibrium also changes the K_{eq} value.

4. Consider the following exothermic reaction at equilibrium.

$$CO_2(g) + 4H_2(g) \rightleftharpoons CH_4(g) + 2H_2O(g)$$

   a. What is the effect of removing some of the CH_4? Explain.

   b. What is the effect of increasing the pressure? Explain.

   c. What is the effect of increasing the temperature? Explain.

5. Predict the shift in equilibrium (to the right or to the left) under the following conditions for the following reaction.

$$CaF_2(s) \rightleftharpoons Ca^{2+}(aq) + 2F^-(aq) \quad \Delta H < 0$$

   a. CaF_2 solid is added.

   b. Ca^{2+} aqueous ions are added.

   c. F^- aqueous ions are removed.

   d. Temperature is increased.

6. How can the chemical manufacturing industry utilize Le Châtelier's principle?

17.3 SOLUTION EQUILIBRIUM

QUESTIONS
» What does a solubility product constant tell us about the solubility of a substance?
» Can I predict whether a mixture of two solutions will produce a precipitate?

TERMS
solubility product constant • common-ion effect

Is it possible to calculate how much of a substance will dissolve?

barium sulfate

Ionic Equilibria

Saturated solutions are a system of undissolved solids and dissolved ions in equilibrium. Some salts, such as calcium sulfate, are slightly soluble.

$$CaSO_4(s) \rightleftharpoons Ca^{2+}(aq) + SO_4^{2-}(aq)$$

Others, such as iron(III) hydroxide, dissolve so little that for all practical purposes they are insoluble.

The forward process is the dissolving process, and the reverse process is precipitation. As with all equilibria, the rate of the forward process matches the rate of the reverse process at equilibrium. Since this process is at equilibrium, the relative amounts of dissolved ions and undissolved solids can be expressed with equilibrium constants. The equilibrium constant for dissolving of $Pb_3(PO_4)_2$ is expressed as

$$K_{eq} = [Pb^{2+}]^3[PO_4^{3-}]^2.$$

As in other equilibrium expressions, solids are not included; thus, the $[Pb_3(PO_4)_2]$ is omitted in the expression of the equilibrium constant. The resulting equilibrium constant for dissolving a solute into a saturated solution is called the **solubility product constant** (K_{sp}). The equilibrium concentrations of the ions are raised to the appropriate powers as in other equilibrium constant expressions.

Solubility and Equilibria

The K_{sp} constant quantifies salt solubilities. Large numbers of ions in solution will cause the solubility product to be relatively large (much greater than 1), and the material is considered soluble. Insoluble materials will produce small numbers of ions in solution, causing the K_{sp} to be small (less than 1). Thus, the larger its K_{sp} is, the more soluble is the salt. Conversely, the smaller the value for K_{sp}, the less soluble the salt is. The K_{sp} of some slightly soluble salts are listed in Table 17-3 (p. 426).

K_{sp} values can be determined from measured solubilities. For example, barium sulfate has a solubility of 1.1×10^{-5} M. When barium sulfate dissolves, each formula unit releases one Ba^{2+} ion and one SO_4^{2-} ion.

$$1.1 \times 10^{-5} \text{ mol } BaSO_4(s) \rightleftharpoons 1.1 \times 10^{-5} \text{ mol } Ba^{2+}(aq) + 1.1 \times 10^{-5} \text{ mol } SO_4^{2-}(aq)$$

$$\begin{aligned}K_{sp} &= [Ba^{2+}][SO_4^{2-}] \\ &= (1.1 \times 10^{-5})(1.1 \times 10^{-5}) \\ &= 1.2 \times 10^{-10}\end{aligned}$$

The K_{sp} of a substance can be used to determine the solubility in moles per liter. The K_{sp} of lead(II) phosphate is known to be 1.5×10^{-32}. The equations for its dissociation and solubility product constant can be written as follows.

$$Pb_3(PO_4)_2(s) \rightleftharpoons 3Pb^{2+}(aq) + 2PO_4^{3-}(aq)$$

$$K_{sp} = [Pb^{2+}]^3 [PO_4^{3-}]^2$$

$$K_{sp} = [Pb^{2+}]^3 [PO_4^{3-}]^2$$

Notice that when 1 mol of lead(II) phosphate dissolves, 3 mol of Pb^{2+} ions and 2 mol of PO_4^{3-} ions are released. The number of Pb^{2+} ions in a saturated solution must be three times the amount of $Pb_3(PO_4)_2$ units that were dissolved. Likewise, there must be twice as many PO_4^{3-} ions as the number of dissolved formula units. Suppose that S stands for the solubility of the salt, in this case $Pb_3(PO_4)_2$. The concentration of Pb^{2+} ions can be expressed as $3S$, and the concentration of PO_4^{3-} ions can be expressed as $2S$. The solubility product constant can now be expressed in terms of the solubility of the salt.

$$K_{sp} = (3S)^3(2S)^2$$

To find the solubility of the salt, we can substitute the value of K_{sp} and solve for S.

$$1.5 \times 10^{-32} = (3S)^3(2S)^2$$
$$1.5 \times 10^{-32} = (27S^3)(4S^2)$$
$$\frac{1.5 \times 10^{-32}}{108} = \frac{108S^5}{108}$$
$$\sqrt[5]{1.3888 \times 10^{-34}} = \sqrt[5]{S^5}$$
$$S = 1.7 \times 10^{-7}$$

The solubility of lead(II) phosphate is 1.7×10^{-7} mol/L. This amount is so small that the salt is considered insoluble.

salt dissolving

EXAMPLE 17-6: DETERMINING SOLUBILITY PRODUCT CONSTANT

Given that the solubility of copper(II) hydroxide is 1.76×10^{-7} M, what is the K_{sp}?

Write what you know.

$$Cu(OH)_2(s) \rightleftharpoons Cu^{2+}(aq) + 2OH^-(aq)$$

Every mole of $Cu(OH)_2$ yields 1 mol of Cu^{2+} and 2 mol of OH^-

Solubility $Cu(OH)2 = S = 1.76 \times 10^{-7}$ mol/L
$K_{sp} = ?$

Write the formula and solve for the unknown.

$$K_{sp} = [Cu^{2+}][OH^-]^2$$

Plug in known values and evaluate.

$$K_{sp} = (S)(2S)^2$$
$$= (1.76 \times 10^{-7})[(2)(1.76 \times 10^{-7})]^2$$
$$= (1.76 \times 10^{-7})(3.52 \times 10^{-7})^2$$
$$= (1.76 \times 10^{-7})(1.239\,04 \times 10^{-13})$$
$$= 2.18 \times 10^{-20}$$

The solubility product constant is 2.18×10^{-20}.

EXAMPLE 17-7: DETERMINING SOLUBILITY

Given that the K_{sp} of silver chloride is 1.8×10^{-10}, determine its solubility.

Write what you know.

$AgCl(s) \rightleftharpoons Ag^+(aq) + Cl^-(aq)$
$K_{sp} = 1.8 \times 10^{-10}$
$[Ag^+] = [Cl^-] = [AgCl]$
$[AgCl] = ?$

Write the formula and solve for the unknown.

$K_{sp} = [Ag^+][Cl^-]$
$K_{sp} = [AgCl][AgCl]$
$\sqrt{K_{sp}} = \sqrt{[AgCl]^2}$
$[AgCl] = \sqrt{K_{sp}}$

Plug in known values and evaluate.

$[AgCl] = \sqrt{1.8 \times 10^{-10}}$
$= 1.3 \times 10^{-5}$

The solubility of silver chloride is 1.3×10^{-5} mol/L.

TABLE 17-3 K_{sp} for Minimally Soluble Substances

Salt	Product	K_{sp} (at 25 °C)	Molar Mass (g)	Solubility (g/L)
aluminum hydroxide, $Al(OH)_3$	$[Al^{3+}][OH^-]^3$	3.0×10^{-34}	78.0	1.28×10^{-7}
barium carbonate, $BaCO_3$	$[Ba^{2+}][CO_3^{2-}]$	2.6×10^{-9}	197.34	1.0×10^{-2}
barium sulfate, $BaSO_4$	$[Ba^{2+}][SO_4^{2-}]$	1.1×10^{-10}	233.39	2.4×10^{-3}
calcium carbonate, $CaCO_3$	$[Ca^{2+}][CO_3^{2-}]$	3.3×10^{-9}	100.09	5.7×10^{-3}
calcium sulfate, $CaSO_4$	$[Ca^{2+}][SO_4^{2-}]$	4.9×10^{-5}	136.14	0.95
iron(II) hydroxide, $Fe(OH)_2$	$[Fe^{2+}][OH^-]^2$	8.0×10^{-16}	89.87	5.3×10^{-4}
iron(III) hydroxide, $Fe(OH)_3$	$[Fe^{3+}][OH^-]^3$	3.0×10^{-39}	106.88	1.1×10^{-8}
lead(II) carbonate, $PbCO_3$	$[Pb^{2+}][CO_3^{2-}]$	1.5×10^{-13}	267.25	1.0×10^{-4}
lead(II) phosphate, $Pb_3(PO_4)_2$	$[Pb^{2+}]^3[PO_4^{3-}]^2$	1.5×10^{-32}	811.66	1.4×10^{-4}
lead(II) sulfate, $PbSO_4$	$[Pb^{2+}][SO_4^{2-}]$	2.1×10^{-8}	303.30	4.4×10^{-2}
magnesium carbonate, $MgCO_3$	$[Mg^{2+}][CO_3^{2-}]$	2.7×10^{-6}	84.32	0.139
magnesium hydroxide, $Mg(OH)_2$	$[Mg^{2+}][OH^-]^2$	5.6×10^{-12}	58.33	6.5×10^{-4}
silver carbonate, Ag_2CO_3	$[Ag^+]^2[CO_3^{2-}]$	8.5×10^{-12}	275.75	3.5×10^{-2}
silver chloride, $AgCl$	$[Ag^+][Cl^-]$	1.8×10^{-10}	143.32	1.9×10^{-3}
silver phosphate, Ag_3PO_4	$[Ag^+]^3[PO_4^{3-}]$	1.5×10^{-18}	418.58	6.4×10^{-3}
zinc hydroxide, $Zn(OH)_2$	$[Zn^{2+}][OH^-]^2$	3×10^{-17}	99.40	1.9×10^{-4}

The K_{sp} can be used to see whether a solution is unsaturated, saturated, or supersaturated. If the product of ion concentrations equals the K_{sp}, the solution is saturated. If the product of ion concentrations is less than the K_{sp}, the solution is unsaturated. A product greater than the K_{sp} indicates a supersaturated solution. Another way to determine whether a solution is unsaturated, saturated, or supersaturated is to compare the solubility of the solute to the concentration of the solution. Be sure, however, to compare it to the concentration in mol/L, not in g/L.

EXAMPLE 17-8: DETERMINING SATURATION STATE

Calculate the product of the ion concentrations of a 2.5×10^{-6} M zinc hydroxide solution. Determine whether the solution is unsaturated, saturated, or supersaturated.

Write what you know.

$Zn(OH)_2(s) \rightleftharpoons Zn^{2+}(aq) + 2OH^-(aq)$
$[Zn(OH)_2] = 2.5 \times 10^{-6}$ M
$[Zn^{2+}] = [Zn(OH)_2] = 2.5 \times 10^{-6}$ M
$[OH^-] = 2[Zn(OH)_2] = 5.0 \times 10^{-6}$ M
$K_{sp} = 3 \times 10^{-17}$ from Table 17-3
$[Zn^{2+}][OH^-]^2 = ?$

Write the formula and solve for the unknown.

$[Zn^{2+}][OH^-]^2$

Plug in known values and evaluate.

$(2.5 \times 10^{-6})(5.0 \times 10^{-6})^2 = (2.5 \times 10^{-6})(2.5 \times 10^{-11})$
$= 6.3 \times 10^{-17}$

This product is greater than the K_{sp} of 3×10^{-17}, so the solution must be supersaturated.

Common-Ion Effect

Since salts dissociate when they dissolve, solutes that have a common ion can affect the equilibrium. For example, an unsaturated silver chloride solution has a concentration less than 1.33×10^{-5} M and an ion concentration product less than the K_{sp} of a saturated solution. What happens when sodium chloride is added to the silver chloride solution, since they both have a chloride ion in common?

Sodium chloride is very soluble. It dissolves and dissociates into sodium and chloride ions. Though the sodium ions have little or no effect on the original solution, the chloride ions do. The chloride ion is called the *common ion* because it is contained in both silver chloride and sodium chloride. The additional chloride ions from the sodium chloride cause the equilibrium to shift to the left in the reaction

$$AgCl(s) \rightleftharpoons Ag^+(aq) + Cl^-(aq).$$

In this example, AgCl is less soluble than NaCl. Adding NaCl shifts the reaction to produce more solid silver chloride. The increased concentration of chloride ions raises the ion concentration product above its K_{sp} value. The result of the **common-ion effect** is that the less soluble salt precipitates out of the solution until the K_{sp} value is reached. Remember that the K_{sp}, like any other equilibrium constant, is a particular value at a given temperature. If that value is exceeded, precipitation will occur to reestablish equilibrium conditions.

Precipitation Reactions

silver chloride precipitate

As we have seen, mixing two solutions can sometimes cause a precipitate to form. For instance, suppose that hydrochloric acid (HCl) and silver nitrate ($AgNO_3$) are mixed together. The resulting solution contains H^+, Cl^-, Ag^+, and NO_3^- ions. Hydrogen chloride and silver nitrate are quite soluble. But an alternate combination of ions, AgCl, is not. A silver chloride precipitate will form if enough silver and chloride ions are in the solution. How much is enough? That depends on the K_{sp} of silver chloride. If the value of $[Ag^+][Cl^-]$ in the solution exceeds the K_{sp} for silver chloride, a precipitate will form. How much precipitate will be deposited? Enough precipitate will form to lower the value of $[Ag^+][Cl^-]$ to the K_{sp} for silver chloride. Thus, the K_{sp} of a salt can be used to predict whether a precipitate will form and how much will form.

EXAMPLE 17-9: PREDICTING PRECIPITATION

Will a precipitate form when equal volumes of a 5.0×10^{-5} M barium nitrate solution and a 1.0×10^{-3} M sodium carbonate solution are mixed?

Write what you know.

Both barium nitrate and sodium carbonate are soluble salts. At this low concentration, they will both be completely dissolved. The first step is to calculate the theoretical concentration of each ion after both salts completely dissolve. Divide their molarities in half since the combined solution has twice the volume.

$$Ba(NO_3)_2(s) \longrightarrow Ba^{2+}(aq) + 2NO_3^-(aq)$$
$$[Ba^{2+}] = 5.0 \times 10^{-5} \text{ M}/2 = 2.5 \times 10^{-5} \text{ M}$$
$$[NO_3^-] = 2(5.0 \times 10^{-5} \text{ M})/2 = 5.0 \times 10^{-5} \text{ M}$$

$$Na_2CO_3(s) \longrightarrow 2Na^+(aq) + CO_3^{2-}(aq)$$
$$[Na^+] = 2(1.0 \times 10^{-3} \text{ M})/2 = 1.0 \times 10^{-3} \text{ M}$$
$$[CO_3^{2-}] = 1.0 \times 10^{-3} \text{ M}/2 = 5.0 \times 10^{-4} \text{ M}$$

Plug in known values and evaluate.

The next step is to check Table 17-3 to determine which combination of ions is most likely to precipitate. Both original combinations are very soluble, and so is sodium nitrate. Barium carbonate is the only salt formed by a combination of the ions in this solution that appears in Table 17-3 ($K_{sp} = 2.6 \times 10^{-9}$). The final step is to determine whether the value of $[Ba^{2+}][CO_3^{2-}]$ exceeds the given K_{sp}.

$$[Ba^{2+}][CO_3^{2-}] = (2.5 \times 10^{-5})(5.0 \times 10^{-4}) = 1.25 \times 10^{-8}$$

Since 1.25×10^{-8} is larger than the K_{sp} of 2.6×10^{-9}, a precipitate will form.

EXAMPLE 17-10: PREDICTING THE AMOUNT OF PRECIPITATE

In Example 17-9, a precipitate forms. If the total volume in the reaction vessel is 1.00 L, how many grams of barium carbonate will precipitate out of solution?

Write what you know.

$[Ba^{2+}] = 2.5 \times 10^{-5}$ M
$[CO_3^{2-}] = 5.0 \times 10^{-4}$ M
$K_{sp} = 2.6 \times 10^{-9}$

$Ba^{2+}(aq) + CO_3^{2-}(aq) \longrightarrow BaCO_3(s)$

Plug in known values and evaluate.

The ion concentrations must be reduced so that the ion concentration product equals the K_{sp}. Since one barium ion combines with one carbonate ion, each concentration will be reduced by the same amount (x).

$[Ba^{2+}][CO_3^{2-}] = 2.6 \times 10^{-9}$

$(2.5 \times 10^{-5} - x)(5.0 \times 10^{-4} - x) = 2.6 \times 10^{-9}$
$1.25 \times 10^{-8} - 5.25 \times 10^{-4}x + x^2 - 2.6 \times 10^{-9} = 0$
$9.9 \times 10^{-9} - 5.25 \times 10^{-4}x + x^2 = 0$

$x = 5.05 \times 10^{-4}$ or 1.95×10^{-5}

Exclude the first solution because it is more than the starting amount.

Convert volume of solution to mass of precipitate.

$1.00 \, \cancel{L} \left(\dfrac{1.95 \times 10^{-5} \, \cancel{\text{mol BaCO}_3}}{1 \, \cancel{L}} \right) \left(\dfrac{197.34 \text{ g BaCO}_3}{1 \, \cancel{\text{mol BaCO}_3}} \right)$
$= 3.85 \times 10^{-3}$ g $BaCO_3$

Therefore, 3.85×10^{-3} g of $BaCO_3$ will precipitate.

17.3 SECTION REVIEW

1. Define *solubility product constant*.

2. The solubility of copper(I) chloride is 1.08×10^{-2} g per 100 mL of solution. Find the K_{sp} value.

3. What is the solubility of lithium phosphate in mol/L if its K_{sp} value is 3.48×10^{-9}?

4. An experimental solution has an ion concentration product of 2.5×10^{-12}. That same solution has a known K_{sp} of 3.2×10^{-10}. Is the experimental solution saturated, unsaturated, or supersaturated? Explain your answer.

5. Is a 4.72×10^{-6} M $Fe(OH)_2$ solution unsaturated, saturated, or supersaturated? ($K_{sp} = 8.0 \times 10^{-16}$)

6. What is the common ion in a mixture of Na_2CO_3 and $MgCO_3$?

7. Define *common-ion effect*.

8. Suppose you mix equal volumes of 3.50×10^{-3} M sodium hydroxide with 2.76×10^{-4} M magnesium nitrate. Does a precipitate form? If so, what is it?

17 CHAPTER REVIEW

Chapter Summary

17.1 EQUILIBRIUM

- Many chemical reactions are reversible. The reactants and products can reach a state of dynamic equilibrium at which point the reaction rates of the forward and reverse reactions are equal.

- At equilibrium, the concentration of each chemical substance reaches a constant value even though both the forward and reverse reactions continue.

- An equilibrium constant (K_{eq}) is a numerical expression of the ratio of products to reactants and is constant for a specific temperature.

- An equilibrium-constant expression is determined from the balanced equilibrium reaction. The equilibrium constant equals the product of the concentrations of the products divided by the product of the concentrations of the reactants, all raised to the power of their stoichiometric coefficients.

- The law of chemical equilibrium states that at a fixed temperature, a chemical system may reach a point where the ratio between the concentration of the products and reactants has a constant value.

- In a homogeneous equilibrium, all the reactants and products are in the same state. In a heterogeneous equilibrium, two or more states are represented.

- The numerical value of an equilibrium constant at a specific temperature must be determined experimentally and is usually expressed without units.

TERMS
reversible reaction
chemical equilibrium
equilibrium position
equilibrium constant
law of chemical equilibrium

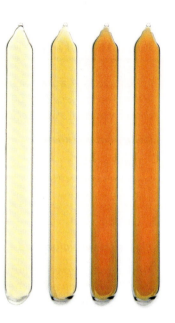

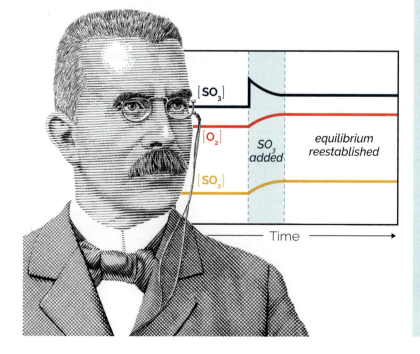

TERMS
Le Châtelier's principle

17.2 LE CHÂTELIER'S PRINCIPLE

- Le Châtelier's principle states that when a system in equilibrium is disturbed, it will shift in the direction that relieves the stress. In chemistry, the stress may be a change in concentration, pressure, or temperature.

- Pressure changes affect the equilibrium only in reactions involving gases that have unequal numbers of moles on each side of the equation.

- A catalyst speeds up both the forward and the reverse reactions, establishing equilibrium sooner. Catalysts do not affect equilibrium concentrations or the value of K_{eq}.

TERMS
solubility product constant
common-ion effect

17.3 SOLUTION EQUILIBRIUM

- The solubility product constant (K_{sp}) is a specific K_{eq} that expresses the solubility of a salt at a specific temperature. The value of K_{sp} for a salt is for a saturated solution and can be calculated from measured solubilities.

- The K_{sp} can be compared with experimental concentration data to determine a solution's state of saturation.

- When two salts contain a common ion in the same solution, the common-ion effect can cause the less soluble salt to precipitate out of solution.

Chapter Review Questions

RECALLING FACTS

1. Define *chemical equilibrium*.
2. What is the difference between a dynamic equilibrium and a static one?
3. Define *equilibrium position*.
4. Which states of matter are included in equilibrium constant expressions? Explain.
5. What generalizations about an equilibrium constant can be made if the value for K_{eq} is large? if the value is small?
6. Name three factors that can shift equilibria.
7. Which factor(s) will change the K_{eq} as well as shift equilibria?
8. Summarize how changes to temperature affect a system in equilibrium.

UNDERSTANDING CONCEPTS

9. Your study partner says that according to the law of chemical equilibrium, in a reaction at equilibrium, the ratio between reactants and products is 1:1. Is she correct? Explain.
10. Write the equilibrium constant expression for each of the following reactions.

    a. Laughing gas can decompose into nitrogen and oxygen.

    $2N_2O(g) \rightleftharpoons 2N_2(g) + O_2(g)$

    b. Carbon monoxide can be converted into methane in one of the processes involved in converting coal into a gas.

    $CO(g) + 3H_2(g) \rightleftharpoons CH_4(g) + H_2O(g)$

    c. Methanol can be synthesized from carbon monoxide and hydrogen.

    $CO(g) + 2H_2(g) \rightleftharpoons CH_3OH(g)$

    d. Baking soda can extinguish fires as it decomposes to produce water vapor and carbon dioxide.

    $2NaHCO_3(s) \rightleftharpoons Na_2CO_3(s) + H_2O(g) + CO_2(g)$

11. A scientist experimenting with the Haber process,

    $N_2(g) + 3H_2(g) \rightleftharpoons 2NH_3(g) + heat$,

    seeks to determine the equilibrium constant at 450 °C.

    a. After permitting a mixture of N_2 and H_2 to react and reach equilibrium, he finds that $[N_2]$ = 1.00 M, $[H_2]$ = 0.0769 M, and $[NH_3]$ = 0.000 200 M. Calculate K_{eq} at this temperature.

    b. The scientist repeats the experiment under the same conditions with different initial amounts of gas in the reaction vessel. The scientist finds that $[H_2]$ = 0.0375 M and $[NH_3]$ = 0.000 318 M. Use the value of K_{eq} already determined to calculate $[N_2]$.

Chemical Equilibrium 431

17 CHAPTER REVIEW

12. Acetic acid ($HC_2H_3O_2$) is the compound that gives vinegar its distinctive smell and taste. When dissolved in water, it ionizes according to the reaction below.

    $$HC_2H_3O_2(aq) + H_2O(l) \rightleftharpoons C_2H_3O_2^-(aq) + H_3O^+(aq)$$

    a. A chemist working in a clinical research laboratory dissolves some acetic acid in water (at 25 °C) and finds that
    $[H_3O^+] = 1.01 \times 10^{-5}$ M,
    $[C_2H_3O_2^-] = 1.01 \times 10^{-5}$ M, and
    $[HC_2H_3O_2] = 5.67 \times 10^{-6}$ M.
    What is the value of K_{eq}?

    b. If $[H_3O^+] = 3.6 \times 10^{-3}$ and $[C_2H_3O_2^-] = 3.6 \times 10^{-3}$, what is the concentration of un-ionized acetic acid at the same temperature?

    c. What is the value of the equilibrium constant if the equation were written reversed?

13. Oxalic acid ($H_2C_2O_4$) experiences a two-step dissociation in an aqueous solution:

    $$H_2C_2O_4 \rightleftharpoons H^+ + HC_2O_4^-,$$
    where $K_{eq} = 5.90 \times 10^{-2}$ at 25 °C,

    and

    $$HC_2O_4^- \rightleftharpoons H^+ + C_2O_4^{2-},$$
    where $K_{eq} = 6.40 \times 10^{-5}$ at 25 °C.

    The equilibrium concentration of $[H^+] = 1.95 \times 10^{-3}$ M and $[HC_2O_4^-] = 1.95 \times 10^{-3}$.

    a. What is the equilibrium concentration of $H_2C_2O_4$?

    b. What is the equilibrium concentration of $C_2O_4^{2-}$?

    c. What is the equilibrium constant for the total reaction?

    $$H_2C_2O_4 \rightleftharpoons 2H^+ + C_2O_4^{2-}$$

14. Explain why a change in pressure shifts a reversible reaction that is in equilibrium as it does.

15. Predict the effect of increasing the pressure on each of the four equilibria given in Question 10.

16. In an equilibrium reaction that is endothermic, will the addition of thermal energy cause the equilibrium to shift toward the reactants or toward the products? Explain.

17. How does adding more catalyst affect reversible reactions?

18. Consider the following exothermic reaction at equilibrium.

    $$CO_2(g) + 4H_2(g) \rightleftharpoons CH_4(g) + 2H_2O(g)$$

    a. What is the effect of adding more carbon dioxide? Explain.

    b. What is the effect of decreasing the pressure? Explain.

    c. What is the effect of decreasing the temperature? Explain.

19. Lithium carbonate (Li_2CO_3) is less soluble in hot water than in cold water.

    a. Do you think the dissolution of Li_2CO_3 is exothermic or endothermic? Explain.

    b. What will happen if a saturated solution of Li_2CO_3 at 25 °C is heated to 100 °C? Explain.

20. Write a K_{sp} expression for the dissolution of each of the following.

    a. AgBr
    b. MgF_2
    c. Ag_2S
    d. $Ba_3(PO_4)_2$
    e. $PbCl_2$

432 Chapter 17

21. A scientist measures the solubility of $Mg(OH)_2$. The concentration of the Mg^{2+} ions in a saturated solution at 18 °C is 1.44×10^{-4} M, and the concentration of OH^- in the same solution is 2.88×10^{-4} M. What is the value of the K_{sp} at this temperature?

22. Using the solubilities and molar masses in Table 17-3 (p. 426), calculate the K_{sp} for each of the following.

    a. Ag_2CO_3
    b. $BaCO_3$
    c. $Fe(OH)_3$
    d. $PbSO_4$

23. Calculate the solubilities (mol/L) of each of the following.

    a. barium fluoride, BaF_2 ($K_{sp} = 1.7 \times 10^{-6}$)
    b. cadmium carbonate, $CdCO_3$ ($K_{sp} = 2.5 \times 10^{-14}$)
    c. lanthanum iodate, $La(IO_3)_3$ ($K_{sp} = 6.0 \times 10^{-10}$)
    d. silver arsenate, Ag_3AsO_4 ($K_{sp} = 1.0 \times 10^{-22}$)

24. A saturated solution of Li_2CO_3 has both Li^+ and CO_3^{2-} ions in it. What will happen if some solid LiCl is added to the solution? (LiCl is much more soluble than Li_2CO_3.)

25. Will a precipitate form if equal volumes of the following solutions are mixed? If so, what substance(s) will precipitate? (*Hint*: First determine whether each substance is soluble or only minimally soluble. If a salt is minimally soluble, refer to its K_{sp}.)

    a. 0.05 M NaCl and 0.05 M LiCl
    b. 6.3×10^{-4} M $MgCO_3$ and 1.0×10^{-4} M $Mg(OH)_2$
    c. 0.5 M NaOH and 2.0×10^{-5} M $Mg(OH)_2$

26. Summarize the process for determining whether a precipitate forms in a double-replacement reaction.

CRITICAL THINKING

27. The reaction $N_2(g) + 3H_2(g) \rightleftharpoons 2NH_3(g)$ starts with the following concentrations: $[N_2] = 0.251$ M, $[H_2] = 0.653$ M, and $[NH_3] = 0.674$ M. If the K_{eq} for this reaction at the current temperature is 62.5, which reaction—the forward or the reverse—must occur to reach equilibrium?

28. The oxygen that you breathe attaches to a protein called hemoglobin (Hb) in red blood cells to form oxyhemoglobin as shown in the reaction below. This carries oxygen from your lungs to other parts of your body that need it.

    $$Hb(aq) + O_2(g) \rightleftharpoons HbO_2(aq)$$

    a. Is the equilibrium in the production of oxyhemoglobin homogeneous or heterogeneous? Explain.

    b. Write the equilibrium constant for the production of hemoglobin.

    c. Why do you think people who live at high elevations don't experience hypoxia, a lack of oxygen in the bloodstream?

    d. Why do you think people with severe hypoxia receive oxygen and not hemoglobin?

29. Sickle-cell anemia is a disease in which a person's hemoglobin has a mutation that often results in misshapen blood cells. The mutation affects the blood cells' ability to deliver oxyhemoglobin and reduces their ability to carry proper hemoglobin. What kind of serious effects could this have?

30. Barium sulfate ($BaSO_4$) is administered to people when x-rays of their digestive systems are taken, despite the fact that the Ba^{2+} ion is toxic to humans. Suggest a method by which $BaSO_4$ could be administered without the danger from Ba^{2+} ions. Base your method on the common-ion effect.

Chemical Equilibrium

Chapter 18

ACIDS, BASES, AND SALTS

Everyone knows that driving under the influence is a real problem on our highways. Deaths caused by driving under the influence are a significant percentage of all vehicular deaths. In 2016 the Centers for Disease Control reported that 10,497 deaths—28% of all auto accident fatalities—were linked to alcohol impairment. According to Federal Bureau of Investigation statistics, over a million drivers were arrested for driving under the influence in 2016. While that number seems like a lot, it is less than 1% of the 111 million people who reported that they had driven while impaired during that year. How can we keep drunk drivers off the road?

Police can measure the blood alcohol content (BAC) of people suspected of drinking and driving by using a Breathalyzer. This device works by reacting the alcohol in a person's breath with sulfuric acid. Throughout this chapter, we will learn about acids and bases and the reactions that they undergo.

18.1 Defining Acids and Bases *435*
18.2 Acid-Base Equilibria *440*
18.3 Neutralization *453*

18.1 DEFINING ACIDS AND BASES

We live in a world full of chemicals, and many of these chemicals are compounds called acids and bases. How can we know whether a chemical is an acid or a base? Since ancient times, people have identified acids and bases by their properties. As early as 1663, Robert Boyle classified many common substances as acids or bases on the basis of their physical properties. But it would be another two centuries before scientists began to explain what causes these properties.

Observable Properties of Acids

Aqueous solutions of acids typically have certain physical properties in common, one of which is a sour taste. For example, the characteristic sour taste of citrus fruits is caused by citric acid. Acetic acid puts the pucker in pickles. Aspirin tastes sour because of acetylsalicylic acid. Lactic acid's smell and flavor help us recognize sour milk. In fact, sourness is so closely tied to acids that their name comes from the Latin *acidus*, meaning "sour." Besides tasting sour, acidic solutions can also conduct an electric current because the acid dissociates or ionizes. For example, hydrochloric acid (HCl) ionizes into hydrogen ions and chloride ions. Any substance that produces ions, which conduct electricity in a solution, is called an **electrolyte**.

Why do we have to test the pH of pool water?

Acids can also be identified by their chemical properties. Many metals react with acids, with the metal replacing the hydrogen in the solution. Often the products of these reactions are hydrogen gas and a salt, though other products are possible. When copper metal reacts with nitric acid (see below), the copper replaces the hydrogen to produce copper(II) nitrate. A toxic reddish-brown gas—nitrogen dioxide gas—and water are also products.

$$Cu(s) + 4HNO_3(aq) \longrightarrow Cu(NO_3)_2(aq) + 2NO_2(g) + 2H_2O(l)$$

A simple test using litmus paper makes one of the chemical properties of acids observable. As litmus paper is produced, it is mixed with a chemical extracted from certain lichens. So litmus paper can be used to identify acids—an acid will turn blue litmus paper red.

QUESTIONS
» What are the characteristics of acids and bases?
» Why are there different definitions for acids and bases?
» What are conjugate acid-base pairs?

TERMS
electrolyte • neutralization reaction • Arrhenius acid • Arrhenius base • Brønsted-Lowry acid • Brønsted-Lowry base • conjugate base • conjugate acid • Lewis acid • Lewis base

TABLE 18-1
Common Acids

Name	Source	Formula
acetic acid	vinegar	$HC_2H_3O_2$
ascorbic acid	vegetables, vitamin C	$H_2C_6H_6O_6$
carbonic acid	soft drinks	H_2CO_3
citric acid	citrus fruit	$H_3C_6H_5O_7$
formic acid	bee venom	HCO_2H
lactic acid	sour milk, sore muscles	$H_2C_3H_4O_3$
oxalic acid	rhubarb	$H_2C_2O_4$
sulfuric acid	battery acid	H_2SO_4

TABLE 18-2 *Common Bases*

Name	Source	Formula
ammonium hydroxide	household ammonia	NH_4OH
calcium hydroxide	lime	$Ca(OH)_2$
magnesium hydroxide	milk of magnesia	$Mg(OH)_2$
sodium hydroxide	lye	$NaOH$

Observable Properties of Bases

Bases, like acids, can be identified by their physical properties. Most bases have a bitter taste, which you might have experienced by accidentally tasting soap. This explains why very few foods contain strong bases. Bases also feel slippery. The soapy feeling associated with bases results from a reaction between the base and the fatty acids and oils in your skin. Bases in solution also dissociate or ionize, allowing them to conduct electric current, signifying that they are also electrolytes.

Chemical properties are also helpful in identifying bases. Litmus paper can be used to identify bases—they turn red litmus paper blue.

Acids and bases have another chemical property—they can neutralize each other in a chemical reaction. A **neutralization reaction** between an acid and a base in an aqueous solution produces a salt and water. For example, the reaction of sulfuric acid with sodium hydroxide produces sodium sulfate and water.

$$H_2SO_4(aq) + 2NaOH(aq) \longrightarrow Na_2SO_4(aq) + 2H_2O(l)$$

Svante Arrhenius

THE Arrhenius MODEL

We have described acids and bases, but how do we determine whether a compound is an acid or a base? Since the physical properties of acids and bases can vary so much, scientists have focused on the chemical properties of acids and bases to define them.

Acid definition: *releases* H^+
Base definition: *releases* OH^-

The earliest of the modern acid-base definitions was proposed in the 1880s by Swedish chemist Svante Arrhenius, who classified acids and bases by the ions that they produce in aqueous solution. **Arrhenius acids** ionize to release hydrogen ions (H^+) in aqueous solutions. Note that a hydrogen ion is just a proton or a hydrogen nucleus. This hydrogen ion is not stable in solution, so it bonds to a water molecule to form H_3O^+—a hydronium ion. In this and many other textbooks, the H^+ and H_3O^+ symbols are considered interchangeable. **Arrhenius bases** are substances that dissociate to release hydroxide ions (OH^-) in aqueous solutions.

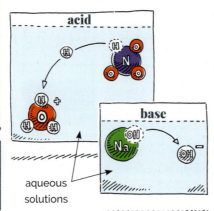

aqueous solutions

Hydrogen-containing compounds are not always acids. For example, methane (CH_4) has four hydrogen atoms, but none of these hydrogen atoms will ionize. Thus, methane is not considered an acid. Likewise, compounds with OH groups in their formulas are not always bases. Methanol (CH_3OH) does not normally release its OH group.

The Arrhenius model is useful but limited because it deals only with compounds in aqueous solutions. Normally this is not a significant limitation, but modern chemistry is pioneering research in acid-base reactions that occur in nonaqueous solvents or even without solvents. The Arrhenius definition of bases does not recognize compounds such as ammonia as being bases because these compounds do not contain an OH group. Nevertheless, ammonia exhibits properties long associated with bases. Although the Arrhenius model has some use, it is not as functional as other definitions. As we have seen often in science, continued research requires that we refine our models to explain new data.

THE *Brønsted-Lowry* MODEL

Some scientists felt that the Arrhenius model was too restrictive in that it dealt only with substances in aqueous solutions. Working independently, Danish chemist Johannes Brønsted and British chemist Thomas Lowry proposed new definitions for acids and bases in 1923. Their definitions, instead of focusing on the release of ions, were based on the transfer of a proton between substances.

Acid definition: *donates protons (H^+)*

Base definition: *accepts protons (H^+)*

A **Brønsted-Lowry acid** is a substance that donates a proton, and a **Brønsted-Lowry base** is a substance that accepts a proton. The process of losing a proton is called *deprotonation*, and the process of gaining a proton is called *protonation*.

As Brønsted-Lowry acids and bases interact, they produce what are called conjugates. The **conjugate base** of a Brønsted-Lowry acid is the acid minus the proton. The **conjugate acid** of a Brønsted-Lowry base is the base with the addition of one proton. For example, when acetic acid mixes with ammonia, it loses a proton (deprotonates) to its conjugate base (acetate ion). The ammonia (base) accepts the proton, forming an ammonium ion (the conjugate acid). The ammonium ion is an acid because it can lose a proton that would be accepted by the acetate ion (a base).

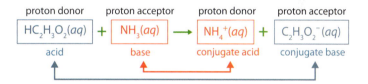

$$HC_2H_3O_2(aq) + NH_3(aq) \rightarrow NH_4^+(aq) + C_2H_3O_2^-(aq)$$

acid — base — conjugate acid — conjugate base

Notice that these conjugate pairs differ only by a single proton (H^+). The molecules in blue are one conjugate pair, and the molecules in red are another.

The Brønsted-Lowry model encompasses all Arrhenius acids and bases plus many others that do not fit the Arrhenius definition. The Brønsted-Lowry definition of acids is essentially the same as the Arrhenius definition because a hydrogen ion is a proton, but the definition of bases differs greatly. The Brønsted-Lowry definition expands the number of substances called bases because it includes many substances that lack the hydroxide ions.

TABLE 18-3 *Conjugate Pairs of Brønsted-Lowry Acids and Bases*

Conjugate Acid		Conjugate Base	
Name	Formula	Formula	Name
perchloric acid	$HClO_4$	ClO_4^-	perchlorate ion
hydrochloric acid	HCl	Cl^-	chloride ion
sulfuric acid	H_2SO_4	HSO_4^-	hydrogen sulfate ion
hydrogen sulfate ion	HSO_4^-	SO_4^{2-}	sulfate ion
acetic acid	$HC_2H_3O_2$	$C_2H_3O_2^-$	acetate ion
ammonium ion	NH_4^+	NH_3	ammonia
water	H_2O	OH^-	hydroxide ion

↑ increasing acid strength

Gilbert Lewis

THE *Lewis* MODEL

American chemist Gilbert Lewis, who developed Lewis structures, published his ideas about acids and bases in 1923, the same year that Brønsted and Lowry reported their findings. His model, though not as commonly used as the Brønsted-Lowry model, is based on bonding and molecular orbital structure and is the broadest definition of acids and bases. It includes many more substances than either the Arrhenius or Brønsted-Lowry model.

Acid definition: *accepts electron pairs*

Base definition: *donates electron pairs*

A **Lewis acid** is any substance that can accept a pair of electrons, and a **Lewis base** is a substance that can donate a pair of electrons. A Lewis acid has at least one empty orbital or bonding site, and a base has at least one unbonded pair of electrons.

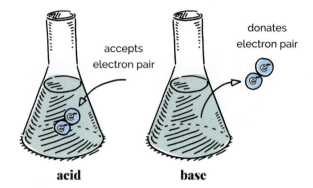

The formation of ammonium ions (NH_4^+) in water, is a reaction between a Lewis acid and a Lewis base.

$$H:\underset{H}{\overset{H}{N}}:H + H:\overset{..}{\underset{..}{O}}:H \longrightarrow \left[H:\underset{H}{\overset{H}{N}}:H\right]^+ + \left[H:\overset{..}{\underset{..}{O}}:\right]^-$$

Lewis base   Lewis acid

Ammonia is a Lewis base because it can donate a pair of electrons in the formation of a covalent bond. The water molecule acts as a Lewis acid by accepting an electron pair to form a hydroxide ion.

Another example of a reaction between a Lewis acid and a Lewis base is ammonia reacting with boron trifluoride.

$$H-\underset{H}{\overset{H}{N}}: + \underset{F}{\overset{F}{B}}-F \longrightarrow H-\underset{H}{\overset{H}{N}}-\underset{F}{\overset{F}{B}}-F$$

Lewis base   Lewis acid

Again, ammonia acts as a Lewis base because it can donate a pair of electrons. Boron trifluoride acts as the Lewis acid by accepting the electron pair, forming a Lewis *adduct*—the product of combining two molecules.

CASE STUDY

ROYAL ACID TO THE RESCUE

In the spring of 1940, as the German army was moving through Denmark, Danish chemist Niels Bohr was in trouble. On a shelf in his laboratory sat two gold Nobel Prize medals, which had been smuggled out of Germany. Bohr was holding them for two German scientists, Max von Laue and James Franck, for safekeeping. What should he do with the medals?

George de Hevesy, a Hungarian chemist, had an idea—use chemistry. Hevesy spent the rest of the day slowly dissolving the medals in *aqua regia*, a very caustic mixture of nitric acid and hydrochloric acid. *Aqua regia* means "royal acid" and is so named because of its ability to dissolve precious metals such as gold and platinum.

When the soldiers arrived, they ransacked the laboratory. Finding nothing of value, they left, leaving behind an innocent-looking orange solution high on a shelf.

After the war, de Hevesy returned to Copenhagen and precipitated the gold out of the solution. Bohr sent the metal back to the Nobel Foundation where the medals were recast. Max von Laue and James Franck were again presented with their Nobel medals in 1952.

18.1 SECTION REVIEW

1. List four common properties of acids.

2. Formic acid (HCOOH) ionizes in water to form formate and hydronium ions.
    a. Write the equation for the ionization of formic acid in water.
    b. What is the acid in the forward reaction?
    c. What is the conjugate base of this acid?
    d. What is the name of the process by which formic acid loses a proton?

3. Define *acid* and *base* according to each model.
    a. Arrhenius model
    b. Brønsted-Lowry model
    c. Lewis model

4. What substance produced in aqueous solutions by H_2SO_4 makes it react strongly with many other substances? What ion is responsible for the corrosiveness of strong bases like NaOH?

5. Summarize the categories of acids and bases, especially how they are related to each other.

6. Identify each of the following as an Arrhenius acid or base, a Brønsted-Lowry acid or base, or a Lewis acid or base. Some substances may fit into more than one model.
    a. HCl
    b. Br^-
    c. CN^-

7. Write the words that complete the statements.
    a. H_2SO_4 is an Arrhenius _____ and therefore must be a Brønsted-Lowry _____.
    b. The Cl^- ion can be classified as a(n) _____ base or a(n) _____ base, but not as a(n) _____ base.

Use the case study above to answer Questions 8–9.

8. What property of acids was demonstrated in the case study?

9. Why did de Hevesy have to use the very caustic *aqua regia* to dissolve the medals?

18.2 ACID-BASE EQUILIBRIA

Some acids are safe to eat and drink, such as citric acid in orange juice. But other acids are dangerous, even deadly. For example, perchloric acid, which explodes on contact with cloth or wood, would wreak havoc as soon as it touched your mouth! What's the difference? Using the Brønsted-Lowry definition, the difference is the degree to which the acids release protons. Many acid-base reactions are reversible processes. Equilibrium constants describe how readily acids deprotonate and bases protonate. These equilibrium constants give scientists a way to describe the acidity or alkalinity of a solution.

QUESTIONS
» What does it mean to say that water is self-ionizing?
» How do we measure acidity and alkalinity?
» Are all acids and bases equally strong?
» What determines the strength of an acid or base?
» How do acid-base indicators work?

TERMS
self-ionization of water • pH • acidic solution • basic solution • neutral solution • pOH • strong acid • weak acid • weak base • strong base • acid-ionization constant • base-ionization constant • amphoteric substance • monoprotic acid • polyprotic acid • diprotic acid • triprotic acid • acid-base indicator • transition interval

If vinegar is an acid, why can we put it on foods?

The Self-Ionization of Water

In addition to all its unique physical properties, water has an intriguing chemical property—it can react with itself. The reaction is an acid-base reaction in which one water molecule donates a proton to another water molecule. Electrical conductivity experiments, which show that pure water is a weak electrolyte, verify that the reaction involves acids and bases. The two water molecules combine to form a hydroxide ion and a hydronium ion—a reaction called the **self-ionization of water**. In the reaction, one molecule acts as an acid and the other acts as a base.

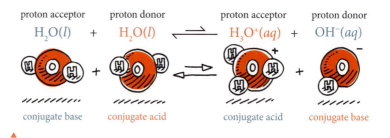

When a proton is transferred from one water molecule to another, self-ionization occurs.

The reaction is an equilibrium reaction in which the reverse reaction predominates. Only a few molecules produce ions. This reaction can be described with an equilibrium constant.

$$K_{eq} = \frac{[H_3O^+][OH^-]}{[H_2O]^2}$$

Since water is the solvent, the water's concentration can't change; it can be considered part of K_{eq}, thus eliminating it from the equation.

$$K_{eq} = K_w = [H_3O^+][OH^-]$$

The resulting constant is called the *ionization constant of water* (K_w). Experimental evidence shows that the concentrations of hydronium and hydroxide ions are both 1.0×10^{-7} mol/L in pure water at 25 °C. Only 1 out of 555 million water molecules is ionized at any given instant. The product of these two concentrations gives K_w a value of 1.0×10^{-14} at 25 °C.

$$K_w = [H_3O^+][OH^-]$$
$$= (1.0 \times 10^{-7} \text{ M})(1.0 \times 10^{-7} \text{ M})$$
$$= 1.0 \times 10^{-14}$$

The value of K_w is a very important number. Whether a solution is acidic, basic, or neutral, the product of the concentration of hydronium ions and hydroxide ions always equals 1.0×10^{-14} at 25 °C.

EXAMPLE 18-1: CALCULATING AMOUNT CONCENTRATION FROM HYDRONIUM ION CONCENTRATION

The concentration of hydronium ions in a mild acid is found to be 5.2×10^{-7} mol/L. What is the concentration of hydroxide ions?

Write what you know.

$[H_3O^+] = 5.2 \times 10^{-7}$
$[OH^-] = ?$

Write the formula and solve for the unknown.

Given that $K_w = [H_3O^+][OH^-]$, solve for $[OH^-]$.

$K_w = [H_3O^+][OH^-]$

$$\frac{K_w}{[H_3O^+]} = \frac{[\cancel{H_3O^+}][OH^-]}{[\cancel{H_3O^+}]}$$

$$[OH^-] = \frac{K_w}{[H_3O^+]}$$

Plug in known values and evaluate.

$$[OH^-] = \frac{1.0 \times 10^{-14}}{5.2 \times 10^{-7}}$$

$$= 1.9 \times 10^{-8}$$

The concentration of OH^- ions is 1.9×10^{-8} M.

Acids, Bases, and Salts

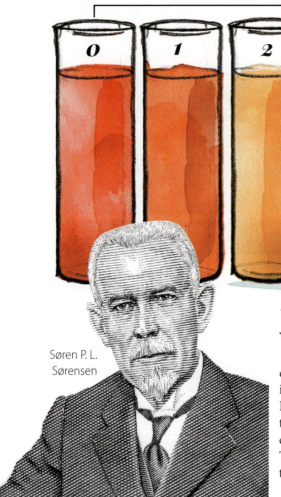

Søren P. L. Sørensen

The pH Scale

You can see that the concentrations of hydronium and hydroxide ions in aqueous solutions are very small, frequently ranging between 1 and 1×10^{-14} mol/L. Scientists find it awkward to deal regularly with such small numbers. Søren P. L. Sørensen, a Danish chemist, proposed the pH scale in 1909 to provide a clear, concise, and convenient way to describe the concentrations of these ions without using scientific notation. The **pH**, potential of hydrogen ions, is related to the concentration of the hydronium ion by the equation

$$pH = -\log[H_3O^+].$$

For example, consider a sample of pure water.

Its hydronium ion concentration is 1.0×10^{-7} mol/L. Let's calculate the pH of a sample of pure water.

$$pH = -\log[H_3O^+]$$
$$= -\log 1.0 \times 10^{-7} \text{ M}$$
$$= -(-7.00) = 7.00$$

Thus, the pH of pure water is 7.00. Significant figures for pH calculations are a special case. The significant figures from the amount concentration determine the number of decimal places to the right of the decimal point because the digits to the left of the decimal point are related to the exponent in the amount concentration. But what does the pH tell us about the solution?

The pH value describes the acidity or alkalinity of a solution. All solutions with a pH less than 7 are called **acidic solutions** because the hydronium ion concentration is greater than the hydroxide ion concentration. All solutions with a pH greater than 7 are called **basic solutions**, or *alkaline solutions*, because the hydroxide ions outnumber the hydronium ions. A solution with a pH of 7 is a **neutral solution**. A sample of pure water is thus a neutral solution.

EXAMPLE 18-2: CALCULATING pH FROM HYDRONIUM ION CONCENTRATION

The hydronium ion concentration in a shampoo is 2.0×10^{-5} M. What is the pH of this shampoo?

Write what you know.

$[H_3O^+] = 2.0 \times 10^{-5}$ M
pH = ?

Write the formula and solve for the unknown.

$pH = -\log[H_3O^+]$

Plug in known values and evaluate.

$pH = -\log 2.0 \times 10^{-5}$ M
$ = -(-4.70)$
$ = 4.70$

The pH of the shampoo is 4.70.

Calculating the concentration from the pH value is the opposite of calculating the pH from the concentration. The inverse of a logarithm is an antilogarithm, or antilog. In this case, you are taking the antilogarithm of the negative pH to get the concentration. For example, if the pH of a solution is known, the concentration of hydronium ions can be calculated. Because pH is a negative exponent, you can find $[H_3O^+]$ by raising 10 to a power given by $-pH$.

$$[H_3O^+] = 10^{-pH}$$

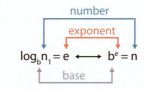

EXAMPLE 18-3: CALCULATING HYDRONIUM ION CONCENTRATION FROM pH

What is the hydronium ion concentration in vinegar with a pH of 2.31?

Write what you know.

pH = 2.31
$[H_3O^+]$ = ?

Write the formula and solve for the unknown.

$pH = -\log[H_3O^+]$
$[H_3O^+] = 10^{-pH}$

Plug in known values and evaluate.

$[H_3O^+] = 10^{-2.31}$
$ = 4.9 \times 10^{-3}$ M

The hydronium ion concentration in the vinegar is 4.9×10^{-3} M.

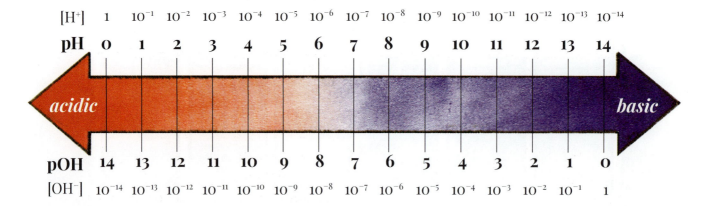

This diagram can help you understand the relationships between pH, pOH, and the concentrations of H^+ and OH^- ions.

The pOH Scale

Just as pH describes hydronium ion concentration, pOH describes hydroxide ion concentration. The **pOH** of a solution is the negative logarithm of the hydroxide ion concentration.

$$pOH = -\log[OH^-]$$

Just like calculating the hydronium ion concentration in a solution from its pH, we can find the hydroxide ion concentration from the pOH value.

$$[OH^-] = 10^{-pOH}$$

Because the hydronium and hydroxide ion concentrations are related, pH and pOH are also related. For a substance at 25 °C, the sum of the pH and the pOH of a solution is 14.

$$pH + pOH = 14.0$$

Thus, $[H_3O^+]$, $[OH^-]$, pH, and pOH are all related.

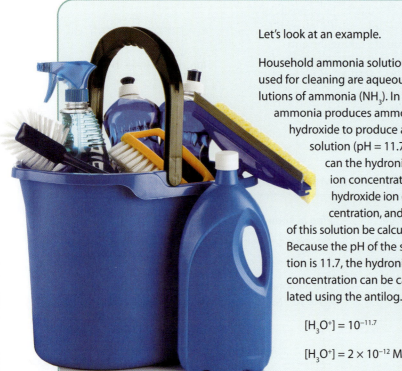

Let's look at an example.

Household ammonia solutions used for cleaning are aqueous solutions of ammonia (NH_3). In water, ammonia produces ammonium hydroxide to produce a basic solution (pH = 11.7). How can the hydronium ion concentration, hydroxide ion concentration, and pOH of this solution be calculated? Because the pH of the solution is 11.7, the hydronium ion concentration can be calculated using the antilog.

$$[H_3O^+] = 10^{-11.7}$$

$$[H_3O^+] = 2 \times 10^{-12} \text{ M}$$

The pOH can be calculated directly from the pH.

$$\begin{aligned} pOH &= 14.0 - pH \\ &= 14.0 - 11.7 \\ &= 2.3 \end{aligned}$$

The hydroxide ion concentration can be calculated from the antilog of the pOH, just as the hydronium ion is calculated from the pH.

$$\begin{aligned} [OH^-] &= 10^{-pOH} \\ &= 10^{-2.3} \\ &= 5 \times 10^{-3} \text{ M} \end{aligned}$$

EXAMPLE 18-4: CALCULATING pH AND pOH FROM CONCENTRATION

A sulfuric acid solution has a hydronium ion concentration of 3.2×10^{-2} mol/L. Calculate the pH, the pOH, and the hydroxide ion concentration of this solution. Identify the solution as acidic or basic.

Write what you know.

$[H_3O^+] = 3.2 \times 10^{-2}$ M
pH = ?
pOH = ?
$[OH^-]$ = ?

Find pH first.

Write the formula and solve for the unknown.

$$pH = -\log[H_3O^+]$$

Plug in known values and evaluate.

$$pH = -\log 3.2 \times 10^{-2}$$
$$= -(-1.49)$$
$$= 1.49$$

Next, find the pOH.

Write the formula and solve for the unknown.

$$pH + pOH = 14.0$$
$$\cancel{pH} + pOH - \cancel{pH} = 14.0 - pH$$
$$pOH = 14.0 - pH$$

Plug in known values and evaluate.

$$pOH = 14.00 - 1.49$$
$$= 12.51$$

The hydroxide ion concentration can now be calculated from the pOH.

Write the formula and solve for the unknown.

$$[OH^-] = 10^{-pOH}$$

Plug in known values and evaluate.

$$[OH^-] = 10^{-12.51}$$
$$= 3.1 \times 10^{-13} \text{ M}$$

The pH of the solution is 1.49, the pOH is 12.51, the hydroxide ion concentration is 3.1×10^{-13} M, and the solution is acidic.

Acid-Base Strength

Though pH describes the concentration of a solution, it does not describe its strength. For solutions, strong is not the same as concentrated and weak is not the same as dilute. The terms *concentrated* and *dilute* refer to the amount of acid in a given volume of aqueous solution. It is possible to have a dilute solution of a strong acid or a concentrated solution of a weak acid. **Strong acids** give up protons easily and ionize completely. The reaction in which the hydronium ions are produced is favored. Surprisingly, only six common acids are classified as strong acids: $HClO_4$, HI, HBr, HCl, H_2SO_4, and HNO_3. Notice below in the equation for the reaction that produces hydronium ions from HCl that the forward reaction is favored.

$$HCl(aq) + H_2O(l) \rightleftharpoons Cl^-(aq) + H_3O^+(aq)$$

Though the solutions look the same, testing with litmus paper demonstrates a significant difference between the two solutions.

Acids, Bases, and Salts

EXAMPLE 18-5: CALCULATING pH FROM MASS OF SOLUTE (STRONG ACID)

What is the pH of 0.500 L of solution that has 0.40 g of hydrogen iodide dissolved in it?

Write what you know.

m_{HI} = 0.40 g
$V_{solution}$ = 0.500 L
n_{HI} = ?
$[H_3O^+]$ = ?
pH = ?

Convert mass HI to moles.

$$0.40 \text{ g HI} \left(\frac{1 \text{ mol HI}}{127.91 \text{ g HI}} \right) = 0.00312 \text{ mol HI}$$

Write the formula and solve for the unknown.

$$[HI] = \frac{n_{HI}}{V_{solution}}$$

Plug in known values and evaluate.

$$[HI] = \frac{0.00312 \text{ mol HI}}{0.500 \text{ L}}$$

$$= 6.25 \times 10^{-3} \frac{\text{mol HI}}{\text{L}}$$

Because HI is a strong acid, we know that every mole of HI will produce a mole of hydronium ions.

Convert [HI] to $[H_3O^+]$

$$[H_3O^+] = 6.25 \times 10^{-3} \frac{\text{mol HI}}{\text{L}} \left(\frac{1 \text{ mol } H_3O^+}{1 \text{ mol HI}} \right)$$

$$= 6.25 \times 10^{-3} \frac{\text{mol } H_3O^+}{\text{L}}$$

Write the formula and solve for the unknown.

$$pH = -\log[H_3O^+]$$

Plug in known values and evaluate.

$$pH = -\log 6.25 \times 10^{-3} \text{ M}$$
$$= -(-2.2)$$
$$= 2.20$$

As electrolytes, both of these acids release ions and conduct electricity. On the basis of this photo, can you tell which one is stronger?

Weak acids do not dissociate completely into ions because only a portion of their molecules loses protons. For example, only 1 in 24 molecules deprotonates in an acetic acid solution. The equilibrium reaction in which they dissociate is reactant-favored.

$$HC_2H_3O_2(aq) + H_2O(l) \rightleftharpoons C_2H_3O_2^-(aq) + H_3O^+(aq)$$

Similarly, the strength of a base depends on how easily it accepts a proton, or protonates. According to the Brønsted-Lowry model, **weak bases**, such as the Cl^- ion, are poor proton acceptors. **Strong bases**, such as the OH^- ion, accept protons readily. The hydroxides of active metals—LiOH, NaOH, KOH, RbOH, CsOH, $Mg(OH)_2$, $Ca(OH)_2$, $Sr(OH)_2$, and $Ba(OH)_2$—are the best sources of the hydroxide ion, so they are commonly called strong bases.

The conjugate of a strong acid or a strong base is always weak. For example, perchloric acid ($HClO_4$) is a strong acid that readily gives up a hydrogen ion. Once the hydrogen and the perchlorate ions are apart, they do not readily rejoin to form perchloric acid. Therefore, the perchlorate ion (ClO_4^-) is a weak conjugate base. Conversely, Cl^- is a weak base with a strong conjugate acid—HCl.

Acid and base strengths are described by the equilibrium constant of their dissociation reactions. The **acid-ionization constant** (K_a) describes the extent of the forward equilibrium reaction in the formation of the hydronium ion. In this generalized formula, HA represents a typical acid where A represents the anion.

$$HA + H_2O \rightleftharpoons H_3O^+ + A^-$$

$$K_a = \frac{[H_3O^+][A^-]}{[HA]}$$

Similarly, the **base-ionization constant** (K_b) describes the extent of the forward equilibrium reaction in the formation of the hydroxide ion. Here, B represents a base. HB^+ represents a base that has captured a hydrogen ion from water.

$$B + H_2O \rightleftharpoons OH^- + HB^+$$

$$K_b = \frac{[OH^-][HB^+]}{[B]}$$

When dealing with acids and bases in aqueous solution, both of these constants are related to the constant for the self-ionization of water, K_w.

$$(K_a)(K_b) = K_w$$

Stronger acids and bases have higher K_a and K_b values because they ionize more readily. In fact, the forward reaction for strong acids and bases is so predominant that the reaction proceeds to completion because they completely dissociate. Table 18-4 lists conjugate pairs and their ionization constants in order from strong to weak. Because the stronger acids have weaker conjugate bases, the bases at the top of the column are the weakest. As in all equilibrium reactions, the values are specified for 25 °C, but their values change as temperature changes.

sodium hydroxide (NaOH) pellets

TABLE 18-4 Relative Strengths of Some Acids and Bases*

Acid		K_a	K_b		Base
perchloric acid	$HClO_4$	large	very small	ClO_4^-	perchlorate ion
hydriodic acid	HI	large	very small	I^-	iodide ion
hydrobromic acid	HBr	large	very small	Br^-	bromide ion
hydrochloric acid	HCl	large	very small	Cl^-	chloride ion
sulfuric acid	H_2SO_4	large	very small	HSO_4^-	hydrogen sulfate ion
nitric acid	HNO_3	large	very small	NO_3^-	nitrate ion
sulfurous acid	H_2SO_3	1.3×10^{-2}	8.3×10^{-13}	HSO_3^-	bisulfate ion
hydrogen sulfate ion	HSO_4^-	1.2×10^{-2}	8.3×10^{-13}	SO_4^{2-}	sulfate ion
phosphoric acid	H_3PO_4	6.9×10^{-3}	1.3×10^{-12}	$H_2PO_4^-$	dihydrogen phosphate ion
hydrofluoric acid	HF	6.8×10^{-4}	1.4×10^{-11}	F^-	fluoride ion
formic acid	HCO_2H	1.8×10^{-4}	5.6×10^{-11}	CO_2H^-	formate ion
acetic acid	$HC_2H_3O_2$	1.7×10^{-5}	5.6×10^{-10}	$C_2H_3O_2^-$	acetate ion
carbonic acid	H_2CO_3	4.3×10^{-7}	2.4×10^{-8}	HCO_3^-	hydrogen carbonate ion
hypochlorous acid	$HClO$	3.0×10^{-8}	2.9×10^{-7}	ClO^-	hypochlorite ion
boric acid	H_3BO_3	5.8×10^{-10}	1.4×10^{-5}	$H_2BO_3^-$	dihydrogen borate ion
ammonium ion	NH_4^+	5.7×10^{-10}	1.8×10^{-5}	NH_3	ammonia
hydrocyanic acid	HCN	4.9×10^{-10}	2.5×10^{-5}	CN^-	cyanide ion
hydrogen carbonate ion	HCO_3^-	4.8×10^{-11}	2.1×10^{-4}	CO_3^{2-}	carbonate ion
hydrogen peroxide	H_2O_2	2.6×10^{-12}	1.3×10^{-2}	HO_2^-	hydroperoxide ion
water	H_2O	1.0×10^{-14}	55.5	OH^-	hydroxide ion
ammonia	NH_3	very small	large	NH_2^-	amide ion
hydrogen	H_2	very small	large	H^-	hydride ion
methane	CH_4	very small	large	CH_3^-	methide ion

← increasing acid strength

increasing base strength →

*First six acids are strong acids and last four bases are strong bases.

Acids, Bases, and Salts

Polarity, Charge, and Acid-Base Strength

Ionization constants *describe* acid-base strengths; they do not *determine* the strengths. The location of shared electron clouds, the availability of a lone pair, and the charge on an acid or base all determine how easily a proton can be released or absorbed. In other words, the strongest acids are more polar and have lower bond energies. The strongest bases are negatively charged and have available bonding sites. Strong bases are often soluble ionic compounds of hydroxides and alkali or alkaline-earth metals, like NaOH and $Mg(OH)_2$. These ionic compounds completely dissociate in water to form hydroxide ions, which are strong proton acceptors.

The polarity of the bond between the hydrogen and the central atom is an important factor in determining acid strength. To be easily ionizable, the nucleus of a hydrogen atom must be exposed. As the polarity of the bond increases, the shared electrons are shifted away from the hydrogen toward a more electronegative atom in the center of the molecule. This creates a positive charge around the hydrogen and weakens the bond. For example, hydrofluoric acid, with its highly polar bond, is a strong acid because of the high electronegativity of fluorine. Methane on the other hand has only slightly polar bonds, making it a weak acid.

The electrical charge of a molecule also determines the strengths of acids. The sulfuric acid molecule (H_2SO_4) has a neutral charge and is a strong acid. As sulfuric acid deprotonates, it forms a hydrogen sulfate ion (HSO_4^-), which has another proton that it can donate. But due to its negative charge, it has a much lower tendency to release this proton. Similarly, H_3PO_4, $H_2PO_4^-$, and HPO_4^{2-} have decreasing acidic strengths. Anions hold protons more strongly than neutral molecules do.

TABLE 18-5
Acid-Base Strength Summary

Strong Acid	Strong Base
has exposed hydrogen nuclei	has exposed electron pairs
has minimal negative charges	has possible strong negative charges

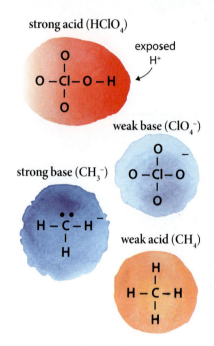

Breathalyzer

The active ingredient in alcoholic beverages is ethanol, also known as ethyl alcohol or grain alcohol (CH_3CH_2OH). Ethanol is unique because it is absorbed unchanged into the bloodstream by the intestines, stomach, esophagus, and even the mouth and does not readily dissociate in the presence of digestive acids. This is why police can tell that a person has been drinking by checking the smell of his breath.

The alcohol in a person's breath is directly related to the amount in his blood—his blood alcohol content (BAC). Is there an accurate way for police officers to determine a person's BAC by measuring the ethanol in his breath? Police officers can now estimate a driver's BAC right on the side of the road using a device that tests a person's breath—a Breathalyzer. The device contains a fuel cell in which the ethanol reacts to form acetic acid and also produces an electric current. The current is proportional to the BAC of the driver. The more current produced, the greater the BAC of the driver and the more impaired he is. Through the chemistry of acids, scientists and police work to keep our roads safe.

Amphoteric Substances

An **amphoteric substance** can act as both a Brønsted-Lowry acid and a Brønsted-Lowry base. Its behavior depends on reaction conditions. For example, the hydrogen carbonate ion accepts a proton (H^+ ion) when an acid is added to the solution. Under those circumstances, the hydrogen carbonate ion is a base.

$$\underset{\text{base}}{HCO_3^-(aq)} + \underset{\text{acid}}{H_3O^+(aq)} \rightleftharpoons \underset{\text{conjugate acid}}{H_2CO_3(aq)} + \underset{\text{conjugate base}}{H_2O(l)}$$

But when a base is added to the solution, the hydrogen carbonate ion acts as an acid by donating a proton.

$$\underset{\text{acid}}{HCO_3^-(aq)} + \underset{\text{base}}{OH^-(aq)} \rightleftharpoons \underset{\text{conjugate base}}{CO_3^{2-}(aq)} + \underset{\text{conjugate acid}}{H_2O(l)}$$

Water is also an amphoteric substance, as demonstrated by its reactions with ammonia and perchloric acid. In a reaction with ammonia, water acts as a Brønsted-Lowry acid, donating a proton.

$$\underset{\text{acid}}{H_2O(l)} + \underset{\text{base}}{NH_3(aq)} \rightleftharpoons \underset{\text{conjugate base}}{OH^-(aq)} + \underset{\text{conjugate acid}}{NH_4^+(aq)}$$

When reacting with perchloric acid, water acts as a base by receiving a proton.

$$\underset{\text{base}}{H_2O(l)} + \underset{\text{acid}}{HClO_4(aq)} \rightleftharpoons \underset{\text{conjugate acid}}{H_3O^+(aq)} + \underset{\text{conjugate base}}{ClO_4^-(aq)}$$

Polyprotic Acids

Some Brønsted-Lowry acids—such as perchloric acid ($HClO_4$), hydrochloric acid (HCl), and nitric acid (HNO_3)—can donate only one proton; these are called **monoprotic acids**. On the other hand, each molecule of a **polyprotic acid** can donate more than one proton. Each stage of ionization has a different equilibrium constant. A **diprotic acid** can donate two protons. For example, carbonic acid (H_2CO_3) is diprotic. The K_a for the first ionization is called K_{a1}, and that for the second ionization is called K_{a2}. (See Table 18-4 on page 447 for the K_a values of carbonic acid and the hydrogen carbonate ion.)

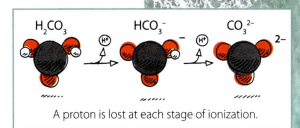

A proton is lost at each stage of ionization.

$$H_2CO_3(aq) + H_2O(l) \rightleftharpoons HCO_3^-(aq) + H_3O^+(aq)$$

$$K_{a1} = \frac{[H_3O^+][HCO_3^-]}{[H_2CO_3]}$$

$$= 4.3 \times 10^{-7}$$

$$HCO_3^-(aq) + H_2O(l) \rightleftharpoons CO_3^{2-}(aq) + H_3O^+(aq)$$

$$K_{a2} = \frac{[H_3O^+][CO_3^{2-}]}{[HCO_3^-]}$$

$$= 4.8 \times 10^{-11}$$

TABLE 18-6
Ionization Constants of Polyprotic Acids

Polyprotic Acid	Acid-Base Reaction	K_a
chromic acid	$H_2CrO_4 \rightleftharpoons H^+ + HCrO_4^-$	1.8×10^{-1}
	$HCrO_4^- \rightleftharpoons H^+ + CrO_4^{2-}$	3.2×10^{-7}
hydrosulfuric acid	$H_2S \rightleftharpoons H^+ + HS^-$	1.1×10^{-7}
	$HS^- \rightleftharpoons H^+ + S^{2-}$	1.0×10^{-14}
phosphoric acid	$H_3PO_4 \rightleftharpoons H^+ + H_2PO_4^-$	6.9×10^{-3}
	$H_2PO_4^- \rightleftharpoons H^+ + HPO_4^{2-}$	6.2×10^{-8}
	$HPO_4^{2-} \rightleftharpoons H^+ + PO_4^{3-}$	4.8×10^{-13}
sulfurous acid	$H_2SO_3 \rightleftharpoons H^+ + HSO_3^-$	1.3×10^{-2}
	$HSO_3^- \rightleftharpoons H^+ + SO_3^{2-}$	6.3×10^{-8}

For any polyprotic acid, all the possible stages of ionization occur in the same solution. For instance, a solution of phosphoric acid (H_3PO_4) contains the ions $H_2PO_4^-$, HPO_4^{2-}, and PO_4^{3-} in addition to the hydronium ion H_3O^+.

All polyprotic acids have at least one amphoteric substance in their series of ionizations. In carbonic acid, the hydrogen carbonate ion is amphoteric. Furthermore, the K_{a2} is smaller than the K_{a1}. The first ionization involves separation of a proton from an uncharged carbonic acid (H_2CO_3) molecule, and the second involves separation of a proton from the negatively charged hydrogen carbonate ion (HCO_3^-). Because the strength of attraction between opposite charges depends on the magnitude of the charges, the second ionization is less favorable than the first. In general, $K_{a1} > K_{a2} > K_{a3}$. Acids with three ionizable protons are called **triprotic acids**. Can you name a triprotic acid? Several more examples of polyprotic acids are shown in Table 18-6.

Measuring pH

Acid-base indicators are substances whose colors are sensitive to pH. As a result, they change colors when the pH of a solution changes. At the beginning of this chapter you were introduced to litmus paper, a type of indicator. Indicators may be weak acids or bases whose conjugates are different colors. Typically, the anion has one color when it is combined with the hydrogen atom and a different color after ionization. Some indicators change colors at low pH values, some at high pH values. Those that change at low pH values are the stronger acids because they ionize easily. The indicators that do not change color until at the higher pH range are weaker acids. Some are polyprotic, so they may exhibit more than one color change.

Examples of indicators, their colors, and the pH range over which their colors change are given below. Notice that each indicator has a range of pH values over which the color change occurs. This is known as the **transition interval**.

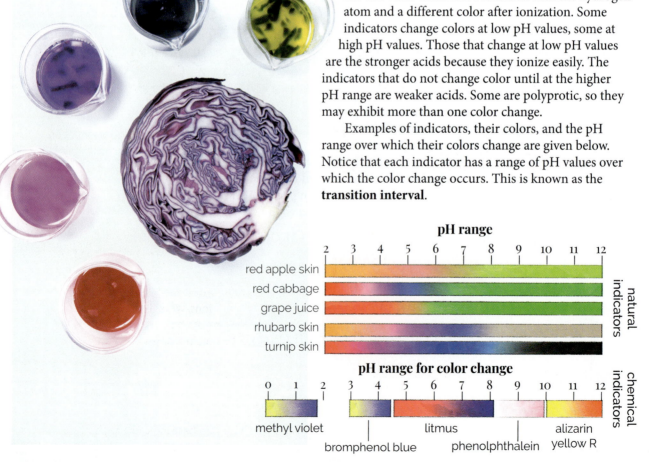

Indicators give rough estimates of the pH value. For instance, if a solution is yellow when tested with both bromthymol blue and methyl orange, then the pH must be between 4.4 and 6.0. Methyl orange has a transition interval from a pH of 3.1 to 4.4. Once the pH is past 4.4, the color remains essentially unchanged. For example, methyl orange cannot indicate the alkalinity of a solution. This problem may be overcome by using carefully chosen combinations of indicators called *universal indicators*. More accurate values can be obtained when large numbers of indicators are used, but this process is tedious.

thymol blue indicator

pH probe

Indicators are subjective because they depend on a person's color perception and comparison. More objective pH measurements can be obtained by using a *pH meter*. A pH meter indirectly determines the concentration of the hydronium ions by measuring the voltage that they produce between two points and comparing the voltage to a standard. The instrument uses a *pH probe*, which is a sensor that responds only to hydronium ions. The meter also contains a reference probe and usually a temperature sensor or adjustment control since the measurement of pH is temperature dependent. The voltage produced by the probe is directly proportional to pH. The slope of the pH probe's output is determined by temperature. The pH instrument converts the voltage and temperature information into a digital display of pH. Before measurements are taken, these meters are standardized, using solutions of known pH, to ensure accuracy.

18.2 SECTION REVIEW

1. What does it mean that water is self-ionizing?

2. What particles are always present in pure water?

3. Using the information given, find the [H_3O^+], [OH^-], pH, and pOH for each solution below. Also identify the solution as acidic, neutral, or basic.

    a. orange juice, pH = 4.13

    b. [H_3O^+] = 9.5×10^{-8}

    c. black coffee, pOH = 9.05

    d. [OH^-] = 3.55×10^{-6}

4. Calculate the [H_3O^+], pH, [OH^-], and pOH of each solution. Assume that the acid completely dissociates.

    a. 1.3×10^{-5} M HCl

    b. 2.71×10^{-2} M H_2SO_4

5. Using Table 18-4, indicate the stronger acid in each pair.

    a. $HClO_4$, H_3PO_4

    b. formic acid, acetic acid

    c. hydrocyanic acid, formic acid

    d. NH_4^+, HSO_4^-

    e. NH_4^+, H_2O

6. Is the HCO_3^- ion an acid, a base, both, or neither? Explain.

7. All the following substances can undergo deprotonation reactions. Write the equations for their ionization constants. If they go through a series of ionizations, include the equation for each.

    a. H_2SO_4

    b. H_3PO_4

8. Explain how an acid-base indicator works.

MINILAB

ACID OR BASE?

Throughout this chapter, we have been discussing acids and bases. How prevalent are these substances? What is the pH of some common substances?

Are there acids and bases around me every day?

1. What are the physical properties of acids and bases?
2. What is an acid-base indicator?

Procedure

A Copy the table at right onto a piece of paper.

B Consider the substances that you will be testing and predict whether each is an acid or base. Record your predictions in your table.

C In each beaker, use a small (3 cm) strip of indicator paper to test the pH of each solution. Test each solution by dipping one end of the paper in the solution and then compare the color of the paper to the indicator paper legend. Record the color, approximate pH, and whether the solution is an acid or base in the table.

Conclusion

3. Which substances did you correctly predict? What was it about the substance that enabled you to predict whether it was an acid or base?

4. Which substances did you not correctly predict? What was it about the substance that caused you to predict it incorrectly?

Going Further

5. What property of acids and bases do you think a pH meter uses? Explain?

EQUIPMENT
- beakers, 50 mL (6)
- universal pH indicator paper
- acids and bases (6), 10 mL

Substance	Prediction	Indicator Color	pH	Conclusion: Acid or Base?

18.3 NEUTRALIZATION

The pain and discomfort of heartburn are usually associated with meals that were either too large or too spicy. But the burning sensation comes not from oregano or green peppers but from excess hydrochloric acid. This acid, which is normally confined to the stomach, irritates the tissues of the esophagus if it rises past the upper part of the stomach. To relieve the pain, a person can neutralize the acid with an antacid. Antacids react with the excess stomach acid to produce harmless products.

Salts

As mentioned in Section 18.1, a neutralization reaction is the reaction of an acid and a base to produce water and a salt. Antacids that contain magnesium hydroxide react to neutralize stomach acid as described by the following equation.

$$Mg(OH)_2(aq) + 2HCl(aq) \longrightarrow MgCl_2(aq) + 2H_2O(l)$$

Sodium hydroxide is much too corrosive to be used as an antacid, but it too neutralizes acids.

$$2NaOH(aq) + H_2SO_4(aq) \longrightarrow Na_2SO_4(aq) + 2H_2O(l)$$

A **salt** is a substance formed when the anion of an acid and the cation of a base combine. In the first reaction shown above, Cl^- (the acid's anion) combines with Mg^{2+} (the base's cation) to form $MgCl_2$ (the salt). Many different salts can be produced in neutralization reactions. Sodium chloride, zinc bromide, and potassium fluoride are all examples of salts. Most salts, like these, are composed of a metal and a nonmetal, but this is not always true. Ammonium chloride (NH_4Cl), potassium sulfate (K_2SO_4), and aluminum phosphate ($AlPO_4$) all contain polyatomic ions.

According to the Arrhenius definitions of acids and bases, neutralization reactions form water from hydronium and hydroxide ions. We can easily see this from the net ionic reaction of a neutralization reaction. The hydronium ions of acids and the hydroxide ions of bases combine to form two water molecules.

$$H_3O^+(aq) + OH^-(aq) \longrightarrow 2H_2O(l)$$

Metal cations and nonmetal anions are present, but they are not involved in the neutralization reaction. Because these ions are found in the solution of the reactant and the product, they are called *spectator ions*. If the resulting solution evaporates, the ions will crystallize into a salt.

QUESTIONS
» What are the products of a neutralization reaction?
» What is titration?
» How can I experimentally determine the concentration of an acid or base?
» How do buffers work?

TERMS
salt • acid-base titration • equivalence point • end point • buffer

How do antacids work?

▲

Antacid reacting with HCl acid. Antacids are effective because they react to reduce the excess hydrochloric acid in the stomach.

EXAMPLE 18-6:
WRITING A NEUTRALIZATION REACTION

Write a neutralization reaction to produce zinc chloride ($ZnCl_2$), a salt.

Solution
Salts are made from the anions of acids and the cations of bases. In this case, the most common acid that releases the chloride ion is hydrochloric acid. Zinc hydroxide is a likely source for the zinc ion.

$$2HCl(aq) + Zn(OH)_2(aq) \longrightarrow ZnCl_2(aq) + 2H_2O(l)$$

EXAMPLE 18-7: DETERMINING SALTS

What salt is produced from the neutralization of barium hydroxide by acetic acid ($HC_2H_3O_2$)?

Solution

$$2HC_2H_3O_2(aq) + Ba(OH)_2(aq) \longrightarrow Ba(C_2H_3O_2)_2(aq) + 2H_2O(l)$$

Barium acetate forms from the barium and acetate ions.

Acid-Base Titrations

Acid-base titrations are controlled neutralization reactions. Scientists use these controlled reactions to determine the unknown concentration of a solution. Scientists use a measured volume of a solution of known concentration to react with a precisely measured volume of solution of unknown concentration. The solution with a known concentration is sometimes called the *standard solution* or *titrant*. This is all based on the concept of stoichiometry, which you studied in Chapter 11.

A burette is a piece of glassware that releases precise amounts of solution to allow titration calculations. Notice that the solution has just turned a slight shade of pink, indicating that the titration is at its end point.

Suppose that a chemist has 100.0 mL of a hydrochloric acid solution, but he does not know its concentration. Because he knows that NaOH neutralizes HCl, he can find the concentration by slowly adding precisely measured volumes of the sodium hydroxide solution until the pH of the solution rises to 7.0. At this point, there are as many moles of H^+ ions from the acid (HCl) as there are OH^- ions from the base (NaOH).

A graph called a *titration curve* shows how pH changes when an acid or base is added to a solution. The graph on page 456 shows how the pH of a solution of unknown concentration changes with the addition of sodium hydroxide solution.

The chemist found that adding 50 mL of 1.00 M sodium hydroxide raised the pH of the acid to 7. So 50.0 mL of 1.00 M sodium hydroxide is chemically equivalent to 100 mL of the acid. With this information, he knows that the 1.00 M sodium hydroxide is twice as concentrated as the hydrochloric acid. He can then calculate that the hydrochloric acid has a concentration of 0.50 M. Lets see how the mathematics works on this problem.

EXAMPLE 18-8: DETERMINING A CONCENTRATION

A volume of 50.0 mL of a 1.00 M sodium hydroxide solution can neutralize 100.0 mL of a hydrochloric acid solution. What is the concentration of the hydrochloric acid?

Write what you know.

$V_{NaOH} = 50.0$ mL $= 0.0500$ L

$c_{NaOH} = 1.00 \dfrac{\text{mol NaOH}}{\text{L}}$

$V_{HCl} = 100.0$ mL $= 0.1000$ L

NaOH(*aq*) + HCl(*aq*) ⟶ NaCl(*aq*) + H$_2$O(*l*)

Unknown: c_{HCl}

Use stoichiometry to determine moles of HCl.

$$0.0500 \text{ L} \left(\dfrac{1.00 \text{ mol NaOH}}{1 \text{ L}} \right) \left(\dfrac{1 \text{ mol HCl}}{1 \text{ mol NaOH}} \right)$$

$= 0.0500$ mol HCl

Write the formula and solve for the unknown.

$$c_{solute} = \dfrac{n_{solute}}{V_{solution}}$$

Plug in known values and evaluate.

$$c_{HCl} = \dfrac{0.0500 \text{ mol HCl}}{0.1000 \text{ L}}$$

$= 0.500$ M HCl

The concentration of the hydrochloric acid is 0.500 molar.

WORLDVIEW INVESTIGATION

INFLUENCING OTHERS

DUI, DWI, OUI are all acronyms for driving or operating a motor vehicle "under the influence." Looking back to the discussion in the Chapter opener, we often think of these terms as being related to driving a car, but they also apply to operating a boat or an aircraft. The federal government continues to work toward a uniform standard for what "under the influence" means.

Task

A recent accident in your community has prompted your classmates to ask you what you know about this issue. Realizing that you don't understand the topic well at all, you have decided to research the topic and write a one page summary of your findings.

Procedure

1. Research the topic of operating under the influence by doing an internet search using the keywords "driving under the influence," "driving while intoxicated," and "operating under the influence." Consider conducting research at the National Traffic Safety Board, National Highway Traffic and Safety Administration, and National Highway Institute websites.

2. Consider how biblical principles, outcomes, and motivations would apply to this issue.

3. Write your summary, citing your sources properly. Have a classmate review your summary and give you feedback.

4. Complete your summary and submit it by the deadline.

Conclusion

The number of alcohol-related accidents and deaths have been decreasing. Much of this improvement has been driven by two factors: awareness and enforcement. In the 1980s, two groups, Mothers Against Drunk Driving (MADD) and Students Against Drunk Driving (SADD), brought this issue to the public's attention. But we still have a significant way to go. Many people continue to work to protect the lives of people on the road.

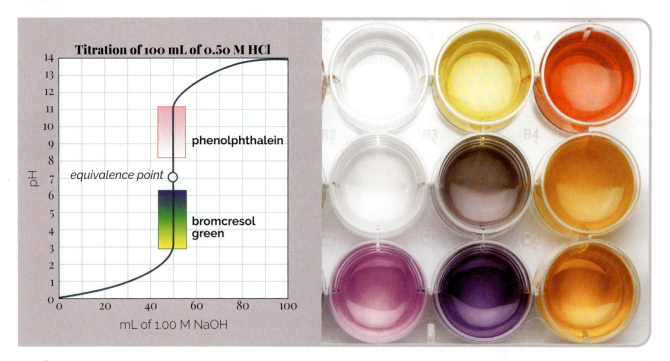

Chemical indicators denote a range of values. The end point is a specific value.

The **equivalence point** is the point in a titration at which an equivalent of titrant is added, that is, the point at which the number of H_3O^+ ions equals the number of OH^- ions. The equivalence point is not always at a pH of 7. If a strong acid reacts with a strong base, the equivalence point is at a pH of 7. If a strong acid reacts with a weak base, the equivalence point will be at a pH below 7. If a strong base reacts with a weak acid, the equivalence point will be at a pH greater than 7.

The equivalence point should not be confused with the end point of a titration. The **end point** is the point at which some change in a property of the solution is detected. For example, if phenolphthalein indicator is added to a solution, it remains clear until it reaches a pH reading of 8.3, at which time a pink color appears. If a 0.1 M HCl solution is titrated with a 0.1 M NaOH solution, the equivalence point will occur at pH 7.0, but the end point will be detected at pH 8.3—the first noticeable pink tinge.

Let's look at an example of a titration problem.

EXAMPLE 18-9: DETERMINING A CONCENTRATION

A volume of 58.3 mL of a 5.00×10^{-3} M sodium hydroxide solution can neutralize 36.2 mL of a nitric acid solution. What is the concentration of the nitric acid?

Write what you know.

$V_{NaOH} = 58.3$ mL $= 0.0583$ L

$c_{NaOH} = 5.00 \times 10^{-3} \dfrac{\text{mol NaOH}}{\text{L}}$

$V_{HNO_3} = 36.2$ mL $= 0.0362$ L

$NaOH(aq) + HNO_3(aq) \longrightarrow NaNO_3(aq) + H_2O(l)$

Unknown: c_{HNO_3}

Use stoichiometry to determine moles of NO_3.

$0.0583 \text{ L} \left(\dfrac{5.00 \times 10^{-3} \text{ mol NaOH}}{1 \text{ L}} \right) \left(\dfrac{1 \text{ mol HNO}_3}{1 \text{ mol NaOH}} \right)$

$= 2.915 \times 10^{-4}$ mol HNO_3

Write the formula and solve for the unknown.

$c_{solute} = \dfrac{n_{solute}}{V_{solution}}$

Plug in known values and evaluate.

$c_{solute} = \dfrac{2.915 \times 10^{-4} \text{ mol HNO}_3}{0.0362 \text{ L}}$

$= 8.05 \times 10^{-3}$ M HNO_3

The concentration of the nitric acid is 8.05×10^{-3} molar.

The next example differs only in the fact that the base provides more hydroxide ions than the acid provides hydronium ions.

EXAMPLE 18-10: DETERMINING A CONCENTRATION

Adding 38.7 mL of 1.50×10^{-2} M barium hydroxide solution can neutralize 21.5 mL of a hydrochloric acid solution. What is the concentration of the hydrochloric acid?

Write what you know.

$V_{Ba(OH)_2} = 38.7$ mL $= 0.0387$ L

$c_{Ba(OH)_2} = 1.50 \times 10^{-2} \dfrac{\text{mol Ba(OH)}_2}{\text{L}}$

$V_{HCl} = 21.5$ mL $= 0.0215$ L

$Ba(OH)_2(aq) + 2HCl(aq) \longrightarrow BaCl_2(aq) + 2H_2O(l)$

Unknown: c_{HCl}

Use stoichiometry to determine moles of NO_3.

$0.0387 \text{ L} \left(\dfrac{1.50 \times 10^{-2} \text{ mol Ba(OH)}_2}{1 \text{ L}} \right) \left(\dfrac{2 \text{ mol HCl}}{1 \text{ mol Ba(OH)}_2} \right)$

$= 1.16 \times 10^{-3}$ mol HCl

Write the formula and solve for the unknown.

$c_{solute} = \dfrac{n_{solute}}{V_{solution}}$

Plug in known values and evaluate.

$c_{solute} = \dfrac{1.161 \times 10^{-3} \text{ mol HCl}}{0.0215 \text{ L}}$

$= 5.40 \times 10^{-2}$ M HCl

The concentration of the unknown acid is 5.40×10^{-2} molar.

Buffers

In the course of a single day, you probably eat foods with very different pH values. It is obvious that the human stomach can handle this wide range of pH values, but the bloodstream cannot. Its pH must stay between 7.35 and 7.45. God has designed the bloodstream with an ingenious protective feature—a buffer system.

Buffers are solutions that resist pH changes when acids or bases are added. Buffer systems usually consist of a weak acid and its conjugate base or a weak base and its conjugate acid. Water is not a buffer; a small amount of acid or base changes its pH dramatically. Adding 0.01 mol of hydrochloric acid or sodium hydroxide to 1 L of pure water changes its pH by 5 units. Due to the blood buffers, the same substances added to 1 L of blood change its pH by only 0.1 unit.

Buffers can be found regulating the pH in many living things. Scientists have also harnessed the benefits of buffers to control chemical reactions for industry and to make safer medicines, such as buffered aspirin. By properly selecting the acid-base pair, buffers can be used in almost any pH range. A given buffer acts to keep the pH relatively constant within its working range. You can see several examples of buffers and their usable ranges in Table 18-7.

TABLE 18-7 *Common Buffer Systems*

Components	Usable pH Range
formic acid + sodium formate	2.6–4.8
citric acid + sodium citrate	3.0–6.2
acetic acid + sodium acetate	3.4–5.9
sodium bicarbonate + sodium carbonate	9.2–10.6
sodium bicarbonate + sodium hydroxide	9.6–11.0

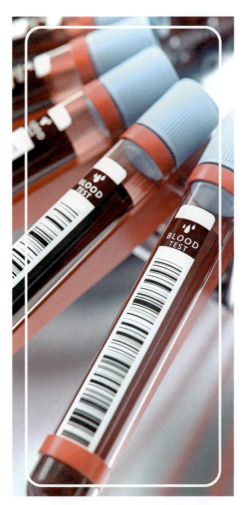

Buffers are most effective in regulating slight pH changes. If an enormous amount of acid or base is added to a buffered solution, the buffer will be depleted and the pH will change drastically. Keep in mind that the buffer does not keep the solution neutral (pH of 7.0) but rather maintains a constant pH for the given solution. For example, suppose that the pH of a given solution containing a buffer is 4.7. As a base is added to this solution, the buffer will maintain the pH at 4.7 until the buffer has been depleted. Only then will the pH begin to rise as more of the base is added.

For example, blood is buffered by a mixture of carbonic acid (H_2CO_3), which is a weak acid, and hydrogen carbonate (HCO_3^-), which is its conjugate base. This combination of solutes keeps the pH relatively constant by reacting with hydronium or hydroxide ions. When an acid intrudes on this blood buffer system, hydrogen carbonate ions snare the hydronium ions.

$$H_3O^+(aq) + HCO_3^-(aq) \rightleftharpoons H_2CO_3(aq) + H_2O(l)$$

When a base disturbs the equilibrium of the same buffer system, carbonic acid molecules spring into action and remove the hydroxide ions.

$$H_2CO_3(aq) + OH^-(aq) \rightleftharpoons H_2O(l) + HCO_3^-(aq)$$

Blood has a second buffer system that utilizes hydrogen phosphate ions (HPO_4^{2-}) and dihydrogen phosphate ions ($H_2PO_4^-$). If excess acid enters the bloodstream, it reacts with HPO_4^{2-} to form $H_2PO_4^-$. Bases react with $H_2PO_4^-$ to form HPO_4^{2-}. In this reaction, the H^+ ions released by $H_2PO_4^-$ react with hydroxide from the base. Thus, the pH is stabilized in either direction.

18.3 SECTION REVIEW

1. Write the chemical equation for each of the following neutralization reactions.

   a. An acid and a base react to produce K_2SO_4 salt.

   b. Calcium hydroxide is neutralized by hydrobromic acid.

2. Define *titration*.

3. Summarize the acid-base titration process.

4. In a titration reaction, 33.6 mL of 0.302 M RbOH neutralizes 20.0 mL of HF. Calculate the amount concentration of the acid solution.

5. A 1.000 molar potassium hydroxide solution is used to neutralize phosphoric acid. If 25.00 mL of the base neutralizes 27.00 mL of the acid, what is the amount concentration of the acid?

6. A solution has a buffer that is effective in the 3.0–6.2 pH range. If the system is overloaded with acid, what will be the effect on the buffer system?

7. What happens to excess H_3O^+ ions when an acid is added to a solution buffered by a combination of acetic acid and sodium acetate?

8. Write the two equations for the buffer reactions involving carbonic acid and the hydrogen carbonate ion in our blood. Identify the conjugate acid-base pairs and explain how it manages the pH in the blood.

18 CHAPTER REVIEW

Chapter Summary

TERMS
electrolyte
neutralization reaction
Arrhenius acid
Arrhenius base
Brønsted-Lowry acid
Brønsted-Lowry base
conjugate base
conjugate acid
Lewis acid
Lewis base

18.1 DEFINING ACIDS AND BASES

- Acids taste sour, turn blue litmus paper red, react with active metals to produce hydrogen and a salt, conduct an electrical current in solution, and neutralize bases.

- Bases taste bitter, turn red litmus paper blue, neutralize acids to form a salt and water, conduct an electrical current in solution, and feel slippery.

- Arrhenius defined acids as substances that ionize to release hydrogen ions into solutions and bases as those that release hydroxide ions into solutions.

- Brønsted and Lowry defined acids as substances that donate protons and bases as those that accept protons. These substances exist as conjugate pairs. A conjugate acid is formed when the base protonates, and the conjugate base forms when the acid deprotonates.

- Lewis defined acids as substances that can accept a pair of electrons. Bases can donate a pair of electrons. His is the broadest definition of acids and bases.

18.2 ACID-BASE EQUILIBRIA

- In self-ionization, water molecules react to form hydronium and hydroxide ions. The equilibrium constant of water (K_w) is the product of the two concentrations of these ions. It is 1.0×10^{-14} at 25 °C.

- The pH of a solution is the negative logarithm of the hydronium ion concentration. The pOH is the negative logarithm of the hydroxide ion concentration. Solutions can be acidic (pH < 7), basic (pH > 7), or neutral (pH = 7).

- The strength of an acid or base is not related to its concentration but instead to its ability to ionize in solution. Ionization constants quantify the degree to which these reactions proceed in the forward direction.

- Amphoteric substances can act as either Brønsted-Lowry acids or bases, depending on the reaction.

- Monoprotic Brønsted-Lowry acids can donate only one proton; polyprotic Brønsted-Lowry acids can donate more than one proton. Donations occur in steps of ionization, but all ions are found in the same solution.

- Indicators are substances whose color is sensitive to pH changes.

TERMS
self-ionization of water
pH
acidic solution
basic solution
neutral solution
pOH
strong acid
weak acid
weak base
strong base
acid-ionization constant
base-ionization constant
amphoteric substance
monoprotic acid
polyprotic acid
diprotic acid
triprotic acid
acid-base indicator
transition interval

18 CHAPTER REVIEW

Chapter Summary

TERMS
- salt
- acid-base titration
- equivalence point
- end point
- buffer

18.3 NEUTRALIZATION

- A strong acid and a strong base neutralize to form a salt and water.

- Titrations are controlled reactions in which the unknown concentration of a measured volume of a substance is determined by adding a known volume of a reactive substance of known concentration.

- In a titration, the system is neutralized at the equivalence point, when the concentration of hydronium and hydroxide ions are equal. The titration continues until the end point, when there is a distinct change in a physical property, such as color.

- Buffers are solutions that resist pH changes despite the addition of acids or bases. Many reactions in living things are possible because of buffer systems.

Chapter Review Questions

RECALLING FACTS

1. List four common properties of bases.
2. What is a neutralization reaction?
3. What is the conjugate base of a Brønsted-Lowry acid?
4. Write the chemical equation for self-ionization of water.
5. Write the equation for the equilibrium constant that describes the reaction of water with itself. What is the numerical value of this constant?
6. Choose the correct answer for each statement.
   a. (Strong/Weak) acids ionize incompletely.
   b. (Strong/Weak) acids have large K_a values.
   c. Strong (acids/bases) accept protons easily.
   d. The conjugate base of a strong acid is a (strong/weak) base.
7. What is a buffer?

UNDERSTANDING CONCEPTS

8. Explain why the Arrhenius model was limited.
9. A classmate tells you that H_2SO_4 and SO_4^{2-} are a conjugate acid-base pair. Is he correct? Explain.
10. Two reactions describe the two-step ionization of H_2SO_4.

    $H_2SO_4(aq) + H_2O(l) \rightleftharpoons HSO_4^-(aq) + H_3O^+(aq)$

    $HSO_4^-(aq) + H_2O(l) \rightleftharpoons SO_4^{2-}(aq) + H_3O^+(aq)$

    a. What is the acid in the forward direction of the first reaction?
    b. What is the conjugate base of this acid?
    c. What is the acid in the forward direction of the second reaction?
    d. What is the conjugate base of this acid?

11. For each of the following reactions, label which substance is the Brønsted-Lowry acid and which is the Brønsted-Lowry base. Also indicate the conjugate pairs.

    a. $H_2C_2O_4 + ClO^- \rightleftharpoons HC_2O_4^- + HClO$
    b. $HPO_4^{2-} + NH_4^+ \rightleftharpoons NH_3 + H_2PO_4^-$
    c. $SO_4^{2-} + H_2O \rightleftharpoons HSO_4^- + OH^-$

12. Identify each of the following as an Arrhenius acid or base, a Brønsted-Lowry acid or base, or a Lewis acid or base. Some substances may fit into more than one model.

    a. $Ca(OH)_2$
    b. H_2O
    c. NH_4^+

13. Write the words that complete each statement.

    a. $Al(OH)_3$ is an Arrhenius _____ and therefore must be a Brønsted-Lowry _____.
    b. NH_3 can be classified as a(n) _____ base and a(n) _____ base, but not as a(n) _____ base.

460 Chapter 18

14. Summarize the different definitions of bases according to the three models in Section 18.1.

15. Create a graphic organizer using the following terms as a minimum: *acid, Arrhenius model, base, Brønsted-Lowry model, conjugate acid, conjugate base, hydronium ion, hydroxide ion, Lewis model*.

16. A young chemist measures [H_3O^+] and [OH^-] in an aqueous solution at 25 °C. She reports that [H_3O^+] = 1 × 10^{-8} M and [OH^-] = 1 × 10^{-8} M. Her lab instructor tells her to go back to the laboratory and make the measurements again. Why?

17. Describe acidic, basic, and neutral solutions on the basis of concentration of hydronium ion, concentration of hydroxide ions, pH, and pOH.

18. Given the information below, find the [H_3O^+], [OH^-], pH, and pOH. Also identify the solution as acidic, neutral, or basic.

    a. seawater, [H_3O^+] = 1.1 × 10^{-8}

    b. pH = 3.52

    c. phosphate detergent solution, pOH = 4.51

    d. [OH^-] = 1.6 × 10^{-3}

19. Calculate the [H_3O^+], pH, [OH^-], and pOH of each solution. Assume that the acid completely dissociates.

    a. 6.54 × 10^{-2} M HCl

    b. 3.9 × 10^{-5} M H_2CO_3

    c. 9.54 × 10^{-1} M H_3PO_4

20. Using Table 18-4, indicate the stronger acid in each pair.

    a. HI, H_2SO_4

    b. HF, HCN

    c. H_2O_2, $HCHO_2$

    d. H_3BO_3, H_3PO_4

    e. HCl, $HClO_4$

21. From each pair, choose the stronger acid and then write a brief justification for your choice.

    a. $HBrO_4$, HBr

    b. PH_3, H_2S

22. All the following substances undergo deprotonation reactions. Write the equations for their ionization constants. If they go through a series of ionizations, include the equation for each.

    a. $HClO_4$

    b. H_3BO_3

23. Compare monoprotic, diprotic, and triprotic acids.

24. A water sample shows a blue color when tested with bromphenol blue and a red color when tested with litmus. What is the approximate pH of the sample? Is it acidic, basic, or neutral?

25. Write the chemical equation for the neutralization reaction

    a. showing how ions from an acid act to neutralize ions in a base.

    b. showing chromic acid neutralizing potassium hydroxide.

26. How could you determine the point at which a chemical reaction raises the pH of a solution past 8.8?

27. In titrations, what is the difference between the equivalence point and the end point?

28. The addition of 37.9 mL of 4.21 × 10^{-2} M HCl solution neutralizes 65.0 mL of NaOH solution. What is the concentration of NaOH?

29. How many milliliters of 2.00 M H_2SO_4 are required to neutralize 45.0 mL of 3.00 M NaOH?

30. A salad-dressing manufacturer wants to make a tangy salad dressing by using vinegar with an acetic acid concentration of at least 1 molar. The quality control department examines a sample of vinegar and determines that 295.7 mL of 0.1000 M sodium hydroxide neutralizes the acetic acid in 25.00 mL of the vinegar. Calculate the amount concentration of the acetic acid to determine whether the vinegar meets the manufacturer's requirements.

31. What happens to excess OH^- ions when a base is added to a solution buffered by a combination of acetic acid and sodium acetate?

32. Explain the purpose of the buffer systems in our blood.

CRITICAL THINKING

33. K_w depends on temperature. We know that since K_w is 1.0 × 10^{-14} at 25 °C, the pH of pure water is 7 (neutral). If the K_w is 2.0 × 10^{-13} at 75 °C, what is the pH of pure water at that temperature?

34. Starting with K_w at 25 °C, derive the equation pH + pOH = 14.

35. Explain the relationship between the self-ionization of water and the pH and pOH scales.

36. How many grams of calcium carbonate would be needed to neutralize 75.7 mL of 0.200 molar sulfuric acid?

Chapter 19

REDOX REACTIONS

When someone mentions the Nobel prizes for science, people think of cutting-edge research. So when the 2019 prize in chemistry was announced, people were mystified that work on batteries could be the winner. Batteries have been around for over two hundred years! What is cutting-edge about them?

The current interest in batteries stems from how many devices we power with them every day. Many people today think that all of today's fields of technology will be limited by how far we can advance battery technology. Can we make them smaller, lighter, more powerful, and longer-lasting?

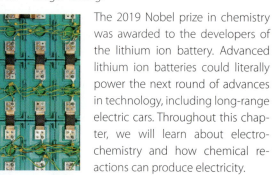

The 2019 Nobel prize in chemistry was awarded to the developers of the lithium ion battery. Advanced lithium ion batteries could literally power the next round of advances in technology, including long-range electric cars. Throughout this chapter, we will learn about electrochemistry and how chemical reactions can produce electricity.

19.1 Redox Reactions *463*
19.2 Electrochemical Reactions *471*

19.1 REDOX REACTIONS

In Chapter 10 you learned about different types of reactions, such as synthesis, decomposition, composition, single-replacement, and double-replacement. Some of these can be grouped together because they are reactions that involve the transfer or shifting of electrons from one atom to another. All single-replacement reactions and combustion reactions as well as some synthesis and decomposition reactions fall into this group. Reactions that involve transferring electrons are called **oxidation-reduction reactions**, or *redox reactions*.

Oxidation: Loss of Electrons

Redox reactions occur in the formation of both ionic and covalent compounds. For example, when magnesium burns to form magnesium oxide, two magnesium atoms each transfer two electrons to the two oxygen atoms that compose an oxygen molecule. Reactions between covalent compounds often end up shifting electrons toward or away from atoms, depending on their electronegativities. When sulfur dioxide forms from sulfur and oxygen, oxygen attracts the shared electrons closer to itself than sulfur does, thus gaining electrons. In both of these cases, one of the reactants loses electrons and the other gains electrons.

Oxidation occurs when an atom loses electrons. In the reaction between magnesium and oxygen previously mentioned, magnesium atoms are oxidized.

$$Mg \longrightarrow Mg^{2+} + 2e^-$$

Lost electrons (e^-) in the reaction above are shown as products. Because neutral magnesium loses electrons, it becomes a cation. When an atom is oxidized, its oxidation number always becomes more positive. Notice the small number above the magnesium in the equation shown below, which is a common system of notation to keep track of the oxidation numbers of the elements (see Chapter 9).

$$2\overset{0}{Mg} + O_2 \longrightarrow 2\overset{+2}{Mg}O$$

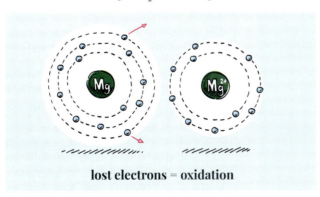

lost electrons = oxidation

At this point it may seem logical to assume that all oxidation reactions involve oxygen. While oxygen usually causes atoms to oxidize because of its high electronegativity, other atoms can do the same thing. For example, when sodium reacts with chlorine, chlorine causes the sodium atoms to oxidize.

$$2\overset{0}{Na} + \overset{0}{Cl_2} \longrightarrow 2\overset{+1}{Na^+} + 2\overset{-1}{Cl^-} \longrightarrow 2\overset{+1}{Na}\overset{-1}{Cl}$$

QUESTIONS
» What is a redox reaction?
» How are oxidation and reduction related to electrons?
» How can I tell what was oxidized or reduced in a redox reaction?
» How can I balance a redox reaction equation?

TERMS
oxidation-reduction reaction • oxidation • reduction • reducing agent • oxidizing agent

How can I prevent my car from rusting?

magnesium burning

Reduction: Gain of Electrons

Electrons lost by atoms must go somewhere; in fact, they are gained by other atoms. An atom that gains electrons undergoes a process called **reduction**. In the reaction between magnesium and oxygen, for example, each oxygen atom is reduced because it gains two electrons. Reduction makes oxidation numbers smaller (*reducing* the oxidation number) as more negative charges join atoms. The oxidation number of oxygen reduces from 0 to −2.

$$2Mg + \overset{0}{O_2} \longrightarrow 2\overset{-2}{MgO}$$

Is the reaction that forms magnesium oxide a reduction reaction? Or is it an oxidation reaction? Actually, it is both. Oxygen atoms gain electrons and are reduced, while the magnesium atoms lose electrons and are oxidized. Oxidation and reduction *always* occur simultaneously. The electrons that leave one atom join another.

gained electrons = reduction

> Recall the rules for oxidation numbers from Chapter 9. (See page 199 for the oxidation numbers of some common atoms.)
>
> **Rule 1** free elements = 0
>
> **Rule 2** monatomic ion charge = oxidation number
>
> **Rule 3** compound sum = 0
>
> **Rule 4A** Group 1 = +1
>
> **Rule 4B** Group 2 = +2
>
> **Rule 4C** hydrogen = +1 or −1 (if with a metal)
>
> **Rule 4D** oxygen = −2, −1 (in a peroxide), or +2 (when with fluorine)
>
> **Rule 4E** Group 17 = −1 (when with a metal)
>
> **Rule 5** sum of oxidation numbers in polyatomic ion = charge

Today we are all used to the instant results of digital photography. In the not too distant past, however, we had to wait as film was developed before we could see our pictures. The process of film developing is still currently used to produce x-rays. The processing of x-ray film involves a redox reaction in which silver ions are reduced to silver atoms. The unexposed x-ray film contains grains of a silver halide such as silver bromide (AgBr). When exposed to x-rays, silver ions in the silver bromide grains are energized. The activated ions are more prone to being reduced during the developing process than those in areas that were not struck by light.

$$\overset{+1}{Ag^+} + e^- \longrightarrow \overset{0}{Ag}$$

▲ Redox reactions make film developing possible.

The film is developed in a solution that contains substances that cause the sensitized silver ions to gain electrons.

Only those reactions in which oxidation numbers change are considered redox reactions. If oxidation numbers do not change, neither oxidation nor reduction has occurred.

EXAMPLE 19-1: IDENTIFYING REDOX REACTIONS

Is either of the following reactions a redox reaction?

a. $HCl + H_2O \rightleftharpoons H_3O^+ + Cl^-$

b. $2KClO_3 \longrightarrow 2KCl + 3O_2$

Solution

a. After assigning oxidation numbers to each atom, we notice that the oxidation numbers of all the atoms remain the same before and after the reaction. Therefore, the reaction is not a redox reaction.

$$\overset{+1\,-1}{HCl} + \overset{+1\,-2}{H_2O} \rightleftharpoons \overset{+1\,-2}{H_3O^+} + \overset{-1}{Cl^-}$$

b. The oxidation numbers for both oxygen and chlorine change during the reaction. This reaction involves both oxidation and reduction.

$$\overset{+1\,+5\,-2}{2KClO_3} \longrightarrow \overset{+1\,-1}{2K\,Cl} + \overset{0}{3O_2}$$

CAREERS

SERVING AS AN ELECTROCHEMIST: *POWER THE FUTURE*

How many battery-powered devices have you used today? Our modern lifestyle relies on batteries, such as those found in our digital gadgets and cars. The future of technologies such as fuel cell vehicles and lighter, smaller, and better digital devices depends on improving battery technology. Scientists today are reaching into the future, developing the batteries that you will be carrying around in your pocket in the decades ahead. These scientists are electrochemists doing battery research. At this point, digital technology is growing faster than battery technology, making innovation in this field even more crucial.

Some electrochemists, such as 2019 Nobel prize winners John B. Goodenough, M. Stanley Whittingham, and Akira Yoshino, focus on battery research to find ways to improve the components of batteries so that they can be smaller, lighter, more efficient, longer-lasting, more powerful, and more environmentally friendly. These scientists focus on specific parts of the battery such as the anode or cathode. They work with materials scientists to pick the best materials for batteries. They create working prototypes, then test and analyze them in the laboratory. They study and analyze oxidation-reduction reactions that could be used in batteries. Through their work, some believe that they can solve many of our future energy needs.

Oxidizing and Reducing Agents

In a redox reaction, one substance is reduced by another substance, called a **reducing agent**. In the film-developing process, one or more reducing agents cause silver ions to be reduced to silver atoms. Because reduction cannot occur without oxidation, reducing agents contain atoms that donate electrons. Since magnesium donates electrons, it is a reducing agent in the reaction between magnesium and oxygen.

$$Mg \longrightarrow Mg^{2+} + 2e^-$$

Remember that oxidation and reduction always occur simultaneously. If magnesium acts as a reducing agent, it donates electrons and is oxidized. *Reducing agents are always oxidized in redox reactions.*

An **oxidizing agent** is a substance that oxidizes other substances. Oxidizing agents are commonly used as cleaners and disinfectants. Chlorine, for instance, is an oxidizing agent added to swimming pools to kill bacteria, algae, and fungi. Chlorine acts as an oxidizing agent by taking electrons from these microbes. In the process, chlorine atoms are reduced to chloride ions.

$$Cl_2 + 2e^- \longrightarrow 2Cl^-$$

Because chlorine gains electrons in this reaction, it is reduced. Oxidizing agents are always reduced in redox reactions.

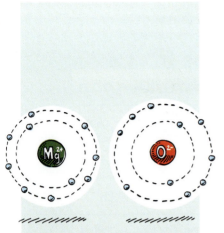

Reducing agents are always oxidized and oxidizing agents are always reduced.

EXAMPLE 19-2: IDENTIFYING OXIDIZING AND REDUCING AGENTS

Which substance in the following reaction is a reducing agent? Which is an oxidizing agent?

$$2Na + 2H_2O \longrightarrow 2NaOH + H_2$$

Solution

$$2\overset{0}{Na} + 2\overset{+1}{H}_2O \longrightarrow 2\overset{+1}{Na}OH + \overset{0}{H}_2$$

The oxidation number of sodium goes from 0 in sodium to +1 in sodium hydroxide. The oxidation number of hydrogen goes from +1 in water to 0 in hydrogen gas. Sodium was oxidized (lost electrons), and hydrogen was reduced (gained electrons). The hydrogen in water, therefore, is acting as an oxidizing agent, and sodium is acting as a reducing agent.

When considering a redox reaction, it is sometimes easier to look at the oxidation number on an entire ion. See the example below.

EXAMPLE 19-3: IDENTIFYING OXIDIZING AND REDUCING AGENTS

Identify which substance is oxidized and which is reduced. Also indicate which substance is the reducing agent and which is the oxidizing agent.

$$2S_2O_3^{2-} + OCl^- + 2H^+ \longrightarrow Cl^- + S_4O_6^{2-} + H_2O$$

Solution
The charge on thiosulfate ($S_2O_3^{2-}$) is 2− and the charge on tetrathionate ($S_4O_6^{2-}$) is also 2−. Since there are two thiosulfate ions on the left (total charge of 4−), one of these elements must be oxidized to keep the charge at 2−. The chlorine in the hypochlorite changes from an oxidation number of +1 to a −1 in the chloride ion.

$$2S_2\overset{-2}{O}_3^{2-} + O\overset{+1}{Cl}^- + 2H^+ \longrightarrow \overset{-1}{Cl}^- + S_4\overset{-2}{O}_6^{2-} + H_2O$$

The thiosulfate is oxidized and the chlorine in the hypochlorite ion is reduced. This means that hypochlorite is the oxidizing agent and thiosulfate is the reducing agent.

Household bleach is a 5% aqueous solution of sodium hypochlorite (NaOCl). Sodium hypochlorite is an effective oxidizing agent. It is reduced in redox reactions according to the equation below.

$$NaOCl + H_2O + 2e^- \longrightarrow NaCl + 2OH^-$$

Bleach is normally used to eliminate stains and make dirty white clothes clean and white again. The molecules in stains possess very mobile electrons. These molecules are colored because they absorb visible light energy, energizing their electrons. When a dirty article of clothing is bleached, the sodium hypochlorite in the bleach grabs the less-tightly-bonded electrons from the molecules in the stains. When these electrons are removed, the molecules no longer absorb visible light. Consequently, the color disappears and the cloth is once again white. In reality, bleach does not remove stains; it merely decolorizes them.

LEO the lion goes **GER**

Lost **E**lectrons = **O**xidation;

Gained **E**lectrons = **R**eduction.

This mnemonic device can help you remember the relationship between reduction, oxidation, and the movement of electrons.

Using Oxidation
TO SOLVE PROBLEMS

We all understand that many metals, especially iron, rust. Rusting is another redox reaction, albeit a complex one. Water reacts with the iron in steel, causing the iron to lose two electrons. The hydrogen from the water gains those two electrons as it forms hydrogen gas and hydroxide ions, which combine with iron(II) ions to form iron(II) hydroxide.

$$Fe(s) + 2H_2O(l) \longrightarrow H_2(g) + Fe(OH)_2(s)$$

The iron(II) hydroxide then reacts in a second redox reaction with oxygen to produce the more familiar rust, or iron(III) oxide (Fe_2O_3).

$$4Fe(OH)_2(s) + O_2(g) \longrightarrow 2Fe_2O_3(s) + 4H_2O(l)$$

The formation of rust is much slower—about one hundred times slower—if no oxygen is present. Anything that pits or creates cracks, divots, and bends in the iron can aid the corrosion process. Another contributing factor is the presence of chloride ions, such as those found in salt. Knowing the conditions that encourage rusting can lead to its prevention. If iron can be coated to avoid contact with water and oxygen, rusting can be prevented.

The process of rust formation occurs in the presence of iron, water, and oxygen. It is very similar to the processes occurring in a battery.

Stainless steel is an iron-chromium alloy that is rust-resistant. It contains 10.5%–28% chromium, which oxidizes in atmospheric oxygen to form a very thin layer of chromium(III) oxide (Cr_2O_3) on the surface of the alloy. This layer, which is thin enough to maintain the luster of the alloy, is impervious to water and oxygen. The layer is added in a process known as *passivation*. Stainless steel has many uses in jewelry, cookware, surgical instruments, and building materials for skyscrapers and industrial machinery. However, even stainless steel can rust in extreme conditions, such as when it is gouged, pitted, stressed, or exposed to high concentrations of chloride ions.

Balancing Redox Reactions

In Chapter 10 you learned how to balance chemical equations. Do you remember why chemical reactions need to be balanced? It is necessary for reactions to have the same number of each kind of atom on both sides of the chemical equation in order to maintain the law of the conservation of mass. Similarly, redox reactions must be balanced. Since electrons are being moved around, both the *number of each kind of atom* and the *overall charge* on each side of the reaction must be balanced. The number of electrons on each side of the reaction must be conserved.

Some redox reactions can be balanced like other reactions in the traditional way of identifying reactants and products, counting atoms of each element, and adjusting coefficients. For example, in Section 10.1 we considered the balanced equation below.

$$2KClO_3(s) \longrightarrow 2KCl(s) + 3O_2(g)$$

This redox reaction can be balanced with little difficulty. There are two potassium atoms, two chlorine atoms, and six oxygen atoms on both sides of the equation. The overall charge is neutral on both sides of the equation.

In reactions where the same element appears in multiple reactants or products, however, a different method sometimes simplifies the process. Consider the unbalanced equation below.

$$\overbrace{NH_3(g) + \underbrace{NO_2(g)}_{\text{reduction}}}^{\text{oxidation}} \longrightarrow N_2(g) + H_2O(l)$$

With nitrogen occurring in both of the reactants, balancing this equation is more difficult, though with enough time and adjustment, you may be able to balance it by inspection. The use of a different method will allow you to balance it more quickly.

Potassium chlorate ($KClO_3$) is found on matches and in fireworks and other explosives.

Steps TO BALANCE A REDOX REACTION

Step 1

Assign oxidation numbers to all atoms, and determine which atoms are oxidized and which ones are reduced.

$$\overset{-3\ +1}{NH_3} + \overset{+4\ -2}{NO_2} \longrightarrow \overset{0}{N_2} + \overset{+1\ -2}{H_2O}$$

Since the oxidation numbers of hydrogen and oxygen remain unchanged in this particular reaction, ignore them for now and focus on the other atoms.

$$\overset{-3}{NH_3} + \overset{+4}{NO_2} \longrightarrow \overset{0}{N_2} + H_2O$$

Step 2

Determine which atoms have been oxidized and which have been reduced. In this case, the nitrogen from the reactants is both oxidized and reduced. Nitrogen from the ammonia (NH_3) is oxidized because its oxidation number increases from −3 to 0. The nitrogen in nitrogen dioxide (NO_2) is reduced from +4 to 0. The oxidation numbers for the hydrogen and oxygen atoms do not change.

$$\overset{oxidation}{\overset{-3}{NH_3} + \overset{+4}{NO_2} \longrightarrow \overset{0}{N_2} + H_2O}$$
(reduction)

Step 3

To help keep track of the oxidation and reduction halves of the reaction, draw a line connecting the atoms involved in reduction and those involved in oxidation. You should also write the net change in oxidation number along each line.

$$NH_3 + NO_2 \longrightarrow N_2 + H_2O$$
(+3 / −4)

Step 4

Balance the magnitude of the oxidation number changes by multiplying both reactant-product pairs by a single coefficient for the reduction change and another coefficient for the oxidation change. In some cases, you may need to multiply only one of the changes to balance the change in oxidation number. In this reaction, both changes must be multiplied to reach the least common multiple of 12. Notice that the N_2 on the product side has been multiplied by both 4 *and* 3. We add these multiples to get a coefficient of 7 on the right side of the equation.

$$4NH_3 + 3NO_2 \longrightarrow (4+3)N_2 + H_2O$$
$+3 \times 4 = 12$
$-4 \times 3 = -12$

Step 5

Now you can use the conventional inspection method to finish balancing the equation. Notice that you have 14 nitrogen atoms on the right. There are half as many nitrogen atoms on the reactant side.

$$4NH_3 + 3NO_2 \longrightarrow 7N_2 + H_2O$$

Doubling the coefficients of each of the reactants would give you the right number of nitrogen atoms. But now you should recognize that the hydrogen and oxygen are still unbalanced.

$$8NH_3 + 6NO_2 \longrightarrow 7N_2 + H_2O$$

Since there are 24 hydrogen atoms and 12 oxygen atoms on the reactant side, the equation balances by simply changing the coefficient on water to a 12.

$$8NH_3 + 6NO_2 \longrightarrow 7N_2 + 12H_2O$$

EXAMPLE 19-4: BALANCING REDOX REACTIONS

Balance the equation for the redox reaction below.

$$HCl + HNO_3 \longrightarrow HOCl + NO + H_2O$$

Solution
Step 1. Assign oxidation numbers to all atoms whose charges have changed. In this reaction, hydrogen and oxygen atoms do not undergo oxidation or reduction.

$$H\overset{-1}{Cl} + H\overset{+5}{N}O_3 \longrightarrow HO\overset{+1}{Cl} + \overset{+2}{N}O + H_2O$$

Step 2. Determine which atoms have been oxidized and which have been reduced.

$$\underbrace{H\overset{-1}{Cl} + H\overset{+5}{N}O_3 \longrightarrow HO\overset{+1}{Cl} + \overset{+2}{N}O + H_2O}_{\text{reduction}}^{\text{oxidation}}$$

Step 3. Write the net change in oxidation number along each line.

$$H\overset{-1}{Cl} + H\overset{+5}{N}O_3 \longrightarrow HO\overset{+1}{Cl} + \overset{+2}{N}O + H_2O$$

(+2 above, −3 below)

Step 4. Balance the magnitude of the oxidation number changes by multiplying both reactant-product pairs by an appropriate coefficient for each oxidation number change.

$$3H\overset{-1}{Cl} + 2H\overset{+5}{N}O_3 \longrightarrow 3HO\overset{+1}{Cl} + 2\overset{+2}{N}O + H_2O$$

(+2 × 3 = +6 above, −3 × 2 = −6 below)

Step 5. Now you can use the conventional inspection method to finish balancing the equation. In this case, the equation is balanced, so no further coefficient changes are necessary.

$$3HCl + 2HNO_3 \longrightarrow 3HOCl + 2NO + H_2O$$

19.1 SECTION REVIEW

1. What is a redox reaction?
2. Define *oxidation*.
3. Write the oxidation number for each element or polyatomic ion in each of the following compounds.

   a. F_2

   b. $Cu(NO_3)_2$

   c. Cr^-

4. Write the oxidation number for each element or polyatomic ion in each of the following compounds.

   a. $Fe(OH)_2$

   b. H_2S

   c. CH_2O_2

5. (True or False) Oxidation always occurs before reduction.

6. For each of the following equations, determine whether the reaction is a redox reaction. For each redox reaction, tell which element is oxidized and which is reduced.

   a. $2KBr(aq) + Cl_2(g) \longrightarrow 2KCl(aq) + Br_2(l)$

   b. $2NaOH(aq) + CuCl_2(aq) \longrightarrow Cu(OH)_2(s) + 2NaCl(aq)$

   c. $Mg(s) + I_2(g) \longrightarrow MgI_2(s)$

   d. $CH_4(g) + 2O_2(g) \longrightarrow CO_2(g) + 2H_2O(g)$

7. In the redox equation $2K + Br_2 \longrightarrow 2KBr$,

   a. which element is reduced?

   b. which element is oxidized?

   c. what is the reducing agent?

   d. what is the oxidizing agent?

   e. which element loses an electron or electrons?

   f. which element gains an electron or electrons?

   g. how many electrons are exchanged between each potassium and bromine atom? how many in total for the reaction?

8. Balance each of these equations using the oxidation-number method.

   a. $SnCl_4 + Fe \longrightarrow SnCl_2 + FeCl_3$

   b. $SO_2 + Br_2 + H_2O \longrightarrow HBr + H_2SO_4$

   c. $CO + I_2O_5 \longrightarrow I_2 + CO_2$

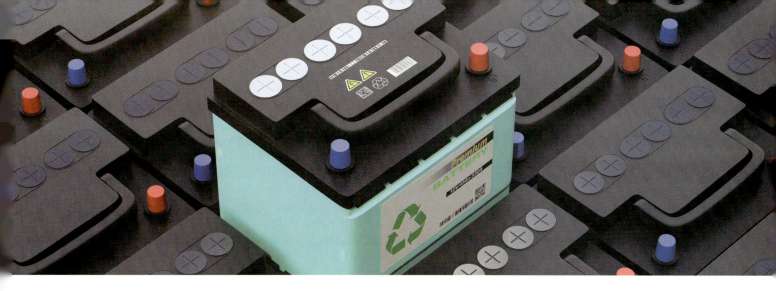

19.2 ELECTROCHEMICAL REACTIONS

With the movement of electrons being required in redox reactions, we can see a definite connection between these chemical reactions and electricity. **Electrochemistry** is the field of chemistry that studies how electricity and redox reactions are related. Today we use redox reactions to produce electricity, as in batteries, or we use electricity to cause a redox reaction, as in the electroplating of metal objects.

Electrochemical Cells

Metals can conduct electricity because their bonds allow electrons to be mobile in a metal lattice. Some solutions can conduct electricity, but for a different reason. Water conducts electricity only when a sufficient number of ions are present. An electrolyte is any substance that, when dissolved in water, allows the resulting solution to conduct electricity. When an electrolyte dissolves in water, anions and cations form. The ions move freely in the solution and therefore may carry a charge.

A substance's ability to act as an electrolyte depends on its ability to form ions. Solutions of strong electrolytes conduct electricity well because they produce many ions in the solution. Most salts, strong acids, and strong bases are strong electrolytes (e.g., NaCl, HCl, H_2SO_4, HNO_3, and NaOH). Substances that do not readily form ions, such as weak acids, weak bases, and less soluble salts, are weak electrolytes. Solutions of weak electrolytes do not conduct electricity as well as those of strong electrolytes. Covalent substances such as sugar, alcohol, and oxygen might dissolve in water, but they cannot conduct electricity. For this reason, such substances are called *nonelectrolytes*.

All electrochemical techniques rely on the electrical conductivity of metals and electrolyte solutions. But to be useful, these substances must be assembled in just the right way. The **electrochemical cell** is the fundamental apparatus used in electrochemistry and either uses a spontaneous chemical reaction to produce electricity or uses electricity to drive a nonspontaneous chemical reaction.

There are two types of electrochemical cells—electrolytic and voltaic. **Electrolytic cells** use electrical energy to force a nonspontaneous chemical reaction to occur. A **voltaic cell** uses a spontaneous redox reaction to produce electrical energy.

QUESTIONS
» What is electrochemistry?
» What different types of electrochemical cells are there?
» How can we use electrochemistry?
» How do different electrochemical cells compare?

TERMS
electrochemistry • electrochemical cell • electrolytic cell • voltaic cell • electrolysis • cathode • anode • electroplating • battery • half-cell • half-reaction • salt bridge • fuel cell

How do batteries transform chemical energy to electrical energy?

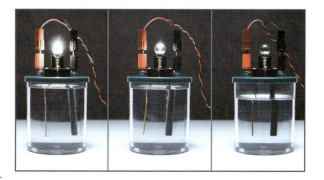

▲
Solutions of a strong electrolyte (HCl), a weak electrolyte ($HC_2H_3O_2$), and a nonelectrolyte ($C_{12}H_{22}O_{11}$) have different electrical conductivities.

Redox Reactions   **471**

Using Electrolytic Cells

Electrolysis is the process of forcing an otherwise nonspontaneous redox reaction to occur with the aid of an electrical current in an electrochemical cell. For example, normally stable water molecules can be decomposed by an electric current to produce hydrogen and oxygen gases. Current can be passed through an electrolytic cell when a source of electricity is connected to two electrodes immersed in an electrolyte solution. Electrodes are nothing more than rods or wires. They are commonly made of metals such as zinc, platinum, or copper.

What happens when electrons flow through the cell? First, electrons flow from the source of the electrical energy into one of the electrodes. These electrons are available to cause reduction reactions in the solution. The electrode where reduction occurs is called the **cathode**, and in the case of an electrolytic cell this electrode is negatively charged. Remember that a redox reaction cannot occur unless oxidation and reduction reactions occur simultaneously. The oxidation occurs at the **anode**, which in electrolytic cells is the positively charged electrode. The electrons freed in the oxidation flow back to the source of electrical energy. Since any anions in the solution are negatively charged, they migrate to the positively charged anode. Similarly, positively charged cations migrate to the negatively charged cathode. As long as the electric potential difference (voltage) between the two electrodes is large enough, the redox reaction will occur.

▲ Reduction occurs at the cathode. Oxidation occurs at the anode.

Electrolytic Cells

Electrodeposition

Sterling silver is at least 92.5% silver. Less expensive silverware is made of some common metal that is covered with a thin layer of silver. How is the thin silver layer put onto the inexpensive metal? You could pound, melt, or glue the silver onto the metal, but there is a better way. Metallic ions in solution can be forced to cling to another metal through an electrochemical process called electroplating, which is one form of *electrodeposition*—any process that deposits one substance onto another by an electrochemical reaction.

Electroplating is a type of electrolysis in which the metal object to be plated is the cathode and the metal that is to cover the object is the anode. For example, we use a silver anode to do silver plating. The electrolyte solution contains silver ions. When current flows in the cell, electrons leave the anode, liberating more silver ions to maintain an adequate level of silver ions in the electrolyte solution. These silver ions migrate to the cathode and are reduced to metallic silver with the electrons from within the circuit. The silver then plates onto the item being electroplated.

$$Ag^+ + e^- \longrightarrow Ag$$

We make galvanized steel by electroplating zinc onto steel to prevent the steel from rusting. While zinc also corrodes, the product of this reaction actually protects the zinc from further oxidation. Therefore, the layer of zinc also protects the iron in the steel from being oxidized. Even if a small crack forms in the zinc plating, the iron is still protected because zinc is more easily oxidized than iron. Any oxidation that occurs will be the oxidation of zinc, which serves as a *sacrificial anode*.

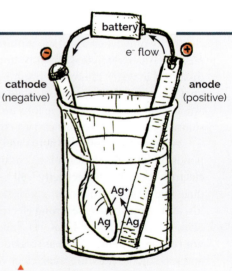

▲ An electrolytic cell is used to electroplate metals.

Nonmetals can also be electrodeposited onto metal surfaces. The automotive industry uses an electrodeposition process to paint car bodies. Positively charged paints are deposited onto the negatively charged sheet metal. This technique more effectively covers the metal and gives better corrosion resistance than conventional spray painting.

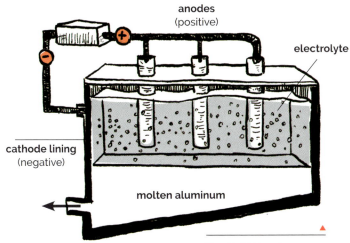

The Hall-Héroult process

Aluminum Smelting

Metals such as copper, tin, and iron, which are low on the activity series (see table on page 232), can be freed from their naturally occurring compounds by chemical means. But active metals such as sodium, lithium, and aluminum bond too strongly for those "mild" techniques to work, so engineers must separate their ores in electrolytic cells. In nature, aluminum atoms are oxidized by oxygen atoms in an ore commonly known as *bauxite*. To get pure aluminum metal from the very stable aluminum oxide (Al_2O_3), the natural oxidation reaction must be reversed—electrons must be *forced* back into the aluminum ions.

The electrolysis process used to produce aluminum in industry is called the *Hall-Héroult process*. This process is named after American chemist Charles Hall and French chemist Paul Héroult, who each developed the process independently in 1886. Until the discovery of this process, aluminum was more precious than gold because there was no practical way to extract the otherwise abundant metal from its ores. In the Hall-Héroult process, aluminum oxide is dissolved in molten cryolite (Na_3AlF_6) and electrolyzed with carbon anodes.

$$3C + 2Al_2O_3 \longrightarrow 4Al + 3CO_2$$

The molten elemental aluminum falls to the bottom and is extracted. The oxygen reacts with the carbon of the anode and forms carbon dioxide.

Melting cryolite and bauxite requires great amounts of electricity. Although less than 6 volts of electricity are needed, the electric current must be between 100 000 and 150 000 amps. To lower operating costs, aluminum refineries are often located close to hydroelectric power stations, where energy is abundant and less expensive. (It's worth noting that it takes only one-twentieth as much energy to recycle aluminum as it does to extract it from its ores, which is why aluminum is commonly recycled, despite the fact that it is abundant in the earth's crust.)

Electrolysis of Brine

Sodium hydroxide, chlorine gas, and hydrogen gas are produced by an electrolytic process from a concentrated sodium chloride solution called *brine*. These chemicals are valuable in industry. The most common type of cell used for this process uses a graphite anode and a steel cathode placed in the brine that are separated by a membrane. Although the membrane will allow cations to pass through it, it prevents substances produced at each electrode to mix. Oxidation at the anode produces chlorine gas from chloride ions.

$$2Cl^-(aq) \longrightarrow Cl_2(g) + 2e^-$$

Reduction at the cathode produces pure sodium from sodium ions and hydrogen gas from water molecules.

$$2H_2O(l) + 2e^- \longrightarrow H_2(g) + 2OH^-(aq)$$

The membrane allows the sodium cations to move toward the cathode side of the device where there are hydroxide anions.

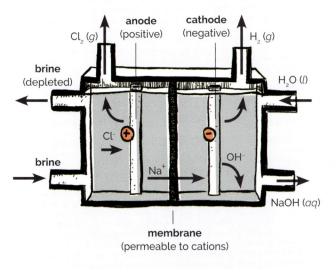

Many very useful elements and compounds can be separated from brine by electrolysis.

Redox Reactions 473

WORLDVIEW INVESTIGATION

BATTERY RECYCLING

Your school is replacing all its laptops. You notice the old laptops being placed in cardboard boxes, and so you ask where they are going. When you are told that they are going to the landfill, you are concerned because you recently read about chemicals from batteries leaking and causing pollution. You ask the principal about the disposal plan. Inspired by your concern, she asks you to research the issue and decides to delay disposing of the laptops until you return with an answer.

Task

You are tasked to research and report back to the principal about the importance of recycling batteries. You must explain why they should be recycled, how they are recycled, how much recycling would cost, and why a Christian should be concerned about this issue. Finally, the principal wants your recommendation about recycling these batteries.

Procedure

1. Research the issue by doing an internet search using the keywords "battery recycling," "why should I recycle batteries," and "how are batteries recycled."
2. Consider how biblical principles, outcomes, and motivations would apply to this issue.
3. Write your summary, citing your sources properly. Have a classmate review your summary and give you feedback.
4. Complete your summary and submit it by the deadline.

Conclusion

Many of our modern electronics run on batteries. If batteries are hazardous to both the environment and people, then we have an obligation to reduce the threat.

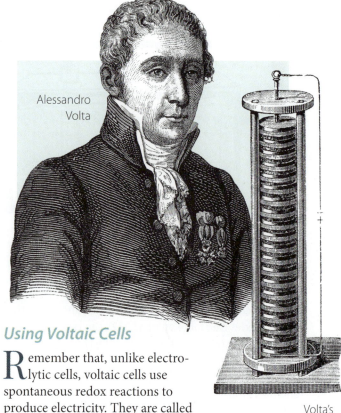

Alessandro Volta

Volta's battery

Using Voltaic Cells

Remember that, unlike electrolytic cells, voltaic cells use spontaneous redox reactions to produce electricity. They are called voltaic cells in honor of Alessandro Volta, who invented the first battery in 1800. A **battery** is comprised of one or more voltaic cells.

In voltaic cells, just as in electrolytic cells, reduction occurs at the cathode and oxidation at the anode. Because the electrochemical process is reversed in the voltaic cell, the anode is now negatively charged and the cathode is positively charged. While the charge on each anode changes, the chemical reaction occurring at each remains the same; in both cases, oxidation occurs at the anode and reduction at the cathode. All voltaic cells contain an anode that loses electrons, a cathode that gains electrons, and an electrolyte between them. The electrolyte may be a liquid, such as the sulfuric acid solution in a car battery, or a moist paste, such as that found in most dry cell batteries.

In the voltaic cell diagrammed at the top of the facing page, the reaction occurring at the anode is the oxidation of zinc, and that at the cathode is the reduction of copper. Every voltaic cell has two **half-cells**, one where oxidation takes place and the other where reduction takes place. The reactions that occur in each half-cell are called the **half-reactions**—together they form a complete redox reaction.

$$\text{anode: } Zn \longrightarrow Zn^{2+} + 2e^-$$

$$\text{cathode: } Cu^{2+} + 2e^- \longrightarrow Cu$$

The electrons from the oxidation half-reaction at the anode travel along the wire to the cathode.

At the cathode, copper(II) ions are reduced to metallic copper by the electrons from the anode. The steady flow of electrons from the anode to the cathode can be harnessed to produce electrical energy. As the reactions in each half-cell continue, the zinc solution builds up a positive charge from the accumulation of zinc ions. As the copper ions plate out as copper, that solution builds up a negative charge. The reactions will stop unless there is a mechanism to prevent the buildup of charges in the solutions. Electrochemists accomplish this with a **salt bridge**—a tube of electrolytic gel that connects the two half-cells of a voltaic cell. The salt bridge allows the flow of ions but prevents the mixing of the solutions.

A salt bridge is illustrated in the image at right. Current flows between the solutions as ions migrate. As zinc is oxidized, excess zinc ions accumulate in the solution around the anode. Chloride ions migrate from the salt bridge toward the concentration of positive charges and keep the solution close to neutral. As copper(II) is reduced to metallic copper at the cathode, positive charges are removed from the solution around the cathode. Potassium ions migrate from the salt bridge into the solution in order to keep the solution electrically neutral. If the salt bridge were removed, current would stop.

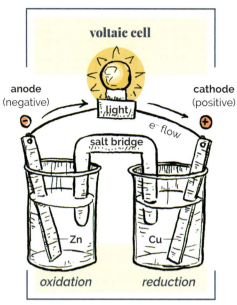

A voltaic cell. The salt bridge allows K^+ and Cl^- ions to flow between the beakers, propagating the redox reaction that generates an electrical current.

TABLE 19-1 *Electrode Conventions*

Properties	Cathode	Anode
ions attracted	cations	anions
electron movement	into cell	out of cell
half-reaction	reduction	oxidation
sign in electrolytic cell	negative	positive
sign in voltaic cell	positive	negative

Sign conventions of the electrodes are a common source of confusion. The confusion can be avoided if we remember that the electrode that emits the electrons is always negative. Recall that oxidation always occurs at the anode. In a voltaic cell, the anode is negative (−) because it is where a metal is *spontaneously* losing electrons. But in an electrolytic cell, the anode is positive (+) because its electrons are being *forcibly* removed by an external power supply. The cathode in a voltaic cell is positive as the reduction of the metal uses electrons. The cathode in an electrolytic cell is negative as the electrons move into solution to reduce ions in the solution. Reduction always occurs at the cathode.

EXAMPLE 19-5: WRITING HALF-REACTIONS

Write the half-reactions and the redox reaction for the oxidation of calcium and the reduction of iron(III) ions into iron atoms.

Solution

LEO GER reminds us that oxidation is the losing of electrons. Calcium in Group 2 has two valence electrons, which it will lose.

$$Ca \longrightarrow Ca^{2+} + 2e^-$$

LEO GER reminds us that reduction is the gaining of electrons. Iron(III) needs to gain three electrons to become neutral iron atoms.

$$Fe^{3+} + 3e^- \longrightarrow Fe$$

To write the balanced redox reaction we have to balance the electrons on either side of the equation.

$$3(Ca \longrightarrow Ca^{2+} + 2e^-)$$
$$2(Fe^{3+} + 3e^- \longrightarrow Fe)$$

We can now cancel the electrons and add the two equations together.

$$3Ca \longrightarrow 3Ca^{2+} + \cancel{6e^-}$$
$$2Fe^{3+} + \cancel{6e^-} \longrightarrow 2Fe$$
$$3Ca + 2Fe^{3+} \longrightarrow 3Ca^{2+} + 2Fe$$

Voltaic Cells

Rechargeable Batteries

We all know the frustration of turning on a battery-operated device only to find that the batteries are dead. Happily, many devices have rechargeable batteries, which means that when they get low on power, we can plug them in and recharge the battery. These batteries can be used multiple times before they need to be replaced. Reversing the polarity of the electrodes recharges batteries. This reverses the redox reaction and regenerates the cell.

The most common uses for rechargeable batteries, also known as *secondary batteries*, are cars, laptops, tablets, and cellphones. Interestingly, as we use a rechargeable battery, it acts as a voltaic cell, but when it is recharging, it acts as an electrolytic cell.

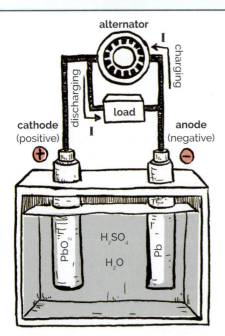

LEAD-ACID STORAGE BATTERIES

Cars and trucks start by cranking the engine from energy stored in a series of six lead storage cells linked together so that their voltages add to each other. These are called *lead-acid storage batteries*. The cathode is a series of lead-antimony alloy plates permeated with lead(IV) oxide (PbO_2). The anode is a series of lead-antimony alloy plates filled with spongy lead. The cathode and anode are immersed in sulfuric acid.

The oxidation half-reaction is

$$Pb + SO_4^{2-} \longrightarrow PbSO_4 + 2e^-.$$

The reduction half-reaction is

$$PbO_2 + 4H^+ + SO_4^{2-} + 2e^- \longrightarrow PbSO_4 + 2H_2O.$$

As strange as it seems, lead atoms are oxidized from 0 to +2 on one plate and reduced from +4 to +2 on another. As the battery discharges, the concentration of the sulfuric acid decreases as H_2SO_4 is consumed at both electrodes.

LITHIUM ION BATTERIES

As mentioned in the Chapter opener, *lithium ion batteries* are the cutting edge of battery technology. They are the newer, lighter, longer-lasting, more powerful batteries that many feel will power our technology-based society for decades to come. These batteries power most of our laptop computers, cellphones, tablets, and electric vehicles.

The specific design of a lithium ion battery depends on the specific application for which it will be used. In general, one electrode is made of carbon and the other is a metal oxide. The electrolyte is a lithium compound in an organic solvent. The lithium ions move between the electrodes during the operation of the battery.

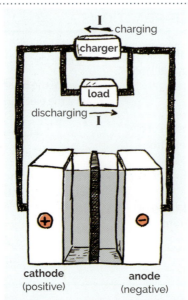

The oxidation half-reaction is
$$LiC_6 \longrightarrow C_6 + Li^+ + e^-.$$

The reduction half-reaction is
$$CoO_2 + Li^+ + e^- \longrightarrow LiCoO_2.$$

One of the most important aspects of the lithium ion battery is that it is rechargeable.

Nonrechargeable Batteries

Some cells cannot or should not be recharged. Manufacturers will often put a label on their alkaline batteries and other *primary* (non-rechargeable) batteries to warn customers against recharging the batteries. It's not that alkaline batteries *cannot* be recharged; rather, they *should not* be recharged because to do so would be dangerous. Even though the redox reaction in an alkaline cell is reversible, recharging produces some gases. Since many alkaline cells have no vents to release gases, they could explode if an attempt is made to recharge them.

ZINC-CARBON DRY CELLS

One common type of battery is the *zinc-carbon dry cell*. A dry cell consists of a zinc container filled with an electrolyte paste made of MnO_2 and $ZnCl_2$ or NH_4Cl in a binder that keeps it all together. The zinc cylinder acts as the anode and loses electrons. Inserted into the electrolyte paste is a graphite rod that acts as the cathode. A coated paper separator acts as an insulator and prevents a short circuit.

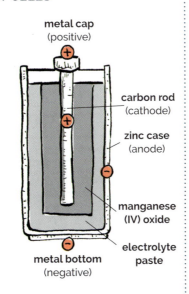

Although the reactions occurring at these electrodes are complicated, they can be summarized as shown here.

$$\text{anode: } Zn + 2Cl^- \longrightarrow ZnCl_2 + 2e^-$$

$$\text{cathode: } 2NH_4Cl + 2MnO_2 + 2e^- \longrightarrow Mn_2O_3 + H_2O + 2NH_3 + 2Cl^-$$

In these half-reactions, zinc is oxidized from a charge of 0 to +2, and manganese is reduced from +4 to +3.

ALKALINE BATTERIES

The most popular batteries in consumer products today are *alkaline batteries*, which are based on the alkaline substance potassium hydroxide. These batteries are a type of dry cell battery but have some chemical and structural differences. During the operation of an alkaline battery, a reaction takes place in a paste of zinc metal and potassium hydroxide. The battery does not have a carbon rod cathode, nor does it have a solid metal liner serving as an anode.

The following reactions occur at each electrode.

$$\text{anode: } Zn + 2OH^- \longrightarrow ZnO + H_2O + 2e^-$$

$$\text{cathode: } 2MnO_2 + H_2O + 2e^- \longrightarrow Mn_2O_3 + 2OH^-$$

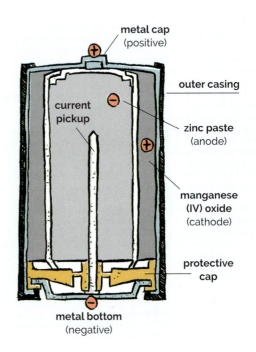

Table 19-2 summarizes information on some of the most common types of batteries. These all produce electricity through redox reactions.

TABLE 19-2 *Types of Batteries in Common Use*

Nonrechargeable (Primary) Batteries				
Type	**Anode**	**Cathode**	**Electrolyte**	**Uses**
zinc-carbon dry cell	Zn (oxidizes Zn)	carbon (reduces MnO_2)	NH_4Cl, $ZnCl_2$	low-cost batteries used in remote controls, flashlights, and calculators
alkaline cells	Zn	MnO_2	KOH	
Rechargeable (Secondary) Batteries				
Type	**Electrodes**		**Electrolyte**	**Uses**
lead-acid	Pb	PbO_2	H_2SO_4	automotive and industrial uses
lithium ion	C	metal oxide ($LiCoO_2$)	lithium salts in organic solvent	electronics such as smartphones, laptops, and vehicles
nickel-cadmium	Cd	NiOOH	KOH	rechargeable batteries
nickel-metal hydride	nickel-metal hydride	NiOOH	KOH	cellphones, cameras, laptops
silver-cadmium	CdO	Ag_2O	KOH	satellites
silver-zinc	Zn	Ag_2O	KOH	military applications

FUEL CELLS

Fuel cells are a special class of very efficient batteries. A fuel cell resembles a voltaic cell in that redox reactions occur, releasing electrons from one electrode to flow through a circuit to another electrode. But there is one major difference. In the voltaic cell, the active ingredients are included within the cell and are depleted as the redox reactions occur. In a fuel cell, a gas or liquid fuel is supplied to one electrode and oxygen or air to the other from an external source. Fuel cells have a long life and have been used extensively in space vehicles since the 1960s. Current fuel-cell research is targeted at powering vehicles with hydrogen as a fuel.

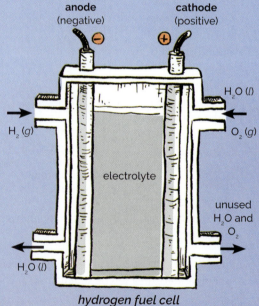

hydrogen fuel cell

MINI LAB

OBSERVING A VOLTAIC CELL

We use voltaic cells in the form of batteries all the time. Voltaic cells work as two half-reactions (oxidation and reduction) as part of an overall redox reaction. This lab activity will allow you to observe each half-cell of the reaction. Zinc metal will be oxidized in one half-cell, while copper(II) ions are reduced in the other half-cell.

How can I make a simple battery?

1. What is oxidation? What is reduction?

Procedure

A Fill one beaker with approximately 75 mL of copper(II) sulfate solution and the other with approximately 75 mL of zinc sulfate solution.

B Connect one of the wires between one end of the copper strip and the positive terminal of the voltmeter. Connect the other wire between one end of the zinc strip and the negative terminal of the voltmeter.

C Place the end of the copper strip in the copper(II) sulfate solution and the end of the zinc strip in the zinc sulfate.

D Turn on the voltmeter.

2. What is the voltmeter reading? Explain why you think it is what it is.

E Put the two ends of the salt bridge strip in each one of the solutions.

3. What is the voltmeter reading? Explain why you think this is what it is.

4. Write the half-reaction occurring in each half-cell. Write the redox reaction equation from the two half-reactions.

Going Further

5. How could we produce a different voltage?

EQUIPMENT

- voltmeter
- beakers, 100 mL (2)
- 1 M copper(II) sulfate solution, 75 mL
- 1 M zinc sulfate solution, 75 mL
- connecting wires (2)
- copper strip, 6 cm
- zinc strip, 6 cm
- salt bridge strip

Redox Reactions

CASE STUDY

SMALLER, SAFER BATTERIES

In 2016 there were numerous news reports of electronic devices igniting. We learned from these incidents that lithium ion batteries had a significant drawback—the electrolyte was a flammable liquid. To address this issue, scientists at both Penn State and Cornell Universities worked on developing batteries with a solid electrolyte.

Both teams worked through the challenges of establishing a reliable interface between the electrodes and the electrolyte. The fluid property of liquids naturally allows them to change shape and remain in contact with electrodes. Researchers worked to establish the same level of contact between the electrodes and a solid electrolyte. Through nanotechnology and polymer chemistry, they achieved their goal.

In addition to solving the safety issue, both teams identified other advantages of a solid electrolyte.

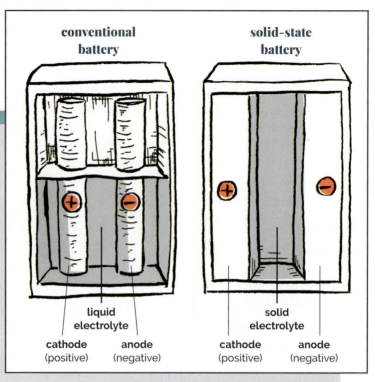

They found that batteries with a solid-state electrolyte also have a high energy density (they can store more energy), they are longer-lasting, and they can be made smaller than their liquid-electrolyte counterparts. Scientists love it when they can solve multiple problems at the same time!

19.2 SECTION REVIEW

1. What is electrochemistry?
2. Name the two types of electrochemical cells and explain the difference between them.
3. Why is a special process necessary to purify active metals such as aluminum and sodium when other metals can be more easily purified from their ores?
4. In a voltaic cell, what is the electric charge on the anode? What type of ion migrates toward the anode? What process (oxidation or reduction) occurs at the surface of the anode?
5. Why is a salt bridge used in voltaic cells?
6. Would the electrodes in the half-reactions represented below make a functional battery? Explain your answer.

   $$Zn \longrightarrow Zn^{2+} + 2e^-$$
   $$Cd \longrightarrow Cd^{2+} + 2e^-$$

7. Write the half reactions and redox reaction for lithium being oxidized and gold(III) ions being reduced to gold atoms.

Use the case study above to answer Questions 8–10.

8. Summarize how a lithium ion battery works.
9. What was the primary issue that the research teams were trying to solve?
10. What is the challenge of using a solid-state electrolyte?

19 CHAPTER REVIEW

Chapter Summary

TERMS
oxidation-reduction reaction
oxidation
reduction
reducing agent
oxidizing agent

19.1 REDOX REACTIONS

- Oxidation-reduction, or redox, reactions are chemical changes in which electrons are transferred from one atom or molecule to another.

- Oxidation involves the loss of one or more electrons (LEO). Reduction involves the gain of one or more electrons (GER). These processes occur simultaneously.

- A reducing agent causes another substance to be reduced and is itself oxidized in the process. An oxidizing agent causes another substance to be oxidized and is itself reduced in the process.

- Equations of redox reactions must be balanced to maintain the laws of the conservation of mass and charge.

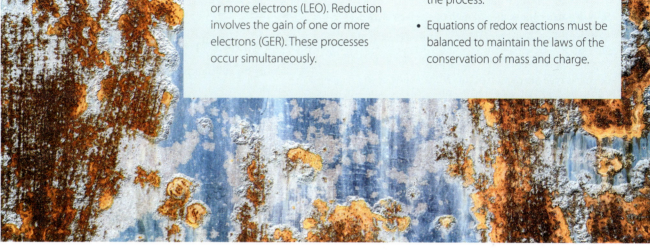

19.2 ELECTROCHEMICAL REACTIONS

- Electrochemistry is the study of chemical reactions that can produce electricity and reactions that require electricity to occur.

- An electrolyte is any substance that releases ions when dissolved in water, allowing the resulting solution to conduct electricity.

- An electrochemical cell has two electrodes (electrical contacts) immersed in an electrolyte, with a wire joining the electrodes. The two main groups of electrochemical cells are electrolytic cells and voltaic cells.

- Electrolysis, a process occurring in electrolytic cells, requires electrical current to cause an otherwise nonspontaneous redox reaction to occur.

- Electrolytic cells use electricity to cause a nonspontaneous chemical reaction to occur. Reduction occurs at the negatively charged cathode and oxidation occurs at the positively charged anode.

- In voltaic cells, spontaneous redox reactions produce electricity by causing electrons to flow. Reduction occurs at the positively charged cathode and oxidation occurs at the negatively charged anode. Batteries are collections of voltaic cells.

TERMS
electrochemistry
electrochemical cell
electrolytic cell
voltaic cell
electrolysis
cathode
anode
electroplating
battery
half-cell
half-reaction
salt bridge
fuel cell

Redox Reactions

19 CHAPTER REVIEW

Chapter Review Questions

RECALLING FACTS

1. Which types of reactions are or can be redox reactions?
2. Does oxidation require that oxygen atoms be present? Explain your answer.
3. Define *reduction*.
4. (True or False) In some redox reactions, the same element is both reduced and oxidized.
5. What determines the strength of an electrolyte?
6. Why is pure water a poor conductor of electricity?
7. Define *electrochemical cell*.
8. What type of reaction occurs at the anode of an electrolytic cell? of a voltaic cell?
9. (True or False) In every type of electrochemical cell, oxidation always takes place at the anode and reduction always takes place at the cathode.
10. What are some useful applications of electrolytic cells?
11. What is the main difference between a voltaic cell and a fuel cell?

UNDERSTANDING CONCEPTS

12. Indicate whether each reaction is a redox reaction.

    a. $2Fe(s) + 3Cl_2(g) \longrightarrow 2FeCl_3(s)$

    b. $CaO(s) + 2HCl(g) \longrightarrow CaCl_2(s) + H_2O(l)$

    c. $2C_2H_6(g) + 7O_2(g) \longrightarrow 4CO_2(g) + 6H_2O(l)$

    d. $Zn(s) + CuSO_4(aq) \longrightarrow Cu(s) + ZnSO_4(aq)$

    e. $H_2SO_4(aq) + 2KOH(aq) \longrightarrow K_2SO_4(s) + 2H_2O(l)$

    f. $Pb(s) + H_2SO_4(aq) \longrightarrow PbSO_4(s) + H_2(g)$

    g. $AgNO_3(aq) + HCl(aq) \longrightarrow AgCl(s) + HNO_3(aq)$

13. Create a graphic organizer using the terms *electrons, oxidation, oxidizing agent, redox reaction, reducing agent,* and *reduction*.

For Questions 14–17, refer to the following redox reactions.

   a. $2KClO_3(s) \longrightarrow 2KCl(s) + 3O_2(g)$

   b. $Zn(s) + 2HCl(aq) \longrightarrow ZnCl_2(aq) + H_2(g)$

   c. $Fe_2O_3(s) + 3CO(g) \longrightarrow 2Fe(s) + 3CO_2(g)$

   d. $Mg(s) + Cu_2SO_4(aq) \longrightarrow 2Cu(s) + MgSO_4(aq)$

   e. $H_2(g) + Cl_2(g) \longrightarrow 2HCl(g)$

   f. $4NH_3(g) + 3O_2(g) \longrightarrow 2N_2(g) + 6H_2O(g)$

   g. $Cu(s) + 4HNO_3(aq) \longrightarrow Cu(NO_3)_2(aq) + 2NO_2(g) + 2H_2O(l)$

   h. $MnO_4^- + 5Fe^{2+} + 8H^+ \longrightarrow Mn^{2+} + 5Fe^{3+} + 4H_2O$

   i. $2SO_4^{2-} + C + 4H^+ \longrightarrow CO_2 + 2SO_2 + 2H_2O$

14. Identify the substance being oxidized in each reaction.
15. Identify the substance being reduced in each reaction.
16. Identify the oxidizing agent in each reaction.
17. Identify the reducing agent in each reaction.
18. Identify what is being oxidized and reduced in the reaction below.

    $2H_2O_2(l) \longrightarrow 2H_2O(l) + O_2(g)$

19. Balance each of the following equations.

    a. $Cu^+(aq) + Fe(s) \longrightarrow Fe^{3+}(aq) + Cu(s)$

    b. $Fe(OH)_2 + H_2O_2 \longrightarrow Fe(OH)_3$

    c. $Cu + HNO_3 \longrightarrow Cu(NO_3)_2 + NO_2 + H_2O$

20. Explain why zinc plating will protect steel from corroding even if the plating is cracked.
21. Describe how the Hall-Héroult process uses electrons to free aluminum atoms from bauxite.
22. Compare electrolytic cells and voltaic cells.
23. What do we mean when we speak of half-reactions and half-cells?
24. Write the half-reactions and redox reaction for aluminum being oxidized and copper(II) being reduced to copper atoms.
25. In a voltaic cell, what is the electrical charge on the cathode? What type of ion migrates toward the cathode? What process (oxidation or reduction) occurs at the surface of the cathode?

26. Why do battery manufacturers recommend not charging some batteries even though they are based on the same reversible redox reactions as those in rechargeable batteries?

27. In the lead-acid storage battery of a vehicle, the degree to which the power has been discharged can be determined by using a hydrometer to measure the density of the sulfuric acid solution. Explain why this is possible.

28. Compare the types of batteries listed in the chapter (zinc-carbon, alkaline, lead acid, and lithium ion).

29. Describe how rechargeable batteries are both voltaic and electrolytic cells.

CRITICAL THINKING

30. Oxidation and reduction involve the losing or gaining of electrons. While this is easily understood in reactions with ionic compounds, how does oxidation and reduction work with covalently bonded substances?

31. Explain why sodium is a strong reducer and fluorine is a strong oxidizer.

32. Consider the list of nitrogen compounds below. Determine the oxidation number for nitrogen in each one. How does this explain the variety of nitrogen-containing compounds?

    NO, NO_2, N_2O, N_2O_5, HNO_2, HNO_3, HNO_4

Use the ethics box below to answer Question 33.

33. How should a Christian respond to the claim that a good steward of our planet's resources and environment should drive only electric vehicles?

ethics

ELECTRIC CARS

Watch almost any news program and you are likely to hear a segment on global warming, air pollution, and fossil fuels. Many people are concerned about our dependence on fossil fuels, the condition in which we are leaving our environment, and even whether we are going to destroy the planet through global warming. Some claim that electric cars are the answer to these issues. Electric cars do not burn fossil fuels and do not produce gases that pollute or contribute to global warming. Others say that electric cars solve none of these issues because the electricity required to operate them still needs to be produced, and it may be produced in a fossil fuel plant that emits polluting greenhouse gases.

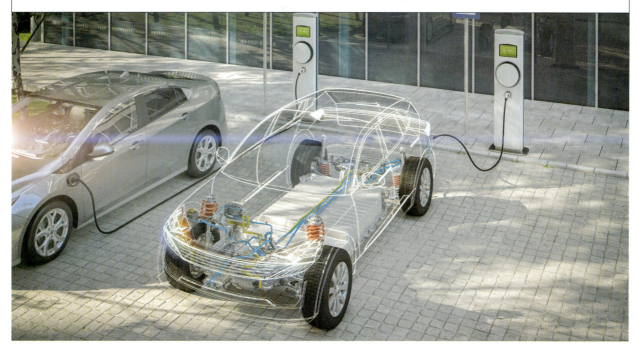

Chapter 20

ORGANIC CHEMISTRY

Jasmine. Lavender. Sandalwood. Rose. Tea Tree. Peppermint. If these sound familiar to you, then perhaps you have had some experience with essential oils. Some people argue that smelling fragrant essential oils offers therapeutic benefits, while others simply enjoy the ambiance created by a few drops of essential oil in a diffuser. But what is not up for debate is what essential oils are—they are mixtures of volatile organic compounds. That's a fancy way of saying that they contain carbon-based molecules that vaporize easily.

Organic compounds are responsible for many of the tastes and smells that humans enjoy. Do you like the taste of fresh butter? You can thank butane-2,3-dione for that. Or how about the smell of a freshly peeled tangerine? That's due to 1-methyl-4-(prop-1-en-2-yl)cyclohex-1-ene. Thankfully, we don't have to know the tongue-twisting names of these compounds in order to enjoy them. But as you'll see in this chapter, the names of organic compounds aren't just gibberish—each is packed with information about the chemical compound that it identifies.

20.1 Organic Compounds *485*
20.2 Hydrocarbons *488*
20.3 Substituted Hydrocarbons *497*
20.4 Organic Reactions *506*

20.1 ORGANIC COMPOUNDS

Carbon is a common element that is found in nature both as a pure element and in compounds. It can be found in all living matter. In fact, without carbon, there would be no life. Carbon is also often in the news these days since it is a major component of both fossil fuels, such as petroleum and natural gas, and the carbon dioxide—a greenhouse gas—that is produced by the combustion of those fuels.

Organic compounds were so named because for many years it was believed that only living things could produce them. But in 1828, Friedrich Wöhler, a German chemist, unintentionally synthesized urea (NH_2CONH_2) in the laboratory. This compound was previously known only from the urine of humans and animals—a compound clearly linked to living things. Beginning with this breakthrough, a new field of study emerged—organic chemistry. Scientists can now synthesize many **organic compounds**, which they define as covalently bonded carbon compounds, with the exception of carbonates, carbon oxides, and carbides. The branch of chemistry that focuses specifically on these carbon compounds is called **organic chemistry**.

QUESTIONS
- When did organic chemistry become a recognized field?
- What makes carbon so special compared with other elements?
- What information can be seen in a structural formula?
- What is the difference between aromatic and aliphatic compounds?

TERMS
organic compound • organic chemistry • condensed structural formula • aromatic compound • aliphatic compound

How does organic chemistry fit within the broader study of chemistry in general?

THE UNIQUE Carbon Atom

Carbon is an element unparalleled in its ability to form a variety of compounds. Other elements can form several hundred thousand compounds, but carbon can form more than fifty million compounds! What makes the carbon atom so versatile? Carbon atoms have unique bonding abilities.

Three important properties of carbon enable it to form large, stable molecules.

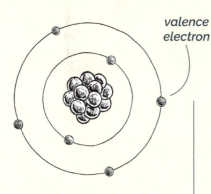
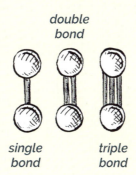
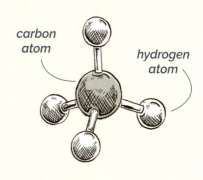

❶ Carbon has four valence electrons, allowing it to form four bonds to obtain a full valence shell.

❷ Carbon forms strong chemical bonds with other carbon atoms.

❸ Carbon forms stable, essentially nonpolar bonds with hydrogen. The electronegativity difference between carbon (2.6) and hydrogen (2.2) is small.

Carbon atoms can bond to a wide variety of other atoms, including hydrogen, phosphorus, oxygen, nitrogen, sulfur, the halogens, and even metal atoms. Carbon atoms can bond with one another to form linear or branched chains of various lengths, and they can even form rings. Carbon atoms can form double and triple bonds, further adding to the almost endless possibilities.

Structural Formulas AND MODELS

Organic chemists study more than just the types of atoms in molecules. They also study how these atoms are arranged. If an analytical chemist were to announce that he had isolated a compound with the molecular formula C_2H_6O, there might be some confusion about what exactly the compound was. Let's look at why this is so.

Ethanol is an alcohol with the formula C_2H_6O. Its two carbon atoms are bonded together in a short chain with the oxygen atom on one end. Shown below is ethanol's structural formula and a ball-and-stick model. Sometimes you might also see a **condensed structural formula**, which is a way of modeling a compound's molecular structure in a single line of text. Ethanol's condensed structural formula is CH_3CH_2OH.

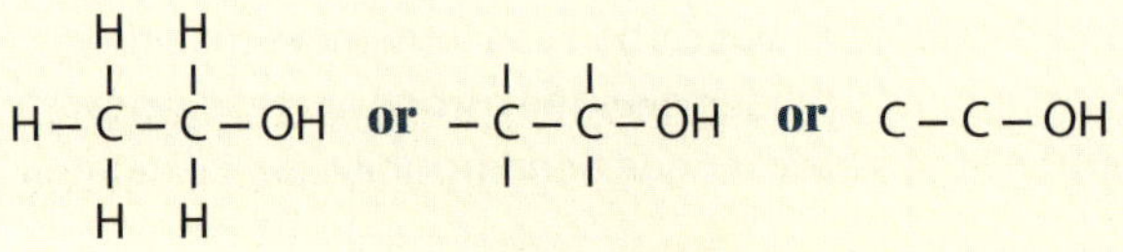

Methoxymethane has the same molecular formula, C_2H_6O, but as shown below, its structural formula and model look quite different. The arrangement of the atoms in the molecule makes the difference. In methoxymethane's case, the oxygen atom is located between the two carbon atoms. It might seem like a minor difference, but this change in structure causes methoxymethane and ethanol to have very different physical and chemical properties. Methoxymethane's condensed structural formula is CH_3OCH_3.

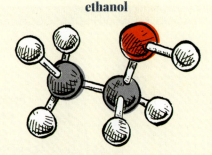

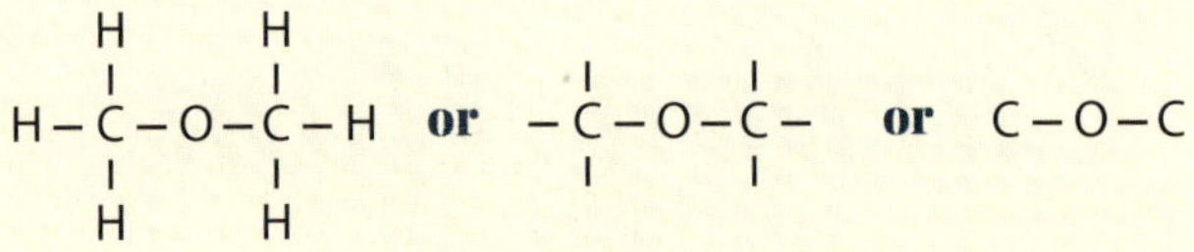

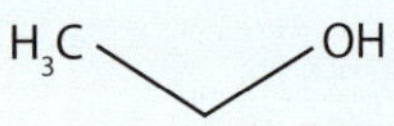

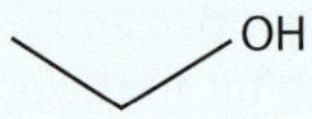

ethanol

methoxymethane

By now you may have noticed that some organic compounds contain a lot of hydrogen atoms. It's commonplace to see these hydrogen atoms omitted in structural formulas. In such instances, it's understood by consensus that the "missing" atoms are hydrogens. In some cases where hydrogen forms part of a functional group, such as a hydroxyl group (OH), it remains shown.

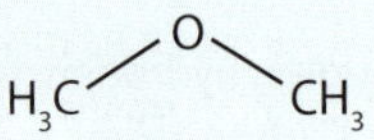

To further reduce the visual clutter in a structural formula, chemists will often use a form known as a *skeletal formula* that omits the carbon atoms as well. In skeletal formulas, it is understood that each unlabeled bend in the structure, along with any "empty" end of a bond, is occupied by a carbon atom and its bonded hydrogens, if any.

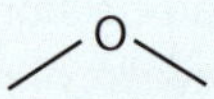

Classification

Every year, many new organic compounds are synthesized for the first time. If a person wants to find his way through this ever-growing field, he must use some guidelines. What's needed is a classification scheme that organizes compounds into easily identifiable groups.

Organic compounds can be divided into two large groups: aromatic compounds and aliphatic compounds. **Aromatic compounds** are organic compounds that contain highly stable ringed structures. The highly stable nature of these rings is due to their electrons not being associated with a specific bond or atom—they are delocalized. This is similar to the way that metal atoms bound by metallic bonds share electrons, as you learned in Chapter 7. Not surprisingly, aromatic compounds are so named because many of the first ones to be identified had distinctive smells. Today, though, the word is applied to compounds with a particular kind of chemical structure without regard to odor. You'll learn more about aromatic compounds and their special structure in the next section. **Aliphatic compounds** consist of open-chain compounds, either branched or unbranched, such as ethanol, or rings that can be formed by bending a straight chain and connecting its ends together. Aliphatic rings do not have delocalized electrons and are less stable than aromatic rings. Cyclohexane is an example of a ring-shaped aliphatic compound.

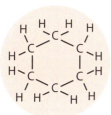
cyclohexane

CAREERS

SERVING AS AN ODOR TESTER:
IMPROVING LIFE ONE SMELL AT A TIME

Some people are more sensitive to smells than others. Some of those sensitive people make their livings with their discerning sniffers. Think about the many everyday products that have distinctive scents—everything from foodstuffs to personal hygiene products. The makers of those products know that pleasant smells make potential customers more likely to buy their wares. The opposite is also true—foul smells are far less likely to tempt buyers! That's where odor testers come in. An odor tester is a person with a keen sense of smell who is able to detect minute differences in smells that most people wouldn't notice. Chances are, your favorite coffee blend, body spray, or air freshener has been approved by a panel of odor testers. Because people are made in God's image, they can use their creative talents to devise a vast array of products that improve people's quality of life. Odor testers help ensure that pleasant smells contribute to that improvement.

20.1 SECTION REVIEW

1. What is an organic compound?
2. What characteristics of carbon enable it to be found in millions of compounds?
3. Why are structural formulas important to organic chemists?
4. List and define the two large groups into which organic compounds are divided.
5. Cyclohexane has a distinctive smell (like that of detergent), just as many aromatic compounds do. Why is cyclohexane not classified as an aromatic hydrocarbon?

20.2 HYDROCARBONS

If you have ever taken the time to read the list of ingredients in your shampoo, you were probably mystified by the list of unpronounceable names. Take *methylisothiazolinone* (MIT), for example. It's used to prevent the growth of germs in cosmetics and personal care products. After all, no one wants to wash with germs, right? The IUPAC name of MIT is 2-methyl-1,2-thiazol-3(2H)-one, and as we have hinted, this name tells a trained chemist about the structure of the compound. In this textbook, we won't handle anything as complex as MIT, but we will examine how to name simpler organic compounds, starting with the simplest of all—hydrocarbons.

As their name implies, **hydrocarbons** are organic compounds that contain only hydrogen and carbon. Hydrocarbons can contain single, double, or triple carbon-carbon bonds. The classification and naming of hydrocarbons is based on the number of carbons in the compound, the number and type(s) of bonds present, and the arrangement of the carbons within a molecule of the compound.

QUESTIONS
» What are the different kinds of hydrocarbons?
» What do the names of organic compounds tell us about their structures?
» How does isomerism affect the physical and chemical properties of compounds?

TERMS
hydrocarbon • alkane • saturated hydrocarbon • alkyl group • isomer • alkene • unsaturated hydrocarbon • alkyne • cyclic aliphatic compound • benzene

What chemicals are in my shampoo?

Aliphatic Hydrocarbons

ALKANES

The simplest aliphatic hydrocarbons are **alkanes**, which contain only single bonds between carbon atoms. Just keep in mind that the word "simple" in this case is relative. Some alkanes consist of many carbon and hydrogen atoms and can have relatively complex structures, but they're still made from only two elements. Their general formula is C_nH_{2n+2}. Because they are energy-dense, our society relies on alkanes for fuel. Natural gas—used to heat homes—contains mainly methane (CH_4). Portable gas barbecue grills use bottles of pressurized propane, and gasoline-powered vehicles burn a mixture composed mostly of alkanes. All these compounds contain only carbon and hydrogen. They have structural formulas that resemble open chains, and they contain only single bonds between their carbons.

Methane (CH_4) is the simplest alkane. Its one carbon atom is bonded to four hydrogen atoms. Other alkanes are formed as additional carbons lengthen the chain. Ethane (C_2H_6) has two carbons, propane (C_3H_8) has three, and butane (C_4H_{10}) has four. Each carbon atom has only single bonds and is therefore surrounded by four other atoms—the maximum number possible. For this reason, an alkane molecule is said to be a **saturated hydrocarbon**.

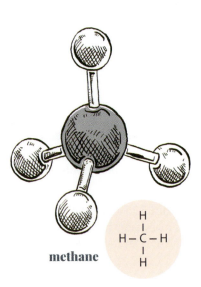

methane

IUPAC has devised and manages a system to accurately name organic compounds. This system relies on a series of prefixes (shown in Table 20-1) to indicate the number of carbon atoms present in the compound's longest chain. A substance's suffix identifies the type of compound it is. All alkanes have an *-ane* ending. The names and some properties of the first five alkanes are listed in Table 20-2 (p. 492).

The simplest hydrocarbons have a single continuous chain of carbons atoms from end to end and are called *straight* or *linear chains*. *Branched hydrocarbons*, as their name implies, have carbon chains that branch off a main chain. Branched alkanes are named on the basis of the longest carbon chain in the molecule. For example, the compound commonly known as isobutane has the structure shown below. Though there are four carbons in this molecule, the longest possible *chain* of carbons is only three, so its IUPAC name is based on a three-carbon alkane—propane. The additional hydrocarbon side chain is an **alkyl group,** an alkane with one hydrogen missing, allowing it to bond to a larger molecule. An alkyl group takes its name from its parent alkane molecule with the *-ane* ending changed to *–yl*. Because this alkyl group has only one carbon atom, its parent molecule is methane, and it is referred to as a methyl group (CH_3). Note that the methyl group is attached to the second carbon of the carbon chain, which makes the IUPAC name of this molecule *2-methylpropane*.

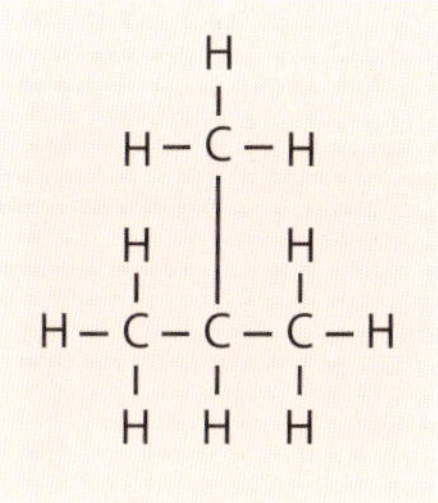

isobutane
(2-methylpropane)

We can also do this process in reverse to derive a compound's structural formula from its name. A close examination of the name 2-methylpropane tells us how a molecule of the compound is configured. Propane is the base structure of the hydrocarbon. Since the name *propane* contains the prefix *prop-*, the main hydrocarbon structure is identified as a three-carbon chain. The *-ane* ending tells us that all the bonds between carbons are single bonds. The *2-methyl-* indicates that there is a branch, or side chain, with the *2-* indicating that the branch is attached to the second carbon of the propane base. *Methyl* is made up of two parts—the *-yl* ending tells us that the branch is an alkyl group and *meth-* indicates that the branch has one carbon.

TABLE 20-1
Numerical Prefixes

Number of Carbon Atoms	Prefix
1	meth-
2	eth-
3	prop-
4	but-
5	pent-
6	hex-
7	hept-
8	oct-
9	non-
10	dec-

Organic Chemistry

From these examples, it should be apparent that once you understand how the IUPAC names of organic compounds are put together, you can quickly deduce the structural formula of an organic compound from its name and vice versa.

EXAMPLE 20-1: WRITING A STRUCTURAL FORMULA FROM A COMPOUND'S NAME

Write the structural formulas for each of the following.

a. 2-methylhexane

b. 2,3-dimethylpentane

c. 3-ethyl-2-methylheptane

Solution:

a. The root *hexane* tells us that the molecule is based on a six-carbon chain containing only single bonds (identified by the *-ane* ending). The next part, *2-methyl-*, tells us that a one-carbon alkyl group is bonded to the second carbon of the hexane molecule; the carbons are numbered from the end nearer the branch. The structural formula is shown below.

```
        H
        |
    H - C - H
H       |       H  H  H  H
|       |       |  |  |  |
H - C - C - C - C - C - C - H
|       |       |  |  |  |
H       H       H  H  H  H
```

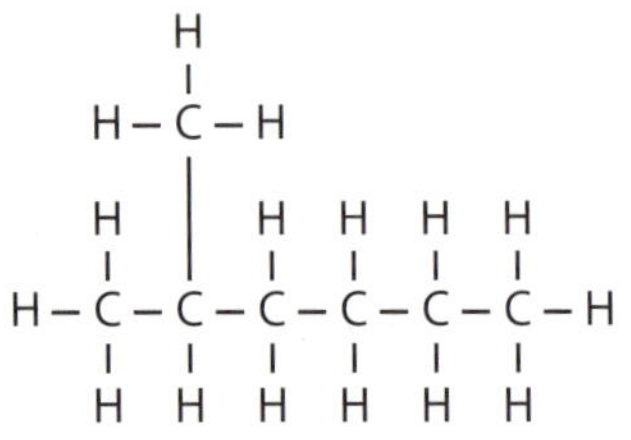

b. When we look at the end of the compound's name, we see that *pentane*, a five-carbon chain with only single bonds, is the parent molecule. *Dimethyl-* indicates that two methyl groups are connected to the pentane molecule. The *2,3-* tells us that one methyl group is bonded to the second carbon, while the other methyl group is bonded to the third carbon in the longest chain.

The structural formula is shown below.

```
            H
            |
        H - C - H
H           |       H  H  H
|           |       |  |  |
H - C - C - C - C - C - H
|   |       |       |  |
H   H       |       H  H
        H - C - H
            |
            H
```

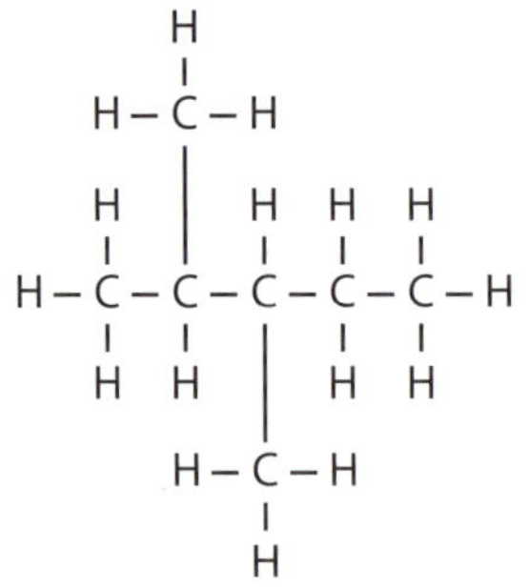

c. Again, look to the end of the name to identify the parent molecule, in this case *heptane*, a seven-carbon chain. Our molecule has two alkyl groups bonded to it, a one-carbon group (*methyl*) and a two-carbon group (*ethyl*). The alkyl groups are placed in alphabetical order, so ethyl is named first, and *3-* tells us that it is bonded to the third carbon of the chain. The methyl group is bonded to the second carbon. The finished structural formula is shown below.

```
                        H
                        |
                    H - C - H
H   H   H   H   H       |   H
 7   6   5   4   3       2   1
H - C - C - C - C - C - C - C - H
|   |   |   |   |       |   |
H   H   H   H   |       H   H
                H - C - H
                    |
                H - C - H
                    |
                    H
```

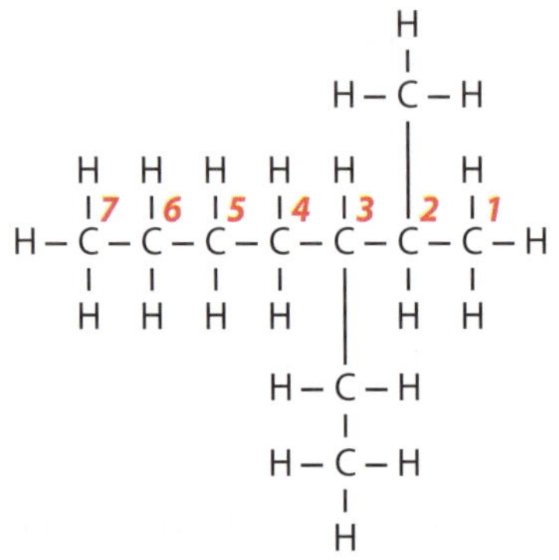

Take note that even though the alkyl groups are placed in alphabetical order in the name, the carbons in the chain are still numbered starting at the end closer to the nearest branching.

490 Chapter 20

EXAMPLE 20-2: NAMING A HYDROCARBON FROM ITS STRUCTURAL FORMULA

Name each of the following branched hydrocarbons.

a.

```
    H   H   H   H   H
    |   |   |   |   |
H − C − C − C − C − C − H
    |   |   |   |   |
    H   H   H   |   H
                |
            H − C − H
                |
                H
```

b.

```
            H               H
            |               |
        H − C − H       H − C − H
    H   H   |   H   H   H   |
    |   |   |   |   |   |   |
H − C − C − C − C − C − C − C − H
    |   |   |   |   |   |   |
    H   H   H   |   H   H   H
                |
            H − C − H
                |
            H − C − H
                |
            H − C − H
                |
                H
```

c.

```
                            H
                            |
                        H − C − H
    H   H   H   H   H       |           H
    |   |   |   |   |       |           |
H − C − C − C − C − C ───── C ───────── C − H
    |   |   |   |   |       |           |
    H   H   H   H   |   H − C − H       H
                    |       |
                H − C − H   H
                    |
                H − C − H
                    |
                    H
```

Solution:

a. Locate the longest chain and number the carbons in the chain, starting from the end closer to a branch.

```
    H   H   H   H   H
    |5  |4  |3  |2  |1
H − C − C − C − C − C − H
    |   |   |   |   |
    H   H   H   |   H
                |
            H − C − H
                |
                H
```

Name the parent molecule. A five-carbon chain with only single bonds is pentane. Add prefixes to indicate any branches, including their locations. In this case, a methyl group is bonded to the second carbon, so it is identified using the prefix 2-methyl. The name is therefore *2-methylpentane*.

b. Locate the longest chain and number the carbons in the chain, starting from the end closer to a branch.

```
            H                       H
            |                       |8
        H − C − H               H − C − H
    H   H   |   H   H   H       |
    |1  |2  |3  |4  |5  |6  |7
H − C − C − C − C − C − C − C − H
    |   |   |   |   |   |   |
    H   H   H   |   H   H   H
                |
            H − C − H
                |
            H − C − H
                |
            H − C − H
                |
                H
```

Name the parent molecule. An eight-carbon chain with only single bonds is octane. Add prefixes for any branches, including their locations. The branches on this molecule consist of a one-carbon group (methyl) and a three-carbon-group (propyl). Remember, the alkyl groups are listed in alphabetical order, so this is *3-methyl-4-propyloctane*.

c. Locate the longest chain and number the carbons in the chain, starting from the end closer to a branch.

```
                            H
                            |
                        H − C − H
    H   H   H   H   H       |           H
    |7  |6  |5  |4  |3      |2          |1
H − C − C − C − C − C ───── C ───────── C − H
    |   |   |   |   |       |           |
    H   H   H   H   |   H − C − H       H
                    |       |
                H − C − H   H
                    |
                H − C − H
                    |
                    H
```

Name the parent molecule. A seven-carbon chain with only single bonds is heptane. Add prefixes for any branches, including their locations. This molecule includes an ethyl group and two methyl groups, both of which are bonded to the number two carbon. Each alkyl group requires its own position identifier, so the two methyl groups will be identified together as 2,2-dimethyl. Listing the alkyl groups in alphabetical order makes this *3-ethyl-2,2-dimethylheptane*. Note that the numerical prefix *di-* is not considered when alphabetizing the alkyl groups.

Organic Chemistry

Are alkanes polar or nonpolar? Bonds between two carbon atoms are not polar because both atoms have the same electronegativity. Bonds between carbon and hydrogen are slightly polar, but these bonds are arranged so that the resulting positive charge of the hydrogen atoms is evenly distributed over the exterior of the molecule. As a result, the molecule has no localized accumulation of charge and is therefore nonpolar. Alkanes dissolve well in nonpolar solvents such as carbon tetrachloride. But in polar substances, such as water, liquid alkanes are immiscible and form a layer on top of water.

Crude oil contains several types of alkanes. Chains from four carbons up to twenty carbons form naturally and are combined into one seemingly inseparable mixture. Chemical engineers take advantage of the properties of these alkanes to separate them so that they can be used as lubricating oils, gasoline, and kerosene. How do chemical engineers separate the individual compounds? The boiling points of alkanes rise as additional carbons are added to the chain, so petroleum engineers can separate the alkanes in a process called *fractional distillation*.

TABLE 20-2 *Linear Alkanes*

Name	Structural Formula	Melting Point (°C)	Boiling Point (°C)
methane	CH_4	−183	−162
ethane	CH_3-CH_3	−183	−89
propane	$CH_3-CH_2-CH_3$	−188	−42
butane	$CH_3-CH_2-CH_2-CH_3$	−138	−1
pentane	$CH_3-CH_2-CH_2-CH_2-CH_3$	−130	36

ISOMERS

You may recall from a course in physical science that there is an issue that arises with longer hydrocarbon chains. Hydrocarbons with four or more carbons can be configured in multiple ways yet still have the same chemical formula. The number of possible configurations increases as the number of carbon atoms increases. These compounds with the same molecular formula but different structural formulas are called **isomers**. Normal butane (*n*-butane, the linear form) and its isomer, 2-methylpropane, have slightly different physical properties. Their melting and boiling points, density, and solubility reflect their different structures. Such differences in physical properties are typical of isomers.

TABLE 20-3 *Comparison of Butane Isomers*

Isomer	Melting Point (°C)	Density (g/mL)	Solubility (mg/L water at 25 °C)
n-butane	−138	0.002 48	61.2
2-methylpropane (isobutane)	−159	0.002 51	48.9

EXAMPLE 20-3: DRAWING ISOMERS

Draw structural formulas for all the isomers of pentane (C_5H_{12}).

```
    H   H   H   H   H
    |   |   |   |   |
H — C — C — C — C — C — H
    |   |   |   |   |
    H   H   H   H   H
```

Solution

For each isomer, we need to move one or more of pentane's end carbons to an internal carbon on the chain. By moving one end carbon to an internal carbon, we arrive at 2-methylbutane (a.). Note that there is no "3-methylbutane" possible; moving an end carbon to butane's "third" carbon makes it necessary to renumber the carbons from the other end of the chain, the end nearer the branch, resulting in that carbon being identified as number two again.

The only other possible isomer of pentane results from moving both end carbons to the internal carbon of a propane chain, producing 2,2-dimethylpropane (b.).

a.
```
        H   H   H   H
        |   |   |   |
    H — C — C — C — C — H
        |   |   |
        H   |   H
            |
        H — C — H
            |
            H
```

b.
```
            H
            |
        H — C — H
        H   |   H
        |   |   |
    H — C — C — C — H
        |   |   |
        H   |   H
            |
        H — C — H
            |
            H
```

ALKENES

Carbons can form double bonds quite easily. Hydrocarbons that contain double bonds between carbon atoms are called **alkenes**. Alkenes use the same prefixes as alkanes but end with the suffix *-ene*. The simplest alkene is ethene. Because double bonds reduce the number of hydrogen atoms in their structures, all alkenes are said to be **unsaturated hydrocarbons**. Ethene contains only two carbons, but when the carbon chain is longer than three carbons, the double bond could be in several locations. As in branched alkanes, an alkene's name pinpoints the location of the double bond by specifying which carbon in the parent chain is the first carbon that is doubly bonded. The number is not placed at the front of the name, but instead is located between the numbering prefix and the *-ene* suffix as shown in Table 20-4.

The physical properties of alkenes are very much like those of alkanes. The first few are gases at room temperature. Pentene and larger compounds are liquids at room temperature because of their greater intermolecular attractions. Alkenes are relatively nonpolar. Alkenes are slightly more reactive than alkanes.

TABLE 20-4 Alkenes

Name	Structural Formula	Melting Point (°C)	Boiling Point (°C)
ethene	H−C=C−H (with H's on each C)	−169	−104
propene	H−C=C−C−H (with H's)	−185	−48
but-1-ene	H−C=C−C−C−H (with H's)	−185	−6.5
but-2-ene	H−C−C=C−C−H (with H's)	−139	3.7

Ethene speeds up ripening. These bananas were picked and shipped green. Before sale, they were exposed to ethene for quick ripening.

Organic Chemistry

TABLE 20-5 Alkynes

Name	Structural Formula	Melting Point (°C)	Boiling Point (°C)
ethyne	H–C≡C–H	−84	−81
propyne	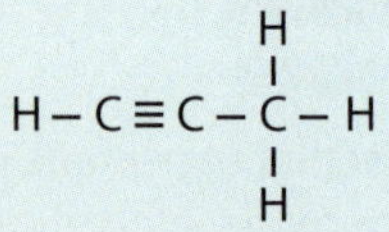 H–C≡C–CH₃	−102.7	−23
but-1-yne	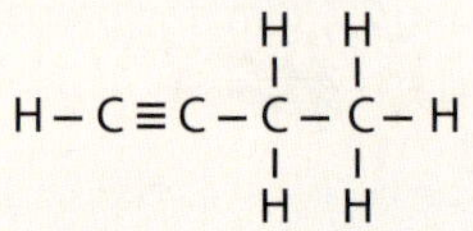 H–C≡C–CH₂–CH₃	−126	8
but-2-yne	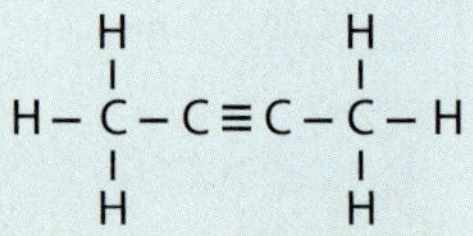 CH₃–C≡C–CH₃	−32	27
pent-1-yne	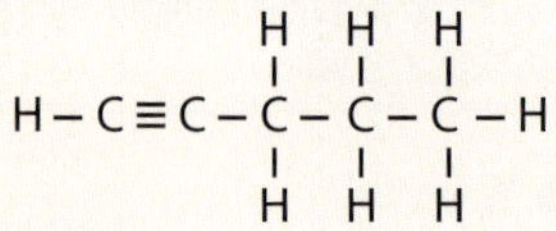 H–C≡C–CH₂–CH₂–CH₃	−106	40
pent-2-yne	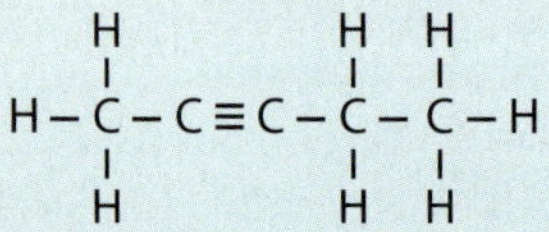 CH₃–C≡C–CH₂–CH₃	−109	56

ALKYNES

A triple bond between two carbon atoms identifies a member of the **alkyne** family. Alkynes use prefixes to indicate the number of carbons and end with the suffix -*yne*. In the same manner as you saw for alkenes, a number is used to tell where the triple bond occurs within the carbon chain, when necessary. The most common alkyne is also the simplest. Ethyne, commonly called *acetylene*, consists of two carbons joined by a triple bond. This compound is often used as a fuel for welding torches and as an ingredient for plastics.

Physically, alkynes are similar to other hydrocarbons. They are practically nonpolar, so they are insoluble in water and very soluble in nonpolar solvents. Their boiling points rise as the carbon chains get longer.

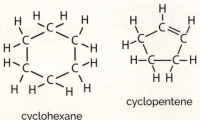

cyclohexane

cyclopentene

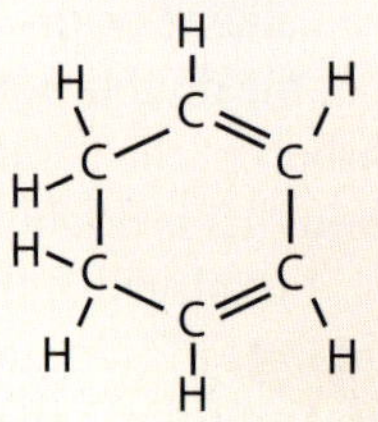

cyclohexa-1,3-diene

Cyclic Aliphatic Hydrocarbons

As stated in Section 20.1, not all aliphatic hydrocarbons consist only of open chains of carbon atoms. Some form rings. Such compounds are called **cyclic aliphatic compounds**. Though many rings are possible, five- and six-carbon rings are the most abundant. Simple alkenes and alkenes with more than one double bond multiply the number of possible structures. Three structures and their names are shown at left.

Some very unusual structures are possible when several rings combine as shown below.

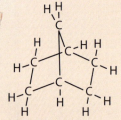

bicyclo[2.2.1] heptane

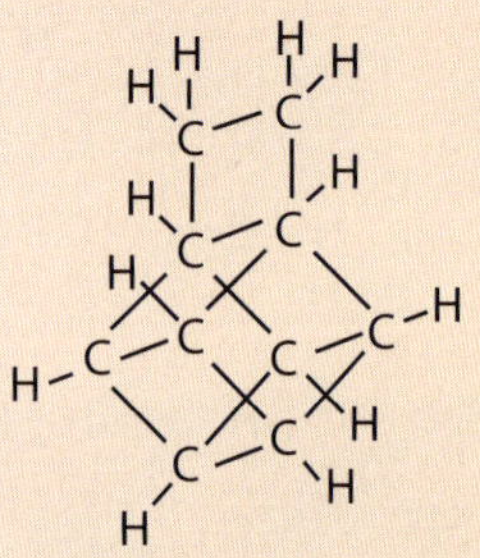

pentacyclo[4.4.0.0^{2,5}.0^{3,8}.0^{4,7}] decane (basketane)

The chemical activity of cyclic compounds is about the same as that of other members of their parent families. *Cycloalkanes* act like alkanes; *cycloalkenes* act like alkenes. Cyclic compounds have several unique uses and are found in anesthetics, car maintenance products, perfumes, and paint removers.

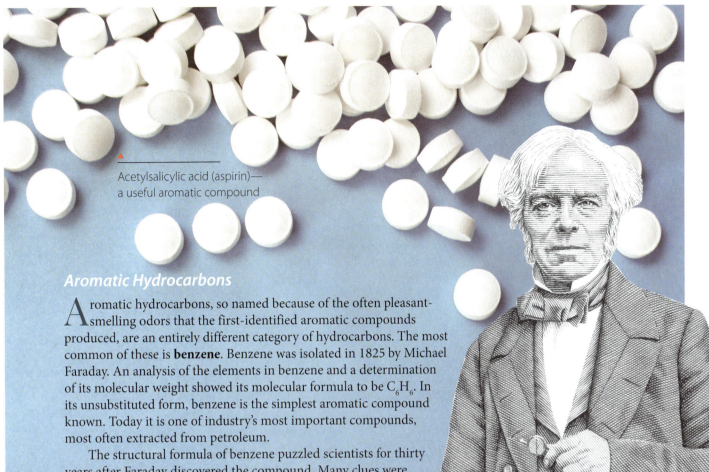

Acetylsalicylic acid (aspirin)—a useful aromatic compound

Michael Faraday

August Kekulé

benzene

Aromatic Hydrocarbons

Aromatic hydrocarbons, so named because of the often pleasant-smelling odors that the first-identified aromatic compounds produced, are an entirely different category of hydrocarbons. The most common of these is **benzene**. Benzene was isolated in 1825 by Michael Faraday. An analysis of the elements in benzene and a determination of its molecular weight showed its molecular formula to be C_6H_6. In its unsubstituted form, benzene is the simplest aromatic compound known. Today it is one of industry's most important compounds, most often extracted from petroleum.

The structural formula of benzene puzzled scientists for thirty years after Faraday discovered the compound. Many clues were gathered, but they did not seem to fit together. The molecular formula C_6H_6 led chemists to believe that the molecule must have several double or triple bonds. Yet the chemical reactions of benzene did not support this idea; it behaved like an alkane, not an alkene or alkyne. When scientists determined the bond lengths between the carbon atoms, they found that the distances were not those of single or double bonds—they were in between. It was as if benzene used one-and-a-half bonds. Furthermore, scientists found that the carbons were arranged in a ring and that all the carbon atoms had identical bonds.

In 1865 August Kekulé proposed a structure that could account for most of the observations. He described benzene in terms of two symmetrical resonance structures.

The actual structure of a benzene molecule was thought to be a dynamic equilibrium of these two structures. Thus, the electrons in the double bonds were mobile and not tied down to specific locations. But the development of the valence bond theory modified this concept. The single bonds between the carbon atoms—sigma bonds—were due to head-on overlap of the carbon atom's hybridized orbitals. The double bond included the additional components of pi bonds—side-to-side overlap of additional orbitals between pairs of carbon atoms.

Organic Chemistry **495**

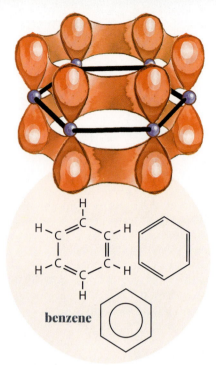

benzene

These pi clouds overlap in the molecule to form doughnut-shaped areas of electron concentration above and below the plane of the ring. Therefore, the electrons are free to move throughout the entire "doughnut." For this reason, they are called *delocalized electrons*. All aromatic compounds have cyclic clouds of delocalized electrons.

Because the pi electrons in the double bonds are not bound between any two carbon atoms, chemists often draw the structure of benzene as shown at left, either with or without the single-bonded hydrogens. They are understood to be there if they are not shown.

Today other non-benzene-based aromatics are known. One example is *furan*, which contains one oxygen atom bonded in a ring with four carbon atoms. One of the oxygen's unbonded pairs of electrons is delocalized into the ring structure. Furan serves as the starting point for the synthesis of a host of other compounds, called *furans*, many of which are used as medicines to treat a variety of conditions. One naturally occurring furan, 1-(3-hydroxyfuran-2-yl) ethan-1-one, contributes to the characteristic taste of bread crust.

20.2 SECTION REVIEW

1. What do hydrocarbons contain? Name some sources of hydrocarbons.

2. A chemist is given two samples of clear liquid, each labeled as C_6H_{14}. When tested, one sample boils at 68.5 °C, while the other does so at 62.9 °C. Explain how this is possible.

3. Why are hydrocarbons mainly insoluble in water?

4. Explain the differences between saturated and unsaturated hydrocarbons.

5. Compare the structures of alkanes, alkenes, and alkynes.

6. Identify each molecular formula as an alkane, alkene, or alkyne.

   a. C_2H_2
   b. $C_{10}H_{22}$
   c. C_5H_8
   d. C_4H_6
   e. C_2H_4

7. What term describes aliphatic compounds whose chains have been bonded into rings?

8. Describe the nature of the bonds between carbon atoms in aromatic compounds.

9. Draw structural formulas for each of the following.

   a. hexane
   b. hex-1-ene
   c. 2-methylhexane
   d. hex-3-yne

10. Name the compounds indicated by each of the following structural formulas.

    a. 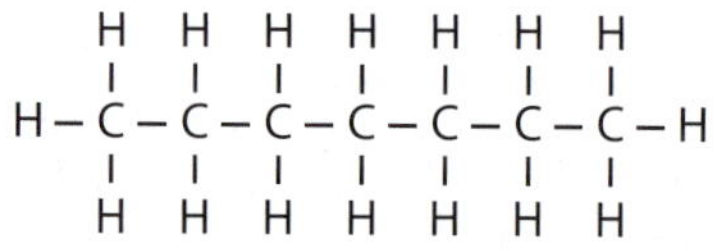 H–C–C–C–C–C–C–C–H (seven carbons, all single bonds, fully hydrogenated)

    b. H–C=C–C–C–C–C–C–H (seven carbons with double bond at position 1)

    c. H–C–C=C–C–C–C–C–H (seven carbons with double bond at position 2)

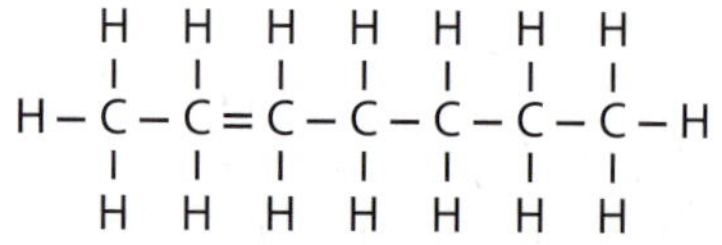

    d. 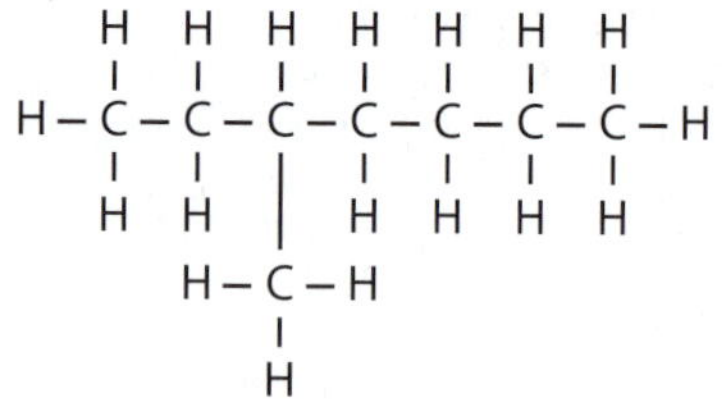 H–C–C–C–C–C–C–C–H (seven carbons with a –CH₃ branch on the third carbon)

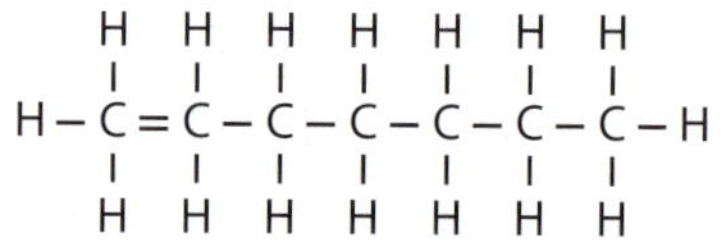

11. Your study partner says that the following structural formulas are both acceptable forms for showing toluene (C_7H_8). Is he correct? Explain.

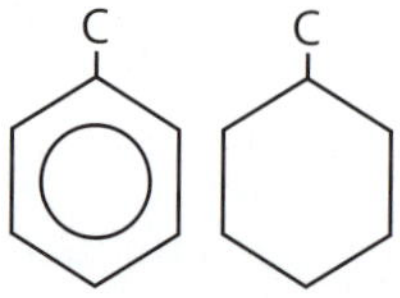

12. Since the two compounds shown in Question 11 both have six-carbon rings, would you expect them to have similar physical and chemical properties? Explain.

20.3 SUBSTITUTED HYDROCARBONS

Functional groups are the hot spots of chemical activity on an organic molecule. A **functional group**, also called a *substituent*, is an atom or group of atoms that can substitute for a hydrogen atom in a hydrocarbon. It greatly modifies the behavior of the hydrocarbon. Hydrocarbons that contain a functional group are often called *substituted hydrocarbons*.

There are many kinds of functional groups. Most functional groups in organic chemistry contain arrangements of oxygen or nitrogen atoms. Some even contain sulfur or halogens. Since the composition and structure of a molecule determine its physical and chemical properties, functional groups partly determine a molecule's properties.

Does substituting one kind of atom for another in a hydrocarbon really make that much of a difference?

QUESTIONS
» What are substituted hydrocarbons?
» Do all substituted hydrocarbons have similar properties?
» What makes one group of substituted hydrocarbons distinct from another?
» How do I name and write structural formulas for substituted hydrocarbons?

TERMS
functional group • halocarbon • haloaromatic • hydroxyl group • alcohol • ether • carbonyl group • aldehyde • ketone • carboxylic acid • carboxyl group • fatty acid • ester • amine • amide

Halocarbons

When a halogen and a hydrocarbon combine, a **halocarbon** forms. The simplest halocarbons are haloalkanes, which are a combination of an alkane and one or more fluorine, chlorine, bromine, or iodine atoms in place of one or more of the alkane's hydrogen atoms. The general formula for haloalkanes is R–X, where R represents an alkyl group (derived from the parent alkane) and X represents a halide. *Haloalkenes* and *haloalkynes* are also possible. When a halogen is attached to an aromatic ring, it forms a **haloaromatic**. Polytetrafluoroethylene (a nonstick coating for pans) and polyvinyl chloride (PVC plastic) are two halocarbon polymers that you have likely used. Many other halides serve as intermediates in the synthesis of other compounds.

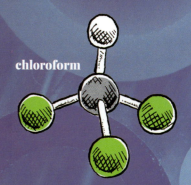
chloroform

TABLE 20-6 *Industrial Uses of Halocarbons*

Name	Structural Formula	Application
trichloromethane (chloroform)	H–C(Cl)(Cl)–Cl	early anesthetic
tetrachloroethene	Cl₂C=CCl₂	nonpolar solvent (dry-cleaning agent)
triiodomethane (iodoform)	H–C(I)(I)–I	veterinary antiseptic

MINI LAB

ISOMERISM IN SUBSTITUTED HYDROCARBONS

Modeling is an essential skill in science. Three-dimensional models of molecules give us a much better understanding of their structures. In this lab activity, you will model isomers and think about how their differences in structure might affect their physical properties.

How does adding substituents to a hydrocarbon affect isomerism?

Procedure

A Build a model of ethene. Refer back to Table 20-4 if you need help.

1. Are any isomers of ethene possible?

B Create a model of chloroethene by removing one hydrogen atom from your ethene model and replacing it with an atom of chlorine.

2. Are any isomers of chloroethene possible? Explain. (*Hint*: To check your answer, move the chlorine atom from one carbon atom to the other. Does that affect the structure of the molecule in any chemically significant way?)

C Create a model of 1,2-dichloroethene by adding a chlorine atom to the other carbon atom in your chloroethene molecule. You should now have one chlorine atom bonded to each carbon atom.

D Examine your 1,2-dichloroethene model by imagining a plane running through both carbon atoms in a way that causes the hydrogen and chlorine atoms to lie either above or below the plane.

3. Are any isomers of 1,2-dichloroethene possible? Explain.

What you have just observed is an example of what chemists call *cis-trans isomerism*. The prefixes *cis-* and *trans-* come from Latin and mean "this side of" and "the other side of."

4. Draw and label the structural formulas for the *cis* and *trans* versions of 1,2-dichlorethene.

Conclusion

5. Would you expect the two isomers of 1,2-dichloroethene to have the same or different physical properties? Explain.

Going Further

6. Is it possible to create a third isomer of dichloroethene? If so, draw and label its structural formula.

Amino acids, the building blocks of proteins, exhibit a characteristic called *chirality* that is similar to cis-trans isomerism. Chirality results when a carbon atom is bonded to four different groups; the molecule and its mirror-image form cannot be superimposed on one another. The two forms, known as *enantiomers*, are often designated as either left- or right-handed.

7. Go to the website of Creation Ministries International and do a search using the keyword "chirality." Why is chirality significant when evaluating theories of the origin of life?

EQUIPMENT

- molecular modeling set

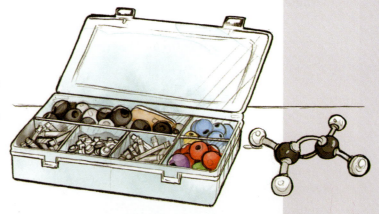

Alcohols

Compounds that have a covalently bonded OH group, called a **hydroxyl group**, attached to an alkyl group are classified as **alcohols**. The general formula for the whole family is R–OH, where R represents an alkyl group. The simplest alcohol is methanol. In this case the R group is the smallest one possible, which is CH_3 (methyl).

Alcohol names consist of the standard prefixes that tell how long the carbon chain is plus an *-ol* ending in place of the *-e* ending. For instance, what is commonly called *rubbing alcohol* is a three-carbon chain with the OH group attached to the middle carbon. The IUPAC name is propan-2-ol. Note that the numeric indicator is inserted between the name of the parent alkane and the *–ol* suffix. This is done even if the hydroxyl group is bonded to an end carbon, as for instance for propan-1-ol.

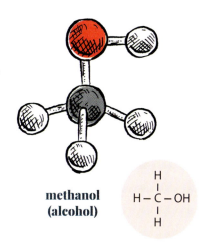

methanol (alcohol)

The physical properties of alcohols are a result of two factors—a polar OH group and a nonpolar alkyl group. The combination of these two opposites determines the behavior of each specific molecule. If the hydrocarbon chain is relatively short, the OH group dominates the molecule. As a result, it behaves as a polar molecule. As the hydrocarbon chain becomes longer, the chain begins to dominate and imparts nonpolar characteristics to the molecule. Alcohols, under the influence of one or more OH groups, form hydrogen bonds and thus have higher boiling points than their parent molecules. Small alcohols, with their polar nature, are soluble in water. Larger alcohols, under the influence of their hydrocarbon chains, are insoluble in water and soluble in nonpolar solvents.

Alcohols that contain more than one OH group are called *polyhydroxy alcohols*. One common polyhydroxy alcohol is ethane-1,2-diol (commonly called ethylene glycol), which is used as an antifreeze in car radiators. Another common alcohol—propane-1,2,3-triol (glycerol)—serves as a moisturizer in cosmetics.

Industries use alcohols as solvents, paint thinners, antifreezes, and ingredients in aftershave lotions. In the United States, most gasoline is 10%–15% ethanol. Some vehicle engines are designed to run on E85 fuel, which is 85% ethanol and 15% gasoline.

TABLE 20-7 Alcohols

Name	Structural Formula	Application
methanol (wood alcohol)	H–C(H)(H)–OH	solvent, fuel
ethanol (grain alcohol)	H–C(H)(H)–C(H)(H)–OH	alcoholic beverages, engine fuel, solvent
propan-1-ol	H–C(H)(H)–C(H)(H)–C(H)(H)–OH	solvent for making pharmaceuticals, resins
propan-2-ol (rubbing alcohol)	H–C(H)(H)–C(OH)(H)–C(H)(H)–H	sterilizing pads, solvent, gasoline additive
butan-1-ol	H–C(H)(H)–C(H)(H)–C(H)(H)–C(H)(H)–OH	solvent, shellac, varnish

Ethers

Compounds that have the general formula R–O–R′ are called **ethers**. Ethers are distinguished by an oxygen bridge between two carbon chains, R and R′ (R-prime). The second alkyl group may be the same as or different from the first group. The name of an ether includes the names of the alkyl groups on each side of the oxygen. If one is smaller than the other, it is named first, using its alkyl form, followed by the root –*oxy*-, and finally the name of the larger alkyl group.

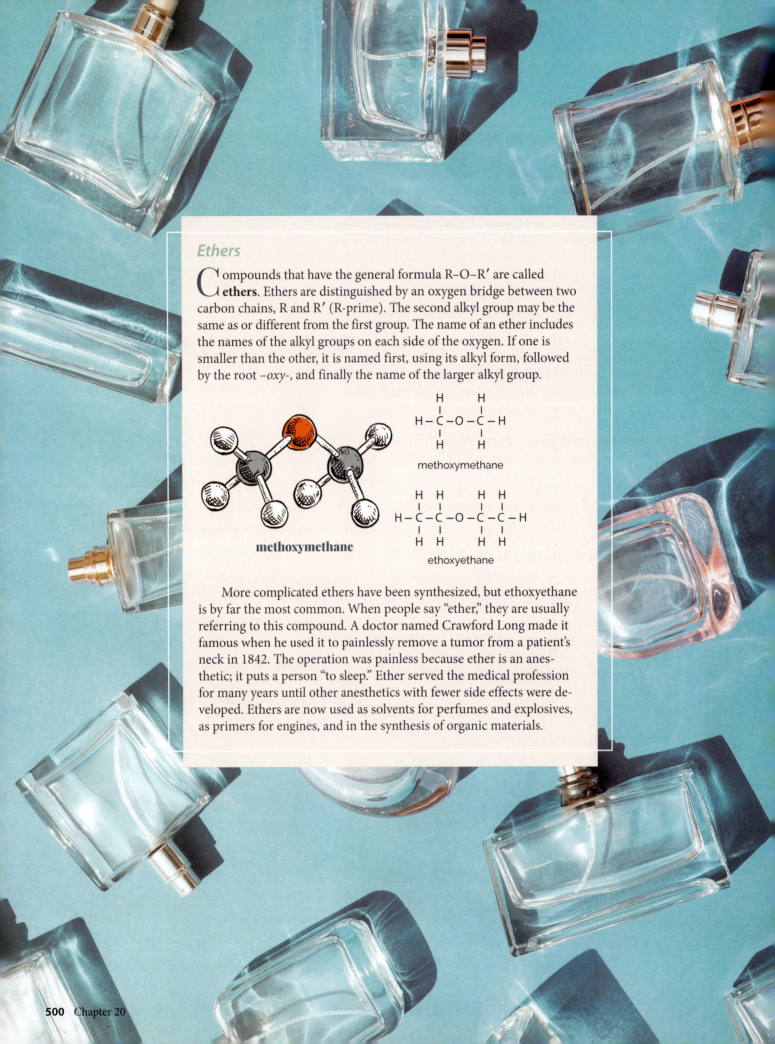

methoxymethane

$$H-\overset{\overset{H}{|}}{\underset{\underset{H}{|}}{C}}-O-\overset{\overset{H}{|}}{\underset{\underset{H}{|}}{C}}-H$$

methoxymethane

$$H-\overset{\overset{H}{|}}{\underset{\underset{H}{|}}{C}}-\overset{\overset{H}{|}}{\underset{\underset{H}{|}}{C}}-O-\overset{\overset{H}{|}}{\underset{\underset{H}{|}}{C}}-\overset{\overset{H}{|}}{\underset{\underset{H}{|}}{C}}-H$$

ethoxyethane

More complicated ethers have been synthesized, but ethoxyethane is by far the most common. When people say "ether," they are usually referring to this compound. A doctor named Crawford Long made it famous when he used it to painlessly remove a tumor from a patient's neck in 1842. The operation was painless because ether is an anesthetic; it puts a person "to sleep." Ether served the medical profession for many years until other anesthetics with fewer side effects were developed. Ethers are now used as solvents for perfumes and explosives, as primers for engines, and in the synthesis of organic materials.

Aldehydes and Ketones

When an oxygen atom forms a double bond with a carbon atom, it forms a **carbonyl group**. (C=O) The location of carbonyl groups is a key feature for distinguishing between two major groups of organic compounds.

ALDEHYDES

Aldehydes are organic compounds that contain an oxygen held with a double bond to an end carbon. Their general structure looks like the structural formula shown below.

$$R-\overset{\overset{O}{\|}}{C}-H$$

According to the IUPAC rules, the name of an aldehyde is formed with an *-al* ending on the name of the corresponding alkane. Other substituents may be bonded to the aldehyde as well, as is shown by 2-methylpentanal.

propanal   2-methylpentanal

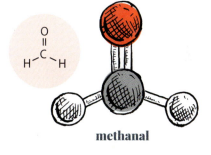

methanal

The simplest aldehyde and one of the most important ones in industry is known by its common name of formaldehyde. Its systematic IUPAC name is methanal. This colorless gas with a piercing odor is often dissolved in water to make a 37% solution called *formalin*. It has been used in the past to preserve frogs, fetal pigs, and other creatures for dissection in biology laboratories. But because of health concerns about its possible link to cancer and allergic reactions, the use of formalin as a preserving fluid has declined. But not all aldehydes are potentially harmful. Some are responsible for the unique smells and tastes associated with certain foods. The pleasant aroma and taste of cinnamon is produced by (2E)-3-phenylprop-2-enal, better known as cinnamaldehyde. The root *phenyl* in the name indicates that a benzene ring is attached to the propene chain.

cinnamaldehyde

Organic Chemistry

KETONES

A carbonyl group within a carbon chain—not on an end carbon—is a key feature of compounds called **ketones**. The general formula of ketones is shown here.

$$R-\overset{\overset{O}{\|}}{C}-R'$$

The simplest ketone is commonly called acetone. The word *ketone* itself is derived from the German word for acetone. The IUPAC name for acetone is *propanone*—the *-e* ending of the alkane is changed to *-one*. For chains over four carbons long, the location of the carbonyl group is indicated by a number placed between the name of the parent alkyl group and the *–one* suffix as shown below for pentan-2-one.

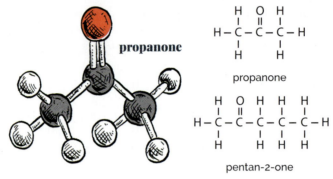

Acetone is an excellent solvent. While the molecule's methyl groups are nonpolar, the carbonyl group is polar. As a result, acetone dissolves most organic compounds but still mixes well with water. Acetone is widely used as a solvent for lacquers, paint removers, explosives, plastics, and disinfectants. It is also the active ingredient in some nail polish removers. Remember MIT from the introduction to Section 20.2? Both its common name and its IUPAC name end in *–one*, so it's a ketone too. Perhaps now it will seem less alien to you when you see it in a list of ingredients.

Since both ketones and aldehydes have carbonyl groups, it is not surprising that they have similar physical and chemical properties. As a rule, though, aldehydes are more reactive because their functional group is exposed on the end of the carbon chain.

Carboxylic Acids

Because of their abundance in nature, carboxylic acids were among the first organic compounds to be studied in detail. As a result, many of these acids acquired common names from their most familiar sources. For example, the Latin word for vinegar is *acetum*, so the acid in vinegar was called *acetic acid*. The Latin word for ant is *formica*. Consequently, the stinging acid of ants was called *formic acid*.

Carboxylic acids derive their name from the presence of a **carboxyl group**, which consists of both a carbonyl group and a hydroxyl group. The chemical formula for a carboxyl group is written as COOH. The generalized structural formula for a carboxylic acid is shown here.

$$R-C\overset{\overset{O}{\|}}{\underset{OH}{}}$$

According to IUPAC nomenclature, the *-e* ending of a carboxylic acid's corresponding alkane is changed to *-oic*, and the word *acid* is added to form the name of a carboxylic acid. Thus, formic acid, which is the simplest carboxylic acid, is called *methanoic acid* in the IUPAC system. Acetic acid's IUPAC name is *ethanoic acid*.

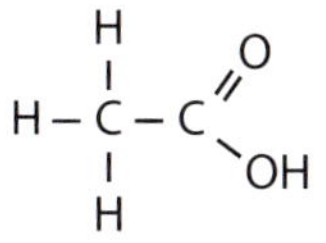

ethanoic acid

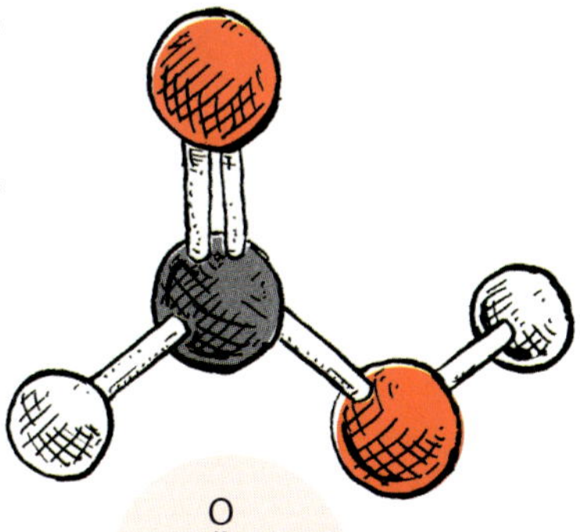

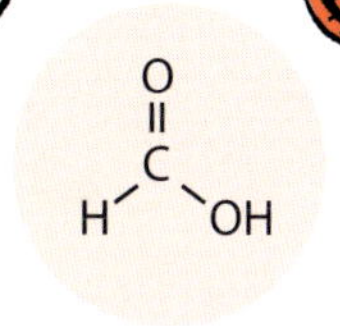

methanoic acid

Carboxylic acids with shorter alkyl groups are liquids at room temperature and have sharp or unpleasant odors. Acids with longer carbon chains are usually waxy solids. When the carbon chains are between twelve and twenty carbon atoms long, they are often called **fatty acids**. They can form hydrogen bonds between themselves and other molecules. These hydrogen bonds keep even the smallest fatty acids solid at room temperature. One such acid, octadecanoic acid (commonly known as stearic acid) is a significant ingredient in a popular food item—chocolate!

Carboxylic acids are called acids because they contain an ionizable hydrogen. The ionizable hydrogen in these acids is the hydrogen in the carboxyl group. Since only a small fraction of carboxylic acid molecules ionize (a mere 4%), these acids are weak. Although they are weak acids, carboxylic acids react quickly with strong bases to form salts. Many of these salts are commercially useful, such as sodium or potassium salts of fatty acids used to make soaps.

Esters

If the hydrogen of a carboxylic acid's carboxyl group is replaced with an alkyl group, an **ester** forms. Esters have the general formula shown below.

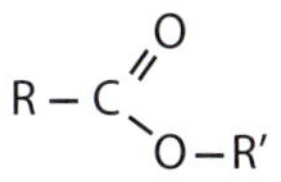

Unlike their cousins the carboxylic acids, esters generally have appealing smells. These compounds are responsible for the flavors of many fruits and the fragrances of many flowers. In the naming of an ester, the R′ group is indicated with its alkyl name and the carboxylic acid part is given an *-oate* ending.

TABLE 20-8: *Esters*

Name	Structural Formula	Flavor or Scent
ethyl butanoate	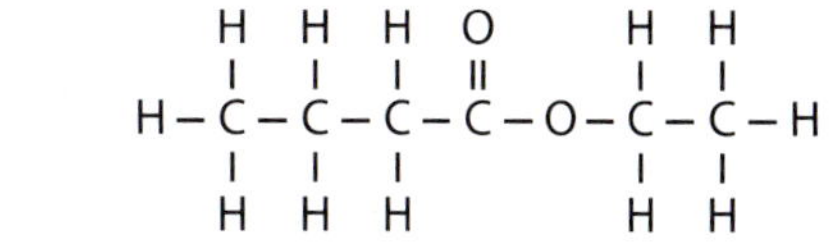	pineapple
pentyl ethanoate	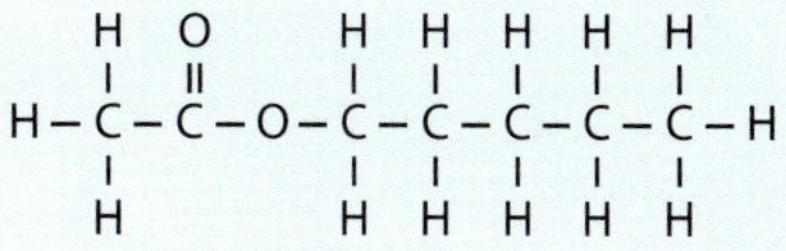	banana
ethyl heptanoate	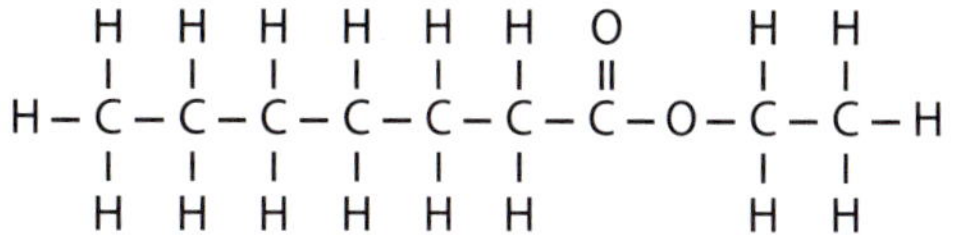	grape
2-methylpropyl formate	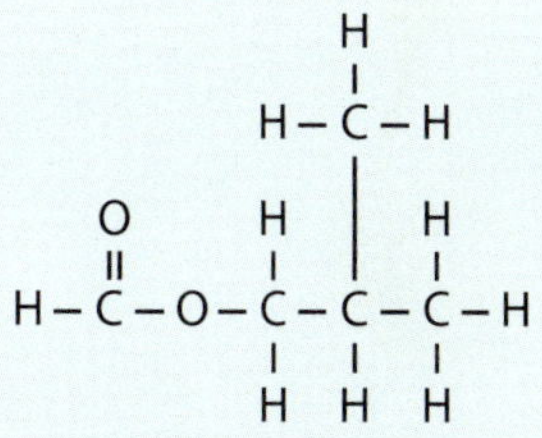	raspberry

Organic Chemistry

Amines and Amides

Nitrogen can bond in organic molecules in several different ways. **Amines** are a family of organic compounds with ammonia (NH_3) as the parent molecule. Derivatives are formed when one or more of the hydrogen atoms in the ammonia molecule is replaced with an organic group, such as an alkyl group. The names of the compounds that result commonly have the word *amine* as a suffix after the names of the alkyl groups.

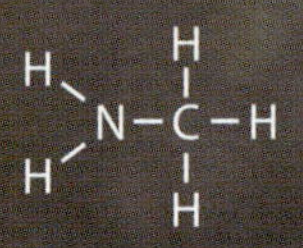

methanamine

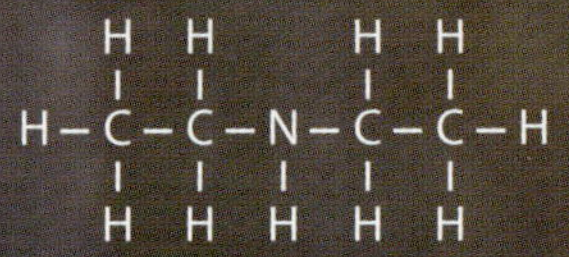

N-ethylethanamine

Another group of compounds that contain nitrogen is called **amides**. Amides are produced when an amine group takes the place of an OH group in a carboxylic acid. All the members of this group have the structure shown below in common. This structure is especially important because it holds the amino acids in proteins together.

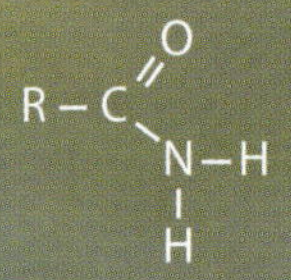

An amide compound that you might be familiar with is *N,N*-diethyl-3-methylbenzamide, better known as DEET. It's the active ingredient in mosquito repellents. Appendix J presents a summary of the families of organic compounds that we have discussed in this chapter.

CASE STUDY

BUGS, BE GONE!

DEET has been used in mosquito repellents for decades. Although studies have indicated that it is generally safe to use, it is not entirely without risks. In rare instances, it can cause skin reactions, breathing difficulty, headaches, seizures, and even death. Extensive use has been linked to insomnia and mood disturbances. But DEET is undeniably effective at what it does—deterring mosquitoes from biting— and this is something that can't be ignored in areas where the risk of contracting serious mosquito-borne diseases, such as West Nile virus and malaria, is high.

20.3 SECTION REVIEW

1. What is a functional group?

2. What is the main difference between a haloalkane and a haloaromatic?

3. The physical properties of alcohols are a result of what two factors?

4. How does the length of an alcohol's hydrocarbon chain influence its properties?

5. Which compounds are characterized by an oxygen bridge between two carbon chains?

6. What is the main difference between an aldehyde and a ketone?

7. What is the main difference between an amine and an amide?

8. In each instance below, identify the organic compound group that has the given IUPAC ending to its name. Then write the structural formula for and name the simplest example of that group that is based on a two-carbon chain. If no two-carbon example is possible, write *none*.

    a. -one

    b. -ol

    c. -al

    d. -oate

    e. -oic acid

9. Identify the class of organic compound to which the molecule indicated by each of the following structural formulas belongs. Then name the compound.

    a.
    $$\text{Cl}-\underset{\underset{\text{H}}{|}}{\overset{\overset{\text{H}}{|}}{\text{C}}}-\underset{\underset{\text{Cl}}{|}}{\overset{\overset{\text{H}}{|}}{\text{C}}}-\underset{\underset{\text{H}}{|}}{\overset{\overset{\text{H}}{|}}{\text{C}}}-\underset{\underset{\text{H}}{|}}{\overset{\overset{\text{H}}{|}}{\text{C}}}-\underset{\underset{\text{H}}{|}}{\overset{\overset{\text{H}}{|}}{\text{C}}}-\text{H}$$

    b.
    $$\text{H}-\underset{\underset{\text{H}}{|}}{\overset{\overset{\text{H}}{|}}{\text{C}}}-\underset{\underset{\text{H}}{|}}{\overset{\overset{\text{H}}{|}}{\text{C}}}-\underset{\underset{\text{H}}{|}}{\overset{\overset{\text{OH}}{|}}{\text{C}}}-\underset{\underset{\text{H}}{|}}{\overset{\overset{\text{H}}{|}}{\text{C}}}-\text{OH}$$

    c.
    $$\text{H}-\underset{\underset{\text{H}}{|}}{\overset{\overset{\text{H}}{|}}{\text{C}}}-\underset{\underset{\text{H}}{|}}{\overset{\overset{\text{H}}{|}}{\text{C}}}-\underset{\underset{\text{H}}{|}}{\overset{\overset{\text{O}}{\|}}{\text{C}}}-\underset{\underset{\text{H}}{|}}{\overset{\overset{\text{H}}{|}}{\text{C}}}-\underset{\underset{\text{H}}{|}}{\overset{\overset{\text{H}}{|}}{\text{C}}}-\underset{\underset{\text{H}}{|}}{\overset{\overset{\text{H}}{|}}{\text{C}}}-\underset{\underset{\text{H}}{|}}{\overset{\overset{\text{H}}{|}}{\text{C}}}-\text{H}$$

    d.
    $$\text{H}-\underset{\underset{\text{H}}{|}}{\overset{\overset{\text{H}}{|}}{\text{C}}}-\underset{\underset{\text{H}}{|}}{\overset{\overset{\text{H}}{|}}{\text{C}}}-\underset{\underset{\text{H}}{|}}{\overset{\overset{\text{H}}{|}}{\text{C}}}-\overset{\overset{\text{O}}{\|}}{\text{C}}-\text{H}$$

Use the case study above to answer Questions 10–11.

10. Why should Christians care about whether mosquito repellents are safe to use?

11. Describe at least one way in which an organic chemist might use his knowledge to help make repelling mosquitoes safer.

20.4 ORGANIC REACTIONS

The number of synthesized organic compounds is continually growing because these compounds can participate in many kinds of chemical reactions. Most biological processes rely on chemical reactions between organic molecules. Such reactions are responsible for the movement of muscles, the digestion of food, the transmission of nerve impulses, and the sensing of light on the retina. Industrial chemists also manipulate organic molecules to make flavorings, plastics, fuels, synthetic fabrics, and a host of other products. Some of the reactions involving organic compounds are quite complicated. This section will survey a few of the basic kinds of reactions in which organic compounds participate.

Oxidation-Reduction

Think back on what you read in Chapter 19 about redox reactions. When oxygen and carbon atoms bond, carbon atoms are oxidized and oxygen atoms are reduced. Since oxygen atoms have high electronegativities, they pull shared electrons away from carbon. This means that adding oxygen always makes the oxidation number of carbon more positive. Adding hydrogen atoms, on the other hand, reduces carbon atoms.

Recall that all combustion reactions are redox reactions. Combustion oxidizes all the carbon atoms in an organic molecule to form carbon dioxide. All hydrocarbons burn in oxygen to form carbon dioxide and water, releasing thermal energy in the process. The energy from these combustion reactions is used to produce motion, heat, and light. The burning of methane gas, a simple alkane, is a combustion reaction.

$$CH_4 + 2O_2 \longrightarrow CO_2 + 2H_2O$$

We can see in the equation below that the carbon is oxidized in the combustion reaction and oxygen is reduced.

$$\overset{-4}{C}\overset{}{H_4} + 2\overset{0}{O_2} \longrightarrow \overset{+4\ -2}{CO_2} + 2\overset{}{H_2}\overset{-2}{O}$$

Fatty acid chains in food are oxidized by the human body in a way that is similar to how other hydrocarbons are burned. The body is designed to control the oxidation precisely so that only small amounts of energy are released at any instant. This regulation keeps the temperatures during oxidation tolerable and allows the body to capture and use most of the released energy.

QUESTIONS
» What kinds of chemical reactions do organic compounds participate in?
» How can I determine the kind of reaction that an organic compound is involved in?
» Is it possible to predict the products in a reaction involving organic compounds?

TERMS
substitution reaction • addition reaction • elimination reaction • condensation reaction

How are organic reactions classified?

Substitution

Substitution reactions replace one part of a molecule with another part. Typically, most of the reactions of chemically inactive compounds are substitution reactions. Alkanes are not very reactive, but when heated to high temperatures or exposed to energetic ultraviolet light, their hydrogen atoms can be replaced by other atoms. For example, methane and chlorine can react to form a variety of substitution products. In the example below, a chlorine atom replaces a hydrogen atom when a mixture of the two gases is exposed to heat or certain types of light.

$$CH_4 + Cl_2 \xrightarrow{\text{heat or light}} CH_3Cl + HCl$$

Various groups can replace one or more of the hydrogens of aromatic compounds. Benzene can be nitrated, halogenated, or even alkylated. The simplest alkylated benzene compound, consisting of a methyl group bonded to a benzene ring, is toluene. It is a common solvent and produces the characteristic smell of paint thinner.

Addition

Compared with carbon-carbon single bonds, double and triple bonds are very reactive. Consequently, the carbons with double and triple bonds react first and determine the molecule's behavior. An **addition reaction** is a reaction in which a multiple bond of a molecule is broken and two atoms or groups of atoms are added. This is a characteristic reaction of unsaturated molecules. Some of the most common addition reactions involve water (called a *hydration reaction*), hydrogen gas (called a *hydrogenation reaction*), halogens, and hydrogen halides. In the example below, hydrogen converts an alkene (ethene) to an alkane (ethane).

$$C_2H_4 + H_2 \longrightarrow C_2H_6$$

Organic Chemistry

WORLDVIEW INVESTIGATION

AROMATHERAPY

Essential oils have been popularized through the practice of *aromatherapy*—the use of essential oils to help improve mental and physical well-being. Some people firmly believe in the therapeutic value of essential oils; others claim that aromatherapy is pseudoscience. Where does the truth lie?

Task

The manager of the store where you work asks you for an opinion on whether the store should stock essential oils, but you don't know much about them. You want to give your manager an informed opinion, so you'll need to do some research and then get back to her.

Procedure

1. Research aromatherapy by doing an internet search using the keywords "aromatherapy" and "essential oils."
2. To get a biblical perspective, do an internet search using the keywords "biblical view of essential oils."
3. Write an opinion paper to present to your manager, citing your sources properly. Have a classmate review your paper and give you feedback.
4. Complete your paper and submit it by the deadline.

Conclusion

In the world that He created, God has given us many compounds that are useful to humans. At the very least, the pleasant scent of essential oils can improve our mood. But like the claims of therapeutic effectiveness made for other substances, those made on behalf of essential oils should be rigorously and scientifically tested.

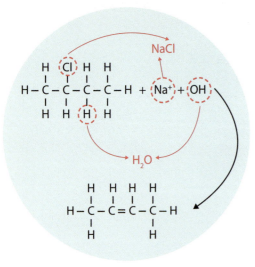

2-chlorobutane reacting with NaOH in solution to form but-2-ene, water, and NaCl

Elimination

We can think of elimination reactions as the opposite of addition reactions. In an **elimination reaction**, two atoms are removed from the organic compound. For example, 2-chlorobutane can react with a sodium hydroxide solution, undergoing an elimination reaction to produce but-2-ene, water, and dissolved sodium chloride.

$$CH_3CClHCH_2CH_3(aq) + NaOH(aq) \longrightarrow$$

$$CH_3CH=CHCH_3(aq) + NaCl(aq) + H_2O(l)$$

This reaction forms unsaturated molecules. Some of the most common elimination reactions involve water (called a *dehydration reaction*), hydrogen (called a *dehydrogenation reaction*), and halogens. The example below shows the dehydration of ethanol to ethylene.

$$CH_3CH_2OH \longrightarrow CH_2=CH_2 + H_2O(l)$$

Condensation

Reactions in which molecules combine to form a larger molecule are called **condensation reactions** and are a type of addition reaction. A water molecule is also produced during some condensation reactions. Under the proper conditions, two identical alcohol molecules can be made to join together to form an ether.

$$\text{R-OH} + \text{R-OH} \xrightarrow{H_2SO_4} \text{R-O-R} + H_2O$$

For example, two methanol molecules can be condensed to form methoxymethane and water. As you can see, this is also a dehydration reaction.

$$2CH_3OH \longrightarrow CH_3OCH_3 + H_2O$$

Esters form when carboxylic acids and alcohols go through a condensation reaction. This condensation reaction is called *esterification*. For example, an artificial banana flavoring, pentyl ethanoate, can be made by mixing ethanoic acid and pentanol.

A condensation reaction is responsible for much of the clothing that people wear. Polymers are substances that consist of huge molecules that have repeating structural units. Polyester, one of the more common polymers, forms when ethylene glycol and terephthalic acid condense. We will discuss polymers more in Chapter 21.

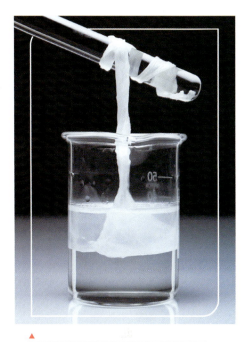

▲
A condensation reaction occurs at the interface between layers of two organic compounds. Above, nylon 6-10 is being formed.

20.4 SECTION REVIEW

For Questions 1–7, determine the most specific organic reaction that is taking place.

1. Hydrogen atoms are replaced by chlorine atoms when an alkane is heated with chlorine.
2. Unsaturated molecules react.
3. Chloroethane reacts with potassium hydroxide to form ethane, potassium chloride, and water.
4. Carbon and oxygen atoms bond.
5. A hydrocarbon reacts with oxygen to form water and carbon dioxide.
6. A reaction between an acid and an alcohol forms an ester and water.
7. Molecules combine and a water molecule is lost.
8. Give one example each of a substitution reaction, an addition reaction, and a condensation reaction.
9. Copy the table below. Fill in the table to summarize and compare different kinds of organic reactions.
10. Predict the compound that will result if propan-1-ol and butan-1-ol undergo a condensation reaction.

Type of Reaction	Key Feature	Example
redox		combustion
	One part of a molecule is replaced with another part.	replacing a hydrogen in a hydrocarbon with a functional group
		breaking a double bond and adding H_2 to convert an alkene to an alkane
	A multiple bond is formed when two groups of atoms are removed.	
condensation		

Organic Chemistry 509

20 CHAPTER REVIEW

Chapter Summary

TERMS
- organic compound
- organic chemistry
- condensed structural formula
- aromatic compound
- aliphatic compound

20.1 ORGANIC COMPOUNDS

- All organic compounds contain carbon. They were thought at first to be produced only by living things, hence the name. Many organic compounds can now be synthesized.

- Carbon has four valence electrons and forms strong bonds with other carbon atoms and atoms of many other elements. It can form chains, rings, and other shapes with single, double, and triple bonds.

- Organic compounds are divided into two main groups: aromatic compounds, which have ringed shapes with delocalized electrons, and aliphatic compounds, which include those with linear chains or ring structures that can be formed by closing a chain.

20.2 HYDROCARBONS

- Hydrocarbons are a class of compounds containing only hydrogen and carbon. Some common hydrocarbon types are alkanes, which have only single bonds; alkenes, which include at least one double bond; and alkynes, which have at least one triple bond.

- The IUPAC names of organic compounds are based on their molecular structures. The structure of an organic compound can be determined from its IUPAC name.

- Isomers are compounds with identical molecular formulas but different structures. Because of their different structures, isomers have different properties.

- The key feature that distinguishes aromatic compounds from similarly shaped cyclic aliphatic compounds is the presence of a highly stable ring structure, often a benzene ring. The electrons in the ring are delocalized, meaning that they are shared by all the carbons in the ring.

TERMS
- hydrocarbon
- alkane
- saturated hydrocarbon
- alkyl group
- isomer
- alkene
- unsaturated hydrocarbon
- alkyne
- cyclic aliphatic compound
- benzene

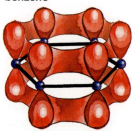

TERMS
- functional group
- halocarbon
- haloaromatic
- hydroxyl group
- alcohol
- ether
- carbonyl group
- aldehyde
- ketone
- carboxylic acid
- carboxyl group
- fatty acid
- ester
- amine
- amide

20.3 SUBSTITUTED HYDROCARBONS

- Functional groups are atoms or groups of atoms that replace hydrogens within organic compounds. The type, number, and location of functional groups partly determine the physical and chemical properties of substituted hydrocarbons. IUPAC names are based on the nature and location of these functional groups.

- Some common functional groups are halides, hydroxyl groups, carbonyl groups, carboxyl groups, amine groups, and amide groups.

- Some of the types of organic compounds that contain functional groups are halocarbons, alcohols, ethers, aldehydes, ketones, carboxylic acids, esters, amines, and amides. Each of these is identified by the presence of one or more specific functional groups.

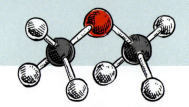

TERMS
substitution reaction
addition reaction
elimination reaction
condensation reaction

20.4 ORGANIC REACTIONS

- Many reactions between organic compounds are redox reactions. Combustion is a common type of hydrocarbon redox reaction.

- Both substitution and addition reactions add atoms to a molecule. In substitution reactions, part of the original molecule is replaced. In addition reactions, a multiple bond is broken and new atoms or groups of atoms are added.

- Elimination reactions remove atoms or groups of atoms from a molecule.

- Condensation reactions join two molecules or two parts of the same molecule, often releasing a water molecule.

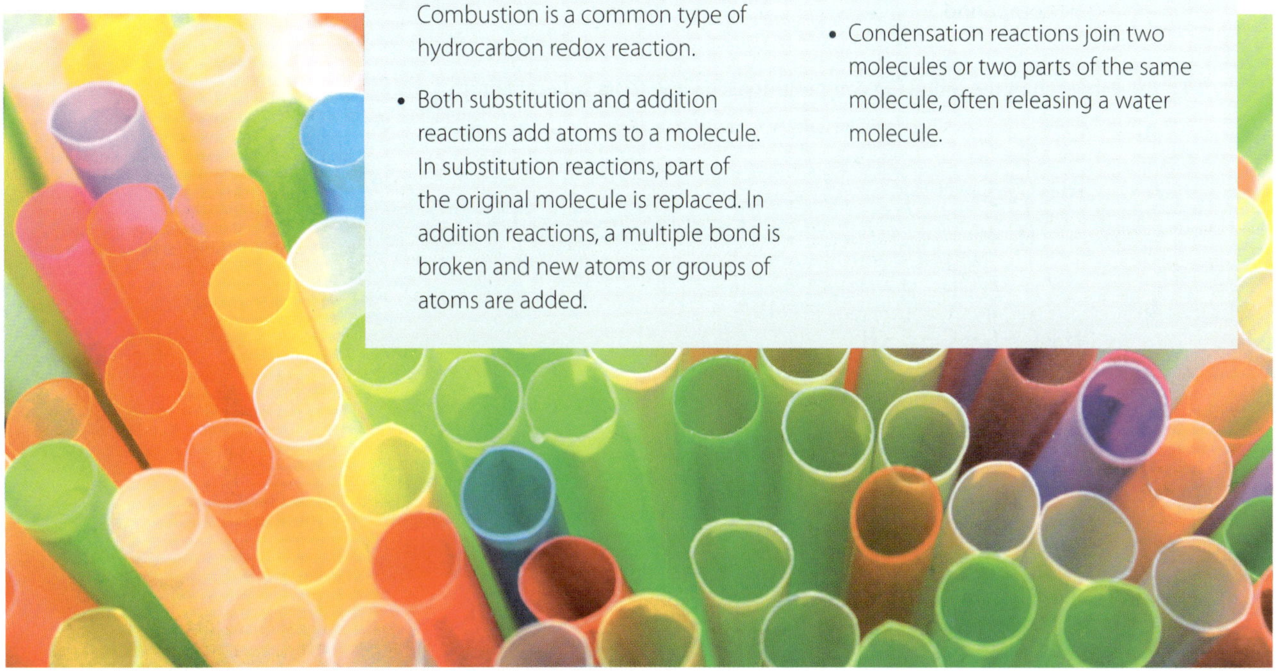

Chapter Review Questions

RECALLING FACTS

1. What is the modern definition of *organic chemistry*?
2. How can you tell when two compounds are isomers? Why is this important?
3. What is unique about the carbon-carbon bonds in benzene and other aromatic compounds?
4. Alcohols and metallic hydroxides both have OH groups in their structural formulas. Explain why metallic hydroxides such as NaOH are caustic but alcohols are not.
5. What is similar about the functions of HCl and CH_3COOH? How are they different?
6. In most substitution reactions of organic compounds, what atom is usually replaced by another atom or group of atoms?

UNDERSTANDING CONCEPTS

7. Compare aliphatic and aromatic compounds.
8. Methane (CH_4) is a gas at room temperature, whereas methanol (CH_3OH) is a liquid. Aside from the difference in molecular masses, suggest an explanation for their different boiling points.
9. Methanol is soluble in water, but larger alcohols such as octanol are not. Why is this?

20 CHAPTER REVIEW

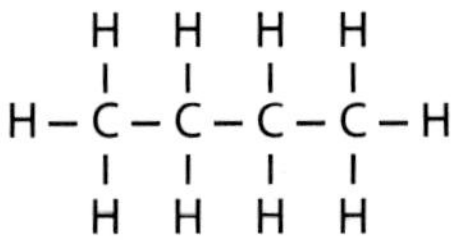

10. Modify the structural formula of butane (left) to create the compounds listed below. Remember that carbon atoms always have four bonds. Add or delete hydrogens when necessary.

   a. but-1-ene
   b. but-1-yne
   c. 2-iodobutane
   d. butan-1-ol
   e. butan-2-one
   f. butoxybutane
   g. butanal
   h. butanoic acid
   i. methyl butanoate

11. Classify each of the following compounds according to its general family.

   a. through l. (structural formulas shown)

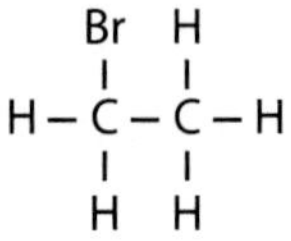

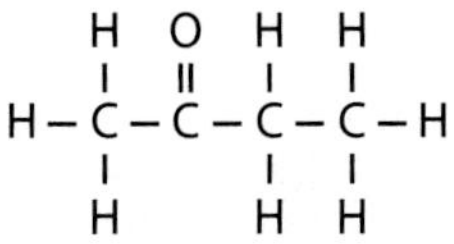

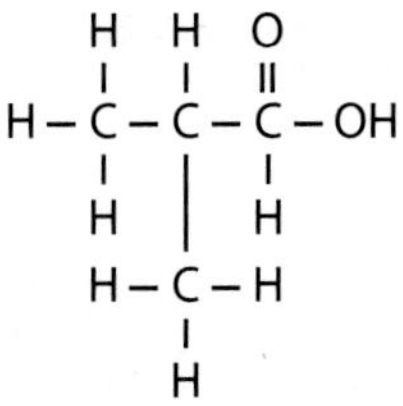

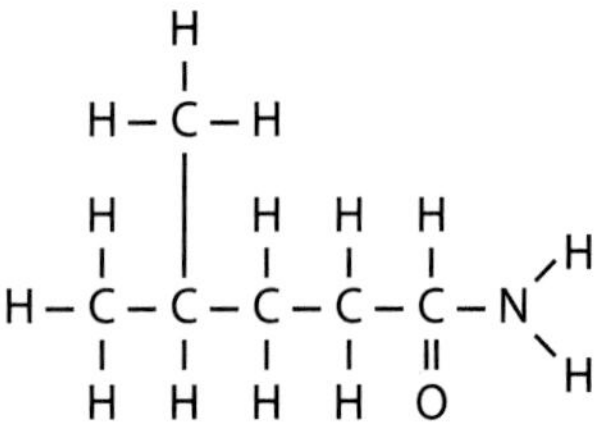

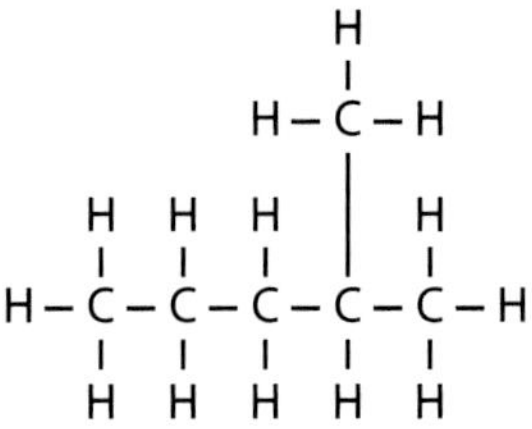

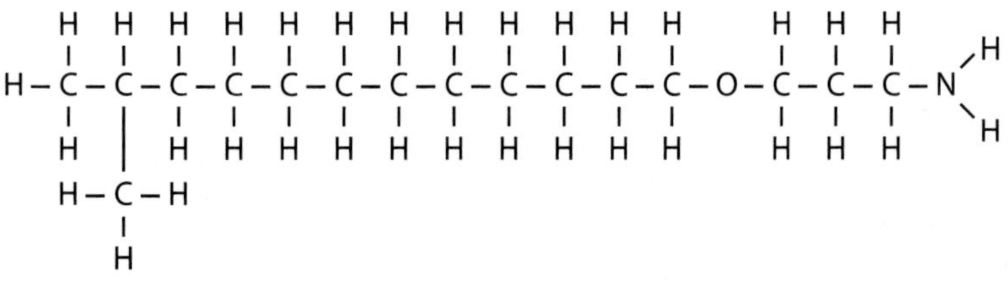

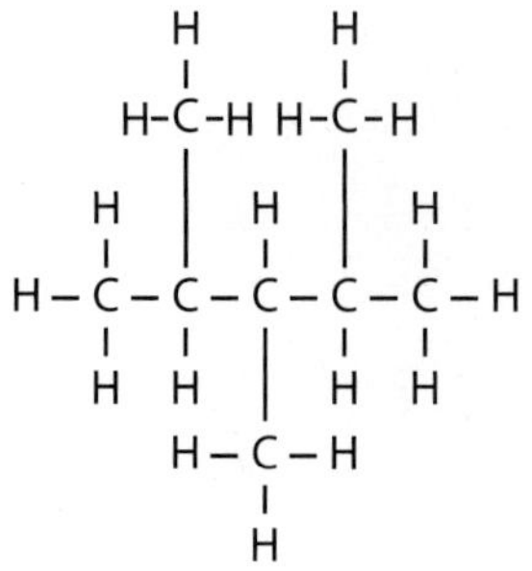

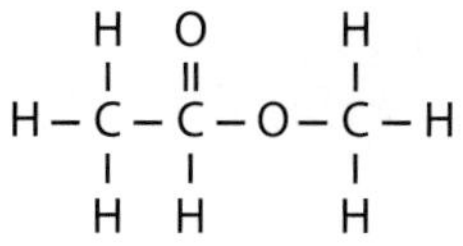

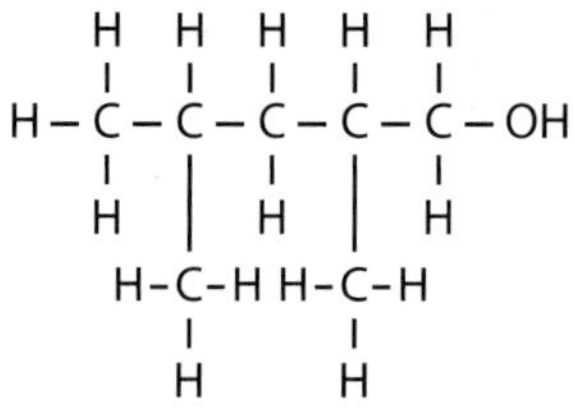

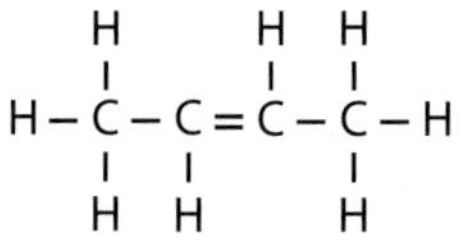

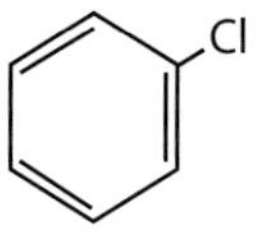

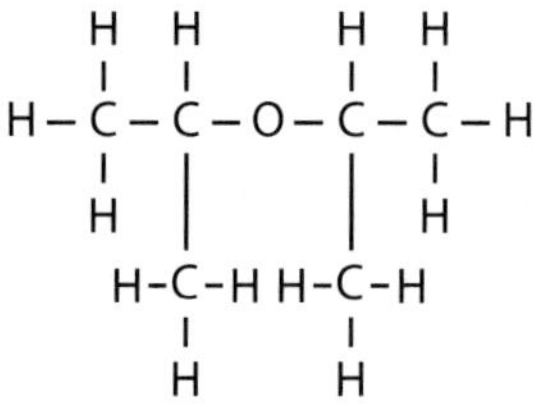

12. Draw the structural formula for each of the following compounds. Assume that the carbon chain is straight in each case.

    a. hexane
    b. hept-1-ene
    c. oct-2-yne
    d. pentan-1-ol
    e. butan-2-ol
    f. 1-chloropropane
    g. ethyl butanoate
    h. hexanoic acid
    i. octan-3-one
    j. 2,2-dimethylbutane

13. Name each of the following compounds.

    a.
    ```
        H H
        | |
    H − C − C − H
        | |
        H H
    ```

    b.
    ```
        H    H H
        |    | |
    H − C = C − C − C − H
             | | |
             H H H
    ```

    c.
    ```
        H
        |
    H − C − OH
        |
        H
    ```

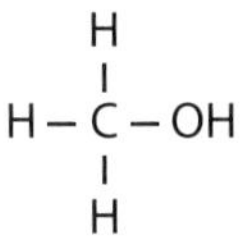

    d.
    ```
        H H H
        | | |
    H − C − C − C − F
        | | |
        H H H
    ```

    e.
    ```
        H OH H H
        | |  | |
    H − C − C − C − C − H
        | |  | |
        H H  H H
    ```

    f.
    ```
        H O    H H
        | ||   | |
    H − C − C − O − C − C − H
        |         | |
        H         H H
    ```

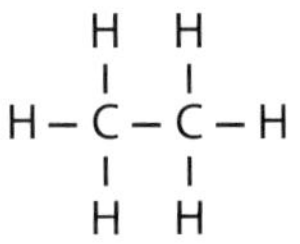

    g.
    ```
        H H H H O
        | | | | ||
    H − C − C − C − C − C − OH
        | | | |
        H H H H
    ```

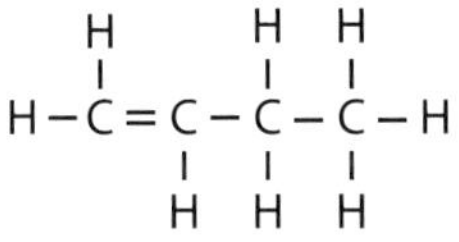

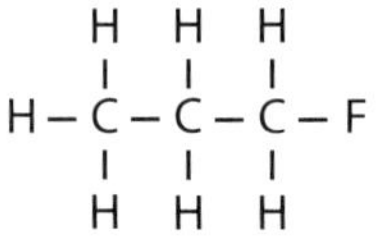

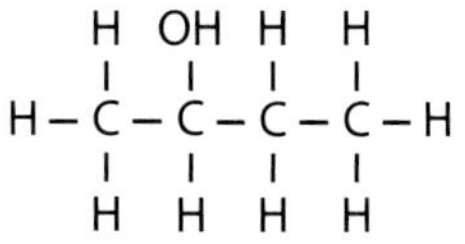

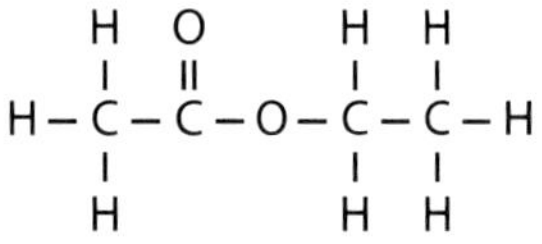

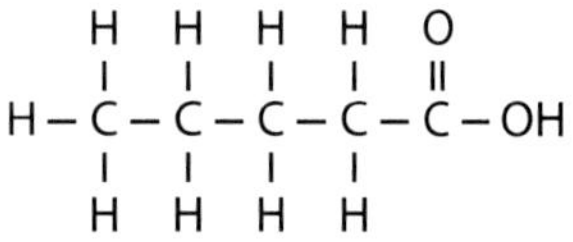

    h.
    ```
                      H
                      |
                  H − C − H
    H H H H H H       | H
    | | | | | |       | |
    H−C−C−C−C−C−C−C−C−H
    | | | | | | | |
    H H H H H | H H
                  H
                  |
              H − C − C − H
                  | |
                  H H
    ```

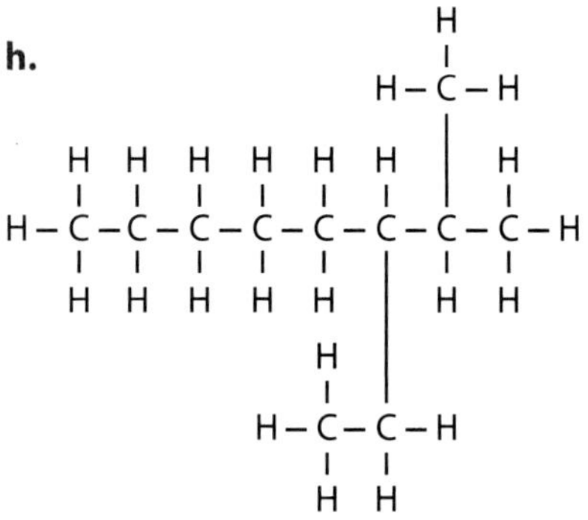

    i.
    ```
        H         H
        |         |
    H − C − C ≡ C − C − H
        |         |
        H         H
    ```

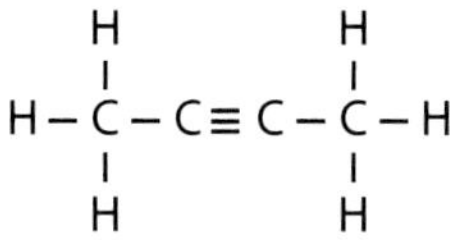

    j.
    ```
        H H H   H H H
        | | |   | | |
    H − C − C − C − O − C − C − C − H
        | | |   | | |
        H H H   H H H
    ```

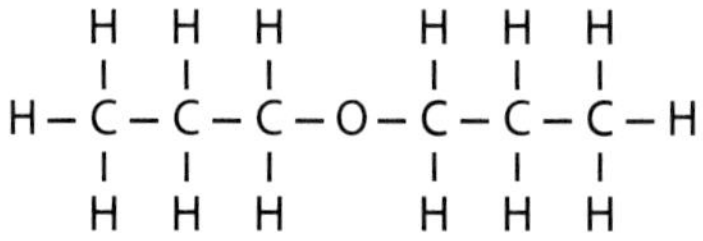

    k.
    ```
        H H H O H H
        | | | || | |
    H − C − C − C − C − C − C − H
        | | | | | |
        H H H H H H
    ```

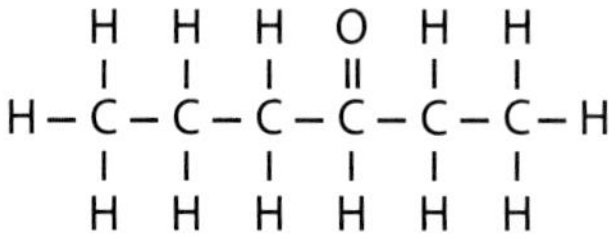

14. Predict the specific products of each of the following reactions. In some cases, more than one product may be possible.

    a. a condensation reaction between butanoic acid and ethanol
    b. the combustion of octane
    c. a reaction between propyne and hydrogen
    d. a reaction between methane and bromine

CRITICAL THINKING

15. On a piece of paper, draw an electron-dot structure for propyne. Would you expect this molecule to be polar or nonpolar? Why?

16. What is the name of the compound shown at right?

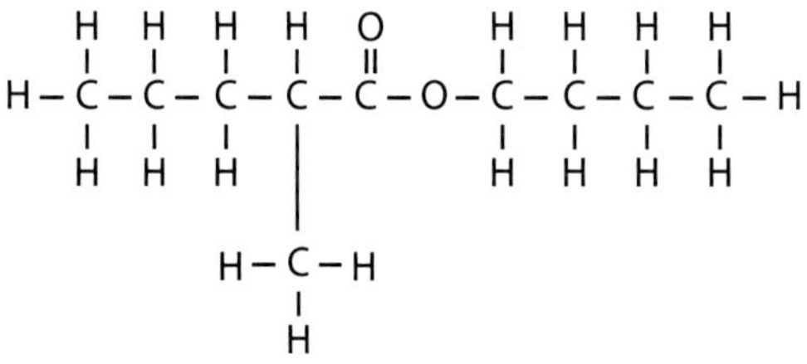

17. Write the structural formula for each of the following compounds.

    a. 3-ethyl-2,2-dimethyl-4-phenyloctane
    b. 3-bromo-2-fluoro-1-methylbenzene

```
    H H H H O     H H H H
    | | | | ||    | | | |
H − C − C − C − C − C − O − C − C − C − C − H
    | | | |                | | | |
    H H H |                H H H H
          H − C − H
              |
              H
```

Organic Chemistry 513

Chapter 21

BIOCHEMISTRY

How did life get here? That is a question that people all over the world debate. Charles Darwin said that life started spontaneously in a "warm little pond" of chemicals on the surface of a developing Earth, and most evolutionists agree. They call this process *abiogenesis*. And if it happened here on Earth, what's to keep it from happening elsewhere in the universe?

But evolutionists needed evidence. In 1952, Harold Urey and his assistant Stanley Miller made a major breakthrough. They discovered what was called "the first step to life in a test tube."

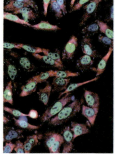

They simulated the conditions that they believed existed in the atmosphere of an early Earth and provided sparks of energy until they produced amino acids, one of the building blocks of life. Scientists all over the world proclaimed a victory for abiogenesis. This experiment has prompted many more in abiogenesis. So does all this prove that life can arise without God?

21.1 Chemistry of Life *515*
21.2 Carbohydrates *517*
21.3 Lipids *522*
21.4 Proteins *525*
21.5 Nucleic Acids *528*

21.1 CHEMISTRY OF LIFE

In Chapter 20 you learned about organic chemistry, which at one time was thought to be the chemistry of living things. Today, though, this narrower focus on living things within the broader context of organic chemistry is known as **biochemistry**. The chemistry of genetics and cellular processes falls within this field of study. Much of this field deals with organic compounds, but living things depend on many inorganic compounds as well. The overlaps between the various fields of chemistry and biology illustrate the fact that the marvelous complexities of life cannot be studied within the confines of a single subject area.

Chemical Reactions in Cells

It's tempting to think of cells as "simple"—they are, after all, the basic building blocks from which living things are made. But individual cells are only *relatively* simple when compared with the extraordinary complexity of multicellular organisms. In truth, every living cell is, by itself, incredibly complex. Each one depends on *thousands* of different chemical processes to live, grow, and reproduce. We use the term **metabolism** as a collective noun for all the chemical reactions that take place within cells. We can break down this very large group of chemical processes into *catabolism* and *anabolism*.

QUESTIONS
» What is biochemistry?
» What chemical reactions occur within cells?

TERMS
biochemistry • metabolism • catabolism • anabolism • polymer

Is biochemistry another name for organic chemistry?

Catabolism & Anabolism

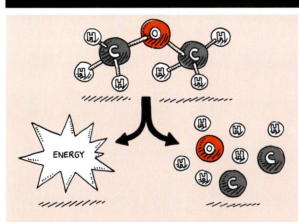

Catabolism

Catabolism includes all the cell's chemical reactions that break molecules into smaller particles. Catabolism provides cells with two very important raw materials—the starting substances needed to build macromolecules and the energy needed for their manufacture.

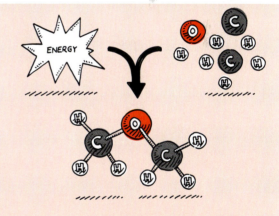

Anabolism

Anabolism can be thought of as the reverse of catabolism—it includes those processes by which cells assemble small particles into larger ones—often *much* larger. Some of the compounds produced by anabolic processes are organic **polymers**, very large molecules built from repeating subunits called *monomers*. The bonding of monomers together is called *polymerization*. Starch, protein, and DNA are examples of organic polymers.

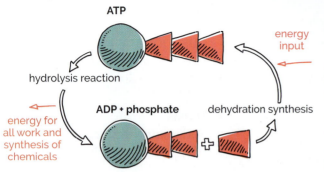

Catabolism and anabolism often go hand in hand. Many cellular processes are cyclical—the products of a catabolic process are used during an anabolic process, whose products are again used by the catabolic process. A good example of this should be familiar to you from biology. Adenosine triphosphate (ATP), the molecule that cells use for energy, is made by bonding a phosphate group to a molecule of adenosine diphosphate (ADP) through an anabolic process. The splitting off of a phosphate group from ATP to release energy is a catabolic process. The product of this process, ADP, is recycled to make ATP once again. Other familiar examples include photosynthesis (anabolic) and cellular respiration (catabolic).

Biochemistry and Ultimate Questions

Because biochemistry deals with life, it particularly leads to philosophical and spiritual issues. As technology develops, scientists encounter perplexing ethical predicaments. Should humans be cloned? Is it right to use stem cells? Is it ever right to harvest human embryos for their stem cells? When does a legal drug become illegal, and vice versa? The Bible, not science, contains the answers to these questions. For this reason, Christians must engage in the field of biochemistry.

Ultimately, the study of biochemistry reveals further levels of complexity in creation, and this shouts out the infinite wisdom of the Creator. Every molecule bears witness to the power, ingenuity, and foresight of God. This flies in the face of what most scientists believe—that life got here through abiogenesis. We'll embark on our study of biochemistry by studying some of the basic molecules of life—carbohydrates, lipids, proteins, and nucleic acids.

21.1 SECTION REVIEW

1. Define *biochemistry* and compare it with organic chemistry.
2. How are catabolism, anabolism, and metabolism related?
3. What is a polymer?
4. Why can catabolism and anabolism not be thought of as entirely separate processes?
5. Considering cellular processes as an example, why do you think that it is misleading to think of cells as "simple" structures?

21.2 CARBOHYDRATES

Low-carb diets are popular these days. Considering how many people that you might know who are trying to cut grains and cereals from their diets, you might think that carbohydrates are bad. As usual, context is the key. At the cellular level, living things absolutely depend on carbohydrates for energy and often for structure as well. Simple sugars make up part of the human diet, as do starches, which consist of long chains of one simple sugar—glucose—bonded together. A slight change to the manner in which individual glucose molecules are bonded together produces cellulose, the stuff of which plant cell walls are made. The exoskeletons of insects and crustaceans are made of chitin, a derivative of glucose.

Are carbohydrates good or bad?

The name *carbohydrate* itself tells us something about the compounds in the group. The word means "water of carbon," since at one time all carbohydrates were thought to be hydrates of carbon, containing only carbon, hydrogen, and oxygen in a ratio of 1:2:1. In biochemistry, the term **carbohydrate** is used to collectively refer to sugars, starches, and cellulose. These organic compounds are also known as *saccharides*, from the Latin word for sugar.

Carbohydrate Structure

When you hear the word *sugar*, you probably think right away of granulated table sugar. But there are actually many different kinds of sugars, some of which can be easily spotted on food labels by their ending in *-ose*. You probably already know something about glucose. Granulated table sugar is sucrose. Other sugars that you may have come across are lactose, fructose, and maltose. How do the structures of these sugars compare? Sugars differ from one another on the basis of how many carbon atoms they contain, how their oxygen atoms are bonded within their structure, and whether they exist as open chains or rings. The simplest sugars are **monosaccharides**, and as their name implies, they consist of a single sugar unit. Let's look at D-altrose to get a better view of carbohydrate structure.

QUESTIONS
» What is a carbohydrate?
» How are carbohydrates classified?
» How are carbohydrates used by living things?

TERMS
carbohydrate • monosaccharide • disaccharide • polysaccharide

Carbohydrates

D-altrose and other carbohydrates are built on a chain of single-bonded carbon atoms. Each internal carbon atom and one end carbon of the chain is bonded to a hydrogen atom and one substituent. The other end carbon is bonded to two hydrogens and a substituent.

The substituent on one end of D-altrose is a double-bonded oxygen atom. Recall from Chapter 20 that this is an identifying feature of aldehydes. This and the presence of multiple hydroxyl groups makes glucose an example of a *polyhydroxy aldehyde*, also referred to as an *aldose*.

D-altrose has six carbon atoms, but not all sugars do. Because it is a six-carbon sugar, D-altrose can be classified as a *hexose*. Because it is also an aldose, these terms are sometimes combined to form *aldohexose*.

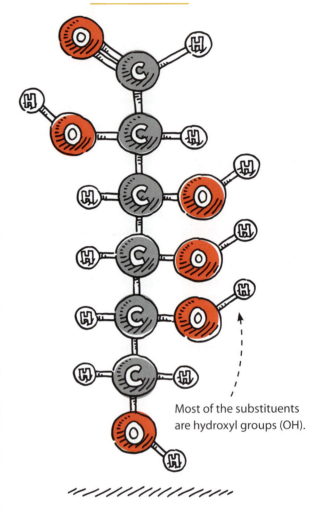

D-ALTROSE

Most of the substituents are hydroxyl groups (OH).

If you did the mini lab in Chapter 20, you may recall that many organic compounds exhibit *chirality*—having left- and right-handed forms called *enantiomers*. D-altrose, as well as many other sugars, exhibit this type of isomerism. The *D-* and *L-* in the names of the enantiomers stand for *dexter* and *laevus*—Latin for "right" and "left."

α-D-altropyranose

β-D-altropyranose

If one of the bonds on D-altrose's double-bonded oxygen atom is broken, the chain can fold back on itself, and the oxygen can bond to the fifth carbon atom of the chain, forming D-altropyranose. The word *pyranose* refers to the six-member ring comprised of five carbons and one oxygen. Because this ring has several possible configurations, two of which can be seen in the *chair conformations* at left, pyranoses have many isomers.

Monosaccharides

Carbohydrates can be classified into three groups on the basis of the number of sugar units that they contain. As we have seen, the smallest possible sugars consist of a single unit. These are the monosaccharides. When purified and dried, monosaccharides—and there are many—are white, water-soluble, crystalline solids. Most have a sweet taste. They rarely occur in nature as free molecules but are usually bonded to a protein, a fat, or another carbohydrate. Two significant exceptions are glucose and fructose.

Glucose is the most abundant sugar in nature. Ripe berries, grapes, and oranges are 20% to 30% glucose in composition. The human body maintains a fairly constant level of 80 to 120 mg of glucose per 100 mL of blood. Hence, it is often called *blood sugar*. Glucose is also the fundamental building block of the most common long-chain carbohydrates: starch and cellulose.

Although monosaccharides are often drawn as open-chain molecules, they most often exist in their ring form. Glucose in an aqueous solution exists in an equilibrium between the ring and open-chain forms. The equilibrium lies strongly in favor of the ring form. Fructose also forms ring structures when it is in solution. In its open-chain form, fructose's carbonyl group is not at the end of the carbon chain. It is thus based on a ketone and is classified as a *polyhydroxy ketone* or *ketose*. Its ring form has only five atoms in its ring, compared with the six in glucose.

fructose

Disaccharides

As their name implies, **disaccharides** consist of two monosaccharide units. Typically, two monosaccharides can join in a condensation reaction to release a water molecule and form a disaccharide. An oxygen bridge between the two monosaccharides holds the two units together. While many monosaccharides exist, an even larger number of disaccharides can be formed from combinations of monosaccharides.

Three disaccharides play an important part in the human diet—maltose, lactose, and sucrose. Maltose consists of two bonded glucose molecules. The bond forms between the first carbon of one molecule and the fourth carbon of the other. Maltose is found in germinating grain and is produced during the digestion of starches. Lactose is a disaccharide found only in milk. It consists of an isomer of glucose (galactose) joined to a glucose molecule. This sugar is not very sweet, but it is an important ingredient in milk. Milk sours when lactose breaks down into lactic acid. A glucose and a fructose molecule can bond in a condensation reaction, releasing a water molecule and forming the disaccharide sucrose (table sugar). Sucrose is abundant in fruits, sugar cane, sugar beets, and nectar.

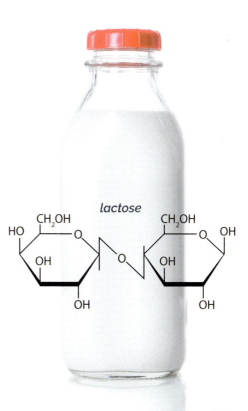
lactose

HOW IT WORKS

New Bone from Sugar?

If you've ever taken a wilderness first aid course, you may have learned how to make a splint for a broken bone by using a tree branch. But medical researchers are taking this idea one step further—they are helping bones heal with help from a surprising source. Researchers in Canada have recently had much success in quickening new bone growth by injecting test subjects with an aerogel derived from plant cellulose. The nanocrystals of the cellulose form a scaffolded structure in which osteoblasts can deposit new bone. And unlike commonly used ceramic implant materials, which are hard and brittle, the new aerogel is soft enough to flex and completely fill all the gaps in a host bone. Those gaps hinder the growth of new bone when using ceramic implants; the new aerogel has demonstrated 50% more bone growth after twelve weeks compared with ceramic. Our God-given ability to think of creative ways to use the resources around us is part of exercising godly dominion over His world.

D-glucose

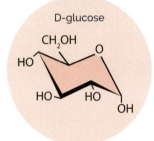

amylose

amylopectin

cellulose

Polysaccharides

Polysaccharides, literally "many sugars," are molecules that contain many sugar units. This is evident from their large molecular masses (up to several million atomic mass units). Multitudes of sugar units are bonded into long chains. Polysaccharides may be built from several different monosaccharides, but we will concentrate on the polymers of glucose.

Plants such as rice, potatoes, wheat, and oats store food in polysaccharide deposits called starch. *Starch* exists as one of two glucose polymers. Most of the mass of starch found in plants exists as the branched isomer called *amylopectin*. But most of the actual starch molecules are the linear form known as *amylose*. Because the amylose molecules are very much smaller than amylopectin molecules, the more numerous amylose molecules make up only about a quarter of the mass of starch. Together, these two polysaccharides supply nearly three-fourths of the world's food energy.

Humans and animals also store glucose as a polysaccharide known as *glycogen*. Like amylopectin, glycogen is branched, but to an even greater degree. In humans, most glycogen is stored in the liver and skeletal muscles. It serves as a readily accessible energy reserve since it can be quickly broken down and converted into the molecules needed for cellular respiration.

Humans can't digest much of the plant matter that herbivorous animals can because plants contain lots of *cellulose*, a polysaccharide that is similar to starch and glycogen. In the polysaccharides that humans can digest, such as starch, each glucose unit is oriented in the same direction. But the glucose units in cellulose alternate directions. Thus, the difference between the starch in a tasty potato and the cellulose in grass, cotton, or a splintery piece of wood is the manner in which the glucose units are arranged. This may seem like a minor difference at first, but it isn't. Humans can digest the uniformly oriented starch, but they lack the enzyme needed to digest cellulose. Cows and other herbivores can't actually digest cellulose either—but symbiotic bacteria that live in their guts can. This mutualistic relationship allows both bacteria and herbivores to thrive.

MINILAB

SIMPLE SUGARS?

You've seen that the ring forms of simple sugars have many isomers, but what about the open-chain forms? Can they also exist as many isomers?

Are simple sugars, you know, like, really simple?

Procedure

A Use your molecular modeling set to build a model of D-glucose as shown at right. When finished, you should be able to view your model from above in a manner that places the hydroxyl (–OH) groups on either the left or right side of the carbon backbone. Keep the double-bonded oxygen of the carboxyl group to the right to help you maintain your orientation when viewing.

B Swap the positions of the hydroxyl group and hydrogen atom on any internal carbon. When viewed from above, that hydroxyl group should now appear to be on the opposite side of the carbon backbone from its original position. Make sure that all the other substituents on the model remain in their original locations.

1. Can the affected carbon atom be rotated so that the hydroxyl group and hydrogen atom are returned to their original positions?

C The affected carbon atom in Step B is an example of a *stereocenter*. Isomers created around a stereocenter are called *stereoisomers*. The three remaining internal carbons are also stereocenters. Repeat Step B for each of those carbons. The resulting model is L-glucose.

D Try to make as many different aldohexoses as you can by continuing to swap substituents on the internal carbons, either individually or in various combinations.

2. Write the structural formula for each aldohexose you come up with for Step D.

3. How many different aldohexoses are possible in addition to D-glucose?

Conclusion

4. So, are open-chain simple sugars simple? Defend your answer.

Going Further

5. Search the internet to find the names of the stereoisomers that you created for Question 2. Label each of your structural formulas with the appropriate name.

EQUIPMENT

- molecular modeling set

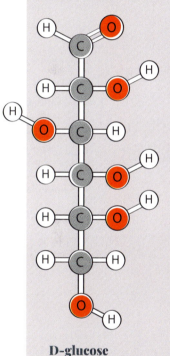

D-glucose

21.2 SECTION REVIEW

1. What are carbohydrates?
2. How does the structure of one kind of monosaccharide differ from another?
3. What can we know about a compound when told that it is an aldohexose?
4. What products result from a condensation reaction between glucose and fructose?
5. How are monosaccharides, disaccharides, and polysaccharides related?
6. Describe at least two roles of carbohydrates in living things.

Biochemistry

21.3 LIPIDS

QUESTIONS
» What are lipids?
» How are saturated fats different from unsaturated fats?
» What functions do lipids serve in living things?

TERMS
lipid • fat • oil • steroid • phospholipid

Why can't I live without fats?

Although water is sometimes called the universal solvent, it does not mix with all the compounds in the human body. An entire class of biomolecules (those molecules necessary for cellular processes) called lipids cannot dissolve in water. They are characterized by a high proportion of C–H bonds, making them nonpolar. As energy storage molecules and as an integral part of cell membranes, **lipids** serve vital roles in living things. Fatty acids, fats, oils, waxes, and steroids are examples of lipids.

Fats and Oils

Propane-1,2,3-triol, commonly known as *glycerol*, is a three-carbon molecule that has three hydroxyl groups on separate carbons. Fats and oils are esters of glycerol and fatty acids, which in turn are carboxylic acids with long hydrocarbon tails. Ester linkages join the carboxyl (COOH) groups of fatty acids to the hydroxyl groups of glycerol to construct a fat or oil molecule and water. Fatty acids rarely occur alone in living things. Typically, three of them bond with a glycerol molecule to form a *triglyceride*.

The fatty acids that bond to glycerol usually contain a large hydrocarbon chain (chains of twelve to eighteen carbons are common) that most often contains an even number of carbons (the chains are made in two-carbon units). Some chains contain only single carbon-carbon bonds, and some contain a few double bonds. This difference leads to the distinction between fats and oils.

Fats are triglycerides that often exist as solids at room temperature. Most of their fatty acid chains are saturated. Saturated chains tend to extend in straight lines from the glycerol backbone. This particular structure allows saturated fat molecules to pack closely together. Moreover, saturated fats have stronger intermolecular forces, so their melting points are higher.

▲ Three fatty acids and a glycerol molecule can combine to form a triglyceride.

Triglycerides that have one double bond somewhere within their fatty acid chains are said to be *monounsaturated*. Those with multiple double bonds are classified as *polyunsaturated*. The greater a triglyceride's degree of unsaturation, the more bent its chains become. The molecules can't pack as tightly, reducing their intermolecular forces, which lowers their melting point. Unsaturated fats that remain liquid at room temperature are referred to as **oils**.

Dietary fat is a vital nutrient for a healthful lifestyle, especially as a source of energy for the body. It is the most concentrated source of energy in one's diet, providing nine calories per gram compared with only four calories per gram from either carbohydrates or protein. Fats also carry vitamins, cushion the body, insulate nerves, act as signaling molecules, regulate immune response, and can act as hormones or serve as precursors for the production of hormones.

Steroids

Steroids are lipids, but their structures do not resemble triglycerides in any way. They are the non-ester lipids. The basic structure of a steroid (shown at right) is a combination of three six-carbon rings and one five-carbon ring. Functional groups attached to various points on the rings result in a wide variety of steroids. Cholesterol, vitamin D, cortisone, testosterone, and estrogen all use this common "chicken wire" frame that you see at right. In living things, steroids serve as important components of cell membranes and as chemical-signaling molecules.

The study of steroids ranks as one of the most active areas of biochemical research. One steroid that has been studied much in recent years is *cholesterol*—a compound that, although vital to good health, some believe can increase the risk of heart disease if excessive amounts are present in the blood. Routine blood panels, such as you might have done for an annual physical, will test for the concentration of cholesterol. Contrary to popular belief that cholesterol levels are controlled by diet, most of the cholesterol found in the body is manufactured by the body itself. The average total blood cholesterol in Americans is 203 milligrams per deciliter (mg/dL). The National Institutes for Health considers less than 200 mg/dL desirable and greater than 240 mg/dL to be too high.

CAREERS

SERVING AS A BIOMEDICAL RESEARCHER: MAKING A MIRACLE MOLECULE

In 1929 Dr. Philip Hench of the Mayo Clinic discovered that a chemical called *cortisone*, a type of steroid, had an amazing ability to relieve the pain of rheumatoid arthritis. It was a miracle molecule. But producing cortisone was extraordinarily difficult. The only known way to make it was from cattle bile using a chemical process that required thirty-six separate steps. The bile from a whopping 14,600 cattle was needed to produce enough cortisone to treat a single patient for one year! Not surprisingly, cortisone was also expensive. The cost for a single dose in 1949 was $200—equivalent to $2162 in 2019.

A solution to producing large quantities of cortisone inexpensively was discovered by Dr. Percy Julian, the grandson of slaves and a chemist working for a chemical company. Julian had experience in synthesizing steroids from chemicals found in plants. In 1949 Julian used his knowledge of steroids to improve the process for making cortisone. His process did not require the use of osmium tetroxide, the rare and expensive compound that made the earlier production method so costly. Other chemists later discovered that a mold could produce cortisone from progesterone, a hormone readily synthesized from a chemical found in a plant, the Mexican yam. Julian eventually decided to work full time for his own company that made progesterone for other businesses to produce cortisone.

Born and raised in Alabama, Dr. Julian lived and worked in an era of institutionalized racism, but he didn't allow the problems he faced to keep him from helping people. Through his work, he was able to save and improve the lives of millions of people around the world. He never lost sight of the potential of chemistry to help people.

Dr. Percy Julian

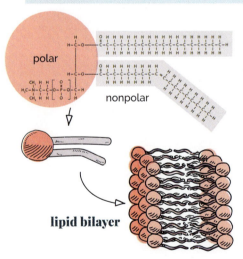

Phospholipids

Phospholipids are structurally similar to triglycerides. But instead of a glycerol head, the head of a phospholipid is a phosphate group. Instead of three fatty acid tails, it has only two. Most importantly though, unlike triglycerides, phospholipids are not completely nonpolar. The fatty acid tails are nonpolar, but the phosphate head is polar. This makes phospholipids *amphiphilic*, meaning that one end (the phosphate head) is attracted to water, but the other (the fatty acid tails) is not. This peculiar physical characteristic is what allows phospholipids to form lipid bilayer cell membranes (see left). The hydrophilic ("water loving") polar heads face outward from both sides of the bilayer, while the hydrophobic ("water fearing") tails face in toward the interior of the bilayer. As you can see, even something as "simple" as the boundary that separates a cell's insides from its environment is incredibly complex.

21.3 SECTION REVIEW

1. What are lipids?
2. Explain the difference between saturated and unsaturated fats.
3. How does the difference in saturation affect physical properties?
4. What is the basic structure of a steroid?
5. Describe at least one function for each of the three kinds of lipids described in this section.
6. Define *amphiphilic*.

21.4 PROTEINS

Protein powders, protein bars, protein shakes—these are products that you can easily find in the health food section of your local grocery. But what exactly is protein, and why is it important? Proteins are versatile molecules in all living organisms. Various proteins form the basis for muscle, hair, blood cells, skin, spider webs, silk, enzymes, insulin, and even snake venom. They are essential nutrients, and they have many life-sustaining functions. Each protein polymer exhibits a unique and intricate three-dimensional architecture custom-made for its function.

What do proteins do for me?

QUESTIONS
» What is a protein?
» What is tertiary structure?
» What is quaternary structure?
» Why is the shape of a protein important?

TERMS
protein • amino acid • peptide bond

Amino Acids

Proteins are organic polymers made from hundreds, even thousands, of amino acids. Each **amino acid** molecule consists of a carbon atom bonded to an amine group (NH_2), a carboxyl group (COOH), a hydrogen atom, and a side chain, or *R-group*. It is the side chain that determines the identity of each amino acid. Though there are about five hundred naturally occurring amino acids, fewer than 5% have so far been found to make proteins.

The human body can synthesize twelve of the twenty amino acids that it needs. The other eight must come from a proper diet. The amino acids that cannot be manufactured by the body are called *essential amino acids*.

amino acid

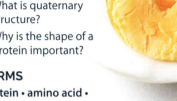

Polypeptide Chains

Amino acids join together when a hydrogen in the amine group of one amino acid reacts with the carboxyl group in another. It is a dehydration reaction and releases water. The bond that links the two amino acids is called a **peptide bond** (see below). Two amino acids bond to form a *dipeptide*. Aspartame (aspartyl-phenylalanine), a popular artificial sweetener, is a dipeptide. *Polypeptides* occur when three or more amino acids bond together in an unbranched chain.

A finished protein molecule is much more than simply a chain of amino acids. In addition to this basic level of complexity, polypeptide chains can be folded, twisted, and combined with other chains to produce a fantastic variety of forms. Let's examine how this variety is produced.

Biochemistry

PROTEIN STRUCTURE

Primary Structure

Any particular protein may consist of one or more polypeptide chains. The basic order of the amino acids in a polypeptide chain is called its *primary structure*.

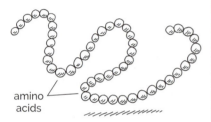

amino acids

Even with a comparatively short chain, the number of possible different proteins is staggering. Using only the twenty amino acids typically found in living things, 1.1×10^{65} unique sequences of fifty amino acids can be formed.

Secondary Structure

The *secondary structure* of a protein is determined by the shape and polarity of its amino acids. Amino acids in one part of the chain can bond with other amino acids in another part. The most common secondary structures are *turns* (the chain bends back on itself), *alpha helixes* (α-helix), and *beta-sheets* (β-sheet).

An α-helix is a tight spiral of amino acids held in place by hydrogen bonds. A hydrogen in one amino acid forms a hydrogen bond with an oxygen of an amino acid four positions away. The side chains of each amino acid extend like spokes from the surface of the helix.

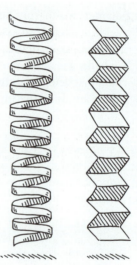

Another form of secondary structure, the β-sheet, results when hydrogen bonding occurs between adjacent polypeptide strands. These sheets have a slightly folded appearance, so they are sometimes referred to as β-pleated sheets.

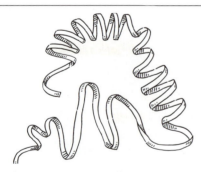

Tertiary Structure

An individual polypeptide may consist of many turns, helixes, and sheets, as well as regions with no specific structure, known as *random coils*, giving it a complex three-dimensional shape. This is known as its *tertiary structure*. At this point, the protein may be complete, or it may only be part of a larger structure, in which case it is called *a protein subunit*.

Quaternary Structure

Many proteins consist of two or more protein subunits fitted together to form a *protein complex*. Such a complex exhibits *quaternary structure*.

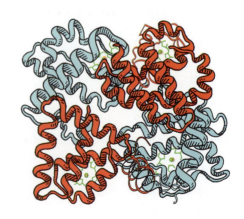

It is amazing that these intricate molecules derive most of their form from ordinary, but carefully arranged, hydrogen bonds. The structure truly is a masterpiece of art and engineering. Even more amazing, each protein's distinctive shape equips it to perform a unique function. For example, hemoglobin, the protein that carries oxygen in the bloodstream, consists of four interwoven chains. The shape of the protein provides perfectly formed crevices for oxygen molecules to nestle into.

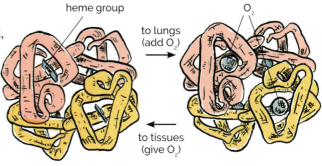

hemoglobin   **hemoglobin with oxygen added**

Enzymes

We've already mentioned the importance of proteins as structural molecules. Another critical role of proteins in living things is their function as organic catalysts called *enzymes*. Most cellular processes involve chemical reactions that would happen far too slowly without the assistance of enzymes. The exact shape of each enzyme normally allows it to catalyze only one specific reaction, so every cell requires an enormous number of enzymes just to maintain everyday function. Scientists know of thousands of enzyme-catalyzed reactions in cells. Without the assistance of these intricately designed chemical machines, life could not exist.

Denaturing

Considering their complexity and the relative weakness of hydrogen bonds compared with either covalent or ionic bonds, you might not be surprised to learn that proteins are not especially durable. Like radioactive elements, the duration of their existence is measured in half-lives, which can range anywhere from a few minutes to many years but averages only a few days. In living things, old, worn-out proteins are constantly being broken down and replaced in a process called *protein turnover*. A decrease in the rate of protein turnover is thought to be a contributing factor of the aging process.

Proteins can also "unwind" in the face of stresses, such as high heat, radiation, extreme salt concentrations, or the presence of particular chemicals. These stress agents cause proteins to lose their proper structure and thus the function that goes with it. Such a loss of protein structure is called *denaturing*. You can see the effect of denaturing when you grill a steak. As the meat cooks, its proteins unravel and its color changes from bright red to brown. As the steak cooks, its denaturing proteins absorb and reflect different wavelengths of visible light compared with the raw meat. Denaturing affects living things too. Stress agents can cause cellular proteins to work inefficiently or not at all. Left untreated, such conditions can lead to the death of an organism.

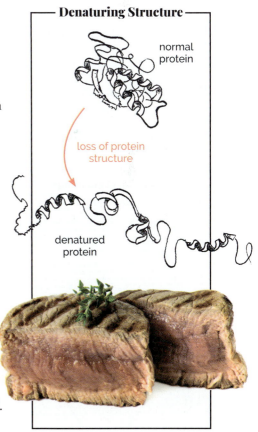

Denaturing Structure

21.4 SECTION REVIEW

1. Define *protein*.
2. How are amino acids bonded together? What is this bond called?
3. In order from least to most complex, describe the types of structure found in proteins.
4. Why is the shape of a protein important?

Biochemistry

21.5 NUCLEIC ACIDS

QUESTIONS
» What are nucleotides?
» How do nucleic acids store information?
» How does a cell's nucleus transmit information to other parts of the cell?

TERMS
nucleic acid • nucleotide

How do cells know how to do things?

When cells reproduce, genetic information is passed from the parent cell to the daughter cells on *chromosomes*. Human body cells contain forty-six of these long structures in their nuclei. A segment of a chromosome that carries the information that directs the production of specific polypeptide chain is called a *gene*. Each gene, in turn, is constructed of many *nucleotides*. A chain of nucleotides forms a **nucleic acid**.

Each **nucleotide** in a nucleic acid consists of three parts: a five-carbon sugar, a phosphate group, and a ring-shaped, nitrogenous (nitrogen-containing) base (see left, below). Sugars and phosphate groups alternate to form a long chain that supports the bases much like the sides of a ladder support its rungs. In RNA (ribonucleic acid), the sugar is ribose, while in DNA (deoxyribonucleic acid), it is deoxyribose, which is a ribose with one less hydroxyl group.

There are five possible nitrogenous bases: adenine, cytosine, guanine, thymine, and uracil (often abbreviated as A, C, G, T, and U). DNA strands contain only adenine, cytosine, guanine, and thymine bases, while RNA strands have adenine, cytosine, guanine, and uracil. Of these bases, cytosine, thymine, and uracil are small, single-ringed structures; adenine and guanine are larger, double-ringed structures. It is the sequence of nucleotides in a nucleic acid that encodes information. Each amino acid in a protein is coded for by a sequence of three nucleotides known as a *codon*.

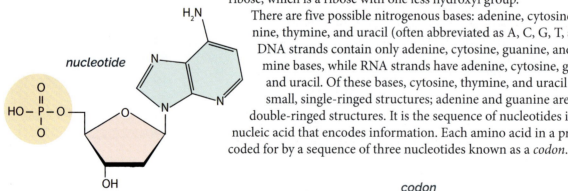

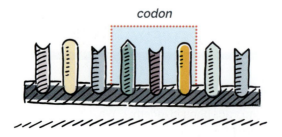

528   Chapter 21

double helix

The nucleic acid DNA has been designed so that two strands coil around each other in a double helix. The backbones circle around the outside, and the bases pair together inside the coil. Large, double-ringed adenine bases pair with smaller thymine bases, while large guanine bases fit neatly next to smaller cytosine bases. Hydrogen bonds hold the complementary bases snugly in position.

When cells divide, a DNA strand reproduces by first unraveling with the help of enzymes. Each of the two resulting strands serves as a template, or pattern, for a new complementary chain. Complementary bases on the newly forming strand match the now-exposed bases of the old strand. When the process of replication is completed, two identical double-stranded DNA molecules result. One strand goes to each half of the dividing cell.

replication

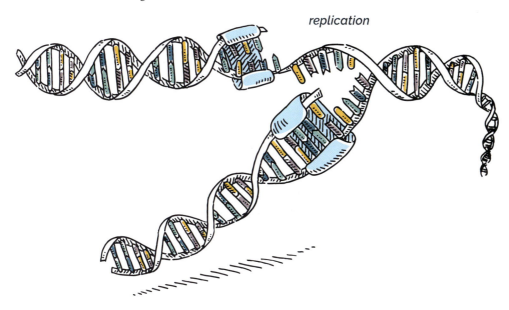

DNA and RNA molecules instruct cells on how to make proteins and regulate a number of other cell functions (such as turning certain metabolic pathways on or off). Because proteins are the building blocks of living things as well as the controllers of all biological processes, these nucleic acids can be described as sets of "blueprints" that make each living thing unique. The amount of DNA—and the information that it encodes—in an average adult human is truly astonishing. Laid out end-to-end, a person's DNA is roughly 2.0×10^{13} m in length—enough to make almost sixty-seven round trips between the earth and sun. During *mitosis*, a cell's DNA condenses into a series of spirals, producing the familiar chromosome structure seen in micrographs.

Of course, the information in DNA is not useful to a cell if it remains stored away in a cell's nucleus. The information must get from the nucleus to the *ribosomes*, where proteins are actually built. To do this, a different kind of copying process, called *transcription*, takes place in the nucleus. A DNA strand is once again "unzipped" by enzymes, but only temporarily. The unzipped portion is read by other enzymes and used as a template for constructing RNA. RNA is different from DNA in several ways. First, the thymine of DNA is replaced by uracil in RNA. Second, RNA consists of a single strand of nucleotides, unlike DNA's double helix, and it is not a copy of an entire chromosome. Instead, a strand of RNA carries a single gene—the information for building a particular protein. Third, RNA exists in multiple forms, and each form has a different function. *Messenger RNA* (mRNA) is the form that is transcribed in a cell's nucleus. It's a working copy of a gene. The mRNA passes through the nuclear membrane into the *cytoplasm* where it can be "read" by a ribosome. The protein-building ribosomes themselves are made of *ribosomal RNA* (rRNA). And molecules of *transfer RNA* (tRNA) transport amino acids to the ribosomes where they will be bonded to form a polypeptide.

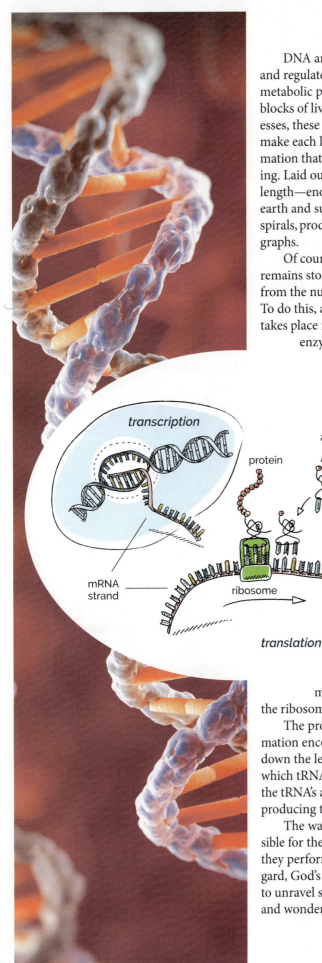

The process of building a polypeptide on the basis of the information encoded in mRNA is called *translation*. As a ribosome moves down the length of an mRNA strand, its codons direct the order in which tRNA molecules dock with the ribosome. This ensures that the tRNA's amino acid cargoes are bonded in the correct sequence for producing the desired protein.

The way that nucleic acids work is marvelous. They are responsible for the faithful duplication of an organism's characteristics, and they perform this task with great precision and accuracy. In this regard, God's creation is astonishingly complex. We have been privileged to unravel some of its workings, further verifying that we are "fearfully and wonderfully made" (Ps. 139:14).

Worldview Conflict in Biochemistry

Scientists have conducted experiments in which they were able to produce amino acids under conditions thought to be those that existed on an ancient Earth. But does that prove that life got here without God? It takes much more than amino acids to make something alive. In fact, some people say, with good reason, that such experiments only show that intelligence and purposeful design are needed to synthesize even the most basic organic compounds.

The origin of the first cell is one of the biggest obstacles to the story of abiogenesis—the creation of living things from nonliving substances. As you read earlier, cells are incredibly complex. Each contains many thousands of complex chemicals, structures, and processes. The early experiments that produced amino acids did not, for example, produce nucleotides, and all life on Earth stores information in nucleic acids. Amino acids alone don't equal life.

Through the years, the difficulties inherent in the prevailing abiogenesis story caused some scientists to seek alternatives. One competing idea to help explain the origins of life is called *panspermia*. The panspermia hypothesis suggests that organic compounds are common throughout the universe and can travel between planets on comets and other space objects. You might be surprised to learn that some of the famous scientists mentioned in this textbook, such as Jöns Jacob Berzelius, Svante Arrhenius, and even Lord Kelvin have put some stock in this idea. But speculation about life arising from somewhere else other than Earth doesn't answer the question of ultimate origins—it just pushes it further back into "deep time."

The Bible tells us that God's creation of the world was a miracle. How would you answer someone who claims that experiments have proven that life got here without God? Use the worldview investigation box on the next page to gather some information and form your own response.

WORLDVIEW INVESTIGATION

ABIOGENESIS

You have a subscription to a popular science magazine. When you get this month's issue, the cover article is all about the Miller-Urey experiment. You read the article. It claims that this experiment was key in proving the validity of the theory of abiogenesis. You decide to write a letter to the magazine to protest this article. But what will you say?

Task

Craft an answer to someone who would tell you that the Miller-Urey experiment proves that life evolved on Earth rather than being specially created. Write a two-hundred-word response, which you could later post.

Process

1. Learn more about the theory of abiogenesis and how it relates to evolution by doing an internet search using the keywords "evolution abiogenesis."
2. Also search online to learn more about the Miller-Urey experiment and what it did by doing an internet search using the keywords "Miller-Urey experiment."
3. To get a Christian perspective on this experiment, do an internet search using the keywords "Miller-Urey experiment Answers in Genesis."
4. Research the Maillard reaction and its link to the Miller-Urey experiment by doing an internet search using the keywords "Maillard reaction Miller-Urey experiment."
5. Write an outline for your response incorporating what you have learned in your research. Be sure to carefully think through the information that you have gathered and use information only from reputable websites and online journals.
6. Write your response to the article, being careful to cite your sources. Point out the problems with trying to use the Miller-Urey experiment as proof of abiogenesis. Consider discussing the types of amino acids found. Above all, be objective—use the information that you find to formulate your response.

Conclusion

Consider posting your rebuttal in an appropriate place, such as on a science blog or forum or on your school or class website. You may want to work with your teacher or a parent to find a good place to post your response. Technology makes it possible for you to reach out to the world and have an influence for Christ, and doing good research and writing about it is an avenue for you to do just that.

21.5 SECTION REVIEW

1. Describe the structure of nucleotides and how they are arranged in a molecule of DNA.
2. How do the nucleotides of DNA and RNA differ?
3. How is information encoded in nucleic acids?
4. How is the information in DNA transferred to ribosomes where proteins are actually built?
5. Defend the idea that a Christian can serve God by working as a biomedical researcher.

21 CHAPTER REVIEW
Chapter Summary

TERMS
biochemistry
metabolism
catabolism
anabolism
polymer

21.1 CHEMISTRY OF LIFE

- Biochemistry is an area of study within general chemistry that focuses on the chemistry of life, including genetics, metabolism, cellular respiration, and other cellular functions.

- Metabolism includes all the chemical processes that take place within cells, both catabolic and anabolic.

- Catabolic processes break larger molecules into smaller particles. Anabolic processes build larger molecules from smaller particles. Often these both occur in cyclical cellular processes.

- Some compounds produced by anabolic processes are polymers, large molecules made of repeating smaller subunits.

21.2 CARBOHYDRATES

- Carbohydrates are macromolecules consisting of carbon, hydrogen, and oxygen and are categorized as sugar, starch, or fiber.

- Carbohydrates are sources of energy and structure in living things.

- Carbohydrates may be simple sugars (monosaccharides), two simple sugars combined (disaccharides), or polymers of many sugars (polysaccharides).

- Starch and glycogen are important polysaccharides found in plants and animals, respectively.

TERMS
carbohydrate
monosaccharide
disaccharide
polysaccharide

TERMS
lipid
fat
oil
steroid
phospholipid

21.3 LIPIDS

- Lipids are a group of compounds that are insoluble in water, including fats, oils, waxes, and steroids.

- Fatty acids are the simplest lipids. Three of these long-chain carboxylic acids can link to a glycerol backbone to form a triglyceride.

- Fats are saturated lipids with high intermolecular forces that exist as solids at room temperature. Oils remain in the liquid state because they are unsaturated and have lower intermolecular forces than saturated fats.

- Steroids are lipids that contain carbon ring structures. Cholesterol is a well-known steroid.

- Phospholipids are amphiphilic lipids with two fatty acid tails. The lipid bilayer of cell membranes is built in large measure from phospholipids.

21 CHAPTER REVIEW

Chapter Summary

21.4 AMINO ACIDS AND PROTEINS

TERMS
protein
amino acid
peptide bond

- Proteins are large, organic polymers built from amino acids. They are essential for life because they provide structure, act as hormones, and serve as enzymes to control the rate of biochemical reactions.

- Amino acids link through peptide bonds to form polypeptide chains. A finished protein consists of one or more of these chains.

- The function of a protein is determined by its composition and resulting shape, which is determined by bonding between different amino acids within its polypeptide chains. A protein subunit normally contains turns, alpha helixes, beta sheets, and random coils.

- Heat, radiation, toxic chemicals, or other stresses can cause proteins to lose their shape, or denature.

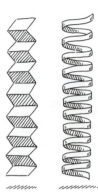

TERMS
nucleic acid
nucleotide

21.5 NUCLEIC ACIDS

- The nucleic acids DNA and RNA are similar in structure and play related but unique roles in living things. DNA stores hereditary information, and RNA uses that code to assemble proteins.

- A chromosome is a long strand of DNA within a cell's nucleus that contains genes. Each gene, constructed of many nucleotides, directs the production of a particular polypeptide.

- RNA exists in multiple forms: mRNA carries genetic information from the nucleus to the cytoplasm, rRNA forms the structure of ribosomes, and tRNA transfers amino acids to the ribosome-mRNA complex for the purpose of synthesizing proteins.

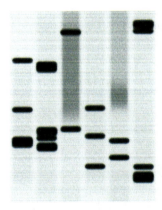

Chapter Review Questions

RECALLING FACTS

1. Which type of metabolic process breaks down substances? Which process builds larger molecules from smaller substances?
2. What does the term *chirality* describe about molecules?
3. List three major functions of carbohydrates.
4. List three functions of lipids in living things.
5. What are the three parts of an amino acid bonded to a central carbon atom?
6. What are enzymes, and why are they important?
7. What is a nucleotide, and what is it composed of?
8. What is the function of DNA?

UNDERSTANDING CONCEPTS

9. A scientist is studying the synthesis of a particular protein in laboratory mice. Is this an example of organic chemistry or biochemistry? Explain.
10. Like many alkanes, carbohydrates serve as an important fuel source. Why are alkanes and carbohydrates not classified together?
11. Describe what is meant by each of the following terms, and give one example of each.
    a. monosaccharide
    b. disaccharide
    c. polysaccharide

12. How are cellulose and glycogen similar? How are they different? How does the difference between cellulose and glycogen affect their nutritional value to humans?

13. Briefly describe the structure and function of glycogen.

14. Why can termites eat wood but humans cannot?

15. What is the difference between a fat and an oil molecule?

16. Explain how fats and oils are esters.

17. What distinguishes steroids from other lipids?

18. Describe how a peptide bond forms.

19. Polysaccharides and polypeptides are both polymers. What is the difference between them?

20. Rank the following in order from least to most complex: alpha helix, amino acid, protein subunit.

21. Can a protein consist of more than one polypeptide? Explain.

22. You're proofreading a friend's report on nucleic acids. The report states that nucleic acids contain the nucleotides adenine, thymine, guanine, and cytosine. Does the report need correction? Defend your answer.

23. A gene consists of 999 nucleotides, not including those that tell a ribosome when to start and stop synthesizing a polypeptide. What is the maximum number of amino acids that could be found in the corresponding polypeptide? Explain.

24. Classify each of the following compounds as either a carbohydrate, lipid, protein, or nucleic acid.

    a. cholesterol

    b. gulose

    c. hemoglobin

    d. tRNA

    e. starch

CRITICAL THINKING

25. Your study partner says that photosynthesis and digestion are both part of a metabolic cycle. Is he correct? Explain.

26. Which of the following is the best source for carbohydrates? Defend your answer.

    a. sugar beet

    b. lean cut of beef

    c. stalk of wheat

    d. mature oak tree

27. Why would you expect the complexity of a protein's shape to be a function of the length(s) of its subunits?

28. As in humans, fish transport oxygen in their blood via a carrier molecule called hemoglobin—a protein. Well-oxygenated fish blood appears bright red. In the presence of elevated levels of toxic nitrite, fish blood takes on a dark-brown color. The condition is known as "brown blood disease" in the aquaculture industry. On the basis of what you have learned in this chapter, suggest a mechanism by which brown blood disease occurs.

29. During DNA replication, several nucleotides are removed from a particular sequence. Assuming that this mutation is passed on to an organism's offspring, how might the organism be affected?

Use the ethics box below to answer Question 30.

30. Use the ethics triad from Chapter 1 and the strategy modeled for you in Chapter 3 to formulate a four-paragraph biblical position on the question of paleo diets. Be sure to address each leg of the triad: biblical principles, biblical outcomes, and biblical motivations. In the last paragraph, formulate your own position on the issue.

ethics

PALEO DIETS—ANCIENT KEY TO MODERN HEALTH?

Paleo dieting is currently very popular. Paleo diets are typically based primarily on eating lean meat, fish, fruits, and vegetables while excluding carbohydrates, dairy, and processed foods. But why exactly is a paleo diet called *paleo* in the first place? And is there a biblical justification for eating paleo?

Chapter 22

NUCLEAR CHEMISTRY

Over the last thirty years, China's economy has grown tremendously. This growth has created a huge demand for electricity. Since 2011 China has been the world's leader in energy consumption. One of the growing pains that China has dealt with has been how to generate all the energy they need without also producing harmful pollution.

Currently the nation produces over half its energy with coal, and, of course, coal has an associated pollution problem. China is looking for alternatives to coal-powered electrical power generation. In this chapter, we will learn about nuclear chemistry. Is nuclear power the answer to China's energy and pollution problem?

22.1 Inside the Nucleus *537*
22.2 Nuclear Decay *545*
22.3 Using Nuclear Chemistry *556*

22.1 INSIDE THE NUCLEUS

Discovery in the Dark

In the previous chapters, the chemistry you studied involved only the atom's valence electrons. But *nuclear chemistry* is concerned with changes in an atom's nucleus—its protons and neutrons—that affect the atom's identity and release radiation. But what causes radiation? How are nuclear changes different from the ones that happen in chemical reactions? Could these mysterious rays hold the key to addressing China's energy and pollution problems? The answers to all these questions lie inside the nucleus.

Why do some atoms decay?

QUESTIONS
» What is radioactivity?
» Why are some atoms stable, but not others?
» How can we measure amounts of radiation?

TERMS
nuclear radiation • radioactivity • nuclide • nucleon • radioisotope • strong nuclear force • band of stability • magic number • mass defect • nuclear binding energy

THE DISCOVERY OF *Radioactivity*

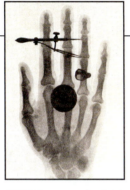

Wilhelm Röntgen

In 1895 German physicist Wilhelm Röntgen discovered x-rays while working with vacuum tubes. Röntgen was awarded the 1901 Nobel Prize in Physics for his discovery.

Henri Becquerel

In 1896 French physicist Henri Becquerel made an unexpected and remarkable discovery. While researching how certain crystals phosphoresce, or glow, after being exposed to light, he wondered whether uranium salts would emit x-rays after being exposed to sunlight. He discovered that uranium salts naturally emitted x-rays even without being exposed to sunlight. Other scientists went on to discover that these atoms emitted energy and particles in order to become more stable. The energy and particles emitted from unstable nuclei are called **nuclear radiation**.

THE DISCOVERY OF *Radioactivity*

Pierre Curie

Marie Curie

Polish physicist Marie Sklodowska Curie and her husband Pierre Curie, a French chemist, used Becquerel's findings to explore radiation. All three shared a Nobel Prize in Physics in 1903 for the discovery of radioactivity. It was the Curies who later coined the term **radioactivity** to describe the process by which unstable nuclei emit nuclear radiation to become more stable. Radioactivity is also referred to as *radioactive decay*. The Curies worked with uranium ore to discover the elements polonium and radium. Marie Curie was awarded a second Nobel Prize, this time in Chemistry for this work, making her the first person to earn two Nobel Prizes.

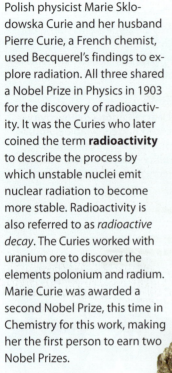
uranium ore

Ernest Rutherford

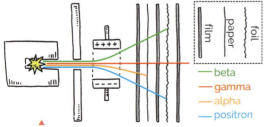

— beta
— gamma
— alpha
— positron

Experiments displayed the charges and relative masses of the particles produced by nuclear radiation.

British physicist Ernest Rutherford continued the research in the field by analyzing the energy and particles emitted during radioactivity. He passed these emissions through powerful electric and magnetic fields. By charting the deflections of the radiation, he came to conclusions about the masses and charges of the different types of radiation. He and later scientists who investigated radiation discovered four types: two that consisted of positively charged particles (alpha particles and positrons), one with a negative charge (beta particles), and one with no charge at all (gamma rays).

Nuclear Stability

As you learned in Chapter 4, an atom's incredibly dense nucleus contains most of the atom's mass. It is so dense that a sphere of nuclear material the size of a golf ball would have a mass of over 9.3 *trillion* kilograms!

Also in Chapter 4 you learned that atoms with identical atomic numbers but different mass numbers are called isotopes. When discussing nuclear characteristics of isotopes, we call them nuclides. Isotopes and nuclides are identical, but the term *nuclide* emphasizes the composition of the nucleus, while *isotope* emphasizes the entire atom. Therefore a **nuclide** is a unique atom of an element with a specific set of protons and neutrons and a specific energy state. Since they are found in the nucleus, protons and neutrons are collectively called **nucleons**. We represent nuclides using isotope notation, which you used in Chapter 4. An isotope that emits nuclear radiation is called a **radioisotope**.

Nuclei contain protons packed close together. Since protons are positively charged, they repel each other, and one would expect them to fly apart. Yet most nuclei are stable and exist for millennia without changing. At the same time, other nuclei are so unstable that they last for only a fraction of a second before changing into another element entirely.

So what determines whether a nucleus will stay together or not? During the development of the quantum model, physicists discovered that the **strong nuclear force** holds nucleons together. This force is a field force that is effective only over small distances (less than 10^{-15} m). Neutrons, having a neutral charge, do not repel each other, nor do they repel the positively charged protons. The strong nuclear force allows neutrons and protons to interact in a way that balances the electrostatic repulsion between the protons. If the strong nuclear force is strong enough, the nucleus will be stable.

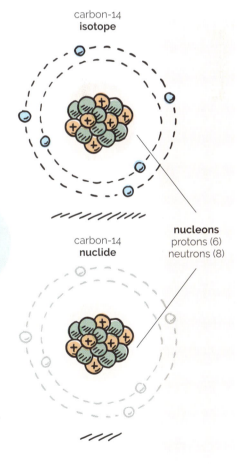

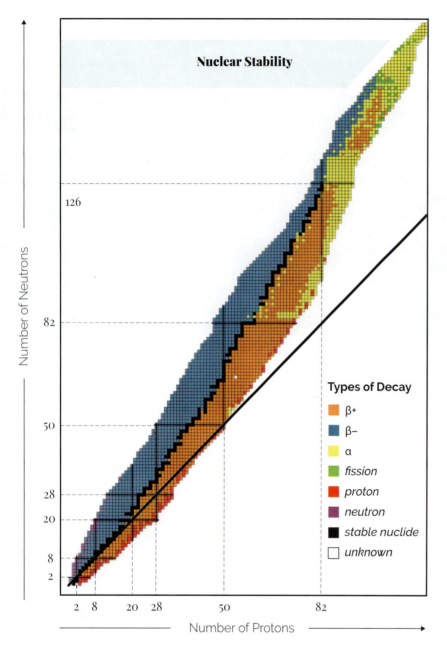

This graph relates the stability of all known nuclides to the numbers of protons and neutrons. The colors represent the types of decay that will occur.

To a certain extent, we can predict nuclear stability on the basis of the ratio of neutrons to protons within the nucleus. For smaller atoms ($Z < 20$), the most stable nuclei have a 1:1 ratio between neutrons and protons. As the number of nucleons increases, additional neutrons, with their strong nuclear forces, are required to balance the repulsive forces between the protons. In some very large nuclei, the ratio may be as high as 1.5:1. The nuclide $^{206}_{82}Pb$ is one such example. With 124 neutrons and 82 protons, its ratio is 1.5:1. The graph at left relates the stability of all known nuclides to their neutron-to-proton ratio. The band of black points is called the **band of stability** because it identifies the most stable nuclides.

The graph of nuclear stability can be used to predict how an unstable nuclide will spontaneously change to form a more stable one—radioactive decay. Nuclides with atomic numbers less than 83 that fall above the band of stability have more neutrons than their stable isotopes. It is thus reasonable that they will decay in a way that will decrease the number of neutrons and increase the number of protons. Nuclides with 83 or more protons frequently decay in a way that makes them lose both protons and neutrons. All the most common isotopes of elements beyond bismuth on the periodic table are radioactive. We will look more closely at the types of radioactive decay in Section 22.2.

EXAMPLE 22-1: NUCLEAR STABILITY PREDICTION

Are $^{60}_{27}Co$ atoms radioactive or stable?

Solution
Cobalt-60 atoms have 27 protons and 33 neutrons. According to the graph of nuclear stability, the point on the graph corresponding to the intersection of 27 protons and 33 neutrons does not represent a stable atom, meaning that that cobalt-60 is an unstable, radioactive nuclide.

The protons and neutrons in a nucleus seem to have energy levels similar to the energy levels in the electron shells. The *nuclear shell model* is a nuclear model in which nucleons exist in levels, or shells, that are analogous to the energy levels that exist for electrons. Just as "full" electron shells are more stable than those that are not full, there are "full" nuclear shells that appear to be more stable. The number of nucleons in a full shell is called the **magic number** for that shell. For protons, the magic numbers are 2, 8, 20, 28, 50, and 82. That explains why tin ($Z = 50$) has ten different stable nuclides, while the elements on either side of it, indium and antimony, have only one and two, respectively. Neutrons have the magic numbers 2, 8, 20, 28, 50, 82, and 126. The nuclide iron-54 is stable because it has a magic number of 28 neutrons, even though it doesn't have a magic number of protons. A few nuclides have stable numbers of both neutrons and protons and are considered very stable, such as calcium-40 with 20 protons and 20 neutrons. Scientists have more complicated rules that govern magic numbers to help predict the stability of nuclides, but these give you an idea of how they work.

If we were to extend the graph of nuclear stability to 180 neutrons or more, we would see that there is a theoretical *island of stability* that is separate from the band of stability. This island could be a future destination for nuclear chemistry research and experimentation. In Chapter 6, we discussed the fact that most of the elements created artificially don't last very long. It is possible that in the future, chemists and physicists may be able to produce a new element with magic numbers that will allow it to be very stable. This could produce exciting new opportunities for exploration and application in nuclear chemistry.

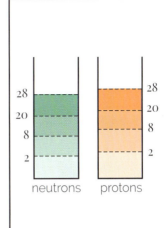

Nuclear Shell Model

- Similar to atomic shells
- Nuclei are particularly stable if either protons or neutrons are a magic number and very stable if both are magic numbers.
- Shells for protons are more widely spaced due to the repulsive force between protons.

Nitrogen-15 has 7 protons and 8 neutrons. The 8 neutrons is a magic number, so nitrogen-15 is stable.

Nitrogen-13 has 7 protons and 6 neutrons. Since neither is a magic number, nitrogen-13 is unstable.

Oxygen-16 has 8 protons and 8 neutrons. Since both are magic numbers, oxygen-16 is very stable.

CAREERS

SERVING AS A NUCLEAR ENGINEER: *HARNESSING THE POWER OF THE ATOM*

Many scientists have spent their careers learning about atomic nuclei. They have learned about the structure of the nucleus and how nuclei can change. As with most discoveries in science, technologists have sought to use these discoveries to help people. Almost immediately following the discovery of x-rays, doctors began using them to view bones within the body.

Today, nuclear engineers continue to investigate how to harness the energy and particles that can be released from the nuclei of atoms. A nuclear engineer develops new processes and devices that will put nuclear energy to good use. As you can imagine, a major task for nuclear engineers is to harness this energy while avoiding the risks of nuclear radiation. If you like the challenge of working on cutting-edge applications of science and energy, you too may be interested in a career as a nuclear engineer.

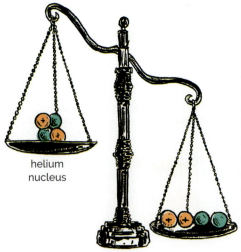

When the nucleons combine to form a nucleus some of their mass gets converted to binding energy. The mass of the nucleus is less than the mass of the nucleons that make it up. We call this the *mass defect*.

Energy Changes and Nuclear Changes

A nucleus has slightly less mass than the protons and neutrons that make up that nucleus. The difference between the mass of an atom's nucleus and the total mass of its nucleons in a free state is called the **mass defect** of the nucleus. The idea that energy and mass are interconvertible explains this phenomenon. According to Albert Einstein's theory of special relativity, mass and energy are equivalent, as indicated by the equation $E = mc^2$, where E is energy, m is mass, and c is the speed of light. Since the speed of light is an astounding 3×10^8 m/s, the energy released by the conversion of even a small amount of mass into energy is huge.

If it were possible, a certain amount of energy would have to be absorbed to separate all the protons and neutrons in a nucleus far from each other. You can think of this like the heat of vaporization when boiling water, which is the energy needed to move liquid particles far away from each other. The energy to separate the nucleons is the same as the energy that is released when a nucleus forms from its nucleons. Scientists call this energy the **nuclear binding energy**. The unit for the nuclear binding energy is a *megaelectronvolt* (MeV). For the purpose of comparing the stability of nuclides, researchers have determined the nuclear binding energies of various nuclei. They then divide these values by the total number of nucleons to obtain the *binding energy per nucleon*. The graph below shows the binding energy per nucleon plotted as a function of mass number. It shows that nuclei with mass numbers near 56 have the most binding energy per nuclear particle.

Binding energy per nucleon peaks near an atomic mass of 56 u.

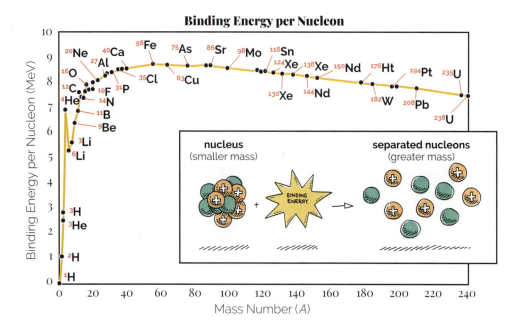

Nuclei with high binding energies per nucleon are more stable than those with lower binding energies per nucleon. As expected, the observed mass defects for the various elements match the observed binding energies. Nuclei with mass numbers near 56 have the most mass defect per nuclear particle. The mass that has "disappeared" (transformed into energy) serves as the nuclear binding energy. These elements have more energy per nucleon holding their nuclei together, so they are more stable.

Geiger counters detect radiation by the tendency of radiation to ionize gases, allowing them to conduct electricity. This change in electrical conductivity indicates the level of radioactivity.

Measuring Radiation

SI UNITS FOR RADIATION

Soon after the discovery of radiation—the matter and energy spontaneously emitted by unstable nuclides—scientists recognized that they could use this technology to "look" inside the human body. Within a year of the first use of x-rays on a human being came reports of injuries from the radiation exposure. Today we have a good understanding of how damaging this radiation can be, and scientists have established safe limits for our exposure.

Scientists recognized that they could measure different aspects of radiation. They could measure radioactivity itself, the absorption of radiation, or the biological effect caused by radiation. Physical radiation units measure the activity of the radiation source, that is, the number of particles or rays produced per unit of time. This measurement is usually expressed using an SI unit called a *becquerel* (Bq). A radiation source with an activity of 1 Bq has one nuclear transformation per second. Another unit for activity which is older but still used is the *curie* (Ci), which is a non-SI unit. Radiation equal to 1 Ci has 3.7×10^{10} transformations per second, or 3.7×10^{10} Bq.

While it is important to understand how much radiation is emitted by its source, it may be more vital in certain instances to understand how much radiation is absorbed by objects. Scientists use an SI unit called the *gray* (Gy) to measure the absorption of *ionizing radiation*. A measure of 1 Gy is equivalent to 1 joule of energy absorbed per kilogram of material. This unit is important in radiotherapy and food irradiation.

These units are important for scientists and technicians, but most of us are more concerned with the effect that the radiation has on living things, especially people. Doctors have devised a unit that factors in the differences in energy, penetrating ability, and ionization capability of differing types of radiation. This unit, which considers both the absorbed dose and its impact on the tissue, is called a *sievert* (Sv). This SI unit is named for Rolf Sievert, a Swedish medical physicist who studied the biological effects of radiation. Table 22-1 shows the health effects of various doses of radiation measured in millisieverts (mSv). But how do we measure how much radiation is present, not just the amount that is absorbed?

TABLE 22-1 *Health Effects of Radiation*

Dose (mSv)	Health Effect
0–250	no detectable clinical effect in humans
250–1000	slight short-term reduction in the number of some types of red blood cells; severe sickness uncommon
1000–2000	fatigue, nausea; vomiting if > 1500 mSv; longer-term decrease in some types of blood cells
2000–3000	nausea, vomiting first day of exposure; a two-week latent period followed by malaise, appetite loss, sore throat, and diarrhea; recovery usually in three months unless complicated by infection
3000–6000	nausea, vomiting, and diarrhea in first few hours; after one week, fever, appetite loss, and malaise; inflammation of the mouth and throat, diarrhea, and emaciation follow, some deaths in two to six weeks; death likely for 50% if exposure was above 4500 mSv; recovery for others in about six months
> 6000	nausea, vomiting, and diarrhea in the first few hours, followed by rapid emaciation and death, possibly within two weeks; death probable in all cases

DEVICES FOR MEASURING RADIATION

A number of different devices can measure radiation. The *Geiger counter* is one. This instrument contains a tube filled with gas that is ionized when exposed to certain kinds of radiation. The ions cause an electric current that is displayed as a series of pulses on a meter. The meter is often connected to an amplifier so that the user can hear the pulses. Another kind of radiation detector, called a *scintillation counter*, works similarly but produces flashes of light. These counters indicate the amount of radiation by the frequency and intensity of pulses in the Geiger counter or flashes in the scintillation counter. People who do work that might expose them to radioactive sources must be able to monitor their exposure to radiation. They can carry pocket instruments called *dosimeters* that monitor their exposure. Modern day dosimeters work on the premise that ionizing radiation alters some characteristic of a material in the dosimeter. The internal circuit monitors both the rate and the total number of those alterations. The dosimeter can then warn the user when either the rate or cumulative limit is approached.

22.1 SECTION REVIEW

1. What is nuclear radiation?
2. What is a nucleon?
3. Compare nuclides and isotopes.
4. What seems to be the main function of neutrons in the nucleus? Explain.
5. Use the nuclear stability graph on page 540 to determine whether the following nuclides are stable. If the nuclide is unstable, explain why this is so.
    a. $^{36}_{18}Ar$
    b. $^{122}_{55}Cs$
    c. $^{166}_{68}Er$
    d. $^{184}_{72}Hf$
6. Using the nuclear binding energy per nucleon graph on page 542, indicate which isotope in each pair will be more stable.
    a. xenon-124, tungsten-182
    b. magnesium-24, carbon-12
    c. calcium-40, lead-208
    d. boron-11, copper-63
7. Why are there different SI units related to radioactivity?
8. Name two types of radiation detectors.

22.2 NUCLEAR DECAY

Radioactive Decay

Recall that in Section 22.1 we defined radioactivity as the spontaneous process that emitted nuclear radiation—particles and energy—from the nuclei of unstable atoms. This process is a naturally occurring nuclear change in which an original unstable nuclide, called the *parent nuclide*, changes to become a more stable *daughter nuclide*. But what do we know about these particles and energy that are emitted?

The particles and waves that a parent nuclide emits have enough kinetic energy to knock electrons out of neighboring atoms. Those atoms could be in any substance. When this happens, the affected atom ionizes, so these kinds of particles and waves are called **ionizing radiation**. Radiation that doesn't have enough kinetic energy to knock out electrons is called *nonionizing radiation*. We will learn about five forms of radioactive decay below, including three that you might remember from a physical science course: alpha decay, beta decay, and gamma decay.

Nuclear equations describe nuclear changes that release radiation during radioactive decay processes. In some cases, the number of protons changes, so the atom becomes a different element. Nuclear equations are similar to chemical equations with reactants on the left and products on the right. But nuclear equations are not written with chemical formulas; rather, they are written with each element's isotope notation. By using isotope notation, we can track changes to the nucleons during nuclear transformations. To balance the nuclear equation, both the mass number and atomic number have to be balanced across the equation.

How long will radioactive waste be around?

QUESTIONS
» Is all radioactivity the same?
» How can I determine the type of decay that an isotope will undergo?
» How can I represent decay events?
» How can I solve half-life problems?

TERMS
ionizing radiation • nuclear equation • alpha decay • beta decay • positron emission • electron capture • gamma decay • radioactive decay series • half-life

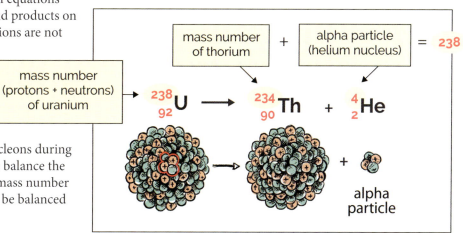

$$^{238}_{92}U \longrightarrow {}^{234}_{90}Th + {}^{4}_{2}He$$

mass number (protons + neutrons) of uranium

mass number of thorium + alpha particle (helium nucleus) = 238

alpha particle

Nuclear Chemistry 545

TYPES OF *Radioactive Decay*

ALPHA DECAY

Alpha decay is radioactivity that produces positively charged *alpha particles* (α, or $^4_2$He). Ernest Rutherford detected alpha particles as a stream of positively charged particles deflected slightly toward a negative electrical plate (see image below). The small deflection indicated that they must be more massive than other types of radiation. It was later determined that alpha particles contained two protons and two neutrons. Nuclides with 83 or more protons frequently decay in a way that releases alpha particles.

For example, uranium-238 emits alpha particles and forms thorium.

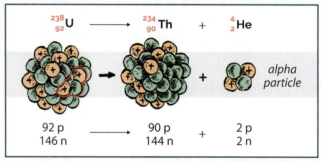

Notice how the mass and atomic numbers are balanced in the equation.

Alpha particles have the greatest ability to ionize atoms with which they collide. But due to their relatively large size and low speed, alpha particles are easily stopped by matter and don't penetrate deeply. For instance, alpha particles cannot go through a piece of paper.

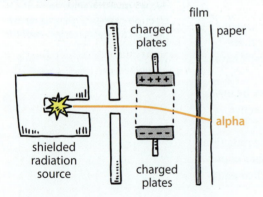

EXAMPLE 22-2: ALPHA DECAY

Radon-222 is a radioactive gas found in buildings. Write the nuclear equation that describes the alpha decay of radon-222.

Solution
Alpha decay implies that an alpha particle is one of the products. Radon-222 is the parent nuclide.

$$^{222}_{86}\text{Rn} \longrightarrow {}^{A}_{Z}X + {}^{4}_{2}\text{He}$$

X is the unknown element, A is its mass number, and Z is its atomic number.

To balance the equation, the mass numbers and the atomic numbers of the products must equal the mass number and the atomic number of the parent nuclide. From the equation above, 222 = A + 4 and 86 = Z + 2. Therefore, the mass number must be 218 and the atomic number must be 84. With an atomic number of 84, the daughter nuclide must be polonium, or $^{218}_{84}$Po. Thus, the final equation is

$$^{222}_{86}\text{Rn} \longrightarrow {}^{218}_{84}\text{Po} + {}^{4}_{2}\text{He}.$$

BETA DECAY

Beta decay is a form of radioactive decay that emits *beta particles* (β, or $_{-1}^{0}e$). Rutherford noted that beta particles acted much like electrons when they swerved sharply toward a positively charged plate. Further studies indicated that beta particles are electrons that are emitted from a nucleus.

As you know, a nucleus contains protons and neutrons, so how can electrons be emitted from it? Scientists theorize that neutrons produce beta radiation by breaking apart to form a proton and an electron. The proton remains in the nucleus and the electron (beta particle) leaves the nucleus. The $_{-1}^{0}e$ symbol used for a beta particle shows both its negligible mass and its effect on the atomic number (−1) of the parent nuclide. You can remember the −1 by recalling that a beta particle is an electron with a negative charge. Notice that when the neutral neutron broke apart, both a positive (proton) and negative (electron) particle had to form to conserve the charge.

Beta decay will most likely occur when the neutron-to-proton ratio is too high. Beta decay reduces the neutron-to-proton ratio by changing a neutron into a proton. This usually happens to nuclides that have an atomic number less than 83, but there are exceptions. Thorium-234 undergoes this kind of decay.

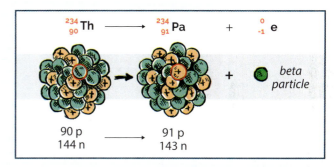

Like alpha decay, this second type of radioactivity also produces ionizing radiation. Because of its higher speed, beta radiation penetrates deeper than alpha radiation. Beta particles can zip through sheets of paper, but wood, glass, or thin metal can stop them. Like alpha particles, they ionize atoms and molecules in the matter that they hit, but not as readily as alpha particles do.

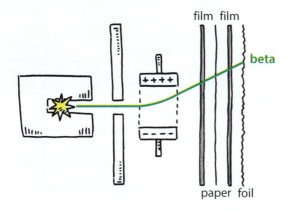

EXAMPLE 22-3: BETA DECAY

Write the nuclear equation for the beta decay of cobalt-60.

Solution
Since beta decay involves the emission of a beta particle ($_{-1}^{0}e$) from the nucleus, we use the equation below.

$$_{27}^{60}\text{Co} \longrightarrow {}_{Z}^{A}X + {}_{-1}^{0}e$$

We know that $60 = A + 0$ and $27 = Z + (-1)$; therefore, $A = 60$ and $Z = 28$. The unknown element is nickel. The final equation is shown below.

$$_{27}^{60}\text{Co} \longrightarrow {}_{28}^{60}\text{Ni} + {}_{-1}^{0}e$$

Nuclear Chemistry

TYPES OF
Radioactive Decay

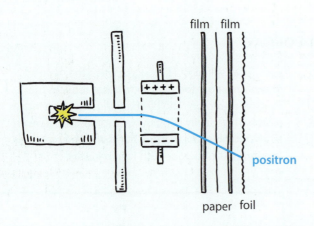

POSITRON EMISSION

In some cases, a nuclide may be unstable because the neutron-to-proton ratio is too low. Positron emission is one decay process that can alleviate this problem by reducing the number of protons in a nucleus by one. During **positron emission**, a proton changes into a neutron and a *positron* ($^{0}_{+1}e$), a particle with positive charge but the same low mass as an electron. Again you can see that charge is conserved as the positive proton breaks into a neutron and a positron. Positron emission is also known as *beta plus decay*. You can remember the +1 by recalling that a positron is like an electron but with a positive charge.

For example, potassium-38 undergoes positron emission as follows.

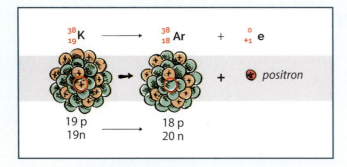

Notice that the atomic number decreased by one, but the mass number didn't change. This occurred because a proton changed into a neutron.

EXAMPLE 22-4:
POSITRON EMISSION

Write the nuclear equation for the positron emission of magnesium-23.

Solution
Since positron emission involves the emission of a positron ($^{0}_{+1}e$) from a nucleus, we use the equation below.

$$^{23}_{12}Mg \longrightarrow\ ^{A}_{Z}X + ^{0}_{+1}e$$

We know that $23 = A + 0$ and $12 = Z + (+1)$; therefore, $A = 23$ and $Z = 11$. The unknown element is sodium. The final equation is shown below.

$$^{23}_{12}Mg \longrightarrow\ ^{23}_{11}Na + ^{0}_{+1}e$$

ELECTRON CAPTURE

Another process that can lower the number of protons in the nucleus is electron capture. **Electron capture** is a nuclear change that occurs when a nucleus pulls in one of its closest electrons (usually from the 1s sublevel) and combines it with a proton to form a neutron. The resulting daughter isotope is in an excited state and may emit a gamma ray as electrons fall into now partially vacated lower energy levels. Silver-106 undergoes electron capture in the following equation.

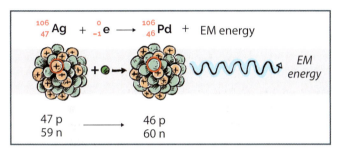

In electron capture, the mass number again remains the same, but the atomic number is reduced by one as a proton becomes a neutron. This process increases the neutron-to-proton ratio to make the daughter nuclide more stable.

EXAMPLE 22-5:
ELECTRON CAPTURE

Write the nuclear equation for potassium-40 undergoing electron capture.

Solution
Since electron capture involves the absorption of an electron ($_{-1}^{0}e$) by a nucleus, we use the equation below.

$$^{40}_{19}K + ^{0}_{-1}e \longrightarrow ^{A}_{Z}X$$

We know that $40 + 0 = A$ and $19 + (-1) = Z$; therefore, $A = 40$ and $Z = 18$. The unknown element is argon. The final equation is shown below.

$$^{40}_{19}K + ^{0}_{-1}e \longrightarrow ^{40}_{18}Ar$$

GAMMA DECAY

Unlike the decay processes outlined above, gamma decay doesn't emit particles. **Gamma decay** is a process that emits high-energy electromagnetic waves called *gamma rays* ($^{0}_{0}\gamma$). These rays are uncharged and thus not deflected by electrical fields, as you can see in the image below. Any nuclear change can produce gamma radiation and most do.

If gamma emission produces no particles, how then are gamma rays produced? As a nucleus changes from an excited state to a lower energy state, it emits gamma rays, similar to electrons changing from higher energy levels to lower energy levels. In an example of gamma decay, an excited technetium-99 nucleus emits a gamma ray to form an unexcited technetium-99 nucleus.

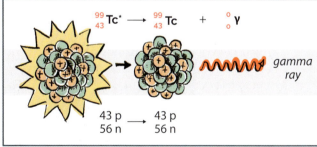

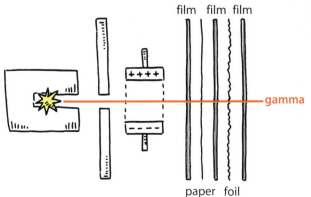

The asterisk on the technetium on the reactant side is used to indicate an isotope in an excited state. As an excited nucleus returns to its ground state, it releases energy in the form of gamma rays.

Nuclear Chemistry **549**

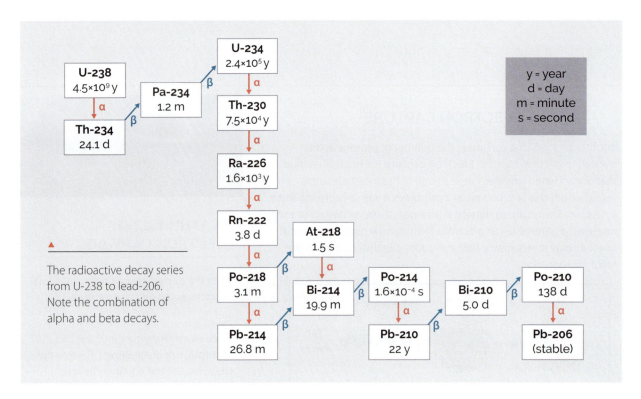

The radioactive decay series from U-238 to lead-206. Note the combination of alpha and beta decays.

Predicting Types of Decay

The nuclear stability graph on page 540 can be used to predict the type of decay that a nuclide might experience. Keep in mind that nuclear stability depends on the neutron-to-proton ratio. Let's look at a few examples.

EXAMPLE 22-6: RADIOACTIVE DECAY PREDICTIONS

Use the graph on page 540 to predict the type of decay that the following nuclides are most likely to undergo: alpha decay, beta decay, or positron emission. Then write the equation for the decay.

a. $^{35}_{13}Al$

b. $^{138}_{60}Nd$

c. $^{220}_{90}Th$

Solution

a. Beta decay is most likely because the nuclide has less than 82 protons, and it falls above the band of stability.

$$^{35}_{13}Al \longrightarrow {}^{35}_{14}Si + {}^{0}_{-1}e$$

b. Positron emission is most likely because the nuclide has less than 82 protons, and it falls below the band of stability.

$$^{138}_{60}Nd \longrightarrow {}^{138}_{59}Pr + {}^{0}_{+1}e$$

c. Alpha decay is most likely because the atomic number is greater than 82.

$$^{220}_{90}Th \longrightarrow {}^{216}_{88}Ra + {}^{4}_{2}He$$

Radioactive Decay Series

Unstable nuclides become more stable by emitting various types of radiation. Some release alpha particles, and some release beta particles or positrons. Some radioactive nuclides, such as uranium-238, are not stable after just one alpha emission.

$$^{234}_{90}Th \longrightarrow {}^{234}_{91}Pa + {}^{0}_{-1}e$$

Thorium's daughter nuclide protactinium-234 is still unstable, and it also decays by beta emission.

$$^{234}_{91}\text{Pa} \longrightarrow {}^{234}_{92}\text{U} + {}^{0}_{-1}\text{e}$$

Uranium-234 is *also* unstable, and it too decays, emitting an alpha particle in this case.

$$^{234}_{92}\text{U} \longrightarrow {}^{230}_{90}\text{Th} + {}^{4}_{2}\text{He}$$

Do these decay reactions ever stop? Yes, but not until a stable nuclide forms! In each step, the daughter nuclide from one decay becomes the parent nuclide of the next decay.

When the product of nuclear decay is at last stable, the radiation emissions stop for that specific nuclide, although other atoms in the same sample will continue to decay until they also reach a stable product. Nuclear scientists have identified several series of predictable alpha and beta decays for a given parent nuclide that result in a single stable daughter nuclide. Sequential alpha and beta emissions often form a long series of nuclear decay events called **radioactive decay series**, though they are statistical models of nuclear decay behavior and not definite decay paths. When uranium-238 decays, nuclear decay reactions proceed until stable lead-206 forms. The image below shows the possible radioactive decay series of uranium-238 in red. The other colors represent the other decay paths for different nuclides. Depending on the pathway of the decay series, as many as fourteen decays may occur involving nine different elements.

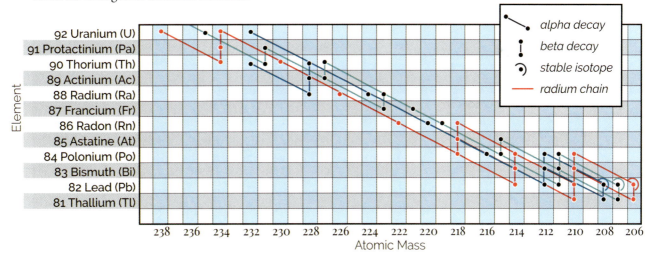

All nuclides for elements with more than 83 protons are radioactive, though every element has some radioisotopes. They all decay according to one of four radioactive decay series: the uranium (or radium) series, the thorium series, the actinium series, or the neptunium series. The first three series can occur naturally. The uranium series begins with the parent nuclide uranium-238 and ends with lead-206. The actinium series begins with plutonium-239 and ends with lead-207. The thorium series begins with uranium-232 or thorium-232 and ends with lead-208. The fourth series, the neptunium series, occurs artificially. It begins with plutonium-241 and ends with thallium-205. Scientists refer to this as the neptunium series instead of plutonium since neptunium is the longest-lived member of the series.

Large Hadron Collider in Geneva, Switzerland

Half-Life

How quickly does nuclear decay occur? Scientists describe the rate of nuclear decay by measuring the **half-life** ($t_{1/2}$) of a parent nuclide, that is, the time it takes for half the atoms of a parent nuclide to decay to a daughter nuclide. Each radioactive nuclide has a unique half-life. For any specific atom, there is a fifty-fifty chance that it will decay in one half-life, similar to your chance of getting heads when you flip a coin. Shorter half-lives indicate that the nuclei tend to quickly decay, while longer half-lives show that radioactive decay usually proceeds more slowly.

For example, the half-life of thorium-234 is 24.1 d. If you had a sample of 40 g of thorium-234, in 24.1 d, that is, in one half-life, half of the thorium, or 20 g, will decay into protactinium-234. How much thorium-234 remains after two half-lives? It might seem logical to say none, but that is not correct. During each half-life, half of the *remaining* thorium-234 nuclides decay. There will be 20 g of Th-234 when the second half-life begins, of which 10 g will decay during the second half-life, leaving 10 g after 48.2 d. After

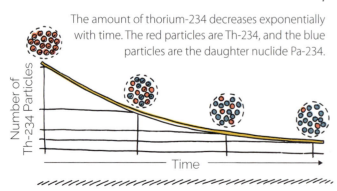

The amount of thorium-234 decreases exponentially with time. The red particles are Th-234, and the blue particles are the daughter nuclide Pa-234.

552 Chapter 22

another 24.1 d, half of the 10 g will decay, leaving 5 g. In this example, the conservation of mass requires that there will always be 40 g of matter—it's just changing from one nuclide to another.

The above example is a great starting point, but we were only considering the sample at the end of each half-life. Real-world science is never so clear-cut. Scientists need to be able to analyze decay after any amount of elapsed time. To accomplish this, scientists model half-life with the exponential decay equation

$$N = N_0(0.5)^{\frac{t}{t_{1/2}}},$$

where N is the remaining amount of a radioisotope in grams or particles, N_0 is the original amount, t is the elapsed time, and $t_{1/2}$ is the radioisotope's half-life. Let's look at a few example problems dealing with radioactive decay and half-life.

TABLE 22-2 *Half-Lives*

Nucleus	Half-Life
oxygen-13	8.9 ms
bromine-80	17.7 min
magnesium-28	21.0 h
radon-222	3.8 d
thorium-234	24.1 d
hydrogen-3	12.33 y
carbon-14	5.73×10^3 y
uranium-238	4.47×10^9 y

EXAMPLE 22-7: CALCULATING AMOUNT REMAINING

A 192 g sample of thorium-234 decays for 17.9 d. At the end of this time, how much thorium-234 remains?

Write what you know.

$N_0 = 192$ g
$t = 17.9$ d
$t_{1/2} = 24.1$ d (from Table 22-2)
$N = ?$

Write the formula and solve for the unknown.

$$N = N_0(0.5)^{\frac{t}{t_{1/2}}}$$

Plug in known values and evaluate.

$N = 192 \text{ g}(0.5)^{\frac{17.9 \text{ d}}{24.1 \text{ d}}}$
$= 192 \text{ g}(0.5)^{0.7427}$
$= (192 \text{ g})(0.5976)$
$= 115$ g

There would still be 115 g of thorium-234 remaining after 17.9 d.

EXAMPLE 22-8: CALCULATING INITIAL AMOUNT

You order a sample that contains magnesium-24. When you receive the sample, it contains 34.5 g of magnesium-24. If the sample was shipped 45.3 h ago, what was the mass of magnesium-24 in the sample when it was shipped?

Write what you know.

$N = 34.5$ g
$t = 45.3$ h
$t_{1/2} = 21.0$ h (from Table 22-2)
$N_0 = ?$

Write the formula and solve for the unknown.

$$N = N_0(0.5)^{\frac{t}{t_{1/2}}}$$

$$\frac{N}{(0.5)^{\frac{t}{t_{1/2}}}} = \frac{N_0 \cancel{(0.5)^{\frac{t}{t_{1/2}}}}}{\cancel{(0.5)^{\frac{t}{t_{1/2}}}}}$$

$$N_0 = \frac{N}{(0.5)^{\frac{t}{t_{1/2}}}}$$

Plug in known values and evaluate.

$$N_0 = \frac{34.5 \text{ g}}{(0.5)^{\frac{45.3 \text{ h}}{21.0 \text{ h}}}}$$

$$= \frac{34.5 \text{ g}}{(0.5)^{2.157}}$$

$$= \frac{34.5 \text{ g}}{0.2242}$$

$$= 154 \text{ g}$$

The sample originally contained 154 g of magnesium-24.

EXAMPLE 22-9: CALCULATING DECAY TIME

A scientist starts with a 15.4 mg sample of radon-222. How many days will it take for the sample to decay so that there is less than 1.00 mg remaining?

Write what you know.

$N_0 = 15.4$ mg
$N = 1.00$ mg
$t_{1/2} = 3.8$ d (from Table 22-2)
$t = ?$

Write the formula and solve for the unknown.

$$N = N_0(0.5)^{\frac{t}{t_{1/2}}}$$

$$\frac{N}{N_0} = \frac{N_0(0.5)^{\frac{t}{t_{1/2}}}}{N_0}$$

$$\log_{0.5}\left(\frac{N}{N_0}\right) = \log_{0.5}(0.5)^{\frac{t}{t_{1/2}}}$$

$$t_{1/2}\frac{t}{t_{1/2}} = t_{1/2}\left(\log_{0.5}\left(\frac{N}{N_0}\right)\right)$$

$$t = t_{1/2}\left(\log_{0.5}\left(\frac{N}{N_0}\right)\right)$$

Plug in known values and evaluate.

$$t = (3.8 \text{ d})\left(\log_{0.5}\left(\frac{1.00 \text{ mg}}{15.4 \text{ mg}}\right)\right)$$

$$= (3.8 \text{ d})(\log_{0.5}(0.064\underline{9}3))$$

$$= (3.8 \text{ d})(3.9\underline{4}4)$$

$$= 15 \text{ d}$$

It will take 15 d for the sample to decay to the point of having less than a milligram of radon-222 remaining.

22.2 SECTION REVIEW

1. What is ionizing radiation?

2. Rank alpha, beta, and gamma decay according to decreasing penetrating power of their products.

3. Use the nuclear stability graph on page 540 to predict the type of decay that each of the following nuclides will undergo.

    a. $^{154}_{66}\text{Dy}$
    b. $^{83}_{38}\text{Sr}$
    c. $^{61}_{26}\text{Fe}$
    d. $^{232}_{94}\text{Pu}$
    e. $^{11}_{6}\text{C}$

4. Write the equation for each of the following.

    a. alpha decay of $^{217}_{86}\text{Rn}$
    b. alpha decay of $^{238}_{94}\text{Pu}$
    c. beta decay of $^{214}_{83}\text{Bi}$
    d. beta decay of $^{234}_{91}\text{Pa}$
    e. positron emission of $^{18}_{9}\text{F}$
    f. positron emission of $^{22}_{11}\text{Na}$
    g. electron capture of $^{26}_{13}\text{Al}$
    h. electron capture of $^{59}_{28}\text{Ni}$
    i. gamma decay of $^{152}_{63}\text{Eu}$
    j. gamma decay of $^{165}_{66}\text{Dy}$

5. Use the uranium decay series shown on page 550 to answer the following questions.

    a. What two kinds of decays does this chain show?
    b. What is the most short-lived nuclide on the chart? Give its half-life.
    c. What is the most stable nuclide on the chart? Give its half-life.

6. How could your worldview affect the way that you would date rocks containing U-238, given its half-life?

7. Solve the following half-life problems.

    a. Oxygen-15 has a half-life of 2.04 min. If you prepare for an experiment by obtaining a 301.75 g sample of pure oxygen-15 and it takes 15.3 min for additional preparations, how much oxygen-15 remains when the experiment begins?

    b. The half-life of hydrogen-3 is 12.32 y. If you have 35.86 g of hydrogen-3 after it is allowed to decay for 19.78 y, with what mass of hydrogen-3 did you start?

    c. The half-life of thorium-230 is 7.538×10^4 y. How long will it take for a 16.2 g sample to decay to less than 1.00 g?

WORLDVIEW INVESTIGATION

RADIOMETRIC DATING

Radiometric dating is a scientific technique in which the radioactive isotopes in a sample are compared with amounts of the decay products in the sample. On the basis of the half-life of the specific isotope, the age of the sample can be estimated. There are many different isotopes that are used for radiometric dating. Radiocarbon dating uses the decay of carbon-14 to estimate the date of formerly living material such as trees.

On the day that you learned about radiometric dating in chemistry, your history teacher told you about the remains of a wooden ship that was said to be 15,000 years old. As a young-earth creationist, you wonder about the age of that ship.

The Oseberg Ship in the Viking Ship Museum, Bygdøy, Oslo, Norway

Task

You have decided to learn more about radiometric dating so that you will have a better understanding of the topic. Prepare a one-page outline of the current research from both secular and creationist sources on the subject of radiometric dating.

Process

1. Conduct research on a biblical view at the website for the Institute for Creation Research and Answers in Genesis. Use the keywords "RATE Project," "Carbon Dating Undercuts Evolution's Long Ages," and "Measurable 14C in Fossilized Organic Materials."

2. Conduct research on a secular view at the website for the American Chemical Society. Use the keywords "Carbon-14 dating," "Carbon Dating Undercuts Evolution's Long Ages," and "Discovery of Radiocarbon Dating."

3. Prepare your outline of radiometric dating and share it with a classmate for review.

4. Finalize your outline and turn it in.

Conclusion

Radiocarbon dating is a scientific tool. As with all things in science, it is affected by worldview. We must assess scientific models and theories with the correct lens of a biblical worldview.

QUESTIONS

» How is a nuclear reaction different from nuclear decay or chemical reactions?

» What is the difference between fission and fusion?

» How can nuclear reactions sustain themselves?

» How much energy is released in a nuclear reaction?

TERMS

transmutation • nuclear fission • chain reaction • critical mass • nuclear fusion

Are nuclear power plants worth the risk?

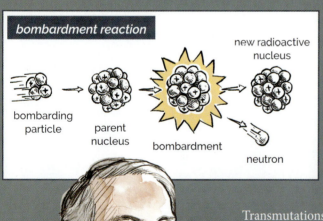

Ernest Rutherford

22.3 USING NUCLEAR CHEMISTRY

Nuclear Reactions

Alchemists dreamed of changing base metals such as lead into gold. We've seen that radioactive decay can naturally change parent nuclides into very different daughter nuclides. But what about changing lead into gold? This can't happen spontaneously through a nuclear decay process—it requires a nuclear reaction. Nuclear reactions are different from radioactive decay because they typically do not occur spontaneously and usually involve more than one particle. One such kind of nuclear reaction is called a *nuclear bombardment reaction*, in which a nucleus is struck by a high-energy particle.

So what are some differences between a nuclear reaction and a chemical reaction? A few of the differences are obvious. In chemical reactions, bonds are broken and formed between atoms, but the atomic nuclei remain unchanged. Chemical reactions involve only electrons, while nuclear reactions affect entire atoms, including their nuclei.

The rates of chemical reactions can be affected by pressure, concentration, temperature, and catalysts, and such reactions usually involve relatively modest changes in energy. But that all changes in a nuclear reaction, where atoms can send out particles and rays, changing their identity in the process and releasing huge amounts of energy. Nuclear reactions are generally unaffected by pressures and temperatures associated with chemical reactions, but some can occur only at extreme pressures and temperatures. A **transmutation** is the process by which an atom of one element changes into another element by changing the number of protons in the atom's nucleus. Transmutations can occur naturally, as in nuclear decay or solar fusion, but can also be caused by humans, as in some nuclear reactions. Scientists have used this process to create artificial elements.

Before the atomic age began, scientists knew of no elements with atomic numbers higher than that of uranium ($Z = 92$). In 1919 Ernest Rutherford, who theorized the existence of atomic nuclei from his gold foil experiment, bombarded nitrogen with alpha particles and found that it produced protons and an isotope of oxygen.

$$^{14}_{7}N + ^{4}_{2}He \longrightarrow ^{17}_{8}O + ^{1}_{1}P$$

His experiment was the first known induced transmutation.

In 1934 Italian physicist Enrico Fermi sent shock waves through the scientific world. Theorizing that if uranium could be forced to emit an electron from one of its neutrons, then the remaining part of the neutron would become a proton. If this were to happen, he could produce an element beyond uranium. Fermi succeeded in producing several elements with atomic numbers higher than uranium, though he couldn't verify their existence. Neptunium was first produced in 1940 when uranium-239 atoms underwent beta decay.

Enrico Fermi

$$^{239}_{92}\text{U} \longrightarrow {}^{239}_{93}\text{Np} + {}^{0}_{-1}\text{e}$$

Presently, scientists have produced elements up to atomic number 118. The elements following uranium on the periodic table were all produced with nuclear bombardment reactions. As you learned back in Chapter 6, scientists call elements with atomic numbers higher than 92 transuranium elements.

Fission

As scientists continued their work to produce heavier elements, some were surprised when their bombardment reactions produced lighter elements. Austrian physicist Lise Meitner was the first to correctly identify the process of fission after observing its products. **Nuclear fission** is the process in which some very heavy nuclei are split apart into smaller, more stable nuclei. Fission reactions are usually triggered by bombardment of heavy nuclei with high-energy particles, though some fission reactions occur spontaneously. For instance, uranium-235 can be bombarded with neutrons ($^{1}_{0}$n) to split the nucleus and release various fragments, some free neutrons, and a tremendous amount of energy. There may be more than one way for the nucleus to split. The equations below show two of the ways that uranium-235 can split.

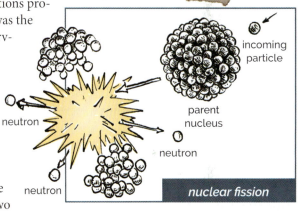

nuclear fission

$$^{235}_{92}\text{U} + {}^{1}_{0}\text{n} \longrightarrow {}^{141}_{56}\text{Ba} + {}^{92}_{36}\text{Kr} + 3{}^{1}_{0}\text{n} + \text{energy}$$

$$^{235}_{92}\text{U} + {}^{1}_{0}\text{n} \longrightarrow {}^{139}_{52}\text{Te} + {}^{94}_{40}\text{Zr} + 3{}^{1}_{0}\text{n} + \text{energy}$$

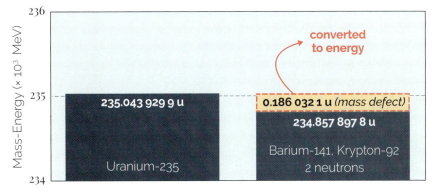

The difference between the binding energy per nucleon of the parent nuclide U-235 and the daughter nuclides Ba-141 and Kr-92 is the energy released in the nuclear reaction, or the mass defect.

Writing equations for nuclear reactions is similar to writing an equation for nuclear decay. Once again, the mass numbers and atomic numbers must balance on each side of the equation. Let's look at an example.

Nuclear Chemistry

EXAMPLE 22-10:
WRITING THE EQUATION FOR A NUCLEAR REACTION

By bombarding a plutonium-239 nucleus with a neutron, we can trigger a fission reaction. The reaction produces two nuclei and three neutrons. One of the nuclei is zirconium-103. Write the nuclear equation.

Solution
To finish the equation, we need to balance the mass numbers and atomic numbers.

$$239 + 1 = A + 103 + 3$$
$$A = 134$$
$$94 + 0 = Z + 40 + 0$$
$$Z = 54$$

The element with atomic number 54 is xenon. The nuclear equation is

$$^{239}_{94}\text{Pu} + ^{1}_{0}\text{n} \longrightarrow ^{134}_{54}\text{Xe} + ^{103}_{40}\text{Zr} + 3\,^{1}_{0}\text{n} + \text{energy}.$$

Even though the atomic and mass numbers balance, empirical measurement of the products and reactants reveals that there is a small change in mass between the two sides of the equation. Recall from Section 22.1 that mass defect is the difference in the mass of a nucleus and the sum of the masses of all the nucleons. A nucleus with a high mass defect needs more energy to hold it together. In the example above, the plutonium nucleus has a higher mass defect than the combined mass defects of the products; in other words the products are more stable than the reactant. The change in energy required to hold the nuclei together is equal to the energy released in the nuclear reaction.

The energy released in a nuclear reaction can also be calculated by calculating the overall change in mass between the two sides of the equation. This "missing" mass can then be converted to its energy equivalence by using Einstein's equation, $E = mc^2$. The nuclear reaction converts the "missing" mass to energy. So how much energy does the fission of uranium-235 into barium-141 and krypton-92 produce?

EXAMPLE 22-11: CALCULATING ENERGY FROM A NUCLEAR REACTION

Given the isotope masses of uranium-235 (235.053 929 9 u), barium-141 (140.914 411 9 u), krypton-92 (91.926 156 1 u), and a free neutron (1.008 664 9 u), calculate the energy released by the fission of uranium-235.

$$^{235}_{92}U + ^{1}_{0}n \longrightarrow ^{141}_{56}Ba + ^{92}_{36}Kr + 3^{1}_{0}n + \text{energy}$$

Calculate the mass of the reactants.

235.043 929 9 u + 1.008 664 9 u = 236.052 594 8 u

Calculate the mass of the products.

140.914 411 9 u + 91.926 156 1 u + 3(1.008 664 9 u)
= 235.866 562 7 u

Calculate the mass defect.

$$m_{defect} = m_{product} - m_{reactants}$$
$$= 235.866\,562\,7 \text{ u} - 236.052\,594\,8 \text{ u}$$
$$= -0.186\,032\,1 \text{ u}$$

Convert the mass defect from daltons to kilograms.

$$-0.186\,032\,1 \text{ u} \left(\frac{1 \text{ g}}{6.022 \times 10^{23} \text{ u}} \right) \left(\frac{1 \text{ kg}}{1000 \text{ g}} \right)$$
$$= -3.089\,207\,9 \times 10^{-28} \text{ kg}$$

Write the formula and solve for the unknown.

$$E = mc^2$$

Plug in known values and evaluate.

$$E = (-3.089\,207\,9 \times 10^{-28} \text{ kg})(3.0 \times 10^8 \text{ m/s})^2$$
$$= -2.78 \times 10^{-11} \text{ J for every nucleus.}$$

To put the energy released in a nuclear reaction in perspective, the energy released in the fusion of 1 kg of uranium-235 would require burning over 2 *billion* kilograms of coal!

In the binding energy per nucleon graph (p. 542), as expected, the observed mass defects for the various elements match the observed binding energies. Nuclei with mass numbers near 56 have the highest binding energy per nuclear particle. This means that they are the most stable. The energy released is the result of the mass defect produced when these more stable nuclei form.

EXAMPLE 22-12: NUCLEAR BINDING ENERGY COMPARISON

Using the binding energy per nucleon graph, determine which of the following nuclei is the most stable on the basis of its nuclear binding energy: hydrogen-2, arsenic-75, or uranium-235.

Solution
From the binding energy per nucleon graph, you can see that of the three nuclei mentioned, arsenic-75 has the highest binding energy per nucleon. Therefore, it is the most stable—it is closest to 56 u.

This picture taken in 1970 shows a French nuclear test at Mururoa, French Polynesia.

Nuclear Chemistry

CHAIN REACTIONS

The fission of a uranium-235 nuclide produces, among other things, an average of between 2 and 3 neutrons. What happens to these neutrons? They can either escape from the sample of uranium or they can collide with other uranium-235 atoms. If three neutrons are released and hit three other uranium-235 nuclei, the impacted nuclei may immediately undergo fission themselves, releasing up to 9 more neutrons. These neutrons can, in turn, initiate more nuclear fission reactions, and so on. This ongoing fission process, in which the neutrons from one reaction trigger additional reactions, is called a **chain reaction**. It can occur only if the sample of uranium is large and dense enough to intercept many of the released neutrons. If the mass of uranium is too small or too spread out, a condition known as a *subcritical mass* exists and many neutrons will escape without initiating more nuclear fission reactions—and a chain reaction does not occur. The smallest mass of a fissionable substance that can sustain a chain reaction is called the **critical mass** of the substance. If the mass of fissionable material is larger or more compact than the critical mass, the mass produces many neutrons and the chain reaction proceeds out of control, as happens in the detonation of a nuclear bomb.

While each reaction converts only a small amount of mass into energy, chain reactions release vast amounts of thermal energy because they involve unimaginably large numbers of nuclei. This is the energy that technology has enabled us to harness to generate electricity in nuclear power plants. But the chain reaction in a power plant has to be closely monitored and controlled. Perhaps nuclear power plants could be one way that nuclear chemistry helps people in China who live in unhealthy environments caused by coal-fired power plants.

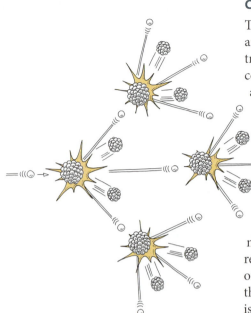

▲ The fission of U-235 can trigger a chain reaction if a critical mass of uranium is present.

MINILAB

INQUIRING INTO CHAIN REACTIONS

Nuclear reactions convert mass into energy. Each individual reaction generates only a small amount of energy, but nuclear power plants produce vast amounts of energy because many fission reactions occur in a short amount of time. This is possible because nuclear engineers design the reactors to take advantage of chain reactions. In this inquiry lab activity, you will try to model a chain reaction and then see whether you can modify the process to speed it up.

How can we alter a chain reaction?

EQUIPMENT
- dominoes (30)
- stopwatch

1. What is a chain reaction?

Procedure

A Standing each domino on its narrow end, arrange all thirty in such a way as to make all the dominoes fall by pushing on only one domino.

B Using the stopwatch, time how long it takes for all thirty dominoes to fall when you push over the first one.

2. What does a domino represent?

3. What does the domino falling represent?

C Rearrange the dominoes to decrease the time needed for all the dominoes to fall.

Conclusion

4. How does this exercise model a chain reaction?

5. What arrangement resulted in the shortest time for all the dominoes to fall?

Going Further

6. How does your answer to Question 5 represent the concept of critical mass?

Fusion

Nuclear fusion is the transmutation process of combining several light nuclei to form more massive nuclei. You can think of it as the opposite of fission. You might expect fusion to absorb energy, but instead, it releases even more energy than fission does. To understand why this is, look back at the binding energy per nucleon graph on the opposite page. Notice how the left side of the graph is very steep, especially at low mass numbers. Fusion reactions will move us toward the right on the graph, and the change in nuclear binding energy is released as we move toward the most stable atomic mass (56). Both fission and fusion result in more stable materials and are *exergonic*, meaning that they release energy.

Fusion in the sun is a multistep process. It starts with two hydrogen (H-1) nuclei colliding to form deuterium (H-2). The deuterium nucleus collides with another hydrogen nucleus to form helium-3. Two helium-3 nuclei collide to form a helium-4 nucleus and two hydrogen atoms that can continue the process.

$$^{1}_{1}H + ^{1}_{1}H \longrightarrow ^{2}_{1}H + ^{0}_{+1}e \text{ (occurs twice)}$$

$$^{1}_{1}H + ^{2}_{1}H \longrightarrow ^{3}_{2}He \text{ (occurs twice)}$$

$$^{3}_{2}He + ^{3}_{2}He \longrightarrow ^{4}_{2}He + 2^{1}_{1}H \text{ (occurs once)}$$

The net reaction is shown below.

$$4^{1}_{1}H \longrightarrow ^{4}_{2}He + 2^{0}_{+1}e$$

As with fission, there is a discrepancy of mass between the parent and daughter nuclides of fusion. The nuclei that form a larger nucleus have more mass than the product of the reaction. The nuclear mass of helium is 4.003 u. However, the combined masses of the two protons (2 × 1.0073 u) and the two neutrons (2 × 1.0087 u) equal 4.0320 u—a 0.029 u difference. Just as in fission, this lost mass is transformed into an enormous amount of energy.

Again refer to the graph of binding energy per nucleon on the facing page. Nuclides to the left of the peak could potentially undergo fusion, and nuclides to the right of the peak could potentially undergo fission. The graph peaks at a mass number of 56, demonstrating that the greater the binding energy per nucleon, the more stable the nucleus will be and the less likely that it will experience fusion or fission. Any element with a mass number of 56 will achieve the greatest stability

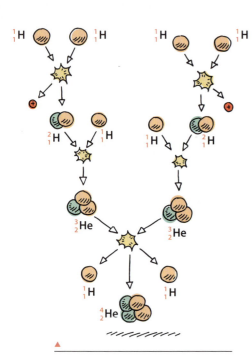

Nuclear fusion is important to all living things because it fuels our sun and other stars. The reaction forms a helium nucleus and an immense amount of energy. This is the most common form of fusion.

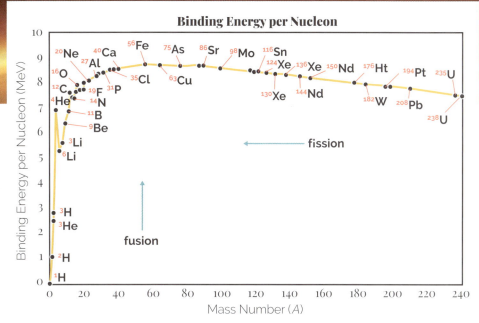

possible. Thus, when bombarded by nuclear particles, elements higher than 56 can fission into smaller elements to become more stable. Elements lower than 56 can fuse to become more stable. Both processes produce vast amounts of energy.

Binding energy per nucleon peaks near an atomic mass of 56 u. Notice how fission and fusion processes are related to binding energy in this diagram.

EXAMPLE 22-13: FUSION OR FISSION PREDICTION

Using the graph above, predict which of the following candidates are more likely to undergo fusion or fission: hydrogen-3, technetium-97, beryllium-7, or plutonium-239.

Solution
Because hydrogen-3 and beryllium-7 are to the left of iron-56 on the graph, they are more likely to undergo fusion. The nuclides technetium-97 and plutonium-239 are strong candidates for fission because they are to the right of iron-56.

   Although the concept of fusing two atoms may appear simple, light nuclei must collide with enough kinetic energy to overcome the natural repulsion between nuclei before nuclear fusion can occur. Temperatures close to 100 000 000 K are required before nuclei have enough energy for fusion to proceed. Because so much thermal energy is required, scientists call fusion reactions *thermonuclear reactions*. Large amounts of energy must be expended to heat atoms to this temperature. In fact, in the first fusion reactions (hydrogen bombs), atomic fission bombs had to be used to supply the vast amount of energy required to initiate the fusion process. Such extreme conditions also exist within the sun, where fusion is an ongoing process. There, hydrogen nuclei are fused together to form helium nuclei. The vast amount of energy produced is emitted from the sun as electromagnetic radiation. Most of the energy that we use on Earth originates in the nuclear reactions within the sun.

Nuclear Chemistry

Aerial view of Grenoble's polygon scientifique, syncrotron ring, and nuclear reactor

Using Nuclear Chemistry to Solve Problems

Over 3 million people in China will be diagnosed with cancer this year. Many of their cancers can be traced back to polluted water and air. And much of the air pollution is from coal-powered electric plants throughout the country. As China continues to develop, its need for electrical power increases. How can they provide this additional power without exacerbating the pollution problem that they already have? Could further decreasing their reliance on coal-burning power plants and increasing their already significant use of nuclear-powered plants solve some of these problems?

Nuclear chemistry can be used to irradiate water by exposing it to ionizing radiation in wastewater treatment plants, removing toxins such as hydrocarbons, nitrogen oxides, heavy metals, and pesticides. Also, many industrialized countries have mastered the technology to harness the fission energy released in nuclear power plants. Using nuclear power plants eliminates the toxic and sometimes radioactive pollutants created by coal-fired power plants. Some reactors even use repurposed nuclear weapons as fuel for generating electricity. Nuclear power produces almost 20% of the electricity generated in the United States. Other countries make even greater use of nuclear power for their energy needs. For example, France provides almost 80% of its electricity using nuclear reactors.

Nuclear chemists and physicists also hope to one day control nuclear fusion reactions and to convert their energy into electricity. This would have the added benefit of avoiding the high-level radioactive nuclear waste products that fission nuclear power plants generate. But the promise of fusion power is not without its challenges. Scientists are currently trying to address the greatest challenge to Earth-based fusion: the necessary containment for the enormous temperatures required for fusion. Researchers are experimenting with various methods of containment, such as magnetic, inertial, and electrostatic confinement. Many nations are spending huge sums of money for the research and development of this almost limitless energy source.

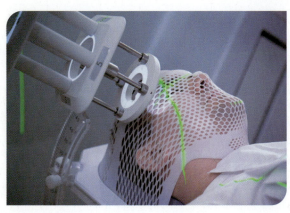

A cancer patient receiving radiotherapy

Nuclear chemistry offers further hope for China and other countries to help treat cancer. We've seen that cobalt-60 undergoes beta decay to release beta particles and gamma rays that can be used to treat cancerous brain tumors. Radiation treatment in a variety of forms is used to treat other cancers as well.

But nuclear chemistry is not limited to producing electricity and treating cancer. Nuclear chemistry helps us explore the outer reaches of space and the depths of the ocean, thanks to the energy density of nuclear fuels. Radiation is used in agriculture as well. It has helped researchers develop new seeds with higher production rates and pest and disease resistance. Radiation is used to control pests by sterilizing them, thereby reducing their numbers. It is a useful tool for sterilizing medical instruments. Radiation also lets us see the invisible by giving us the ability to image the body's systems, defects in bridges, the levels of liquids in tanks and oil wells, and even the atmosphere of Jupiter.

But, as with any area in science, we need to consider the risks along with the benefits of nuclear chemistry. Newsworthy accidents, such as the ones that occurred at Chernobyl and Fukushima, highlight the risks of nuclear power. In the Nuclear Power ethics box on page 569, you'll be tasked with developing a position on the use of nuclear energy for power production. Part of your decision making must include assessing the risks of nuclear power. Using God's world in ways that please Him involves minimizing the risks and maximizing the benefits of our efforts to exercise dominion. This means that we should develop a mindset of safety that evaluates and minimizes risk.

In your journey through chemistry this year, you've seen how the Bible's account of Creation, the Fall, and Redemption encapsulates a biblical worldview that gives us a framework for all of life. As you looked at chemistry through the lens of Scripture, you have found that the best reasons to study and practice chemistry are biblical ones. We can use the tool of chemistry to help find solutions to the challenges of malaria, cancer villages, electricity generation, and so much more for God's glory and mankind's good.

Inner Radiation Belts of Jupiter

22.3 SECTION REVIEW

1. What is a bombardment reaction?

2. State two ways in which nuclear fission and nuclear fusion are similar and one way in which they are dissimilar.

3. Complete the nuclear equation for the fission of uranium-233 using the partial equation below.

    $$^{233}_{92}U + ^{1}_{0}n \longrightarrow ^{A}_{Z}X + ^{137}_{54}Xe + 3^{1}_{0}n + energy$$

4. What happens to the missing mass in a nuclear reaction represented by the mass defect?

5. Explain the relationship between the terms *critical mass* and *chain reaction*.

6. Given the isotope masses of plutonium-239 (239.052 163 4 u), xenon-134 (133.905 394 5 u), zirconium-103 (102.926 601 2 u), and a free neutron (1.008 664 9 u), calculate the energy released by the fission of plutonium-239.

    $$^{239}_{94}Pu + ^{1}_{0}n \longrightarrow ^{134}_{54}Xe + ^{103}_{40}Zr + 3^{1}_{0}n + energy$$

7. Would each of the following nuclides be more likely to undergo fusion or fission? (Use the graph on page 563.)

    a. carbon-12

    b. uranium-235

    c. neon-20

22 CHAPTER REVIEW

Chapter Summary

TERMS
nuclear radiation
radioactivity
nuclide
nucleon
radioisotope
strong nuclear force
band of stability
magic number
mass defect
nuclear binding energy

22.1 INSIDE THE NUCLEUS

- Nuclear chemistry studies the changes that can occur in an atom's nucleus.

- Nuclear radiation, or radioactivity, is the spontaneous emission of particles and energy from an unstable nucleus in order to become more stable.

- The stability of the nucleus depends on the ratio of neutrons to protons. Nuclear shells model nuclear stability well, including the specific numbers of neutrons or protons, or magic numbers, associated with stable nuclei.

- The difference between the mass of an atom's nucleus and the total mass of all the individual nucleons is called the mass defect. Mass defect is related to the energy that binds the nucleus together.

- The SI unit for measuring emitted radiation is the becquerel. The gray is the SI unit that quantifies radiation absorption. A sievert is the measure of the biological effects of absorbed radiation.

22.2 NUCLEAR DECAY

- Nuclear equations use isotope notation to represent the nuclides involved in a nuclear event. Nuclear equations are balanced by accounting for both mass numbers and atomic numbers.

- Alpha decay, beta decay, positron emission, electron capture, and gamma emission all cause predictable changes to the charge and mass of an atom.

- Alpha particles are composed of two protons and two neutrons, giving them a +2 charge. They easily ionize atoms that they collide with and have little penetrating power.

- Beta particles are electrons that form in and are released from the nucleus. They have lower ionizing potential than alpha particles but greater penetrating ability. Beta decay changes a neutron to a proton.

- Positron emission is a form of beta decay, except that a positron (positive electron) is emitted from the nucleus. Positron emission changes a proton to a neutron.

- Electron capture occurs when an inner orbital electron is absorbed by the nucleus. Electron capture changes a proton to a neutron.

- Gamma rays are the most energetic type of radiation and are uncharged electromagnetic waves. They have the greatest power of penetration.

- Many nuclides undergo a series of nuclear reactions that culminate in a stable form. These changes can be represented as a radioactive decay series.

- Half-lives measure the amount of time that it takes for half of a sample of a nuclide to decay.

TERMS
ionizing radiation
nuclear equation
alpha decay
beta decay
positron emission
electron capture
gamma decay
radioactive decay series
half-life

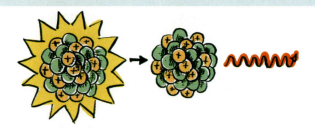

22.3 USING NUCLEAR CHEMISTRY

- Nuclear reactions can be triggered through nuclear bombardment with particles or even other nuclei. This process can be used to create synthetic transuranium elements.

- In nuclear fission, a nucleus splits to produce more stable nuclides.

- A chain reaction of controlled fission is produced when particles emitted from one reaction trigger additional reactions. There must be sufficient mass—the critical mass—to sustain a chain reaction.

- In nuclear fusion, nuclei are fused to form larger, more stable nuclei. Both fission and fusion release great amounts of energy.

TERMS
transmutation
nuclear fission
chain reaction
critical mass
nuclear fusion

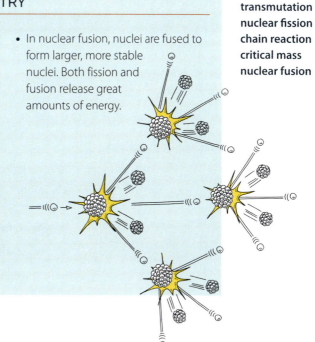

Chapter Review Questions

RECALLING FACTS

1. What is radioactivity?
2. What is the band of stability?
3. Define *binding energy per nucleon*.
4. What are the three SI units related to radioactivity?
5. How are nuclear equations balanced?
6. (True or False) Alpha particles have more mass but less penetrating power than beta particles.
7. Which types of radiation consist of particles? Which consist of only waves?
8. Explain the relationship between ionizing power and the depth of penetration of particles emitted as nuclear radiation.
9. Make a table for the five different types of radioactive decay. Make columns for the name, symbol, nuclear notation, charge, mass number, effect on atomic number, and penetration power. Rate penetration power as low, medium, or high.
10. Make a table that compares nuclear decay, a nuclear reaction, and a chemical reaction.
11. (True or False) The mass needed to sustain a nuclear chain reaction is known as the critical mass.
12. Where does the energy released by fission and fusion come from?
13. Why do some nuclei undergo fission but not fusion?

UNDERSTANDING CONCEPTS

14. Summarize the discoveries of scientists and their impact on our understanding of radioactivity.
15. What determines the stability of a nucleus? Explain.
16. Use the nuclear stability graph on page 540 to predict whether the following nuclides are stable. Explain each prediction.

    a. $^{160}_{70}Yb$ c. $^{55}_{25}Mn$

    b. $^{200}_{80}Hg$ d. $^{107}_{42}Mo$

17. Using the nuclear binding energy per nucleon graph on page 542, determine which nuclide in each pair will be more stable.

    a. oxygen-16, tin-119

    b. helium-4, lithium-6

    c. uranium-238, neon-20

    d. tin-119, helium-4

18. Gamma radiation, visible light, and radio waves are all forms of electromagnetic radiation. Why are gamma rays much more dangerous than visible light and radio waves?

22 CHAPTER REVIEW

19. Using the nuclear stability graph on page 540, predict the type of decay that each nuclide will undergo.

   a. $^{143}_{58}Ce$

   b. $^{37}_{18}Ar$

   c. $^{15}_{8}O$

   d. $^{286}_{111}Rg$

   e. $^{106}_{44}Ru$

20. Write the nuclear equation for each of the following.

   a. alpha decay of $^{180}_{74}W$

   b. alpha decay of $^{223}_{88}Ra$

   c. beta decay of $^{210}_{83}Bi$

   d. beta decay of $^{225}_{88}Ra$

   e. positron emission of $^{126}_{53}I$

   f. positron emission of $^{13}_{7}N$

   g. electron capture of $^{7}_{4}Be$

   h. electron capture of $^{53}_{25}Mn$

   i. gamma decay of $^{60}_{28}Ni$

   j. gamma decay of $^{237}_{93}Np$

21. Why do the decay processes in a radioactive decay series not continue indefinitely?

22. Using Table 22-2, determine which nuclide decays more quickly: bromine-80 or magnesium-28.

23. Solve the following half-life problems.

   a. Iodine-131 is used to treat a hyperactive thyroid gland. The half-life of iodine-131 is 8.02 d. If a hospital receives a 215.25 g shipment of iodine-131, how much will be left after 21.3 d?

   b. The half-life of copper-62 is 9.67 min. If you need 10.0 g to conduct an experiment and it will take 65 min to set up the experiment once the sample arrives, how much must be in the sample when it arrives?

   c. Carbon-10 has a half-life of 19.29 s. If you have 1.53 kg, how long will it take until you have less than 10.0 g?

24. Explain the difference between nuclear fission and fusion.

25. Given the partial equation below, complete the nuclear equation for the fission of plutonium-239.

   $^{239}_{94}Pu + ^{1}_{0}n \longrightarrow ^{A}_{Z}X + ^{140}_{51}Sb + 3^{1}_{0}n + energy$

26. Why do nuclear fission and nuclear fusion both release energy?

27. Given the isotope masses of plutonium-239 (239.052 163 4 u), niobium-104 (103.922 464 7 u), iodine-134 (133.909 744 9 u), and a free neutron (1.008 664 9 u), calculate the energy released by the fission of plutonium-239.

   $^{239}_{94}Pu + ^{1}_{0}n \longrightarrow ^{104}_{41}Nb + ^{134}_{53}I + 2^{1}_{0}n + energy$

28. In the following pairs, determine which nucleus is more stable by referring to the graph on page 563.

   a. lithium-6 or magnesium-24

   b. uranium-235 or iron-56

   c. helium-4 or uranium-235

29. Write a paragraph about another area in chemistry in which safety and risk are an important part of using this field for God's glory and mankind's good. Include in your paragraph uses, risks, and safety considerations in this field of chemistry.

CRITICAL THINKING

30. Why do scientists suspect that there is an island of stability beyond the band of stability?

31. As you are working on a nuclear decay lab activity, you get to the point of having one atom remaining. Your lab partner says, "Just one more half-life and they will all have decayed." Is he correct? Explain.

32. You are working in the laboratory with carbon-11. Your sample has gone from 13.58 g to 8.074 g in 15.25 min. What is the half-life of carbon-11?

33. Iodine-131 experiences beta decay with a half-life of 8.02 d. You start an experiment with a 10.0 g sample. Answer the following questions.

   a. How many beta particles will be emitted in the first half-life? Iodine-131 has a molar mass of 130.906 g/mol.

   b. Is the number of beta particles emitted per second a constant? Explain.

   c. How many beta particles will be emitted per second, on average?

34. Suppose that a uranium-238 nucleus were bombarded with a high-energy neutron to fission into iodine-135 and krypton-96, also emitting both neutrons and protons in the process. Write the nuclear equation for this event and determine how many protons and neutrons would be emitted.

$$^{238}_{92}U + ^{1}_{0}n \longrightarrow ^{135}_{53}I + ^{96}_{36}Kr + X^{1}_{0}n + Y^{1}_{1} + \text{energy}$$

35. Given the isotope masses of uranium-233 (233.039 u), strontium-94 (93.915 361 u), xenon-137 (136.911 562 1 u), and a free neutron (1.008 664 9 u), calculate the energy released by the fission of 1.75 g of uranium-233.

$$^{233}_{92}U + ^{1}_{0}n \longrightarrow ^{94}_{38}Sr + ^{137}_{54}Xe + 3^{1}_{0}n + \text{energy}$$

ethics

NUCLEAR POWER

Introduction

We live in a technology-driven society, and most technology requires electrical energy—lots of electrical energy. We are also living in a time when the environment is a major concern. Many are advocating decreasing our dependence on fossil fuels and increasing our electricity generation from clean sources. Many green technologies are expensive. So how do we balance our energy needs, our obligation to protect the world that God gave us, and the economic reality that solutions are expensive and we have limited resources?

Task

You are a representative in your state legislature. Your state is facing a significant increase in the demand for electrical power. At the same time, numerous groups have approached you about pollution and the environment. Advocates for green power have proposed new, cleaner sources for electrical power generation. One of the proposals is to increase power production from nuclear sources. In the upcoming weeks, you will have to decide on a proposal to build nuclear power plants in the state.

Process

1. Search online for information on nuclear power generation. Include research on safety and security, benefits and risks, cleanliness of nuclear power, and nuclear waste issues.

2. Research biblical principles, outcomes, and motivations for electrical power generation from nuclear sources.

3. Organize your research into evidence that would argue either for or against the use of nuclear power.

4. Prepare for a debate on the topic of whether your state should increase electrical power production by building five new nuclear power plants. Be prepared to argue for or against the proposal.

5. Participate in a debate with your classmates, arguing the side assigned by your teacher.

Conclusion

We constantly face the challenge of deciding between conflicting demands. As a society, we want cheap, clean, and safe electrical power. We also want to leave our descendants a livable world. We have to be knowledgeable about the issues around us so that we can make informed decisions. We must always strive for our decisions to glorify God and benefit other people.

Appendix A
UNDERSTANDING SCIENTIFIC TERMS

You may find science a little intimidating because of all the long, unfamiliar words that scientists use. But you can often unravel the meanings of these words by breaking them down into simple parts that do have meaning to you. When you see a difficult scientific word, look at the entries in this appendix to help you understand that term. These word parts may come at the beginning, at the end, or in the middle of the term, depending on their meaning.

For example, if you ran into the word *heterogeneous equilibrium*, you could separate it into four parts—*hetero-*, *-gen*, *equi-*, and *-libra*. *Hetero-* means "other, different," *-gen* means "kind," *equi-* means "equal," and *-libra* means "balance." So heterogeneous equilibrium has to do with different kinds of substances being in equal balance.

a, an (Gk.)—not, without
ab (L.)—away from
ac, ad, ag (L.)—to, toward
acou (Gk.)—hearing
aer, aero (Gk.)—air
alter (L.)—change
amal (Gk.)—soft
amphi, ampho (Gk.)—on both sides
ant, anti (Gk.)—opposite, against
ante (L.)—before
aqua (L.)—water
ar (L.)—of or related to
audio (L.)—hear
aut, auto (Gk.)—self
bar, baro (Gk.)—weight, pressure
bi (L.)—two, twice, double
bio, bios, biot (Gk.)—life
calc, calci (L.)—calcium
calor (L.)—heat
centi (L.)—a hundred
centr, centri, centro (Gk.)—center
chem, chemi, chemo (Gk.)—chemistry

chrom, chromo (Gk.)—color
chron, chrono (Gk.)—time
cline (Gk.)—sloping
co, com, con (L.)—with, together
cupr (Gk.)—copper
cycl, cyclo (Gk.)—circle, wheel
de (L.)—loss, removal
deci (L.)—tenth
di (Gk.)—two
div (Gk.)—apart
duce, duct (L.)—to lead
dyna (Gk.)—power
eco (Gk.)—house
electro (L.)—electricity
en, end, endo (Gk.)—within, inner
equ, equa, equi (L.)—equal
ex, exo (Gk.)—out, outside, without
extra (L.)—outside, more, beyond, besides
fissi (L.)—split, divide
flam (L.)—fire
fund (L.)—basis
fusi (L.)—to join together

gen (L.)—kind
grad (L.)—step, walk, slope
graph, grapho, graphy (Gk.)—to write
grav (L.)—heavy
gyro (Gk.)—spinning
halo (Gk.)—salt
hemi (Gk.)—half
hetero (Gk.)—other, different
homeo, homio, homo (Gk.)—like, same, resembling
hydr, hydra, hydro (Gk.)—water, fluid
hyper (Gk.)—over, beyond
hypo (Gk.)—under, beneath
ic (Gk.)—of, relating to
inter (L.)—between
is, iso (Gk.)—equal
ism (Gk.)—belief, process of
kine, kinema, kinemato, kines, kinesi, kinet, kineto (Gk.)—move, moving, movement
libra (L.)—balance
log, logo, logus, logy (Gk.)—word, study of

macr, **macro** (Gk.)—large

magneto (Gk.)—magnetic

medi, **media**, **medio** (L.)—middle

met, **meta** (Gk.)—between, with, after, change

meter, **metry** (Gk.)—measure

micro (Gk.)—small

mill, **mille**, **milli**, **millo** (L.)—one thousand

mit (L.)—to send

mono (Gk.)—one

morph, **morpha**, **morpho** (Gk.)—form, shape

multi (L.)—many

nan, **nani**, **nano**, **nanus** (Gk.)—dwarf, one billionth

nomy (Gk.)—the science of

nuc, **nucle**, **nucleo** (L.)—central part

ocul, **oculi**, **oculo**, **oculus** (L.)—eye

opt, **opti**, **opto** (Gk.)—eye, vision

organ (Gk.)—living

orth, **ortho** (Gk.)—upright, perpendicular

ox, **oxy** (Gk.)—oxygen

par, **para** (Gk.)—beside

pause (Gk.)—to stop

pend (L.)—hanging

peri (Gk.)—around, near

phon, **phono** (Gk.)—sound

phos, **phot**, **photo** (Gk.)—light

phyt, **phyto**, **phytum** (Gk.)—plant

poly (Gk.)—many

post (L.)—after

pre (L.), **pro** (Gk.)—before, in front of

prot, **prote**, **proto** (Gk.)—first, original

pyro (Gk.)—fire

radi, **radia**, **radio** (L.)—spoke, ray

retro (L.)—backward

sal (L.)—salt

scientia (L.)—knowledge

scope, **scopy** (Gk.)—to see, watch

sect (L.)—to cut

seism (Gk.)—earthquake

semi (L.)—half

sol (L.)—sun

son (L.)—sound

spec (L.)—to see, look at

sphere (Gk.)—ball, globe

stasis (Gk.)—to stand still

sub (L.)—below, under

super (L.)—above, over

syn (Gk.)—together

tele (Gk.)—distant

terra (L.)—earth

tetr, **tetra** (Gk.)—four

therm, **thermo** (Gk.)—heat

top, **topo** (Gk.)—place

tran, **trans** (L.)—across, through

trop, **tropae**, **trope**, **tropo** (Gk.)—turn, change

uni (L.)—single, one

vacu (L.)—empty

vari, **vario** (L.)—difference

vect (L.)—to carry

volu (L.)—bulk, amount

Appendix B
MATH HELPS

This appendix provides you the necessary helps for solving the many kinds of chemistry problems found in this textbook. It's not intended to be all-inclusive. We assume that you understand or are learning the methods for solving single-variable algebraic equations, including using the order of operations and solving equations for an unknown quantity.

ROUNDING RULES

We round numbers all the time. Sometimes it's convenient to round, for example, the number of residents in your town to the nearest 100 people or the distance around the earth to the nearest 1000 kilometers. Rounding is useful for several reasons. It helps us grasp large numbers when the specific number is not needed but the approximate size is. Rounding also allows us to report measured scientific data correctly.

For the purposes of measuring scientific data and performing calculations, scientists have agreed on certain rounding rules. These ensure that calculation solutions are properly rounded to show the needed precision of numerical quantities. We suggest that you use the following rules for problems in this textbook to closely imitate those used by scientists.

1. Identify the place value that you are going to round to. This is the rounded place value.

2. If the digit to the right of the rounded place value is 0–4, the digit in the rounded place value remains unchanged ("rounded down"). If the digit to the right is 5–9, the digit in the rounded place value is increased by 1 ("rounded up").

   a. If the rounded place value is to the right of the decimal point, all digits to the right of the rounded place value will be dropped after rounding.

   Examples: 56.739 mL → 56.7 mL

   39.47 mm → 39.5 mm

   74.95 s → 75.0 s

   b. If the rounded place value is to the left of the decimal point (or assumed decimal point), all digits between the rounded place value and the decimal point become zeros and the decimal point (if one is shown) is omitted.

   Examples: 1694 m → 1700 m

   19.6 °C → 20 °C

3. Notice in two of the examples above that rounding to a particular place value may affect other digits to the left of the rounding place value. When a 9 is rounded up, the 9 rounds up to 10, so the 9 becomes a 0 and 1 is added to the digit to the left.

MATHEMATICS AND MEASUREMENT

When measured quantities appear in equations as terms or factors, follow the same arithmetic principles that you have learned for solving any equation. But acting as a scientist, you have to think about some extra things before arriving at a solution to a problem.

Measured quantities include units. Units, such as meters (m), seconds (s), and meters per second (m/s), are treated as parts of a measurement. So any arithmetic operation applies to both the number and the unit factors in a measurement.

Measured quantities also have precision (see page 60). Precision is the exactness of a measurement. Scientists exert a lot of effort to ensure that their data properly represents the exactness that their instruments can read. Scientists alert other scientists to the precision of their measurements by the number of significant figures (SF) they include in their reported data.

On the following pages you will find some rules for doing basic arithmetic operations using measurements.

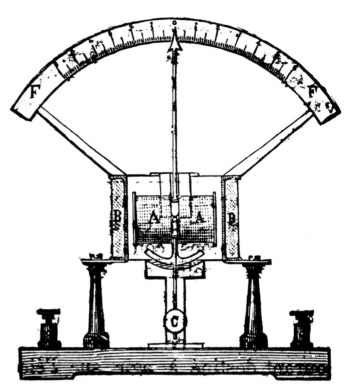

DETERMINING SIGNIFICANT FIGURES

When we are reading measured data collected by someone else, we have to be able to identify the significant figures. (Significant figures do not apply to counts or definitions—only to measured data.) Scientists follow an established set of rules to determine which digits in a measurement are significant.

Rule 1: All nonzero digits are significant.

Rule 2: All zeros between significant figures are significant.

Rule 3: All ending zeros to the right of the decimal point are significant.

Examples:

a. 1.694 mL → 4 SF (all four digits are nonzero—Rule 1)

b. 51.021 m → 5 SF (four nonzero digits and a zero between SF—Rule 2)

c. 22.7500 g → 6 SF (four nonzero digits and two ending zeros to the right of the decimal point—Rule 3)

d. 0.097 mL → 2 SF (Rule 1 [Since the zero in the tenths place is not an ending zero, Rule 3 does not apply.])

The fourth example can be confusing, but it is much easier to understand if you change the number to scientific notation.

$$0.097 \text{ N} = 9.7 \times 10^{-2} \text{ N} \rightarrow \text{only 2 SF}$$

One other case in which determining SF can be difficult is when ending zeros occur to the left of the decimal point.

Example:

How many SF does the measurement 7600 m have? You know that two of the digits must be significant, but you as the reader cannot determine which, if either, of the zeros are significant. Does this number have two, three, or four SF? Again, the person reporting the value can use scientific notation to avoid confusion.

7600 m → 7.6×10^3 m (two SF)

7600 m → 7.60×10^3 m (three SF)

7600 m → 7.600×10^3 m (four SF)

Appendix B
MATH HELPS

ADDING AND SUBTRACTING CHEMISTRY DATA

Math Rule 1: Added or subtracted data must have the same units. Even if the kinds of data are the same, if their units are not the same, then you are adding apples and oranges. For example, you can't add the lengths 3.1 m and 45 cm together without first converting one of the measurements to the other's units.

Math Rule 2: The result of adding or subtracting data can't be more precise than the least precisely measured data used in the calculation. After adding or subtracting, round the result to the decimal place of the estimated digit in the least precise measurement.

EXAMPLE 1: ADDING DATA: LENGTH

Add 3.1 m and 45 cm.

Unit Conversion:

$$45 \text{ cm} \left(\frac{1 \text{ m}}{100 \text{ cm}} \right) = 0.45 \text{ m}$$

Add data:

$$3.\underline{1} \text{ m} + 0.4\underline{5} \text{ m} = 3.\underline{55} \text{ m} = 3.6 \text{ m}$$

The estimated digits in each piece of data and the sum are underlined. The sum contains two underlined digits that result from the summing operation. These must be rounded to the place having the least precise estimated value, in this case the 0.1 m place.

MULTIPLYING AND DIVIDING CHEMISTRY DATA

Math Rule 3: The result of multiplying or dividing data can't have more SF than the data with the fewest SF used in the calculation. After multiplying or dividing, round the results to the same number of significant figures as the data with the fewest SF.

EXAMPLE 2: DIVIDING DATA: DENSITY

Calculate the density of a sample of quartz with a mass of 27.55 g and a volume of 10.4 cm³.

Given data: $m = 27.55$ g (four SF), $V = 10.4$ cm³ (three SF)
Calculation:

$$d = \frac{m}{V}$$

$$= \frac{27.55 \text{ g}}{10.4 \text{ cm}^3}$$

$$= 2.6\underline{4}9 \; \frac{\text{g}}{\text{cm}^3}$$

$$= 2.65 \; \frac{\text{g}}{\text{cm}^3} \text{ (three SF allowed)}$$

Since 10.4 cm³ has only three SF, only three SF are allowed in the quotient.

Math Rule 4: The result of multiplying or dividing a measured quantity and a pure number has the same number of decimal places, or the same precision, as the measured quantity used in the calculation.

Examples:

a. (7)(2.35 cm) = 16.45 cm (not 16.5 cm)

b. 2.63 cm ÷ 5 = 0.53 cm (not 0.526 cm)

Notice that the number of SF in the measured data do not determine the number of SF in the result when multiplying or dividing by pure numbers. Math Rule 4 ensures that you preserve the precision of the original measurement, even if you change the number by multiplying or dividing by a pure number.

PROPORTIONALITIES IN CHEMISTRY

As you study chemistry, you will discover that many measurable quantities change or vary in a reliable, predictable way with other quantities. For example, the pressure of a gas in a rigid container changes as the absolute temperature of the gas varies. Double its temperature and the pressure doubles too. Similarly, if you double the volume of the container (at a constant temperature), the pressure will be cut in half. These kinds of relationships are called *proportionalities*.

Direct Proportionality

If two numbers, A and B, are directly proportional, they can be set equal to each other by including a proportionality constant factor—k. The equation is

$$A = kB,$$

which means that

$$\frac{A}{B} = k$$

and k is a constant for all ratios of these two quantities. For an example, let A equal time in minutes and B time in seconds. Then the proportionality constant k is equal to 1 min/60 s as shown below.

$$A = \left(\frac{1 \text{ min}}{60 \text{ s}}\right)B$$

In this textbook, you will use direct proportions in Charles's law, Gay-Lussac's law, and Avogadro's law.

Inverse Proportionality

You'll also find that other pairs of quantities are *inversely proportional*. If the value of one increases, the other decreases in proportion. The equations for these proportions set up as shown below.

$$A = \frac{k}{B}$$

If you multiply both sides of the equation by B, you end up with

$$AB = k.$$

So if k is a constant, as A increases, B must decrease in the same proportion to maintain the equality.

In chemistry, Boyle's law uses the inverse proportion concept.

Other Proportionalities

Lastly, there are other kinds of proportionalities that are neither entirely direct nor inverse relationships. For example, the combined gas law is a joint variation with some variables that vary directly and others that vary inversely. In another case, the kinetic energy of particles changes directly with the mass of the particles but according to a square of the particles' velocity.

Exponentials and Logarithms

Two other relationships that are common in chemistry are exponential and logarithmic relationships. These two functions are inverses of each other and are needed when studying acids, bases, and nuclear chemistry. The independent variable in an exponential function is an exponent. The general form is

$$y = Ab^x,$$

where A is a coefficient and b is the base of a power. In chemistry, we see this function used in three different calculations. To calculate the concentration of hydronium from pH, we use the equation $[H_3O^+] = 10^{-pH}$. When calculating the concentration of hydroxide from pOH, the equation is $[OH^-] = 10^{-pOH}$. In nuclear chemistry, the half-life formula involves an exponential.

$$N = N_0(0.5)^{\frac{t}{t_{\frac{1}{2}}}}$$

As mentioned above, the logarithmic function is the inverse of the exponential. The general form of a logarithmic function is $y = A\log_b x$. Again A is the coefficient and b is the base of the logarithm. In chemistry, we use logarithmic functions to solve for pH and pOH. We also use logarithms to solve half-life problems to determine elapsed time or the half-life of an isotope.

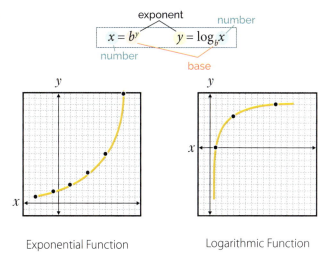

Exponential Function Logarithmic Function

Appendix B
GRAPHING

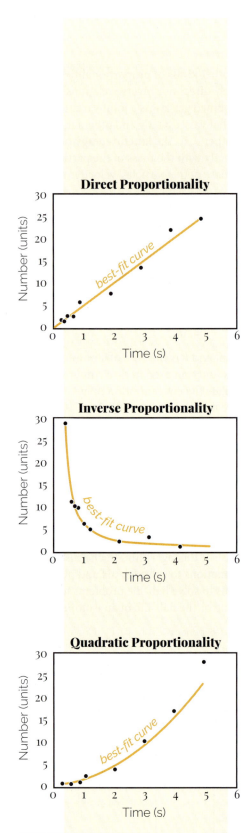

Scientists often like to compare two or more groups of numbers visually to see whether they are related in some way. When scientists plot data to compare different quantities, they make a graph. The simplest graphs compare two changing quantities, called *variables*. Usually, one of the variables is determined by the scientist or else changes in a regular way (e.g., time). This is called an *independent variable*. Its value doesn't depend on anything in the data. The other variable is expected to change in some way related to the independent variable. Its value depends on the first variable, so it is called the *dependent variable*. The values of the independent and dependent variables are called the *coordinates of the data*. To plot the data, an ordered pair of coordinates is used. You have graphed ordered pairs in the form (*x*, *y*) in a math class.

Scientists usually plot the independent variable on the horizontal axis (*x*-axis), with increasing values to the right. The dependent variable is typically plotted on the vertical axis (*y*-axis), increasing upward. These are not hard and fast rules, and many graphs are arranged differently to more clearly see the relationships in the plotted data. Useful graphs include a title describing the graph's purpose and labels identifying the quantities and units used on each axis. Graphs are often depicted on a scaled grid with the *x*- and *y*-axis scales shown so that estimates of the values of the plotted variables can be made.

SCATTERPLOTS

Graphs come in different forms. A simple plot of points on a graph is called a *scatterplot*. This is the starting point for many graphs. Scientists like to detect trends in the data and then create a mathematical equation (model) that describes the trend. They draw a *best-fit curve* through the pattern of dots. The kind of curve drawn depends on how the data changes. We still call it a best-fit curve even if it doesn't curve. Curves that are straight lines are called *linear graphs*. These are fairly rare in nature. Most trends in nature range from slightly curved to really wavy! These graphs, logically enough, are called *nonlinear graphs*.

SLOPE

One of the most important things to be learned from a graph is the rate at which the dependent variable changes in comparison to the independent variable. This rate is indicated by the *slope* of the graph. A graph with a steep slope shows that things are changing quickly. One with a horizontal slope shows that the dependent variable is not changing at all. A graph that rises from left to right has a positive slope; a dropping curve has a negative slope.

To calculate the slope (*m*) of a curve, use the slope formula,

$$m = \frac{\Delta y}{\Delta x} = \frac{y_f - y_i}{x_f - x_i},$$

where *m* is the slope, Δy is the change in the *y* variable, and Δx is the change in the *x* variable.

ESTIMATING DATA

You can sometimes use scatterplots and trend lines to obtain information that is not in the measured data set. If you follow the trend line between two data points, these values are estimated—not measured. Obtaining unmeasured data this way is called *interpolation*. Scientists also often try to predict the values of the dependent variable beyond the range of the measured independent variable data points. This method of analysis is called *extrapolation*. Extrapolating data assumes that the trend will continue in the same fashion as observed within the measured data.

OTHER TYPES OF GRAPHS

In a bar graph, dependent data is plotted as vertical bars at each independent data value. Area graphs fill in the area of a graph. When several dependent variables are plotted on a single graph, the areas can be compared to show their relationships visually. Pie charts are another type of area graph. They are especially useful for showing percentages of a whole.

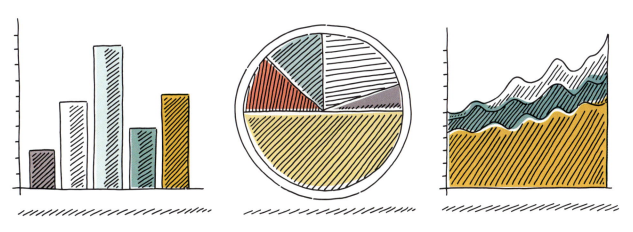

A bar graph (left), pie chart (center), and an area graph (right)

Appendix B   577

Appendix C
FUNDAMENTAL AND DERIVED UNITS OF THE SI

FUNDAMENTAL UNITS

Dimension	Name	Symbol	Definition
length	meter	m	The meter is the length of the path traveled by light in a vacuum during a time interval of 1/299 792 458 of a second.
mass	kilogram	kg	The kilogram is the unit of mass. It is defined on the basis of the Planck constant ($6.626\,070\,15 \times 10^{-34}$ J·s), the meter, and the second.
time	second	s	The second is the duration of 9 192 631 770 cycles of the radiation associated with a specific transition of the cesium-133 atom.
electric current	ampere	A	The ampere is the current required for one coulomb of charge to pass through a conductor every second. The coulomb is defined on the basis of the elementary charge, the electric charge carried by a single proton.
temperature	kelvin	K	The kelvin, the unit of thermodynamic temperature, is defined on the basis of the Boltzmann constant ($1.380\,649 \times 10^{-23}$ J/K), the kilogram, the meter, and the second.
amount of substance	mole	mol	A mole is the amount of material that contains $6.022\,140\,76 \times 10^{23}$ entities. This value is the Avogadro constant.
luminous intensity	candela	cd	The candela is the luminous intensity, in a given direction, of a source that emits monochromatic radiation of frequency 540×10^{12} Hz and that has a radiant intensity in that direction of 1/683 watts per steradian (W/sr).

Note: The unit definitions are a contribution of the National Institute of Standards and Technology.

COMMONLY USED DERIVED UNITS

Dimension	Name	Symbol	Derivation
force	newton	N	$\dfrac{kg \cdot m}{s^2}$
energy, work, heat	joule	J	$\dfrac{kg \cdot m^2}{s^2}$
power	watt	W	$\dfrac{J}{s} = \dfrac{kg \cdot m^2}{s^3}$
pressure	pascal	Pa	$\dfrac{N}{m^2} = \dfrac{kg}{m \cdot s^2}$
electric charge	coulomb	C	A·s
voltage	volt	V	$\dfrac{J}{C} = \dfrac{kg \cdot m^2}{A \cdot s^3}$
electrical resistance	ohm	Ω	$\dfrac{V}{A} = \dfrac{kg \cdot m^2}{A^2 \cdot s^3}$
frequency	hertz	Hz	$\dfrac{cycles}{s} = s^{-1}$
particle mass	atomic mass unit	u	1/12 of a carbon-12 nuclide

Appendix D
METRIC PREFIXES

Prefix	Meaning (Origin)	Symbol	Factor	Example	Application
exa-	six (Gk.)	E	10^{18}	exabyte (EB)	amount of data created every second
peta-	five (Gk.)	P	10^{15}	petahertz (PHz)	frequency of ultraviolet radiation
tera-	monster (Gk.)	T	10^{12}	terawatt (TW)	average annual power generation in the United States
giga-	giant (Gk.)	G	10^{9}	gigabyte (GB)	hard-drive storage capacity
mega-	great (Gk.)	M	10^{6}	megajoule (MJ)	energy to heat 10 L of water from 0 °C to 100 °C
kilo-	thousand (Gk.)	k	10^{3}	kilometer (km)	distances on Earth
hecto-	hundred (Gk.)	h	10^{2}	hectometer (hm)	wavelength of radio waves
deca-	ten (Gk.)	da	10^{1}	decapascal (daPa)	pressure of sound
(base)			10^{0}		
deci-	tenth (L.)	d	10^{-1}	decibel (dB)	sound loudness
centi-	hundredth (L.)	c	10^{-2}	centimeter (cm)	distances in a laboratory
milli-	thousandth (L.)	m	10^{-3}	millivolt (mV)	EKG signal from heart at skin
micro-	small (Gk.)	μ	10^{-6}	microsecond (μs)	time for light to travel 1 km
nano-	dwarf (Gk.)	n	10^{-9}	nanometer (nm)	size of atoms
pico-	small (Sp.)	p	10^{-12}	picometer (pm)	long gamma ray wavelength
femto-	fifteen (Dan.)	f	10^{-15}	femtogram (fg)	mass of a virus
atto-	eighteen (Dan.)	a	10^{-18}	attosecond (as)	period required to image an orbiting electron

Appendix E
PHYSICAL CONSTANTS

Quantity	Symbol	SI Units
atomic mass unit	u	$1.660\,539 \times 10^{-27}$ kg
Avogadro's number	N_A	$6.022\,140\,76 \times 10^{23}$ particles/mol
fundamental electric charge	e	$1.602\,176\,634 \times 10^{-19}$ C
ionization constant of water	K_w	1.0×10^{-14} mol/L at 298.15 K
molar volume (STP)	V_{STP}	22.4 L/mol
normal boiling point of water	T_b	373.15 K = 100 °C
normal freezing point of water	T_f	273.15 K = 0.00 °C
speed of light	c	$2.997\,924\,58 \times 10^8$ m/s
universal gas constant	R	$8.314 \dfrac{\text{J}}{\text{mol} \cdot \text{K}}$

Appendix F
COMMON ABBREVIATIONS AND SYMBOLS

Unit	Abbreviation	Dimension	Unit	Abbreviation	Dimension
becquerel	Bq	nuclear activity	liter	L	volume
candela	cd	light intensity	meter	m	length/distance
coulomb	C	electric charge	millibar	mbar	pressure
dalton	Da	atomic mass	minute	min	time
day	d	time	molal	m	molality
decibel	dB	relative power	molar	M	amount concentration
degree	°	temperature or angle	mole	mol	quantity (particles)
degree (Celsius)	°C	temperature	newton	N	force
degree (Fahrenheit)	°F	temperature	ohm	Ω	electrical resistance
gram	g	mass	pascal	Pa	pressure
gray	Gy	radiation absorption	percent	%	ratio (per hundred)
hertz	Hz, s^{-1}	frequency	second	s	time
hour	h	time	sievert	Sv	biological effect of radioactivity
joule	J	energy or work	volt	V	electric potential difference
kelvin	K	temperature	watt	W	power
kilogram	kg	mass	year	y	time
kilowatt-hour	kW·h	energy			

Appendix F
FORMULA QUANTITIES AND SYMBOLS

Quantity	Symbol	Quantity	Symbol
acid-ionization constant	K_a	heat	Q
activation energy	E_a	ionization constant of water	K_w
amount concentration	c_{solute}	Kelvin temperature	T_K
amount of substance (moles)	n	kinetic energy	KE
atomic number	Z	mass	m
base-ionization constant	K_b	mass fraction	w_{solute}
boiling-point elevation	ΔT_b	mass number	A
Celsius temperature	T_C	molal boiling-point constant	K_b
constant (unspecified)	k	molal freezing-point constant	K_f
critical pressure	P_c	molality	b_{solute}
critical temperature	T_c	Planck's constant	h
density	ρ	potential energy	PE
enthalpy	H	pressure	P
enthalpy of fusion	ΔH_{fus}	principal quantum number (energy level)	n
enthalpy of reaction	ΔH_r	solubility	S
enthalpy of solution	ΔH_{soln}	solubility product	K_{sp}
enthalpy of vaporization	ΔH_{vap}	specific heat	c_{sp}
entropy	S	speed of light	c
equilibrium constant	K_{eq}	standard molar enthalpy of combustion	$\Delta H°_c$
Fahrenheit temperature	T_F	standard molar enthalpy of formation	$\Delta H°_f$
freezing-point depression	ΔT_f	time interval	Δt
frequency	f	universal gas constant	R
fundamental electric charge	e	volume	V
Gibbs free energy	G	volume fraction	φ_{solute}
Gibbs free-energy change	ΔG	wavelength	λ
half-life	$t_{1/2}$		

Appendix F
OTHER ABBREVIATIONS

Term	Abbreviation
acidity (potential of hydrogen)	pH
alkalinity	pOH
alpha particle	α or ^4_2He
beta particle	β or $^0_{-1}e$
chemical state, aqueous	*(aq)*
chemical state, gas	*(g)*
chemical state, liquid	*(l)*
chemical state, solid	*(s)*
deoxyribonucleic acid	DNA
electron	e^-
fundamental electric charge	*e*
gamma ray	γ or $^0_0\gamma$
International Union of Pure and Applied Chemistry	IUPAC
kinetic energy	KE
National Institute of Standards and Technology	NIST
neutron	n or $^1_0 n$
potential energy	PE
proton	p or $^1_1 p$
ribonucleic acid	RNA
significant figure	SF
standard temperature and pressure	STP
Système International d'Unités	SI
valence shell electron pair repulsion theory	VSEPR

Appendix G
COMMON IONS

1−	2−	3−	1+
acetate, $C_2H_3O_2^-$	carbonate, CO_3^{2-}	arsenate, AsO_4^{3-}	ammonium, NH_4^+
amide, NH_2^-	chromate, CrO_4^{2-}	arsenite, AsO_3^{3-}	hydronium, H_3O^+
azide, N_3^-	dichromate, $Cr_2O_7^{2-}$	phosphate, PO_4^{3-}	
bromate, BrO_3^-	hydrogen phosphate, HPO_4^{2-}	phosphite, PO_3^{3-}	
bromite, BrO_2^-	oxalate, $C_2O_4^{2-}$		
chlorate, ClO_3^-	peroxide, O_2^{2-}		
chlorite, ClO_2^-	sulfate, SO_4^{2-}		
cyanate, OCN^-	sulfite, SO_3^{2-}		
cyanide, CN^-	thiosulfate, $S_2O_3^{2-}$		
dihydrogen phosphate, $H_2PO_4^-$			
hydrogen carbonate, HCO_3^- (bicarbonate)			2+
hydrogen sulfate, HSO_4^- (bisulfate)			mercury(I), Hg_2^{2+}
hydrogen sulfide, HS^-			
hydrogen sulfite, HSO_3^- (bisulfite)			
hydroxide, OH^-			
hypobromite, BrO^-			
hypochlorite, ClO^-			
hypoiodite, IO^-			
iodate, IO_3^-			
iodite, IO_2^-			
nitrate, NO_3^-			
nitrite, NO_2^-			
perbromate, BrO_4^-			
perchlorate, ClO_4^-			
periodate, IO_4^-			
permanganate, MnO_4^-			
thiocyanate, SCN^-			

Appendix H
STANDARD THERMODYNAMIC PROPERTY VALUES

Compound	$\Delta H°_f$ (kJ/mol)	$S°$ [J/(mol·K)]	$\Delta G°$ (kJ/mol)
Ag (s)	0	42.6	0
AgBr (s)	−99.5	107.1	−96.9
AgCl (s)	−127.0	96.3	−109.8
AgI (s)	−61.8	115.5	−66.2
Ag_2O (s)	−31.1	121.3	−11.2
Ag_2S (s)	−32.6	144.0	−40.7
Al (s)	0	28.3	0
Al_2O_3 (s)	−1675.7	50.9	−1582.3
Ba (s)	0	62.5	0
$BaCl_2$ (s)	−855.0	123.7	−806.7
$BaCO_3$ (s)	−1213.0	112.1	−1134.4
$BaSO_4$ (s)	−1473.2	132.2	−1362.2
Br_2 (l)	0	152.2	0
C (s)	0	5.7	0
Ca (s)	0	41.6	0
$CaCl_2$ (s)	−795.4	108.4	−748.8
$CaCO_3$ (s)	−1207.6	917	−1129.1
CaO (s)	−634.9	38.1	−603.3
$Ca(OH)_2$ (s)	−985.2	83.4	−897.5
$CaSO_4$ (s)	−1434.5	56.5	−477.4
CCl_4 (l)	−128.2	214.4	−65.3
CH_4 (g)	−79.6	186.3	−50.5
C_2H_2 (g)	227.4	200.9	209.9
C_2H_4 (g)	52.4	219.3	68.4
C_2H_6 (g)	−84.0	229.2	−32.0
C_3H_8 (g)	−103.8	270.3	−23.4
C_6H_6 (l)	49.1	173.4	124.5
CH_3OH (l)	−239.2	126.8	−166.6
C_2H_5OH (l)	−277.6	160.7	−174.8
Cl_2 (g)	0	165.2	0
CO (g)	−110.5	197.7	−137.2
CO_2 (g)	−393.5	213.8	−394.4
Co (s)	0	30.1	0
Cr (s)	0	23.8	0
Cr_2O_3 (s)	−1139.7	81.2	−1058.1

Compound	$\Delta H°_f$ (kJ/mol)	$S°$ [J/(mol·K)]	$\Delta G°$ (kJ/mol)
Cu (s)	0	33.2	0
CuO (s)	−157.3	42.6	−129.7
Cu_2O (s)	−168.6	93.1	−146
CuS (s)	−48.5	120.9	−86.2
$CuSO_4$ (s)	−771.4	109.2	−662.2
F_2 (g)	0	158.8	0
Fe (s)	0	27.3	0
Fe_2O_3 (s)	−824.2	87.4	−742.2
H_2 (g)	0	130.7	0
HBr (g)	−36.3	198.7	−53.4
HCl (g)	−92.3	186.9	−95.3
HF (g)	−273.3	173.8	−275.4
Hg (l)	0	75.90	0
HgO (s)	−90.8	70.25	−58.5
HgS (s)	−58.2	82.4	−50.6
HI (g)	26.50	206.6	1.7
HNO_3 (l)	−174.1	155.6	−80.7
H_2O (g)	−241.8	188.8	−228.6
H_2O (l)	−285.8	70.0	−237.1
H_2S (g)	−20.6	205.8	−33.4
H_2SO_4 (l)	−814.0	156.9	−690.0
I_2 (s)	0	116.1	0
K (s)	0	64.7	0
KBr (s)	−393.8	95.9	−380.7
KCl (s)	−436.5	82.6	−408.5
$KClO_3$ (s)	−397.7	143.1	−296.3
KF (s)	−567.3	66.6	−537.8
KOH (s)	−424.6	81.2	−379.4
Mg (s)	0	32.7	0
$MgCl_2$ (s)	−641.3	89.6	−591.8
$MgCO_3$ (s)	−1095.8	65.7	−1012.1
MgO (s)	−601.6	27.0	−569.3
$Mg(OH)_2$ (s)	−924.5	63.2	−833.5
$MgSO_4$ (s)	−1284.9	91.6	−1170.6
Mn (s)	0	32.0	0

Compound	$\Delta H°_f$ (kJ/mol)	$S°$ [J/(mol·K)]	$\Delta G°$ (kJ/mol)
MnO (s)	−385.2	59.7	−362.9
MnO_2 (s)	−520	53.1	−465.1
N_2 (g)	0	191.6	0
Na (s)	0	51.3	0
NaBr (s)	−361.1	86.8	−349.0
NaCl (s)	−411.2	72.1	−384.1
NaF (s)	−576.6	51.1	−546.3
NaI (s)	−287.8	98.5	−286.1
NaOH (s)	−425.8	64.4	−379.7
NH_3 (g)	−45.9	192.8	−16.4
NH_4Cl (s)	−314.4	94.6	−262.9
NH_4NO_3 (s)	−365.6	151.1	−183.9
NO (g)	91.3	210.8	87.6
NO_2 (g)	33.2	240.4	51.3
O_2 (g)	0	205.2	0
O_3 (g)	142.7	238.9	163.2
Pb (s)	0	51.9	0
$PbBr_2$ (s)	−278.7	161.5	−261.9
$PbCl_2$ (s)	−359.4	136.0	−314.1
PbO (s)	−217.3	66.5	−187.9
PbO_2 (s)	−277.4	68.6	−217.3
Pb_3O_4 (s)	−718.4	211.3	−601.2
PCl_3 (l)	−319.7	217.1	−272.3
S (s)	0	32.1	0
Si (s)	0	18.8	0
SiO_2 (s)	−859.4	41.5	−856.3
Sn (s)	0	51.2	0
$SnCl_4$ (l)	−511.3	258.6	−440.1
SnO (s)	−280.7	57.2	−251.9
SnO_2 (s)	−577.6	49.0	−515.8
SO_2 (g)	−296.8	248.2	−300.1
SO_3 (g)	−395.7	256.8	−371.1
Zn (s)	0	41.6	0
ZnO (s)	−350.5	43.7	−320.5
ZnS (s)	−206.0	57.7	−201.3

Appendix I
K_{sp} FOR MINIMALLY SOLUBLE SUBSTANCES

Salt	Product	K_{sp} (at 25 °C)	Molar Mass (g)	Solubility (g/L)
aluminum hydroxide, $Al(OH)_3$	$[Al^{3+}][OH^-]^3$	3.0×10^{-34}	78.0	1.28×10^{-7}
barium carbonate, $BaCO_3$	$[Ba^{2+}][CO_3^{2-}]$	2.6×10^{-9}	197.34	1.0×10^{-2}
barium sulfate, $BaSO_4$	$[Ba^{2+}][SO_4^{2-}]$	1.1×10^{-10}	233.39	2.4×10^{-3}
calcium carbonate, $CaCO_3$	$[Ca^{2+}][CO_3^{2-}]$	3.3×10^{-9}	100.09	5.7×10^{-3}
calcium sulfate, $CaSO_4$	$[Ca^{2+}][SO_4^{2-}]$	4.9×10^{-5}	136.14	0.95
iron(II) hydroxide, $Fe(OH)_2$	$[Fe^{2+}][OH^-]^2$	8.0×10^{-16}	89.87	5.3×10^{-4}
iron(III) hydroxide, $Fe(OH)_3$	$[Fe^{3+}][OH^-]^3$	3.0×10^{-39}	106.88	1.1×10^{-8}
lead(II) carbonate, $PbCO_3$	$[Pb^{2+}][CO_3^{2-}]$	1.5×10^{-13}	267.25	1.0×10^{-4}
lead(II) phosphate, $Pb_3(PO_4)_2$	$[Pb^{2+}]^3[PO_4^{3-}]^2$	1.5×10^{-32}	811.66	1.4×10^{-4}
lead(II) sulfate, $PbSO_4$	$[Pb^{2+}][SO_4^{2-}]$	2.1×10^{-8}	303.30	4.4×10^{-2}
magnesium carbonate, $MgCO_3$	$[Mg^{2+}][CO_3^{2-}]$	2.7×10^{-6}	84.32	0.139
magnesium hydroxide, $Mg(OH)_2$	$[Mg^{2+}][OH^-]^2$	5.6×10^{-12}	58.33	6.5×10^{-4}
silver carbonate, Ag_2CO_3	$[Ag^+]^2[CO_3^{2-}]$	8.5×10^{-12}	275.75	3.6×10^{-2}
silver chloride, $AgCl$	$[Ag^+][Cl^-]$	1.8×10^{-10}	143.32	1.9×10^{-3}
silver phosphate, Ag_3PO_4	$[Ag^+]^3[PO_4^{3-}]$	1.5×10^{-18}	418.58	6.4×10^{-3}
zinc hydroxide, $Zn(OH)_2$	$[Zn^{2+}][OH^{-2}]$	3×10^{-17}	99.40	1.9×10^{-4}

Appendix J
FAMILIES OF ORGANIC COMPOUNDS

Hydrocarbons

alkane	alkene	alkyne	cyclic	aromatic
propane	propene	propyne	cyclopropane	benzene

Substituted Hydrocarbons

Family	General Structural Formula	Prefix, Suffix, or Group Name	Typical Compound	
halide	R–CH₂–X (R–C(H)(H)–X)	fluoro- chloro- bromo- iodo-	1-fluoropropane	F–CH₂–CH₂–CH₃
alcohol	R–CH₂–OH	-ol	propan-1-ol	H₃C–CH₂–CH₂–OH
ether	R–O–R'	-oxy-	methoxymethane	H₃C–O–CH₃
aldehyde	R–CH=O	-al	propanal	CH₃–CH₂–CH=O
ketone	R–C(=O)–R'	-one	propan-2-one	CH₃–C(=O)–CH₃
carboxylic acid	R–C(=O)–OH	-oic acid	propanoic acid	CH₃–CH₂–C(=O)–OH
ester	R–C(=O)–O–R'	-oate	methyl propanoate	CH₃–CH₂–C(=O)–O–CH₃
amine	R–NH–R'	-amine	N-methylethan-1-amine	CH₃–CH₂–NH–CH₃
amide	R–C(=O)–NH₂	-amide	propanamide	CH₃–CH₂–C(=O)–NH₂

Glossary

A

absorption. The ability of a fluid substance to seep into the pores or interstitial spaces between the molecules of a solid.

accuracy. An evaluation of how close a measured value is to the expected or accepted value of a dimension being measured.

acid-base indicator. A substance that will change colors when it reacts with an acid or base, making it helpful in identifying acidic or basic solutions.

acid-base titration. A controlled neutralization reaction that scientists use to determine the unknown concentration of a solution.

acidic solution. A solution with a pH less than 7 and a pOH greater than 7.

acid-ionization constant (K_a). The equilibrium constant for the dissociation of an acid. Higher values indicate stronger acids.

acoustic energy. The energy carried in a sound wave.

actinoid series. A set of elements typically placed at the bottom of the periodic table that consists of the *f*-block elements from Period 7.

activated complex. An intermediate substance that forms during a chemical reaction as reactants transition into products.

activation energy (E_a). The minimum energy necessary for a chemical reaction to occur when reactants collide in the proper orientation.

actual yield. The amount of product produced when a chemical reaction is done in a laboratory.

addition reaction. A reaction in organic chemistry in which a multiple bond of a molecule is broken and two atoms or groups of atoms are added.

adhesion. The attraction between the particles of a liquid and the particles of other materials.

adsorption. A process similar to absorption but in which one substance collects on the surface of another and does not enter into the pores or interstitial space of the other substance.

alchemy. An ancient philosophy that blended astrology and mysticism with observation and experimentation and was the precursor of chemistry.

alcohol. A substituted hydrocarbon in which a hydroxyl group has replaced at least one of the hydrogen atoms.

aldehyde. A substituted hydrocarbon that contains a carbonyl group at one or both ends of a carbon chain.

aliphatic compound. An organic compound, either branched, unbranched, or a ring that does not contain delocalized electrons.

alkali metal. An element of Group 1 on the periodic table. It is a soft, highly reactive metal that has one valence electron that it easily loses to form a 1+ cation.

alkaline battery. A type of primary battery that contains a hydroxide electrolyte.

alkaline-earth metal. An element of Group 2 on the periodic table. It is a reactive metal, having two valence electrons that it easily loses to form a 2+ cation.

alkaline solution. See *basic solution*.

alkane. The simplest class of hydrocarbons consisting of substances that contain only single bonds.

alkene. A hydrocarbon that contains at least one double bond between two of its carbon atoms.

alkyl group. A hydrocarbon branch connected to another hydrocarbon.

alkyne. A hydrocarbon that contains at least one triple bond between two of its carbon atoms.

allotrope. One of the different forms that an element can have.

allotropic element. An element that can be found in different forms.

alloy. A homogenous mixture of a metal with another element, often a second metal.

alpha decay. Radioactivity that emits an alpha particle and energy, causing a decrease to both the mass number and the atomic number of the parent nuclide.

alpha helix. A secondary structure of polypeptides consisting of a tight spiral of amino acids held in place by hydrogen bonds.

alpha particle (α or ^4_2He). A particle consisting of two protons and two neutrons and having a 2+ charge.

amide. A substituted hydrocarbon in which an ammonia molecule replaces the hydroxyl group in a carboxylic acid.

amine. A substituted hydrocarbon in which an ammonia molecule combines with a hydrocarbon.

amino acid. A biochemical molecule made up of a carbon atom with an amine group, a carboxyl group, a hydrogen atom, and a side chain; a monomer of a protein.

amorphous solid. A solid that has no distinct shape or underlying pattern.

amount concentration (c_{solute}). A measure of concentration that compares the moles of a solute with the volume of an entire solution in liters; also known as *molar concentration* or *molarity*.

amphiphilic. A term used to describe a molecule that has both a polar end and a nonpolar end. The polar end of the molecule is water soluble, while the nonpolar end is not water soluble.

amphoteric substance. A substance that can act as both a Brønsted-Lowry acid and a Brønsted-Lowry base.

anabolism. Any biochemical process that combines small molecules or particles into larger molecules.

anhydrous compound. A compound without water in its crystalline structure; used to describe a hydrate with its water removed.

anion. A charged atom or group of atoms with a net negative charge due to the gain of electrons.

anode. The electrode at which oxidation occurs in an electrochemical cell. It is positively charged in an electrolytic cell and negatively charged in a voltaic cell.

applied science. The use of science to solve real-world problems.

aromatic compound. An organic compound that contains highly stable ring structures consisting in part of delocalized electrons.

Arrhenius acid. A chemical substance that when dissolved in water will ionize to form hydrogen ions (H^+) or hydronium (H_3O^+) ions; the earliest and most restrictive definition of acids.

Arrhenius base. A chemical substance that when dissolved in water will dissociate to form hydroxide ions (OH^-); the earliest and most restrictive definition of bases.

atom. The smallest particle that makes up an element and is capable of chemical interactions.

atomic crystal. A crystalline solid that forms from atoms of noble gases when they freeze.

atomic mass. The weighted average of the masses of all the naturally occurring isotopes of an element.

atomic number (Z). The unique number that identifies a specific element. It is equal to the number of protons in an atom of that element.

atomic radius. The distance from the center of an atom's nucleus to its outermost electron.

atomism. An ancient Greek model of matter that stated that matter consisted of discrete particles, called *atoms*, which established a limit on how far matter could be divided.

aufbau principle. A principle that states that electrons will fill the lowest available energy sublevels before any can occupy higher-energy sublevels.

Avogadro's law. The law that states that the volume of a gas at a constant temperature and pressure is directly proportional to the number of moles of the gas.

Avogadro's number (N). The number of particles in one mole: $6.022\ 140\ 76 \times 10^{23}$ particles.

B

band of stability. The group of stable nuclides plotted on a graph of the number of neutrons versus the number of protons.

base-ionization constant (K_b). The equilibrium constant for the dissociation of a base. Higher values indicate stronger bases.

basic solution. A solution with a pH greater than 7 and a pOH less than 7; also known as an *alkaline solution*.

battery. A device that contains one or more voltaic cells, storing chemical energy to provide an on-demand electrical energy supply.

becquerel (Bq). The SI unit for measuring the radioactivity emitted by a source.

bent molecule. A molecule shape characterized by three or four regions of electrons and two outer atoms surrounding a central atom.

benzene. A six-carbon, ring-structured hydrocarbon with delocalized electrons forming its carbon-carbon bonds; a common component of many aromatic hydrocarbons.

beta decay. Radioactivity that emits a beta particle (electron) and energy. The electron forms as a neutron changes to a proton, increasing the parent nuclide's atomic number but not its mass number.

beta particle (β or $_{-1}^{0}e$). An electron emitted from a nucleus during beta decay.

beta plus decay. See *positron emission*.

beta sheet. A secondary structure of polypeptides consisting of strands of amino acids held in place by hydrogen bonds.

Big Bang model. The most popular evolutionary theory for the origin of the universe. It suggests that all matter in the universe began in a highly dense, high-energy state and expanded rapidly to form the universe over the past 13.8 billion years.

binary acid. An acid containing hydrogen and a nonmetal.

binary compound. A compound made of only two elements.

biochemistry. The field of chemistry that focuses on the chemistry of living things: genetics, cellular respiration, and other cellular functions.

Glossary

Bohr model. A model of atoms proposed by Niels Bohr in which electrons move around the nucleus in well-defined orbits. Each orbit represents a specific energy level for electrons. Also known as the *planetary model*.

boiling. The process of vaporization caused by the vapor pressure in regions of a liquid exceeding the atmospheric pressure, allowing liquid particles to transition to the gaseous phase.

boiling point. The temperature at which the vapor pressure of a liquid is equal to the current atmospheric pressure.

boiling point elevation (ΔT_b). The increase of the boiling point of a solvent due to the presence of dissolved solute particles.

bond axis. The line formed between two atoms involved in bonding.

bonding pair. The two shared valence electrons participating in a covalent bond.

Boyle's law. The law that states that the pressure and volume of a sample of gas at a constant temperature are inversely related; $PV = k$.

branched hydrocarbon. A hydrocarbon that has alkyl groups attached such that no single chain of carbon atoms can be traced.

brine. A concentrated sodium chloride solution.

Brønsted-Lowry acid. A chemical substance that donates a proton.

Brønsted-Lowry base. A chemical substance that accepts a proton.

Brownian motion. The constant motion of fluid particles in random directions due to their kinetic energy.

buffer. A solution that resists pH changes when acids or bases are added; usually consists of a weak acid and its conjugate base or a weak base and its conjugate acid.

C

calorimeter. An insulated container, similar to a thermos, in which a thermometer detects the temperature change that occurs during a chemical reaction.

capillary action. The process of a liquid in a container moving upward against gravity due to the adhesive force between the liquid and container molecules.

capillary tube. A narrow glass tube that will draw liquids through the action of adhesion between the glass and liquid molecules.

carbohydrate. An organic compound that consists of carbon, hydrogen, and oxygen; collectively used to refer to sugars, starches, and cellulose.

carbonyl group. A functional group consisting of a carbon atom double bonded to an oxygen atom (C = O); a key feature of aldehydes and ketones.

carboxyl group. A functional group consisting of a carbon atom double bonded to an oxygen atom and to a hydroxyl group (C = OOH).

carboxylic acid. A substituted hydrocarbon containing a carboxyl group at one end.

catabolism. Any biochemical process that breaks large molecules down into smaller particles, providing building materials and energy for other processes.

catalyst. A substance that changes a reaction rate without being consumed in the chemical reaction.

catalyzed reaction. A reaction that occurs in the presence of a catalyst. The reaction rate of a catalyzed reaction is faster than the rate of the uncatalyzed reaction.

cathode. The electrode at which reduction occurs in an electrochemical cell. It is negatively charged in an electrolytic cell and positively charged in a voltaic cell.

cathode ray. The stream of electrons emitted from the cathode in a cathode-ray tube.

cation. A charged atom or group of atoms with a net positive charge due to the loss of electrons.

cellulose. A polysaccharide that forms the basis for structural components in plants.

Celsius scale. A temperature scale with one hundred degrees between the normal boiling point (100 °C) and freezing point (0 °C) of water.

chain reaction. An ongoing fission process in which the neutrons released by one reaction trigger additional reactions.

Charles's law. The law that states that the Kelvin temperature and volume of a sample of gas at a constant pressure are directly related; $\frac{V}{T} = k$.

chemical bond. A durable electrostatic attraction that forms between atoms.

chemical change. Any change that causes a substance to change its chemical identity; also known as a *chemical reaction*.

chemical energy. The potential energy stored in chemical substances.

chemical equation. A model of a chemical reaction that shows the chemicals used (reactants) and the chemicals produced (products).

chemical equilibrium. The state in a reversible chemical reaction in which the forward and reverse reaction rates are equal and the concentration of each substance is stable.

chemical formula. A combination of chemical symbols, coefficients, and subscripts that identifies the elements and number of atoms of each element in a compound.

chemical property. Any property that describes the way that matter acts in the presence of other materials or the way that it changes composition under certain conditions.

chemical reaction. Any change that causes a substance to change its chemical identity; also known as a *chemical change*.

chemical symbol. A one- or two-letter symbol that identifies an element.

chemistry. The study of matter and the changes that it undergoes.

chirality. The characteristic of a molecule that indicates that it can have either a left- or right-handed arrangement.

chromosome. A strand of DNA containing some or all of the genetic code of a cell.

coagulation. The clumping together of particles. In solutions, this refers to the clumping together of impurities in the solution that are to be removed.

coefficient. A number placed before a chemical substance in a chemical equation that indicates the number of particles of that substance in the reaction.

cohesion. The attraction between the particles of a liquid.

colligative property. A property of a solvent that changes as the number of solute particles in a solution changes.

collision theory. A theory explaining how chemical reactions occur that states that for a reaction to occur the substances (1) must collide (2) in the proper orientation and (3) with sufficient energy, called the *activation energy*.

colloid. A heterogeneous mixture formed when medium-sized particles (> 1 nm and < 1 μm) are dispersed in a dispersing medium.

combination reaction. See *synthesis reaction*.

combined gas law. The law that states that the pressure, volume, and Kelvin temperature of a sample of a gas vary jointly; $\frac{PV}{T} = k$.

combustion reaction. A chemical reaction that combines a reactant with oxygen and releases energy in various forms.

common ion. The ion that exists in two different solutes when mixed together. The concentration of both solutes must be considered when dealing with a common ion.

common-ion effect. The effect when there is a common ion in a mixture of solutions that causes the ion concentration product to exceed the K_{sp} of the less soluble salt, which will then precipitate out.

complete iconic equation. An expanded form of the chemical equation for a double-replacement reaction showing all the dissociated ions.

complex reaction. A chemical reaction that occurs in two or more elementary steps.

compound. A pure substance that consists of two or more elements chemically combined in a fixed ratio.

compressibility. The ability of a substance to be squeezed into a smaller volume by applying pressure.

concentration. A measure of the amount of a solute in a certain volume or mass of either a solvent alone or of an entire solution.

condensation. The phase change from a vapor to a liquid due to a loss of thermal energy.

condensation reaction. A reaction in organic chemistry in which molecules combine to form a larger molecule.

condensed structure formula. An abbreviated form of a structural formula that omits some components of a molecule and their bonds.

conductivity. The property of a material that enables it to transfer thermal energy or electricity.

conjugate acid. A Brønsted-Lowry base that has gained a proton.

conjugate acid-base pair. A Brønsted-Lowry acid and base that have exchanged a proton.

conjugate base. A Brønsted-Lowry acid that has given up a proton.

continuous spectrum. A complete set of all the items along a continuum.

continuous theory of matter. An ancient Greek model of matter that stated that matter could be divided infinitely into smaller portions.

conversion factor. The ratio of two equivalent measures, used to perform unit conversions.

covalent bond. A chemical bond characterized by the sharing of valence electrons between two atoms.

covalent molecular crystal. A molecular solid that is made of covalently bonded atoms held in a crystal lattice by its intermolecular forces.

covalent network solid. A nonmetallic element that is able to bond covalently to form a large network of atoms, as in diamond.

Creation Mandate. God's commandment to mankind to have dominion over the world that He created.

Glossary

critical mass. The minimum mass required to maintain a chain reaction.

critical pressure. The pressure that is needed to liquefy a gas at just below its critical temperature.

critical temperature. The temperature at which a gas can not be liquefied regardless of pressure.

crystal lattice. The orderly repeated arrangement of cations and anions in an ionic compound.

crystalline solid. A solid that has a naturally orderly shape, forming a regular three-dimensional pattern with distinct edges and sharp angles.

cyclic aliphatic compound. Any aliphatic organic compound that exists in a ring form rather than in an open chain.

D

Dalton's law of partial pressures. The law that states that the total pressure of a mixture of gases equals the sum of the pressures of each individual gas, called the *partial pressures*; $P_{total} = \Sigma P_i$.

daughter nuclide. The nuclide that is produced by any nuclear transmutation.

de Broglie hypothesis. The idea that if waves can behave like particles, then particles can behave like waves.

decanting. A process for separating mixtures in which a less dense medium is poured off the top of a more dense medium.

decomposition reaction. A chemical reaction that splits a single reactant into two or more products, having the general form $AB \longrightarrow A + B$.

deductive reasoning. The process of logically proceeding from general statements to a specific conclusion.

dehydration reaction. An elimination reaction in which water is removed from a molecule.

dehydrogenation reaction. An elimination reaction in which hydrogen atoms are removed from a molecule.

delocalized electron. A bonding electron that is associated with a group of atoms rather than forming part of a particular bond.

denaturing. The loss of protein structure caused by high heat, radiation, or toxins; results in a loss of a protein's function.

density (ρ)**.** The amount of matter in a unit of volume.

deoxyribonucleic acid. See *DNA*.

dependent variable. The variable in a controlled experiment that changes in response to variations in the independent variable.

deposition. The change of state directly from vapor to solid without transitioning through liquid due to a loss of thermal energy.

derived unit. Units in the SI that are mathematical combinations of fundamental units.

descriptive chemistry. The study of elements and the compounds they form.

diagonal rule. A visual aid to help remember the order of energy sublevels in the quantum-mechanical model of an atom.

diatomic element. An element that naturally occurs as molecules of two atoms.

diffusibility. The ability of a fluid substance to spread throughout and mix with another fluid.

diffusion. The process by which fluid matter and energy spread from a region of higher concentration to a region of lower concentration.

dilution. The process of making a solution less concentrated.

dimension. A measurable property of a phenomenon.

dipeptide. A molecule formed by the bonding of two amino acids.

dipole-dipole force. An intermolecular force between the positive end of one polar molecule and the negative end of another.

diprotic acid. An acid that can donate two protons.

disaccharide. A complex carbohydrate that is comprised of two sugar units.

dispersed phase. The medium-sized particles (> 1 nm and < 1 μm) that are spread through the dispersing medium of a colloid.

dispersing medium. The medium in which the medium-sized particles (> 1 nm and < 1 μm) of a colloid are spread.

dispersion force. See *London dispersion force*.

dissociation. The process by which ionic solutes dissolve when their ions are physically separated from each other by a solvent.

distillation. A process for separating a mixture on the basis of the different boiling points of its components.

DNA. A nucleic acid that contains deoxyribose as the sugar in its nucleotides and which encodes genetic information; the abbreviation for *deoxyribonucleic acid*.

dosimeter. A wearable device that collects data about the amount of radiation that a wearer has been exposed to.

double bond. A covalent bond that forms when two pairs of valence electrons are shared between atoms.

double-displacement reaction. See *double-replacement reaction*.

double-replacement reaction. A chemical reaction between two compounds in which the cations or the anions switch places. It has the general form $AX + BY \longrightarrow BX + AY$ (cation switching) or $AX + BY \longrightarrow AY + BX$ (anion switching). Also known as a *double-displacement reaction*.

ductility. The property of a material that enables it to be drawn into wire.

E

effusion. The process by which particles pass through a tiny opening into an evacuated chamber.

electrical energy. The potential energy of a charged particle due to its position in an electric field.

electrochemical cell. A device that uses either a spontaneous chemical reaction to produce electricity or electricity to drive a nonspontaneous chemical reaction.

electrochemistry. The field of chemistry that studies how electricity and redox reactions are related.

electrodepositing. The general term for processes that apply layers of material on an electrode through electrochemistry.

electrolysis. The process of forcing an otherwise nonspontaneous redox reaction to occur with the aid of an electrical current in an electrochemical cell.

electrolyte. Any substance that produces ions, which conduct electricity in a solution.

electrolytic cell. An electrochemical cell that use electrical energy to force a nonspontaneous chemical reaction to occur.

electromagnetic energy. The energy carried in electromagnetic waves that is produced by the vibration of charged particles and that disrupts the electromagnetic field.

electron ($e-$). A subatomic particle located in the electron cloud around the nucleus of an atom. It has a negative charge and very little mass—less than 1/1800 the mass of a proton or neutron.

electron affinity. The change in energy when an electron is added to a neutral atom to form an anion.

electron capture. A nuclear transformation caused when a low-energy-level electron is absorbed by a nucleus. The transformation changes a proton to a neutron, causing a decrease in the parent nuclide's atomic number but not to its mass number.

electron cloud. The combination of all the orbitals of an atom that forms a roughly spherical region containing all the electrons of an atom.

electron configuration. In general, the arrangement of electrons in an atom of an element in its ground state. Specifically, the notation of an atom's electron configuration, consisting of its energy sublevels and superscripts that indicate the number of electrons in each.

electron deficiency. The state of a molecule that forms with one or more of the atoms having less than eight valence electrons.

electron dot notation. A shorthand method of representing atoms and their valence electrons, consisting of the atom's chemical symbol with surrounding dots representing its valence electrons.

electron sea model. A model for explaining bonding in metals and the properties of metals that states that metals consist of an extensive crystal lattice of metal cations surrounded by and submerged in a "sea" of mobile electrons; also known as the *free-electron model*.

electronegativity. The measure of the attraction between the nucleus of an atom and shared electrons in chemical bonds.

electroplating. A type of electrolysis in which the cathode is a metal object to be plated and the anode is the metal that is to cover the object.

electrostatic force. The force between electric charges. Opposite charges attract and like charges repel.

element. A pure substance that consists of only one kind of atom.

elementary step. An individual step in a reaction mechanism.

elimination reaction. A reaction in organic chemistry in which two atoms or groups of atoms are removed and a multiple bond fulfills the octet rule for the remaining atoms.

emission spectrum. An incomplete spectrum, in the form of a series of colored lines, that is unique for each element.

empirical formula. A formula that indicates the smallest whole-number ratio of atoms in a compound.

empirical science. Science that is based on data gathered from observations or an experiment.

enantiomer. An isomer based on chirality, being either right- or left-handed.

endothermic process. A process that absorbs more energy than it releases.

end point. The point in a titration at which some change in a property of the solution is detected. Typically the end point is when an indicator changes color.

energy. The ability to do work.

Glossary

energy diagram. A graph of the energy in reactants and products during a chemical reaction.

energy shell. See *principal energy level*.

energy sublevel. Divisions of principal energy levels within an atom that are further divided into orbitals.

enthalpy (ΔH). The thermal energy content of a system at a constant pressure.

enthalpy of reaction (ΔH_r). The change in enthalpy that occurs during a reaction. Positive values indicate an endothermic reaction and negative values indicate an exothermic reaction. Also known as the *heat of reaction*.

enthalpy of solution. The net energy change during the dissolving process; the energy required to form a solution; also known as the *heat of solution*.

entropy (S). A mathematical measure of the dispersal of energy in a system.

enzyme. A naturally occurring biochemical catalyst.

equilibrium. In general, any condition in which competing influences are balanced. In chemistry, any condition in which reactions or processes are in balance.

equilibrium constant (K_{eq}). A mathematical representation of the equilibrium position for a chemical reaction at a particular temperature.

equilibrium position. The concentrations of reactants and products at equilibrium in a reversible chemical reaction.

equivalence point. The point in a titration at which an equivalent of titrant is added; that is, the point at which the number of H_3O^+ ions equals the number of OH^- ions.

essential amino acid. One of the eight amino acids that are necessary for human health but which cannot be produced by the body and must be obtained from diet.

essential oil. A mixture of volatile organic compounds, that is, carbon-based molecules that vaporize easily.

ester. A substituted hydrocarbon in which a carboxyl group connects two hydrocarbon chains; usually formed by the reaction of a carboxylic acid with an alcohol.

ether. A substituted hydrocarbon in which two hydrocarbon chains are connected by an oxygen atom.

ethics. The application of moral principles to how a person lives, including the use of chemistry.

evaporation. The process of vaporization caused by random liquid particles close to the surface of a liquid obtaining sufficient kinetic energy to leave the liquid phase.

excess reactant. A reactant that remains after a limiting reactant is consumed.

excited state. Any energy state, other than the ground state, resulting from an electron absorbing energy.

exothermic process. A process that releases more energy than it absorbs.

expansibility. The ability of a substance to spontaneously fill its available space.

experiment. An orderly process to observe a phenomenon, ideally under controlled conditions.

F

family. See *group*.

fat. A saturated ester of fatty acids and glycerol having relatively high intermolecular forces and which is typically solid at room temperature.

fatty acid. A carboxylic acid found in plant and animal fats with an even number of carbon atoms (normally twelve to eighteen).

filtration. A process for separating mixtures on the basis of particle size.

first law of thermodynamics. See *law of conservation of mass-energy*.

flammability. The property of a substance that enables it to burn in the presence of oxygen.

flocculation. The process by which coagulated material precipitates out of a solution as part of a water purification system.

fluid. Any substance that can flow.

formula unit. The chemical formula that represents the smallest whole number ratio of the elements in an ionic compound.

forward reaction. The chemical reaction depicted in an equation for a reversible reaction with the reactants on the left forming the products on the right.

fractional distillation. A process for separating mixtures in which the components have significantly different boiling points.

free-electron model. See *electron sea model*.

freezing. The phase change from a liquid to a solid due to a loss of thermal energy.

freezing point depression (ΔT_f). The decrease in the freezing point of a solvent due to the presence of dissolved solute particles.

fuel cell. A highly efficient voltaic cell in which a fuel is provided to drive a chemical reaction that produces electrical energy.

functional group. An atom or group of atoms that replaces one or more hydrogen atoms in a hydrocarbon. Also called a *substituent*.

fundamental unit. One of seven units in the SI that is based on a physical constant. The seven fundamental units are the basis for almost all the derived units in the SI.

G

gamma decay. Radioactivity that emits energy, producing an isotope that differs only in having a lower energy state.

gamma ray. The high-energy electromagnetic radiation emitted during gamma decay.

gas. A fluid state of matter characterized by widely spaced particles that interact only in momentary collisions, resulting in a changeable shape and volume; also known as *vapor*.

Gay-Lussac's law. The law that states that the Kelvin temperature and pressure of a sample of gas at a constant volume are directly related; $\frac{P}{T} = k$.

Geiger counter. A device used to measure ionizing radiation by measuring the current produced by ionized gases in the counter.

gene. A subunit on a chromosome that contains information for the sequencing of amino acids in a protein.

Gibbs free energy (ΔG). A mathematical measure of a chemical system that incorporates changes in enthalpy and entropy. A negative value for Gibbs free energy indicates a spontaneous reaction; a positive value indicates a reaction that is nonspontaneous.

glass. See *supercooled liquid*.

glycerol. A three-carbon alcohol with a hydroxyl group on each carbon; forms the backbone of triglycerides.

glycogen. A polysaccharide that stores energy in humans and animals.

Graham's law of diffusion. See *Graham's law of effusion*.

Graham's law of effusion. The law that states that the rate of effusion for a gas is inversely proportional to the square root of its molar mass; also known as *Graham's law of diffusion*.

gray (Gy). The SI unit for measuring the radioactivity absorbed by an object.

Greek prefix system. A system used to name covalently bonded compounds that includes prefixes to indicate the number of atoms of each element in the compound.

greenhouse gas. Any atmospheric gas that transmits incident radiant energy but absorbs reflected energy, retaining the energy as thermal energy in the atmosphere.

ground state. The lowest energy level of an electron in the Bohr model.

group. A column of the periodic table that includes elements that have similar properties because they have similar electron configurations. Sometimes called a *family*.

H

Haber process. A chemical process that converts nitrogen gas into ammonia, which is vital to producing artificial fertilizers. Also known as the *Haber-Bosch process*.

half-cell. One of the two halves of a voltaic cell in which either the oxidation or the reduction occurs.

half-life. The amount of time required for half of a radioactive sample to decay.

half-reaction. Either of the separate oxidation and reduction reactions that occurs in a half-cell of a voltaic cell.

halide. Compounds that contain a halogen, often a halogen combined with one other element.

Hall-Héroult process. An electrolysis process used to produce aluminum.

haloaromatic. A substituted aromatic hydrocarbon in which a halogen has replaced at least one of the hydrogen atoms.

halocarbon. A substituted hydrocarbon in which a halogen has replaced at least one of the hydrogen atoms.

halogen. An element of Group 17 on the periodic table. It is a highly reactive nonmetal with seven valence electrons. It easily gains an additional electron to make a 1– anion.

heat. The movement of thermal energy due to a temperature difference.

heat of reaction. See *enthalpy of reaction*.

heat of solution. See *enthalpy of solution*.

Heisenberg uncertainty principle. A fundamental property of all submicroscopic systems described as the impossibility of simultaneously knowing both the energy, or momentum, and the exact position of a particle.

Henry's law. The law that states that the solubility of a gas is directly proportional to the partial pressure of that gas above the solution.

Hess's law. The law that states that the enthalpy change of a reaction equals the sum of the enthalpy changes for each step of the process.

heterogeneous catalyst. A catalyst that is in a different phase, or state of matter, as the substances in a chemical reaction.

heterogeneous equilibrium. The state of equilibrium between substances that are in different states of matter.

Glossary

heterogeneous mixture. A mixture that has two or more distinct regions, which give it a nonuniform appearance.

historical science. The body of scientific knowledge acquired by making inferences about events and processes in the past on the basis of observations of evidence in the present.

HIV. Two species of retrovirus that cause AIDS. Acronym for *human immunodeficiency virus*.

homogeneous catalyst. A catalyst that is in the same phase, or state of matter, as the substances in a chemical reaction.

homogeneous equilibrium. The state of equilibrium between substances that are all in the same state of matter.

homogeneous mixture. A mixture that has only one phase, which gives it a uniform appearance.

human immunodeficiency virus. See *HIV*.

Hund's rule. The rule stating that electrons fill a sublevel by a single electron occupying each orbital before a second electron can occupy any orbital.

hydrate. Compounds that hold a certain amount of water within their crystalline structures.

hydride. Compounds that contain hydrogen acting as an anion, often hydrogen combined with one metallic element.

hydrocarbon. An organic compound that contains only carbon and hydrogen.

hydrogen bond. An especially strong type of dipole-dipole force that forms between hydrogen in one molecule and nitrogen, oxygen, or fluorine in another molecule.

hydroxide. Compounds, often ionic, that contain a hydroxide ion (OH⁻).

hydroxyl group. A functional group having the formula OH; a key feature of alcohols.

hypervalent molecule. A molecule that forms with one or more of the atoms having more than eight valence electrons.

hypothesis. A testable explanation for a phenomenon that forms the basis for scientific inquiry.

I

ideal gas. A gas whose behavior is perfectly predicted by the kinetic-molecular theory.

ideal gas law. The law that relates the pressure, temperature, volume, and the number of moles in any sample of a gas; $PV = nRT$.

immiscibility. The tendency of two liquids to not mutually dissolve. They remain in distinct layers.

independent variable. The variable in a controlled experiment that is manipulated by a scientist.

inductive reasoning. The process of logically proceeding from specific statements to a general conclusion.

inhibitor. A substance that reduces the effectiveness of a catalyst, slowing its action.

inner transition metal. An element from the *f* block of the periodic table (lanthanoid and actinoid series).

intermediate. A chemical substance that is formed during one elementary step of a reaction mechanism and is consumed in a later elementary step.

intermolecular force. A type of electrostatic attraction that forms between molecules and explains many bulk properties of matter. Also known as a *van der Waals force*.

International Union of Pure and Applied Chemistry (IUPAC). An international organization responsible for standardization in chemistry.

ion. A charged atom or group of atoms due to the gain or loss of electrons.

ionic bond. A chemical bond due to the attraction between ions formed by the transfer of valence electrons.

ionic crystal. An ionic solid made of a repeating network of ions defined by a unit cell.

ionization. The process by which covalent solutes dissolve as a solvent separates out ions from a solute.

ionization constant of water (K_w). The equilibrium constant for the self-ionization of water. The value at 25 °C is 1.0×10^{-14}.

ionization energy. The amount of energy required to remove an electron from an atom.

ionizing radiation. Any radioactivity of sufficient energy to ionize atoms with which it collides.

irreversible reaction. A reaction that can occur only in the forward direction.

island of stability. A theoretical group of stable nuclides with atomic numbers higher than those for presently known stable nuclides.

isomer. A substance whose molecular formula is identical to that of another substance but which has a different structural formula.

isotope. One of two or more variants of atoms of the same element, having the same number of protons but different numbers of neutrons, resulting in them having different mass numbers.

isotope notation. A special notation, used to identify important information about an isotope, consisting of a chemical symbol, atomic number, and mass number.

J

joule (J). The SI derived unit for work, energy, and heat; $1\text{ J} = 1\text{ N} \cdot \text{m} = 1\frac{\text{kg} \cdot \text{m}^2}{\text{s}^2}$.

K

Kelvin scale. An absolute temperature scale with 100 kelvins between the normal boiling point (373.15 K) and freezing point (273.15 K) of water.

ketone. A substituted hydrocarbon in which a carbonyl group exists anywhere between the end carbons of a carbon chain.

kinetic energy. The energy that an object has due to its motion.

kinetic-molecular theory. The model of matter that states that properties of matter can be explained by the interrelationships between the forces between particles and the particles' kinetic energy.

kinetics. The study of the rates of chemical reactions and the steps by which they occur.

L

lanthanoid series. A set of elements typically placed at the bottom of the periodic table that consists of the f-block elements from Period 6.

latent heat. The thermal energy that produces a phase change when transferred to or from a substance.

lattice energy. The energy released when gaseous particles form crystals.

law. A scientific model that describe a phenomenon, often in mathematical terms.

law of charges. The law that states that opposite charges attract and like charges repel.

law of chemical equilibrium. The law that states that at a specific temperature a chemical system may reach a point at which the ratio of the concentration of the products to the reactants is constant.

law of combining volumes. The law that states that under equivalent conditions, the volumes of reacting gases and their gaseous products can be expressed in ratios of small whole numbers.

law of conservation of mass-energy. The law that states that matter and energy cannot be created or destroyed but can change from one form to another, including changes between the two; also known as the *first law of thermodynamics*.

law of definite proportion. The law that states that every compound is formed of elements combined in specific ratios by mass that are unique for that compound.

law of energy conservation. The law that states that energy cannot be created or destroyed but can change from one form to another.

law of entropy. The law that states that the entropy of an isolated system will always increase; that is, its energy will become more spread out; also known as the *second law of thermodynamics*.

law of multiple proportions. The law that states that whenever a fixed amount of one element can combine with different masses of a second element, the ratio of masses can be reduced to small, whole numbers.

law of octaves. An early form of the periodic law devised by John Newlands as he noticed that every eighth known element had similar properties when ordered by increasing atomic mass.

Le Châtelier's principle. The principle that states that when a stress is put on a system in chemical equilibrium, the system will adjust in order to reduce the effect of the stress.

Lewis acid. Any substance that can accept a pair of electrons; the broadest definition of an acid.

Lewis base. A substance that can donate a pair of electrons; the broadest definition of a base.

Lewis structure. A visual, two-dimensional representation of a molecule that shows its atoms, bonds, and unbonded electron pairs.

limiting reactant. The reactant that is completely consumed in a chemical reaction.

linear hydrocarbon. A hydrocarbon without any alkyl groups attached; a hydrocarbon with only a single chain of carbon atoms.

linear molecule. A molecule shape characterized by two regions of electrons and two outer atoms on either side of a central atom.

lipid. A nonpolar biochemical molecule that is insoluble in water and is vital for energy storage; includes fats, oils, waxes, and steroids.

liquid. A fluid state of matter characterized by closely spaced particles that can slide past each other, resulting in a fixed volume but changeable shape.

litmus paper. Paper that has been treated with an acid-base indicator that changes color in the presence of an acid or a base. Blue litmus paper turns red in the presence of an acid, and red litmus paper turns blue in the presence of a base.

localized electron theory. See *valence bond theory*.

London dispersion force. An intermolecular force that forms between both polar and nonpolar molecules due to temporary dipoles formed from the random motion of the valence electrons in the participating molecules; also known as a *dispersion force* or *London force*.

Glossary

London force. See *London dispersion force*.

lone pair. A pair of electrons in a bonded atom that are not involved in bonding; also known as a *nonbonding pair*.

M

magic number. The number of protons or neutrons associated with a full nuclear shell. Nuclides with a magic number of protons or neutrons tend to be stable.

malleability. The property of a material that enables it to be hammered or pressed into a sheet.

mass defect. The mass difference between a nucleus and the sum of the masses of the components that make up that nucleus.

mass fraction (w_{solute}). A measure of concentration that compares the mass of a solute with the mass of an entire solution and is reported as a decimal or percent; also known as *percent by mass*.

mass number (A). The sum of the protons and neutrons in an atom that is used to identify specific isotopes of an element.

matter. Anything that takes up space and has mass.

measurement. A comparison of an unknown quantity to a known standard unit.

melting. The phase change from a solid to a liquid due to a gain of thermal energy.

melting point. The temperature at which a solid changes to a liquid.

meniscus. The curved surface of a liquid caused by the interplay between cohesion and adhesion.

messenger RNA. RNA that carries protein synthesis instructions from the nucleus to the ribosomes; also known as *mRNA*.

metabolism. A collective term for all the chemical reactions that take place within cells.

metabolite. An intermediate in a chemical reaction that carries out metabolism.

metal. One of a general group of elements with similar characteristics, such as being solid at room temperature, lustrous, malleable, ductile, and a good conductor of heat and electricity.

metallic bond. A chemical bond characterized by the attraction between all the metal ions in a sample and their shared valence electrons.

metallic crystal. A network of positive metal ions surrounded by a sea of valence electrons.

metalloid. One of a general group of elements with characteristics that are between those of metals and nonmetals; also known as a *semiconductor*.

miscibility. A measure of the tendency of two liquids to mutually dissolve and mix in a solution.

mixture. The combination of two or more substances that are physically combined in a changeable ratio.

model. A workable explanation, description, or representation of a phenomenon.

molal. A term used to indicate the molality of a solution.

molal boiling point constant (K_b). A constant for a solvent that indicates the change to the boiling point of a solution effected by the presence of solute particles.

molal freezing point constant (K_f). A constant for a solvent that indicates the change to the freezing point of a solution affected by the presence of solute particles.

molality (b_{solute}). A measure of concentration that compares the moles of a solute with the mass of a solvent in kilograms.

molar. A term used to indicate the amount concentration of a solution.

molar concentration. See *amount concentration*.

molar enthalpy of fusion (ΔH_{fus}). The quantity of energy required to melt one mole of a solid to a liquid with no temperature change; also known as the *molar heat of fusion*.

molar enthalpy of vaporization (ΔH_{vap}). The quantity of energy required to vaporize one mole of a liquid to a gas with no temperature change; also known as the *molar heat of vaporization*.

molar heat of fusion. See *molar enthalpy of fusion*.

molar heat of vaporization. See *molar enthalpy of vaporization*.

molar mass. The mass of one mole of any pure substance.

molarity. See *amount concentration*.

mole. The amount of any substance that contains exactly $6.02214076 \times 10^{23}$ particles.

mole ratio. The ratio of coefficients in a balanced chemical equation that is used in stoichiometric calculations.

molecular dipole moment. The measure of the polar nature of a molecule.

molecular formula. The formula that indicates the types and numbers of atoms found in a molecule, ion, or formula unit.

molecular orbital theory. A model of chemical bonding that states that atoms obtain full valence shells by the formation of new molecular orbitals.

molecule. A distinct group of atoms that are covalently bonded together.

monatomic element. An element that can naturally occur as an individual atom.

monomer. A small molecule-like structure that forms the basis for creating a polymer.

monoprotic acid. An acid that can donate only one proton.

monosaccharide. A simple carbohydrate consisting of a single sugar unit.

monounsaturated fat. A lipid that contains one carbon-carbon double bond.

mRNA. See *messenger RNA*.

N

naturalism. A worldview that is based on a belief that matter is all that exists and that human reasoning informed by science is the only reliable path to truth.

net ionic equation. An abbreviated form of the chemical equation for a double-replacement reaction showing only the ions involved in the chemical change. Spectator ions are removed.

network covalent substance. A covalent substance that forms into a crystalline structure instead of forming distinct molecules (e.g., diamond).

neutralization reaction. A chemical reaction in which an acid and a base react to form water and a salt.

neutral solution. A solution with a pH and pOH both equal to 7.

neutron (n^0 or n). A subatomic particle located in the nucleus of most atoms. It has no charge and a mass of approximately 1 u.

noble gas. An element of Group 18 on the periodic table. It is a nonreactive (inert) nonmetallic gas with a full valence shell.

noble gas notation. An abbreviated form of electron configuration consisting of the last noble gas in the configuration in brackets followed by any additional electron configuration for that element.

nomenclature. A system of naming that follows a standardized set of rules. The International Union of Pure and Applied Chemistry (IUPAC) has developed and oversees a nomenclature for compounds.

nonbonding pair. A pair of electrons in a bonded atom that are not involved in bonding; also known as a *lone pair*.

nonelectrolyte. Any substance that will not conduct electricity when dissolved.

nonmetal. One of a general group of elements with similar characteristics, such as generally being a gas or a soft, crumbly solid at room temperature, and being a poor conductor of heat and electricity.

nonrechargeable battery. A single-use battery that is not designed to be rechargeable. Its electrochemical process cannot be safely reversed. Also known as a *primary battery*.

normal boiling point. The temperature at which the vapor pressure of a liquid is equal to the standard atmospheric pressure of 101.325 kPa.

nuclear binding energy. The energy that holds a nucleus together. It is equivalent to the energy released as the nucleus is formed and is related to the mass defect of the nucleus.

nuclear bombardment reaction. Any type of nuclear transformation that is triggered by striking a nucleus with another particle.

nuclear chemistry. The study of the nuclei of elements, particularly their composition as they undergo nuclear changes.

nuclear energy. The potential energy in the nucleus of an atom due to the interaction of the subatomic particles and the electromagnetic, weak nuclear, and strong nuclear forces.

nuclear equation. An equation used to model nuclear reactions. Nuclear equations are balanced by balancing both the mass numbers and the atomic numbers across the equation.

nuclear fission. The nuclear process in which a heavy nucleus is split apart into smaller, more stable nuclei.

nuclear fusion. The transmutation process of combining light nuclei to form a more massive nucleus.

nuclear radiation. The energy and particles emitted from unstable nuclei as they move toward stability.

nucleic acid. A chain of nucleotides that encodes the information needed for the production of proteins.

nucleon. Any particle found in a nucleus, whether proton or neutron.

nucleotide. A monomer of a nucleic acid that consists of a five-carbon sugar, a phosphate group, and a nitrogenous base.

nucleus. The extremely small and dense center of an atom that contains the protons and neutrons and is surrounded by an electron cloud.

nuclide. A unique atom of an element with a specific set of protons and neutrons and a specific energy state.

O

octet rule. A rule that states that elements are most chemically stable when they have a full valence electron shell, typically with eight valence electrons.

oil. An unsaturated ester of a fatty acid and glycerol having relatively low intermolecular forces and which is typically liquid at room temperature.

Glossary

operational science. The body of scientific knowledge acquired by methods of investigation that involve real-time observations of present-day phenomena or first-person observations of events in the past.

orbital. A three-dimensional region representing the most probable position for an electron according to the quantum-mechanical model.

orbital hybridization. The formation of a new orbital due to the overlapping of atomic orbitals.

orbital notation. A notation or diagram used to illustrate the electron configuration of an atom, consisting of horizontal lines, representing the orbitals and labeled by sublevel, and arrows, representing the electrons in that orbital.

organic chemistry. The study of covalently bonded carbon compounds, with the exception of carbonates, carbon oxides, and carbides.

organic compound. Any covalently bonded carbon compound, with the exception of carbonates, carbon oxides, and carbides.

osmosis. The movement of a solvent across a semipermeable membrane due to differences in the solution concentration on each side of the membrane.

osmotic pressure. The amount of pressure required to prevent osmosis from occurring.

oxidation. A chemical process in which an element loses electrons, increasing the oxidation number of the element.

oxidation number. The number of electrons that an atom loses or gains when it bonds; also known as *oxidation state*.

oxidation-reduction reaction. A class of chemical reactions that involves the transfer of electrons; also known as a *redox reaction*.

oxidation state. See *oxidation number*.

oxide. A compound that contains oxygen, often oxygen combined with one other element.

oxidizing agent. A chemical substance that causes the oxidation of another substance in a redox reaction. The oxidizing agent is reduced in a redox reaction.

oxyanion. An anion that contains oxygen and one other element.

P

paramagnetism. The attraction of a molecule to a magnetic field.

parent nuclide. The starting nuclide in any nuclear transmutation.

partial pressure. The pressure contributed by an individual gas to the total pressure of a mixture of gases.

pascal (Pa). The SI unit of pressure; $1 \text{ Pa} = 1 \text{ N/m}^2$.

Pauli exclusion principle. A principle that states that the two electrons in an orbital must have opposite spins.

peer review. The process of having other scientists review scientific work by which the research is evaluated.

peptide bond. The bond formed between the carboxyl group of one amino acid and the amine group of another amino acid.

percent by mass. See *mass fraction*.

percent by volume. See *volume fraction*.

percent composition. The percent ratio of the mass of a component of a compound to the total mass of the compound.

percent yield. The percent ratio of an actual yield to its corresponding theoretical yield.

period. A row in the periodic table that includes elements that have their valence electrons in the same energy level. The period number represents the valence shell. Also known as a *series*.

periodic law. The law that states that the properties of elements vary periodically with their atomic numbers.

periodic table. A table of all the known elements ordered by increasing atomic number and arranged in a manner that models the atomic structure of the elements.

periodic trend. The pattern of change in particular properties as one moves along periods and groups in the periodic table.

periodicity. A repeating pattern in a characteristic; demonstrated in the periodic table as elements in the same family having similar properties.

pH. The potential of hydrogen ions is a simple indication of the acidity of a solution ($pH = -\log[H_3O^+]$); related to pOH by the equation $pH + pOH = 14$.

pH meter. An electronic device that reports a numerical value of the pH of a solution.

pharmacokinetics. The study of how the human body processes medications.

phase change. The transition between different states, or phases, of matter due to the transfer of thermal energy.

phase diagram. A graph of temperature and pressure that shows the states of matter and phase changes for a particular substance.

phospholipid. A lipid consisting of two fatty acid chains bonded to a phosphate group instead of to glycerol.

photon. A massless particle of light consisting of a bundle of wave energy.

physical change. Any change in the appearance, shape, or state of matter of a material that does not cause a change in its chemical composition.

physical property. Any property of matter that can be observed or measured without altering a substance's chemical composition.

pi (π) bond. A bond that forms when side by side orbitals overlap, forming two separate regions of the same bond.

planetary model. See *Bohr model*.

plum pudding model. The atomic model suggested by J. J. Thomson that viewed an atom as a sphere of positive charge in which numerous negative electrons are embedded.

pOH. A simple indication of the alkalinity of a solution (pOH = −log[OH⁻]).

polar covalent bond. A covalent bond with positive and negative ends caused by the unequal sharing of valence electrons.

polar molecule. A molecule with both a positive end and a negative end due to the polar nature of its bonds and the shape of the molecule.

polarity. The tendency of an object to form two localized regions of opposite character or charge.

polyatomic element. An element that naturally occurs as molecules of two or more atoms.

polyatomic ion. A charged group of covalently bonded atoms that act as a single ion.

polymer. A large molecule made of regularly repeated smaller subunits called *monomers*.

polymerization. Any process that combines monomers into polymers.

polymorph. A crystalline solid that can form different crystal lattices.

polypeptide. A molecule formed by the bonding of three or more amino acids.

polyprotic acid. An acid that can donate more than one proton.

polysaccharide. A complex carbohydrate that is composed of many sugar units.

polyunsaturated fat. A lipid, typically an oil, that contains two or more carbon-carbon double bonds.

positron ($_{+1}^{0}e$). A subatomic particle identical to an electron, but having a positive charge. It is emitted from a nucleus during positron emission.

positron emission. Radioactivity that emits a positron (positive electron, 1+ charge) and energy. The positron forms as a proton changes to a neutron, causing a decrease in the parent nuclide's atomic number but not its mass number. Also known as *beta plus decay*.

post-transition metal. A metallic element in Groups 13–16. It easily loses electrons to form a 2+ cation.

potential energy. The energy that an object has due to its position or condition.

precipitate. A solid that settles out of a solution as the result of physical or chemical changes.

precision. An evaluation of the exactness of a measurement or a measuring instrument. It indicates how repeatable a measurement is or how exactly one can make a measurement.

premise. A statement that forms the basis of a logical argument.

pressure (P). The force per unit area caused by the collisions of fluid particles with their container.

presupposition. An assumption about the world. The sum total of a person's presuppositions forms the basis of his worldview.

primary battery. A single-use battery that is not designed to be rechargeable. Its electrochemical process cannot be safely reversed. Also known as a *nonrechargeable battery*.

primary structure. The basic order of amino acids in polypeptide chains.

principal energy level (n). The orbits or energy levels in atomic models; also known as *energy shells*.

principal quantum number. The quantum number that identifies the principal or main energy level of an electron.

product. A chemical substance that is formed in a chemical reaction. It is found on the right side of a chemical equation.

protein. An organic polymer comprised of amino acids.

protein complex. A large biomolecule formed by the combination of two or more protein subunits.

protein subunit. A finished polypeptide that combines with other protein subunits to form a protein complex.

protein turnover. The process in living things of breaking down and replacing worn-out proteins.

proton (p^+ or p). A subatomic particle located in the nucleus of every atom. It has a positive charge and a mass of approximately 1 u. The number of protons in an atom is equal to the atomic number of that element.

pure science. The effort to understand how and why things work the way that they do.

pure substance. Any material that consists of only one type of matter (an element or a compound).

Glossary

Q

qualitative data. Data in the form of words that is used to describe something.

quantitative data. Data in the form of numbers that is determined through measuring.

quantized. The condition of a phenomenon having only discrete values for a particular dimension.

quantum-mechanical model. The current model of the atom in which protons and neutrons are located in the nucleus with electrons in orbitals around the nucleus. Orbitals represent the regions of highest probability for the locations of electrons.

quantum mechanics. The study of the mechanics of matter at the atomic and subatomic levels, with an emphasis on the quantized nature of their energies and momentums.

quantum numbers. A series of four numbers that uniquely identifies the placement of an electron in an atom according to the quantum-mechanical model.

R

radioactive decay. See *radioactivity*.

radioactive decay series. A sequence of radioactive decays through which an isotope proceeds in order to become a stable nuclide.

radioactivity. The spontaneous emission of nuclear radiation from the nuclei of unstable atoms as they endeavor to become more stable; also known as *radioactive decay*.

radioisotope. Any isotope that emits nuclear radiation.

rate-determining step. The elementary step in a reaction mechanism that determines how fast the overall reaction can occur. The reaction can go no faster than the slowest elementary step.

rate law. A mathematical law relating the concentration of reactants to the rate of a reaction. Rate laws can be determined only empirically.

reactant. A chemical substance that is consumed in a chemical reaction. It is found on the left side of a chemical equation.

reaction mechanism. The series of steps that make up a chemical reaction.

reaction order. An indication of the effect that the concentration of a particular reactant has on a reaction rate.

reaction rate. A measure of how quickly reactants change into products.

reactivity. The property of a substance that enables it to interact with another specific chemical.

rechargeable battery. A battery that is designed so that its electrochemical process can be reversed, allowing the battery to be recharged; also known as a *secondary battery*.

redox reaction. A class of chemical reactions that involves the transfer of electrons; also known as an *oxidation-reduction reaction*.

reducing agent. A chemical substance that causes the reduction of another substance in a redox reaction. The reducing agent is oxidized in a redox reaction.

reduction. A chemical process in which an element gains electrons, decreasing the oxidation number of the element.

resonance. The state of molecules that can have two or more possible Lewis structures. The observed state of the molecule is intermediate to the possible structures.

reverse osmosis. The process by which a solvent is forced to move "backward" across a semipermeable membrane by applying a force to the system.

reverse reaction. A chemical reaction that occurs when the concentration of products increases and the products react to form the original reactants. A reverse reaction is the opposite of the corresponding forward reaction.

reversible reaction. A reaction that can occur in both the forward and the reverse directions.

ribonucleic acid. See *RNA*.

ribosomal RNA. RNA of which the ribosomes in a cell are made; also known as *rRNA*.

RNA. A nucleic acid that contains ribose as the sugar in its nucleotides and which encodes the instructions needed for protein synthesis in cells; the abbreviation for *ribonucleic acid*.

roman numeral system. See *Stock system*.

rRNA. See *ribosomal RNA*.

S

salt. A substance formed when the anion of an acid and the cation of a base combine.

salt bridge. A tube of electrolytic gel that connects the two half-cells of a voltaic cell in order to permit the movement of ions to complete a circuit.

saturated hydrocarbon. A hydrocarbon in which all the carbons are surrounded by four other atoms, the maximum number possible. Saturated hydrocarbons have only single bonds between their carbon atoms.

saturated solution. A solution that contains the maximum amount of solute that it can hold at current conditions.

scientific inquiry. The process by which scientists answer scientific questions; also known as the *scientific method*.

scientific survey. A process that involves randomly selected representative samples from a larger population.

scintillation counter. A device used to measure ionizing radiation by measuring the light produced by ionized gases in the counter.

secondary battery. A battery that is designed so that its electrochemical process can be reversed, allowing the battery to be recharged; also known as a *rechargeable battery*.

secondary structure. A more complex structure of polypeptides in which the amino acids in one part of a chain bond with amino acids in other parts of the chain. Alpha helixes and beta sheets are the most common secondary structures.

second law of thermodynamics. See *law of entropy*.

self-ionization of water. A reaction that occurs in water in which two water molecules react to form a hydronium ion and a hydroxide ion.

semiconductor. See *metalloid*.

sensible heat. The transfer of thermal energy that produces a temperature change in a substance.

series. See *period*.

SI. The modern metric system used for science and adopted almost worldwide for daily usage. An abbreviation for the French *système international d'unités* for "international system of units."

sievert (Sv). The SI unit for measuring the biological effect of radioactivity on an organism.

sigma (σ) bond. A bond that forms when orbitals overlap along the bond axis.

significant figure. The digits in a measurement that include those that are known with certainty plus one estimated digit. An indication of the precision of a measurement.

single-displacement reaction. See *single-replacement reaction*.

single-replacement reaction. A chemical reaction in which a more reactive element replaces a less reactive element in a compound. It has the general form $A + BX \longrightarrow AX + B$ (cation being replaced) or $X + AY \longrightarrow AX + Y$ (anion being replaced). Also known as a *single-displacement reaction*.

smelting. The process of melting an ore in order to separate a metal from its ore.

solid. A state of matter characterized by closely spaced particles vibrating in fixed positions, which give a material a fixed shape and volume.

solubility. The maximum amount of a solute that can dissolve in a specific solvent under specific conditions, such as temperature and pressure.

solubility product constant (K_{sp}). The equilibrium constant for the process of dissolving a solute.

solute. The substance that is dissolved by a solvent; the substance(s) in lower abundance in a solution.

solution. A homogeneous mixture of variable composition in a single phase.

solvation. The dissolving process in solid-in-liquid solutions in which the solvent particles separate the solute particles and then surround them.

solvent. The most abundant substance in a solution; the substance that dissolves the solute.

specific heat (c_{sp}). The amount of thermal energy required to raise the temperature of 1 g of a substance by 1 °C.

specific rate constant. The empirically determined value for the rate law for a particular chemical reaction.

spectator ion. An ion in a double-replacement reaction that is not involved in the chemical reaction.

spectroscopy. The analysis of light emitted or absorbed by matter.

spin. A measure of the angular moment of an electron in an orbital. Each of the two electrons in an orbital must have opposite spins.

spontaneous combustion. A combustion reaction that starts without an ignition source due to existing conditions, typically high temperatures.

standard molar enthalpy of combustion ($\Delta H°_c$). The thermal energy released by the complete burning of 1 mol of a substance at standard conditions; also known as the *standard molar heat of combustion*.

standard molar enthalpy of formation ($\Delta H°_f$). The enthalpy change to produce 1 mol of a compound in its standard state from its elements in their standard states; also known as the *standard molar heat of formation*.

standard molar heat of combustion. See *standard molar enthalpy of combustion*.

standard molar heat of formation. See *standard molar enthalpy of formation*.

standard molar volume. The property of gases that states that a mole of a gas at standard temperature and pressure occupies a volume of 22.4 L.

standard temperature and pressure (STP). A temperature and pressure agreed upon by chemists for comparison of research. STP is 0 °C (273.15 K) and 101.325 kPa.

starch. A complex carbohydrate that contains multiple glucose molecules.

Glossary

steroid. A lipid comprised of three six-carbon rings and a five-carbon ring.

Stock system. A system for naming ionic compounds with cations that can have more than one oxidation number; also known as the *roman numeral system*.

stoichiometry. The mathematical relationship between the quantities of substances involved in a chemical reaction.

strong acid. A substance that ionizes easily, freely giving up at least one proton.

strong base. A substance that easily accepts protons.

strong nuclear force. The short-range force that holds protons and neutrons together in a nucleus.

subcritical mass. A mass less than the minimum mass required to maintain a chain reaction.

sublimation. The change of state directly from solid to vapor without transitioning through liquid due to a gain of thermal energy.

substituent. See *functional group*.

substituted hydrocarbon. Any hydrocarbon in which a hydrogen has been removed and replaced by a functional group.

substitution reaction. A reaction in organic chemistry in which one part of a molecule is replaced with a new element or functional group.

sugar. A generic term for sweet-tasting saccharides.

sulfide. A compound that contains sulfur, often sulfur combined with one other element.

supercooled liquid. A term used for some amorphous solids that are cooled so fast that they cannot form a crystalline structure; also known as *glass*.

supersaturated solution. A solution that contains more than the maximum amount of solute that it can normally hold at current conditions.

surface tension. The elastic "skin" that forms on the surface of liquids, caused by the imbalance of intermolecular forces at the liquid's surface.

surfactant. Any substance that breaks down normal surface tension by interfering with hydrogen bonds.

suspension. A heterogeneous mixture formed when large particles (> 1 μm) are suspended in another substance. The suspended particles are so large that they will eventually settle out.

synthesis reaction. A chemical reaction that combines two or more reactants into a single product, having the general form A + B ⟶ AB; also known as a *combination reaction*.

system. Everything within a set of boundaries, defined by a scientist, that is being studied.

T

temperature. A measure of the average kinetic energy of all the particles within a substance.

ternary acid. An acid containing more than two elements, namely hydrogen, oxygen, and at least one other nonmetal.

tertiary structure. The three-dimensional shape of polypeptides resulting from combinations of turns, helixes, sheets, and random coils.

tetrahedral molecule. A molecule shape characterized by four regions of electrons and four outer atoms surrounding a central atom. An outer atom is at each vertex of a tetrahedron.

theoretical yield. The predicted amount of product produced by a chemical reaction as determined by a stoichiometric calculation.

theory. A scientific model that explains a phenomenon.

thermal energy. The sum of the kinetic energy of all the particles within a substance.

thermochemical equation. A chemical equation that shows the reactants, products, and amount of energy that is released or absorbed as heat.

thermochemistry. The branch of science that studies the transfer of energy during phase or chemical changes.

thermodynamics. The study of the movement and conversion of energy, especially thermal energy.

thermometric property. A physical property that changes in a predictable way with changes in temperature.

third law of thermodynamics. The law that states that we can never decrease a system's entropy to an absolute minimum.

titrant. The solution of known concentration used in a titration.

titration curve. A graph showing the pH of a system during titration as an acid and base are combined.

toxicity. The property of a substance that makes it harmful to living organisms.

transfer RNA. RNA that transports appropriate amino acids to ribosomes for protein synthesis; also known as *tRNA*.

transition interval. The range of pH values over which an indicator transitions from one color to another.

transition metal. An element from the *d* block of the periodic table (Groups 3–12). It is a metal that typically has one or two valence electrons that it easily loses to form a 1+ or 2+ cation.

translation. The process of building a polypeptide on the basis of information encoded in mRNA.

transmutation. Any process by which an atom of one kind of element changes into another kind of element as a result of a change in the number of protons in the atom's nucleus.

transuranium element. An element with an atomic number greater than 92, the atomic number of uranium.

triad. A group of three elements with similar characteristics that demonstrates periodicity.

triglyceride. A biochemical molecule formed by three fatty acids bonded to a glycerol molecule.

trigonal planar molecule. A molecule shape characterized by three regions of electrons and three outer atoms surrounding a central atom.

trigonal pyramidal molecule. A molecule shape characterized by four regions of electrons and three outer atoms surrounding a central atom. Three outer atoms and one unbonded pair of electrons are at the four vertices of a tetrahedron.

triple bond. A covalent bond that forms when three pairs of valence electrons are shared between atoms.

triple point. The specific pressure and temperature at which a substance will be in the solid, liquid, and vapor forms in equilibrium.

triprotic acid. An acid that can donate three protons.

tRNA. See *transfer RNA*.

Tyndall effect. An effect demonstrated by colloids in which a beam of light is visible due to the scattering of light by the dispersed phase.

U

uncatalyzed reaction. A reaction that occurs without a catalyst.

unified atomic mass unit (u). The unit used to measure the mass of an atom; also known as a *dalton* (Da).

unit cell. The smaller structure within a crystalline solid identified by scientists as that which is repeated to form a crystal lattice.

unit conversion. The mathematical process used to change a measurement in one unit into an equivalent measurement using a different unit.

universal indicator. A substance that will change colors when it reacts with an acid or base and works throughout the entire range of pH values.

unsaturated hydrocarbon. A hydrocarbon that has fewer than four atoms surrounding at least some of its carbon atoms. These hydrocarbons contain at least one double or triple bond.

unsaturated solution. A solution that contains less than the maximum amount of solute that it could hold at current conditions.

V

valence bond theory. A model of chemical bonding that states that atoms fill their valence shells by filling vacancies in particular orbitals; also known as the *localized electron theory*.

valence electron. The electrons in the outermost energy level of an atom that are the ones most likely to be involved in chemical bonding.

valence shell electron pair repulsion (VSEPR) theory. A model for explaining the shapes of molecules that states that the atoms in a molecule will be arranged in such a way as to minimize the repulsive force between bonded and unbonded electrons.

van der Waals force. A collective term for the various intermolecular forces. See *intermolecular force*.

vapor. See *gas*.

vapor pressure depression. The lowering of the vapor pressure of a solvent due to the presence of dissolved solute particles.

vaporization. The phase change from liquid to vapor due to a gain of thermal energy that can occur by evaporation or boiling.

viscosity. A liquid's ability to resist flowing.

voltaic cell. An electrochemical cell that uses a spontaneous chemical reaction to produce electrical energy.

volume fraction (φ_{solute}). A measure of concentration that compares the volume of a solute with the volume of an entire solution and is reported as a decimal or percent; also known as *percent by volume*.

W

wave-particle duality. The current description of both light and matter as having characteristics of waves and particles simultaneously.

weak acid. A substance that doesn't ionize easily, releasing relatively few protons.

weak base. A substance that doesn't easily accept protons.

worldview. The perspective from which a person sees and interprets all of life. It is made up of what one believes about the most important things in life.

Index

A **boldface** page number denotes the page on which the vocabulary term is defined.

A

abiogenesis, 514, 516, 531, 532
absolute percent error, 59
absolute zero, **37**, 277
absorption, 315, 394
accuracy, **58**–62, 69
acetone, 502
acid, 213–14, 435–58
    common, 435
    properties of, 435
    strong, **445**, 447–48
    weak, **446**, 448, 457–58
acid-base equilibria, 440–51
acid-base indicator, **450**
acid-base strength, 445–48
acid-base titration, **454**
acidic solution, **442**
acid-ionization constant, **446**
acoustic energy, 32
actinoid series, 124, **139**
activated complex, **390**–91
activation energy, **390**–91, 393, 394
actual yield, **261**
addition reaction, **507**, 509
adenosine diphosphate, 516
adhesion, **314**–15
ADP, 516
adsorption, 394
aerobic cellular respiration, 374
airbag, 268, 290, 297
alchemist, 117
alchemy, 3
alcohol, **499**
aldehyde, **501**
aliphatic compound, **487**, 488–94
alkali metal, **136**
alkaline-earth metal, **137**
alkaline solution, 442
alkane, **488**–92
alkene, **493**
alkyl group, **489**–91
alkyne, **494**
allotrope, **309**
allotropic element, 309
alloy, **168**

alpha decay, **546**
alpha helix, 526
alpha particle, 82–83, 538, 546
amalgam, 328
amide, **504**
amine, **504**
amino acid, **525**, 528, 530, 531
    essential, 525
amorphous solid, **305**–6
amount concentration, **339**–41
amphoteric substance, **449**–50
amylopectin, 520
amylose, 520
anabolism, **515**–16
analytical chemist, 293
anesthesiologist, 208
anhydrous compound, **209**
anion, **112**, 128
anode, **472**–73, 474–75, 476–78, 480
antibonding orbital, 181
antimatter, **94**
apothecary, 2
applied science, 1
aqua regia, 439
Aristotle, 77
aromatherapy, 508
aromatic compound, **487**, 495–96
aromatic hydrocarbon, 495–96
Arrhenius acid, **436**
Arrhenius base, **436**
Arrhenius model, 436–37
Arrhenius, Svante, 436
Artemisia annua, 16
artesunate, 16
atmospheric pressure, 271
atom, **27**–28, 30, 76–83, 85–89
    early models of, 77–79
    Greek philosophy of, 77
atomic mass, **88**–89, 121, 242
atomic model, 2, 3, 77–83
    of Ernest Rutherford, 82–83
    of J. J. Thomson, 80–81, 82
    of John Dalton, 78–79
    of Niels Bohr, **96**–97, 100, 102, 104
    quantum-mechanical, **100**–105, 123

atomic number (Z), **83**, 85–87, 120, 121
atomic radius, **127**, 128, 129
atomism, 77, 94
atomos, 77
ATP, 516
aufbau principle, **108**–110
Avogadro, Amedeo, 241, 284–85
Avogadro's law, **284**
Avogadro's number, **241**–42, 243

B

Baekeland, Leo, 152
Bakelite, 152
balancing chemical equations, 222–27
    limitations of, 227
balancing redox reactions, 468–70
band of stability, **540**
base, 435–58
    common, 436
    properties of, 436
    strong, **446**, 447–48
    weak, **446**, 448, 457
base-ionization constant, **447**
base unit, 51
basic solution, **442**
battery, 462, **474**–78, 480
    alkaline, 477
    lead-acid storage, 476
    lithium ion, 476
    nonrechargeable, 477
    primary, 477
    rechargeable, 476
    secondary, 476
    solid-state, 480
battery recycling, 474
becquerel, 543
Becquerel, Henri, 537
bent molecule, **183**
benzene, **495**–96
beta decay, **547**–49
beta particle, 538, 547
beta plus decay, 548
beta sheet, 526
biblical worldview, 4, 6–7
Big Bang theory, 32, 44

binary acid, **213**
binary compound, **200**, 204
binary covalent compound
    chemical formula, 211
binary ionic compound, 204
    chemical formula, 211
    naming, 204
binding energy per nucleon, 542, 557, 559, 562–63
binding force, in a crystal, 309–10
biochemistry, 142, **515**–20, 522–31
blood alcohol content, 434, 448
Bohr, Niels, 95, 96, 439
Bohr model, **96**–97, 100, 102, 104
boiling, **317**
boiling point, **317**–18, 344–345, 346, 347
boiling point elevation, **344**–45
bond
    chemical, **153**–69
    covalent, **155**–57, 158–62
    double, **160**, 176, 177
    hydrogen, **300**–302, 303, 305
    ionic, **155**, 157, 163, 303
    metallic, **155**, 165–66, 303
    nonpolar, 155–56
    pi (π), 176
    polar, 186, 188, 190
    polar covalent, **156**
    sigma (σ), 176
    triple, **160**, 176
bond angle, 182–83
bond axis, 176
bonding orbital, 181
bonding pair, 158
boron group, 141
Bosch, Karl, 408
Boyle, Robert, 274, 435
Boyle's law, **274**–75
Brahe, Tycho, 20
branched hydrocarbon, 489
Breathalyzer, 434, 448
Briggs-Rauscher reaction, 415
Broecker, Wallace Smith, 70
Broglie, Louis de, 100
Brønsted, Johannes, 437

Brønsted-Lowry acid, **437**, 449
Brønsted-Lowry base, **437**, 449
Brønsted-Lowry model, 437
Brown, Robert, 35, 350
Brownian motion, 35, **350**
buckminsterfullerene, 15
buffer, **457**–58

C

caloric theory, 35
calorie, 39
calorimeter, **357**
capillary action, 315
carbohydrate, **517**–21
    structure of, 517-18
carbon group, 142
carbon monoxide, 240, 267
    detector, 242, 267
    poisoning, 240
carbonyl group, **501**–2
carboxyl group, **502**
carboxylic acid, **502**–3, 504
Carson, Rachel, 19
catabolism, **515**–16
catalyst, 225, **393**–94, 420
    heterogeneous, **394**
    homogeneous, **394**
cathode, **472**–73, 474–75, 476–78, 480
cathode ray, 80
cation, **112**, 128
cellular respiration, 516
cellulose, 520
Celsius, Anders, 37
Celsius scale, **37**–39
Chadwick, James, 83, 94
chain reaction, **560**, 561
chalcogen, 144
Charles, Jacques, 276–**77**
Charles's law, 276–**77**
chemical abatement specialist, 255
chemical bond, **153**–69
    properties of, 165
chemical change, 23
chemical energy, 32, 135
chemical engineer, 147

chemical equation, 220–27 (**221**)
    as a model, 221
chemical equilibrium, **409**
chemical evolution, 356
chemical formula, **29**–30
    and Stock system, 205–6
    for acid, 213–14
    for binary covalent compound, 211
    for binary ionic compound, 211
chemical property, **23**
chemical reaction, 219–36
    combustion, **230**
    decomposition, **229**
    double-replacement, **232**–36
    evidence of, 219
    single-replacement, **231**–32
    synthesis, **228**
chemical subscript, 29
chemical symbol, **27**, 29
chemistry, **1**, 3
    and ancient philosophy, 2
    and worldview, 6–7
    history of, 2, 3
chirality, 498, 518
cholesterol, 523
chromatography, 26
chromosome, 528, 530
climate change, 70
climatologist, 70
coagulation, 351
codon, 528, 530
cohesion, **314**–15
collection by water displacement, 281
colligative property, **343**–47
collision, effective, 388–89
collision theory, **388**–89
colloid, 24, 329, **349**–50
combination reaction, 228
combined gas law, **279**
combustion reaction, **230**
common ion, 427
common ion effect, **427**
complete ionic equation, **235**
complex reaction, **396**
composite particle, 94

Index

compound, 24, **29**
    properties of, 167–69
compressibility, **271**
concentration, 339
    of a solution, **338**
condensation, **42**, 316, 318
condensation reaction, **509**
condensed structural formula, **486**
conductivity, 22
conjugate acid, **437**, 449, 457
conjugate base, **437**, 449, 457–58
conjugate pair, 437
control group, 13
conversion factor, **55**
cortisone, 524
covalent bond, **155**–57, 158–62
covalent compound, 210–11
    naming, 211
    properties of, 167
Creation Mandate, **8**, 49, 68, 74
crisscross method, 200
critical mass, **560**
critical pressure, **319**
critical temperature, **319**
cryogenic engineer, 307
cryogenics, 307, 325
cryonics, 325
cryosurgery, 307
crystal lattice, **164**, 165–66, 308, 310
crystalline solid, **305**, 310–11
crystalline structure, 30, 308–9
curie, 543
Curie, Marie Sklodowska, 538
Curie, Pierre, 538
cyclic aliphatic compound, **494**
cycloalkane, 494
cycloalkene, 494

D

Dalton, John, 78–79, 117, 280
Dalton's atomic model, 103, 117
Dalton's law of partial pressures, **280**–81
Darwin, Charles, 514

data, 10–12
    qualitative, **10**
    quantitative, **10**, 49
daughter nuclide, 545
DDT, 19
de Broglie's hypothesis, 100
debye (D), 189
decanting, 25
decay
    nuclear, 545–54
    radioactive, 545–54
decomposition reaction, **229**
deductive reasoning, 11
DEET, 505
de Hevesy, George, 439
delocalized electron, **165**, 168, 496
Democritus, 77
density, 22
    of a liquid, 312
    of a solid, 304–5
dependent variable, 13
deposition, 42, **43**, 307
deprotonation, 437
derived unit, **52**
descriptive chemistry, 133
deuterium, 87
dew point, 42
diagonal rule, 105
diatomic element, 28, 158
diffusibility, 315
diffusion, 34, **270**, 331, 341
dimension, 49
dimensional analysis, 56, 66, 67, 69
dimensional unit, 49
dipeptide, 525
dipole-dipole force, **299**, 303
diprotic acid, **449**
disaccharide, **519**
dispersed phase, **349**–50
dispersing medium, **349**
dispersion force, 301–3
dissociation, **330**
distillation, 25, **318**
DNA, 529–30
Döbereiner, Johann, 118, 119
dosimeter, 544

double bond, **160**, 176, 177
double-displacement reaction, 232–36
double helix, 529, 530
double-replacement reaction, **232**–36
driving under the influence, 434, 455
drug testing, 217
ductility, 22
dynamic equilibrium, 316, 334–35
dynamite, 218, 230

E

effervescence, 337
effusion, **270**
Einstein, Albert, 31
electric car, 483
electrical energy, 32, 135
electrochemical cell, **471**–72
electrochemist, 465
electrochemistry, **471**–78
electrodeposition, 472
electrolysis, **472**, 473
electrolyte, **435**–36, 440, 471, 472–73, 476, 477, 478
electrolytic cell, **471**–73, 475
electromagnetic energy, 32, 96, 98
electromagnetic radiation, 98
electron, 27, **81**–83, 85, 88, 94–97
    valence, 122, 127–28
electron affinity, **130**, 154
electron capture, **549**
electron cloud, 103, 127
electron configuration, **106**–112, 121–23, 133, 153–54
    and chemical bonding, 155
electron deficiency, 179
electron dot notation, **111**–12
electronegativity, **131**, 132, 155–56
    and chemical bonding, 156, 157, 163–64
electron-sea model, **165**
electroplating, **472**
electrostatic force, 158
element, 24, **27**–29, 85–86, 88, 117, 118, 119, 120–23
    diatomic, 28
    isolating, 130

monatomic, 28
naming of, 29
origin of, 20, 21
polyatomic, 28
symbol, 121
element origins, 126
elementary particle, 94
elementary step, **396**
elimination reaction, **508**
emission spectrum, **95**–96
empirical formula, **245**
    calculations with, 248–51
end point, **456**
endergonic process, 36
endothermic process, **36**, 371–72
endothermic reaction, 366, 371
energy, **31**–33, 100–101
    acoustic, 32
    chemical, 32, 135
    electrical, 32, 135
    electromagnetic, 32, 96, 98
    in chemical bonding, 153
    ionization, **129**
    kinetic, **32**, 36, 273, 312, 316
    mechanical, 32
    nuclear, 32
    potential, **32**
    sound, 32
    thermal, **32**, 35, 36, 42–43, 98, 316, 357–63, 364, 419
energy diagram, 388
energy shell, 96
energy sublevel, **102**, 104, 106–7, 123
enthalpy, **357**–59, 365–66
    of reaction, **365**–66, 368–70
    of condensation, 359
    of solution, **330**
entropy, 34, **372**–74
    and life, 374
environmental scientist, 351
enzyme, **394**, 402
equilibrium, 409–15, 416–20, 421–23
    and catalysts, 420
    and concentration, 416–17, 420
    and ethanol, 422
    and pressure, 418, 420
    and temperature, 419, 420
    heterogeneous, 412–13
    homogeneous, 412
equilibrium constant, **412**–14, 416–20, 440
equilibrium position, 261, **411**
equivalence point, **456**
essential oil, 484, 508
ester, 422, **503**, 522
esterification, 509
ether, **500**
ethics, 8–9
ethics triad, 8, 74
evaporation, **316**–17
excess reactant, **259**–60
excited state, **96**
exergonic process, 36
exoplanet, 97
exothermic process, **36**, 371
exothermic reaction, 366, 371
expanded octet, 179
expansibility, **271**
experiment, 13
    natural, 13
experimental group, 13
explosive ordnance disposal technician, 233
extensive property, 22

F

Fahrenheit scale, 38
family, 122
Faraday, Michael, 495
fat, **522**
    monounsaturated, 523
    polyunsaturated, 523
fatty acid, **503**, 522
Fermi, Enrico, 557
Feynman, Richard, 2
fiduciary point, 37
field ion microscope, 76
field research, 12
field work, 12
filtration, 25
first law of thermodynamics, 33
fission, nuclear, **557**–59, 560, 561, 564
flammability, 23
flocculation, 351
fluid, 41, **271**
fluidity, 271
formalin, 501
formula, chemical, **29**–30
formula unit, 30, **163**, 200
forward reaction, 409
fossil fuel, 152, 485
fraction, 318
fractional distillation, 492
Franck, James, 439
free energy. *See* Gibbs free energy
free radical, 178
free-electron model, 165
free-electron theory, 168
freezing, 42, **43**, 306
freezing point, 43, 306, 346–47
freezing point depression, **346**–47
fuel cell, **478**
functional group, **497**, 499–504
fundamental unit, **51**
fusion, nuclear, **562**–63, 564

G

gamma decay, **549**
gamma ray, 538, 549
gas, 40–43 (**41**)
    density, 291
    ideal, **286**
    properties, 269–73
gas discharge tube, 80
gas laws, 274–82
gas stoichiometry, 283–291, 293
Gay-Lussac, Joseph Louis, 278, 283
Gay-Lussac's law, **278**
Geiger, Hans, 82
Geiger counter, 544
gene, 528
Gibbs, J. Willard, 375
Gibbs free energy, **375**–79, 387
global warming, 70, 292
glucose, 519
glycerol, 522
glycogen, 520
Graham, Thomas, 270

Index

Graham's law of diffusion, 270
Graham's law of effusion, **270**
Greek prefix system, **211**
greenhouse effect, 292, 301
greenhouse gas, 292
ground state, **96**, 102–3, 104–5, 106, 108, 110, 153
group, **122**

H

Haber, Fritz, 408, 421, 423
Haber process, 421
half-cell, **474–75**
half-life, **552**–54, 555
half-reaction, **474–75**
Hall, Charles, 473
Hall-Héroult process, 473
haloalkane, 497
haloalkene, 497
haloalkyne, 497
haloaromatic, **497**
halocarbon, **497**
halogen, **145**
heat, 32, **35**–36, 357
heat death, 380
heat of fusion, 359
heat of solution, 330
heat of vaporization, 359, 361
heating curve, 358
Heisenberg, Werner, 101
Heisenberg uncertainty principle, **101**
Hench, Philip, 524
Henry's law, **337**
Héroult, Paul, 473
Hess's law, **369**
heterogeneous catalyst, **394**
heterogeneous mixture, **24**, 25, 349
historical science, 12
HIV, 401
homogeneous catalyst, **394**
homogeneous mixture, **24**, 327–28
Hund, Friedrich, 109
Hund's rule, **109**–10, 111
hydrate, **209**
hydration, 330

hydrocarbon, 134, **488**–96
   aromatic, 495–96
   branched, 489
   saturated, **488**
   substituted, 497, 499–504
   unsaturated, **493**
hydrogen 7, 158
hydrogen bond, **300**–302, 303, 305
hydrogen fuel cell, 135
hydrogen group, 134
Hydrogen Seven, 28
hydroxyl group, **499**, 502
hypervalent molecule, 179
hypothesis, **12**–13

I

ideal gas, **286**
ideal gas law, **287**–89
Iijima, Sumio, 15
image-bearer, 8
independent variable, 13
inductive reasoning, 11, 14
inhibitor, **394**
inner transition metal, 122, **139**
intensive property, 22
intermediate, **396–97**
intermolecular force, 167, **299**–303, 312, 314–15, 316, 317
   and chemical bonds, 303
   and melting point, 304
   and polarity, 299
   strength of, 301
International Space Station, 13
International Union of Pure and Applied Chemistry (IUPAC), 29, 130
ion, 27, **111**–12, 128
ionic bond, **155**, 157, 163, 303
ionic compound, 163–64
   and oxidation number, 204–5
   chemical formula, 200, 202–8, 211
   naming, 202–8
   properties of, 168
ionic equation, complete, **235**
ionic radius, 128
ionization, **330**
ionization constant of water, 441

ionization energy, **129**, 154
ionizing radiation, 543, 544, **545**, 547, 564
irreversible reaction, 409
island of stability, 541
isomer, **492**–93, 521
isomerism, 498
isotope, **86**–89, 539–40
isotope notation, **86**
ISS, 13
IUPAC, 29, 130

J

joule (J), **39**
Joule, James Prescott, 39
Julian, Percy, 524

K

Kekulé, August, 495
Kelvin, Lord, 37
Kelvin scale, **37**–39
ketone, **502**
kilogram, redefinition, 56
kinetic energy, **32**, 36, 273, 312, 316
kinetic-molecular theory, **40**, 269, 270, 271–72, 274, 286, 304, 312
kinetics, **387**–400, 402
Kolesnikov, Alexander, 15

L

lanthanoid series, 124, **139**, 151
latent heat, **358**–59
lattice energy, **310**
Laviosier, Antoine, 117
law, **14**
   of chemical equilibrium, **412**
   of combining volumes, **283**, 284
   of conservation of mass-energy, **33**
   of conservation of matter, 220, 221–22
   of definite composition, 247
   of definite proportion, **78**, 283
   of energy conservation, 33
   of multiple proportions, **78**
   of octaves, 118
   scientific, 14

Le Châtelier, Henri, 416
Le Châtelier's principle, **416**–20, 421
Lewis, Gilbert, 159, 438
Lewis acid, **438**
Lewis adduct, 438
Lewis base, **438**
Lewis model, 438
Lewis structure, **159**–62, 164–65, 182
    limitations of, 175, 177–78
limiting reactant, 259–60
limnic eruption, 293
linear chain, 489
linear molecule, **183**
lipid, **522**–24
liquid, 40–43 (**41**), 312–21
literal equation, 69
localized electron theory. *See* valence bond theory
London dispersion force, **301**–3
London force, 301–3
lone pair, 158
Lord Kelvin, 37
Lowry, Thomas, 437

M

magic number, **541**
magnetic property
    and molecular orbital theory, 180–81
malaria, xiv, 1, 10, 11, 15, 16, 19
malleability, 22
Mars Climate Orbiter, 48
Marsden, Ernest, 82
mass composition, 246, 249, 250
mass defect, **542**, 557, 558–59
mass fraction, **338**
mass number (A), 85–88 (**86**)
mass spectrograph, 120
mass-to-mass conversion, 257–58
mass-to-mole conversion, 254–55
materials scientist, 43
matter, **1**, 31, 77–79, 116
    classification of, 21–30
    continuous theory of, 77
    origin of, 20, 21, 44
    particle model, 35, 40, 78
    state of, 40–44

measurement, **49**–69
    integrity in, 61
    limitations of, 58
    uncertainty in, 58
mechanical energy, 32
medical marijuana, 407
medical testing, 75
megaelectronvolt (MeV), 542
Meitner, Lise, 557
melting, 42, **43**, 306
melting point, 43, **306**
Mendeleev, Dmitri, 119
meniscus, **315**
messenger RNA, 530
metabolism, **515**
metabolite, 402
metal, **124**
    properties of, 168
metallic bond, **155**, 165–66, 303
metallic hydride, 199
metalloid, **125**, 140, 141
meteorologist, 70
metric system, 50–53
microscope, field ion, 76
Miller, Stanley, 514
miscibility, **328**
mitosis, 530
mixture, 21, **24**–26
    heterogeneous, **24**, 25, 349
    homogeneous, **24**, 327–28
    separating, 25
model, 1–2, 4
    atomic, 77–83
    Bohr, **96**–97, 100, 102, 104
    nuclear shell, 541
    quantum-mechanical, **100**–105, 123
molal boiling point constant, 344
molal freezing point constant, 346
molality, 338, **341**
molar concentration, 339–41
molar enthalpy of fusion, **359**
molar enthalpy of vaporization, **359**, 361
molar mass, **242**–43, 244
molar volume, 284
    standard, **285**

molarity, 339–41
mole (mol), **241**–44, 252–54, 255, 527–28
mole ratio, 248–51, **253**–55, 257
mole to mole conversion, 252–54
molecular dipole moment, **189**
molecular formula, **245**, 251
molecular geometry, 182–92
molecular orbital model, 192
molecular orbital theory, 175–76, **180**–81
molecule, **28**, 29, 30
monatomic element, 28
monomer, 515
monoprotic acid, **449**
monosaccharide, **517**, 519
Moseley, Henry, 120
Mosquirix, 15
mRNA, 530
Müller, Erwin, 76

N

nanoscience, 15
NASA, 48
natural experiment, 13
naturalism, 4, **7**, 44
net electrical charge, 111
net ionic equation, **235**–36, 453
net reaction, 396
network covalent substance, **167**
neutralization, 453–58
neutralization reaction, 235–36, **436**
neutral solution, **442**
neutron, 27, **83**, 85–88, 94
Newlands, John, 118, 119
Newton, Isaac, 101
nitrogen narcosis, 280
nitroglycerin, 218, 230, 239
Nobel, Alfred, 218, 230, 239
noble gas, **146**
noble gas notation, **109**, 110
nomenclature, **197**
nonbonding pair, 158
nonelectrolyte, 471
nonionizing radiation, 545
nonmetal, **125**, 140
nonpolar bond, 155–56

Index

nonpolar solute, 332
normal boiling point, 317
nuclear binding energy, **542**, 557, 559, 562–63
nuclear bombardment reaction, 556–57
nuclear change, and energy, 542
nuclear chemistry, 537–65
    use in solving problems, 564–65
nuclear decay, 545–54
nuclear energy, 32
nuclear engineer, 541
nuclear equation, **545**
nuclear fission, **557**–59, 560, 561, 564
nuclear fusion, **562**–63, 564
nuclear power, 569
nuclear radiation, **537**–41, 545
nuclear reaction, 556–63
nuclear shell model, 541
nuclear stability, 539–41
nucleic acid, **528**–30
nucleon, **539**–41
nucleotide, **528**
nucleus, 82–**83**, 95–96
nuclide, **539**–541, 551

O

octet rule, **154**
odor tester, 487
oil, **523**
operational science, 12
orbital, **101**, 102–3, 104–5, 106–7, 123
orbital hybridization, 184–85
orbital notation, **108**–10
ordnance disposal technician, 233
organ transplant, 298, 320–21
organic chemistry, 142, **485**–509
organic compound, 484, **485**–87
    classification, 487
organic reaction, 506–9
osmosis, **347**, 348
osmotic pressure, **347**, 348
oxidation, **463**, 466, 468–70, 472–73, 474–75, 506
oxidation number, **197**, 198–202, 204–5, 210, 469–70
oxidation-reduction reaction, **463**–70
oxidation state, 197
oxide, **144**
oxidizing agent, **466**–67
oxyanion, **206**–7
oxygen toxicity, 280

P

paleo diet, 535
paper chromatography, 26
paramagnetism, 181
parent nuclide, 545
partial pressures, 280
    Dalton's law of, **280**
particle
    composite, 94
    elementary, 94
particle model of matter, 35, 40, 78
pascal (Pa), 271
Pascal, Blaise, 271
passivation, 467
patent attorney, 179
Pauli, Wolfgang, 108
Pauli exclusion principle, **108**
Pauling, Linus, 131, 156
peer review, 14
peptide bond, **525**
percent composition, **246**–47, 249
percent error, 59
percent yield, 260–**61**
period, **122**, 123
periodicity, 118, 119, 127
periodic law, 118, 119, **120**
periodic table, **117**, 119, 120–25, 126–34, 136–46
periodic trends, 127–32
pesticide, 19
pH, **442**–43, 444–46, 452
    measuring, 450–51
pH meter, 451
pH probe, 451
pharmaceutical pollution, 329
pharmacokinetics, 402, 403
pharmacologist, 401, 403
phase, 24
phase change, **42**
phase diagram, **319**
phospholipid, **524**
photon, **100**–101
photosynthesis, 516
physical change, **22**
physical property, **22**, 23, 25
pi (π) bond, **176**
planetary model. *See* Bohr model
plasma, 41
plastic, 152, 157, 169, 173
    biodegradable, 169
plum pudding model, 80–81, 82
pnictogen, 143
pOH, **444**–45
polar bond, 186, 188, 190
polar covalent bond, **156**
polar molecule, **186**, 188, 191
polar solute, 332
polar solvent, 332
polarity, **155**–57, 186–87, 189
pollution, pharmaceutical, 329
polyatomic element, 28
polyatomic ion, **164**–65, 207, 208
polyatomic ionic compound, naming, 206–8
polyhydroxyl alcohol, 499
polymer, 153, **515**
polymerization, 515
polymorph, 309
polypeptide, 525, 530
polyprotic acid, **449**–50
    ionization constants, 450
polysaccharide, **520**
Popper, Karl, 6
positron, 94, 538
positron emission, **548**
post-transition metal, 140, 141
potential energy, **32**
precipitate, **219**, 234, 428–29
precipitation, 334
precision, **60**–62, 65–66, 69
pressure, **271**
    of a gas, 272–73
    in a liquid, 316–18
presupposition, **7**

principal energy level, **96**, 103, 104, 106–7, 122, 123
principal quantum number, **102**, 103, 110
prism spectroscope, 98
probability plot, 102
problem solving, 69
product, **221**
protein, **525**–27
    as an enzyme, 527
    denaturing, 527
    primary structure, 526
    quaternary structure, 526
    secondary structure, 526
    tertiary structure, 526
protein complex, 526
protein subunit, 526
protein turnover, 527
proton, 27, **83**, 85–88, 94
protonation, 437
Proust, Joseph, 78
pure science, 1
pure substance, 21, **24**, 29, 30

Q

qualitative data, **10**
quantitative data, **10**, 49
quantum-mechanical model, **100**–105, 123
quantum mechanics, 100
quantum number, 102
quinine, 1, 16

R

radiation
    electromagnetic, 98
    measuring, 543–44
radioactive decay, 545–54
radioactive decay series, 550–**51**
radioactivity, **538**–41, 545
    discovery of, 537–38
radioisotope, **539**, 551, 553
radiometric dating, 555
Radium Girls, 93
Ramsay, William, 146
random uncertainty, 59

rare-earth element. *See* lanthanoid series
rate-determining step, **396**, 399
rate law, **398**–400
rate of solution, factors affecting, 335
ray, cathode, 80
Rayleigh, Lord, 146
reactant, **221**
reaction
    addition, **507**, 509
    biochemical, 394
    combustion, **230**
    condensation, **509**
    decomposition, **229**
    dehydration, 508
    dehydrogenation, 508
    double-replacement, 235–36
    elimination, **508**
    endothermic, 366, 371
    exothermic, 366, 371
    favorable, 387–89
    forward, 409
    homogeneous, 397
    hydration, 507
    hydrogenation, 507
    irreversible, 409
    net ionic, 453
    neutralization, 235–36, 453–54
    nuclear, 556–63
    organic, 506–9
    oxidation-reduction, **463**–70
    redox, 463–70
    reverse, 409
    reversible, 224, **409**, 410
    single-replacement, **231**–32
    substitution, **507**
    synthesis, **228**
    thermonuclear, 563
    types of, 228–33
reaction mechanism, 393, **396**–400
reaction order, **398**
reaction rate, 387–94 (**391**), 398–400
    factors affecting, 392–93
reaction tendency, 371–79
reactivity, 23
reactivity series, 232

reasoning
    deductive, 11
    inductive, 11, 14
recycling, 157, 169
redox reaction, 463–70, 506
reducing agent, **466**
reduction, **464**, 466, 468–70, 472–73, 474, 506
replication, 529
resonance, **177**–78
reverse osmosis, 348
reverse reaction, 409
reversible reaction, 224, **409**, 410
Rf value, 26
ribosomal RNA, 530
RNA, 530
Röentgen, Wilhelm, 82, 537
roman numeral system, 205
Ross, Ronald, 11
rRNA, 530
Rush, Benjamin, 3
Rutherford, Ernest, 82–83, 538, 556

S

saccharide, 517
Sagan, Carl, **7**
salt, **453**, 454
salt bridge, 472, **475**
saturated hydrocarbon, **488**
saturated solution, **334**
scanning tunneling microscopy (STM), 77
science
    applied, 1
    historical, 12
    operational, 12
    pure, 1
scientific inquiry, **12**–14
scientific method, 12–14
scientific model, 65
scientific notation, 63, 69
scientific survey, 14
scintillation counter, 544
Seaborg, Glenn, 124
second law of thermodynamics, 34, 372, 374

Index

self-ionization of water, **440**
semiconductor. *See* metalloid
semipermeable membrane, 347, 348
sensible heat, **358**–59
series, 122, 123
SI, **50**–53
SI unit prefix, 53
sievert, 543
Sievert, Rolf, 543
sigma (σ) bond, **176**
significant figure, 62–64 (**63**)
    math rules, 66–67
    rules for determining, 63
single-displacement reaction, 231–32
single-replacement reaction, **231**–32
skeletal formula, 486
smelting, 473
solid, **40**, 304–11
solidification, 43
solubility, **332**–33, 425–27
    factors affecting, 336–37
solubility product constant, **424**–27, 429
solubility rules, 234, 333
solute, **327**–28, 329, 330, 332–33, 339
    nonpolar, 332
    polar, 332
solution, 24, **327**–30, 339, 349
    acidic, **442**
    basic, **442**
    concentration, **338**
    dissolving process, 327–42
    saturated, 334
    supersaturated, 334
    unsaturated, 334
solution equilibrium, 424–29
solvation, **330**
solvent, **327**–30, 332–33
    polar, 332
    selectivity, 332
Sørensen, Søren P. L., 442
sound energy, 32
specific heat, **360**–63
specific rate constant, **398**
spectator ion, **235**, 453
spectroscopy, 95, 97, 98, 99

spin, **103**
spontaneous combustion, 395
standard molar enthalpy of combustion, **367**
standard molar enthalpy of formation, 366–70 (**367**)
standard molar entropy, 372
standard molar volume, **285**
standard solution, 454
standard state, 366
standard temperature and pressure (STP), **274**
starch, 515, 517, 519, 520
state of matter, 40–44
stereocenter, 521
stereoisomer, 521
steroid, **523**, 524
Stock, Alfred, 205
Stock system, **205**, 206
stoichiometry, 241, **252**–55, 257–61
    gas, 283–91, 293
Stoney, George Johnstone, 81
straight chain hydrocarbon, 489
strong acid, **445**, 447–48
strong base, **446**, 447–48
strong nuclear force, **539**–40
structural formula, 245, 486, 490–91
subcritical mass, 560
sublimation, 42, **43**, 306–7
subscript, chemical, 29
substance, pure, 21, **24**, 29, 30
substituent, 497
substituted hydrocarbon, 497, 499–504
substitution reaction, **507**
sulfide, **144**
supercooled liquid, 305
supernova, 20
supersaturated solution, **334**
surface tension, 188, 190, 191, **314**–15
surface-active agent, 314
surfactant, **314**
suspension, 24, 329, **349**
sustained-release medication, 403
symbol, chemical, **27**, 29
synthesis reaction, **228**
system, **34**

Système International d'Unités, 50–53
systemic uncertainty, 59

T

temperature, **36**–39, 357
    absolute thermodynamic, 37
ternary acid, **213**
tetrahedral molecule, **182**
theoretical yield, **260**
theory, **14**
    kinetic-molecular, 35, **40**, 78
    molecular orbital, 175–76, **180**–81
    of special relativity, 542
    valence bond, **175**–76, 180
    valence shell electron pair repulsion (VSEPR), **182**
thermal energy, **32**, 35, 36, 42–43, 98, 316, 357–63, 364, 419
thermochemical equation, **365**, 369
thermochemistry, **357**
thermodynamics, **33**, 35–36, 357–70, 387, 388
    and worldview conflict, 380–81
    first law of, 33
    second law of, 34, 372, 374
    third law of, 37
thermometer, 37
thermometric property, 37
thermonuclear reaction, 563
third law of thermodynamics, 37
Thomson, J. J., 80–81, 94
time-release medication, 402
titrant, 454
titration, acid-base, **454**
titration curve, 454
toxicity, 23
transcription, 530
transfer RNA, 530
transition interval, **450**–51
transition metal, 119, **138**
translation, 530
transmutation, **556**, 562
transuranium element, **121**
triad, 118
triglyceride, 522–23
trigonal planar molecule, **183**

trigonal pyramidal molecule, **182**–83
trihydrogen, 29
triple bond, **160**, 176
triple point, 37, **319**
triprotic acid, **450**
tritium, 87
tRNA, 530
Tyndall, John, 350
Tyndall effect, **350**

U

uncertainty
    random, 59
    systemic, 59
unified atomic mass unit (u), **88**
unit, 49
unit cell, **308**–9
unit conversion, **55**–57
universal gas constant, 287
universal indicator, 451
unsaturated hydrocarbon, **493**
unsaturated solution, **334**
Urey, Harold, 514

V

vaccine, xiv, 10, 15
valence bond model, 192
valence bond theory, **175**–76, 180
    limitations of, 184
valence electron, **110**, 111, 122, 127–28, 154, 155
valence shell, 154
valence shell electron pair repulsion (VSEPR) theory, **182**
van der Waals force. *See* intermolecular force
van't Hoff factor, 344–46
vapor, 41, 42
vapor pressure, 316–18
vapor pressure depression, **343**
vaporization, **42**, 316–18
variable, 13
    dependent, 13
    independent, 13
viscosity, **314**–15
volatility, 343

Volta, Alessandro, 80, 474
voltaic cell, **471**, 474–76, 478, 479
volume fraction, **339**
von Laue, Max, 439
VSEPR theory, **182**

W

wastewater management, 186, 355
water
    and hydrogen bond, 300, 305
    as a resource, 195, 326, 351
water displacement, collection by, 281
water molecule, 174, 175, 188, 190, 191
    and design, 191, 301
water of hydration, 209
water strider, 188
water vapor pressure, 281
wave-particle duality, **101**
weak acid, **446**, 448, 457–58
weak base, **446**, 448, 457
Wöhler, Friedrich, 485
work, 31
worldview, **1**, 6, 44, 68
    and biochemistry, 516, 531
    and equilibrium, 423
    and half-life, 555
    biblical, 4, 6–7
    naturalistic. *See* naturalism

X

x-ray, 82

Z

zinc-carbon dry cell, 477

Photo Credits

Key: (t) top; (c) center; (b) bottom; (l) left; (r) right; (bg) background

COVER

Nick Poon/Moment/Getty Images

FRONT MATTER

i real444/iStock/Getty Images Plus/Getty Images; **vii**tl dotted zebra/Alamy Stock Photo; **vii**tr UPI/Alamy Stock Photo; **vii**rc LWM/NASA/LANDSAT/Alamy Stock Photo; **vii**b LightField Studios Inc./Alamy Stock Photo; **viii**l © DETLEV VAN RAVENSWAAY/SCIENCE SOURCE; **viii**r ivan-96/Digital Vision Vectors/Getty Images; **ix**t NASA/Handout/Getty Images; **ix**c © USTC Institute of Advanced Technology & Tsinghua University Press/SCIENCE SOURCE; **ix**bl Roongrote Amnuaysook/EyeEm/Getty Images; **ix**br Michael Marfell/Moment/Getty Images; **x**tl Aerogel hand by Courtesy NASA/JPL-Caltech/Wikimedia Commons/Public Domain/Cropped; **x**tr ranasu/iStock/Getty Images; **x**b Ashley Cooper/Alamy Stock Photo; **xi**t Jose A. Bernat Bacete/Moment/Getty Images; **xi**c Floortje/iStock/Getty Images; **xi**b MARTIN BERNETTI/AFP/Getty Images

CHAPTER 1

xiv Paul Starosta/Stone/Getty Images; **2** (philosophers) "Sanzio 01 Plato Aristotle" by Raphael/Wikimedia Commons/Public Domain/cropped; **2–3** (gutter) ilbusca/DigitalVision Vectors/Getty Images; **3**t benoitb/Digital Vision Vectors/Getty Images; **3**b anyaivanova/iStock/Getty Images Plus/Getty Images; **4–5**bg TommoT/Shutterstock.com; **5**l Public Domain; **6–7** georgeclerk/E+/Getty Images; **8**r Klaus Vedfelt/DigitalVision/Getty Images; **8–9** (gutter), **17**l Neil Thomas/Corbis News/Getty Images; **10–11** dra_schwartz/E+/Getty Images; **12**bg Tommy Trenchard/Alamy Stock Photo; **12**r Edwin Remsberg/Alamy Stock Photo; **13** AFP/Getty Images; **14** Worldwide Picture Library/Alamy Stock Photo; **15, 17**r Hailshadow/iStock/Getty Images Plus/Getty Images; **16** Florilegius/Alamy Stock Photo; **19** Everett Collection Inc/Alamy Stock Photo

CHAPTER 2

20–21, 34l blackphobos/iStock/Getty Images; **21** studiocasper/iStock/Getty Images; **22**l PHOTO RESEARCHERS, INC./Science Source/Getty Images; **22**tr Jennifer A Smith/Moment/Getty Images; **22**br, **45**t FactoryTh/iStock/Getty Images; **23**tr Fuse/Corbis/Getty Images; **23**l Jose A. Bernat Bacete/Moment/Getty Images; **23**br jarun011/iStock/Getty Images; **24**t Sebastian Janicki/Shutterstock.com; **24**c Mandy Disher Photography/Moment/Getty Images; **24**b pederk/iStock Unreleased/Getty images; **25**t Spiderstock/iStock/Getty Images; **25**b gameover2012/iStock/Getty Images; **27** © Vincent Boisvert, all right reserved/Moment/Getty Images; **28** Andrew Brookes/Cultura/Getty Images; **29** NASA/Handout/Getty Images; **31** Mimadeo/iStock/Getty Images; **32–33**t Jose A. Bernat Bacete/Moment/Getty Images; **32**b Floortje/iStock/Getty Images; **33**b, **45**b duncan1890/Digital Vision Vectors/Getty Images; **34**r emilyhayward/iStock/Getty Images; **35** © Richard J. Green/SCIENCE SOURCE; **36–37** JTSorrell/E+/Getty Images; **39** UniversalImagesGroup/Getty Images; **40, 46** jonnysek/iStock/Getty Images; **41**t videophoto/E+/Getty Images; **41**b E.R. Degginger/Alamy Stock Photo; **42** Allan Wallberg/EyeEm/Getty Images; **43** Instants/E+/Getty Images; **44** MARIIA VASILEVA/iStock/Getty Images

CHAPTER 3

48 © DETLEV VAN RAVENSWAAY/SCIENCE SOURCE; **49** ivan-96/Digital Vision Vectors/Getty Images; **50**t, **71**tr studiocasper/iStock/Getty Images; **50**b INTERFOTO/Alamy Stock Photo; **51** (Ampere's Stand) World History Archive/Alamy Stock Photo; **51** (Candle) ilbusca/Digital Vision Vectors/Getty Images; **51** (thermometer) benoitb/Digital Vision Vectors/Getty Images; **51** (weights) ivan-96/Digital Vision Vectors/Getty Images; **51** (measuring tape) ivan-96/Digital Vision Vectors/Getty Images; **51** (molecule) Book end/Alamy Stock Photo; **51** (watch) pictore/Digital Vision Vectors/Getty Images; **51** (scale) Mark Weiss/Photodisc/Getty Images; **51** (gauge) ChuckSchugPhotography/iStock/Getty images; **52** Bryan Mullennix/Getty Images; **53** HHakim/E+/Getty Images; **54**tl macroworld/E+/Getty Images; **54**tr Universal History Archive/Universal Images Group/Getty Images; **54**b lucato/iStock/Getty Images; **55** malerapaso/iStock/Getty Images; **56** picture alliance/Getty Images; **57** agafapaperiapunta/iStock/Getty Images; **58**t tdub303/E+/Getty Images; **58**b, **60**c, **71**cl Rawpixel/iStock/Getty Images; **59** Jenny E. Ross/Corbis Documentary/Getty Images; **60**t TEK IMAGE/SCIENCE PHOTO LIBRARY/Getty Images; **61** ilbusca/Digital Vision Vectors/Getty Images; **62** Tomas Ragina/iStock/Getty images; **63** Mailson Pignata/EyeEm/Getty Images; **64** AlexLMX/iStock/Getty Images; **65**t, **71**br shotsstudio/iStock/Getty Images; **65**b channarongsds/iStock/Getty Images; **66–67** Virojt Changyencham/Moment/Getty Images; **68** Casarsa Guru/E+/Getty Images; **69** EmirMemedovski/E+/Getty Images; **70** Kativ/E+/Getty Images; **72** tdub303/E+/Getty Images; **74** Plan Shoot/Multi-bits/Taxi/Getty Images

CHAPTER 4

76 Prof. Erwin Mueller/Science Source; **79**bg MARK GARLICK/Science Photo Library/Getty Images Plus/Getty Images; **79**c Dalton's Element List by John Dalton/Wikimedia Commons/Public Domain/Edited; **80**bl ZU_09/Getty Images/DigitalVision Vectors/Getty Images; **80**br Artwork by Ina Stanimirova Chronicle/Alamy Stock Photo Archive/Alamy Stock Photo; **82** Artwork by Frank Ramspott Universal Images Group North America LLC/Alamy Stock Photo; **83**t Artwork by Ina Stanimirova Universal Images Group North America LLC/Alamy Stock Photo; **83**bl Artwork by Ina Stanimirova derivative of "James Chadwick"/Los Alamos National Laboratory. Unless otherwise indicated, this information has been authored by an employee or employees of the Los Alamos National Security, LLC (LANS, operator of the Los Alamos National Laboratory under Contract No. DE-AC52-06NA25396 with the US Department of Energy.)/Wikimedia Commons; **85** Patrice Loiez, CERN/Science

Source; **87** BJU Photo Services; **89** SPL/Science Source; **93**t "Radium Dial"/Arma95/Wikimedia Commons/CC BY-SA 3 /cropped; **93**b Daily Herald Archive/SSPL/Getty Images

CHAPTER 5

94 ATLAS Collaboration/CERN/Science Source; **95, 113** Adobe Stock/Morphart; **100** standing wave electron cloud /CK-12 Foundation/Wikimedia Commons/CC BY SA 3.0 /modified; **101**cl Artwork by Ina Stanimirova derivative of "Werner Heisenberg"/Bundesarchiv, Bild 183-R57262 /Wikimedia Commons/CC-BY-SA 3.0; **109** Artwork by Leib Chigrin derivative of Hund, Friedrich 1920er Göttingen by GFHund/Wikimdia Commons/CC BY 3.0

CHAPTER 6

116 Science History Images/Alamy Stock Photo; **117**cl Artwork by Ina Stanimirova derivative of "Lavoisier explaining to his wife the result of his experiment" by Ernest Board/WelcomeImages/Wikimedia Commons/CC By-SA 4.0; **117**t © SPL/SCIENCE SOURCE; **117**b Photos .com/Getty Images; **118**tl © Turtle Rock Scientific/SCIENCE SOURCE; **118**tr Artwork by Ina Stanimirova INTERFOTO /Alamy Stock Photo; **118**cl © Charles D. Winters/SCIENCE SOURCE; **118**c Newlands periodiska system 1866/John Alexander Reina Newlands/Wikimedia Commons/Public Domain/modified; **118**bl Editorial Image, LLC/Alamy Stock Photo; **119**t Science History Images/Alamy Stock Photo; **119**b Phil Degginger/Alamy Stock Photo; **124** (hydrogen), **134**t © SPL/SCIENCE SOURCE; **124, 136** (lithium) Bjoern Wylezich/Shutterstock.com; **124, 136** (sodium) Andraž Cerar/Shutterstock.com; **124, 136** (potassium) © Turtle Rock Scientific/SCIENCE SOURCE; **124, 136** (rubidium) © SPL/SCIENCE SOURCE; **124, 136** (cesium) © SPL/SCIENCE SOURCE; **124, 137** (beryllium) Bjoern Wylezich/Shutter stock.com; **124, 137** (magnesium) Bjoern Wylezich/Shut terstock.com; **124, 137** (calcium) Bjoern Wylezich/Shutter stock.com; **124, 137** (strontium) © SPL/SCIENCE SOURCE; **124, 137** (barium) © Turtle Rock Scientific/SCIENCE SOURCE; **124, 138** (scandium) © SPL/SCIENCE SOURCE; **124, 138** (titanium) Bjoern Wylezich/Shutterstock.com; **124, 138** (vanadium) Bjoern Wylezich/Shutterstock.com; **124, 138** (chromium) Bjoern Wylezich/Shutterstock.com; **124, 138** (manganese) Bjoern Wylezich/Shutterstock.com; **124, 138** (iron) Iron electrolytic and 1cm3 cube/Alchemist -hp/Wikimedia Commons/CC By-Sa 3.0/modified; **124, 138** (cobalt) ALFRED PASIEKA/SCIENCE PHOTO LIBRARY/Getty Images; **124, 138** (yttrium) Yttrium sublimed dendritic and 1cm3 cube/Alchemist-hp/Wikimedia Commons/CC By SA 3.0/modified; **124, 138** (zirconium) Bjoern Wylezich/Shut terstock.com; **124, 138, 149** (niobium) Bjoern Wylezich /Shutterstock.com; **124, 138** (molybdenum) © Science Stock Photography/SCIENCE SOURCE; **124, 138** (ruthenium) © SPL/SCIENCE SOURCE; **124, 138** (rhodium) © SPL/SCIENCE SOURCE; **124, 138** (lanthanum) Bjoern Wylezich/Shutterstock.com; **124, 138** (hafnium) Bjoern Wylezich/Shutterstock.com; **124, 138** (tantalum) Bjoern Wylezich/Shutterstock.com; **124, 138** (tungsten) Wolfram evaporated crystals and 1cm3 cube/Alchemist-hp /Wikimedia Commons/CC By-SA 1.2/modified; **124, 138** (rhenium) Rhenium single crystal bar and 1cm3 cube /Alchemist-hp/Wikimedia Commons/CC By-SA 3.0/modified; **124, 138** (osmium) LuYago/Shutterstock.com; **124, 138** (iridium) © SPL/SCIENCE SOURCE; **124, 139** (cerium) Bjoern Wylezich/Shutterstock.com; **124, 139** (praseodymium) Bjoern Wylezich/Shutterstock.com; **124, 139** (neodymium) Kim Christensen/Shutterstock.com; **124, 139** (samarium) Bjoern Wylezich/Shutterstock.com; **124, 139** (thorium) Thorium sample 0.1g/Alchemist-hp/Wikimedia Commons/CC By-SA 3.0/modified; **124, 139** (uranium) abadonian/iStock/Getty Images; **125, 138** (nickel) Bjoern Wylezich/Shutterstock.com; **125, 138** (copper) Bjoern Wylezich/Shutterstock.com; **125, 138** (zinc) © Turtle Rock Scientific/SCIENCE SOURCE; **125, 138** (palladium) RHJ-Phtotoandilustration/Shutterstock.com; **125, 138** (silver) Bjoern Wylezich/Shutterstock.com; **125, 138** (cadmium) Bjoern Wylezich/Shutterstock.com; **125, 138** (platinum) Bjoern Wylezich/Shutterstock.com; **125, 138** (gold) studiocasper/iStock/Getty Images; **125, 138** (mercury) MarcelC/iStock/Getty Images; **125, 139** (europium) © SPL /SCIENCE SOURCE; **125, 139** (gadolinium) Bjoern Wylezich /Shutterstock.com; **125, 139** (terbium) Bjoern Wylezich /Shutterstock.com; **125, 139** (dysprosium) Bjoern Wylezich /Shutterstock.com; **125, 139** (holmium) Bjoern Wylezich /Shutterstock.com; **125, 139** (erbium) © SPL/SCIENCE SOURCE; **125, 139** (thulium) © SPL/SCIENCE SOURCE; **125, 139** (ytterbium) Bjoern Wylezich/Shutterstock.com; **125, 139** (lutetium) © SPL/SCIENCE SOURCE; **125, 139** (americium) Americium microscope/Bionerd/Wikimedia Commons/CC By-SA 3.0/cropped; **125, 141** (boron) Kim Christensen/Shutterstock.com; **125, 140**b, **141** (aluminum) Bjoern Wylezich/Shutterstock.com; **125, 141** (gallium) © Turtle Rock Scientific/SCIENCE SOURCE; **125, 141** (indium) Bjoern Wylezich/Shutterstock.com; **125, 141** (thallium) © SPL/SCIENCE SOURCE; **125, 142** (carbon) PjrStudio/Alamy Stock Photo; **125, 142** (silicon) Bjoern Wylezich/Shutter stock.com; **125, 142** (geranium) ALFRED PASIEKA/SCIENCE PHOTO LIBRARY/Getty images; **125, 142** (tin) Bjoern Wylezich/Shutterstock.com; **125, 142** (lead) Lead electrolytic and 1cm3 cube/Alchemist-hp/Wikimedia Commons /CC By-SA 3.0/cropped; **125, 143** (phosphorus) © SPL /SCIENCE SOURCE; **125, 143** (arsenic) Harry Taylor/Dorling Kindersley/Getty Images; **125, 143** (antimony) Bjoern Wylezich/Shutterstock.com; **125, 143** (bismuth) MarcelC /iStock/Getty Images; **125, 144** (sulfur) Bjoern Wylezich /Shutterstock.com; **125, 144** (selenium) Bjoern Wylezich /Shutterstock.com; **125, 144** (tellurium) Bjoern Wylezich /Shutterstock.com; **125, 145** (fluorine) Dorling Kindersley /Getty Images; **125, 145** (chlorine) Unusual Films; **125, 145** (bromine) © Martyn F. Chillmaid/SCIENCE SOURCE; **125, 145** (iodine) Bjoern Wylezich/Shutterstock.com; **125, 146** (helium) Helium discharge tube/Alchemist-hp/Wikimedia Commons/CC By-SA 3.0/modified; **125, 146** (neon) Neon discharge tube/Alchemist-hp/Wikimedia Commons/CC By-SA 3.0/modified; **125, 146** (argon) Argon discharge tube/Alchemist-hp/Wikimedia Commons/CC By-SA 3.0

Photo Credits

/modified; **125, 146** (Krypton) Krypton discharge tube /Alchemist-hp/Wikimedia Commons/CC By-SA 3.0/modified; **125, 146** (xenon) Xenon discharge tube by Alchemist-hp/Wikimedia Commons/CC By-SA 3.0/modified; **126** Steve Allen/DigitalVision/Getty Images; **129** BJU Press artwork derivative of First Ionization Energy/Sponk/Wikimedia Commons/CC By-SA 3.0; **132** Atlantide Phototravel /Corbis Documentary/Getty Images; **133** Bjoern Wylezich /Shutterstock.com; **134** © Martyn F. Chillmaid/SCIENCE SOURCE; **135** GIPhotoStock X/Alamy Stock Photo; **136t** sciencephotos/Alamy Stock Photo; **136b** © Charles D. Winters/SCIENCE SOURCE; **137b** E.R. Degginger/Alamy Stock Photo; **138l** MarcelClemens/Shutterstock.com; **138r** Eddie Jordan Photos/Shutterstock.com; **140t** © Turtle Rock Scientific/SCIENCE SOURCE; **141t** Sean Gallup/Getty Images; **141b** KrimKate/iStock/Getty Images; **142r** sankai/iStock /Getty Images; **143t** jz86/iStock/Getty Images; **143b** © Charles D. Winters/SCIENCE SOURCE; **144t** R.M. Nunes /iStock/Getty Images; **144b** Charles D. Winters/SCIENCE SOURCE; **145l** © Turtle Rock Scientific/SCIENCE SOURCE; **145r** © Charles D. Winters/SCIENCE SOURCE; **146r** hudiemm/iStock/Getty Images; **147l** dpa picture alliance archive/Alamy Stock Photo; **147r** Dr. Yuki Abe; **151** HAYKIRDI/E+/Getty images

CHAPTER 7

152 FotografiaBasica/E+/Getty Images; **153** DOE Photo /Alamy Stock Photo; **156b** Artwork by Ina Stanimirova World History Archive/Alamy Stock Photo; **157** Deepak Aggarwal/Dorling Kindersley/Getty Images; **159t** Artwork by Ina Stanimirova Science History Images/Alamy Stock Photo; **159–61bg** Photo by Alex Tihonov/Moment/Getty images; **162–63** © Turtle Rock Scientific/SCIENCE SOURCE; **164t** Artwork by David Lompe derivative of Xtals combined 2 300ppi/Paburr/Wikimedia Commons/CC By-SA 4.0; **164l** Edwin Remsberg/Photolibrary/Getty Images; **165–66l** Gregory_DUBUS/iStock/Getty images; **166–67r** Science Photo Library/Getty Images; **168t** Bjoern Wylezich/Shutterstock.com; **168l** © Turtle Rock Scientific/SCIENCE SOURCE; **168b** mizikm/iStock/Getty Images; **169** Aryfahmed/iStock /Getty Images; **173** MARTIN BERNETTI/AFP/Getty Images

CHAPTER 8

174 Paul Taylor/Stone/Getty Images; **177** Vadim Petrakov /Shutterstock.com; **178–79bg** Christian Schulze/EyeEm /Getty Images; **179tc** stocknshares/iStock/Getty Images; **180** © Charles D. Winters/SCIENCE SOURCE; **184** Zagursky /iStock/Getty Images; **186** BNBB Studio/Moment/Getty Images; **188l** JanMiko/iStock/Getty Images; **188r** Gregory S. Paulson/Cultura/Getty Images; **191** Xuanyu Han /Moment/Getty Images; **192** Simon Annable/Shutterstock.com; **195** Jason Doiy/iStock/Getty images

CHAPTER 9

196 wektorygrafika/iStock/Getty Images; **197** Alpha Stock /Alamy Stock Photo; **198–99, 215tl** © CMEABG-UCBL /SCIENCE SOURCE; **200–201** hekakoskinen/iStock/Getty images; **202** David Wall Photo/Lonely Planet Images/Getty Images; **203cl** © Turtle Rock Scientific/SCIENCE SOURCE; **203cc** Phil Degginger/Alamy Stock Photo; **203cr** Andy Crawford & Tim Ridley/Dorling Kindersley/Getty Images; **203bl** © Turtle Rock Scientific/SCIENCE SOURCE; **203bc** © Turtle Rock Scientific/SCIENCE SOURCE; **204** (aluminum oxide) © Turtle Rock Scientific/SCIENCE SOURCE; **204** (magnesium oxide) © Turtle Rock Scientific/SCIENCE SOURCE; **204** (potassium sulfide) Potassium sulfide /Leiem/Wikimedia Commons/CC By-SA 4.0/modified; **204** (sodium chloride) mariusFM77/E+/Getty Images; **205** © Turtle Rock Scientific/SCIENCE SOURCE; **206–7** MICHAEL W DAVIDSON/Science Source/Getty Images; **208** BraunS/E+/Getty Images; **210–11, 215cl** THOMAS DEERINCK, NCMIR/Science Photo Library/Getty Images; **213–14, 215bl** © Vincent Boisvert, all right reserved /Moment/Getty Images; **217** RapidEye/E+/Getty Images

CHAPTER 10

218 Juergen Ritterbach/The Image Bank/Getty Images; **219tl** filipfoto/iStock/Getty Images; **219tr** Dorling Kindersley ltd/Alamy Stock Photo; **219bl** © MARTYN F. CHILLMAID /SCIENCE SOURCE; **219br** Scott O'Neill/iStock/Getty Images / babyblueut/iStock/Getty Images; **221** © Turtle Rock Scientific/SCIENCE SOURCE; **222** Bigyy/iStock/Getty Images; **223** Dennis Macdonald/Photolibrary/Getty Images; **224, 237tl** © USTC Institute of Advanced Technology & Tsinghua University Press/SCIENCE SOURCE; **227** © USTC Institute of Advanced Technology & Tsinghua University Press/SCIENCE SOURCE; **228–29tl** unkas_photo/iStock/Getty Images; **228b, 237br** Michael Marfell/Moment/Getty Images; **229r** © Turtle Rock Scientific/SCIENCE SOURCE; **230** Cavan Images/Cavan/Getty Images; **231** © Alexandre Dotta /SCIENCE SOURCE; **233t** © GIPhotoStock/Cultura/Getty Images; **233b** Stocktrek Images, Inc./Alamy Stock Photo; **234** © Turtle Rock Scientific/SCIENCE SOURCE; **235** Xvision /Moment/Getty Images; **236–37l** Roongrote Amnuaysook /EyeEm/Getty Images; **239t** World History Archive/Alamy Stock Photo; **239b** Bob_Eastman/iStock/Getty Images

CHAPTER 11

240 Дерябин Андрей Николаевич/iStock/Getty Images; **242** dehooks/iStock/Getty Images; **243** - - /123RF; **244** TibiP03/iStock/Getty Images; **245** Yon Marsh/Alamy Stock Photo; **246** Robert Daly/Stone/Getty Images; **248** nicolas_ /E+/Getty Images; **250–51** © Jonathan Knowles 2015 /Stone/Getty Images; **252–53tl, 263cr** Sarah Marchant /Shutterstock.com; **252bl** Prostock-Studio/iStock/Getty Images; **253tr** AnthonyRosenberg/E+/Getty Images; **254** membio/iStock/Getty Images; **255** Ashley Cooper /Alamy Stock Photo; **257** gaffera/E+/Getty Images; **259, 263b** grandriver/E+/Getty Images; **260–61** georgeclerk /E+/Getty Images; **262** Muslianshah Masrie/Alamy Stock Photo; **267** AndreyPopov/iStock/Getty Images

CHAPTER 12

268 fStop Images - Caspar Benson/Brand X Pictures/Getty Images; **271** dml5050/iStock/Getty; **272t** Artwork by David Lompe Universal Images Group North America LLC/Alamy Stock Photo; **272cl** Alter_photo/iStock/Getty Images; **272cr** ilbusca/Digital Vision Vectors/Getty Images; **272bl** Alter_photo/iStock/Getty Images; **272br** nicoolay/iStock/Getty Images; **273** Zero Creatives/Image Source/Getty Images; **274–75t** Mr.Nino/Shutterstock.com; **274b** Science History Images/Alamy Stock Photo; **276** Nastasic/Digital Vision Vectors/Getty Images; **278–79, 294c** LeManna/iStock/Getty Images; **280** Luis Javier Sandoval/Stockbyte/Getty Images; **282–83** Kalistratova/iStock/Getty Images; **285** fmajor/E+/Getty Images; **286** sarote pruksachat/Moment/Getty Images; **287** jacquesdurocher/iStock/Getty Images; **290t** B toy Anucha/Shutterstock.com; **290b** Maurice Savage/Alamy Stock Photo; **291** Leon Neal/Getty Images; **292** Westend61/Getty Images; **293l** Eric BOUVET/Gamma-Rapho/Getty Images; **293r** Fabian Plock/iStock/Getty Images; **297** Kohei Hara/Stone/Getty Images

CHAPTER 13

298 Ben Edwards/The Image Bank/Getty Images; **299** ZU_09/Digital Vision Vectors/Getty Images; **301** Maxiphoto/E+/Getty Images; **302–3** Rudi Sebastian/The Image Bank/Getty Images; **304tl** rep0rter/iStock/Getty Images; **304–5tr, 322** Danita Delimont/Gallo Images/Getty Images; **304b** © Charles D. Winters/SCIENCE SOURCE; **305bl** Dafinchi/iStock/Getty Images; **305br** MahirAtes/iStock/Getty Images; **306l** HadelProductions/E+/Getty Images; **306r** Matt Meadows/Photolibrary/Getty Images; **307t** Jay Freis/Stone/Getty Images; **307b** Kubra Cavus/E+/Getty Images; **308** (Fluorite) Epitavi/iStock/Getty Images; **308** (Wulfenite) Coldmoon_photo/iStock/Getty Images; **308** (Tanzanite) Albert Russ/Shutterstock.com; **308** (Azurite) Kerrick/iStock/Getty Images; **308** (Amazonite) MarcelC/iStock/Getty Images; **308** (Emerald) photo-world/Shutterstock.com; **308** (Rhodochrosite) WILDLIFE GmbH/Alamy Stock Photo; **308–9bg** Bkamprath/iStock/Getty Images; **309t** Nikolai Gavriliukov/Shutterstock.com; **309b** Reload_Studio/iStock/Getty Images; **310l** Argon ice 1/Deglr6328 ~commonswiki/Wikimedia Commons/CC By-SA 30/modified; **310r** AlasdairJames/iStock/Getty Images; **311t** J-Palys/iStock/Getty Images; **311cr** Dafinchi/Shutterstock.com; **311cl** ROBERT BROOK/SCIENCE PHOTO LIBRARY/Getty Images; **312–13** Nick Poon/Moment/Getty Images; **314–15bg** Patrick Boelens/iStock/Getty Images; **314b** MarcelC/iStock/Getty Images; **315b** © Sinclair Stammers/SCIENCE SOURCE; **316** Kis Arpad/EyeEm/Getty Images; **317** ssuaphoto/iStock/Getty Images; **318, 323** gameover2012/iStock/Getty Images; **320** imageBROKER/Alamy Stock Photo; **321** uchar/E+/Getty Images; **325l** 1939 Ted Williams/Apex Photo Company/Wikimedia Commons/Public Domain; **325r** Alcor-Dewar2/Alcor Life Extension Foundation/Wikimedia Commons/Public Domain

CHAPTER 14

326 hadynyah/E+/Getty Images; **327l** Shawshots/Alamy Stock Photo; **327r** pederk/iStock Unreleased/Getty images; **328–29tl** Jeff Rotman/The Image Bank/Getty Images; **328c** Cobalt88/iStock/Getty images; **328bl** ChrisEllis85/iStock/Getty Images; **328br** Epitavi/iStock/Getty Images; **329tr** Creative Frame Studio/Moment/Getty Images; **329b** abadonian/iStock/Getty Images; **332–33tbg** rudchenko/iStock/Getty Images; **332cl** © Martyn F. Chillmaid/SCIENCE SOURCE; **333tc** © Turtle Rock Scientific/SCIENCE SOURCE; **334t** © Turtle Rock Scientific/SCIENCE SOURCE; **334c** © Turtle Rock Scientific/SCIENCE SOURCE; **334b** © Turtle Rock Scientific/SCIENCE SOURCE; **335t** VICTOR DE SCHWANBERG/Science Photo Library/Getty Images; **335b** Man_Half-tube/DigitalVision Vectors/Getty Images; **337** © Turtle Rock Scientific/SCIENCE SOURCE; **338tl, 341bl, 352** © Turtle Rock Scientific/SCIENCE SOURCE; **338–39tr** cookelma/iStock/Getty Images; **341tr** rrocio/iStock/Getty Images; **342** © Eye of Science/SCIENCE SOURCE; **343** robcocquyt/iStock/Getty Images; **344** RyersonClark/iStock/Getty Images; **346t** Helen King/The Image Bank/Getty Images; **346b** Steven White/iStock/Getty Images; **347** Pongsak Tawansaeng/EyeEm/Getty Images; **348** Tetiana Garkusha/iStock/Getty Images; **349cl** antonioiacobelli/RooM/Getty Images; **349cr** Ashva/iStock/Getty Images; **349b, 353** Anton Petrus/Moment/Getty Images; **350l** Artwork by Leib Chigrin INTERFOTO/Alamy Stock Photo; **350cr** © Charles D. Winters/SCIENCE SOURCE; **350bl** De Agostini/P. Castano/De Agostini Picture Library/Getty Images; **351** dszc/E+/Getty Images; **355** robas/iStock/Getty Images

CHAPTER 15

356 Dimitri Otis/Stone/Getty Images; **357** benoitb/Digital Vision Vectors/Getty Images; **358–59bg, 382t** Martin Puddy/Stone/Getty Images; **359l** © ADAM HART-DAVIS/SCIENCE SOURCE; **359r** © Turtle Rock Scientific/SCIENCE SOURCE; **360–61l** Jasmin Merdan/Moment/Getty Images; **361r** Lena_Zajchikova/iStock/Getty Images; **362** © Turtle Rock Scientific/SCIENCE SOURCE; **365t** Xvision/Moment/Getty Images; **365b** © SPL/SCIENCE SOURCE; **366tl** Daniel Grill/Getty Images; **366tr** NaturaLight/Alamy Stock Photo; **366b, 382b** © TREVOR CLIFFORD PHOTOGRAPHY/SCIENCE SOURCE; **367** TANGOART/iStock/Getty Images; **368, 368–69** (border) azerberber/iStock/Getty Images; **370** Leonid Ikan/iStock/Getty images; **371bg** Ivan/Moment/Getty Images; **371, 383** duncan1890/Digital Vision Vectors/Getty Images; **372** Bruno Gori/Cultura/Getty Images; **373** Alatom/E+/Getty Images; **374** Trevor Williams/DigitalVision/Getty Images; **375** Artwork by Leib Chigrin derivative of Josiah Willard Gibbs -from MMS-/Wikimedia Commons/Public Domain; **375** ss404045/Alamy Stock Photo; **376, 378** Sunny/Stone/Getty Images; **380** Alina Rosanova/iStock/Getty Images; **381** Melinda Podor/Moment/Getty Images

Photo Credits

CHAPTER 16

386 ljubaphoto/iStock/Getty Images; **387**l Koichi Yajima/EyeEm/Getty Images; **387**c Leonid Ikan/iStock/Getty Images; **387**r © Turtle Rock Scientific/SCIENCE SOURCE; **390–91** changyu lu/Moment/Getty Images; **392**cl © Charles D. Winters/SCIENCE SOURCE; **392**cc © Charles D. Winters/SCIENCE SOURCE; **392**b © Turtle Rock Scientific/SCIENCE SOURCE; **392**r © Turtle Rock Scientific/SCIENCE SOURCE; **393**tl © Turtle Rock Scientific/SCIENCE SOURCE; **393**tr CTK/Alamy Stock Photo; **393**bl © Turtle Rock Scientific/SCIENCE SOURCE; **393**br Xvision/Moment/Getty Images; **394**tr Aleksandr Zubkov/Moment/Getty Images; **394**bl © SPL/SCIENCE SOURCE; **395** SilvanBachmann/Shutterstock.com; **396**bl © Charles D. Winters/SCIENCE SOURCE; **396–97**bg Kitsana1980/Shutterstock.com; **397**tr Editorial Image, LLC/Alamy Stock Photo; **398–99**bg, **404–5**bg rudchenko/iStock/Getty Images; **399**br Dorling Kindersley ltd/Alamy Stock Photo; **400** E.R. Degginger/Alamy Stock Photo; **401**t Motortion Films/Shutterstock.com; **401**b Michael Moloney/Shutterstock.com; **402** Yulia Reznikov/Moment/Getty Images; **403**t ilbusca/Digital Vision Vectors/Getty Images; **403**b KatarzynaZakowska/Shutterstock.com; **407** Sara Stathas/Alamy Stock Photo

CHAPTER 17

408 Jevtic/iStock/Getty Images; **409** Sueddeutsche Zeitung Photo/Alamy Stock Photo; **410**tl, **430**r © Charles D. Winters/SCIENCE SOURCE; **410**tr © ANDREW LAMBERT PHOTOGRAPHY/SCIENCE SOURCE; **410**b Westend61/Getty Images; **411** © Editorial Image/SCIENCE SOURCE; **412** © Turtle Rock Scientific/SCIENCE SOURCE; **413**t © Charles D. Winters/SCIENCE SOURCE; **413**c © Charles D. Winters/SCIENCE SOURCE; **413**b © Charles D. Winters/SCIENCE SOURCE; **416–17** Nik Merkulov/Shutterstock.com; **419** Cultura RF/Joseph Giacomin/Cultura/Getty Images; **420** Nivellen77/iStock/Getty Images; **421**b Artwork by David Lompe derivative of Haber-Bosch-En/by Francis E Williams/Wikimedia Commons CC By-SA 2.5; **422**t Chuanthit Kunlayanamitre/Shutterstock.com; **422**b istetiana/Moment/Getty Images; **423** Science History Images/Alamy Stock Photo; **424–25**t Savany/iStock/Getty Images; **424**b © Turtle Rock Scientific/SCIENCE SOURCE; **425**br durk gardenier/Alamy Stock Photo; **426** Anna Usova/iStock/Getty images; **427**, **428**cl, **431** © Turtle Rock Scientific/SCIENCE SOURCE; **428–29**t NagyDodo/iStock/Getty Images; **432–33** Sergio Ballivian/500px Prime/Getty Images

CHAPTER 18

434 RossHelen/Shutterstock.com; **435**tr channarongsds/iStock/Getty Images; **435**bl © Turtle Rock Scientific/SCIENCE SOURCE; **436–38**bg Savushkin/iStock/Getty Images; **438** Artwork by Leib Chigrin GL Archive/Alamy Stock Photo; **439** KrivoTIFF/iStock/Getty Images; **440**tl Vesna Jovanovic/EyeEm/Getty Images; **440–41**tr borchee/iStock/Getty Images; **442–43** Artwork by Ina Stanimirova © Turtle Rock Scientific/SCIENCE SOURCE; **442** New Africa/Shutterstock.com; **444** fcafotodigital/E+/Getty Images; **445**, **459** © SCIENCE PHOTO LIBRARY/SCIENCE SOURCE; **446** © Turtle Rock Scientific/SCIENCE SOURCE; **447** © Turtle Rock Scientific/SCIENCE SOURCE; **448** kosmos111/iStock/Getty Images; **449** Abstract Aerial Art/Digital Vision/Getty Images; **450** © Andrew Lambert Photography/SCIENCE SOURCE; **451**tr © Turtle Rock Scientific/SCIENCE SOURCE; **451**bl © Turtle Rock Scientific/SCIENCE SOURCE; **453**t © Turtle Rock Scientific/SCIENCE SOURCE; **453**b © GIPhotoStock/SCIENCE SOURCE; **454**, **460** © GIPhotoStock Images/SCIENCE SOURCE; **455** simonkr/E+/Getty Images; **456** © Turtle Rock Scientific/SCIENCE SOURCE; **457** Nazar Abbas Photography/Moment/Getty Images; **458** undefined/iStock/Getty Images

CHAPTER 19

462 Westend61/Getty Images; **463** © Charles D. Winters/SCIENCE SOURCE; **464** JGI/Jamie Grill/Getty Images; **465** ra2studio/iStock/Getty Images; **467**tr benoitb/Digital VisionVectors/Getty Images; **467**br Tsuji/E+/Getty Images; **467**bg, **481**t Xuanyu Han/Moment/Getty Images; **468** onlyyouqj/iStock/Getty Images; **471**t Chesky_W/iStock/Getty Images; **471**br © Turtle Rock Scientific/SCIENCE SOURCE © Turtle Rock Scientific/SCIENCE SOURCE; **474**tl benoitb/DigitalVisionVectors/Getty Images; **474**tr ilbusca/DigitalVision Vectors/Getty Images; **474**bl DAMRONG RATTANAPONG/Shutterstock.com; **476** MileA/iStock/Getty Images; **476**t Artwork by David Lompe derivative of Secondary Cell Diagram by Barrie Lawson/Wikimedia Commons/CC By-SA 3.0; **476**b Artwork by David Lompe © Argonne National Laboratory/SCIENCE SOURCE; **477**t, **481**b Matveev Aleksandr/Shutterstock.com; **477**b Artwork by David Lompe derivative of Zincbattery (1)/by Mcy jerry/Wikimedia Commons/CC By-SA 2.5; **478** Tribune Content Agency LLC/Alamy Stock Photo; **483** Sven Loeffler/iStock/Getty Images

CHAPTER 20

484 Anna Ok/Shutterstock.com; **487** SeanShot/E+/Getty Images; **488–89**t freemixer/E+/Getty Images; **490** wolv/E+/Getty Images; **492** Shelly Still/iStock/Getty Images; **493** Picture Store/Shutterstock.com; **494** Syda Productions/Shutterstock.com; **495** Charles Gatewood/iStock/Getty Images; **497** FactoryTh/iStock/Getty Images; **499** sirawit 99/iStock/Getty Images; **500** Studio Doros/iStock/Getty Images; **501** Alina Yudina/Shutterstock.com; **502**t Isabel Pavia/Moment/Getty Images; **502**b Westend61/Getty Images; **503** Michal Kin/EyeEm/Getty Images; **504–5**bg Tunatura/iStock/Getty Images; **505**t Elizaveta Galitckaia/Shutterstock.com; **506–7**l, **511** Lorna Roberts/Shutterstock.com; **507**r photograph by dorisj/Moment/Getty Images; **508** AmyLv/Shutterstock.com; **509** © Turtle Rock Scientific/SCIENCE SOURCE

CHAPTER 21

514 RGB Ventures/SuperStock/Alamy Stock Photo; **516, 533**t Paul Souders/Stone/Getty Images; **517**t didecs/iStock/Getty Images; **517**b MediaProduction/E+/Getty Images; **518** fcafotodigital/E+/Getty Images; **519**t Vesna Jovanovic/EyeEm/Getty Images; **519**b benoitb/E+/Getty Images; **520**t Aerogel hand by Courtesy NASA/JPL-Caltech/Wikimedia Commons/Public Domain/cropped; **520**r–**21** ranasu/iStock/Getty Images; **522**l Fattyplace/iStock/Getty Images; **522**r–**23**t, **533**b apomares/E+/Getty Images; **524**t Artwork by Ina Stanimirova Science History Images/Alamy Stock Photo; **525** virtustudio/iStock/Getty Images; **526**t Yulia Furman/Shutterstock.com; **526**b Artwork by Frank Ramspott derivative of Oxy-Hemoglobin by Belle2018/Wikimedia Commons/CC By-SA 4.0; **527** robynmac/iStock/Getty Images; **528**tl, **534** Steven Puetzer/Photographer's Choice RF/Getty Images; **528**tr–**29**t Varavin88/Shutterstock.com; **530** Science Picture Co/Alamy Stock Photo; **531** Science Photo Library - NASA/ESA/STSCI/M. ROBBERTO,HST ORION TREASURY TEAM/Brand X Pictures/Getty Images; **532** AvigatorPhotographer/iStock/Getty Images; **535** Fascinadora/Shutterstock.com

CHAPTER 22

536 guowei ying/Moment/Getty Images; **537**tr Artwork by Ina Stanimirova derivative of Würzburg, Germany - DSC04289/by Daderot/Wikimedia Commons/CC By-SA 1.0; **537** UniversalImagesGroup/Getty Images; **537**b Artwork by Ina Stanimirova Sueddeutsche Zeitung Photo/Alamy Stock Photo; **538** abadonian/iStock/Getty Images; **538**b, **546** Artwork by Ina Stanimirova derivative of Ernest Rutherford 1905/Wikimedia Commons/CC By-SA 4.0; **539** Rost9/Shutterstock.com; **540** Table isotopes ru by Sjlegg/Wikimedia Commons/CC By-SA 3.0/modified; **541** H. Mark Weidman Photography/Alamy Stock Photo; **543** wellphoto/iStock/Getty Images; **544–45**t cigdem/Shutterstock.com; **544**b alexkuehni/iStock/Getty Images; **548–49** cokada/iStock/Getty Images; **552** Massimo Dallaglio/Alamy Stock Photo; **555**t Ian Dagnall/Alamy Stock Photo; **555**b Collection Christophel/Alamy Stock Photo; **556** JESPER KLAUSEN/SCIENCE PHOTO LIBRARY/Getty Images; **558** Evdoha_spb/Shutterstock.com; **559** AFP/Getty Images; **560** zhongguo/E+/Getty Images; **562–63** IngaNielsen/iStock/Getty Images; **562** Artwork by Frank Ramspott derivative of FusionintheSun by Borb/Wikimedia Commons/CC By-SA 2.0; **564**t mmac72/E+/Getty Images; **564**b Mark Kostich/E+/Getty Images; **565**t Jupiter radio by NASA Jet Propulsion Laboratory/Wikimedia Commons/Public Domain/Cropped; **565**b Lisa Schaetzle/Moment/Getty Images; **569** DKosig/E+/Getty Images

BACK MATTER

571 ilbusca/DigitalVision Vectors/Getty Images; **572** SolStock/E+/Getty Images; **573** Morphart Creation/Shutterstock.com; **577** Look Studio/Shutterstock.com

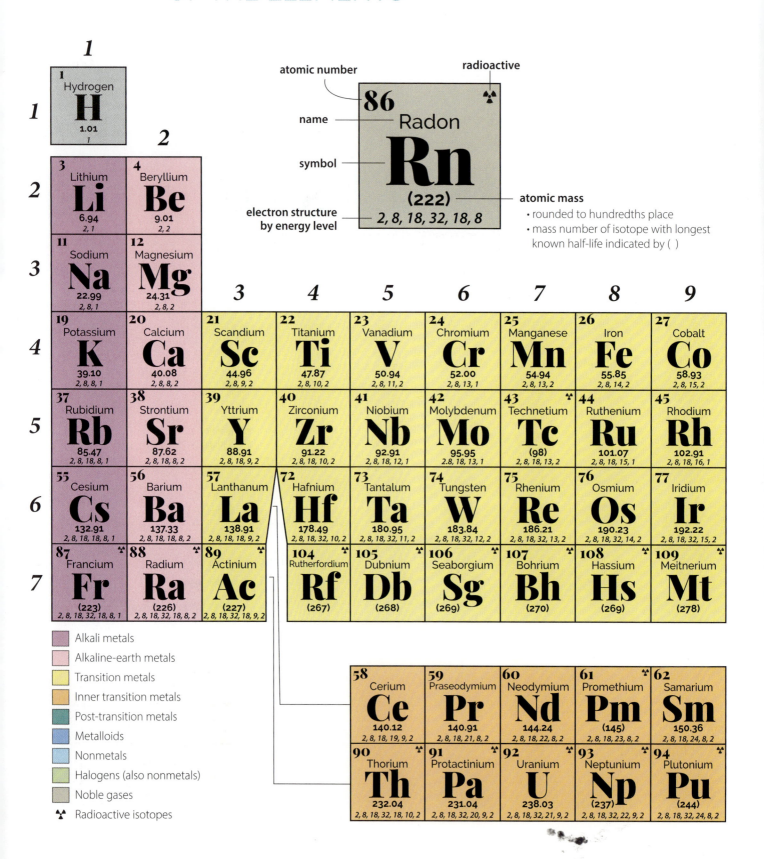

10	11	12	13	14	15	16	17	18
								2 Helium **He** 4.00 2
			5 Boron **B** 10.81 2, 3	6 Carbon **C** 12.01 2, 4	7 Nitrogen **N** 14.01 2, 5	8 Oxygen **O** 16.00 2, 6	9 Fluorine **F** 19.00 2, 7	10 Neon **Ne** 20.18 2, 8
			13 Aluminum **Al** 26.98 2, 8, 3	14 Silicon **Si** 28.09 2, 8, 4	15 Phosphorus **P** 30.97 2, 8, 5	16 Sulfur **S** 32.06 2, 8, 6	17 Chlorine **Cl** 35.45 2, 8, 7	18 Argon **Ar** 39.95 2, 8, 8
28 Nickel **Ni** 58.69 2, 8, 17, 1	29 Copper **Cu** 63.55 2, 8, 18, 1	30 Zinc **Zn** 65.38 2, 8, 18, 2	31 Gallium **Ga** 69.72 2, 8, 18, 3	32 Germanium **Ge** 72.63 2, 8, 18, 4	33 Arsenic **As** 74.92 2, 8, 18, 5	34 Selenium **Se** 78.97 2, 8, 18, 6	35 Bromine **Br** 79.90 2, 8, 18, 7	36 Krypton **Kr** 83.80 2, 8, 18, 8
46 Palladium **Pd** 106.42 2, 8, 18, 18	47 Silver **Ag** 107.87 2, 8, 18, 18, 1	48 Cadmium **Cd** 112.41 2, 8, 18, 18, 2	49 Indium **In** 114.82 2, 8, 18, 18, 3	50 Tin **Sn** 118.71 2, 8, 18, 18, 4	51 Antimony **Sb** 121.76 2, 8, 18, 18, 5	52 Tellurium **Te** 127.60 2, 8, 18, 18, 6	53 Iodine **I** 126.90 2, 8, 18, 18, 7	54 Xenon **Xe** 131.29 2, 8, 18, 18, 8
78 Platinum **Pt** 195.08 2, 8, 18, 32, 17, 1	79 Gold **Au** 196.97 2, 8, 18, 32, 18, 1	80 Mercury **Hg** 200.59 2, 8, 18, 32, 18, 2	81 Thallium **Tl** 204.38 2, 8, 18, 32, 18, 3	82 Lead **Pb** 207.24 2, 8, 18, 32, 18, 4	83 Bismuth **Bi** 208.98 2, 8, 18, 32, 18, 5	84 Polonium **Po** (209) 2, 8, 18, 32, 18, 6	85 Astatine **At** (210) 2, 8, 18, 32, 18, 7	86 Radon **Rn** (222) 2, 8, 18, 32, 18, 8
110 Darmstadtium **Ds** (281)	111 Roentgenium **Rg** (282)	112 Copernicium **Cn** (285)	113 Nihonium **Nh** (286)	114 Flerovium **Fl** (289)	115 Moscovium **Mc** (290)	116 Livermorium **Lv** (293)	117 Tennessine **Ts** (294)	118 Oganesson **Og** (294)

63 Europium **Eu** 151.96 2, 8, 18, 25, 8, 2	64 Gadolinium **Gd** 157.25 2, 8, 18, 25, 9, 2	65 Terbium **Tb** 158.93 2, 8, 18, 27, 8, 2	66 Dysprosium **Dy** 162.50 2, 8, 18, 28, 8, 2	67 Holmium **Ho** 164.93 2, 8, 18, 29, 8, 2	68 Erbium **Er** 167.26 2, 8, 18, 30, 8, 2	69 Thulium **Tm** 168.93 2, 8, 18, 31, 8, 2	70 Ytterbium **Yb** 173.05 2, 8, 18, 32, 8, 2	71 Lutetium **Lu** 174.97 2, 8, 18, 32, 9, 2
95 Americium **Am** (243) 2, 8, 18, 32, 25, 8, 2	96 Curium **Cm** (247) 2, 8, 18, 32, 25, 9, 2	97 Berkelium **Bk** (247) 2, 8, 18, 32, 27, 8, 2	98 Californium **Cf** (251) 2, 8, 18, 32, 28, 8, 2	99 Einsteinium **Es** (252) 2, 8, 18, 32, 29, 8, 2	100 Fermium **Fm** (257) 2, 8, 18, 32, 30, 8, 2	101 Mendelevium **Md** (258) 2, 8, 18, 32, 31, 8, 2	102 Nobelium **No** (259) 2, 8, 18, 32, 32, 8, 2	103 Lawrencium **Lr** (266) 2, 8, 18, 32, 32, 9, 2